Pharmaceutical Computer Systems Validation

Pharmaceutical Computer Systems Validation

Quality Assurance, Risk Management and Regulatory Compliance
Second Edition

Edited by

Guy Wingate
GlaxoSmithKline
Barnard Castle, U.K.

CRC Press
Taylor & Francis Group
Boca Raton London New York

CRC Press is an imprint of the
Taylor & Francis Group, an **informa** business

First published in paperback 2024

First published 2010 by Informa Healthcare

Published 2024
by CRC Press
2385 NW Executive Center Drive, Suite 320, Boca Raton FL 33431

and by CRC Press
4 Park Square, Milton Park, Abingdon, Oxon, OX14 4RN

CRC Press is an imprint of Taylor & Francis Group, LLC

© 2010, 2024 Taylor & Francis Group, LLC

Library of Congress Cataloging-in-Publication Data

Pharmaceutical computer systems validation : quality assurance, risk management and regulatory compliance / edited by Guy Wingate. — 2nd ed.
    p. ; cm.
  Rev. ed. of: Computer systems validation / editor, Guy Wingate. Boca Raton, Fla. : Interpharm/CRC, c2004.
  Includes bibliographical references and index.
  ISBN-13: 978-1-4200-8894-6 (hardcover : alk. paper)
  ISBN-10: 1-4200-8894-7 (hardcover : alk. paper) 1. Pharmaceutical industry—Management. 2. Pharmaceutical industry—Data processing. 3. Health facilities—Risk management. 4. Risk management—Data processing. I. Wingate, Guy. II. Computer systems validation.
  [DNLM: 1. Computer Systems—standards. 2. Drug Industry–standards. 3. Drug Industry—instrumentation. 4. Guideline Adherence. 5. Health Care Sector—standards. 6. Quality Control. 7. Software Validation. QV 26.5 P5352 2010]
  HD9665.5.C664 2010
  615.1068′4—dc22

                                  2009049688

ISBN: 978-1-4200-8894-6 (hbk)
ISBN: 978-1-03-291751-1 (pbk)
ISBN: 978-0-429-13762-4 (ebk)

DOI: 10.3109/9781420088953

Visit the Taylor & Francis Web site at
http://www.taylorandfrancis.com

and the CRC Press Web site at
http://www.crcpress.com

For Sarah, Katherine, Robert, and Edward

Validation should be viewed as an integral part of the overall computer system's life cycle. It should promote improved process control and not bureaucracy. Good practice and common sense should prevail.

Foreword to the Second Edition

It is pleasing to note that the second edition has been significantly enhanced and updated since 2004. It seeks to anticipate some emerging FDA and EU requirements (1–4). It is aligned with the latest fifth edition of the GAMP Guide (5), explains the latest thinking from ASTM (6) and DIA (7), and presents developing industry best practice. Regulatory inspection findings and FDA Warning Letters have been analyzed; risk-based decisions are now considered in more detail; case studies have been revisited to take account of regulatory and technological trends; there are additional case studies on databases and spreadsheets, together with new material on PAT and process control systems to fit with pending new guidance from ISPE/GAMP.

I am happy to repeat the recommendation that I gave in 2004, in support of this book and its editor and team of collaborating authors: "Whether you are looking for the missing piece of the jigsaw puzzle for your project guidance on how to meet the regulations in a practical sense, (whilst making cost-effective use of your investments), this information resource [which puts principles into practice] is a good place to start! "

Anthony J. Trill
Senior Inspector
MHRA (Retired)

REFERENCES

1. Food and Drug Administration. Pharmaceutical CGMPs for the 21st Century: A Risk-Based Approach. Rockville, MD, 2004.
2. International Conference on Harmonisation. Quality Risk Management, Q9 Document, Technical Requirements for Registration of Pharmaceuticals for Human Use, 2005. Available at: www.ich.org.
3. Food and Drug Administration. 21 CFR part 211. Supplementary information. Rockville, MD, 2008.
4. European Commission. Draft Annex 11. Computerised Systems (EU guideline to Good Manufacturing Practices for Medicinal Products for Human and Veterinary Use), Public Consultation Document. Brussels, April 2008.
5. International Society for Pharmaceutical Engineering. GAMP®5: Risk-Based Approach to Compliant GxP Computerised Systems. Tampa, Florida, 2008. Available at: www.ispe.org.
6. American Society for Testing and Materials. E2500-07 standard guide for specification, design and verification of pharmaceutical and biopharmaceutical manufacturing systems and equipment, 2007.
7. Drug Information Association. Computerized Systems used in Non-Clinical Safety Assessment: Current Concepts in Validation and Compliance (known as Red Apple II), 2008.

Editor's Note: Tony Trill has written a personal reflection on his career in the MHRA inspecting computer systems, which can be found online:

"A regulator's perspective of his GAMP experience," Anthony Trill, retired senior MHRA inspector, online exclusive article, Pharmaceutical Engineering. November/December 2008, Vol. 28 No. 6.

Foreword to the First Edition

Computer technology is all pervasive. It hides behind the smallest button on domestic appliances and is in smart cards and security devices, mobile phones, cash dispensers, PCs, integrated networks, process plant, automobiles, "jumbo jets, " and power plants. Automation is everywhere and is gathering complexity, innovation, and momentum, and we have to rely on it more and more in our everyday lives. The inexorable rise of automation is also seen in the corporate strategies of pharmaceutical manufacturers calling for investment in new technology to improve business efficiency and competitive edge. When such technology is associated with high-risk public safety projects or the production and control of life-saving medicines or devices, we (businesses and regulators) need to know that it is reliable, quality assured, and validated. Easy to say, but the technology (and the terminology) is difficult to understand, let alone prove and qualify, if you are not an electronic systems engineer or a latent Einstein.

Pharmaceutical and healthcare companies have historically engineered their businesses to be profitable while ensuring that quality is built into their medicinal products, or devices, through the observance of GxPs (viz GCPS, GLPs, GMPs, etc.) which essentially require computerized systems in the pharmaceutical sector to be fully documented, defined as to functionality, quality assured, and validated. This book considers the requirements of the various international regulations, guides, and codes in historical perspective and leads the reader into business and project life cycle issues and activities. This book is invaluable in bridging the gap between theory and practice and is supported by case studies from experienced professional practitioners and engineers who have had to face up to the challenges of proving the quality, structural integrity, and validation of different systems in the "real world" (process control, utility management, networked control, information and high level business IT, and integrated real-time application, etc.). The case studies are organized hierarchically from low-level instruments and PLCs through integration to higher-level proprietary electronic document and information management systems, and beyond.

Pharmaceutical and healthcare companies that invest in computerized systems need systems that are delivered on time and within budget and that fulfill business functional and performance requirements. In their rush to place new products and versions on the market, however, computer software and systems suppliers rarely delivered error-free products. In fact, some two-thirds of life cycle costs can be incurred after delivery of the software and system to the users. Pharmaceutical and healthcare companies do not want lots of downtime, disruption, and escalating costs once a system has been delivered and implemented (1,2). And, of course, in GxP applications, any deficiencies will be of particular interest during regulatory inspections.

Inspectors and investigators working for the different national regulatory bodies have to apply their national GxPs and regulations when assessing these systems. While these are published, they are not necessarily up to date and, as we all would acknowledge, they are often open to interpretation not only by different inspectors, depending on their background and training, but also depending on the particular computerized system and application. Regulators need to be satisfied that computerized systems installed in pharmaceutical companies are fit for their intended purposes by considering the nature of the application, specifications, quality assurance of the development life cycle activities, qualification, performance validation, in-use controls, accuracy, and reliability in the context of relevant GxPs. The increasing complexity of (integrated) propriety computer systems, critical applications, project validation issues, and inspection findings have been considered before together with the challenge for all parties (as ever) to apply sensible regulations and cost-effective good computer validation practices (1,3,4).

The pharmaceutical and healthcare industries (including software and system suppliers and developers) have also reportedly had some difficulty in ensuring that these projects actually deliver the proposed business benefits and that the systems, as built, actually meet the specifications and are reliable and validated and quite apart from determining just how much and what type of validation evidence is required to satisfy the different regulatory bodies, particularly the FDA. While the GAMP Guide (5) and, to some extent, the PDA 18 report (6) provide the latest interpretation of acceptable development and project guidance in this field (to ensure compliance with the needs of the principle regulatory bodies around the world), and TickIT provides a guide to software quality system construction and certification (using ISA 9001:1994) (7), there has been a lack of papers on practical experiences from pharmaceutical sector project teams seeking to implement new technology.

Today, both the industry and regulators have a much better understanding (8) of the ways and means to develop and validate computerized systems. Regulatory inspections now have more to do with risk-based assessments of what these systems are being used for in the context of broader GxP requirements rather than software and system validation per se. Inspectors (9) now rarely concentrate on "simply" inspecting computerized systems as an entity on sites; they are more often directly concerned with what the systems are being used for and how they are being used and controlled. Risk-based findings for noncompliances associated with computerized systems will often be linked with other chapters of the EU or PIC/S GMP apart from annex 11. However, where a detailed inspection of a computerized system is indicated (from risk assessments or other inspections), it can be arranged as a specific exercise.

It is interesting to note the ongoing collaboration between ISPE and PDA (10,11) to publish guidance on electronic records and management and to influence opinion. It is to be hoped that the technological implementation of electronic records and signature requirements worldwide will not be frustrated by a lack of understanding and agreement by all stakeholders of the real issues. Recognition must be given to the need for regulated users to have robust information security management practices and a risk-based approach applied to important relevant records and inspection compliance.

I believe that this book will be welcomed by novices and experts, suppliers, developers, purchasers, and regulators alike in providing insight into the practical aspects of these complex automation– and life cycle–related projects. Many staff assigned to validation projects could also benefit from sharing the experience of other practitioners. Whether you are looking for the missing piece of the jigsaw for your project or guidance on how to meet the regulations in a practical sense, this information resource (which puts principles into practice) is a good place to start!

Anthony J. Trill
Senior Inspector
MHRA

REFERENCES

1. Stokes T, Branning RC, Chapman KG, et al. Good Computer Validation Practices Common Sense Implementation. Buffalo Grove: Interpharm Press, Inc., 1994.
2. Wingate GAS. Computer systems validation: a historical perspective. Pharmaceutical Engineering, July/August 1995, pp. 8–12.
3. Trill AJ. EU GMP Requirements and the good automated manufacturing practice (GAMP) supplier guide for the validation of automated systems for pharmaceutical manufacturing. Pharmaceutical Engineering, May/June 1995, pp. 56–62.
4. Trill AJ. An EU/MCA view of recent industry guides to computer validation, including GAMP 1996, PDA technical report 18, and the validation toolkit. Proceedings of PIC/S Seminar "Inspection of Computer Systems," Sydney, Australia, September 1996.
5. International Society for Pharmaceutical Engineering. Good Automated Manufacturing Practice Guide for Validation of Automated Systems (known as GAMP®4), 2001. Available at: www.ispe.org.
6. Parenteral Drug Association. The Validation of Computer Related Systems, Technical Report No. 18. J Pharm Sci Technol 1995; 49(1).
7. TickIT Guide. A guide to software quality management system construction and certification using ISO 9001:2000, issue 5.0. DISC/BSI TickIT Office, London, 2000.

8. Pharmaceutical Inspection Co-operation Scheme. Good Practices for Computerised Systems in Regulated GxP Environments, Pharmaceutical Inspection Convention, PI 011-1. Geneva, 2003.

9. Medicines Control Agency. "Top 10 GMP inspection issues," MCA Seminar. Trill AJ. "Computerised systems and GMP—current issues." London, September 24, 2002.

10. International Society for Pharmaceutical Engineering/Parenteral Drug Association. Good Practice and Compliance for Electronic Records and Signatures: Part 1. Good Electronic Record Management (GERM), 2002. Available at: www.ispe.org.

11. International Society for Pharmaceutical Engineering/Parenteral Drug Association. Good Practice and Compliance for Electronic Records and Signatures: Part 2. Complying with 21 CFR Part 11, Electronic Records and Electronic Signatures, 2001. Available at: www.ispe.org.

Preface

This is the second edition of this book, and much has changed in the world of computer systems compliance since it was first published five years ago. The combination of economic pressures and the desire to maximize value for money is driving a new mind set. Bureaucratic practices often associated with validation have been widely challenged rather than accepted through hearsay that they are driven by the regulations. Practitioners feel empowered to justify more efficient and effective ways of working. The role of risk management throughout the life cycle of a computer system is understood not only for the benefit of patient/consumer safety but also in terms of cost-effectiveness. The next step involves making largely subjective risk assessments more scientific through utilizing relevant data. Couple this thinking with the focus on demonstrating that a computer system is fit for purpose through process capability rather than a compliance cross-check, and it is fair to say that a new paradigm has emerged.

The structure of this second edition remains the same, with the first portion providing chapters discussing the new paradigm, organization and management of compliance activities, life cycle requirements stepping through development, verification, operation, and decommissioning of computer systems. In addition, there are chapters covering latest developments in electronic record and signature requirements, new practical guidance on risk-based decisions, handling regulatory inspections, and improving organizational capabilities to deliver highly efficient and effective compliance through six-sigma and lean manufacturing philosophies. Throughout the book, I have included real observations recently made by FDA on the various topics being discussed.

A series of case studies are provided in the second portion of the book showing how the principles explained earlier can be put into practice. The real world is rarely as simple and straightforward as we might hope, and I have been very keen to bring together a group of case studies to share practical solutions to the many and varied challenges that can face practitioners. The case studies cover most types of computer systems used today in R&D, manufacturing, and commercial supply organizations. Original case studies have undergone significant review and new case studies added to keep the information presented current and relevant. I have encouraged the contributors to include pertinent checklists on key aspects of their case studies to complement the more generic guidance provided in the first portion of the book.

Once again, I am hugely indebted to the large number of contributors each of whom was invited to participate on the basis of their recognized expertise (see Biographies). In addition, I would also like to acknowledge friends and colleagues who have provided invaluable discussions and explorations of computer compliance principles and practices over the years. In particular, I would like to thank Sam Brooks (Novartis Consumer Healthcare), Ellis Daw (GlaxoSmithKline), Paul D'Eramo (Johnson & Johnson), Howard Garston-Smith (Gartson Smith Associates), Jerry Hare (Independent Consultant), Niels Holger Hansen (Novo Nordisk), Paige Kane (Wyeth Biotech), Scott Lewis (Eli Lilly), Takayoshi Matsumura (Eisai), Karl-Hienz Menges (German Darmstadt GMP Inspectorate), Gordon Richman (EduQuest), Peter Roberston (AstraZeneca), David Selby (Selby-Hope), and Sion Wyn (Conformity).

Last but by no means least, I would like to thank my wife Sarah and our children (Katherine, Robert, and Edward) for their love, patience, and support during the preparation of this book.

Guy Wingate

DISCLAIMER

The information contained in this book is provided in good faith and reflects the personal views of the contributors. These views do not necessarily reflect the perspectives of the contributor's respective employers. The information provided does not constitute legal advice and no liability can be accepted in any way.

PERMISSIONS

ISPE and PDA are thanked for their kind permission to print the following artwork: Figure 1.6 (New 21st Century Paradigm), Table 3.1 (Characteristics of Various Document Types), Figure 5.4 (Criticality and Impact Analyses), Table 11.1 (Monitoring Plan for Server-Based LIMS), Figure 11.4 (Data Life Cycle), Figure 13.3. (Electronic Record Life Cycle), Figure 13.4 (Example Audit Trail), and Appendix 13.C (Procedural and Technological Controls for 21 CFR Part 11). Specific reference sources are provided in the text where artwork is located. Figure 1.5 is reprinted, with permission, from ASTM E2500-07 Standard Guide for Specification, Design, and Verification of Pharmaceutical and Biopharmaceutical Manufacturing Systems and Equipment, copyright ASTM International, 100 Barr Harbor Drive, West Conshohocken, PA 19428.

Contributor Biographies

JOHN ANDREWS (ANDREWS CONSULTING ENTERPRISES)

John is a Director of Andrews Consulting Enterprises Limited. Prior to this, he was the European IT Consulting Service Manager at KMI, a division of PAREXEL International LLC. He has been a member of the GAMP Forum Special Interest Group on process control systems; he also sat on the editorial board for GAMP®4. His activities include providing consultancy and training on computer systems validation and compliance and quality assurance activities within the pharmaceutical, biopharmaceutical, medical device, and other regulated healthcare industries. He is currently specializing in the application of GAMP principles to the project management and validation of process analytical technologies (PAT), PAT data collection methods, chemometric models, and real-time release strategies. Prior to the above, he held positions as a site computer system validation manager and a supply chain systems project manager with GlaxoSmithKline (GSK). Responsibilities there covered process control systems, business systems, and laboratory system validation. Previous employment includes SmithKline Beecham Pharmaceuticals where he held positions as a senior engineering standards engineer, secondary manufacturing electrical engineer, projects engineer, and electrical supervisor.

Contact: john.m.andrews@gsk.com

PETER BOSSHARD, PHD (HOFFMANN-LA ROCHE)

Peter is global Quality Manager for F. Hoffmann-La Roche Limited based at Basel, Switzerland. He is a pharmacist by education and has spent more than 15 years in the pharmaceutical industry. His experience covers quality assurance of computerized systems validation, establishment of policy and guidelines, training of business and IT departments, audits of the internal implementation of the standards, consultancy for projects and sites, and external representation of the Roche approach to computer system validation. Currently, Peter is responsible for all aspects of computerized systems validation, electronic records and signature interpretation, and change control for global systems within Roche Pharma Technical Operations.

Contact: peter.bosshard@roche.com

ROGER BUCHANAN (ELI LILLY)

Roger is a Computer Quality Consultant for Eli Lilly; before joining Lilly, he was working in the petrochemical industry (Middle East and the United Kingdom). Roger has been with Lilly for the last 20 years; his first 10 years was working as a technical support engineer for biotechnology human growth hormone facilities before moving into IT and computer quality. It was in this first period with Lilly that he obtained his BSc (Hon) degree via the open university program. He has spent the last eight to nine years working in computer systems quality assurance supporting the European and Asia Pacific sites providing training, quality evaluations, and consultation on computer validation for IT, laboratory, and process automation systems. He is actively involved with his U.S. consultation group to look at benchmarking across industry on approaches to computer systems validation. He is an active member of GAMP and has been involved in a number of special interest groups working on building management systems, testing, and, more recently, process control systems. Roger is currently involved in developing guidance on the application of dose calculators in the pharmaceutical industry.

Contact: buchanan_roger@lilly.com

WINNIE CAPPUCCI (BAYER PHARMACEUTICALS)

Winnie is currently Associate Director for Product Supply IT Systems Compliance, North America. She has worked in the pharmaceutical industry for 39 years and in global roles for the last 14 years. Ms Cappucci has worked as a business process owner, an IT professional, and, lastly, as a quality and compliance specialist in a highly regulated environment. In her current role, Ms Cappucci is responsible for developing and implementing Bayer's global standards for computer systems compliance. She is a member of the GAMP Americas Steering Committee, the GAMP Editorial Board, and cochair of the GAMP special interest group working on quality and compliance guidance on the topic of outsourcing. She is also a member of the Parenteral Drug Association (PDA) Industry Advisory Board for Technical Report 32 (Auditing of Suppliers Providing Computer Products and Services for Regulated Operations) and the Audit Resource Center repository.

Contact: winnie.cappucci@bayer.com

MARK CHERRY (ASTRAZENECA)

Mark is the Integrated Assurance Manager for Global Operations at AstraZeneca where he is responsible for computer systems validation, risk management and security. Mr Cherry graduated in Instrumentation and Process Control Engineering in 1987, and worked as a Project Engineer for Sterling Organics until joining Glaxo in 1990. He was involved in a variety of process control projects within Glaxo until 1995 when he became the Engineering Team Manager, responsible for all engineering maintenance activities on a bulk API plant. In 1997 he was appointed Systems and Commissioning Manager for a major capital project within GlaxoWellcome, involving the installation of a large DCS system using the S88.01 approach to batch control. From 1999 to 2001, Mr Cherry was responsible for computer systems validation for bulk API manufacturing sites within GlaxoSmithKline. Mr Cherry is a Chartered Engineer, a member of The Institute of Measurement and Control, the International Society for Pharmaceutical Engineering (ISPE) and of the GAMP European Steering Committee.

Contact: mark.cherry@astrazeneca.com

CHRIS CLARK (NAPP PHARMACEUTICALS)

Leaving Lancaster University with a BSc degree in biochemistry, Chris has gained over 30 years' QA experience within the pharmaceutical and healthcare industries, commencing with Sterling-Winthrop, then moving to Baxter Healthcare Limited, and finally joining NAPP in 1993. As head of quality compliance, he is responsible for aspects of the company quality management system ensuring compliance to current international regulatory requirements for GMP, GCP, GPvP, and GLP. Chris has been involved in a variety of computer-related projects, including the local implementation of Oracle® Applications 11i, an international enterprise document management system and an international implementation of the Oracle clinical data management system. As an active member of the ISPE, having formerly held the chair of the GAMP European Steering Committee and being the current chair of the GAMP Editorial Review Board, Chris regularly speaks at conferences on developing and implementing science-based quality risk approaches to computerized systems compliance.

Contact: chris.clark@napp.co.uk

PETER COADY (PFIZER)

Peter is a computer systems specialist in the global quality operations validation organization of Pfizer, Inc., based in the United Kingdom. He has over 35 years of industrial experience and has been employed at senior level by major British companies in the pharmaceutical, petrochemical, chemical, and water sectors. He is a member of GAMP Europe Forum and a steering committee member and represents the GAMP Forum on the BSI Joint TickIT Industry Steering Committee. He is a registered lead TickIT and quality management system (ISO9000) auditor with the International Register of Certified Auditors (IRCA) and works internationally for Lloyd's Register Quality Assurance (LRQA) as an independent lead assessor. Mr Coady has a BSc (Hon) degree and is a chartered engineer. He is a fellow of the Institution of Mechanical Engineers (FIMechE), a fellow of the Institute of Measurement and Control (FInstMC), an associate of the Chartered Quality Institute (ACQI), and a member of the International Society of Pharmaceutical Engineers (ISPE).

Contact: peter@coadyassociates.com

TONY DE CLAIRE (APDC CONSULTING)

Tony is a Director and Principal Consultant of APDC Consulting Limited. The company provides GxP regulatory compliance and validation consultancy, addressing the broad spectrum of computer system application and related quality assurance in pharmaceuticals, biologicals, and medical devices. Following process control and automation career development in the chemical and steel industries, he moved into the pharmaceutical sector and directed major capital and technical programs for new and replacement GxP automated systems. As an "end user, " he developed and managed the Manufacturing Automation and Information Systems Group worldwide for SmithKline Beecham Corporate Engineering. Moving into computerized system validation consultancy, he worked alongside several notable former FDA inspectors before setting up an independent consultancy. A contributing author for the GAMP®4, he also produced key sections of the associated supplementary GAMP Good Practice Guides for "Validation of Process Control Systems" and "IT Infrastructure Control and Compliance," and to the ISPE Baseline® Guide on "Commissioning and Qualification." He has lectured on "quality management for software" at the University of Brighton (U.K.) and contributed material for the process control module of the "pharmaceutical engineering and advanced training" (PEAT) MSc course at the University of Manchester, U.K. Tony is a member of the Institute of Measurement and Control, a senior member of the International Society of Automation, a member of the ISPE, and a trained TickIT auditor.

Contact: tony_de_claire@btinternet.com

CHRISTOPHER EVANS (GLAXOSMITHKLINE)

Christopher joined GSK on July 1, 1999 following 27 years of service with ICI PLC. He is now GSK's global computer validation manager, responsible for managing resource for compliance oversight of centrally managed IT projects related to IT manufacturing systems. In addition, Christopher provides consultancy and inspection support for central systems implemented at GSK manufacturing sites as part of these IT projects. Christopher has wide international experience in the establishing and managing teams for validation projects in primary and secondary manufacturing facilities working for a number of major pharmaceutical manufacturers in the United Kingdom and Europe including Pfizer, AstraZeneca, Roche, and Napp Laboratories. Christopher was the lead for the two teams developing revised guidance on the topics of software and hardware categories and supplier auditing for GAMP®4. He has also been a member of the GAMP Process Control Special Interest Group.

Contact: christopher.x.evans@GSK.com

JOAN EVANS (ABB)

Joan was originally qualified as a chemical engineer at University College Dublin (Ireland) and has extensive experience in manufacturing industry covering project management, quality management, line management, and consultancy positions. In 1995 she transferred to the life sciences group of Eutech (now ABB), and is now part of ABB's Global Consultancy Group. As a principal consultant with ABB, Joan has responsibility for the management and technical leadership of a range of assignments for blue chip companies, specializing in tailored compliance services for computer systems across the research, manufacturing, and distribution spectrum. She is also internal product champion for ABB's audit and remediation and system obsolescence management consultancy offerings.

Contact: joan.evans@gb.abb.com

MARK FOSS (BOEHRINGER-INGELHEIM)

Mark Foss has been Head of Engineering for Boehringer-Ingelheim in the United Kingdom for the last 10 years. He has worked in both the nuclear and pharmaceutical industries for BNFL, SmithKline Beecham, as well as Roche Products and Glaxo. He is a member of the GAMP Industry Board, GAMP European Steering Committee, and is cochair of the UK GAMP Forum. He chairs the ISPE Good Engineering Practice and Calibration Special Interest Groups, which have produced ISPE good practice guides. He also previously chaired the GAMP special interest group that produced the GAMP Good Practice Guide on Validating Process Control Systems (VPCS).

Contact: mark.foss@boehringer-ingelheim.com

ROBERT FRETZ (HOFFMANN-LA ROCHE)

Robert is Head of Process Automation and MES at F. Hoffmann-La Roche Limited based at Basel, Switzerland. He has a degree in chemical engineering from Swiss Federal Institute of Technology (ETH Zurich) and joined Hoffmann-La Roche more than 30 years ago. Robert is presently responsible for process automation and system integration in all chemical, biotechnical, and galenical manufacturing sites of the Hoffmann-La Roche Pharmaceuticals Division and leads the corporate manufacturing execution systems (MES) program. He has a wide international experience in all levels of control and automation projects from instrumentation to the enterprise level. Many of these projects included computerized system validation and US CFR Part 11 electronic record and signature compliance. He has also coauthored the Hoffmann-La Roche corporate guideline on process automation qualification. More recently, Robert has been concentrating on developing a collaborative system architecture and electronic batch record systems supporting business processes for drug product release.

Contact: robert.fretz@roche.com

LUDWIG HUBER, PHD (LABCOMPLIANCE)

Ludwig is Director and Chief Editor of Labcompliance, the global online resource for validation and compliance. Before, he worked for Agilent Technologies for more than 20 years as compliance program manager. He is author of several computer compliance books including "Validation and Qualification in Analytical Laboratories" and "Validation of Computerized Analytical and Networked Systems." He has published more than 100 papers on validation and compliance and 21 CFR Part 11 and regulatory contributes to international conferences on the topic of computer compliance.

Contact: ludwig_huber@labcompliance.com

LOUISE KILLA (LOGICACMG)

Louise is a senior IT consultant within the pharmaceutical sector at LogicaCMG, specializing in the validation of commercial and distribution supply chain applications. Her expertise covers various aspects of GxP software development and delivery of different computer systems from R&D through to commercial distribution. She joined LogicaCMG in 1997 and has over 10 years of experience in software development, quality systems, and project management. She received a masters degree in transportation from the University of Wales College at Cardiff. She is an ISO 9001 lead auditor, a member of the ISPE, and an active member of GAMP Europe. Louise sadly died as this book was being revised.

BOB MCDOWALL (MCDOWALL CONSULTING)

Bob has over 23 years experience in the validation of computerized systems in regulated environments, was the editor of the first book on LIMS in 1987, and is the author of the book on "validation of chromatography data systems" in 2005. He has written the Questions of Quality column in LC-GC Europe since 1993 and the Focus on Quality column for Spectroscopy magazine since 2000. Bob is also the 1997 LIMS awardee for teaching and advances to the subject. Bob has been principal of McDowall Consulting since 1993, specializing in process improvement and validation of computerized systems in GxP environments. He has a PhD in forensic toxicology and fifteen years of experience of analytical chemistry.

Contact: rdmcdowall@btconnect.com

BARBARA NOLLAU (ABBOTT VASCULAR)

Barbara is a Director of Quality Services at Abbott Vascular, responsible for validations, reliability engineering, supplier quality, microbiology, and document management. She has 26 years of experience and increasing responsibility in the pharmaceutical and medical device industry, spanning the areas of manufacturing, quality assurance and compliance, IT, and information services. She holds a BA in communications from Cabrini College and has completed graduate work in the areas of adult education and pharmaceutical engineering at Penn State University and California State University, respectively. Barbara is an ASQ-certified manager of quality and organizational excellence and a member of PDA and ISPE. She is a member of the GAMP Americas Steering Committee and is serving as a commissioner on the ISPE Professional Certification Commission. She is a member of the Technology Advisory Board for LSIT and of the Editorial Review Board of The Journal of Validation Technology and the Journal of GxP Compliance and has published numerous papers and presented extensively on validation, 21 CFR Part 11, and related topics.

Contact: barbara.nollau@av.abbott.com

ARTHUR (RANDY) PEREZ, PHD (NOVARTIS PHARMACEUTICALS)

Randy Perez currently holds the position of Executive Expert, IT Quality Assurance for Novartis Pharmaceuticals. His responsibilities at Novartis include a wide range of IT compliance issues, such as GxP, Sarbanes-Oxley, and data privacy. He serves on several global Novartis teams dealing with computer systems compliance issues and has authored many of the firm's global GxP compliance policies. During his 26-year tenure at Novartis, he has developed a broad range of experience, working as a chemistry group leader in process research, managing a chemical manufacturing process validation program, and running a QA validation group for pharmaceutical operations. Randy was a member of the PhRMA Computer Systems Validation Committee from 1995 to 1999 and was instrumental in the formation of GAMP Americas when that group started in 2000. From 2002 to 2008, he was chairman of GAMP Americas and has been a member of the global GAMP Council since 2002. He initiated and led the GAMP special interest group who produced the GAMP Good Practice Guide on Global Information Systems in 2005, and was part of the core team that led the development of GAMP®5, published in 2008. Randy has been a speaker and a course leader at numerous conferences in the United States and Europe and has been published in industry journals and textbooks. In 2005 he was elected to the ISPE International Board of Directors and was elected to the office of board secretary in 2008.

Contact: arthur.perez@novartis.com

CHRIS REID (INTEGRITY SOLUTIONS)

Chris has worked in life science industries for over 15 years, prior to which he was a computerized system development engineer. Chris is the Managing Director of Integrity Solutions Limited, providers of quality and compliance services to life science industries. Chris currently works with leading global organizations developing and implementing quality and compliance solutions including defining and implementing strategic quality initiatives, implementing corporate quality policies and standards, skills development, and system validation. Chris works with a variety of organizations and disciplines including IS/IT, engineering, business, QA, and suppliers. Chris has worked across pharmaceuticals, biotechnology, medical devices and cosmetic industries, and all regulatory domains including clinical, laboratories, manufacturing, distribution, and pharmacovigilance. Chris is a current member of the GAMP European Steering Committee and contributed to the development of GAMP®5 and a variety of GAMP good practice guides.

Contact: creid@integrity-solutions.co.uk

TOM RYAN (BOSTON SCIENTIFIC)

Tom is Principal Engineer for validation for Boston Scientific Corporation based in Ireland. He is an engineering graduate of the University of Limerick and joined Boston Scientific in 1996. Tom has been involved in several major validation projects from process to product. He is a member of the ISPE and the Institute of Engineers of Ireland (IEI) and has presented on validation-related forum topics. He holds an MSc in pharmaceutical validation from the European College of Validation and in management and technology from the Open University.

Contact: ryant1@bsci.com

MICHAEL SCHNEIDER, PHD (HOFFMANN-LA ROCHE)

Michael is MES Project Manager for F. Hoffmann-La Roche based at Basel, Switzerland. He is currently managing a project for the introduction of a MES in a newly built aseptic filling plant. Previously, he headed the automation and MES user team in the investment project for a large biotechnology plant. Michael graduated with a PhD degree in chemical engineering of the Swiss Federal Institute of Technology (ETH Zurich) in 2004. In the same year, he joined Roche as technical operations trainee where he was responsible for several quality management projects.

Contact: michael.schneider@roche.com

ROBERT STEPHENSON, PHD (PFIZER)

After successfully completing a PhD in physics, Rob joined the Boots Company Technical Services Group in 1977 and, since then, has worked in various roles within the pharmaceutical and personal product sectors for companies such as Eli Lilly, Unilever and Coty. He joined Pfizer in 2000 as member of their quality unit operating within the IT group where his responsibilities included coordinating the manufacturing site's initiative to achieve 21 CFR Part 11 compliance and authoring their IT quality management system. Robert is currently regulatory systems team leader, Sandwich, U.K. During his career, Rob has had experience with the implementation and operational control of a wide range of applications including MRPII, document management (EDMS), laboratory management (LIMS and CDMS), and CAPA/change management systems. He is a member of the GAMP Europe Steering Committee and has contributed material to GAMP®5 and the new GAMP Good Practice Guide on "A Risk-Based Approach to Operation of GxP Computerised Systems."

Contact: Robert.Stephenson@Pfizer.com

ANTHONY TRILL (RETIRED SENIOR INSPECTOR, MHRA)

Tony Trill served for 24 years with the United Kingdom's Medicines Inspectorate [known now as the Medicines and Healthcare Regulatory products Agency (MHRA)], retiring in July 2008. His career in the Medicines Inspectorate required him to conduct regular inspections of companies against requirements of the U.K. Medicines Act and EU directives related to GMP (manufacturing), GDP (wholesaling and distribution), GMP investigational medicinal products (IMP) (for clinical trials), and "specials" (industry and NHS). Since 1988 he also had leadership responsibility within the agency for GMP standards and inspection guidance relating to computerized systems. Prior to joining the Medicines Inspectorate, Tony had worked for more than 18 years for three multinational pharmaceutical companies in research and development, formulation and process development, manufacturing, quality assurance, and technical services in professional, technical, and management roles. His industrial experience covered a broad spectrum of product types, processes, and responsibilities, and he has regularly spoken at conferences and published many articles on the topic of computer compliance. He is a member the GAMP European Steering Committee and led for a time the Pharmaceutical Inspection Cooperation Scheme's (PICS) Expert Circle for Computerised Systems, which developed a guideline for use by regulatory authorities around the world entitled "Good Practices for Computerised Systems in Regulated 'GXP' Environments" (Ref. PI 011-1) adopted in 2003. Most recently, Tony, who holds a BSc (Hon) in pharmacy from the University of Aston, an MSc in pharmaceutical technology from the University of London (Chelsea, now Kings College), and an OU diploma in digital computing, led the general GMP revision process for the Orange Guide and represented MHRA on the revision panel for the redrafting of GMP chapter 4 (Documentation) and annex 11 (Computerised Systems), which is still ongoing. He also facilitated a recent EU/PICS regulatory benchmarking audit of MHRA's Inspection and Standards Division and assisted with internal quality and systems improvement initiatives. He retains membership of IQA's IRCA lead assessor/auditor panel and is eligible as an EC qualified person.

Contact: quality.assured@ymail.com

GUY WINGATE, PHD (GLAXOSMITHKLINE)

Guy is a Senior Quality Director at GSK and has held a range of operational and corporate roles in GSK's global manufacturing and supply. These include overall responsibility for quality for one of GSK largest manufacturing sites (covering topicals, orals, inhalations, steriles, biopharmaceuticals, and vaccines), responsibility for corporate QA technology strategy, leading a major revision to the GSK corporate GMP Quality Management System, and overall responsibility for computer validation standards and implementation. Guy has worked with a number of regulatory authorities including FDA and MHRA since the early 1990s to help them develop regulatory requirements and inspection strategy (e.g., coauthor of highly influential ISPE white paper on risk-based approach to electronic records and signatures, which lead to the proposed revision of U.S. 21 CFR Part 11 regulation). He was the task team leader for GAMP®4 and GAMP®5 and has chaired the governing GAMP Council since 2000. A well-known speaker on computer validation, Guy has previously published three popular validation books with Interpharm Press. He is also visiting lecturer for the University of Manchester's MSc course on pharmaceutical engineering advanced technology (PEAT) and Institute of Dublin's MSc course on validation sciences. He is a chartered engineer, a member of the ISPE, a member of the PDA, and a member of the Institution of Engineering and Technology (IET). Guy holds BSc, MSc, and PhD from University of Durham in computing, advanced electronics, and engineering science, respectively. He is widely published in journals and books and regularly chairs and speaks at validation conferences in the United Kingdom and Europe. In 2008 Guy was elected to become a member of the International Board of Directors of the ISPE.

Contact: guy.wingate@gsk.com

Abbreviations

4GL	Fourth Generation Language
ABAP	Advanced Business Application Program (SAP R/3)
ABB	Asea Brown Boveri
ABO	Blood Groups: A, AB, B, O
ABPI	Association of the British Pharmaceutical Industry
ACDM	Association for Clinical Data Management
ACRPI	Association for Clinical Research in the Pharmaceutical Industry
ACS	Application Configuration Specification
A/D	Analog to Digital
ADE	Application Development Environment
AGV	Automated Guided Vehicle
AIX	Advanced Interactive eXecutive, a version of UNIX produced by IBM
ALARP	As Low As Reasonably Practical
ANSI	American National Standards Institute
API	Active Pharmaceutical Ingredient
APV	Arbeitsgemeinschaft für Pharmazeutische Verfahrenstechnik
AQAP	Association of Quality Assurance Professionals
ASAP	Accelerated SAP R/3 application development methodology
ASCII	American Standard Code for Information Interchange
ASTM	American Society for Testing and Materials
AUI	Application User Interface
BARQA	British Association for Research Quality Assurance
BASEEFA	British Approvals Service for Electrical Equipment in Flammable Atmospheres
BASIC	Beginners All-purpose Symbolic Instruction Code
BCD	Binary Coded Decimal
BCS	British Computer Society
BGA	Bundesgesundheitsamt (German Federal Health Office)
BIOS	Basic Input Output System
BIRA	British Institute of Regulatory Affairs
BMS	Building Management System
BNC	Boyonet Neil Concelman
BOM	Bill of Materials
BPC	Bulk Pharmaceutical Chemicals
BPR	Business Process Re-engineering
BS	British Standard
b/s	bits per second
BSI	British Standards Institution
CA	Certification Agency
CAD	Computer Aided Design
CAE	Computer Aided Engineering
CAGR	Compound Annual Growth Rate
CAM	Computer Aided Manufacturing
CANDA	Computer Assisted NDA (United States)
CAPA	Corrective And Preventative Action
CASE	Computer-Aided Software Engineering
CBER	Center for Biologies Evaluation and Research, FDA
CCTA	Central Computer and Telecommunications Agency

CD	Compact Disk
CDDI	Copper Distributed Data Interface
CDER	Center for Drug Evaluation and Research, FDA
CDMS	Clinical Database Management System
CDRH	Centre for Devices and Radiological Health
CD-ROM	Compact Disk — Read Only Memory
CD(-RW)	Compact Disk — Rewritable
CDS	Chromatography Data System
CE	Communauté Européene (EU Medical Device Mark)
CE	Capillary Electrophoresis
CEFTC	Chemical European Federation Industry Council
CENELEC	European Committee for Electrotechnical Standardization
CFR	United States Code of Federal Regulation
CGM	Computer Graphics Metafile
cGMP	Current Good Manufacturing Practice
CHAZOP	Computer Hazard and Operability Study
CIM	Computer Integrated Manufacturing
CIP	Clean In Place
CISPR	International Special Committee on Radio Interference (part of IEC)
CMM	Capability Maturity Model
CO	Costing
COBOL	Common Business Oriented Language
COM	Component Object Model
COQ	Cost of Quality
COTS	Commercial Off-The-Shelf
CPG	Compliance Policy Guide (United States)
CPP	Critical Process Parameter
CPU	Central Processing Unit
CQA	Critical Quality Attribute
CRC	Cross Redundancy Check
CRM	Certified Reference Material
CROMERR	Cross-Median Electronic Reporting and Record-Keeping
CSA	Canadian Standards Association
CSV	Computer System Validation
CSVC	Computer Systems Validation Committee (of PhRMA)
CTQ	Critical to Quality
CV	Curriculum Vitae
DAC	Digital to Analog Converter
DACH	German-speaking countries of Germany (D), Austria (A), and Switzerland (CH)
DAD	Diode Array Detector
DAM	Data Acquisition Method
DAT	Digital Audio Tape
DBA	Database Administrator
DBMS	Database Management System
D-COM	Distributed Component Object Model
DCS	Distributed Control System
DDMAC	Division of Drug Marketing, Advertising and Communications
DECnet	Digital Equipment Corporation Network
DIA	Drug Information Association
DLL	Dynamic Link Library
DLT	Digital Linear Tape
DoH	U.K. Department of Health
DOS	Disk Operating System
DPMO	Defects Per Million Opportunities
DQ	Design Qualification
DR	Design Review
DRP	Distribution Requirement Planning
DSL	Digital Subscriber Line

DSP	Digital Signal Processing
DVD	Digital Video Disk
DXF	Data Exchange File
EAM	Engineering Asset Management
EAN	European Article Number
EBRS	Electronic Batch Record System
EC	European Community
EDI	Electronic Data Interchange
EDMS	Electronic Document Management System
EEC	European Economic Community
EEPROM	Electronically Erasable Programmable Read Only Memory
EFPIA	European Federation of Pharmaceutical Industry Association
EFTA	European Free Trade Association
EIA	Electronics Industries Association
EISA	Extended Industry Standard Architecture
ELA	Establishment License Application
ELD	Engineering Line Diagram
EMC	Electro-Magnetic Compatibility
EMEA	European Medicines Evaluation Agency
EMI	Electro-Magnetic Interference
EMS	Engineering Management System
ENCRESS	European Network of Clubs for Reliability and Safety of Software
EOLC	Environmental/Operation Life Cycle
EPA	U.S. Environmental Protection Agency
EPROM	Electronic Programmable Read Only Memory
ERD	Entity Relationship Diagram
ERES	Electronic Records, Electronic Signatures
ERP	Enterprise Resource Planning
ESD	Electro-Static Discharge
ESD	Emergency Shutdown
EU	European Union
FAT	Factory Acceptance Testing
FATS	Factory Acceptance Test Specification
FAX	Facsimile Transmission
FDA	U.S. Food and Drug Administration
FD&C	U.S. Food, Drug, and Cosmetics Act
FDDI	Fiber Distributed Data Interface
FDS	Functional Design Specification
FEFO	First Expired First Out
FFT	Fast Fourier Transform
FI	Finance
FIFO	First In-First Out
FM	Factory Mutual Research Corporation
FMEA	Failure Mode Effect Analysis
FORTRAN	Formula Translator
FS	Functional Specification
FTE	Full-Time Employee
FT-IR	Fourier Transform — Infrared
FTP/IP	File Transfer Protocol/Internet Protocol
GALP	Good Automated Laboratory Practice
GAMP	Good Automated Manufacturing Practice
GB	Giga-Byte
GC	Gas Chromatography
GCP	Good Clinical Practice
GDP	Good Distribution Practice
GEP	Good Engineering Practice
GERM	Good Electronic Record Management

GIGO	Garbage In, Garbage Out
GLP	Good Laboratory Practice
GMA	Gesellschaft Me Meβ- und Automatisierungstechnik
GMP	Good Manufacturing Practice
GPIB	General Purpose Interface Bus
GPP	Good Programming Practice
GUI	Graphical User Interface
GxP	GCP/GDP/GLP/GMP
HACCP	Hazard Analysis and Critical Control Point
HATS	Hardware Acceptance Test Specification
HAZOP	Hazard and Operability Study
HDS	Hardware Design Specification
HIV	Human Immunodeficiency Virus
HMI	Human Machine Interface
HP	Hewlett-Packard
HPB	Canadian Health Products Branch Inspectorate
HPLC	High Performance Liquid Chromatography
HPUX	Hewlett-Packard UNIX
HSE	U.K. Health and Safety Executive
HTML	Hyper Text Markup Language
HVAC	Heating, Ventilation, and Air Conditioning
IAPP	Information Asset Protection Policies
IBM	International Business Machines
ICH	International Conference on Harmonization
IChemE	U.K. Institution of Chemical Engineers
ICI	Imperial Chemical Industries
ICS	Integrated Control System
ICSE	U.K. Interdepartmental Committee on Software Engineering
ICT	Information and Communications Technologies
ID	Identification
IEC	International Electrotechnical Commission
IEE	U.K. Institution for Electrical Engineers
IEEE	Institute of Electrical and Electronic Engineers
IETF	Internet Engineering Task Force
IIP	Investors in People
IKS	Swiss Agency for Therapeutic Products (also known as SwissMedic)
IMechE	U.K. Institution for Mechanical Engineers
INS	Instrument File Format
InstMC	U.K. Institution for Measurement and Control
InterNIC	Internet Network Information Center
I/O	Input/Output
IP	Index of Protection
IP	Ingress Protection
IP	Internet Protocol
IPC	Industrial Personal Computer
IPC	In-Process Control
IPng	IP Next Generation
IPR	Intellectual Property Rights
IPSE	Integrated Project Support Environment
IPv4	Internet Protocol version 4
IPv6	Internet Protocol version 6
IPX	Internet Packet eXchange
IQ	Installation Qualification
IQA	U.K. Institute of Quality Assurance
IRCA	International Register of Certificated Auditors
IS	Intrinsically Safe
ISA	Industry Standard Architecture bus (also known as AT bus)
ISA	Instrument Society of America

ISM	Industrial, Scientific, and Medical
ISO	International Standards Organization
ISP	Internet Service Provider
ISPE	International Society for Pharmaceutical Engineering
IT	Information Technology
ITIL	Information Technology Infrastructure Library
ITT	Invitation to Tender
IVRS	Interactive Voice Recognition System
IVT	Institute of Validation Technology
JAD	Joint Application Development
JETT	North American Joint Equipment Transition Team
JIT	Just In Time
JPEG	Joint Photographic Experts Group
JPMA	Japanese Pharmaceutical Managers Association
JSD	Jackson Development Method
KOSEISHO	Ministry of Health and Welfare (Japan)
KPI	Key Performance Indicator
KT	Kepner Tregoe
LAN	Local Area Network
LAT	Local Area Transport, a DEC proprietary Ethernet protocol
LC	Liquid Chromatography
LIMS	Laboratory Information Management System
L/R	Inductance/Resistance Ration
MAU	Media Attachment Unit
MASCOT	Modular Approach to Software Construction, Operation, and Test
MB	Mega-Byte
Mb/s	Mega bits per second
MC	Main cross connect room
MCA	Micro Channel Architecture
MCA	U.K. Medicines Control Agency
MCC	Motor Control Center
MD	Message Digital, an algorithm to verify data integrity
MDA	U.K. Medical Device Agency
MDAC	Microsoft Data Access Components
MES	Manufacturing Execution System
MHLW	Japanese Ministry for Health, Labor, and Welfare
MHRA	U.K. Medicines and Healthcare products Regulatory Authority
MHW	Japanese Ministry for Health and Welfare
MIME	Multipurpose Internet Mail Extension
MIS	Management Information System
MM	Materials Management
MMI	Man Machine Interface (see HMI)
MMS	Maintenance Management System
MODEM	Modulator-Demodulator Units
MPA	Swedish Medical Products Agency
MPI	Manufacturing Performance Improvement
MPS	Master Production Schedule
MRA	Mutual Recognition Agreement
MRP	Materials Requirements Planning
MRP II	Manufacturing Resource Planning
MRM	Multiple Reaction Monitoring
MSAU/MAU	IBM's Multi-Station Access Unit (Token Ring hubs)
MTTF	Mean Time To Failure
NAMAS	U.K. National Measurement Accreditation Service
NAMUR	Normenarbeitsgemeinschaft fiir Me β- und Regelungstechnik
NATO	North Atlantic Treaty Organization
NDA	U.S. New Drug Application
NetBEUI	NetBIOS Extended User Interface

NetBIOS	Network Basic Input/Output System
NIC	Network Interface Card
NIST	National Institute of Standards and Technology
NIR	Near Infra-Red
NMR	Nuclear Magnetic Resonance
NOS	Network Operating System
NPL	National Physics Laboratory
NSA	U.S. National Security Agency
NT	New Technology
NTL	National Testing Laboratory
OCR	Optical Character Recognition
OCS	Open Control System
OECD	Organisation for Economic Co-operation and Development
OEM	Original Equipment Manufacturer
OICM	Swiss Office Intercantonal de Controle des Medicaments
OLE	Object Linking and Embedding
O&M	Operation and Maintenance
OMM	Object Management Mechanism
OOS	Out Of Specification
OQ	Operational Qualification
OS	Operating System
OSI	Open System Interconnect
OTC	Over The Counter
OTS	Off The Shelf
OWASP	Open Web Application Security Project
PAI	Pre-Approval Inspection
PAR	Proven Acceptable Range
PAT	Process Analytical Technology
PC	Personal Computer
PCI	Peripheral Component Interconnect
PCX	Graphics File Format
PDA	Parenteral Drug Association
PDA	Personal Digital Assistant
PDF	Portable Document Format
PDI	Pre-Delivery Inspection
PhRMA	Pharmaceutical Research and Manufacturing Association
PIC	Pharmaceutical Inspection Convention
PIC/S	Pharmaceutical Inspection Co-operation Scheme
PICSVF	U.K. Pharmaceutical Industry Computer System Validation Forum
PID	Proportional, Integral, Derivative (Loop)
P&ID	Process Instrumentation Diagram
PIR	Purchase Item Receipt
PKI	Public Key Infrastructure
PLC	Programmable Logic Controller
PMA	Pharmaceutical Manufacturers Association
POD	Proof of Delivery
PP-PI	Production Planning — Process Industries
PQ	Performance Qualification
PQG	Pharmaceutical Quality Group (part of IQA)
PRINCE2	Projects In Controlled Environments 2
PRM	Process Route Maps
PSI	Statisticians in the Pharmaceutical Industry
PSU	Power Supply Unit
PTB	Physikalische-Technische Bundesanstalt
PTT	Public Telephone and Telecommunications
PV	Performance Verification
QA	Quality Assurance
QC	Quality Control

QM	Quality Management
QMS	Quality Management System
QP	European Union Qualified Person
QS	Quality System
QSIT	FDA Quality System Inspection Technique
QTS	Quality Tracking System
RAD	Rapid Application Development
RAD	Role Activity Diagram
RAID	Redundant Array of Inexpensive Disks
RAM	Random Access Memory
RCCP	Rough Cut Capacity Planning
RCM	Reliability Centered Maintenance
R&D	Research and Development
RBE	Review By Exception
RDB	Relational Database
RDT	Radio Data Terminal
RF	Radio Frequency
RFI	Radio Frequency Interference
RFID	Radio Frequency Identification
RFP	Request for Proposal
RH	Relative Humidity
ROM	Read Only Memory
RP	German Federal Ministry for Health
RPharmS	U.K. Royal Pharmacy Society
RPN	Risk Priority Number
RSA	Rivest, Shamir, Adleman Public-Key Cryptosystem
RSC	U.K. Royal Society of Chemists
RTD	Radio Data Terminal
RTF	Rich Text Format
RTL/2	Real-Time Language, Version 2
RTM	Requirements Traceability Matrix
RTSASD	Real-Time System-Analysis System-Design
SAA	Standards Association of Australia
SAM	Software Assessment Method
SAP	Systems, Applications, Products in Data Processing (Company)
SAP R/3	An ERP system developed by SAP
SaRS	U.K. Safety and Reliability Society
SAS	Statistical Analysis System
SAT	Site Acceptance Testing
SATS	System Acceptance Test Specification
SCADA	Supervisory Control and Data Acquisition
SCR	Source Code Review
SD	Sales and Distribution
SDLC	Software Development Life Cycle
SDS	Software Design Specification
SEI	Carnegie Mellon University's Software Engineering Institute
SFC	Sequential Function Chart
SGML	Standard Generalized MarkUp Language
SHA	Secure Hash Algorithm
SHE	Safety, Health & Environment
SIP	Sterilization In Place
SKU	Stock Keeping Unit
SLA	Service Level Agreement
SLC	System Life Cycle
SM	Section Manager
SMART	Specific, Measurable, Achievable, Recorded, Traceable
SMDS	Software Module Design Specification
S/MIME	Simple Multipurpose Internet Mail Extension

SMS	Microsoft's System Management Server
SMTP	Simple Mail Transfer Protocol
SNA	Systems Network Architecture
SOP	Standard Operating Procedure
S&OP	Sales and Operations Planning
SOUP	Software Of Unknown Pedigree
SPC	Statistical Process Control
SPICE	Software Process Improvement Capability d'Etermination
SPIN	Software Process Improvement Network
SPSS	Statistical Product and Service Solutions
SQA	Society of Quality Assurance
SQAP	Software Quality and Productivity Analysis
SQL	Software Query Language
STARTS	Software Tools for Large Real-Time Systems
STD	Software Technology Diagnosis
STEP	STandard for Exchange of Product model data in ISO 10303
STP	Shielded Twisted Pair
StRD	Statistical Reference Dataset
SWEBOK	Software Engineering Body of Knowledge
TC	Terminal Cross connect room
T&C	Threats and Controls
TCP	Transmission Control Protocol
TCP/IP	Internet Protocol/Transmission Control Protocol
TCU	Temperature Control Unit
TIA	Telecommunications Industry Association
TIFF	Tagged Image File Format
TIR	Test Incident Report
TGA	Australian Therapeutic Goods Administration
TÜV	Technischer Überwachungs-Verein
UAT	User Acceptance Testing
UCITA	U.S. Uniform Computer Information Transactions Act
U.K.	United Kingdom
UL	Underwriters Laboratories Inc.
ULD	Utility Line Diagrams
UPC	Universal Product Code
UPS	Uninterruptible Power Supply
URL	Universal Resource Locator
URS	User Requirement Specification
U.S.	United States (of America)
U.S.A.	United States of America
USD	United States Dollars
UTP	Unshielded Twisted Pair
UV	Ultra Violet
VBA	Visual Basic
VDS	Validation Determination Statement
VDU	Visual Display Unit
VMP	Validation Master Plan
VMS	Virtual Memory System
VP	Validation Plan
VPN	Virtual Private Network
VR	Validation Report
VSR	Validation Summary Report
V-MAN	Validation Management
WAN	Wide Area Network
WAO	Work Station Area Outlet
WAP	Wireless Application Protocol
WFI	Water For Injection
WHA	World Health Agreement

WHO	World Health Organisation
WIFF	Waveform Interchange File Format
WIP	Work In Progress
WMF	Windows Metafile Format
WML	Wireless Markup Language
WORM	Write Once, Read Many
WWW	World Wide Web
WYSIWYG	What You See Is What You Get
XML	Extensible Markup Language
Y2K	Year 2000

Contents

Contributors

John Andrews   Andrews Consulting Enterprises Ltd.

Peter Bosshard   Hoffmann-La Roche

Roger Buchanan   Eli Lilly

Winnie Cappucci   Bayer Pharmaceuticals

Mark Cherry   AstraZeneca

Chris Clark   NAPP Pharmaceuticals

Peter Coady   Pfizer

Tony de Claire   APDC Consulting Ltd.

Christopher Evans   GlaxoSmithKline

Joan Evans   ABB

Mark Foss   Boehringer-Ingelheim

Robert Fretz   Hoffmann-La Roche

Ludwig Huber   Labcompliance

Louise Killa   LogicaCMG

Bob McDowall   McDowall Consulting

Barbara Nollau   Abbott Vascular

Arthur (Randy) Perez   Novartis Pharmaceuticals

Chris Reid   Integrity Solutions Limited

Tom Ryan   Boston Scientific

Michael Schneider   Hoffmann-La Roche

Robert Stephenson   Pfizer

Guy Wingate   GlaxoSmithKline

1 | Introduction

Computer systems support billions of dollars of pharmaceutical and healthcare sales revenues. The pharmaceutical and healthcare industry has increasingly used computers to support the development and manufacturing of their products. Within research environments, computer systems are used to speed up product development, reducing the time between the registration of a patent and product approval and, hence, optimizing the time available to manufacture a product under a patent. Computer systems are also used to support routine supply of medicinal products to improve manufacturing performance, reduce production costs, and improve product quality. It is important that these systems are fit for purpose from a business and regulatory perspective. Regulatory authorities treat a lack of regulatory computer system compliance as a serious GxP deviation. Pharmaceutical and healthcare companies need a balanced, proactive, and coordinated strategy that addresses short-, medium-, and long-term and internal and external needs and priorities. This book aims to provide practical advice and guidance on how to do this on the basis of extensive industry experience and with reference to the latest regulatory developments and industry trends.

REGULATED COMPUTER SYSTEMS

Computer systems that are used to support regulated processes in the pharmaceutical, biological, and healthcare industry need to comply with the expectations of the regulatory authorities overseeing development, manufacture, and supply. Applicable regulations are commonly referred to collectively as *GxP regulations* and include but are not limited to

- Good Clinical Practices (GCP)—dealing with clinical trials,
- Good Laboratory Practices (GLP)—dealing with laboratory operations,
- Good Manufacturing Practices (GMP)—dealing with manufacturing of products,
- Good Distribution Practices (GDP)—dealing with distribution and onward warehousing, and
- Electronic Records and Electronic Signatures.

The scope of regulated computer systems includes systems used to manage data or support decisions that are submitted or subject to review by regulated authorities whether they are being submitted because they are required or used as supporting information.

The list of potential computer system applications that must comply with regulatory requirements is extensive. Some examples of computer system applications impacted are illustrated in Figure 1.1. Such diversity, coupled with the increasing scope of applications, has led to some individual regulators to suggest that a simpler approach would be to declare that *all* computer systems used within a manufacturing environment, whatever their application be, must comply with regulatory requirements.

Business Case

Investments in computer systems are made on the basis of realizing a benefit. Claimed benefits in the business case might include the following:

- Building in quality controls to ensure that the process is followed correctly, reducing human error and the need to conduct manual checks.
- Standardization of practices to build consistent ways of working. This is increasingly important for large multisite manufacturing organizations that are rationalizing their operations.

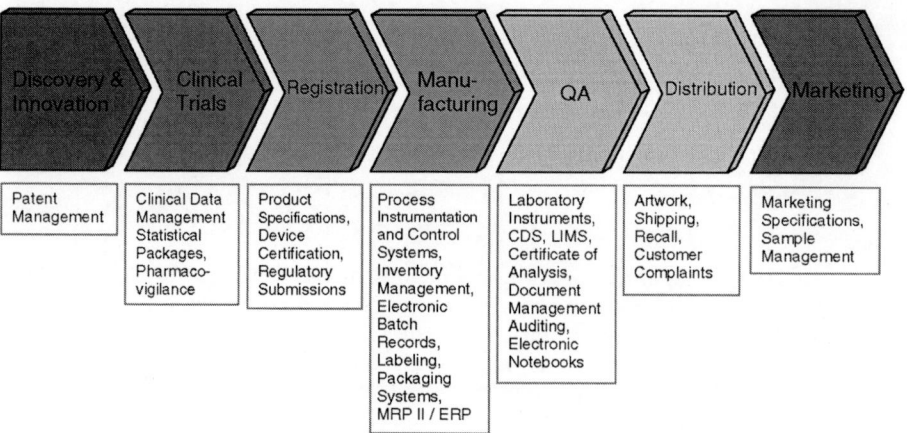

Figure 1.1 Regulated computer system applications.

- Reducing the cost of sales by removing non-value-added activities (e.g., quality inspections, exception handling, rework, and scrap).
- Speed-up of process cycle times by reducing wait times and by improved scheduling.
- Elimination of duplicate effort by establishing electronic master records and thus avoiding the need for the presentation of information in various paper formats, each of which must be controlled.
- Replacing systems that are no longer supported by their suppliers.
- Addressing regulatory deficiencies in processes or existing computer systems.

More than one benefit should normally be delivered. Computer systems should not be implemented solely for regulatory compliance; operational benefits should always be explored as well.

Regulatory Compliance

The pharmaceutical industry is subject to GxP regulations such as the World Health Organization's (WHO) resolution WHA 22.50; the European Union's (EU) Directive 2003/94/EC; the Japanese Manual on Computer Systems in Drug Manufacturing; the U.S. Code of Federal Regulations Title 21, Parts 210 and 211; and Medicinal Products, Part 1 of the Australian Code of Good Manufacturing for Therapeutic Goods. Other countries have their own regulatory requirements, which tend to have aligned expectations, although some specific differences do exists.

GMP is enforced on the ground by the national regulatory authorities. Well-known GMP regulatory authorities in the pharmaceutical industry include the U.S. Food and Drug Administration (FDA), the U.K. Medicines and Healthcare products Regulatory Agency (MHRA), the Australian Therapeutic Goods Administration (TGA), Health Canada, and the Japanese Ministry of Health, Labor and Welfare (MHLW). The regulatory authorities can prevent the sale of any product in their respective country if they consider its manufacture not to be GxP compliant. To pharmaceutical and healthcare companies, GxP is nothing less than a license-to-operate matter.

The process of demonstrating GxP has become known as *validation* and involves

> "establishing documented evidence which provides a high degree of assurance that a specific process will consistently produce a product meeting its pre-determined specifications and quality attributes (1)."

This definition has been widely adopted, albeit with minor modifications, by the various GxP regulatory authorities around the world and embraces facilities, equipment, systems, and processes.

A general awareness within both the industry and the regulatory community of the need to validate computer systems began to emerge formally in 1979, when the United States introduced GxP regulatory legislation that specifically referred to automated equipment (2). In 1983 the FDA issued what became known as the "Blue Book" (because of the color of its cover) guidance to inspectors on what was reasonable to accept as evidence for computer systems compliance (3). The first widely publicized FDA citation (a formal written regulatory criticism of a perceived noncompliance with the regulations) for computer system noncompliance was issued in 1985.

The issue of computer systems validation assumed a high profile in Europe in 1990 when two European pharmaceutical manufacturers did not satisfy computer compliance expectations and were temporarily prohibited from exporting their products to the United States. EU requirements for computer systems compliance were issued a few years later in 1993 and can be found in Annex 11 in the EU GMPs (4). Once formal requirements had been defined, European regulators started to make significant noncompliance findings at various pharmaceutical companies.

Australia and Canada have subsequently adopted Annex 11. Meanwhile, Japan issued its first computer compliance guideline in 1993 (5). Computer compliance, however, did not receive the industry profile in Japan that it did in the United States and Europe until the need for electronic record and electronic signature controls emerged.

The United States was the first country to issue specific requirements for electronic records and electronic signatures—21 CFR Part 11 was published in 1997 (6). It caused much debate in terms of whether its expectations were reasonable and practical. The U.S. FDA subsequently published industry guidance that applied a risk-based approach to controls (7). Similar expectations were published by PIC/S for use by Australian, Canadian, and EU regulatory authorities (8). The Japanese MHLW issued electronic record/signature requirements in 2005 on the basis of various discussions with industry (9). Both U.S. and EU regulations are currently being updated to address these developments.

Table 1.1 provides a general summary of regulatory expectations for computer validation together with prominent regulations for different sectors of pharmaceutical and healthcare industry.

Cost of Compliance

Computer systems can account for significant capital and operating costs. Such assets deserve careful management. The cost of compliance will depend on the breadth and depth of work undertaken.

Current industry good practice suggests that compliance activities should account for not more than 10% of project delivery costs (10). Anecdotal stories still circulate relating to examples of excessive cost of validation. The root cause behind such high costs tends to be poor project management, and when things go bad, people tend to want to find a scapegoat.

The primary factor that has typically driven the amount of validation has been the nature of the computer system involved [whether standard application, configurable package, or custom-built (bespoke) application]. This approach has been driven by ISPE/GAMP®, specifically the use of software categories to derive basic compliance approach. Other factors have had a limited influence on the amount of compliance work conducted, although risk-based thinking has started to have an impact.

While current best practice represents a huge improvement in industry practice from the 1990s, there is still much that can be done to reduce costs further. As discussed later in this chapter, it should be possible to reduce the cost of compliance below 5% of project costs while at the same time improving the standard of compliance.

Operational Dividend

Compliant computer systems should yield an operational dividend (10). A survey of over three hundred applications by Weinberg Associates suggests that maintenance savings within four years generally offset the investment in validation. An example of such a maintenance dividend is illustrated by a production planning system at ICI that adopted the principles of validation for about half of its 800 computer programs. Management, halfway through the

Table 1.1   Regulatory Expectations Concerning Computer Validation

Industry sector	Validation expectation	Prominent regulations
Food	Requirement for process validation on effective controls at critical points, and equipment maintenance. Computer validation is recommended.	U.S. Code of Federal Regulations 21 CFR Part 110, and EU Directive 93/43/EEC
Dietary supplements	Requirement for process characterization and equipment maintenance. Computer validation is recommended.	U.S. Code of Federal Regulation 21 CFR Part 111, 113 and 114), and EU Directive 2002/46/EC
Cosmetics	Requirement for computer validation.	COLIPA and U.S. Code of Federal Regulation 21 CFR Part 700, 701, 710, 720, and 740)
OTC medicines	Process and computer validation is required.	U.S. Code of Federal Regulation 21 CFR Part 210 and 211, and EU Directive 2003/94/EEC
Prescription pharmaceuticals	Process and computer validation is required.	U.S. Code of Federal Regulation 21 CFR Part 210 and 211, and EU Directive 2003/94/EEC
Biological and biopharmaceuticals	Process and computer validation is required.	U.S. Code of Federal Regulation 21 CFR Part 600, 606, and 610, and EU Directive 2003/94/EEC
Blood processing and products	Process and computer validation is required.	U.S. Code of Federal Regulation 21 CFR Parts 800 and 820, EU Directive 2005/02/EEC
Medical devices	Process and computer validation is required for medical device and its production.	U.S. Code of Federal Regulations 21 CFR Part 820 (Quality Systems Regulation), and EU 2007/47EEC coupled with CE marking.

Abbreviations: COLIPA, European Cosmetic, Toiletry, and Perfumery Association; EU, European Union.

project, abandoned the quality approach because there was no perceived project benefit. The total operational life of the system was later examined. It was found that maintenance costs for the software adopting the principles of validation were about 90% *less* than the comparable costs for the remainder of the software. Similar data has been found for MRP II/ERP systems. With poor-quality software typically accounting for 50% to 60% of maintenance costs, validation really does make good business sense.

Cost of Failure

The failure to comply with regulatory expectations can have significant financial implications. Noncompliance incidents may lead to delays in the issue of a license or its withdrawal and thus an embargo on the distribution of their product in the relevant marketplace (e.g., the United States).

The financial consequences of correcting deficient validation might, at first sight, seem small compared with the typical investment of US$800,000,000 to bring a new drug to market (11). The real financial impact is the loss in sales revenue arising from a prohibition to market the product. For top-selling drugs in production, citations for noncompliance by GxP regulatory authorities can cost their owner upward of US$2,000,000/day in lost sales revenue. One FDA Warning Letter cost the pharmaceutical manufacturer concerned over US $200,000,000 to replace and validate a multisite networked computer system.

The trick is to cost-effectively conduct sufficient work to ensure GxP compliance, but, as illustrated in Figure 1.2, there is always debate over how much is sufficient to fulfill the regulator's expectations. Excessive validation may increase confidence in regulatory compliance, but it does not come cheap. Inadequate validation may actually be cheaper, but, in the long term, the cost of regulatory noncompliance could be devastating. This book aims to clarify how much validation is enough and to suggest how it can be cost-effectively organized and also to discuss areas of debate.

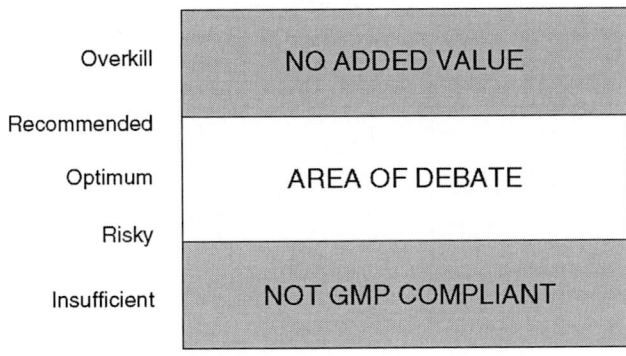

Figure 1.2 How much validation is enough?

There are numerous stakeholders with an interest in successful GxP inspection outcome. GxP noncompliance is likely to reduce public confidence in the pharmaceutical and healthcare industry and the offending company. Political pressures may result in improved industry practices, influence the inspection approaches and methods of regulatory authorities, and/or review the acceptability of compliance standards and guides. The standing of regulatory authorities may be affected if they fail to notice noncompliance that lead directly to substandard drug products being distributed and used. Associated legal liabilities may arise for both the regulator and offending company. The company's corporate reputation may take years to recover. Drug sales are likely to fall as the consumers of the products, the prescribers, and their patients, become uneasy about the quality and/or consistency of supply. Market confidence in the offending company will be reduced, and the brand image will be tarnished. The reputation of distributors may also be undermined through "guilt by association" with the offending company. Insurance premiums for the company are likely to increase. As an overall consequence, the jobs of all those working for the company and associated suppliers will be less secure.

Pharmaceutical and healthcare companies should beware of persistent regulatory noncompliance. Repeated failure to address specific topics will deeply concern regulatory authorities. It is likely that they will escalate the severity of regulatory action taken. Company reputation will suffer—not just with the regulatory authority concerned but also with industry, and reputation can take a long time to reestablish. There may be fines too that need to be considered. In exceptional circumstances, U.S. law allows *"disgorgement,"* whereby a percentage of sales revenue can be taken as part of a penalty for persistent and serious noncompliance that impacts public health.

TRENDS IN TODAY'S COMPUTING ENVIRONMENT

Computer systems share some basic hardware and software characteristics that must be understood to appreciate the quality and compliance issues discussed in this book.

First, it is important to grasp that the proportion of hardware costs is, on the whole, reducing as a percentage of the lifetime cost of a computer system, as illustrated in Figure 1.3. Computer systems are now less reliant on custom (bespoke) hardware than was the case and now largely consist of an assembly of standard components that are then configured to meet their business objective. Standard software products are more readily available than ever before, although these products are often customized with bespoke interfaces to enable them to link into other computer systems. Software products are also becoming larger and more sophisticated. With the use of ever larger and more complex software applications, the task of maintenance has also increased, especially as many vendors of commercial software have acquired the habit of releasing their products to market while significant numbers of known errors still remain. The effective subsequent management of defect correction patch installations and other code changes can be challenging.

Second, while software shares many of the same engineering tasks as hardware, it is nevertheless different (2). The quality of hardware is highly dependent on design,

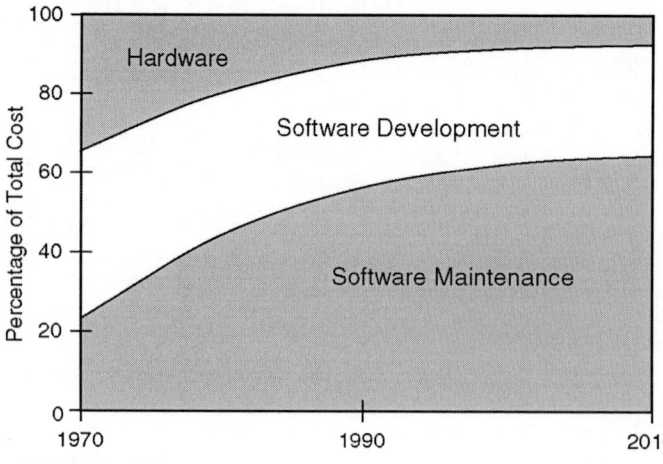

Figure 1.3 Changing proportions of software and hardware costs.

development, and manufacture. The quality of software is also highly dependent on design and development, but its manufacturer consists merely of replication, a process whose validity can be easily verified. For software, the hardest part is not replicating identical copies but rather the design and development of software being copied to predetermined specifications. Then again, software does not wear out like hardware. On the contrary, it often improves over time as defects are discovered and corrected.

Third, another related characteristic of software is the speed and ease with which it can be changed. Most applications today are based on configurable networked packages. Changes are made by super users resetting configuration parameters. This characteristic often gives users the impression that software problems can be easily corrected. This is true at one level, but what appear simple changes to address a problem in one part of the application can also create unwanted changes to the functionality elsewhere in the application. A similar issue exists with custom-built (also known as bespoke) software applications. These include macros developed for configurable packages. The complexity of custom software means that even professional programmers can find modifications to code in one place causing unexpected and very significant defects to mysteriously arise elsewhere in the program.

Fourth, the scope of different types of computer systems is changing. For instance, more analytical equipment is being moved from the laboratory to be in-line or at-line with manufacturing operations. With this move, responsibility has also changed. Typically, production and engineering are responsible for these systems once colocated instead of QA when the analytical equipment was located in a separate laboratory facility. Meanwhile, many process control applications are now residing on IT platforms and as a consequence IT is responsible rather than engineering for their upkeep. At the same time, some functionality once commonly found as an integral part of IT applications has shifted and is now supported by IT infrastructure. A good example is security access and password control. Many organizations now have a single user log-on and shared password control for multiple applications, which is facilitated through the desktop environment.

Finally, focus is now being put on the capability of an automated process rather than the computer system itself. This emphasis promotes a better understanding and control of business processes, which, in terms of developing, manufacturing, and supplying drug and healthcare products, also supports the principal aim of GxP regulations to ensure product safety, quality, and efficacy.

PROBLEMS IMPLEMENTING COMPUTER SYSTEMS
Industry Experience

The Standish Group have been reviewing trends in success rates of computer projects for almost 15 years. Currently, about one-third of computer system projects are on-time, without

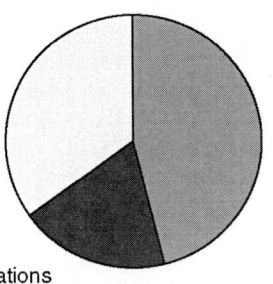

35% of Applications
are completed on time,
to budget, and met
user requirements

46% of Applications
are late, overspent,
and/or did not meet
user requirements

19% of Applications
were not delivered

Figure 1.4 Project outcomes. *Source*: From Ref. 12.

overspend, with all planned functionality present, and almost one-fifth of applications are never delivered at all (Fig. 1.4). Even if a project appears superficially to have been successful, that does not imply that the application it delivered will be used long enough to repay the investment. Many business cases for new software-related products require a return on investment within three years, but in practice, a high proportion of systems have a shorter life than this. Computer technology and business IT strategies tend to be very dynamic. In such a changing environment, applications tend to be quickly labeled as inflexible and/or redundant, requiring their replacement by new and more sophisticated systems long before they have paid for themselves. And so the challenge to more robustly implement and exploit computer systems continues.

Quality management systems (QMSs) must be mature and robust to mitigate the risk of unsuccessful projects. Factors that critically determine the likelihood of success of computer projects are summarized in Table 1.2. Lack of user input can almost be guaranteed to result in an incomplete user requirement specification, a foundation of sand on which only the shakiest edifice can be built. Those with only general skills should not be deployed on critical tasks such as quality assurance, testing, and project management. Specific technical expertise and proven competence are also required for handling new technology. Good team communications are also vital. Ineffective supplier management and poor relationships with subcontractors aggravate an already weak technical base. Software upgrades are often conceived to rectify hardware deficiencies rather than seeking a more appropriate hardware solution. Teams often mistakenly focus on innovation rather than on containing cost and risk.

Gaining acceptance of quality management is vital (13). Both the heart and the mind of senior management must be behind the use and benefits of QMSs. There is more than enough evidence around to make the case that QMSs work. Without clear leadership at an executive level, however, it will be almost impossible to overcome statements like "We do not have time for paperwork," "We cannot afford the luxury of procedures," "Too structured approach will undermine flexibility, slow projects down, and increase costs," and "The concept is good, and we hope to use it some day but not just yet."

Simply monitoring quality performance is just not adequate. The effectiveness of QMSs should be actively managed, and performance improvement opportunities should be seized. Business benefits should more than compensate for any investment in quality. Senior

Table 1.2 Factors That Affect Project Success Based on Standish 2006 Data

Successful project	Unsuccessful project
User involvement	Lack of user input
Executive management support	Lack of executive support
	Unrealistic schedule pressure
Clear statement of requirements	Changing requirements
Proper planning	Lack of project management expertise
Routine use of standard tools and infrastructure	Lack of structured methodology
Assignment of competent staff	Lack of resources
	Technological incompetence

management, system owners, project managers, and anyone else involved with computer projects need to appreciate this. This book will help explain what needs to be done to successfully achieve quality and compliance of computer systems in the pharmaceutical and healthcare industries.

Regulatory Observations

There has been a steadily decreasing number of computer noncompliance observations during regulatory inspections since the turn of the millennium. Indeed, in recent years, computer systems–related issues are no longer even in top ten issues cited by regulators as compliance deficiencies (14). This is a complete reversal of the trend seen in the 1990s when concern around computer compliance was stoked by anxiety regarding electronic records, electronic signatures, and Y2K millennium bugs.

It has been suggested that the steady decline in computer noncompliance observations is in part due to the loss of experienced inspectors with expert knowledge in computer systems from the major regulatory authorities. This might be true in part, but probably a bigger impact is down to a change in approach and focus during inspections.

Computer systems rarely come under targeted and detailed inspection these days without a specific reason (sometimes referred to as a "for cause" inspection). This is because FDA and other regulatory authorities are adopting "quality systems"–based inspections that are focused on examining processes rather starting from the bottom up looking at particular SOPs, pieces of equipment, and physical systems. A recent legal challenge has affirmed that noncompliant computer systems do not automatically compromise the safety of medical products so long as an effective quality approach is taken (15). Similar feedback had been given to FDA in response to a computer compliance–related Warning Letter (16).

Nowadays, there is also a general acceptance of a risk-based approach to better understand the potential risk posed to patients and consumers. While some regulatory authorities have been operating this way for many years, it is a much newer concept for others. The real risk posed by computer systems may have been overestimated in the past because due account was not taken of other controls that would detect and stop any adverse impact reaching patients and consumers.

Finally, there is much more conscious attempt to link noncompliance observations on computer systems back to actual regulatory requirements (sometimes referred to as predicate rules). This has stopped personal interpretation by individual inspectors and encouraged a much more consistent approach to identifying noncompliances.

Current noncompliance observations for computer systems tend to be made in one of four following areas:

- Access control: lack of security to prevent unauthorized access or use by untrained personnel
- Data integrity: lack of controls to preserve data integrity including, where appropriate, use of audit trails for electronic records
- Change control: during projects but also perhaps more importantly during operation and maintenance of computer systems
- General validation: missing or deficient specification and verification of computer systems

These reflect basic good practice deficiencies. These regulatory requirements are not new, and due diligence should ensure that they are addressed. Senior management have no excuse not to provide sufficient resource and time to validate computer systems that need it.

LATEST INDUSTRY EXPECTATIONS

A significant shift came into effect in 2003 with the prominence given to the acceptability of risk-based approach to quality and compliance. The aim was to improve the effectiveness and efficiency of activities by ensuring that controls were commensurate with the risk posed to patients. It was also hoped that this new emphasis would promote early adoption of new

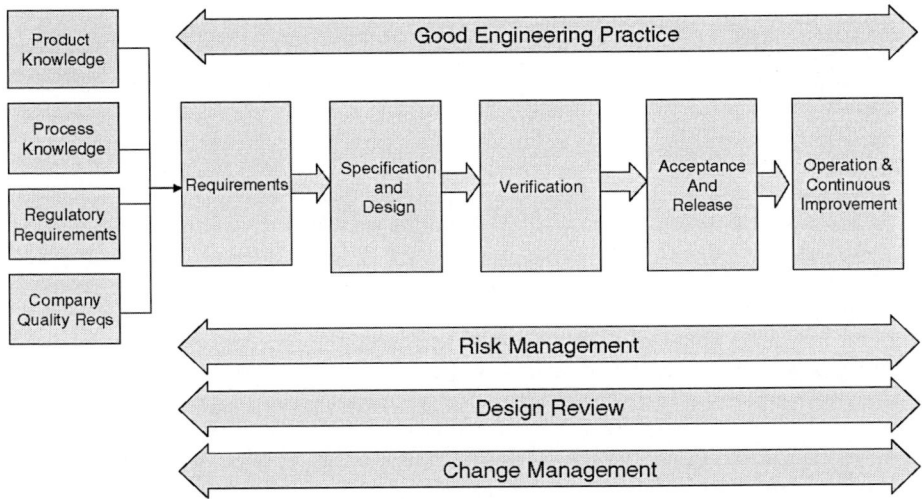

Figure 1.5 Specification and verification model. *Source*: From Ref. 19.

technological advances (17). Early examples of specific regulatory guidance mentioning risk management for computer systems include

- FDA Final Guidance on Scope and Application for 21 CFR Part 11 Electronic Records; Electronic Signatures (8) and
- PIC/S Good Practices for Computerized Systems in Regulated GxP Environments (9).

Early attempts at risk management were not particularly sophisticated. Risks were often oversimplified, and their management was based on opinion rather than scientific knowledge. Perhaps most crucially, risk management did not typically take into consideration the ultimate impact risks posed to patients. The need for further regulatory guidance was clear. In response, in 2005 the ICH published the consensus expectations from U.S. FDA, European EMEA, and Japanese regulatory authorities for quality risk management (18). The general principles and process presented were consistent with medical device practices and directly applicable to computer systems. Q9 was incorporated into U.S., EU, and Japanese GMPs in 2006.

The impact of taking a risk-based approach to facilities, equipment, and systems was subsequently explored by ASTM and resulted in the publication of high-level guidance in 2007 (19). ASTM promoted the model in Figure 1.5 that emphasized risk management as occurring throughout the life cycle and not just as a discrete activity. The work also drew attention to the need to remove non-value-added activities (such as repeating supplier testing and creating unnecessary documentation) that had emerged over the past decades as part of industry custom and practice.

Although GAMP®4 Guide (20) provided some basic guidance on risk management for computer systems, it was not until 2008, when GAMP®5 was published (21), that detailed practical guidance became available, showing how to adopt and exploit a risk-based approach. GAMP has emerged through its various editions as the unofficial consensus standard for computer compliance between regulators, pharmaceutical companies, and suppliers around the world. GAMP guidance covers IT applications, analytical laboratory applications, process control systems, and IT infrastructure (networks and desktop environments) governed by GxP regulations. The main GAMP®5 Guide is complemented by a number of Good Practice Guides providing further guidance on particular topics:

- Electronic data archiving (22)
- Testing GxP systems (23)
- Global information system and compliance (24)

- IT infrastructure control and compliance (25)
- Laboratory computerized systems (26)
- Risk-based approach to electronic records and signatures (27)
- Legacy systems (28)
- Process control systems (29)
- Calibration management (30)

Another key industry update to occur in 2008 was the publication of the second edition of the Red Apple Report first published 20 years earlier (31). Although the document is titled for computerized systems used in nonclinical assessments, it effectively provides general GCP/GLP guidance. The contents cover full life cycle requirements including electronic records management but does not deal with risk management.

A number of regulatory updates are in progress to remove inadvertent dependencies that constrain use of new technology, ensure that controls for electronic records and signatures are practical, and facilitate the use of a risk-based approach. U.S. GMPs (21 CFR 211) for finished pharmaceutical were updated in 2008 to allow the second person checks to be replaced by automated verification. The U.S. regulation on electronic records and signatures (21 CFR 11) is expected to be updated to align to FDA industry guidance issued in 2003 (32). Amendments have also been proposed to EU GMPs to clarify requirements for handling records and documentation and the introduction of particular requirements for electronic records (33).

A summary of regulatory authority expectations and industry guidance is presented in appendixes 1 and 2, respectively.

NEW 21ST CENTURY PARADIGM

This book explains how to implement the new paradigm that has emerged over recent years that will lead to a step change reduction in the cost of computer compliance while at the same time improving compliance (35–37). The paradigm is based on a combination of reinforcing well-established principles along with the introduction of new principles that may have been implied in the past but never formally stated and exploited. GAMP®5 endeavors to define these principles (21), and these are summarized in Figure 1.6.

The focus of regulated organizations should be on having a fundamental understanding of their products and the pharmaceutical science behind them. They cannot be experts in every aspect of their operations including the use of computer systems. QMSs should exist to govern supply chain activities for medicinal products and suppliers. The management and control of these activities can be scaled using science-based risk management appropriate to the medicinal product concerned. Regulated organizations need to leverage supplier expertise in

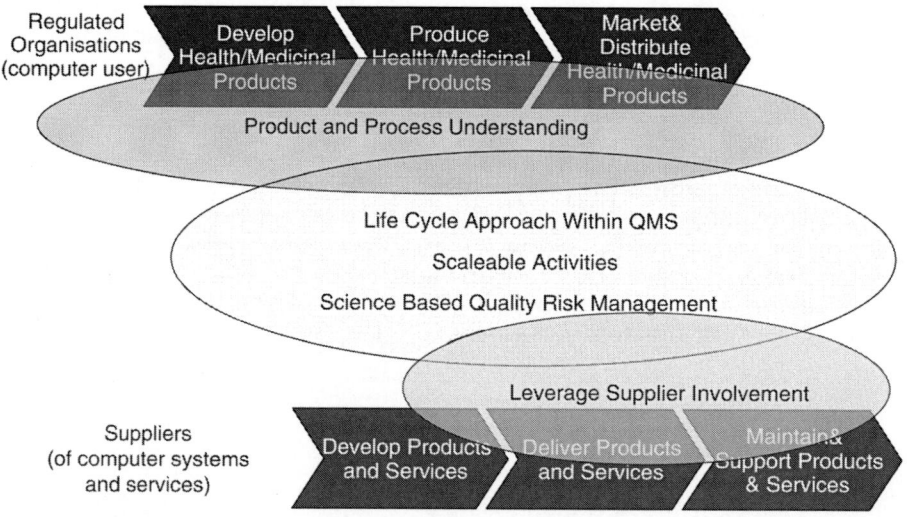

Figure 1.6 New 21st century paradigm. *Source*: From Ref. 21.

the products and services used. After all, the focus of supplier organizations is to have a fundamental understanding of the computer systems and associated services they offer to customers.

Process and Product Understanding

Efforts to ensure a computer system is fit for its intended purpose should focus on criticality to patient/consumer safety, product quality, and data integrity. An understanding of the process being automated is fundamental to the determining system requirements. Manufacturing this will involve the identification of critical quality attributes (CQAs) and critical process parameters (CPPs). It is only through this understanding that we can have complete confidence in identifying critical functionality and data. Incomplete process understanding will hinder effective and efficient compliance. Less critical computer systems, or aspects of those systems, will have to be treated as critical (and incur otherwise unnecessary attention) because it is not possible to distinguish between them from more critical systems.

The GAMP®5 Guide suggests that effort is focused on computer systems that support the following critical processes (21):

- Generate, manipulate, or control data supporting regulatory submissions
- Control critical parameters for preclinical and clinical, development, and manufacturing
- Control or provide information for product release
- Control information required in case of product recall
- Control adverse event or complaint handling
- Support pharmacovigilence

It is important to appreciate the context in which a computer system is automating a business process. For instance, manufacturing control systems are more critical in sterile manufacturing compared with production of topical ointments. Both are important in their own right, but one has a much greater potential patient impact.

Science-Based Quality Risk Management

Quality risk management should be based on clear understanding and potential impact on patient safety, product quality, and data integrity. Risks are relative and very difficult to quantify in an absolute sense. High risks might be defined as having the potential for severe harm to users of medicinal products (e.g., death, hospitalization, long-term effects impeding daily life). Low risks might be defined as having potential for inconsequential impact on patient safety. A computer system cannot pose a greater risk than the business process being automated.

Risk assessments need to use structured and repeatable processes that are data driven. It is no longer acceptable to just organize a team or committee meeting to discuss and announce a risk assessment outcome. Such meetings were acceptable but could not be relied on to reproduce the same decisions. Reliance on good judgement in the absence of any data is not acceptable.

Risk management must reduce risks to an acceptable level. Quantitative and qualitative techniques may be used. It is important to realize that it may not be possible or practical to eliminate a risk. This is quite acceptable so long as the efficacy, safety, and quality of medicinal products are not compromised. Implemented controls must be monitored to verify their sustained effectiveness.

Life Cycle Approach Within Quality Management System

The complete life cycle of a computer system from conception to system retirement should be subject to management and control. It is well known that procedures must be established to ensure that a consistent approach is taken. Implied, but not explicitly stated before, is the expectation that these procedures exist within a QMS. A separate QMS should exist for computer systems, rather a single QMS should exist, which covers all activities of an organization including the development and use of computer systems.

The QMS should promote a holistic approach to compliance. An integrated approach should be taken to computer systems with the equipment and facilities they manage and control. Common processes such as change control and CAPA should not have separate procedures for computer systems but rather be governed by shared procedures. Pharmaceutical and healthcare companies should also ensure that their QMS expects and aligns to a complementary QMS in use within their supplier organizations.

A healthy QMS must promote continuous improvement. For computer systems, this is achieved through determining root cause of incidents and deviations (with emphasis on using data to support investigations), implementing corrective and preventative actions, and conducting regular periodic reviews. Improvements can only be implemented under suitable change management.

Scalable Activities

Compliance activities through a computer system life cycle should be scaled where it makes sense to stop otherwise redundant work. The degree of scaling will depend on the attributes of an individual computer system.

Scaling has conventionally almost always been limited to just considering system novelty. However, there are several other important factors worthy of consideration. Key attributes to take account of include

- Impact of system and its data on quality, safety, and efficacy of medicinal product taking into account severity of potential harm to patient/consumer that could result,
- System complexity (size and architecture),
- System novelty (degree of standardization and number of other organizations actively using the system), and
- Supplier capability (project management, technical, and level of reliance on temporary staff and third-party organizations).

Scaling may not always be practical. The time and effort to determine and implement scaling may outweigh the benefits of just validating the whole system to a higher-than-needed level for simple computer systems. No adverse regulatory comment will be received in these situations provided the computer systems are compliant.

Leveraging Supplier Involvement

Pharmaceutical and healthcare companies should recognize and exploit the capability of supplier organizations. There is no need to duplicate supplier activities and documentation unless they fall short of basic good practice standards.

On the whole, there has been insufficient recognition of supplier competence despite industry guidance on the topic. This has resulted in pharmaceutical and healthcare companies unnecessarily rewriting system development documents provided by the supplier and repeating verification activities conducted by the supplier. The reason given is typically founded on the perceived need for records to conform to their document conventions (e.g., layout, format, style, and approvals).

Planning must clearly define the respective responsibilities of pharmaceutical and healthcare companies and their suppliers and how these responsibilities fit together into an integrated approach. The degree of reliance on supplier activities and documentation must be justified. This is usually done by leveraging the outcome of supplier assessments, which may include supplier audits depending on risk.

Pharmaceutical and healthcare companies cannot abdicate their accountability for compliance. In additional formal contracts, some form of monitoring supplier performance is expected. Integrated plans, shared responsibility for risk mitigation, and common escalation processes for issues will all improve the effectiveness of customer-supplier relationships. If activities and documents are substandard, then the issues must be resolved in a timely manner. It may be necessary in extreme circumstances to replace suppliers and ensure that they are not used again. Incompetent suppliers should, of course, not be selected in the first place. The compliance of a computer system must not be compromised.

Subject Matter Experts

Roles should be performed by those with appropriate skills, training, and experience. In this context, individuals are referred to as subject matter experts (SMEs) because of their knowledge and capabilities. Departmental responsibilities should not take precedence over the capabilities of staff to perform various roles. The only exception is the quality unit, which, according to regulatory requirements, must provide impartial oversight of compliance activities.

Too often in the past, roles have been aligned to departments even if the staff in those departments were not capable of fulfilling those duties. The same situation often existed in relation to suppliers. Suppliers can perform the role of SMEs.

Terminology

The choice of terminology to describe these principles must not interfere with effective implementation of these principles. While it is up to individual companies to agree to their own lexicon, it is important to recognize the benefits of using industry standard terminology to promote common understanding between pharmaceutical companies, suppliers, and regulatory authorities.

This book uses established regulatory definitions for computer validation that are consistent with the key principles behind the 21st century paradigm. GAMP®5 has introduced a framework of specification and verification to act as umbrella terms for the various activities involved in defining a computerized system and the various review and testing activities involved in qualifying a computerized system as fit for its intended purpose. This framework aligns both with established use of qualification and validation terminology and also the use of specification and verification championed by ASTM (19).

GOOD PRACTICE

Quality Assurance

The achievement of quality in a product should be based on the adoption of good practices. Neither these, in relation to computer systems or not, nor the concept of quality were invented by the pharmaceutical and healthcare industry. Good computer practices existed long before pharmaceutical and healthcare industry regulations required their application. The basic underlying premise is that quality cannot be tested into computer systems once developed. On the contrary, it must be built in right from the very start. Defects are much cheaper to correct during the early stages of system development compared to leaving them to be weeded out just before release or, perish the thought, by disaffected customers. The additional cost generated by ensuring that the system is sound at every stage in its development from conception to testing is far less than the cost and effort of fixing the computer system afterward, not forgetting the hidden losses suffered through customer disaffection. So do not wait until the end of the project to put things right!

Sir John Harvey-Jones, Chairman of the former industrial chemicals giant ICI, summed this up pithily, "The nice thing about not planning is that failure then comes as a complete surprise." This said, it is important to appreciate that planning does not come naturally to many people, and the temptation to jump into code development before the groundwork of properly defining requirements has been completed often proves irresistible. This tendency can be exacerbated by managers expecting too much too soon. A degree of self-discipline is required as shortcutting the quality process will almost certainly wreak havoc later on.

Duty of Care

Like other financial governing bodies around the world, the London Stock Exchange requires pharmaceutical and healthcare companies to comply with laws and regulations including those dealing with GxP and consumer protection (38). Collectively, these are often portrayed as the exercise of a "duty of care" through operating in a responsible and reasonable manner. This duty of care embraces the use of computer systems because of the crucial role they play in determining the quality of drug and healthcare products. Failure in this duty of care implies, at best, negligence or incompetence; at worst, it may infer fraud and may subject senior personnel to prosecution and legal penalty. However, the net of responsibility falls wider than the pharmaceutical or healthcare company involved. It may jointly or individually include equipment hardware suppliers, software suppliers, system integrators, and system users. Notwithstanding this, GxP regulators hold pharmaceutical and healthcare companies solely accountable for GxP compliance despite the unavoidable supplier dependencies. Examples of matters where such accountability may be cited include deficient design, defective construction, weak or inadequate inspection, incomplete, ambiguous, or confusing user instructions provided by supplier, software installed on an inappropriate hardware platform, inappropriate use of a system, or neglect of operational instructions.

Buyer Beware
Regulatory authorities do not approve computer systems, neither do they agency certify suppliers or consultants. Pharmaceutical and healthcare companies have always been and remain accountable for compliance of the computer systems they use. Audits and certifications, however rigorously applied and conscientiously implemented, are no substitutes for validation.

WIDER APPLICABILITY
Validation is recommended for all computer, control, and laboratory systems that have the potential to seriously affect safety, health, and environmental (SHE) conformance and adherence to other regulatory requirements (e.g., Data Protection Act and financial control legislation such as Sarbanes-Oxley).

REFERENCES
1. FDA. General Principles of Validation. Rockville, MD: Food and Drug Administration, Center for Drug Evaluation and Research, 1987.
2. FDA. General Principles for Software Validation. Final Guidance for Industry and FDA Staff, Food and Drug Administration, Centre for Devices and Radiological Health, 2002.
3. FDA. Guide to Inspection of Computerised Systems in Drug Processing. Technical Report, Reference Materials and Training Aids for Investigators. Rockville, MD: Food and Drug Administration, 1983.
4. European Union. EU GMP Annex 11 Computerised Systems. Guide to Good Manufacturing Practice for EU Directive 2003/94/EC, Community Code Relating to Medicinal Products for Human Use, Vol 4, 2003.
5. Japanese Ministry of Health and Welfare. Guideline on Control of Computerised Systems in Drug Manufacturing. Manual for Control of Computerised Systems in GMP, Audit Manual for Manufacturers of Pharmaceutical Product with Computer Systems, 1993.
6. FDA. Preamble to Electronic Signatures and Electronic Records. Code of Federal Regulation Title 21, Part 11. Rockville, MD: Food and Drug Administration, 1997.
7. FDA. 21 CFR Part 11 Electronic Records; Electronic Signatures—Scope and Application. Guidance for Industry. Rockville, MD: Food and Drug Administration, 2003.
8. Pharmaceutical Inspection Co-operation Scheme. Good Practices for Computerised Systems in Regulated GxP Environments. Pharmaceutical Inspection Convention, PI 011-1, Geneva, 2003.
9. Japanese Pharmaceutical Manufacturers Association. Guideline for the Application of ERES in Production Control and Quality Control for Human Drug Manufacturing, 2002.
10. Wingate GAS. Computer Systems Validation: Quality Assurance, Risk Management and Regulatory Compliance for Pharmaceutical and Healthcare Companies. Boca Raton, FL: Interpharm Press, 2003.
11. Faiella C. Pharmaceutical Industry Report: the Next Pharmaceutical Agenda. Ernst & Young, 2002.
12. Standish Group. Sixth Chaos Report. Available at: www.standishgroup.com, 2006.
13. Bennatan EM. On Time, Within Budget: Software Project Management Practices and Techniques. New York: Wiley Press, 2000.
14. BioQuality. US Versus EU Inspection Results. 2008; 13(1).
15. BioQuality. Validation and Compliant Handling Court Decision Allows Increased Flexibility. 2005; 10(11).
16. Wingate GAS. Computer Compliance: Expectations & Experiences. FDA Process Analytical Technologies Sub-Committee Meeting, Washington, October 23, 2002.
17. FDA. Pharmaceutical CGMPs for the 21st Century: a Risk-Based Approach. Rockville, MD: Food and Drug Administration, 2004.
18. ICH. Quality Risk Management. Q9 Document, Technical Requirements for Registration of Pharmaceuticals for Human Use, International Conference on Harmonisation (ICH). Available at: www.ich.org, 2005.

19. ASTM. E2500-07 Standard Guide for Specification, Design and Verification of Pharmaceutical and Biopharmaceutical Manufacturing Systems and Equipment. American Society for Testing and Materials, 2007.
20. ISPE. GAMP® Guide for Validation of Automated Systems (known as GAMP®4), Tampa, FL: International Society for Pharmaceutical Engineering. Available at: www.ispe.org, 2001.
21. ISPE. GAMP®5: Risk-Based Approach to Compliant GxP Computerised Systems. Tampa, FL: International Society for Pharmaceutical Engineering. Available at: www.ispe.org, 2008.
22. ISPE. GAMP® Good Practice Guide: Electronic Data Archiving. Tampa, FL: International Society for Pharmaceutical Engineering. Available at: www.ispe.org, 2007.
23. ISPE. GAMP® Good Practice Guide: Testing GxP Systems. Tampa, FL: International Society for Pharmaceutical Engineering. Available at: www.ispe.org, 2005.
24. ISPE. GAMP® Good Practice Guide: Global Information System and Compliance. Tampa, FL: International Society for Pharmaceutical Engineering. Available at: www.ispe.org, 2005.
25. ISPE. GAMP® Good Practice Guide: IT Infrastructure Control and Compliance. Tampa, FL: International Society for Pharmaceutical Engineering. Available at: www.ispe.org, 2005.
26. ISPE. GAMP® Good Practice Guide: Validation of Laboratory Computerized Systems. Tampa, FL: International Society for Pharmaceutical Engineering. Available at: www.ispe.org, 2005.
27. ISPE. GAMP® Good Practice Guide: Risk-Based Approach to Electronic Records and Signatures. Tampa, FL: International Society for Pharmaceutical Engineering. Available at: www.ispe.org, 2005.
28. ISPE. GAMP® Good Practice Guide: Legacy Systems. Tampa, FL: International Society for Pharmaceutical Engineering. Available at: www.ispe.org, 2003.
29. ISPE. GAMP® Good Practice Guide: Validation of Process Control Systems. Tampa, FL: International Society for Pharmaceutical Engineering. Available at: www.ispe.org, 2003.
30. ISPE. GAMP® Good Practice Guide: Calibration Management. Tampa, FL: International Society for Pharmaceutical Engineering. Available at: www.ispe.org, 2001.
31. DIA. Computerized Systems Used in Non-Clinical Safety Assessment: Current Concepts in Validation and Compliance (known as Red Apple II), published by Drug Information Association, 2008.
32. FDA. 21 CFR Part 211 Supplementary Information—Code of Federal Regulation. Rockville, MD: Food and Drug Administration, 2008.
33. European Commission. Draft Annex 11—Computerised Systems (EU Guideline to Good Manufacturing Practices for Medicinal Products for Human and Veterinary Use), Public Consultation Document, Brussels, April 2008.
34. Wingate GAS. GAMP®5 Drivers, Objectives and Benefits. Launch of GAMP®5, ISPE US Conference on Manufacturing Excellence, Florida, February 25–28, 2008.
35. Selby D. Validating Complex Computer Systems. ISPE Turkey Annual Meeting, Istanbul, March 28, 2008.
36. Daw E. Case Study—Effective & Efficient Compliance. Launch of GAMP®5, ISPE European Conference on Innovation, Copenhagen, Denmark, April 7–10, 2008.
37. The Institute of Chartered Accountants in England and Wales. Internal Control: Guidance for Directors on the Combined Code. ISBN 1-84152-010-1, 1999.
38. ISO. ISO 90003: Software Engineering—Guidelines for the Application of ISO 9001:2004 to Computer Software. Geneva, Switzerland: International Organization for Standardization, 2004.

Appendix 1: Body of Knowledge: Regulatory Authority Expectations

Title	Date of Publication	Applies to	Risk Management	Training	Supporting Processes					Project Delivery						Operation & Maintenance											Phase-Out			Electronic Records & Signatures
					Document Management	Change Control	Configuration Management	Self-Inspection	Managing Deviations	Project Initiation & Compliance Determination	User Requirements & Supplier Selection	Design & Development	Coding, Configuration, & Build	Development Testing	User Qualification & Authorisation to Use	Performance Monitoring	Repair & Preventative Maintenance	Upgrades, Bug-Fixes & Patches	Data Maintenance	Backups & Restoration	Archiving & Retention	Business Continuity Planning	Security	Contracts & Service Level Agreements	User Procedures	Periodic Review & Revalidation	Retirement	Replacement	Decommissioning	
FDA Blue Book for Drug Manufacturing	1983	Computerized Systems																													
FDA Software Development Activities	1987	Computerized Systems																													
Australian TGA GMP for Therapeutics Section 900	1990	Computerized Systems																													
EU GMP Guide for Medicinal Products Annex 11	1992	Computerized Systems																													
Japanese MHLW	1993	Computerized Systems																													
Drug Manufacturing Guidance	1993	Computerized Systems																													
OECD GLP for Computerised Systems	1995	Laboratory Systems																													
FDA Computerised Systems for Food Processing	1998	Computerized Systems																													
FDA Medical Device Software Validation	2002	Medical Devices																													
PIC/S Good Practices for GxP Computerised Systems	2003	Computerized Systems																													
FDA Computerised Systems Used in Clinical Investigations	2007	Clinical Data Systems																													

Legend:
- ▒ Mentioned by topic, with or without supplementary guidance
- ☐ Not mentioned

Appendix 2: Body of Knowledge: Industry Guidance

Title	Date of Publication	Applies to	Risk Management	Training	Document Management	Change Control	Configuration Management	Self-Inspection	Managing Deviations	Project Initiation & Compliance Determination	User Requirements & Supplier Selection	Design & Development	Coding, Configuration, & Build	Development Testing	User Qualification & Authorisation to Use	Performance Monitoring	Repair & Preventative Maintenance	Upgrades, Bug-Fixes & Patches	Data Maintenance	Backups & Restoration	Archiving & Retention	Business Continuity Planning	Security	Contracts & Service Level Agreements	User Procedures	Periodic Review & Revalidation	Retirement	Replacement	Decommissioning	Electronic Records & Signatures
			Supporting Processes							Project Delivery						Operation & Maintenance											Phase-Out			
APV Guide to Annex 11	1996	Computerized Systems		▓		▓	▓			▓	▓	▓	▓	▓	▓	▓				▓	▓	▓	▓		▓	▓		▓		
ACDM/PSI Computer Validation in Clinical Research	1997	Clinical Systems	▓	▓		▓	▓			▓	▓	▓	▓	▓	▓			▓				▓	▓		▓	▓				
JPMA GMP ERES Guideline	2002	Laboratory & IT Systems	▓	▓		▓				▓	▓	▓	▓	▓	▓					▓	▓	▓	▓							▓
ISO 90003	2004	Computer Systems	▓		▓	▓	▓		▓	▓	▓	▓	▓	▓			▓		▓											
GAMP® ERS Guide	2005	Computerized Systems	▓	▓	▓	▓	▓			▓	▓	▓	▓	▓	▓		▓	▓	▓	▓	▓	▓	▓	▓	▓				▓	▓
ASTM E2500	2007	Equipment & Systems	▓			▓	▓			▓	▓	▓	▓											▓						
GAMP® 5	2008	Computerized Systems	▓	▓	▓	▓	▓		▓	▓	▓	▓	▓	▓	▓	▓	▓	▓		▓	▓	▓	▓	▓	▓	▓			▓	
DIA Red Apple 2	2008	Clinical Systems		▓	▓	▓						▓	▓					▓		▓	▓	▓	▓		▓	▓	▓			▓

Legend:

☐ Not mentioned   ▓ Mentioned by topic, with or without supplementary guidance

Notes: ACDI/PDI evaluation is based on generic guidance provided and excludes case study examples. ASTM E2500 uses the term 'verification' in preference to 'qualification'. GAMP®5 is complemented by Good Practice Guides covering process control systems, laboratory systems, IT systems and IT infrastructure (see reference list in this chapter). See reference [38] for ISO standard. DIA use the term 'formal testing' in preference to 'development testing'.

2 | Organization and Management

INTRODUCTION

This chapter suggests an approach to the organization and management of computer compliance that satisfies the regulatory accountabilities. However, in presenting this offering, we do not purport to suggest that this is the only acceptable approach. The needs of each pharmaceutical and healthcare company will vary since these depend on many different factors, including company culture, how much work there is to do, and the availability of suitably skilled company and contract staff.

ORGANIZATIONAL RESPONSIBILITIES

Structure and Management

The organizational structures in the enterprise, whatever its size, must be defined and documented. GxP regulatory authorities expect to see an organizational chart with the various roles of different departments clearly specified. Of critical importance in these is the position of quality staff whose role and reporting relationships must not compromise their ability to discharge their responsibilities including the escalation and resolution of quality and compliance issues. Quality departments are typically independent of other functions to ensure impartiality. This will prove crucial if things go awry later on.

Senior management is responsible for ensuring that personnel assigned to compliance work are competent to fulfill their role and for arranging any supplementary training requirements. Evidence that individuals have sufficient education and experience to enable them to undertake their assigned functions will need to be collected and kept ready for presentation when required. Role descriptions for individuals should be prepared and kept up-to-date.

It is not acceptable for senior managers to rely on individuals to fulfill their role without management support. Individuals must be given sufficient authority to fulfill their duties. Duties may be delegated to designated deputies with satisfactory levels of competence. It is recommended that deputies are assigned for critical roles so that working practices are not hamstrung when key staffs are absent.

Quality and Compliance Roles

Pharmaceutical and healthcare companies should appoint a senior management representative with specific responsibility for ensuring that computer compliance requirements are implemented and maintained. This individual, who is often graded as a Director, must wholeheartedly champion the cause of GxP. The authority and responsibility of this senior position should be clearly defined and recorded.

It is very important that this senior manager has the authority to block the release of drug product on the grounds of noncompliance, since this can compromise the quality of drug and healthcare products. Without such a level of authority, the individual will have to rely solely on his or her powers of personal persuasion with others, who may themselves be under acute production or sales pressures to permit the release of drug product. Bitter experience shows that this will seldom be enough. As a result, it has long been taken for granted in the industry that quality managers must have the authority to place an embargo on drug products that they deem to be substandard. The senior manager responsible for GxP must possess a similar level of empowerment; many companies achieve this by placing their GxP personnel within their quality control/assurance management hierarchy.

The senior manager responsible for regulatory compliance is expected to recruit appropriately qualified and experienced staff and ensure that the work conducted is properly and effectively carried out (1). In many organizations, the senior manager responsible for GxP may also have other duties. The company must formally acknowledge and concede that these other duties do not excuse or relieve his or her responsibility for compliance to the GxP regulatory authorities. In the phrase forever associated with President Harry S. Truman, *the buck stops here*!

Sufficient Resources

Senior management must ensure that an adequate number of personnel are available with the necessary qualifications and practical experiences appropriate to their responsibilities. Insufficient personnel will attract regulatory criticism, since this is likely to lead to individuals being burdened with excessive responsibilities and workloads and subjected to inordinate pressure. The consequential risk is that quality will then be compromised in some way. In the worst case situation, senior company executives are subject to potential prosecution if they fail to meet such regulatory expectations (2). Blaming a lack of competent personnel on corporate challenge for minimum staffing levels or reduced training budgets will not be accepted by the regulatory authorities for whom public health comes above company profits.

Governance

The expected role of senior managers is defined by the ISO 9001 standard, which states that management shall

> define and document [the company's] policy and objectives for, and commitment to, quality and ensure that this policy is understood, implemented and maintained at all levels in the organization. (2)

While this is useful, more practical guidance is available. An interpretation of the ISO requirements based on work by Teri Stokes is presented below (3).

1. Establish a work group to define company policy, including a statement of commitment to such policies and objectives, and any associated company plans.
2. Establish a company steering committee to set the company computer compliance strategy, approve the compliance policy, provide oversight, and agree funding models.
3. Establish local site steering committees to prepare an inventory of systems, set priorities, establish site master plans, approve procedures, assign resources, and monitor progress.
4. Develop an awareness/education program for all senior managers and their employees.
5. Monitor progress, priorities, resources, and funding against company objectives and strategy.

Company steering committees should be multidisciplinary teams with representatives from research, production, engineering, quality control/assurance, and business support. Site steering committees should also be multidisciplined with representatives from technical operations, IT, laboratory management, engineering, and quality. Members of both steering committees and supporting work groups should be trained via internal or external courses so that they gain an understanding of the basic principles of computer compliance. Alternatively, members might attend external conference or training event where they could also meet practitioners from other companies. Major conferences are regularly organized by International Society of Pharmaceutical Engineering (ISPE), Parenteral Drug Association (PDA), and Institute of Validation Technology (IVT) throughout Europe and the United States. External consultants with specialist knowledge and experience may also be engaged to support the steering committees and work groups, perhaps assisting with internal training courses.

Recent Inspection Findings

> Inadequate organizational structure to ensure quality system requirements met (Food and Drug Administration (FDA) 483, 2002)
> Failure to have a quality control unit adequate to perform its functions and responsibilities, as required by 21 CFR 211.22, as demonstrated by the number and type of inspectional observations (FDA Warning Letter, 2002)

No standard operating procedure (SOP) delineating QA oversight duties and responsibility for computer systems, networks, and associated servers/workstations (FDA 483, 2006)

COMPLIANCE STRATEGY
Organizational Capability

From the outset the aim must be to achieve compliance in a manner as cost-effective as practicable. Many pharmaceutical companies who have rushed into the remediation of computer systems have subsequently discovered to their cost that inefficient compliance programs are hugely expensive, involving much more work than is really necessary.

Figure 2.1 illustrates three basic compliance strategies by comparing the cost associated with compliance (prospective validation) against the cost associated with noncompliance (the combined impact of retrospective validation and business disruption).

Point *A* in the graph denotes the breakeven point where the cost of noncompliance equals the cost of compliance. This may appear to indicate the ideal amount of compliance effort, an effort that just delivers compliance but constrains cost by going no further. Is this really the ideal that should be aimed at? Compliance requirements enforced by the various regulatory authorities are interpretative, not prescriptive. Further, regulatory inspections are never exhaustive; practical limitations of time and resources mean that inspections can only ever examine a proportion of a company's operation. Therefore, they cannot ensure the exposure of all noncompliances. Assessing where point *A* really is thus inspired guesswork, something of an art rather than a science. Aiming at point *A* but missing it will mean either that compliance is not achieved or that money has been wasted! An alternative strategy that is often adopted after a serious noncompliance has been revealed by a regulatory authority is to aim for point *C*. This point represents an exhaustive (risk-averse) compliance effort in a climate of zero tolerance of any regulatory criticism, however minor. The cost that this point implies is exorbitant and adding little value to the business or indeed to the likelihood of regulatory compliance—most of it is unnecessary overhead. It is useful, then, to consider a third compliance strategy, the *middle way*, whereby a more balanced approach to compliance is adopted. It has been claimed that this type of approach can lead to 40% to 50% cost savings when compared with those incurred at point *C* while still maintaining a sustainable level of regulatory compliance (4). Point *B* can be viewed as a *common sense approach*, erring on the side of caution by being more conservative than that represented by point *A*. There is a broad consensus on the wisdom of setting compliance effort at this point. Its precise location can only be determined by surveying industry practice and monitoring regulatory expectations. Benchmark exercises are not readily available, so a more ad hoc collation of information derived from consultants, new recruits, industry associations, and informal regulatory contacts is often used. As long as the quality of information is adequate, not being tainted by hidden political agendas, it should be possible to arrive at point *B* relatively easily.

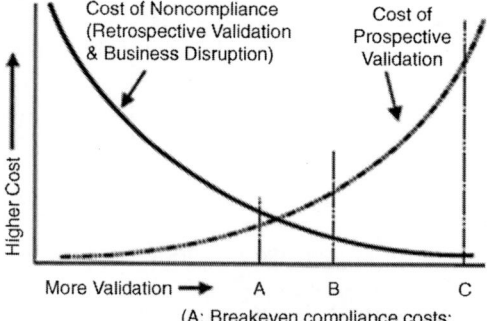

(A: Breakeven compliance costs;
B: Balanced compliance scorecard;
C: Zero tolerance to noncompliance)   **Figure 2.1**   Compliance strategy.

Members of the group establishing a compliance strategy must appreciate the implications of their policy on working practices. For instance, point *B* should not be determined solely on the basis of the cost of compliance but also on its effectiveness, by examining the standards and practices to be employed. Inefficient practice can inflate project overhead costs by up to 30%. This could entirely undermine the potential cost savings associated with adopting a compliance strategy based on point *B*.

Outsourcing and Offshore Partnerships

The outsourcing of systems and services (irrespective of whether it is onshore or offshore) does not change the basic regulatory expectations for computer compliance. Senior management may have an expectation that compliance becomes simpler and less likely to come under inspection. Whether or not this is true is subject to debate. What is clear, however, is that the capability of the outsource partner in terms of appreciating regulatory expectations and implementing robust quality practices will have a significant impact on the level of management and control required by the pharmaceutical company to establish and maintain a state of compliance. The same basic economic model still holds.

POLICY AND PROCEDURES
Establishing Policy Framework

Compliance policies vary greatly between different companies. There are no set rules governing their content or structure, but it will typically cover the following:

- A definition of the overall principles to achieve and sustain compliance
- The scope of policy application
- A statement of commitment
- A definition of who is to be responsible
- A glossary defining the terminology to be used

Both prospective and retrospective validation must be considered for new and existing computer systems, respectively. Figure 2.2 outlines the relationship between prospective and retrospective validation. New systems must be authorized for use, while the continuing use of existing systems must be justified. Once validation is achieved, it must be maintained despite any changes that are made to it. All such changes must be scrutinized through change control and the revised system specifically authorized for use. Periodic reviews must also be performed to ascertain whether an overall revalidation is required, as a result of cumulative system changes over time, regulatory developments, or organizational changes that impact company standards. If the use of a computer system cannot be justified, then it should be decommissioned through a premature retirement. Computer systems will require decommissioning at some time anyway once they reach the end of their useful life.

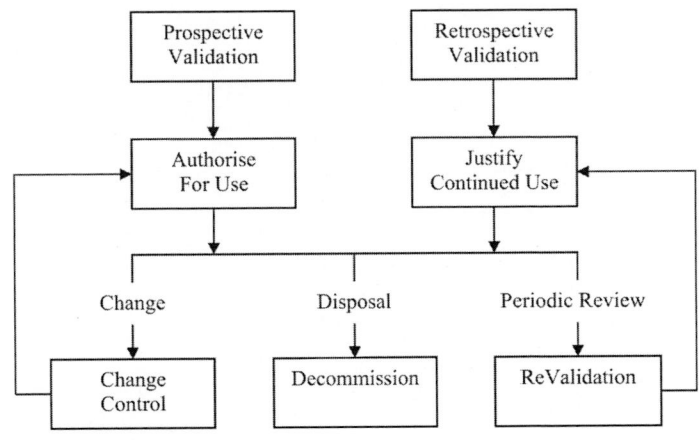

Figure 2.2 Relationship between prospective and retrospective validation.

Key principles for computer systems compliance policy are presented in Appendix 2.A. These principles have been prepared to address the basic requirements of the following regulatory authorities:

- U.S. FDA
- U.K. Medicines Healthcare products Regulatory Authority (MHRA)
- Japanese Ministry of Health, Labour and Welfare (MHLW)
- Australian Therapeutics Goods Administration (TGA).

Experience has shown that a clear, concise compliance policy for computer systems can be achieved within a 5- to 10-page document and produced in a couple of man-months. Typically, most of the effort is spent in consultative meetings to secure a consensus on the content of the policy that should, in any event, confine itself to high-level statements of principle. It should, by its nature, be relatively stable. Nevertheless, it should be periodically reviewed and kept up-to-date.

Standard Operating Procedures

SOPs should be drafted by experienced practitioners who have experience in developing such procedures. The number of SOPs that are required will depend on organizational complexity and the number of the systems that need to be addressed. It may be necessary to hire an external consultant to fulfill this role. He or she should join a team of operational procedure end users as a ghostwriter to aid the development of the procedures. The aim here is not to impose a set of generic procedures that might be in some way foreign to the organization. The goal is rather to tailor the end user's current working practices into compliant procedures with a minimum of change. The end user's personal involvement should ensure that they and their colleagues will readily adopt the procedures without resentment.

About 20 to 25 generic procedures will be needed to cover the life cycle for a computer system. The following list is based on GAMP®5 (5).

Management

- Validation planning
- Supplier audit
- Risk management
- Design review and traceability analysis
- Quality planning
- Validation reporting
- Change control (project and operational)
- Configuration management
- Document management

Project

- User requirements specification (URS)
- Functional specification
- Hardware design specification
- Software design specification
- Software controls
- Testing (qualification/verification)

Operation

- Periodic review
- Service level agreements
- Security
- Performance monitoring
- Record retention, archive, and retrieval

- Backup and recovery
- Business continuity planning
- Decommissioning

Managing electronic records and electronic signatures may be handled with separate SOPs or integrated into the above. An abridged set of procedures will be appropriate for small systems, but supplementary procedures will be needed for larger ones.

Management must approve procedures and any subsequent changes made to them. Approval signatures will be required from at least two individuals representing a quality and technical perspective. Management must then ensure that any deviation from these procedures is properly authorized and documented. Deviations will often be associated with corrective actions arising from internal audit findings; these must also be documented together with the evidence that demonstrates their resolution.

The effort to produce these procedures may require about 40 days from an experienced consultant. Use should be made of industry guidance when developing procedures, for example, GAMP®5 example procedures and IEEE standards. Where existing procedures are being revised to achieve compliance, this estimate of effort could be reduced by about two days per procedure. As recommended above, this should be supplemented with about 20 days of effort across all the procedures, shared among a team of end users. The individuals should contain the core users who are involved in all the procedures, to ensure consistency. Other end users on the team, however, can be seconded for the development of particular procedures in which they have a specific interest, or can contribute a particular skill or competence. For instance, an end user quality representative may wish to be seconded for the development of the security procedure.

To make the preparation of these procedures easier, many organizations are developing document templates and tools to assist practitioners prepare, review, and approve documents in a rapid, quality-conscious fashion.

Many pharmaceutical and healthcare companies find it beneficial to also tailor particular sets of procedures to different types of systems. This may also help ownership and adoption since different types of systems are usually supported by QA/laboratory, engineering, and IT departmental functions. From a technical standpoint too, it is very difficult to make a single set of procedures easy to use while providing a practical level of detail to address the various technical characteristics of different types of computer systems—one size does not readily fit all. As a consequence, typically, there might be four sets of procedures.

- Laboratory applications (e.g., analytical, measurement)
- Control systems (e.g., PLC, SCADA, DCS)
- IT systems (e.g., ERP, MRP II, LIMS, EDMS)
- Computer network infrastructure (e.g., servers, networks, clients)

An additional set for desktop applications (e.g., spreadsheets, databases, and web applications) may be needed, but more typically, these are included within the general scope of IT systems.

There are inevitable interfaces between the application areas of the various sets of procedures and the computer systems to which they apply, as indicated in Figure 2.3. Client-server technology is typically the deciding factor in determining whether control system and laboratory application projects would be better served by IT system procedures. Another example might be that robotic systems used to automate laboratories would be better served by control system procedures.

Recent Inspection Findings

No evidence that the quality policy has been implemented, understood, and maintained by all levels of the organization (FDA Warning Letter, 2001)

Quality system procedures not implemented (FDA 483, 2002)

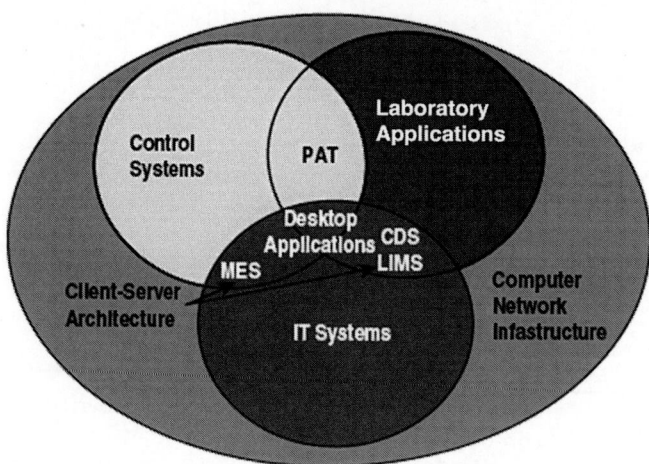

Figure 2.3 Mapping procedures to computer systems.

COMPUTER SYSTEMS INVENTORY

An inventory of computer systems should exist and may be requested during regulatory inspections. All computer systems that are subject to GxP regulatory compliance requirements should be listed. When determining whether or not a computer system has a GxP impact, it is useful to reflect on the advice given by Sam Clark, a former FDA investigator: "If it looks like a duck, flies like a duck, and sounds like a duck, then it's probably a duck"! In other words, use your common sense (Fig. 2.4).

It is important to consider the effort to maintain increasing numbers of fields when designing the system inventory. The greater the number of fields the more maintenance will be required, and the harder it will prove to keep it up-to-date. Remember that there may be many thousands of computer systems in use on larger pharmaceutical and healthcare manufacturing sites.

The inventory should be periodically reviewed and approved by a QA representative. When reviewing the inventory, it is important to remember that the compliance status will also have to be periodically checked to determine whether revalidation is required. Revalidation may be triggered by cumulative effect of changes made to the computer system or by new regulatory requirements.

Recent Inspection Findings

There is no list of all systems that are subject to validation (FDA 483, 2004).
Unvalidated systems have not been identified (FDA 483, 2004).

SYSTEM LIFE CYCLE MANAGEMENT

The management of compliance can be considered as a cyclical process, as shown in Figure 2.5. The cycle begins with GxP assessments surveying the compliance requirements of computer systems in readiness for preparing validation (master) plans. Supplier assessments consider the capabilities of suppliers providing computer systems and associated services. Compliance activities are then conducted according to any prevailing priorities. Once validation is completed for individual computer systems, its operational compliance must be maintained.

Throughout the compliance management process, there should be formal opportunities to review practices. The status of compliance work is likely to change; some existing systems may be decommissioned, new systems may be planned, and the priorities on the current inventory work may vary because of changing company needs. Compliance policy and supporting procedures may change as a result of new regulatory requirements or feedback from project experience.

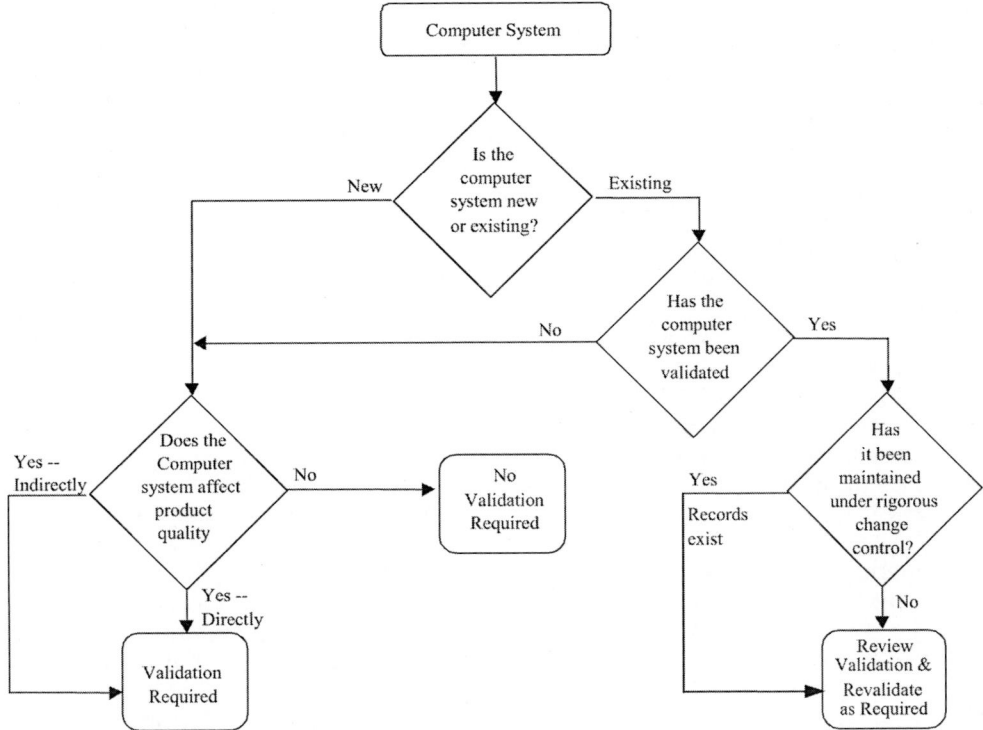

Figure 2.4   Determining a validation requirement.

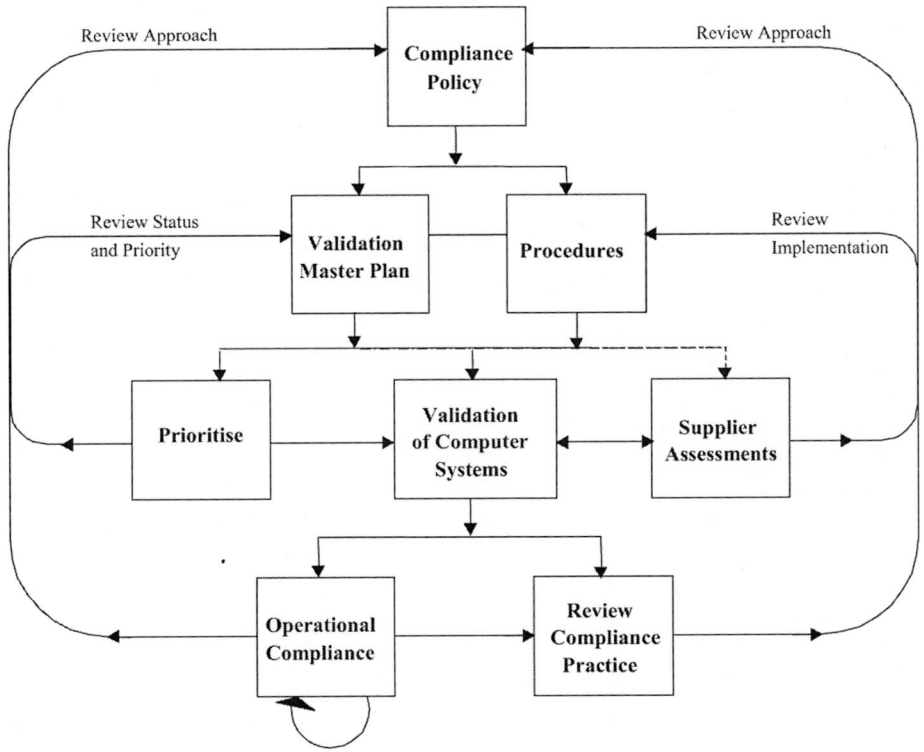

Figure 2.5   Compliance management.

Getting Started

There are two main obstacles to be surmounted here. One is a lack of compliance experience, and the other, an absence of focus and determination. Weak compliance experience is often characterized by questions like "Why are we doing this anyway?" and "What are the fundamental principles we should follow?" Education and training, and managing the learning curve are both key issues. Managers and sponsors, as well as practitioners, need a practical appreciation of compliance requirements (trends, constraints, areas of flexibility, and benefits). It must be understood that initial work will require more effort, because practitioners will not be familiar with the new way of working. It is wise to avoid large work packages as the starting point for adopting the new quality-assured mode of working, since the scale of inefficiency might shock an organization, is tempting it to abandon the quality-assured approach altogether. It is better to begin with smaller work packages and watch costs reduce as practitioners improve their understanding and efficiency. The learning curve will eventually flatten out to a plateau. However, herein lies another danger. If key staffs move on and leaning has not been captured in the corporate memory (policies, procedures, guidance, training, and succession planning), learning and efficiency will be lost. Hopefully avoiding these pitfalls, there is still a need to pace the introduction of the new way of working in case progress is not forthcoming. The best course is to build on success and extend the new ways of working based on a proven track record.

Risk Management

Risk management is very important if appropriate resources are to be deployed in a timely fashion to mitigate or reduce the potential effect of identified risks. It is recommended that a risk map be produced showing where computer systems are used to support the various process streams of operational activity.

Determining which operational aspects are most critical requires an understanding of the potential impact on drug or healthcare product safety, quality, and efficacy. The Canadian Health Products and Food Branch Inspectorate have already identified a number of high-risk issues that are likely to result in noncompliant drug product and present an immediate or latent public health risk (6). A similar identification of high-risk issues can be achieved by applying International Conference on Harmonisation (ICH) Q9 regulatory guidance (7). These high-risk issues are applied here to computer systems and aligned to the six operational areas identified by FDA quality system-based approach (8).

Quality Systems
- Document management
- SOP administration
- Security access controls (e.g., user profiles and password management)
- Change control records
- Customer complaints
- Adverse event reporting
- Review/audit/corrective actions management
- Training records

Facilities and Equipment Systems
- HVAC controls and alarm handling
- Critical equipment and instrumentation (calibration and maintenance)
- Change control records

Materials Systems
- Traceability of material handling
- Raw material inspection/testing/quarantine management
- Storage conditions
- Containers usage and cleaning management
- Distribution records and recall management

Production Systems
- Recipe/formulation management
- Batch manufacturing instruction and records
- In-process testing
- Yield calculation
- Purified water
- Aseptic filling

Packaging and Labeling Systems
- Labeling information

Laboratory Control Systems
- QC raw data
- Stability testing
- Sterility testing
- QC analytical results
- Quality disposition
- Out of specification investigations

Workflow analysis using flowcharts marking critical decision points and information flows is an effective way of pictorially mapping risks. The use of computer systems should be identified on the maps. Other relevant supplementary information can be added as deemed appropriate. A balance has to be struck on the amount of information included so that the process can be mapped in a manageable number of pages. Producing the risk map on A3 size paper can help given a better overview.

The rigor of validation for computer systems supporting these critical operational aspects of the processes should take account of their composite custom (bespoke) software, commercial off-the-shelf (COTS) software, and supporting computer network infrastructure. The risk map and supporting rationales will form the basis of validation master plans that are discussed in more detail later in this book.

Life Cycle Approach
The life cycle approach has attracted broad acceptance across the pharmaceutical and healthcare industry and can be refined to meet the needs of particular applications. Different organizations use variants of the life cycle, but the methodology of dividing a life cycle into phases remains the same. For instance, some companies develop the subphases that are indicated in the phase descriptions above as distinct phases in their own right. The specific life cycle model chosen does not really matter. Its constituent phases must, however, be clearly defined in advance, with entry and exit criteria for each phase, and appropriate verification procedures to ensure the controlled completion of constituent phases.

Figure 2.6 presents a set of life cycle phases that summarizing the approach typically used within the pharmaceutical and healthcare industry. Life cycle phases may be known by alternative names in different organizations. There is no standard glossary throughout the industry relating to naming conventions or groupings of phases. It is important, however, that all the activities covered by this chapter are included in any alternative scheme.

The life cycle is consistent with guidance provided by the Australian, European, Japanese, and U.S. GxP regulatory authorities (9–13). It is also consistent with guidance provided by the GAMP®5 and DIA Red Apple (5,14).

The steps in the life cycle are not necessarily executed in the order indicated. Rather, the steps are usually executed as an iterative process, where various functions may be carried out concurrently. If necessary, steps may be repeated. For instance, the validation master plan may be developed after or concurrently with the URS rather than before as indicated. Equally, a

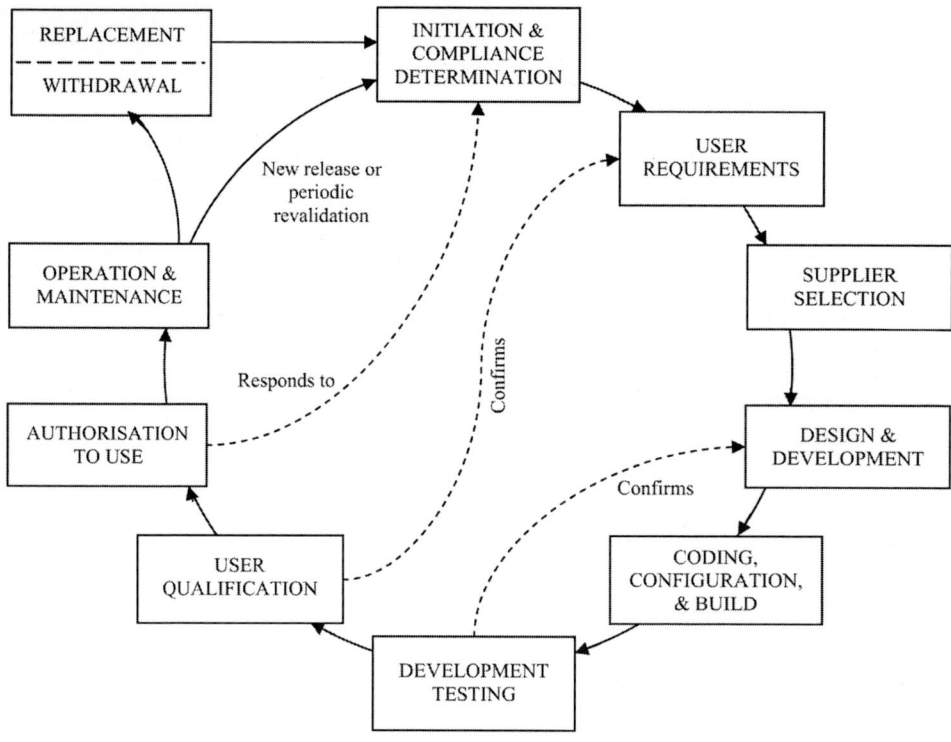

Figure 2.6 Computer systems life cycle.

supplier selection often involves a series of steps beyond assessment to include corrective actions that may not be complete until well into the project.

Some of the steps in the life cycle will not be needed in some projects. For instance, the use of preferred suppliers for software and hardware products or services removes the need for repeated supplier appraisal and selection.

For systems that have been in use for some time, a compilation and review of documentation, and a review of historical data supplemented by a series of functional tests may be adequate to demonstrate that the system performs its intended function. Evidence that the software is structurally sound may be provided by a formal evaluation of the supplier's software development and testing procedures and by an analysis of historical, system-related data. Where historical, system-related data is not available, for whatever reason, additional functional testing may be required.

MANAGEMENT REVIEW

A management review will usually be conducted periodically to draw out lessons from the compliance work conducted to date, to consider the impact of any regulation developments, and to report any recommendations. It should also effectively address any inspection issues raised by regulatory authorities concerning the company's computer systems. The overall aim is the continuous improvement of the company's compliance capability (policy, strategy, procedures, training, etc.).

Projects can provide a central focus for applying and refining policies and procedures. Feedback from practical experience is vital if a cost-effective compliance approach is to be established. An external consultant may be seconded to provide an independent perspective, to comment on current industry practices, and to provide updates on topical regulatory issues. Compliance does not have to be unduly expensive if the issues involved are managed in a timely manner.

It is recognized that senior managements in pharmaceutical and healthcare companies are faced with multiple and changing priorities: for example, customer service, quality, and financial performance. Nevertheless, it is very important that the level of support given to compliance is sustainable. Should it falter, the organization will face pendulum swings of compliance investment and compliance underfunding. Such feast and famine nearly always leads to a serious noncompliance sooner or later, as practices try to adapt to the level of prevailing financial support. Senior management should beware of making arbitrary cuts and rather work on cost-effectiveness improvements. Equally, throwing money at compliance does not necessarily result in it becoming "solved." Senior management must avoid the notion that compliance is a one off project type activity. Inspection readiness results from maintaining the ongoing compliance of legacy systems that must be properly supported; otherwise their compliance will be compromized over time.

Recent Inspection Findings

No management review procedures and no documented management reviews (FDA Warning Letter, 2001).

Quality audits did not verify effectiveness in fulfilling quality system objectives (FDA 483, 2002).

REFERENCES

1. Organisation for Economic Co-operation and Development Environmental Directorate. The application of the principles of GLP to computerised systems. No. 10 OECD Series on Principles of Good Laboratory Practice and Compliance Monitoring, GLP Consensus Document, Environmental Monograph No. 116. Paris, 1995.
2. International Organisation for Standardisation. ISO 9001: International Standard: Quality Management Systems—Requirements. Geneva, 2008.
3. Stokes T. The role of senior management in computer systems validation. In: Stokes T, Branning RC, Chapman KG, et al., eds. Good Computer Validation Practices: Common Sense Implementation. Buffalo Grove: InterPharm Press, 1994.
4. Bruttin F, Dean D. A risk-based approach to reducing the cost of compliance in pharmaceutical manufacturing. Pharm Technol Eur 1999; 36–44.
5. International Society for Pharmaceutical Engineering. GAMP®5—Risk-Based Approach to Compliant GxP Computerised Systems. Available at: www.ispe.org, 2008.
6. Canadian Health Products and Food Branch Inspectorate, Good Manufacturing Practices—Risk Classification for GMP Observations, 2000.
7. International Conference on Harmonisation. Quality risk management. Q9 document, technical requirements for registration of pharmaceuticals for human use (ICH). Available at: www.ich.org, 2005.
8. U.S. Food and Drugs Administration. Quality systems based approach for pharmaceutical current Gmp regulations, guidance for industry. Rockville, MD, 2006.
9. Therapeutic Goods Administration. Australian Code of Good Manufacturing for Therapeutic Goods. Medicinal Products—Part 1, Woden. Australia, 1990.
10. European Union. EU GMP Annex 11 Computerised Systems. Guide to Good Manufacturing Practice for EU Directive 2003/94/EC. Community Code Relating to Medicinal Products for Human Use 2003; 4.
11. Japanese Ministry of Public Welfare. Guideline on computer systems in drug manufacturing in manual on computer systems in drug manufacturing, 1993.
12. U.S. Food and Drugs Administration. Guide to inspection of computerised systems in drug processing. Technical report, Reference Materials and Training Aids for Investigators. Rockville, MD, 1983.
13. Pharmaceutical Inspection Co-operation Scheme. Good Practices for Computerised Systems in Regulated GxP Environments. Pharmaceutical Inspection Convention, PI 011-1, Geneva, 2003.
14. Drug Information Association. Computerized Systems Used in Nonclinical Safety Assessment: Current Concepts in Validation and Compliance (known as "Red Apple II"), 2008.

APPENDIX 2.A: KEY PRINCIPLES FOR COMPUTER SYSTEM COMPLIANCE

These principles apply to computer systems that can affect the quality of drug and healthcare products. Such computer systems include laboratory systems, process control systems, spreadsheet and database applications, business systems, and associated computer network infrastructure.

Founding Principles

1. Validation requires establishing documented evidence that provides a high degree of assurance that a computer system will consistently perform according to its predetermined specifications. A key consideration is the protection of the integrity, authenticity, and security of data relating to the quality of drug and healthcare products and/or supporting regulatory submissions.
2. Decisions on the extent of validation and data integrity controls should be based on a justified and documented risk assessment. Factors to consider include business and regulatory criticality of the process the system is to support, the potential risks to product quality or safety the system could pose, and the source of the system (custom written or COTS).
3. It is essential that there is the closest cooperation between key personnel involved with the development, management, and use of computerized systems such as users, system administrators, quality assurance, and technical staff. Persons performing such roles should have the appropriate and documented training, technical expertise, and experience to carry out their assigned duties.
4. Electronic records may be signed electronically or by applying a handwritten signature to a printed copy of the record. Electronic signatures are expected to be legally equivalent to handwritten signatures, linked to their respective record, and include the time and date they were applied. Hand-signed printed copies should provide indisputable linkage with the electronically held record.

Management Requirements

5. All new systems requiring validation must be validated prospectively. Existing systems that require validation but have not already been validated must be retrospectively validated.
6. Validation must be planned, executed, and reported. Validation encompasses the entire life of the computer system, from planning through development and implementation, and use and operational support to decommissioning. Responsibilities and accountabilities must be defined and documented.
7. Computerized systems must have documented authorization to be used in their operating environments. Any restrictions impacting use must be recorded and authorized.
8. Validation must employ predefined procedures and plans designed to build in quality during all stages of the computer system life cycle. The effectiveness of these procedures must be assessed periodically, and improvements must be made as required.
9. System requirements must be traceable throughout specification, verification, and subsequent change control records.
10. Suppliers of computer systems and associated services must be managed to assure that the software, hardware, and/or related services they supply are fit for purpose.
11. Enhancements and modifications to computer systems and associated documentation must be implemented under change control, and, where appropriate, configuration management. Changes affecting the compliance status of a computer system must be approved before they are implemented.
12. Those involved in the development and implementation, use and operational support, and decommissioning of computer systems must have the documented education, training, and experience to fulfill their duties.
13. Rationales must be developed and documented to justify compliance decisions not supported elsewhere.

Project Requirements

14. User requirements and design must be specified, documented, and approved.
15. Software controls must be used to manage programming.

16. Development testing of computerized systems must be documented and cover structural and functional attributes.
17. Design reviews (also known as design qualification) must be conducted and documented to verify that user and regulatory requirements are satisfied in the wider system context including equipment, processes, and manual interaction.
18. User/site acceptance testing of computerized systems (known as qualification) must cover installation, operation, and performance in the wider system context including equipment, processes, and operator interaction (i.e., installation qualification, operation qualification, and performance qualification).
19. Data migration must preserve the integrity and security of original data. Processes and methods used to load data (manually and automatically) must be defined and validated with supporting documentation before they are used.

Operation Requirements

20. The performance of computer systems must be monitored against predefined requirements to demonstrate acceptable operational service.
21. Preventative maintenance and calibration, where required, must be planned, conducted, and documented.
22. Software changes and upgrades must be conducted according to defined procedures and documented.
23. Records must be established to demonstrate data integrity being maintained.
24. Backups of software, configuration, and data must be planned, conducted, and documented. Backup data must be readable and therefore retrievable.
25. Archiving of software, configuration, data, and associated documentation must be planned, conducted, and documented. Storage media must be retained for a predefined period at a separate and secure location under suitable environmental conditions, be protected against willful or accidental damage, and be periodically checked for durability and restoration. Archived materials must not be destroyed until this retention period has expired.
26. The procedures to be followed if the computerised system breaks down and is unavailable must be documented and periodically verified.
27. Access to computerized systems and associated functionality must be restricted to authorized persons, and documented.
28. Formal agreements regarding operational support (e.g., contracts and service level agreements) must be established by defining responsibilities and accountabilities.
29. User procedures must be established and trained out to ensure that computer systems are consistently used.
30. The compliance status of computerized systems and the cumulative effect of change must be periodically reviewed, and any required revalidation must be conducted.
31. Decommissioning of computer systems must be planned and conducted in accordance with defined procedures.

Responsibilities and Accountabilities

System owners/users are accountable for assuring that computer systems used by them or on their behalf by other organizations are compliant with pharmaceutical regulatory requirements.

Developer and operational support organizations are responsible for the technical delivery and quality of work of their associated compliance activities. These activities must fulfill regulatory expectations. Developer and operational support organizations have a mutual responsibility to ensure that project handover activities are appropriate and completed.

Quality and compliance are responsible for establishing necessary policies and procedures to manage compliance activities and for approving compliance work. They must be able to demonstrate their independence to the system owner/user and to developer and operational support organizations.

3 | Supporting Processes

INTRODUCTION

Successful validation depends on the satisfactory operation of a number of underlying supporting processes. Among these are training, document management, change control, configuration management, requirements traceability, self-inspections, and managing deviations. Compliance is fundamentally flawed without them, and so they are discussed here.

TRAINING

All personnel (permanent staff, contractors, consultants, and temporary staff) developing, supporting, or using a GxP computer system must be trained so that they acquire the necessary level of competency before they may be allowed to perform their designated duties. To this end, all personnel involved in any aspect of compliance should have (1)

- A role description,
- Appropriate qualifications that have been documented, and
- Plans and records for training, both prospective and retrospective.

Qualifications

When recruiting, pharmaceutical and healthcare companies should try to verify the details of the education, training, and experience claimed by the candidates. Copies of their certificates should be requested and retained. Personal references should also be taken up, although their value should be weighed with care, remembering that the commendations given can be presented in a politically correct fashion that carries a hidden, subliminal, and rather less favorable implication! For example, consider the following embarrassingly flattering accolade

> You write to ask me for my opinion of XXXX, who has applied for a position in your department. I cannot recommend him too highly, nor say enough good things about him. The validation he conducts is the sort of work you don't expect to see nowadays. His documentation clearly demonstrates his complete capabilities. His understanding and appreciation of regulatory requirements will surprise you. You will indeed be fortunate if you can get him to work for you.

Curricula vitae (CVs) for permanent staff are usually kept by the Human Resources department. This is not necessarily the case for contractors, consultants, and temporary staff. Their CVs, typically retained by the responsible manager, can be easily lost when individuals move on to new contracts. One way to systematically capture such training records for contractors, consultants, and temporary staff is to attach them as appendixes to validation plans or validation reports.

The profile suggested for a computer compliance practitioner is given in Ref. 2 and comprises

- Technical background involving computer systems,
- Technical qualifications associated with computer systems, and
- Two or more years of Gxp experience, not necessarily involving computer systems.

Compliance practitioners must have a measure of tenacity and self-discipline so that they stay the course, progressing work through to completion without constant supervision. This is not to say, however, that they should be discouraged from seeking advice but rather that they should have sufficient judgment and regulatory knowledge to make some basic decisions themselves. Practitioners should be flexible in their approach, so that when the inevitable problems arise, they do not instinctively resist exploring new solutions.

Managers are likely to be qualified by experience, perhaps supplemented by training. Personal attributes, and especially attitudes, are of crucial importance, but there is insufficient space to develop this theme here. Educational attainment alone does not ensure that a manager has the competencies required to manage effectively.

Training Plans and Records

Training plans should be used to manage the development of staff, and subsequent training should be conducted in accordance with approved procedures. All personnel should be aware of the principles of GxP affecting them, and receiving initial and continuing training relevant to their job responsibilities. This includes those developing, supporting, and using computer systems including electronic records and signatures. It is important to recognize that necessary training must be provided prior to the need for the use of the associated competency arising, rather than when the lack of such competency has already been painfully demonstrated!

Training records must be maintained. Some pharmaceutical and healthcare companies make use of questionnaires to try to verify in a formal way that personnel have understood their training and really acquired the intended competence. Authorized assessors should be engaged to mark such miniature examination papers. If personnel fail to pass such a competency test, some supplementary training is required. Care should be taken not to simply default to repeating the original training and the examination. Perhaps the training materials or delivery were at fault, and they may require improvement. There may be a systematic reason why individuals have not understood what they have been taught.

The performance of personnel should be periodically reviewed to identify any refresher training requirements. Some pharmaceutical and healthcare companies achieve this through an audit or a periodic review of training records, which must be updated to reflect the training received. Marked competency questionnaires or test papers should be attached to training records where possible.

Recent Inspection Findings

No written procedures describing training given to those who use the computer system (FDA 483, 2004).

No records to document that the information technology (IT) service provider staff personnel have received training that includes current good manufacturing practice regulations and written procedures referred to by the regulations (FDA 483, 2000).

No documentation to indicate that users are trained in the software and its applications (FDA Warning Letter, June 2000).

No documentation that a qualified person reviewed the training records (FDA Warning Letter, 2000).

No assurance that adequate training was given on how to use the *(computer)* system software (FDA 483, 2002).

Corrective action to noncompliance with SOPs consists of retraining in the same manner as initially trained. There is no limit to the frequency of retraining (FDA 483, 2000).

There is no evidence that the training provided during 2000 and 2001 to your analysts is adequate as evidenced in the following events. The efficiency and adequacy of the training program are questionable in that numerous training sessions are performed during the same day (a specific list of eight training sessions a particular employee received on the same day was then made) (FDA 483, 2002).

The current training procedure for employees does not determine the proficiency or comprehension at the end of training. Your response to this letter should include (your) plan for establishing a system of training and evaluation to ensure that personnel have the capabilities commensurate with their assigned function (FDA Warning Letter, 2000).

DOCUMENT MANAGEMENT

Documentation of research, development, and manufacturing practice is vital to pharmaceutical and healthcare companies because unless they do this, they have no way of demonstrating compliance to the various GxP regulatory authorities. Examples of documents include policies, procedures, plans, reports, and operational data. Regulatory inspectors will expect to see document management procedures established covering preparation, review, approval, issue, change, withdrawal, and storage. This is especially important for contractual documents and documents endorsed by the pharmaceutical or healthcare company's QA organization.

Document Preparation

Documentation standards should be defined so that there is consistent document layout, style, and reference numbering. Documents should be clearly marked as *draft* until they are formally released. Version control should be apparent. The version identifiers should distinguish documents under development (drafts) from those that have been issued formally. Documents should include a document history section to log the changes made in successive issued versions of the document.

Individual documents should have the following controls:

- Document title
- Document number
- Version number
- Page x of y
- Date of issue
- Copy number

Some organizations include a date for the next routine review of the document. This ensures that even if no changes have occurred, the document will still be examined to verify that it is still relevant and accurate.

Documentation has progressed enormously from the days when word processing represented the apex of efficiency and automation in this arena. Special considerations for more sophisticated document types are given in Table 3.1. In some instances, it is recommended that the document type should be stored with or within the document to make ongoing document maintenance easier. This also applies to some types of embedded documents.

Document Review

Documentation should be subject to review prior to its formal release. Such a review might assume a number of forms ranging from the evaluation of the document and collection of comments through to its inspection within formally convened review meetings. The reviewers should be identified in advance within the validation plan, the project and quality plan, or the document management procedures. Many firms give staff guidance on how to decide the most appropriate reviewers for different documents.

The review can be recorded using a template or by simply recording meeting minutes in the traditional manner. In either case, the date of the review and the names of the reviewers should be noted. It is recommended that a multidisciplinary team that includes technical and quality representatives review documents. Each reviewer, however, does not necessarily have to be an authorizing signatory for the document under inspection as long as his or her comments are included in the review records.

For the review process to be effective, reviewers must come prepared. Copies of the documents under review should be distributed and scrutinized prior to the meeting. A chairperson for the review meeting should be nominated beforehand. Similarly, remote

Table 3.1 Characteristics of Various Document Types

Document type	Characteristics	Examples of application area	Special precautions
ASCII text document	The simplest document type to manage. Typically created in a word processor, it consists of only characters belonging to the ASCII or ANSI sets (text and symbols). Such documents can be viewed by the originating word processor or by file viewer programs commonly available in most computer systems.	Memos, master production and control records, SOPs, deviation reports, validation protocols, batch documentation, etc.	Specify file format, application, version, and language.
Portable format document	A homogeneous document type created from many other types of document but stored in a standard or proprietary file format. The international standard is SGML. Proprietary formats include Adobe's PDF format. A semiportable format is HTML, which is used on the Internet. Files cannot be edited without special editing software.	All sorts of documents including more complex ones such as integrated batch documentation, illustrated SOPs	Documents must be stored in "neutral" file format. Specify file format, application, and version. Special printer drivers may be necessary.
Graphical document	A homogeneous document type stored in a standard graphical file format. Includes scanned paper documents. Many file formats exist, from bit-mapped pictures (e.g., TIFF, PCX, GIF, JPEG) to highly complex vectored drawings in a CAD environment (e.g., CGM, DXF). Many proprietary formats exist. Some CAD formats include product database information.	CAD drawings, SOP illustrations, scanned paper documents, label pictures for batch documentation	Specify file format, application, version, and language.

reviews will require the appointment of an individual to coordinate and collate the review feedback.

The review must systematically cover each section of the document. If a section attracts no comments, this should be indicated in the records. Any corrective actions identified in the review must be assigned to a named individual with a completion date. The progress of individual actions should be tracked through to closure. Care must be taken to ensure that associated documents are also reviewed and updated as necessary.

Sometimes in the absence of any consensus, a compromise on the content of a document will have to be reached. Such compromise positions should be agreed before approval is sought to avoid delaying the approval process. Normally, the document author has the responsibility for resolving these issues. Often this is not simple, especially when reviewers have entrenched and opposing views! An escalation route to resolve any impasse should therefore also be defined and agreed beforehand.

Speakers at conferences have expressed the opinion that in their experience, 80% of review comments might relate only to format and style. In spite of this, over 80% of the problems arising from poor documentation could be attributed to omissions and accuracy! Something is wrong here. Reviewers need to bear these statistics in mind and strive to make their reviews as effective as possible in identifying the defects in documents, defects that will cost time and money later on.

Once the agreed changes have been incorporated into the document, it is ready for approval. The document history should be created to record the changes made. There is no need for the document history to affirm what remained the same. Document histories are usually written in the form of a summary at the beginning or end of a document.

Review records should not be destroyed, at least until the document is approved and formally issued. Many projects have a policy of retaining all document review minutes and records until computer system is handed over and commissioned into use. Even then the project files containing the review records may be retained for a few years, just in case some question or defect arises in the future.

Document Approval

Just like the reviewers, document approvers should be identified in advance within the validation plan, the project and quality plan, or the document management procedures. Again, many firms guide staff on the most appropriate choices of personnel for the sensitive and important task of document approval.

The number of signatures on individual documents should be monitored. There are usually four principal signature roles (all not required for each document).

- Technical approval where relevant
- Regulatory compliance
- Compliance with corporate procedures (including format)
- Authorization to proceed

There is no reason why one individual cannot fulfill more than one role provided he or she has appropriate competencies. There is one prohibition, however; no single individual should represent both quality and technical roles. The minimum requirement is for two signatories, representing quality and technical roles, respectively.

Documents with up to 10 signatures are common where the number of signatories is not controlled. During a survey at one European pharmaceutical company, a document was found bearing no less than 18 signatures! Was this really necessary? Indeed, too many signatories will retard the release of documentation, while some practitioners have argued that many signatures leads to less effective document scrutiny rather than more. It is not hard to see why—human nature being what it is. The temptation to believe that the effort of an effective review is pointless as so many others have already endorsed it becomes almost irresistible (a phenomenon also known as the *rubber stamp effect*)! Furthermore, what personal price will be exacted for questioning the combined wisdom of so many colleagues? There is thus some truth in the cynic's maxim that the quality of a document is inversely proportional to the number of approval signatures.

The rationale for the presence of each approval signature should be unambiguous and documented—signatories should know why they are signing the document! Example approval signatures include technical authority, QA compliance, and user acceptance. Regulatory authorities normally require key compliance documents to have formal signatures from two or more authorized persons. Signatures should be written in black or blue ink as the pigments and dyes in them render the signatures more resistant to fading. Approval signatures should be dated and accompanied by the name of the person signing, as the identities of some signatures are indecipherable (a weakness normally associated with the medical profession and its prescriptions but also all too evident during regulatory compliance activities). Interestingly, in some countries such as Japan, it is legally admissible to use as a signature a mark or stamp that does not identify the person's spelled name.

Document Issue

Document Distribution

Approved documents should be made available in accordance with defined distribution procedures. Many organizations use document management systems including intranet that allow users to download copies of master documents. Such copies may be used for GxP purposes provided users check that they are using the correct version of the document (i.e., verify current version number) before using the copy. It may be necessary for pharmaceutical and healthcare companies to establish confidentiality agreements with suppliers who receive copies of their documents as well as ensure that suitable security access controls are in place so that only authorized persons can access documentation.

Some organizations define and record an "effective date" and "expiry date" on documents. The effective date is usually defined to allow a period for dependent activities, for instance, the distribution and training implied in the associated SOPs. The expiry dates of different document types should take account of their change frequency. SOPs might have an expiry date equivalent to their next planned periodic review. Master documents that have passed their expiry date must be clearly marked as *superseded*. Where effective and expiry dates are deployed, they should be prominently displayed on the document's frontispiece.

Document Changes

All changes to released approved documents must be subject to change control. The revised document should be clearly marked as a draft and managed accordingly, as described above. Modifications to approved documents should be reviewed and approved by the same functions/organizations that performed the original review and approval unless specifically designated otherwise. Despite changes in individual signatories, there should be a consistent allocation of responsible review and approval roles.

Document Withdrawal

From time to time, documents must be withdrawn from use, having perhaps come to the end of their useful life or being superseded. Some firms send an e-mail to relevant portions of their organization notifying them of withdrawn documents. Document keepers can be asked either to notify the central distribution group that they have destroyed their copies of withdrawn documents or to return specially printed copies to the central distribution group for disposal. Safeguards should be instituted to prevent the retention and use of superseded or withdrawn documents. Many firms use audits to this purpose.

Document Administration and Storage

It is wise to keep the organization and administration of documentation as simple as can be conceived. Complex systems are much harder to manage successfully. Centralized versus distributed administration of documentation has advantages and disadvantages. Centralized administration offers economies of scale and easier control of master documents as the latter are held in one location. Distributed administration, meanwhile, offers more "ownership" because it is closer to its users, and it is easier to plan for busier periods. Most organizations tend to centralize administration on a site basis, but either approach can be adopted as long as it is controlled. In theory, the best of both worlds is available through the implementation of an electronic document management system (EDMS). Such systems must of course be validated!

Master copies of documentation should be stored in a safe and secure location according to defined procedures. These master copies should be stored with

- Approval signatories,
- Document history,
- Change control records,
- Document distribution records where applicable,
- Superseded versions, clearly marked as such, and
- Withdrawn documents, clearly marked as such.

Stored documents should be protected against accidental and malicious damage. They must be retrievable in a legible format throughout their predefined retention period. This usually means that a minimum of two copies are retained, each in a separate place, just in case of accidents. Once the retention period has expired, a decision can be taken whether or not to destroy the master copies. A record of destruction, evidence that the document once existed but has since been destroyed, should be made and retained for a further period.

A document index should be maintained to log documentation by reference/title, version, and physical storage location. As the status of a document changes, the document index needs to be updated.

Quality of Documentation

The quality of documentation must be assured. Poor documentation is often marred by rambling, unfocussed, and verbose text, with omissions in some areas and excessive detail in others (Appendix 3.A). This impedes its use as well as undermines the goal of achieving GMP compliance. Those preparing documentation and records should therefore ensure that documents meet the *six virtuous Cs*, that is,

- Concise,
- Complete,
- Consistent,
- Comprehensible,
- Correct, and
- Controlled.

Wherever possible, keep documents short (preferably fewer than twenty pages) and avoid the duplication of information.

Basic regulatory expectations include

- Mistakes being altered correctly: single strike, initialed and dated, with a brief reason for the correction (or a reference to a change control number if appropriate),
- Avoiding use of dittos or arrows, as FDA consider them insufficiently descriptive where actual values with corresponding signatures are needed,
- Avoiding transcriptions even if the original document/record looks messy, and
- Ensuring that white opaque correction fluid is *never* used, as it hides the original information.

Regulatory authorities will search documentation for certain vague words that, in their experience, are often associated with imprecise documentation.

- Calculate: the actual calculation intended to be used specified?
- Automatic: the degree of manual intervention specified?
- Typically, usually: exactly how often is meant here?
- Normally: what is normal, what is abnormal?
- Appropriate: what is appropriate, what is not appropriate?

A simple word search can be used on word processors when a document is being written to identify the use of these words, which should then either be replaced with alternative phrases or be clarified to alleviate the uncertainty.

Recent Inspection Findings

QA has failed to ensure that all appropriate computer validation documents are identified and controlled (FDA 483, 2004).

Lack of appropriate documentation procedures (FDA Warning Letter, 2001).

Lack of procedures to ensure that records are included with validation documentation, maintained, and updated when changes are made (FDA Warning Letter, 2001).

SOPs do not clearly describe who must approve documents or what each type of approval represents (FDA 483, 2002).

Significant deficiencies regarding documentation controls were reported. Documents were not dated, lacked a documentation control number, were missing, or were reported in pencil on uncontrolled pages, or dates were crossed out without initials, dates, or explanation (FDA Warning Letter, 2001).

Errors on batch production, control, and laboratory records must not be erased or overwritten (interpret as no whiteout). A line must be drawn through an incorrect entry and the corrected figure or word written neatly and initialed. Significant data must not be discarded without explanation. To discard significant data, the data must be crossed out and initialed, and a valid reason for discarding the data must be explained (FDA Warning Letter, 2001).

Numerous instances of lack of control of official controlled documents were observed: use of incorrect version of testing forms, incorrect data sheet used because old sheets are not replaced with new, incorrect log sheets were used (FDA 483, 2002).

Several pages were missing in printouts (FDA Warning Letter, 2000).

Two pages of a laboratory notebook were written in pencil and erased. Your abbreviation for . . . could be read on one of the erased pages (FDA Warning Letter, 2000).

Values in at least two *laboratory records* were altered. Altered values were written under computer-generated values . . . . and used in potency calculations. Review of the electronic data confirmed the incorrect values, which were part of your submission to the drug master file (FDA Warning Letter, 2000).

CHANGE CONTROL

The following maxims of change are based on work by Lehman and Belady (2).

- *First maxim of change: change will happen*
  Computer systems do not have static requirements. A system that is being used will almost certainly undergo continuing change either because its requirements were not fully understood in the first place or because the use of the system is changing. Change will only stop when the system's functionality becomes obsolete or it is judged more cost-effective to reengineer the system or replace it by a completely new version.
- *Second maxim of change: change breeds change*
  Programmers often find it hard to resist adding unsolicited functionality.
- *Third maxim of change: change increases complexity*
  Software subject to change becomes less and less structured and thus becomes more complex. Extra effort is required when implementing changes to avoid increasing complexity.
- *Fourth maxim of change: documentation eases change*
  The quality of documentation associated with computer systems is a limiting factor to the ease of implementing change over its operational life. Faster rates of change typically indicate the dominance of developing functionality over documenting the change. Slower rates of change may indicate that system modifications are being hindered by previous changes not being fully documented or that developing functionality is being fully documented.
- *Fifth maxim of change: more resource does not imply faster change*
  There is an optimum level of resource for change. Applying more people to implement a change does not imply that the change will be achieved faster. Indeed, it can quite often add management complexity and slow change down.

These maxims should raise awareness of the need for effective change management. Quality and compliance has to be preserved.

All changes to regulated computerized systems must be reviewed, authorized, documented, tested (if applicable), and approved before implementation. Software cannot be partially verified. When a change (even a small change) is made to a software program, the

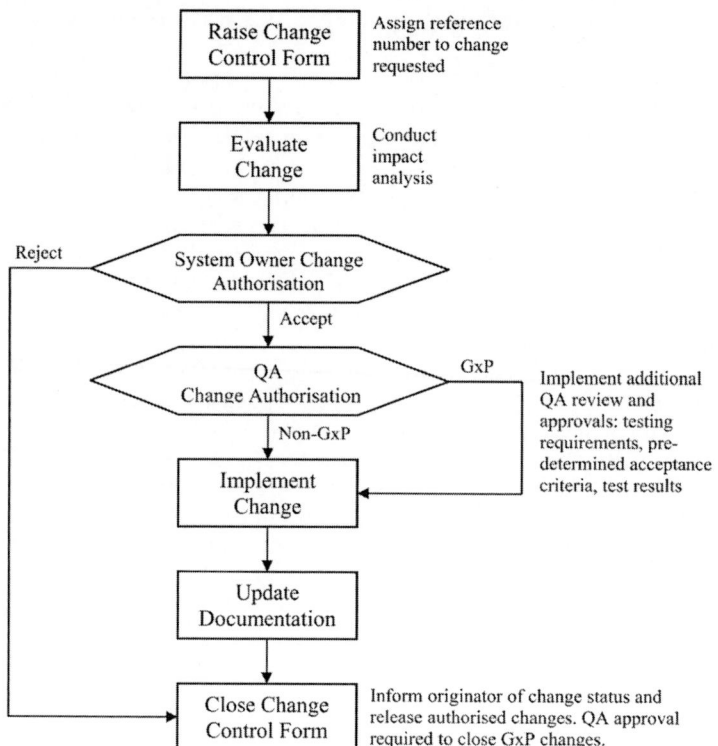

Figure 3.1 Change control process.

compliance status of the entire software system should be reconsidered, not just the individual change (3). Retrospective validation and reverse engineering of existing software are very difficult but may be necessary to properly document and verify changes.

The procedure for change control can be divided into four phases (Fig. 3.1).

- Request for change
- Change evaluation (impact analysis) and authorization
- Testing and implementation of the change
- Change completion and approval

Request for Change

A system owner should be (have been) appointed for every system. This should be laid down in the system documentation (or validation plan). A proposal for a change should be directed first to the system owner, who shall be responsible for ensuring that all changes to the system are reviewed, authorized, documented, tested (if applicable), approved, and implemented in a controlled manner. The system owner may delegate this responsibility if permitted to do so and the delegation is documented.

Any proposed change should be requested and recorded by submitting a *change request form*. An example of such a form is given in Figure 3.2. The change request part should at least include the following items:

- Requester name
- Origination date
- Identification of component or software module to be changed
- Description of the change
- Reason for the change

Unique reference number is to be assigned by the system owner or his delegate using a logging mechanism.

CHANGE CONTROL FORM	Change No.:

Computer System:

Location:

Name/Date of Person Submitting Change Request:

Request for Change

Details of Proposed Change:

Reason for Change:

Change Authorization

Disposition: Accepted/Rejected (delete as appropriate)

Signature: Date: Representing: User

Signature: Date: Representing: QA

Signature: Date: Representing: Technical

Change Details

Comments: (include reasons for rejection if appropriate, details of testing requirements, other relevant information)

Change Completion & Approval

These following approvals signify completed implementation of the change including any updates required to associated documentation.

Signature: Date: Representing: User

Signature: Date: Representing: QA

Signature: Date: Representing: Technical

Figure 3.2   Example of change control form.

Change Evaluation and Authorization

Each change request raised must be reviewed, and a judgment must be made (accepted or rejected). For system changes with a scope wider than that solely of the department that owns it, the change request should be circulated to all the departments affected by the change. These should be identified by the system owner or his delegate, and they must be obliged to review the change request with their appropriate technical, management, QA, and user personnel. Once the impact of the proposed change has been assessed, the system owner or his delegate must decide whether to accept or reject the proposed change.

In principle, for systems used for GxP-related activities, QA should be involved in the decision-making process. Future QA involvement in the rest of the change process will depend on the compliance status of the computer system and the nature of the change being implemented. Authorization by QA is required at several stages when the change is regarded as GxP relevant.

Changes may be implemented separately or collected into bundles for implementation. Consideration should be given in either case as to whether or not revalidation of the whole system is required. As more and more changes are applied, revalidation becomes increasingly appropriate.

Testing and Implementation of the Change

After evaluation and acceptance, the change can be effected, tested (if applicable), and formally commissioned into use. This principle applies equally to hardware and software; in the case of the latter, code redevelopment and testing should follow the same procedure as newly developed software. It is wise if possible to develop and test such changes in an isolated test/ development environment before applying the change to the operational system.

Testing is necessary to determine whether the change(s) work(s) properly and has (have) not compromised the system's functionality. The scope of testing should be based on the impact analysis. Where potential impact on other system functionality or other applications is identified, testing must be extended to include affected areas. This is sometimes referred to as regression testing.

Testing should be performed according to a test plan and all testing should be fully documented (e.g., test description, test items, acceptance criteria, results, date of test, and names and signatures of persons who performed the test). While testing is of course necessary, it is vitally important to understand a critical principle where software changes are concerned. This is that the assurance of the safety of the change should rest far most heavily on a review of the change to the *design* of the software. If reliance is confined to test results alone, serious new flaws consequential on the change but quite unanticipated may be overlooked.

After implementation of the change (in the operation environment), the system owner should formally accept the change. This formal approval can be made on the basis of the test results, or the system owner might decide to perform some separate acceptance test.

Change Completion and Approval

Finally, all the documentations concerning the change and all documents required for operation with the change need to be completed. It is important to identify and satisfy any training needs. The change request form shall be completed and passed to the system owner for final review and approval. Depending of the GxP relevance of the system, or the impact of the change, QA should review and endorse the implementation of the change. QA should always be informed about the change by being sent a copy of the completed change request form. The users should be informed (and trained if applicable) about the change. The system owner gives the final approval of the change and releases the system.

Recent Inspection Findings

The quality unit failed to put in place procedures to ensure that design documentation is updated when program code changes are made (FDA 483, 2007).

Changes were made to macro in process control software but change control process was not used, instructions on how to use the software were not modified, and changes were not validated (FDA 483, 2006).

Firms change control procedure does not include software changes (FDA 483, 2003).

There was no evaluation of impact of software changes on other parts of the program (FDA 483, 2003).

There is no system in place to ensure that parameter adjustments, which are executed in the during production runs, are made by authorized personnel (FDA Warning Letter, 2002).

Changes to in-house written templates used in QC laboratory computer system not controlled (FDA 483, 2003).

Failure to establish test plan/protocol for approved hardware changes (FDA 483, 2001).

Change control records were found signed off by the quality unit that had not been properly annotated in the code (FDA 483, 2001).

The firm failed to document review and approval of test records supporting program modification (FDA 483, 2001).

In managing change, personnel will receive their appropriate XXXX via an e-mail that has been sent from an e-mail distribution list. The firm has failed to implement controls to document that these distribution lists are maintained updated with the current approved listing of users (FDA 483, 2001).

Software "bug" that could result in erroneous release not scheduled for correction (FDA 483, 2002).

Computer program change requested to prevent shipping error has not been addressed (FDA 483, 2003).

The vendor edited the software and configuration, but there is no documentation of the vendor's evaluation (*of the problem*), description of root cause, or recommendation for correction (FDA 483, 2009).

Computer enhancement was identified as needed to correct labeling deviation but not implemented over one year later (FDA 483, 2002).

No record of review of software fix and correction of incorrect electronic records (FDA 483, 2002).

Documentation regarding system and software changes not fully available (FDA 483, 2004).

CONFIGURATION MANAGEMENT

Configuration management refers to the overall task of managing the use of varying versions of the various components (hardware, software, and documentation) that comprise a complex computer system. Both ISO 90003 and GAMP®5 promote configuration management as a recommended and necessary discipline. The level of formality needed is greater for an operating system compared with, say, a system in its early development, but the principles are the same. The use of configuration management tools can considerably ease the effort required here, especially in the case of larger systems where the level of complexity grows exponentially, rather than in a linear fashion.

Configuration management consists of the following activities:

- Configuration identification (what to keep under control)
- Configuration control (how to perform the control)
- Configuration status accounting (how to document the control)
- Configuration evaluation (how to verify that control)

Configuration management should be planned and conducted in accordance with defined procedures. This should include specified roles and responsibilities. Configuration management activities are normally specified with the validation plan or project and quality plan, although the complexity of larger projects implies the desirability of a separate configuration management plan.

Configuration Identification

Configuration management begins with the system assembly. The task here is to identify and document the build configuration by ensuring that the mix of software, hardware, and documentation, all in their various versions, is unambiguously known and coordinated. Clearly, if this is not done, chaos rapidly ensues. It is important to be able to establish the exact composition of a particular system build that can act as a baseline, or known reference point, against which any subsequent changes or behavior can be referred to. Key configuration management records include

- Document index (approved documents including key documents provided by suppliers such as user manuals),
- Hardware unit index (clients, servers, communication interfaces, printers, etc.), and
- Software program index (source code, executables, configuration files, data files, and third-party software such as operating system, library files, and drivers).

Configuration Control
All documents, hardware units, and software programs must be uniquely identified. It is not necessary to violate warranty seals to uniquely identify subcomponents. However, in situations where hardware units and software programs do not have a unique identification, physical labels and software header information can be added retrospectively. Unique identification should include the model number for hardware and the version number for software (e.g., MS Vista Service Pack 2). Current approved versions of source codes must correspond to current approved versions of software documents, object code, and test suites. All source codes should have associated documentation.

Configuration Status Accounting
Documentation showing the status and history of configuration items should be maintained. Such documentation may include details of changes made, with the latest version and release identifiers.

Configuration Evaluation
A disciplined approach must be sustained to maintain the integrity of configuration management. It can be tempting to relax configuration management just to release resources and reduce costs. Consequential problems that often arise include

- Only partial backups/archives of data made in a rush or on insufficient storage media,
- Items labeled unclearly or ambiguously labeled with poorly handwritten labels that cannot be understood by anyone other than author,
- Ambiguous version numbering nomenclatures, particularly for software where media might be labeled by date rather than by the version of the software carried,
- Supplier's notification of serial numbers that do not match the actual serial numbers delivered, and
- Documentation not fully synchronized with system changes, often because documentation was accorded a lower priority than the physical implementation of a change.

It is therefore recommended that the configuration status and practices should be regularly checked. Periodic reviews should include configuration management. It is important to demonstrate that

- The configuration management plan is up to date,
- Recorded configuration is consistent with physical status,
- Naming and labeling conventions are being followed,
- Software version controls are being applied,
- System is in the intended baseline state in accordance with defined milestones (e.g., for supplier testing, at installation, for user acceptance testing, and for use), and
- Change management is effective.

The fundamental challenge to configuration management record keeping is whether it can make a full system rebuild possible by relying solely on these records. A good test is to assess the measure of confidence the responsible person has that the system could be successfully restored on a first attempt. At a practical level, this capability is the foundation of disaster recovery, directed at the effective support of business continuity plans.

Recent Inspection Findings

The firm has failed to adequately control software configuration (FDA 483, 2003).

System configuration documentations were found to be obsolete (FDA 483, 2004).

The firm has failed to establish an overall revision control system for the program throughout its software life cycle (FDA 483, 2001).

The *computer system* is not validated in that ... configuration management. The firm failed to document all sites, departments, or connections on the network ... The firm has failed to document external program interfaces ... The firm has failed to define or describe the various used for the development, test, and production environments (FDA 483, 2001).

Networked system can only support four interfaced systems but had up to five systems attached. There was no validation showing this configuration to be acceptable (FDA 483, 2000).

The vendor edited the software configuration, but there is no documentation of the vendor's evaluation (*of the problem*), description of root cause, or recommendation for correction (FDA 483, 2009).

REQUIREMENTS TRACEABILITY

Traceability is a fundamental in terms of demonstrating regulatory compliance. In ISO terminology, traceability demonstrates that design input is linked to design output and has been verified. For example, does the functional specification fulfill the user requirements specification?

Principles

- It should be possible to explicitly link specifications forward where there is appropriate design, testing and subsequent changes. Likewise, it should be possible to trace from testing backward through to specification.
- The linkage between specification, design, and test is not necessarily limited to a 1:1:1 relationship. Multiple requirements can be covered by a single test. Multiple tests can be required to cover one requirement.
- It should be possible to measure coverage of design and testing in fulfilling user requirements. Focus should be placed on the traceability of critical requirements. Criticality can then be cascaded through design to testing.
- Traceability should be maintained throughout system operation and deal with any changes made to the system.
- Traceability should not force a single stream of sequenced activities. Many activities can be conducted in parallel; only approvals and key dependencies need to be sequential.

Benefits of Traceability

Establishing a robust means of traceability brings many benefits. During the project development, it can be used as a means to

- Identify and control critical aspects of system,
- Identify and challenge user qualification duplicating development tests,
- Demonstrate testing percentage coverage of system specification, and
- Determine scope of testing to address design revisions.

The value of traceability is not limited to project. Maintaining traceability during the operational life of a system is also beneficial.

- Facilitates impact assessment for proposed changes
- Identifies scope of regression testing for proposed changes
- Links compliance documents with supplementary change control records
- Supports inspection readiness by quickly identifying where certain aspects of the system are documented and tested

Traceability Requirements

Typical traceability requirements for requirements traceability matrices (RTMs) and other trace mechanisms are

- URS element links to functional specification element,
- Functional specification element links to design element,
- Design element links to elements of source code,
- Test case links to URS element,
- Test case links to functional specification element, and
- Test case links to design element.

It is not necessary to link test cases to elements of source code as the practicalities of this are generally not manageable.

Special consideration may be required for COTS applications where design, source code, and development testing documentation may not be available without supplier support. Likewise, where there is a combined functional design specification, the need for specific traceability may be alleviated.

Requirements Traceability Matrix

Traceability information is often recorded in a tabular form commonly known as a RTM. The concept of requirements traceability is shown in Figure 3.3 with an example RTM provided in Table 3.2.

The structure and level of granularity of an RTM will have a major impact on its practicality. The following comments should help ensure that a practical approach is adopted.

- Strategy for traceability should be established during validation planning. User requirements should be developed with traceability in mind.
- Level of traceability should stop where user documentation references supplier documentation. The suppliers should have their own traceability for the documentation and testing under their control. This should be verified during supplier audits where these are appropriate.

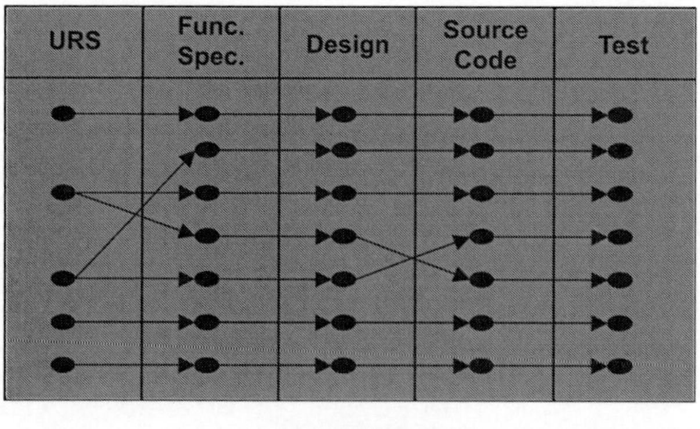

Figure 3.3 Requirements traceability.

Table 3.3 Example of Requirements Traceability Matrix Extract

URS reference	URS topic	Specification	Design		Testing			Change controls not incorporated into documents
		Functional specification reference	Design specification reference	Source code review reference	Development testing reference	User qualification protocol reference	SOP reference	
1	Scope	N/A	N/A	N/A		N/A	N/A	N/A
1.1		1.1	1.1, 1.7	N/A		N/A	N/A	N/A
1.2		1.2, 1.4	2.3	1.5		N/A	N/A	N/A
1.3	Out of scope	7.1	N/A	N/A		N/A	N/A	CC-005
2	System boundary	N/A	N/A	N/A	N/A	N/A	N/A	N/A
2.1	Server	2.1	2.1, 2.3, 2.7	2.2	DT-1-3,4,5	IQ-ABC-001 OQ-ABC-001	N/A	CC-001
2.2	Networks	2.2	2.1, 2.7	4.3	DT-1-1,2	OQ-ABC-002 OQ-ABC-003 PQ-ABC-001	N/A	CC-002
7	Interfaces	N/A	N/A	N/A	DT-1-6	N/A	N/A	N/A
7.1	HMI	7.3, 7.6	N/A	N/A	DT-1-7	IQ-ABC-021	N/A	N/A
7.2	Instrumentation	7.3, 7.6	N/A	N/A	DT-2-1,2,3 DT-2-4,5,6	OQ-ABC-037 PQ-ABC-023	N/A	CC-006
8	System Performance	N/A	N/A	N/A	DT-3-1	N/A	N/A	N/A
8.1	Security	8.7	N/A	N/A	DT-4-1	OQ-ABC-038 PQ-ABC-024	SOP-ABC-07	N/A
8.2	Data integrity	8.7	N/A	5.6	DT4-2	N/A	SOP-ABC-12	N/A

- Requirements need not trace to technical controls in all circumstances. Requirements can trace to procedural controls in which case a cross-reference to identified SOPs is appropriate.
- Adding a column to include a brief description of each requirement may assist verification that matrix contents are referenced correctly. It is also possible to include specific identification of what are critical and noncritical requirements.
- For simple systems, a RTM is not recommended as sufficient traceability can be incorporated as cross-references embedded within individual document.
- For large systems, RTM may exist as one or more tables. Each table may have a slightly different structure depending on any special needs for tracking combination of local and global requirements.

RTMs are usually created immediately after or in parallel with the functional specification. The functional specification represents the contractual system definition. The URS is then retrospectively mapped against the functional specification. Subsequent activities are prospectively mapped.

Manual maintenance of RTM documents can be labor intensive and error prone. It is therefore recommended that tools for managing traceability information be adopted as an alternative. Spreadsheet applications are an obvious choice in simple cases. For more complex document sets, however, a more sophisticated automated solution is much more likely to be up to the job.

Completed RTMs should be retained as part of the compliance package as a controlled document. QA approval is not necessarily required. Tools used to support traceability should themselves be qualified and controls should be implemented where records are maintained electronically.

Recent Inspection Findings

Your response fails to trace back to source code, and the related software development cycle that establishes evidence that all software requirements have been implemented correctly and completely and are traceable to system requirements (FDA Warning Letter, 2001).

Your response did not evaluate requirements to trace changes to determine side effects (FDA Warning Letter, 2001).

SELF-INSPECTIONS (INTERNAL AUDITS)

Self-inspections, also known as internal audits, are a fundamental activity of competent quality assurance. They typically focus on reviewing compliance documents and the SOPs used to generate them. Guidance is not usually audited unless it is effectively being used as the procedure for work.

The PIC/S harmonized computer inspection guide contains aide memoirs that can be used as input to develop a self-inspection checklist (Appendix 3.B) (4). Such checklists can be used to examine any aspect of working practices and assess the level of GxP compliance of processes, including computer systems, with a view to identify poor practices and opportunities for improvement. They can also be used with equal beneficial effect in non-GXP areas.

Personnel independent of the work practices being examined should conduct these audits. Independence implies the ability to demonstrate a measure of impartiality. It does not necessarily mean that the person conducting the self-inspection is from a separate department or function. A peer review is perfectly acceptable.

A report describing the self-inspection and its observations should be produced unless a dispensation has been specifically not required and given within the terms of reference. Observations from internal audits should be precise and objective. Documenting subjective opinions should be avoided. Closure of actions should be tracked.

Internal audits are not usually subject to regulatory scrutiny without due cause, and reports of such should not be presented during a regulatory inspection without good reason. If an inspector does ask to see evidence of self-inspections being conducted, the pharmaceutical or healthcare company should consider sharing with the inspector the schedule of self-inspections recently completed and leave it at that.

MANAGING DEVIATIONS

A nonconformance may be discovered during testing, internal audits, or regulatory inspections, through a customer query, or by chance. It may be caused by failure to follow procedural controls, by a failure of the procedural controls themselves, or by an ambiguity or lack of detail in the documents and records supporting the procedural controls. Examples of nonconformances include the following:

- Computer systems that do not behave in the necessary manner (e.g., electronic record and electronic signature requirements)
- Compliance activities that do not conform to defined procedures
- Compliance documents that do not conform to defined procedures
- Defined procedures that do not fulfill regulatory requirements

When a nonconformance is discovered, it must be reported to the manager responsible for the computer system, service, data, or document as appropriate. Depending on the significance of the nonconformance, QA management may also need to be informed.

Investigation

A thorough investigation should be undertaken to determine the root cause of the deviation (4,5). Sometimes, it will only be possible to determine the most probable root cause. The rigour of the investigation will depend on the nature of the deviation and its criticality. Cursory investigations should be avoided as their outcome will not help anyone in the long term. The nature of root cause can be categorized, and these categories trended over time to determine underlying problems with particular aspects of computer systems, their operation, or failure to determine the real root cause.

Issue Resolution

Resolution of issues can take a variety of forms. The approach taken will depend on the significance of the issue, whether action can be taken to avoid the issue occurring, whether the issue can be mitigated, and whether the issue can be tolerated without mitigation. Options for resolving issues include

- Modifying the design of a computer system,
- Implementing a separate downstream computer system to detect and remedy any potential quality failures,
- Introducing manual procedures in parallel with the computer application to complement the computer system's functionality (e.g., security measures),
- Introducing independent manual procedures in parallel with the computer application to verify the computer's functionality, and
- A combination of the above.

Some issues will need specific and immediate attention to provide assurance of control and to justify the use of the computer system. Issues with a high impact on drug product safety (and hence patient safety), for instance, may benefit from a more detailed design review or testing to determine the extent and severity of the issue before a root cause can be properly identified with confidence and then corrected. Other issues may be accepted as having no or negligible impact without the need for any corrective action. This is quite acceptable so long as the justification is robust and suitably documented.

DEVIATION REPORT	**Deviation No:**

System or Service:

Location:

Name/Date of Person Identifying Nonconformance:

Details of Nonconformance

Date identified:

Description of Nonconformance:

Description of Remedial Action or Concession

Disposition: Remedial Action/Concession (delete as appropriate)

Signature: Date: Representing: User

Signature: Date: Representing: QA

Signature: Date: Representing: Technical

Approval

These following approvals signify a satisfactory outcome to identified non-conformance.

Signature: Date: Representing: User

Signature: Date: Representing: QA

Signature: Date: Representing: Technical

Figure 3.4   Example of deviation report.

Deviation Reports

Deviation Reports should be prepared to describe the nonconformance, analyze the nature of the deviation, and define how the deviation is being addressed. The criticality of the deviation will determine appropriate controls.

- Deviations that have a critical GxP impact, that is, those that affect the quality, efficacy, or safety of pharmaceutical and healthcare products, will require root cause remediation.
- More general deviations affecting GxP processes may be accepted with consequential changes such as processes changes, computer design changes, changes to user procedures (sometimes known as work-arounds), and concessions on the acceptability of validation (e.g., acceptance of unexpected test results). Some deviations may require further investigation (e.g., testing) to scope a problem before a concession can be justified.
- Deviations that do not impact GxP processes, that is, those that cannot impact pharmaceutical and healthcare products, do not necessarily require remediation and can be accepted without corrective action so long as there is no other key operational deficiency.

Deviation reports must document the approval of the remedial actions (or concession to accept the deviation without remedial action) and justify closure of the deviation explaining where appropriate ongoing controls to stop it happening again. An example of deviation report is shown in Figure 3.4. Deviation reports associated with GxP impacting nonconformance require the signature and approval of the QA organization. The reports with supporting evidence should be retained as part of the computer system documentation. How deviations are handled is a good indicator on how well an organization understands compliance requirements, and pharmaceutical and healthcare companies should not be surprised if deviation records are requested for review during regulatory inspections.

The management of deviations and compliance issues is discussed further in chapters 5 and 10 where the use of project compliance issue logs, risk, actions, issues, decisions (RAID) log, and validation reports are introduced.

Recent Inspection Findings

Thirty-seven cases were found in which software changes were made in response to problems but the firm was unable to provide any investigation records (FDA 483, 2003).

No SOP describing use of computer system incident logbook (FDA 483, 2004).

No root cause determined to initiate necessary corrective actions for database software errors (FDA 483, 2005).

No SOP for access, operation, maintenance, and control of system used to track investigations and corrective actions (FDA 483, 2006).

No procedures in place defining or controlling the use of the database used by QA to monitor deviations and investigations (FDA 483, 2007).

REFERENCES

1. ACDM/PSI. Computer Systems Validation in Clinical Research: a Practical Guide, Version 1.1, December 1998.
2. Wingate GAS. Computer Systems Validation: Quality Assurance, Risk Management and Regulatory Compliance for Pharmaceutical and Healthcare Companies. Rockville, MD: Interpharm Press, 2003.
3. Food and Drug Administration. General Principles for Software Validation, Final Guidance for Industry, January 2002.
4. Pharmaceutical Inspection Co-operation Scheme. Good Practices for Computerised Systems in Regulated GxP Environments, Pharmaceutical Inspection Convention, PI 011-1, Geneva, 2003.
5. International Society for Pharmaceutical Engineering. GAMP®5: Risk-Based Approach to Compliant GxP Computerised Systems, Tampa, 2008. Available at: www.ispe.org.

Appendix 3.A Examples of Deficient Documentary Evidence

- Failure to use standardized document formats
- Incomplete definitions
- Constraints not cited
- Formulae inconsistencies
- Inappropriate or inconsistent "<" and ">"
- Legends with inconsistent or misleading scales
- Excessive changes
- Different corrections not distinguished
- Not reporting significant changes
- Standard reports and forms not used (raw data recorded informally)
- Illegible writing
- Raw data records not available
- Inconsistent dates
- Inconsistent units of measure
- Lack of double checking for accuracy
- Not reporting all adverse reactions
- Problems and deviations not fully reported
- Unusual or unexpected recorded results
- Records retained informally
- Inappropriate procedures
- Deficient procedures
- Staff responsibilities not defined
- Training records not up to date
- Poorly organized documents and records
- Hard-to-follow documents and records
- Report conclusions that seem too good to be true
- Reports with exaggerated claims
- Reports with political half-truths
- Reports with incorrect absolute terms (all, every, none, never, etc.)

Appendix 3.B Example of Self-Inspection Checklist

- Determine the critical control points (base investigation on FMEA or other hazard analysis technique). Examples would be
  - Pasteurization,
  - Sterilization,
  - PH control,
  - Temperature control,
  - Cycle timing,
  - Control of microbiological growth,
  - Quality status of materials and products, and
  - Record keeping.
- For those critical control points controlled by computerized systems, determine if failure of the computerized system may cause drug adulteration.
- Identify computerized system components including
  Hardware Inventory
  - Input devices
  - Output devices
  - Signal converters
  - Central processing unit
  - Distribution system
  - Peripheral devices

  Software Inventory
  - Inventory of files (program and data)
  - Documentation:
  - Manuals
  - Operating procedures
- Hardware
  Obtain a simplified drawing of the computerized system (covering major computer components, interface, and associated system/equipment). For computer hardware, determine the manufacturer, make and model number.

Appendix 3.B   Example of Self-Inspection Checklist (*Continued*)

- Software
  For all critical software determine
  - Name,
  - Function,
  - Inputs,
  - Outputs,
  - Set points,
  - Edits,
  - Input manipulation of date,
  - Program overrides, and
  - Version control.
  Who developed the software (standard, configured, customized, bespoke)
  Software security to prevent unauthorized changes
  Checking computerized systems input/outputs
  Obtaining simplified drawing of overall functionality of collective software within
  Computerized systems
- Data
  What data is stored and where?
  Is data distributed over a network, and how is it controlled?
  How is compliance to electronic record regulations achieved?
  How is data integrity verified?
- Personnel
  Type (developer, user, owner)
  Training records
- Observe the system as it operates to determine if
  - Critical processing limits are met,
  - Records are accurate,
  - Input is accurate (sensor or manual input),
  - Time keeping is accurate, and
  - Personnel are trained in systems operations and functions.
- Determine if the operator or management can override computer functions. How is this controlled?
- How does the system handle deviations from set or expected results?
  Check all alarms, calculations, algorithms, and messages
  Alarms
  - Types (visual, audible, etc.)
  - Functions
  - Records
  Messages
  - Types (mandate action?)
  - Functions
  - Records
- Determine the steps used to insure that the computerized system is functioning as designed
- Was the computerized system tested upon installation?
  - Under worst case conditions?
  - Minimum of three test runs?
- Are there procedures for routine maintenance?
  - User manual
  - Vendor-supplied manual
  - Third-party support manual
  - Management manual
- Does the equipment here meet the original specifications?
- Is verification and validation of the computerized system documented?
- How often is system
  - Maintenance performed,
  - Calibrated, and
  - Revalidated?
- Check scope and records of any service level agreements.
- Are there procedures for revalidation? How often is revalidation conducted?
- Are system components located in a hostile environment that may effect their operation (ESD, RFI, EMI, humidity, dust, water, power fluctuations)? Are system components reasonably accessible for maintenance purposes?
- Determine if the computerized system can be operated manually. How is this controlled?

(*Continued*)

Appendix 3.B   Example of Self-Inspection Checklist (*Continued*)

- Automated cleaning in place (CIP).
  How is automated verified?
  Documentation of CIP steps.
- Automated sterilization in place (SIP)
  How is automated sterilization verified?
  Documentation of SIP steps.
- Shutdown procedures
  Does firm use a battery backup system?
  Is computer program retained in control system?
  What is the procedure in the event that power is lost to computer control system?
  Have backup and restore procedure been tested?
- Is there a documented system for making changes to the computerized system?
  Is there more than one change control system (hardware, software, infrastructure, networks)? For each of these, challenge as follows:
  - The reason for the change
  - The date of the change
  - The changes made to the system
  - Who made the changes
  How do they interface? Challenge change history, and verify audit trail.
- What are the auditor's impressions of
  - Presentation of regulatory compliance,
  - State of documentation,
  - State of compliance,
  - Maintaining operational compliance, and
  - Requirements for revalidation?

4 | Prospective Verification and Validation

INTRODUCTION

The various compliance approaches promoted within the pharmaceutical and healthcare industry by GxP regulators, industry initiatives, and individual pharmaceutical and healthcare companies all adopt the same basic approach. This follows the following basic steps:

- Define what is to be done (plan)
- Define how to do it (specification, procedures, and resources)
- Do it, controlling any changes (change control)
- Establish that the end result was what was originally intended (verification)
- Provide evidence demonstrating this (audit trail)

This chapter presents the set of life cycle phases summarizing the project approach typically adopted within the pharmaceutical and healthcare industry. These life cycle phases may be known by alternative names within different organizations, as there are yet no generally accepted naming conventions or groupings of phases throughout the industry. It is important, however, that all the activities covered by this chapter are included in any alternative scheme.

CHARACTER OF APPLICATION

The features of a computer system and hence its compliance requirements can be described in terms of its hardware and software. International Society for Pharmaceutical Engineering (ISPE)/GAMP® have defined categories of software found in computer systems (1). The numbering system for these software categories is not sequential because in the latest copy of GAMP®5 guide, the original category 2 software (firmware) has now been designated to the other categories depending on nature of software involved. The remaining four software categories are intended to be all embracing, so it should not be possible for any software to fall entirely outside of them. The four software categories are as follows:

- **GAMP category 1 software: operating systems**
  This category defines established commercial off-the-shelf (COTS) operating systems. Examples include OS/2, MPE, Microsoft Windows, Fix DMACS, and database management systems. Regrettably, upgrades can be a mixed blessing, for while correcting defects and delivering enhancements, they can, at the same time, have a serious impact on overall system performance and security. Because of this, regression testing of the application using the new version of the operating system cannot be skipped when the latter is upgraded. Operating systems that are new and unusual and thus cannot be regarded as having been adequately market tested should be managed as category 5 software (see below).
- **GAMP category 3 software: standard software packages**
  This category defines COTS software packages. Any configuration to which such software would be subjected in a pharmaceutical and healthcare operation is limited to operating parameters, and configuration of the system environment parameters (e.g., file names, directory/folder structures). Examples include statistics packages and base software for spreadsheets. Commercial software packages that lack wide exposure on the market must not be recognized as market tested and should be managed as category 5 software.

Table 4.1 Applying GAMP® Software Categories to Typical Computer Systems

Type of system	Typical applications based on following computer system[a]	GAMP software categories			
		1	3	4	5
Computerized analytical laboratory equipment[b]	Analytical instrument		▓		
	High performance liquid chromatography			▓	
	Chromatography data system				b
Process control and monitoring systems[b]	Field instrumentation		▓	c	
	Programmable logic controller			▓	▓
	Supervisory control and data acquisition			▓	▓
Spreadsheet and database applications[b]	Spreadsheet application			▓	d
	Database application			▓	d
Corporate computer systems	Laboratory information management system			▓	d
	Manufacturing resource planning II/enterprise resource planning			▓	d
IT infrastructure and services	Desktop environment including personal computer or workstation	▓	▓	▓	▓
	Communication network	▓	▓	▓	▓

[a]The GAMP software categories of the various computer system applications should be verified for individual systems.
[b]Excludes resident personal computer or workstation.
[c]"Fieldbus" instrumentation is online user configurable.
[d]Typically, interfaces and bespoke reports.

- **GAMP category 4 software: custom-configurable software packages**
  This category defines COTS software packages whose configuration makes use of standard calculations and predefined control routines. Configuration of the system environment may also be required. Examples of commercial configurable software packages of this kind include SCADA, DCS, MES, LIMS, MRPII, and some test equipment. The core software package and its configuration should be managed as category 3 and 5 softwares, respectively.
- **GAMP category 5 software: custom (bespoke) application software**
  This category defines software written entirely to meet the exclusive requirements of a single user/company, or small group of users. Because of this, it is likely to be updated frequently. Examples of custom software developments include code for data migration, code for user reports, macros for spreadsheets, database scripts, and code for interfaces. Software from another category that has been customized should also be managed as category 5 software. Regression testing is required when such software is upgraded.

Most computer systems will have a number of software components falling into several of these categories, as illustrated in Table 4.1.

Hardware can also be divided into categories representing different compliance requirements. There are two GAMP categories of hardware: standard hardware and custom (bespoke) hardware.

APPROACH TO VERIFICATION AND VALIDATION

Validation requires that a computer system as a whole meets its predetermined specifications. The specific project life cycle model adopted does not matter as long as it covers planning, requirements and design, implementation, and testing (1–3). The constituent phases making up the life cycle must be clearly defined with entry and exit criteria for each phase. Appropriate verification is needed to ensure the controlled completion of constituent phases.

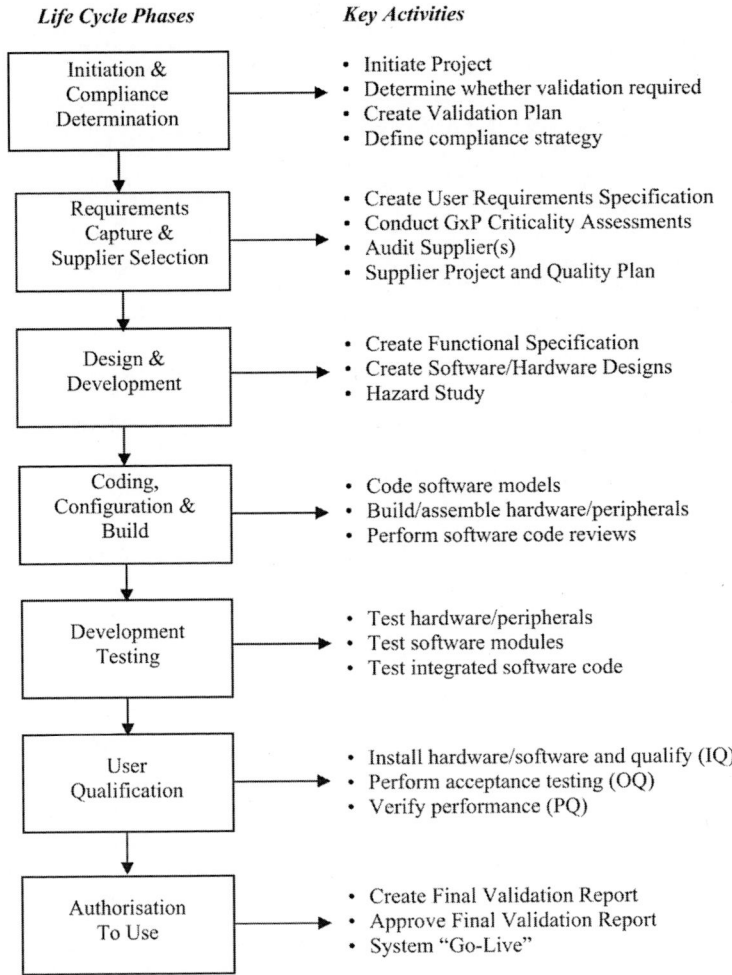

Life Cycle Phases	Key Activities

Initiation & Compliance Determination
- Initiate Project
- Determine whether validation required
- Create Validation Plan
- Define compliance strategy

Requirements Capture & Supplier Selection
- Create User Requirements Specification
- Conduct GxP Criticality Assessments
- Audit Supplier(s)
- Supplier Project and Quality Plan

Design & Development
- Create Functional Specification
- Create Software/Hardware Designs
- Hazard Study

Coding, Configuration & Build
- Code software models
- Build/assemble hardware/peripherals
- Perform software code reviews

Development Testing
- Test hardware/peripherals
- Test software modules
- Test integrated software code

User Qualification
- Install hardware/software and qualify (IQ)
- Perform acceptance testing (OQ)
- Verify performance (PQ)

Authorisation To Use
- Create Final Validation Report
- Approve Final Validation Report
- System "Go-Live"

Figure 4.1   Key compliance activities.

There is therefore a subtle distinction between validation and verification, but they are mutually supportive.

Figure 4.1 illustrates the project life cycle used in this book. Although represented as a sequence of phases, many activities can be conducted in parallel without compromising compliance. Indeed, where appropriate, the parallel execution of activities should be encouraged to help de-bottleneck project critical paths. Documents may also be combined so long as any necessary sequential order of activities is preserved.

The life cycle approach applies to both "in-house"-developed and in-house-purchased computer systems. Supplier responsibilities are indicated later in this chapter. Suppliers include internal development or support groups, external vendors, and outsource organizations.

Emphasis within the life cycle will change depending on whether computer hardware is custom (bespoke) or standard, and the mix of software categories in which the application software falls. Tables 4.2 and 4.3 outline the preferred approach toward different categories of software and hardware.

While this approach may seem simple, it becomes more complex when validating an application containing software in multiple categories. Most computer systems will contain software in multiple software categories, as it is very unusual to find a system that is made up of software falling into only one category. It is not usually practical or desirable to treat each item of software independently. Rather, a holistic approach should be taken that collectively

Table 4.2 Software Categories (Based on GAMP®5)

Category	Software type	Compliance expectations
1	Operating system	Record version (including any service pack). The operating system will be challenged indirectly by the functional testing of the application.
3	Standard software packages	Record version and any configuration of environment. Verify operation against user requirements.
		Consider auditing the supplier for critical and complex applications.
4	Custom-configurable software packages	Record version, any parameter configuration, and any configuration of environment. Verify operation against user requirements.
		Normally, assess (audit) software development capability maturity of the package supplier for complex and critical applications.
		Manage any bespoke programming (e.g., macros) as category 5 software.
5	Custom (bespoke) Application software	Assess (audit) software development capability maturity of supplier, and validate complete computer system.

Table 4.3 Hardware Categories (Based on GAMP®5)

Category	Hardware type	Compliance expectations
1	Standard hardware	Record the model, version number, and, where available, serial number of preassembled hardware. Retain hardware data sheets and other supplier specification material. Document hardware configuration details. Verify installation and performance of hardware components.
2	Custom (bespoke) hardware	Manage standard hardware components as category 1 hardware.
		Prepare design specification including any hardware configuration. Verify installation and performance of hardware components. Assess (audit) hardware development capability maturity of supplier.

addresses the software as a complete application. A well-organized and streamlined approach is necessary.

Commercial Off-the-Shelf Products

Nowadays, most softwares and hardwares are based on purchased COTS products, rather than being bespoke/custom built. Pharmaceutical and healthcare companies (the COTS product user) are accountable to the regulatory authorities for ensuring that the product development methodologies used by the COTS developer are of a sufficient degree of capability maturity and adequate for the intended use of the COTS product.

The key compliance activities presented in Figure 4.1 can be potentially simplified if the supplier can provide information about the development of their COTS product. Figure 4.2 presents key user activities for COTS products, the shared area of the life cycle reflecting supplier responsibilities. In many cases, the necessary documentary evidence to facilitate this can only be obtained through the mechanism of a formal audit of the supplier, as discussed in chapter 6. If an audit is conducted and significant progress with any corrective actions cannot be proven, then suitable alternative suppliers or sources of software should be sought, including custom (bespoke) developments. Any decision not to change supplier when there is no meaningful progress on corrective actions needs to be justified in a documented rationale.

Another issue that can occur with COTS products is that access to product development documentation may not be available. The suppliers may refuse to share their proprietary information, or it may be unavailable for some other reason. In such circumstances, pharmaceutical and healthcare companies are expected by the regulatory authorities to compensate with additional *black box* testing to establish with sufficient confidence that the

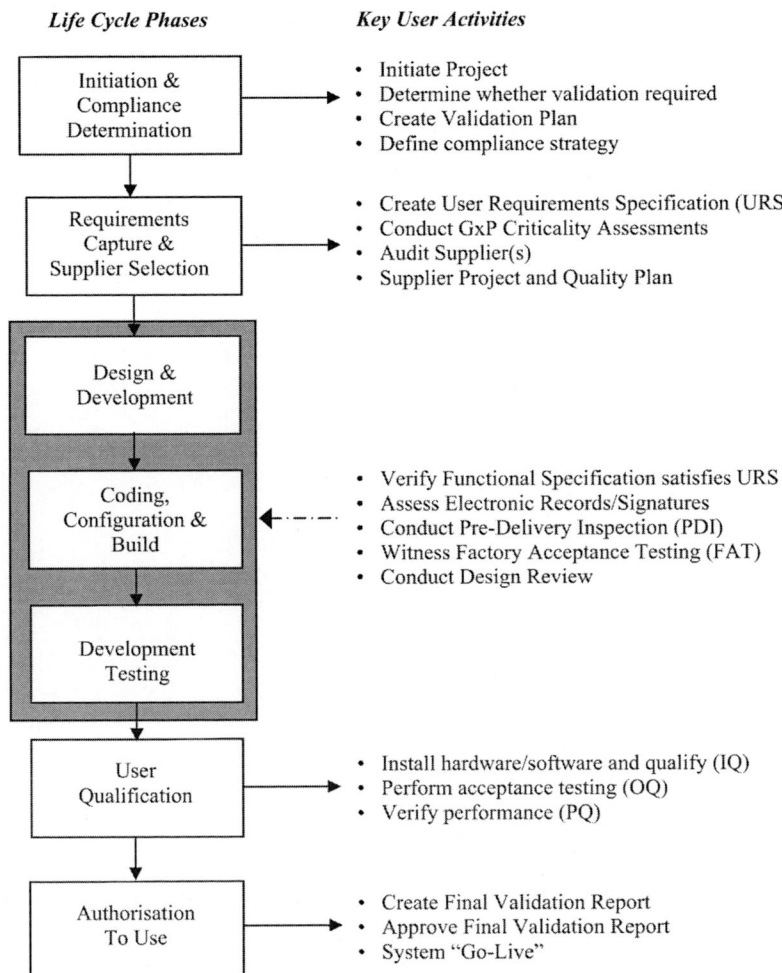

Figure 4.2 Key COTS compliance activities.

COTS products meet user needs. (Black box testing is functional testing of software often based on verifying expected outputs from defined inputs, without having any real idea of how the software does it or how it works. The imaginary box is *black* because it cannot be opened to inspect its inner workings.) Commercial software products may have "bug lists," user manuals, and product specifications that can be compared with the user requirements to structure such black box testing effort.

The FDA recognizes that user black box testing is not practical for some COTS products. The proper operation of such products may be satisfactorily inferred by other means such as independent certification by an accreditation body (2).

Open Source Software

Open source software (OSS) (also known as freeware or shareware) is increasingly being incorporated into COTS products. OSS is developed by informal communities who claim no ownership and refute any accountability for the code. Features and bug fixes emerge out of uncoordinated custom developments of freely available uncontrolled copies of the source code. There is usually no formal quality umbrella for software development and support, and accordingly, it is very difficult to demonstrate that such software is fit for purpose. Reverse engineering development documentation from the code and conducting comprehensive testing is likely to be extremely expensive. The use of OSS should therefore be avoided for

Table 4.4 Tools Verification

Type of software tool	Compliance requirement	Examples
Computer-aided software engineering used in context of supporting creation, ongoing maintenance, and/or retrieval of compliance records	Verification required	Configuration management supporting IQ records. Automated testing tools supporting OQ/PQ records. Tools applying approvals to compliance records. Change management tools supporting change control records.
System software	Verification by inference	Compilers, communications interface driver software, Acrobat Writer.
Incidental use	Verification not required	Word processors, Microsoft Project, network performance, computer-aided software engineering not requiring validation as above (e.g., debuggers, static code analysis, process modeling, and PLC programming).

Abbreviations: IQ, installation qualification; OQ, operational qualification; PQ, performance qualification.

critical applications that have a direct impact on product quality and patient safety. If it cannot be avoided (e.g., embedded in COTS products), then intensive black box testing should be undertaken commensurate with the criticality of the functionality involved.

Software Tools

Software tools may require verification even though they might be considered supplementary to the application software, as summarized in Table 4.4.

Software used to automate any part of the software development process, its maintenance, or any part of the quality system is expected to be verified (2). Examples of such software tools include code-generating tools, quality diagnostic tools, automated testing tools, and configuration management tools. These software tools should be assessed in line with their software category, and their use should be described in the documentation of the application using them. Tools that create electronic records also need to be assessed for compliance with applicable regulatory requirements for electronic records and electronic signatures.

Software tools supporting the functionality of applications should also be verified. Formal activities are not required, however, because verification is inferred by the correct operation of the application itself.

Software tools do not require verification if they are incidental to the creation of paper records that are subsequently maintained in traditional paper-based systems. Examples include word processors, essentially act as a typewriter, and project scheduling tools that do not have a regulatory dimension. Software that provides record storage and retrieval facilities would be treated like traditional "file cabinet." Overall reliability and trustworthiness would derive primarily from well-established and generally accepted procedures and control for paper records.

Managing Change

An important matter when choosing a life cycle is the relative cost of making changes to software as the cycle progresses. The relative cost of correcting errors is shown in Figure 4.3. It is very important to clarify incomplete or ambiguous information as soon as reasonably practical, because the longer it is left, the more expensive it will be to correct if a wrong assumption has crept in in the meantime. The later changes are made, the more they are likely to cost.

Experience suggests that poor design and programming errors account for up to one-third of system malfunctions (5). This emphasizes the importance of not only avoiding these problems but also discovering these problems as early as possible. In particular, the use of design reviews is promoted to catch errors early and thereby reduce the cost of their correction

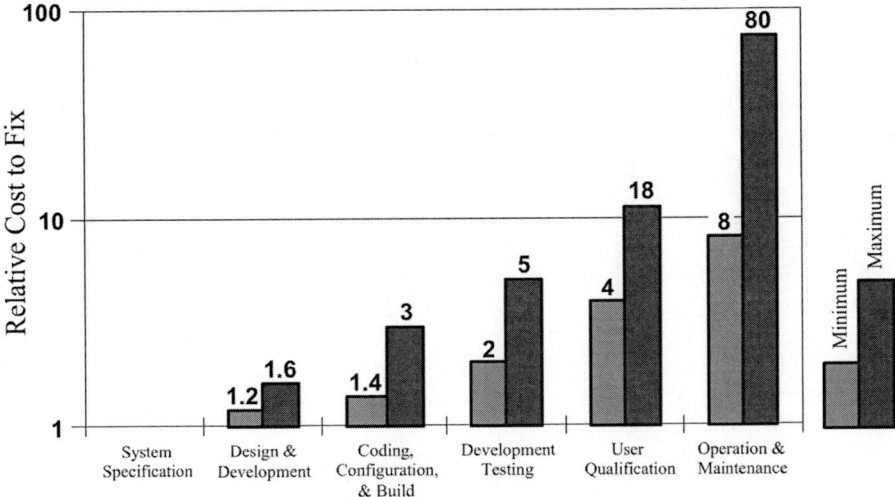

Figure 4.3 Relative cost of error correction. *Source*: From Ref. 4.

when compared with only discovering the errors during testing. The use of software inspections (either conducted formally or via less formal code walkthroughs) should also save money by the early exposure and correction of errors. It is estimated that projects implementing effective design reviews and software inspections can reasonably expect the overall project effort to shrink by about 10% when compared with projects paying little or ineffective attention to these questions (6). Not only is the project cheaper to run, but also the outcome (the computer system) is more predictable and has higher quality, with all the benefits that this implies for compliance.

Another important feature of change is the changing influence between the pharmaceutical and healthcare companies and the supplier during the project. At the beginning, the pharmaceutical and healthcare company has enormous influence, which is right and proper as this is where the definition and the direction of the project originate. However, the pharmaceutical and healthcare company's influence quickly declines as the project progresses because of the increasing relative cost of change. The supplier soon has a significant influence because he or she largely dictates whether or not changes can be implemented within the constraints of the project (time, functionality, and cost). This transition of influence is illustrated in Figure 4.4. Practitioners need to ensure that a project does not inadvertently fall into the position of becoming resistant to the implementation of critical changes, because time and budgets were frittered away earlier in the project in the pursuit of less important changes.

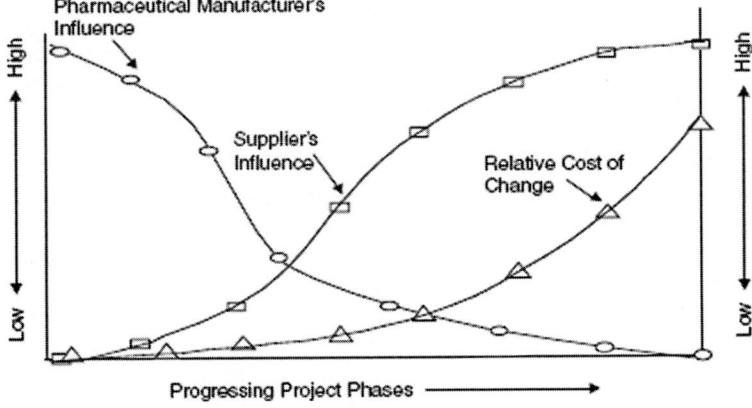

Figure 4.4 Changing project influences.

Project Roles

The diagram for the life cycle presented in Figure 4.1 is complemented in Figure 4.5, with one showing the main roles of system owner/user, developer, and quality and compliance. A summary of each role is described below:

-The *system owner/user* role is responsible for defining and approving requirements. It should be possible to describe in overview the system to be implemented from this information. The system owner/user should then agree with quality and compliance on what functionality within the computer system is GxP critical. This is used to help select the supplier to develop the system. The system owner/user should lead a design review to verify that what is being developed meets the requirements, with feedback to the design and development group(s) as required. The user qualification process should take account of GxP assessment of critical functions, the capability of the supplier from the supplier selection process, and any recommendations from the design review. The system owner/user should approve user qualification and the final validation (summary) report.

- The *developer* role should start with the drafting of an agreed contract of supply. The scope of supply may alter as a result of the supplier selection process. Design and development, system build, coding, and configuration (including software inspection/ source code review) then follow. The computer system is functionally tested (with traceability) to confirm that the design intentions (and in turn the user requirements) were achieved. Satisfactory testing is finally used to authorize release of the system for distribution. The system owner/user will then qualify the system using as much evidence from development testing process as possible to reduce the user qualification effort required.

- The *quality and compliance* role traditionally starts with the preparation and approval of a validation plan and associated specification of a compliance strategy. Both will be influenced by the user requirements specification and the output from the GxP assessment. A supplier audit may be required depending on the level of criticality and degree of custom software involved (discussed further below). Predelivery inspection should be conducted as needed to follow up on any supplier audit issues and to confirm that the supplier is implementing his or her own quality management system as required on the development of the computer system. Quality and compliance should further participate in the design review and verify that what is being developed meets compliance requirements. User qualification should also be reviewed and approved together with authorization of the validation (summary) report concluding project validation.

CHOOSING AN APPROPRIATE LIFE CYCLE METHODOLOGY

The compliance activities presented in this book can be implemented in accordance with a number of life cycle methodologies. The "waterfall model" life cycle is the most rudimentary approach and basically cascades the activities presented in chapters 5 to 12. The ordered sequence of the "waterfall" life cycle works well with tightly defined and understood requirements. Alas, in the real world, although we might like to think otherwise, most projects are not tightly defined and understood. In these situations, an appropriately controlled but more flexible approach is needed.

The "V model" life cycle has found prominence in the pharmaceutical and healthcare industry for computer system projects (1). The compliance activities are presented in a "V" shape, as illustrated in Figure 4.6. The V model was developed to promote planning and designing in anticipation of testing. Phases of the life cycle are conducted in a controlled sequence. In theory, the current phase must be completed before progressing to the next phase. In practice, while the edges may blur a little between phases, the order of finalizing key documents cannot be broken. User qualification (acceptance testing) should only be conducted after coding, configuration, and build. Design should not be concluded without a finalized user requirements specification.

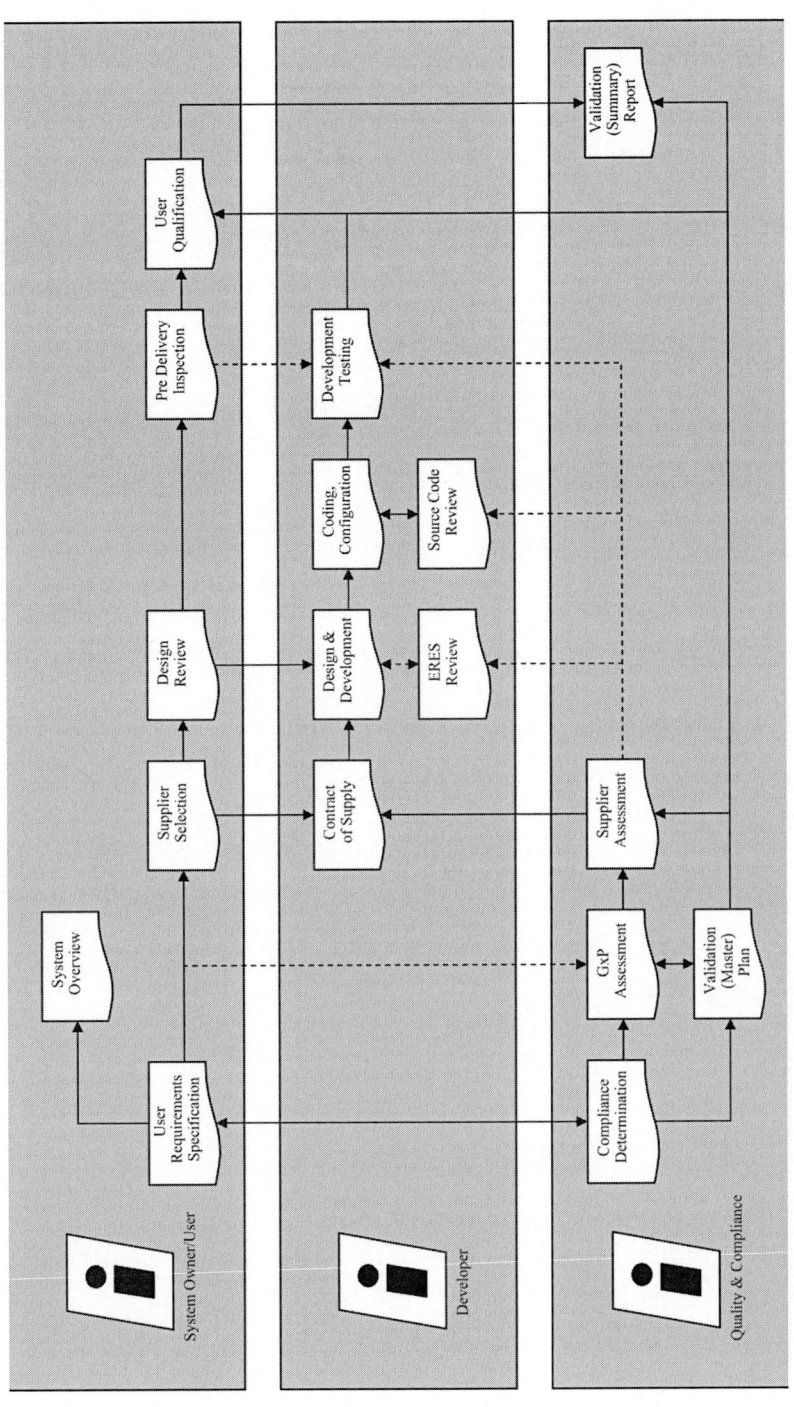

Figure 4.5 Role activity diagram for V model.

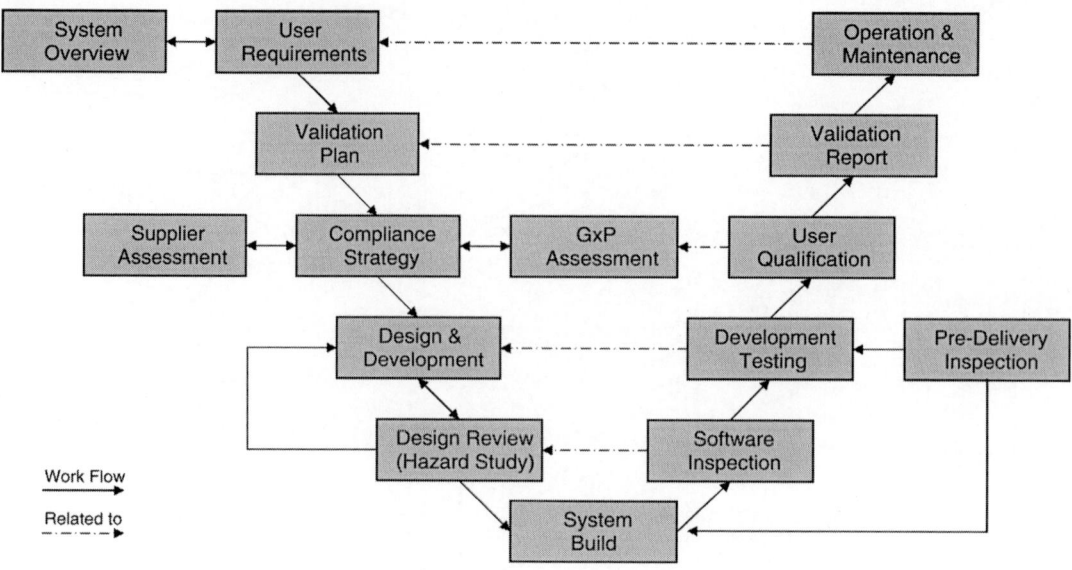

Figure 4.6 V model life cycle for computer systems.

The strengths of the V model have been summarized as follows (7):

- The model emphasizes planning for verification of the computer system in the early stages of the project.
- The model encourages verification of all deliverables, not just the system.
- The V model encourages definition of the requirements before designing the system, and it encourages designing software before building it.
- It defines the computer system that the project development process should deliver; each deliverable must be testable.
- It enables project management to track progress accurately; the progress of the project follows a timeline, and the completion of each phase is a milestone.
- It is easy to use (when applied to a project to which it is suited).

Weaknesses of the V model have also been summarized (7).

- It does not handle concurrent events.
- It does not handle iterations of phases.
- The model is not equipped to handle dynamic changes in requirements throughout the life cycle.
- The requirements are tested too late in the cycle to make changes without affecting the schedule of the project.
- The model does not contain project risk analysis activities.

The structure and control brought by the V model can be quite frustrating to project managers up against challenging deadlines. Accelerated development is often thought of as the solution, but there is no "silver bullet." Rapid application development (RAD) and prototyping allow management and users to "see" the system requirements as they are developed. It is very useful when requirements for a new system are not fully defined at the beginning of a project. Clarifying requirements is especially important "proof of concept" projects. It has been proven well suited for highly interactive systems, first-of-a-kind systems, decision support systems, and medical diagnosis. Such accelerated development can be habit forming and may go on too long, or be applied inappropriately to projects that do not warrant such an approach. Undisciplined programmers can fall into a "code and fix" cycle (akin to relying on breadboards for hardware "build and fix") that does not embody the quality assurance principles expected for GxP applications. A particular approach is not mandated (1–3). What is most important is to select an approach suited to the computer application and then to ensure that approach is appropriately managed and controlled to assure quality and compliance.

PROJECT INITIATION AND COMPLIANCE DETERMINATION

	Deliverables
Inputs	Project scope (outline specification)
Outputs	Compliance determination statement
	Validation (master) plan
	Project and quality plan

Project Scope (Outline Specification)

The Project Scope provides an executive summary justifying the purpose and benefits of the proposed computer system. Particular project management risks will normally be identified at this stage so that key controls to help ensure project success can be planned.

An outline specification is typically produced as part of a project scope to sanction a project. The outline specification should not be specific to hardware and software products but should provide sufficient information for a compliance determination statement to be prepared.

The project scope is not maintained after the project is completed. The outline specification, however, is not lost. It is carried forward to form the basis of a system overview and is maintained during the operational life of the computer system.

Compliance Determination

All computer systems should be assessed as early as possible to determine whether or not they need to comply with regulatory requirements. Compliance determination statements should be periodically reviewed against changes in pharmaceutical and healthcare regulatory requirements, or use of the system, and updated as necessary. Some pharmaceutical and healthcare companies include the compliance determination statement within the validation plan. The benefit of separating this as a separate document is that it can be used to justify why some computer systems are outside the scope of regulatory requirements.

During regulatory inspections, the compliance determination statement can be presented with a validation certificate to demonstrate that a particular computer system has been validated without delving into any detail.

Where pharmaceutical and healthcare regulatory requirements or system use change, the regulatory status of the system in the compliance determination statement must be reviewed and any remedial actions must be identified and implemented through change control and/or validation plans.

Validation (Master) Planning

Validation plans record standards, methods, and personnel to be used and involved to assure quality through the system development life cycle and to establish the adequacy of the performance of the computer system. The term validation master plan is typically used for large or multiple computer system projects. Validation planning should be initiated at the earliest practicable opportunity and may be reviewed and updated through subsequent stages of the project.

The size of the validation plan should be commensurate with the project complexity. Any planned amalgamation or split of documentation to fit the size and complexity of a computer system should also be defined here. Review and approvals should be defined either within referenced procedures or specifically within the plan.

The information contained in user requirements specifications is often used to help determine the basic approach to compliance to be taken for individual projects. Guidance on the scope of compliance activities required for different types of computer system is defined at the beginning of this chapter. A rationale must always be given supporting any compliance decisions made during a project, in the form of a separate document or included within the validation plan.

During regulatory inspections, the validation plan can be presented with a validation report or validation summary report to demonstrate that a particular computer system has been validated. These documents provide a regulator with additional detail to the compliance determination statement and validation certificate without digressing into the supporting evidence.

If the computer system is relatively small and contained entirely within a stand-alone piece of equipment, the validation plan for the computer system may be embodied within the equipment's overall validation plan.

Project and Quality Plans

Project plans are typically based on work schedules. The project life cycle stages should be specified with the project control documentation to be delivered. These disciplines, though perhaps sometimes regarded as irksome, should soon become recognized as being of enormous aid in mitigating project risk. Of all the project documentation, the GANTT charts are of the highest importance, defining deliverables and timetables. They are often created with project management tools under version control to monitor project progress against defined milestones and critical paths. Project phases can overlap as long as compliance with regulatory expectations is not compromised. Project milestones are usually also included in the validation plan to indicate a commitment to planned project implementation and stop project plans becoming a regulatory document.

Separate quality plans are sometimes established to define the procedures, documentation, roles, and responsibilities that will collectively assure the quality of a computer system. However, validation plans are often considered to supersede the need for user quality plans.

REQUIREMENTS CAPTURE AND SUPPLIER (VENDOR) SELECTION

	Deliverables
Inputs	Project scope (outline specification)
Outputs	User requirement specification (URS)
	GxP assessment
	Request for proposal[a]
	Supplier proposal[a]
	Supplier assessment[a]
	Evaluation of proposal[a]
	Purchase orders and contracts of supply[a]

[a]For external rather than internal suppliers.

User Requirements Specification

URS describes the user's functionality requirements, level of user interaction, interfaces with other systems and equipment, the operating environment, and any constraints. Specific regulatory requirements should be included, for example, requirements regarding use of electronic records and electronic signatures. The documentation making up the URS should

- Allow the developer to understand the user's requirements,
- Clearly define any design constraints,
- Provide sufficient detail to facilitate acceptance testing,
- Support operation and maintenance of the computer system, and
- Anticipate and ease phaseout and withdrawal of the computer system.

Emphasis is on the user requirements, not the method of implementation, and should therefore not be product specific. Requirements should be defined such that individual requirements can be identified (with acceptance criteria) for traceability through development and testing. Diagrams should be used to enhance readability.

Because URS are written in conversational English, rather than a stricter, more formal notation, they are notoriously prone to imprecise expression and ambiguity. While this is a feature of everyday conversation, where it is often no handicap, it undermines the very objective for which the URS in intended. Care should be taken through review to try to minimize such source of error wherever possible.

GxP Assessment

Examination of the URS to identify GxP functionality, processes, and/or components is implemented by the computer system. These aspects of the computer system should be the focus of attention during validation, especially during design review and user qualification. Some GxP assessments may be delayed until the functional specification has been issued. The GxP assessment can then be conducted on the functional specification. It is often useful for processes to be mapped showing critical points in the process and which computer system(s) support these critical process points and in what way.

Request for, Receipt of, and Evaluation of Proposals

Requests to potential suppliers for a proposal should include a copy of the URS, with specific reference to regulatory compliance requirements. For instance, such references might be to the *GAMP®5 Guide* and specific paragraphs of the electronic record and electronic signature rule. The proposals received in response to requests need to be evaluated against the request to ascertain any deviations. Once a preferred supplier or more than one has been short-listed, a decision whether or not to conduct a supplier audit can be made.

Supplier Selection

The software development capability maturity of suppliers needs to be assessed to determine the level of assurance that they can provide contracted computer systems, services, and/or documents to meet the pharmaceutical and healthcare company's requirements. These requirements include compliance with regulatory expectations. Regardless of origin, computer systems and their software should

- Be developed according to a defined, documented life cycle of adequate scope and depth that ensures the structural integrity of the delivered software and facilitates its testing and maintenance,
- Undergo appropriate testing and be released according to approved procedures, and
- Be maintained under configuration management (if in the form of multiple source code files) and change control.

User project controls may need to be introduced to compensate for any supplier deficiencies identified.

Supplier assessments are conducted to determine at first hand the supplier's capability. Audits are not normally required for COTS products because they are market tested. The performance of a supplier with the pharmaceutical or healthcare company and/or other users can be used to determine whether an audit is appropriate.

Supplier audits are most beneficial if they are conducted as part of the supplier selection and procurement process, so that any actions arising can be progressed within a project's implementation. More than one supplier audit may be appropriate or necessary for a system where there are multiple subsystem suppliers or subcontractors are used. The validation plan will document which suppliers do and do not require auditing and when these audits should take place. Table 4.2 indicates when supplier audits are required.

Purchase Orders and Contracts of Supply

Copies of purchase orders and contracts of supply should be retained in accordance with pharmaceutical and healthcare regulatory requirements.

DESIGN AND DEVELOPMENT

	Deliverables
Inputs	User requirements specification
	GxP assessment
	Supplier proposal[a]
Outputs	Supplier project and quality plans[a]
	System overview
	Functional specification
	Architectural design
	Software design and program specification
	Hardware design
	Data definition (including configuration)
	Operating manuals[a]
	Design review (including hazard study)

[a]For external rather than internal suppliers.

Supplier Project and Quality Plans

Essentially, it is the same document as described earlier for the project initiation phase but defining the scope of the supplier's role and responsibility. The supplier quality plan should act as an extension to the supplier proposal and contract of supply. The supplier quality plan should be approved before the functional specification is approved. If the supplier is an internal function of the pharmaceutical or healthcare company, then these plans do not need to exist as separate documents but rather integrated within overall project validation plans, project plans, and quality plans.

System Overview

This is a single document (clear, concise, accurate, and complete) describing the purpose and function of the system. System overviews are therefore not necessary for less complex systems where there is a single URS or functional specification document.

The system overview is a useful document to present during an inspection and should be written in nontechnical language so that it is meaningful to an inspector without the need for technical expertise. It should include

- A diagram indicating the physical layout of the system hardware,
- Any automated and manual interfaces, and
- The key functions indicating inputs, outputs, and main data processing.

This information may be contained within the validation plan or compliance determination statement, in which case it need not be duplicated.

Functional Specification

This describes the functionality of the chosen or developed system and how it will fulfill the user requirements. This document is the specification against which the system operability will be tested and maintained. Specific hardware and software products may be referenced. The functional specification should not duplicate information available in standard prepublished supplier documentation if commercial software/hardware is being used as long as it is referenced and managed under change control.

Contents of the functional specification should be cross-referenced to the URS to demonstrate coverage. A requirements traceability matrix (RTM) is one way of confirming that all relevant aspects of the URS have been brought forward into the functional specification.

This includes both positive and negative requirements (the latter is often overlooked), that is, what the software is supposed to do and what it is NOT supposed to do!

The functional specification must not be approved until its corresponding URS has first been approved. After the functional specification is issued, a GxP assessment may be conducted or updated if it has been already prepared in association with the URS.

Larger and more complex computer systems can benefit from an architectural design defined in a separate document. For smaller and simpler computer systems, this information is more readily included in the functional specification.

Architectural Design
Larger systems will typically benefit from a separate architectural design to explain the structure of the computer system and help link functional specification to the hardware and software design. The architectural design will be required for the design review.

Hardware and Software Design
Design specifications define the equipment hardware and/or software system in sufficient detail to enable it to be built. Design considerations include inputs and outputs, error handling, alarm messages, range limits, defaults, and algorithms and calculations. The design specification may make use of formal supplier documentation such as data sheets for commercial software products. Design information will be used to support installation, commissioning, and testing. Software design may itself be supplemented by program specifications depending on the appropriate level of granularity. Diagrams should be used where appropriate to assist readability.

The design specification needs to cross-reference relevant sections of the functional specification. Again, the use of an RTM may prove useful. The design specification must therefore not be approved prior to the approval of the functional specification. Detailed design information can be included within the functional specification, in which case it should be called a functional design specification.

Data Definition (Including Configuration)
Data structures and content must be defined. Actual data to be loaded into tables, files, and databases must be specified by reference to its source. Data dictionaries should be used to describe different data types. Specific data requirements include

- Electronic record and electronic signature compliance,
- Built-in checks for valid data entry,
- That data is only entered or amended by authorized persons, and
- That GxP data is independently checked.

The data definition may stand on its own as a separate document(s) or be incorporated within functional specification or design specification. Data definition needs to be approved before data load can begin.

Operating Manuals
Operating manuals need to be formally reviewed as fit for purpose by the supplier, as they form the basis of user procedures and user qualification. Operating manuals must be kept up to date with developments to the computer systems to which they relate, and they must refer to specific hardware models and software versions making up the computer system being supplied. Recommended ways of working defined by the supplier should be verified as part of the development testing.

Design Review (Including Hazard Study)
The Design review provides a feedback mechanism to the design and development phase to refine the design of the computer system before construction begins in the system build phase.

The design review can also be used to carry forward issues to be tested during development testing and user qualification.

Documentation must be reviewed to ensure that all requirements, functional and design specifications, drawings, and manuals have been produced and updated appropriately. Documentation should be reviewed for the following principal attributes:

- *Clear and concise:* Documentation should conform to document standards and should be readily understandable.
- *Complete:* All requirements are carried forward and are traceable through design and development. An RTM may prove a valuable asset in this regard. It is important to be able to determine which user requirements are and are not addressed.
- *Current:* Verify that the documentation is current and that the necessary change control measures have been applied.
- *Testable:* Criteria to be used for user acceptance testing must be specific, measurable, achievable, realistic, and traceable to the functional specification and design specification.

Hazard studies are conducted to identify potential operational problems with the computer system and how they are or will be managed. They may include confirming functionality required for electronic records and electronic signatures (where appropriate).

Design reviews are sometimes referred to as design qualifications especially when their scope is extended to cover all activities leading to the readiness of systems for user qualification.

CODING, CONFIGURATION, AND BUILD

	Deliverables
Inputs	Hardware design
	Software design and program specification
	Configuration definition
	Software programming standards
Outputs	Hardware platform
	Application software
	Source code review

Hardware Platform
Computer hardware needs to be assembled in accordance with good practice and any manufacturer's instructions. Hardware components should be itemized in the hardware design and checked as part of the installation qualification.

Application Software
Application software includes all of the softwares required to implement a computer system. An inventory of application software should be itemized and checked and installed as part of the installation qualification.

Initial and subsequent COTS product releases should only be used once they have been proven over a period of time by a user community to be fit for purpose. It is recommended that software is not used until it has been commercially available for at least six months. Software products that are released by a supplier pending final evaluation (so-called *β testing* of software) must NOT be used to support the manufacture of drug and healthcare products!

Source Code Review
A source code review must be performed on custom (bespoke) software in critical applications unless there is evidence that the source code has been or will be developed in a quality-assured

manner and subjected to review as part of its development life cycle. The decision and justification not to perform a review must be documented within the validation plan.

The source code review aims to provide confidence in the operability of the system and, as such, has four basic objectives.

- To verify the adoption of good programming practices (e.g., headers, version control, and change control) and compliance with documented, audited programming standards
- To determine the level of assurance by which the code fulfills design specifications including process sequencing, affirming I/O handling, formulae, algorithms, message, and alarm handling
- To detect possible coding errors
- To identify evidence of dead code
- To check that superfluous options are deselected or disabled and cannot accidentally be enabled

DEVELOPMENT TESTING

	Deliverables
Inputs	Functional specification
	Hardware design
	Software design and program specification
	Design review (including hazard study)
Outputs	Unit/module testing
	Integration/system testing
	Predelivery inspection

Unit/Module Testing

Unit testing focuses on verifying the smallest unit of software design—the unit/module or program. Using the software design and program specification, important control paths with worst case conditions are exercised. These activities include data entry operations and also that the limits of internal boundaries have been checked by exercising the ranges of acceptable values. All these tests are intended to expose errors within the unit/module or program. The relative complexity of tests and the level of residual concealed errors are limited by the constrained scope established for unit testing. Unit/module testing is normally white box orientated, in other words, tested by someone who understands exactly how the code works (so the "box" is "white," i.e., you can look inside it). The work can be conducted in parallel for multiple units/modules or programs. Supplier audits and source code reviews may recommend specific unit/module testing activities. Unit/module testing may not be appropriate for instrumentation and control systems that tend to be more self-contained as computer systems.

System/Integration Testing

Integration testing is a systematic technique for constructing a system while conducting tests to reveal errors associated with interfacing. It may be tempting to adopt a nonincremental approach to integration testing—that is, to test the system as a whole without incrementally testing each assembly. The problem with this approach is that when errors are discovered, it is difficult to precisely identify the cause and hence specify effective and error-free corrective action. When corrections are thus implemented, they can introduce new system errors that themselves are hard to be isolated for the same reason. Such uncontrolled testing can become extended, frustrating, and dangerous for ultimate product quality.

Incremental integration testing is the antithesis of the nonincremental approach. The system is tested as subsystems before being tested as collections of subsystems; finally, the complete system is tested. Errors should be easier to be isolated and correct predictably, interfaces should be easier to be tested more comprehensively, and it should be easier to develop a systematic test plan.

System testing should be comprehensive and includes

- Normal system operation including failure mode operation
  - Processing sequences
  - I/O handling
  - Formulae, calculations, and algorithms
- All error conditions (contrivance of error conditions may be needed) and associated error message intelligibility, relevance, and treatment
- All range limits with alarm conditions
- Service continuity throughout defined operating environment
- Recommended testing from source code review

Predelivery Inspection

Pharmaceutical and healthcare companies may consider conducting predelivery checks on their suppliers to verify that supplier project/quality plans have been implemented. Computer systems should not be accepted at their user sites if outstanding and agreed issues from the supplier audit have not been resolved to the satisfaction of the pharmaceutical or healthcare company.

Predelivery checks can significantly reduce overall project timelines if conducted properly. For example, some or all of the documentation destined for delivery with the computer system may be used in lieu of on-site inspection if the appropriate approvals were obtained prior to the predelivery checks. It is recognized, however, that there may be situations where the cost of attending to these checks may outweigh the benefits. The application of predelivery checks should be contractually specified with suppliers, and the outcome of predelivery inspections should be documented.

USER QUALIFICATION AND AUTHORIZATION TO USE

	Deliverables
Inputs	User requirements specification
	GxP assessments
	Functional specification
	Hardware design
	Data definition (including configuration)
	Operating manual
	Design review (including hazard study)
Outputs	Commissioning and calibration
	Data load (including configuration)
	Installation qualification (IQ)
	Operational qualification (OQ)
	Performance qualification (PQ)
	Operation and maintenance prerequisites
	Validation (summary) report

Note: As discussed in chapter 1, some companies refer to IQ, OQ, and PQ as installation verification, operational verification, and performance verification, respectively.

Site Readiness Activities
The installation site must be prepared (commissioned) as required, and necessary calibration of equipment and instruments must be conducted. This activity is usually separated from installation qualification.

Data Load (Including Configuration)
Procedures and protocols must define the data load (entry) process to fulfill the data definition requirements identified during the design and development phase. All GxP data, including configuration, must at least be double-checked to verify that it is correct. Statistical sampling can be used to check whether other data commensurate with the business need to verify data integrity. Rationales justifying sampling regimes must be defined. Automated data load tools should be verified.

Installation Qualification
This records the checks conducted to establish that the installation has been completed in accordance with system specifications. It may contain

- Inventory checks (hardware, software, data, user manuals, and SOPs),
- Operating environment checks (e.g., power supply, RFI, EMI, RH, temperature), and
- Diagnostic checks (installation diagnostics and software launch).

The boundary of the system and hence the scope of the IQ must be defined in the validation plan. The RTM should be updated with IQ cross-references.

Operational Qualification
Tests must be designed to demonstrate that the installed computer system functions as specified under normal operating conditions and, where appropriate, under realistic range conditions. Destructive testing is not required.

OQ testing should only be conducted after the IQ has been successfully concluded. The scope of the OQ should be defined in the validation plan. System testing can be repeated or referenced to reduce the amount of OQ testing required provided supplier documentation standards fulfill user qualification requirements. The OQ should cover

- Confirmation of user functionality,
- Audit trails for electronic records,
- Application of electronic signatures,
- Verification of operation and maintenance SOPs, and
- Verification of business continuity plans.

OQ protocols should define any ordering between individual tests. Specific tests may be recommended by supplier audit, GxP assessments, and design reviews. The RTM should be updated with OQ cross-references. The RTM should specifically identify where GxP functionality identified by the GxP assessment is tested.

Performance Qualification
PQ has often been the least understood phase of user qualification. This is probably because the character of PQ testing can vary considerably between different computer systems.

PQ testing should be designed to demonstrate that the installed computer system's functionality is consistent and reproducible. The PQ may be conducted, in whole or part, immediately after the computer system is brought into use, but it cannot be brought into routine use until the PQ has been successfully completed.

PQ testing should only be conducted after the OQ has been successfully concluded. The scope and timing of the PQ should be defined in the validation plan. The scope of the PQ should cover (as appropriate)

- Product performance qualification to verify that critical items/records are consistent (e.g., batch records, labeling variants) and
- Process performance qualification to verify that critical functionality is reproducible (e.g., coping with operating environment variations, demonstrating integrity and accuracy of data processing, and managing service metrics such as availability, reliability, and probability of failure on demand).

Parallel ways of working are expected to complement PQ in case the validation fails, or a catastrophic failure occurs. A back-out strategy should be developed involving manual ways of working, switching over to an alternative validated system, or a hybrid combination of both. Maintaining data integrity is the most important objective here. The strategy is likely to make use of current business continuity plans.

PQ protocols should define any ordering between individual tests. The RTM should be updated with PQ cross-references.

Operation and Maintenance Prerequisites

It is necessary to ensure that either arrangements for operation and maintenance of the computer system are already established or documented plans have been prepared to ensure that these arrangements are in place by the time the system is authorized for use.

- *Performance monitoring*
  SOPs for collection and analysis of performance data must be approved before data is collected. Statistical analysis should be conducted under the supervision of a professional statistician where results are used to support GxP decisions. Performance monitoring may be linked to preventative maintenance, for example, as part of a reliability-centered maintenance (RCM) program.
- *Repair and preventative maintenance*
  The validation plan may cover the requirements for maintenance planning for the system, in which case they should be verified by the IQ/OQ. Where this is not the case, the following areas will be addressed under the validation report:
  - Recommended spares holding
  - Frequency of routine testing/calibration
  SOPs covering maintenance activities must be approved before the system is used.
- *Upgrades, bug fixes, and patches*
  SOPs for software upgrades, bug fixes, and patches must be approved before such changes can be made.
- *Data maintenance*
  SOPs supporting the management and control of data integrity must be approved before the computer system is used. Procedures will include data change control.
- *Backups and recovery*
  SOPs for backup and restoration of software and data files must be approved and verified before the computer system is used. It is not unknown for some projects to verify the backup process but to forget to verify the restoration process with the adverse results that can occur if it does not work when called upon.
- *Archiving and retrieval*
  SOPs for archiving, retention, and retrieval of software, data, documentation, and electronic records must be specified, tested, and approved before the system is approved for use.

- *Business continuity planning*
  Procedures and plans supporting business continuity (disaster recovery plans and contingency plans) must be specified, tested, and approved before the system is approved for use. Business continuity plans will normally be prepared for a business or operational area rather than individual computer systems. It is likely that the only way to verify the plan is to walk through a variety of disaster scenarios. Topics for consideration should include catastrophic hardware and software failures, fire/flood/lightning strikes, and security breaches. Alternative means of operation must be available in case of failure if critical data is required at short notice (e.g., in case of drug product recalls). Reference to verification of the business continuity plans is appropriate during OQ/PQ.
- *Security management*
  SOPs for managing security access (including adding and removing authorized users, virus management, and physical security measures) must be specified, tested, and approved before the system is approved for use.
- *Contracts and service level agreements*
  Commercial contracts and service level agreements for operation and support of computer systems should be established prior to their use.
- *User procedures*
  User procedures for operating and maintaining the computer systems must be specified, approved, and, where possible, tested before the systems are approved for use. Projects should aim to use and thereby test user procedures within user qualification activities. This approach offers the opportunity to use end users to help conduct testing with the user procedures. This can often be coordinated as a training exercise. User procedures can be refined by end users themselves in readiness for handover of the computer system.
- *Availability of software and reference documentation*
  Application software and relevant development documentation must be available for inspection. Formal access agreements should be established if access to software is restricted, for example, escrow accounts.

The RTM should be updated to link operation and maintenance activities that address system requirements. Operation and maintenance requirements are discussed in more detail in chapter 11.

Validation (Summary) Report

A validation report must be prepared at the conclusion of the activities prescribed in the validation plan. Where there are deviations from the validation plan or unresolved incidents, these should be documented and justified. Where critical unresolved issues remain, the computer system cannot be considered validated.

The validation report for a system must not be approved until all the relevant documents defined within its validation plan have been approved. Approval of the validation report marks the completion of the validation process. The validation report must, therefore, include a clear statement confirming whether or not the whole computer system is validated and authorized for use.

Some pharmaceutical and healthcare companies prepare what is referred to as a validation summary report. This is really the validation report described above, but it ensures that the summary of deviations from the validation plan reported within the validation report are described at a very high level with little detail.

The concept of a validation summary report can be taken a step further in the form of a validation certificate. This merely states that a computer system is validated, and it specifies a review date against this compliance status. While a validation certificate could be presented to a regulatory inspector as evidence of validation, we believe that this would prompt the inspector to request more detailed information. The effort to produce and maintain validation certificates must be carefully weighed.

PROJECT DELIVERY SUPPORTING PROCESSES

The following should be established to support the project and its handover to ongoing use and are discussed in more detail in chapter 3.

Training

Project staff must be trained as required for the project. Users should also receive training in advance of being expected to use a computer system. All personnel using or maintaining a computer system *must* be trained to the correct level of competency before they are allowed to operate the system.

Training should be conducted in a timely manner. A training plan should be developed to achieve this objective. Training must be conducted against approved user procedures. User training records must be updated to reflect training given on the system. Any requirements for refresher training should be audited through periodic reviews.

Contractors and temporary staff should have a CV or staff training records including all relevant details retained by the responsible manager. Staff should be selected to fully exploit their skills, education, and experience.

Document Management

Documentation, records, and raw data must be reviewed prior to approval and maintained under change control. Amendments to documentation, records, and raw data must not obscure the original entry. All such amendments must be signed and dated.

Raw data attached to documents and records or existing in their own right should be clearly itemized, signed, and dated. Sets of raw data should be physically coupled (e.g., bound or stapled). Raw data forming attachments to documents and records must be marked as such (e.g., supplier audit reports and test results). Sets of raw data should have each page marked as belonging to the set, and the front page should be signed and dated.

Change Management

Change control must be established for software, hardware, data, and documentation under development. Procedures for change control may be project specific and vary from the procedure used after handover of the project. Any transition in change control procedures needs to be managed. Software should further be checked for unauthorized changes. Although they should not occur and must be strictly forbidden, it is to be regretted that unauthorized changes often do occur in practice.

Configuration Management

A system should be established to document and control the versions of computer system software and hardware through the development, testing, and use of the system. Configuration management needs to link software and hardware versions to particular data sets and document versions. Changes to software, hardware, data, and documentation may be interdependent.

Requirements Traceability

In the words of ex-FDA investigator Martin Browning, "traceability is the absolute key [to documentation]" through design, development, deployment, all the way through to decommissioning. This is vital when conducting capability assessments of software package suppliers and suppliers of equipment where software is embedded. Specifically, requirements traceability will improve

- Test coverage,
- Impact assessments of change, and
- Inspection support.

A RTM or equivalent mechanism for establishing and maintaining requirement traceability should be established during projects for use during operation and maintenance. This mechanism should provide a method of tracing a requirement from the URS through the

Table 4.5 Compliance Package Contents

Compliance package: documents and records	Software category			Comments
	3	4	5	
System overview	X	X	X	Executive introduction to overall system.
Compliance determination statement	X	X	X	May initiate a detailed electronic record/signature assessment.
Validation (master) plan	X	X	X	The scope of a validation plan need not be limited to one system. For exact replicas of a computer system, the validation plan can be proceduralized. Reference may be made to the project quality plans and supplier quality plans if they are separate documents.
User requirements specification	X	X	X	Typically, a single document covering whole system.
GxP assessment	X	X	X	For overall system.
Supplier audit report		X	X	For bespoke and critical COTS-based applications.
Functional specification	X	X	X	For standard COTS packages reference to product specification is sufficient.
Architectural design		X	X	Only consider for large or complex systems; may be combined with functional specification.
Software design and program specification		X	X	Configuration only for category 4 software.
Hardware design				For bespoke hardware, otherwise reference standard COTS hardware documentation.
Data definition (including configuration)	X	X	X	To cover electronic records and any data configured/bespoke data structures.
Design review (including hazard study)		X	X	Typically, covering whole system (sometimes known as design qualification).
Source code review		X	X	Configuration only for category 4 software.
Unit/module testing			X	May be combined as appropriate.
Integration/system testing			X	
Predelivery inspection		X	X	For bespoke and critical COTS-based applications, follow on from supplier audit.
Site readiness activities	X	X	X	Site preparations, commissioning, and calibration.
Data load (including configuration)	X	X	X	Only for systems that require data population.
Installation qualification	X	X	X	Documents may be combined as long as test plans and test cases are collectively approved before testing. For COTS instrumentation, testing may be based on calibration activities.
Operational qualification	X	X	X	
Performance qualification	X	X	X	
Operation and maintenance prerequisites	X	X	X	Activities include user procedures/manuals, maintenance, backup and restoration, security, training, business continuity plans.
Validation (summary) report	X	X	X	May be in the form of a certificate for very simple systems.

Note: There are no specific compliance activities for category 1 software (standard COTS compilers and operating systems) beyond documenting version details.
X- Element of compliance package is relevant to software category.
Abbreviation: COTS, commercial off-the-shelf.

design and development and user qualification phases. It provides assurance that all system requirements have been included in the design and tested to verify correct operation.

During the operational life of the computer system, requirements traceability will enable impact assessments to be conducted in support of regression testing of changes. It is quite normal for a particular aspect of a computer system to require more than one test. When making a change, it is important to appreciate if this one-to-many situation exists and ensure that all appropriate tests are executed.

Requirements traceability should also ease inspections by facilitating the rapid location of any specification-to-test information requested. For many organizations, this forward tracking of requirements can be quite cumbersome if they rely solely on reverse tracking offered by test documentation referencing requirements and specifications.

Deviation Management

Compliance deviations during the project must be managed. Details of deviations must be recorded with a description of the circumstances under which the deviation was noted (e.g., reference to design review or test case) and the name of the person noting the deviation. A record is also required for the remedial action taken to a deviation (including any testing required) or the justification for not taking action. An index of all deviations should be maintained. Deviations can be prioritized for resolution.

COMPLIANCE PACKAGE

Throughout the verification and validation process, a package of documentation should be collated, which contains documentary evidence of compliance. The contents of a package are indicated in Table 4.5 and should be defined in the validation plan. Change control records must also be included. Example review and approval expectations for key documents are indicated in Figure 4.7.

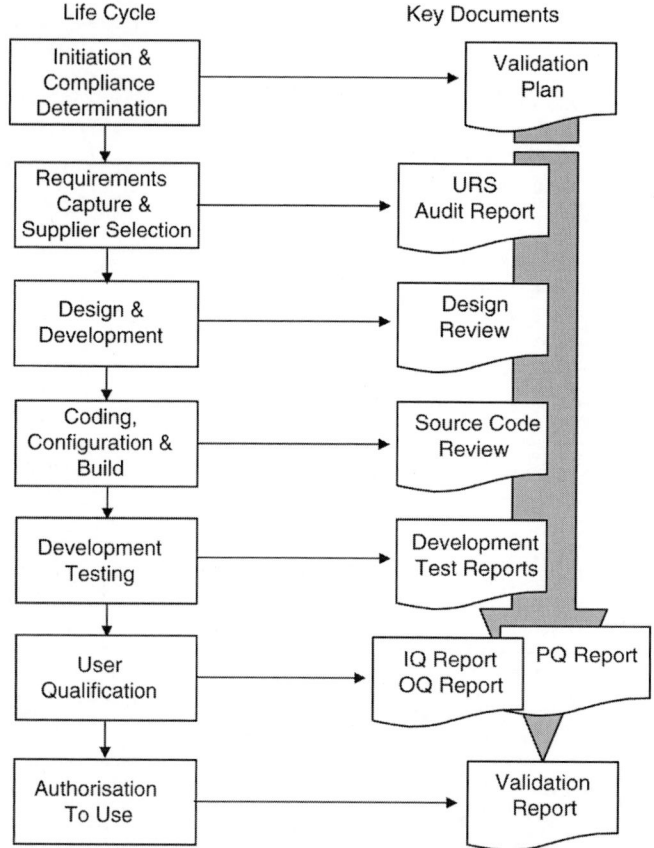

Figure 4.7   Key compliance deliverables.

The compliance package (both formal documents and raw data) must be securely stored in accordance with site procedures and be readily retrievable. The package will normally be stored in a centralized site archive/repository ready for future use.

Reviews and Approvals

Review and approval requirements for documentation can be defined in procedures or specified in the validation plan. Roles and responsibilities will vary between organizations; there are no definitive role models.

Recommended minimum approval levels for compliance documentation generated at each stage and the associated responsibilities for each signatory are suggested in Table 4.6.

Table 4.6 Example Compliance Package Reviews and Approvals

Project document	System owner/user	Developer		Quality and compliance	
		User project manager	Supplier	Compliance oversight	Operational quality
System overview	O/A			Advice, support, and audit as required	R
Compliance determination statement	A				O/A
Validation (master) plan	A	A			O/A
Project and quality plans	A	O/A			
User requirement specification	O/A				A
GxP assessment	A				O/A
Supplier audit report	R	R			O/A
Supplier project and quality plans		R	O/A		R
Functional specification			O/A		R
Architectural design			O/A		
Software design and program specification			O/A		
Hardware design			O/A		
Data definition (including configuration)			O/A		
Operating manual	R		O/A		R
Design review (including hazard study)	A	O/A	R		A
Source code review			O/A		R
Unit/module testing			O/A		
Integration/system testing			O/A		
Predelivery inspection	A	O/A			R
Site readiness (commissioning and calibration)	A	O/A			R
Data load (including configuration)	A	O/A			A
Installation qualification	A	O/A			A
Operational qualification	A	O/A			A
Performance qualification	A	O/A			A
Operation and maintenance prerequisites	A	O/A			R
Validation (summary) report	A	A			O/A

Note: O, originator; R, recommended review; A, approve; shaded, review or approval not mandated.

The "developer" role can be split into the user project manager and supplier. The user project manager is the pharmaceutical or healthcare company's own project manager. The supplier can be an internal group within the pharmaceutical or healthcare company's organization or an external company. Similarly, the "quality and compliance" role can be split into compliance oversight and operational quality. Compliance oversight represents QA staff with specialist computer compliance knowledge. Operational quality represents QC department staff supporting general GxP operations. Other roles may be more appropriate depending on the organizational structure.

Table 4.6 also defines responsibilities. The "originator" responsibility is preparing a document or record, "review" responsibility is confirming technical content and consistency with other compliance activities, and "approve" responsibility is authorizing the activity as complete and correctly documented. The quality unit may, at their discretion, as part of their approval process, review or audit supporting and referenced deliverables. Due account must also be taken where a supplier audit determines that the pharmaceutical or healthcare company should take over a lead responsibility for a supplier's documentation responsibilities. The review and approval roles presented in Table 4.6 are consistent with regulatory expectations.

REFERENCES

1. International Society for Pharmaceutical Engineering. GAMP®5: Risk-Based Approach to Compliant GxP Computerised Systems. Tampa, Florida, 2008. Available at: www.ispe.org.
2. U.S. Food and Drug Administration. General Principles of Software Validation; Final Guidance for Industry and FDA Staff. Rockville, MD, 2002.
3. Food and Drug Administration. Software Development Activities. Technical Report, Reference Materials and Training Aids for Investigators. Rockville, MD, 1987.
4. David Begg Associates. Computer Systems and Automation Quality and Compliance. York, 2001.
5. Wingate GAS, Smith M, Lucas PR. Assuring confidence in pharmaceutical software. safety and reliability of software based systems. First annual ENCRESS conference, Bruges, Belgium, 1995.
6. Wingate GAS. Validating Automated Manufacturing and Laboratory Applications: Putting Principles into Practice. Buffalo Grove, IL: Interpharm Press 1997.
7. Futrell RT, Shafer DF, Shafer LI. Quality software project management. Upper Saddle River, NJ: Prentice Hall, 2002.

5 | Project Initiation and Compliance Determination

INTRODUCTION

The first phase in the project life cycle is that of project initiation. Being the responsibility of the pharmaceutical or healthcare company, it consists of five main activities that may be conducted sequentially or concurrently. These are

- Definition of the project scope,
- Determination of the compliance requirements,
- Drafting of the validation master plan,
- Drafting of other, subordinate validation plan(s), and
- Definition of the compliance strategy.

PROJECT SCOPE

Project resources (financial and human) are always scarce and the subject of conflicting claims, so they are usually at a premium. Quality assurance and compliance measures are often criticized, almost always unfairly, for not delivering benefits commensurate to their cost. For this reason, it is vital to make the very most of the resources available. The compliance strategy adopted ought to be challenged at the outset, and any compliance rationales ought to be rigorously scrutinized for their long-term benefits in mitigating risk to the operation. This is far better than seeking penny-pinching economies under pressure of criticism after the project is under way. Understand the project risks, then manage them proactively. A well-specified and planned project has a good chance of reaching a successful completion.

Understanding the Scope of the Computer System

Computer systems can be thought of as comprising a programmable electronic device connected to actuators and sensors (and possibly other computer systems) via communication links. Closely associated with the computer system are data and/or equipment, people, and procedures. A schematic of a computer-related system based on some work by the U.S. Pharmaceutical Manufacturers Association (PMA) (1) is shown in Figure 5.1. However, let us first define our terms.

- *Computer System* is a group of hardware components and associated software designed and assembled to perform a specific function or group of functions.
- *Operating Procedures* are documents that describe the steps to be taken to implement the authorized processes intended to meet the business objective, processes that are expressed by work operations. These can be complex or mundane. Examples of typical operations requiring procedural controls are the operation of computers and equipment, development of computer systems, maintenance of computers and equipment, and training of personnel.
- *Controlled Process* is a sequence of operations that are controlled or directed by a computer system. It is often described in terms of the physical equipment and/or manual operations required. Such a description also includes physical devices where the computer technology is integrated with control and monitoring equipment (e.g., manufacturing equipment, laboratory equipment, facilities, and

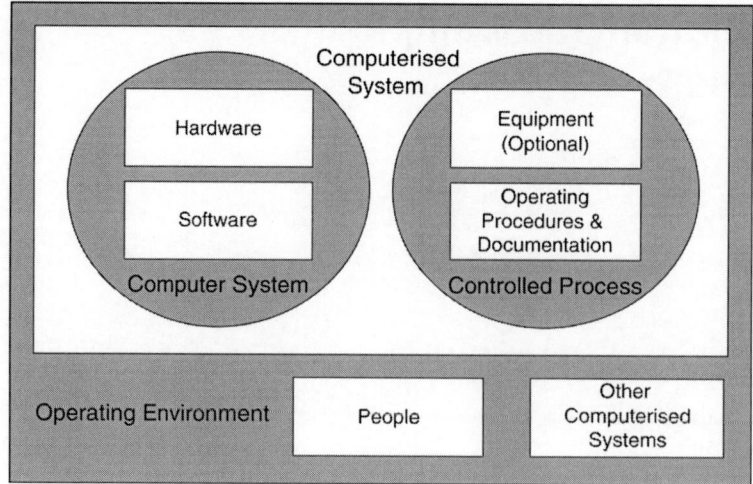

Figure 5.1   The Pharmaceutical Manufacturers Association conception of a computerized system.

equipment). Quality and compliance activities associated with equipment and manual operations are dictated by regulatory requirements. These are intended to ensure the suitability of the equipment, physical processes, and operating procedures.

- *Computerized Systems* encompass processes and/or operations integrated with a computer system. (It is the collected elements of the computer system itself together with the operating procedures and any equipment where applicable.)
- *Operating Environment* includes all outside influences that affect the computer system and the manner in which it is used. It addresses the defined, integrated work flow (procedures) between people and machines (computer system and equipment) needed to accomplish a specific task. It can be used synonymously with the term *computerized system*, but really, it is broader in scope, for it encompasses work environments that utilize more that one computer system to accomplish an overall work operation.

Outline Specification

An outline specification is usually required to sanction a project and authorize the allocation of a budget. It typically presents the business case for the computer system, defining key functionality and compliance requirements. This document may be used later on for developing the user requirements specification (URS), as discussed in chapter 6. Where COTS products are being acquired, this document may be used as the URS. Outline specifications must be reviewed and approved.

Project Risk Management

A league table of top project risks is presented in Table 5.1. If these are relevant in a particular case but in practice treated casually or even ignored, then project budgets, schedules, and/or system functionality will almost certainly be compromised. This in turn is bound to affect the standard of regulatory compliance. It is therefore painfully clear that project risk management is very important, not only in terms of project delivery, but also in terms of the operational compliance that the computer system will be capable of achieving once put into use.

Potential risks to a project must be identified early so that they can be proactively addressed. It is naïve to assume that there are no risks, and therefore, the project will automatically run smoothly—if only life were like that! The amount of effort devoted to identifying risks should correspond to the perceived criticality of the project, so simple projects

Table 5.1 Project Risk Factors

Risk	Description
Lack of competent personnel	Exemplified by inexperience with tools and development techniques to be used, personnel turnover, loss of critical team members, and a failure to match the size of the team to the complexity of the project.
Unrealistic schedule/budget	Too little money, too little time.
Wrong functionality	Causes include failure to understanding customer needs accurately or comprehensively, and a lack of development process capable of implementing these faithfully into code.
Inappropriate user interface	The quality (e.g., ease of use) of the user interface may be critical to system success.
Gold plating	Developers may allow their enthusiasm for technical sophistication to outstrip the features demanded by the customer.
Requirements volatility	Requirements and thus designs altered during development often have a hugely damaging impact on system quality.
Poor external component/tasks	Suppliers may provide inadequate products/services.
Real-time shortfalls	The performance of the system may be inadequate.
Capability shortfalls	An unstable operating environment or new/untried technology poses a risk to the development schedule.

Source: From Ref. 2.

implementing standard commercial computer systems will generally be of a lower risk level than complex projects involving custom (bespoke) software.

Project risks can be mitigated using a four-stage approach (2).

- Identify the risk factors such as those listed in Table 5.1.
- Determine the level of exposure to the identified risks (the seriousness these carry coupled with the probability that they will occur). Formal risk assessment methodologies such as Kepner-Tregoe® (KT) might play a useful part here.
- Develop strategies to minimize and mitigate risk. This would normally be done for risks with the highest exposure level or for risks whose risk probability/seriousness or KT factor exceeds some defined threshold.
- Implement risk management strategies.

Hopefully, the risk management activities implemented will prove sufficient. Some risks, however, may still threaten success despite risk management activities, so the strategy should be to continually keep risks under review and revise activities as necessary to deal with residual and emergent risks. It may prove necessary to invoke contingency plans or even move into crisis management mode if the risks appear to be leading the project to failure.

Project and Quality Plans

Project and quality plans provide a mechanism to control the management of the project. Planning needs to focus on scheduling activities, controlling cost, allocating resources, and managing technical issues. Good project management is a basic business expectation.

Project plans should address

- Establishment of project organization,
- Allocation of resource,
- Identification of all activities, critical path, and key milestones (GANTT charts),
- Recognition and management of project risk,
- Ensuring that project prerequisites are put in place,
- Establishment of progress reporting method,
- Budget management, ensuring business case maintained (beware scope creep),
- Resolution process for open issues, and
- Links to quality plan for standards and quality control.

Quality Plans should address

- Quality responsibilities of project organization,
- Standards for work conducted, including the software development life cycle to be used, and the documents to be delivered by it,
- List of other deliverables to be produced (e.g., project life cycle documents such as control charts) and quality criteria,
- Quality control activities (e.g., software reviews, testing, and document reviews),
- Definition of change control procedures applicable to project,
- Configuration management practices,
- Document management practices, and
- Training requirements.

Successful projects do not happen by accident. They result from all the activities being well managed and being done right first time, but, as Tony Simmons acknowledges, this is an ideal that is seldom realized in practice, where change happens and people make mistakes (3). Chapter 1 reviews some of the most common problems that afflict projects. Project managers might want to consider using established project management methodologies such as PRINCE2 to help engineer predictability into the process and thus ensure a successful project outcome.

Quality planning and, to a lesser extent, project planning overlap validation planning. Project managers should consider how to avoid duplicating information. Project plans and quality plans can be combined into a single document, although there are pitfalls here. The project plan forms the charter for the project managers, who have a *driving* role. The quality plan, on the other hand, is the charter for quality assurance, whose role is to *constrain* the project into the paths of engineered best practice. These roles are always potentially in conflict, but both are essential to project success. Keeping the two plans separate but complementary reinforces this distinction and helps build clarity of understanding of the roles among the project participants, especially the project sponsors. Both the project plan and the quality plan are typically approved by the system owner/user and project manager. Project plans and quality plans are not normally subject to regulatory inspection if validation plans are prepared to meet regulatory expectations.

Glossary of Terms
A glossary of terms and acronyms should be collated as an integral part of the planning exercise. There must be a common understanding of terminology used in a project. It is all too easy for individuals to attribute their own meaning to standard terminology and thereby proliferate endless misunderstandings!

COMPLIANCE DETERMINATION
Early Indication of Compliance Requirement
An early decision as to whether or not a computer system requires validation can help ensure that the necessary support is provided for from the outset of a project. Late identification of compliance requirements could result in significant costs and delays if additional activities and documentation are then required. In addition, many regulatory authorities devalue such *retrospective validation.*

Validation Criteria
Computer systems require validation when an affirmative answer must be given to one or more of the following questions (4):

- Does the application or system directly control, record for use, or monitor product quality?
- Does the application or system directly control, record for use, or monitor laboratory testing or clinical data?
- Does the application or system affect regulatory submission/registration?

- Does the application or system perform calculations/algorithms that will support a regulatory submission/registration?
- Is the application or system an integral part of the equipment, instrumentation, or identification methods used in testing, release, and/or distribution of the product/samples?
- Does the application or system define materials (i.e., raw materials, packaging components, formulations, etc.) to be used?
- Can the application or system be used for product/samples recall, reconciliation, stock tracing, product history, or product-related customer complaints?
- Will data from the application or system be used to support quality control product release?
- Does the application or system deal with coding of materials, formulated products, or package components (i.e., labels or label identification)?
- Does the application or system hold or manipulate stock information, stock status, location, or shelf life information?
- Does the application or system handle data that could affect product quality, strength, efficacy, identity, status, or location?
- Does the application or system employ any electronic signature capabilities and/or provide the sole record of the signature on a document subject to review by a regulatory agency?
- Is the application or system used to automate a manual QC check of data subject to review by a regulatory agency?
- Does the application or system create, update, or store data prior to transferring to another computer system? Is that subsequent system compliant with regulatory requirements?
- Is the application or system the official archive or record of any regulated activity and thus subject to regulatory audit?

An in-depth assessment is not required at this stage. It should be possible to furnish an immediate "yes" or "no" to each bullet point question.

Compliance Determination Statement

A simple form, such as the example given in Appendix 5.A, can be used to document the decision whether or not to validate a system. Compliance determination statements should be completed no later than the moment the URS is released. The basic information on which to base a decision should be available at this time, and in any case once issued, there is nothing to prevent the compliance determination statement from being updated with additional information when available. The important thing is establishing early on whether or not validation is required, so that project planning can take this into account. To give a practical implication here, purchasing departments may need this information to implement contractual clauses relating to compliance expectations with the suppliers.

Recent Inspection Findings

Change control computer database not validated (FDA 483, 2004)

Reports generated (*containing GxP data*) from *computer system* observed in use despite the lack of validation (FDA 483, 2004)

The spreadsheet used by quality unit personnel for XXXX not validated (FDA 483, 2004)

Process control system and monitoring system not validated (FDA 483, 2004)

Software application used for tracking process deviations and complaints not validated (FDA 483, 2003)

Spreadsheets used for calculations not validated (FDA 483, 2003)

Autoclave control system not validated (FDA 483, 2003)

Software program for inventory of released products not validated (FDA 483, 2003)

Firms computerized inventory control system not validated (FDA 483, 2003)

Firm not validated program used to calibrate microbiology bioburden and environmental data (FDA 483, 2003)

Stability management software not validated (FDA 483, Nov 2004)

Change control computer database not validated (FDA 483, 2004)

The facility environmental monitoring computer system not validated (FDA 483, 2004)

Computer system used in documentation of complaints not validated (FDA 483, 2006)

Computer system used to control stability chambers not validated (FDA 483, 2004)

Software used for document management including SOPs, validation reports, and change control records not validated (FDA 483, 2006)

Software used to receive, track, and monitor adverse drug experiences not validated (FDA 483, 2006)

The database used to monitor deviations and investigations not validated (FDA 483, 2007)

VALIDATION MASTER PLAN
Hierarchy of Validation Plans

The term validation master plan usually denotes that the associated project covers large or multiple computer systems. For example, where more than one validation plan exists to cover the implementation of a system, a supervisory plan should be produced to define the overall approach. Similarly, if the computer system is relatively small and contained in its entirety within a stand-alone piece of equipment, then the validation plan may be embodied within the overall validation master plan for the equipment/area in question.

The International Society for Pharmaceutical Engineering (ISPE) *GAMP®5 Guide* suggests a hierarchy that might look like (5)

- Management plan for multisite activities,
- Site validation master plan (for entire site),
- Validation master plans (for complex systems or supervision of multiple systems), and
- Validation plans (for individual systems).

Whatever pattern a supervisory planning takes, it must clearly identify the scope of any subordinate validation plans.

Figure 5.2 shows how the compliance activities for a pair of linked computer systems might be structured within a validation master plan. This can be compared with Figure 5.3, which shows how the compliance activities might be further structured to include a third and entirely independent computer system within the same validation master plan.

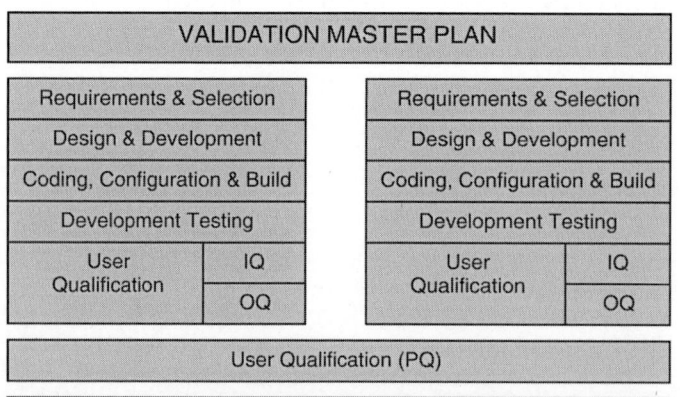

Figure 5.2 Simple validation master plan.

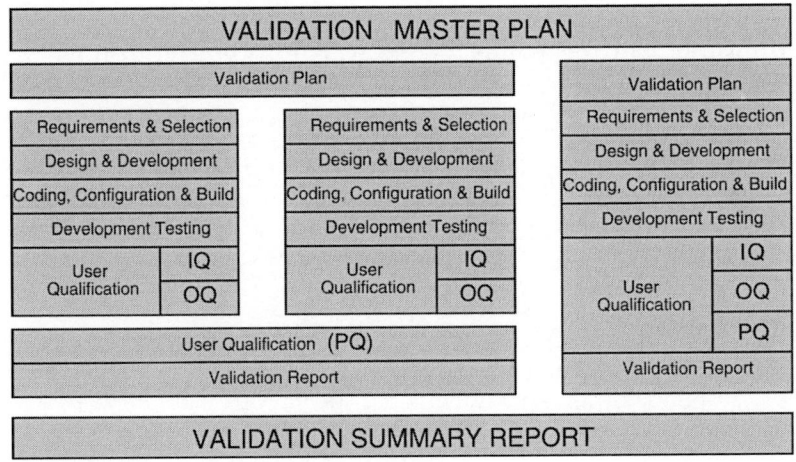

Figure 5.3   Complex validation master plan.

Contents of Validation Master Plan

Validation master plans typically have three main sections. The first section states how the pharmaceutical or healthcare company's own compliance philosophy and policy address regulatory GxP requirements. The second section defines the scope of validation, identifying which computer systems require validation. All computer systems whose malfunction could possibly affect the safety, quality, and efficacy (during development or manufacture) or batch tracking (during distribution) of drug products should be validated. A register or inventory of computer system systems to be validated is sometimes attached to the validation master plan as an appendix. Finally, the third section commits the pharmaceutical or healthcare company to some basic milestones. These milestones are usually associated with the launch of a new drug product but may also be timetabled to satisfy anticipated inspections by GxP regulatory authorities. The milestones demonstrate the pharmaceutical and healthcare company's intent and may be revised during the course of a program of work to reflect changing conditions.

Senior managers should not be unduly worried about delays as long as the pharmaceutical or healthcare company can demonstrate progress to the regulatory authorities and continued commitment to a revised timetable. This said, however, senior managers must expect to be held to any completion dates specifically agreed with regulatory authorities. It is usually a good idea to keep them regularly appraised on progress at suitable intervals. Failure to deliver on agreed dates is likely to be seen as demonstrating a distinct lack of commitment to address compliance requirements.

As much notice as possible should be given to regulatory authorities if delays are expected to agreed completion dates. Care should be taken to explain to the regulatory authority concerned how the delays came about and how they are being addressed.

Structure of Validation Master Plan

The ISPE *GAMP Guide* suggests the following layout for a validation master plan (5):

- Introduction and scope
- Organizational structure
- GxP assessment process
- Compliance strategy
- Change control
- Procedures and training
- Document management
- Timeline and resources

In addition, a glossary may be added as required to aid understanding of the validation master plan. Bear in mind that not all terms routinely used within the organization will be familiar to those outside it.

Introduction and Scope
Here, reference should be made to relevant policies, and where document fits into level of planning should be described. The scope and boundaries of the site/area/systems being addressed should be defined as appropriate. Reference should also be made to any subordinate validation plans, and the period within which this plan will be reviewed should be stated.

Organizational Structure
The roles and responsibilities such as ownership, technical support, and QA should be defined. Depending on level of planning, these may be departmental or corporate roles and responsibilities. Individuals should not be named, but their job titles should be identified.

GxP Assessment Process
The computer systems compliance requirement should be explained, including how any prioritization was determined and how any changes in priority will be managed. Identify procedures used in GxP assessment.

Compliance Strategy
The overall computer compliance strategy within validation master plan as well as the life cycle model to be applied and the procedures to be used should be outlined.

Change Control
The approach to change management referring to relevant change control procedures should be mentioned.

Procedures and Training
The training requirements for new and existing SOPs should be described.

Document Management
The contents of compliance packages to be produced should be defined together with the identification of document management and control procedures to be used. Any special requirements need to be specified and clearly understood.

Timeline and Resources
Timeline and resources indicate planned end dates and appropriate intervening milestones, identify resources to be assigned to various activities, and note any critical dependencies that may impact progress.

Preparation of a Validation Master Plan
The time to prepare a validation master plan depends on the number, type, and use of computer systems. For a facility with 30 computer systems, a validation master plan with an inventory might take about three to five days to draft. The length of the document is likely to be between five and fifteen pages depending on its detail. The layout of an example validation master plan is shown in Appendix 5.B.

When writing a validation master plan, it is important to minimize the duplication of information between it and other documents. Sufficient information, however, must be provided so that personnel not directly involved in the planning can understand the compliance approach in general. Clarity is important, because validation master plans may be examined by GxP regulators.

Validation master plans should be straightforward, easy-to-read documents written in the vernacular and as far as possible free of irritating jargon. Needless complexity usually retards the review process. Beware of demanding too many approval signatories in the interests of political correctness—it can make the review process cumbersome! Indeed, only signatories whose presence lends weight to the technical or quality value of the document should be included, and these persons should then be able to justify the plan to a regulator on request.

The validation master plan must be reviewed and approved before use. They are regulatory documents demonstrating an organization's intention to validate one or more computer systems. At a minimum, a validation master plan should be signed and dated by the site director/area manager and appropriate site/area quality assurance manager.

Recent Inspection Findings
See validation plans.

VALIDATION PLAN
The purpose of the validation plan is to provide a clear compliance strategy for the validation exercise based on risks arising from a number of factors.

- Functional criticality of application
- Size and complexity of application
- Capability of organization (including supplier)
- Customization and configuration (degree of standardization)

The validation plan and validation report are often some of the first documents to be examined by GxP regulatory authorities inspecting a pharmaceutical or healthcare facility (6). Figure 5.4 illustrates this relationship. Validation reports are discussed in more detail in chapter 10. Without a plan to direct and organize work, compliance activities are likely to be chaotic and ineffective. GxP regulatory authorities have frequently found projects where members were incorrectly assuming that somebody else was providing key compliance evidence. (*Everybody thought that somebody was validating it, but actually it was nobody!*)

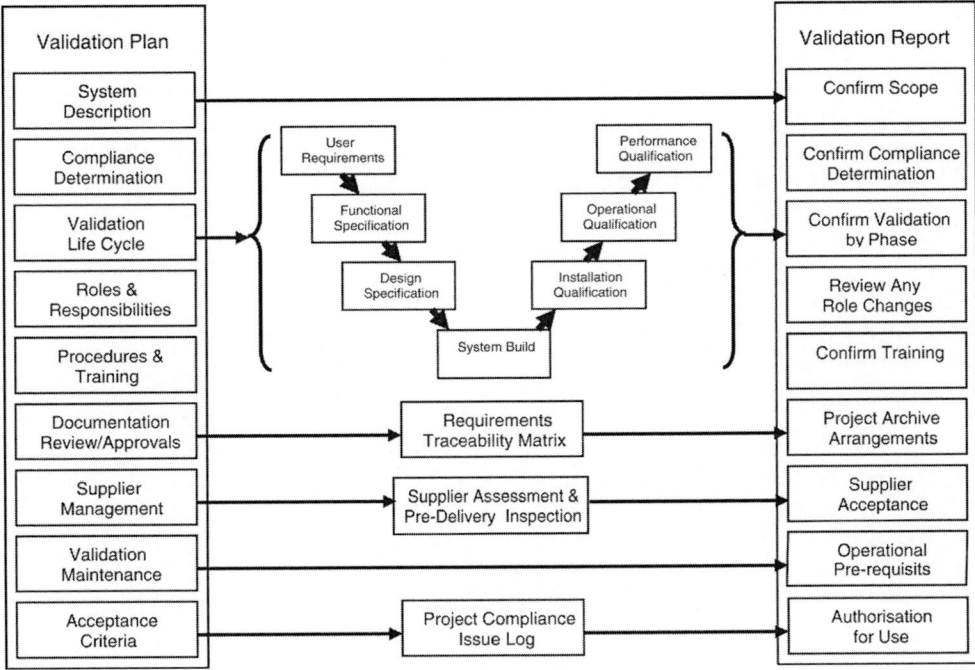

Figure 5.4 Relationship between validation plans and reports.

Contents of a Validation Plan

Validation plans specify how a pharmaceutical or healthcare company intends to satisfy the GxP requirements affecting a single computer system or group of associated computer systems. Validation plans should

- Identify the computer system being validated,
- Give any relevant background information,
- Reference procedures to be used,
- Define compliance package deliverables
- Specify review and approval responsibilities,
- Identify personnel assigned to the project together with an indication of their competency to participate, and any training requirements, and
- Indicate project milestones.

The depth and scope of compliance activities depend on the degree of customization, complexity, and criticality of the computerized application (5,7). The URS can provide useful information when determining the most appropriate approach to compliance.

Structure of a Validation Plan

The following sections are suggested for a validation plan (5):

- Introduction
- System description
- Compliance determination
- Project life cycle
- Acceptance criteria
- Roles and responsibilities
- Procedures and training
- Documentation review and approvals
- Supplier and subcontractor management
- Support program for maintaining operational compliance

In addition, it is often useful to include a glossary so that a terminology used by everyone is unambiguous and clear. Although a common set terminology is now widely used, the precise understanding of these terms can vary dramatically (e.g., use of the term "verification" discussed in chap. 1). Just as for the validation master plan, remember, as we mentioned earlier, that not all terms routinely used within the organization will be familiar to external personnel. Establishing a definitive understanding on key terms provides a point of reference when different perspectives arise.

Introduction
Validation plans often begin by citing the authority under which they have been issued (e.g. a validation master plan or the personal authority of a senior manager). They preview the compliance requirements for the project.

System Description
A brief system description covers the system's history (whether it is an entirely new system, a replacement system, or an existing system) and its business purpose. GxP regulatory authorities expect replacement systems to be as least as reliable as their predecessor manual or automatic system (8). The aim is to provide a management overview of the project boundaries, defining what is and what is not included within the scope of work. The system description may be supplemented later by a separate system overview document especially for larger systems.

Compliance Determination
The validation plan should refer to or include a brief description why a particular computer system is being validated. The role of the compliance determination statement was described

earlier in this chapter. It is useful to the reader to have this placed early in the validation plan to put the validation into context. This information will be supplemented later with a GxP assessment (see chap. 6) in terms of detail once the user requirement specification is available.

Project Life Cycle
Project activities are then laid out with a description of any issues for specific GxP consideration. A typical project life cycle has already been described in chapter 4 and consists of

- Requirements capture and supplier (vendor) selection,
- Design and development,
- Coding, configuration, and build,
- Development testing,
- User qualification and authorization to use,
- Operation and maintenance, and
- Phaseout and withdrawal.

The project life cycle adopted with supporting compliance activities should be customized for each individual project. Often there are opportunities for various phases to be carried out concurrently. However, the phases must be clearly defined in advance, with acceptance criteria and appropriate verification to ensure that each phase is completed in a state of control. This must be specified and documented in the validation plan. The following chapters examine in detail the activities associated with the life cycle outlined above. The resulting process, tailored for the computer system being validated, is in essence the compliance strategy being adopted.

Acceptance Criteria
A clear definition of what criteria must be met for acceptance of the computer system must be defined. Reference to a validation report alone is not sufficient. The criteria to be used by the validation report in measuring the computer system for acceptability as validated together with the manner in which compliance failures logged during the project are resolved must be stated.

Roles and Responsibilities
The roles and responsibilities divided and allocated between the pharmaceutical or healthcare company and any suppliers need to be defined. A simple organizational chart can prove very useful when describing such interrelationships between project staff. Resumes of project staff should be collated and include references to their education and degrees obtained, professional certificates, previous job titles and responsibilities, and, most importantly, their *competency* to fulfill their role. Resumes for temporary staff, such as supplier personnel, are often included as an appendix to the validation plan or associated validation report. Some pharmaceutical and healthcare companies may prefer to use a central management system to collate the resumes of their own permanent staff.

Procedures and Training
The procedures to be adopted and the documentation to be produced must be identified. Planned deviations from procedures should be defined in advance within the validation plan to demonstrate management's acceptance of and continued control over the project. Key procedures, in addition to those required to prepare the life cycle documents, include change control and document management.
Training requirements against project procedures should be identified. Although already stated earlier, it is worth repeating that it is important to recognize that the competency of individuals is critical to a successful validation exercise.

Documentation Reviews and Approvals
Approval signatories for supporting documentation should be specified. An individual's job title may not necessarily reflect the function performed and the responsibility shouldered by a person's

signature. On numerous occasions, GxP regulatory authorities have challenged individuals who have signed documents, discovering to their alarm that the signatories had misunderstood their full responsibilities and performed less thorough reviews than were required.

Supplier and Subcontractor Management

Procedures and documentation to be provided by suppliers should be distinguished from the pharmaceutical or healthcare company's own documentation. The pharmaceutical and healthcare company's dependence on suppliers should be explained, identifying any necessary supplier audits (or reviewing the results of audits already conducted). The awareness that quality can only be built into products, not tested in retrospectively, has led the pharmaceutical and healthcare industry in recent years to place increasing emphasis on this activity. The regulators have endorsed this healthy trend, reflecting their expectation for user companies to ensure the software they use is fit for purpose. An important factor to consider is who approves and accepts work as satisfactorily completed. Any requirements for monitoring a supplier's work should be defined.

Support Program for Maintaining Operational Compliance

A support program to maintain regulatory compliance should be identified. It should include procedures for change control, maintenance and calibration, security practices, contingency planning, operating procedures, training, performance monitoring, and periodic review. Operation and maintenance are discussed further in chapter 11.

Preparation of a Validation Plan

An experienced practitioner can produce a validation plan for a computer system (usually ten to fifteen pages long) in about three days. An example validation plan layout is illustrated in Appendix 5.C. The structure of validation plans must be scaleable to fit the system or software being validated. For large projects, configuration management and functional test planning may be split out into separate documents. Validation plans can also make use of appendixes, for instance, project milestones and resumes of project team members.

The validation plan must be reviewed and approved before issue. They are regulatory documents specifying the quality and compliance controls used to manage the deployment of a computer system. At a minimum, a validation plan should be signed and dated by the system owner as well as by the quality and compliance representative.

The time to review and issue the validation plan will vary between different organizations. It is primarily dependent on the number of people involved in the review process. Validation plans should be easy-to-read clear documents, written in the vernacular, as we mentioned in the context of validation master plans, and as far as possible free of irritating jargon. If review is slow or protracted, it may be because it is too complex or imposes a cumbersome review and approval signatories process. Each signature should genuinely add value to the document. Signatories should not be added solely for political representation (to share the glory) or as a fail-safe (to share the blame). Every signatory must be able to explain and justify the plan to an inspector if required.

Maintenance of Validation Plans

Validation plans should be maintained to reflect changes in project activities. Specific revisions to plans should be considered when transitioning between major project phases.

- Requirements capture and supplier (vendor) selection
- Design and development
- Coding, configuration, and build
- Development testing
- User qualification and authorization to use

Thereafter, validation plans should be updated or superseded by new plans when there is a change to the architecture of the system, a change in use of the system to that validated, or a change in operation and maintenance standards or when a revalidation is initiated.

Supplier Relationships

The ultimate responsibility for the regulatory compliance of a computer system provided by vendors, system integrators, and service providers lies with the user. This said, certain elements remain the province of the supplier, while others will be under control of the user. Validation plans must address the requirements relating to both.

Commercial Off-the-Shelf Computer Systems

The documentation set provided by a supplier of COTS products should be standard, rather than being tailored in advance with the users. Nevertheless, pharmaceutical and healthcare companies should map supplier documentation to compliance requirements to try to satisfy the latter (9). The mapping exercise is typically conducted by pharmaceutical and healthcare companies as a desktop exercise, although a supplier audit should embrace most of these if conducted competently and thoroughly. The auditors chosen to conduct supplier audits should be formally qualified for their role and therefore be able to demonstrate competency, just as GxP regulations require other staff to be competent. Pharmaceutical and healthcare companies are increasingly seeking supplier auditors qualified either to ISO 9001:2008 (10) or, preferably, to the current standard ISO 90003:2004 (11) through the *International Register of Certificated Auditors* (IRCA). Other internal accreditation measures may be sufficient for internal auditors. Compliance strategies are discussed later in chapter 15, specifically the X model for verifying COTS software. If the supplier's document set is insufficient to support the pharmaceutical or healthcare company's validation of the computer system and the deficiencies cannot be remedied, then the system cannot be considered compliant with regulatory requirements and as such must not be used.

Contracted (Commissioned) Systems

Computer systems supplied by third parties (often known as system integrators) should adhere to the same compliance requirements as in-house developments. This, of course, requires that compliance requirements are clearly identified and understood by the supplier at the outset of a project.

The definition of supplier activities and documentation should be embedded in contractual agreements. In addition, suppliers should agree to potential inspection by GxP regulatory agencies and supplier audit by pharmaceutical and healthcare companies. Supplier audits can be conducted by the pharmaceutical or healthcare company's own personnel or, if this would compromise the supplier's commercial interests, by an independent software quality assurance auditor or expert consultant employed by the pharmaceutical or healthcare company. Auditors must be suitably qualified, for example, by independent certification by examination to the quality system standards such as. ISO 9001:2008. Supplier audits are discussed in more detail in chapter 6.

Availability of Software and Reference Documentation

Another important aspect to take into consideration is access to proprietary source code and associated documentation (12). All custom (bespoke) application-specific software and reference documentation should be available. COTS software and development documentation, however, is not always available. It is generally accepted that formal access agreements for operating systems, compilers, and system software are unnecessary, since the ubiquity of such systems provides adequate implied evidence of their fitness for purpose.

Regulators now expect formal access agreements to be established (e.g., escrow accounts) where access to application software and associated reference documentation is restricted. It is important to keep any access agreement up-to-date; current and historical software versions should be covered with their respective documents and revisions.

Copies of the software and documentation must be retained within safe and secure areas, protected within fireproof safes. The duration of storage of legacy software and documentation needs to be defined. Guidance on this is given in chapter 12. However, it must be borne in mind that should a supplier cease to trade and the escrow agreement then invoked to secure access to the code, it is highly unlikely that this will prove an acceptable basis for the ongoing long-term use of the computer system. Support for code in these circumstances is fraught with difficulty. Pharmaceutical and healthcare companies have hitherto rightly concluded that there is far less risk shouldered by premature retirement and replacement of such systems, rather than attempting to support them in the face of the insolvency and lack of support from their developers.

Recent Inspection Findings

Completion dates in validation master plan inadequate to assure equipment/systems appropriate to intended use (FDA 483, 2002).

Validation master plan allowed only a single signature for both validation and quality control (FDA 483, 2002).

Validation plan for XXXX not performed in accordance with cGMPs: plan lacked number of qualified personnel, completion dates were inadequate to assure valid performance of manufacturing processes, and validation strategy did not contain procedures, standard protocols, and specific requirements (FDA 483, 2002).

Validation plan contained no instructions pertaining to what constitutes acceptable test results (FDA 483, 2002).

No information on qualifications of those reviewing and approving protocols/reports (FDA 483, 2002).

COMPLIANCE STRATEGY

The magnitude of compliance effort should be commensurate with the risk associated with the computer system, the computer's dependence on software for potentially hazardous or critical functions, and the role of the specific software modules in higher-risk functions (13). For example, while all software modules should be verified, modules that are critical should be subjected to a more thorough and detailed design, development, and testing. Such assurance is derived far more securely from the methodology by which such code was developed and is supported then from functional testing by the user. Likewise, size and complexity are an important factor in establishing the appropriate level of effort and associated documentation for the software. The larger the system, the more extensive procedures and management controls are required. Small and large firms alike cannot justify a superficial validation on the basis of scarce resources.

Compliance approaches for both the hardware and software of computer systems need to be considered.

- Is hardware/software bespoke, commercial off-the-shelf (COTS), or a combination?
- Is there any hardware configuration?

The fitness for purpose of a proposed solution should be assessed, and any history of usage in similar applications should be considered when determining the compliance strategy. A holistic perspective should be taken with computer systems comprising components found in multiple categories of software and hardware. It is worth noting here that some companies are now starting to use the term 'verification' to describe qualification activities (see chap. 10 for more details).

Approach to Hardware

The approach to validation must reflect whether the associated computer system hardware is a unique combination of components put together by the pharmaceutical or healthcare company or preassembled as a standard product by the original equipment manufacturer (e.g., PCs packaged with laboratory analytical equipment, or production equipment containing an embedded control system). It must also reflect any hardware configuration (e.g., switch settings). Bespoke items of hardware must have a design specification and be subjected to acceptance testing. A supplier audit should be performed for custom (bespoke) hardware development. The design of standard hardware products must also be documented. In this regard, reference may be made the hardware manufacturer's data sheet or other specification material as long as its source is recorded, typically the supplier's name and address. It should also include details of any hardware configuration. Complete systems consisting of a unique assemblage of hardware from various sources must be checked to ensure that interconnected hardware components are compatible with one another. Any hardware configuration must be defined in the design documentation and verified in the installation qualification (IQ). The model number, version number, and, where available, serial number of each component of the assemblage must be recorded. Preassembled hardware that is sealed does not have to be dismantled as such action often invalidates the maker's warranty. Rather, the hardware details should be obtained from the system data sheet or other specification document.

Approach to Software

GAMP Category 1 Software: Operating Systems

It is not normally necessary to attempt to verify the operability of established, commercially available operating systems. Because of their ubiquity and because they are exercised every time the applications installed upon them are used, these should be considered to be already verified. Only the name and version number need to be recorded in the hardware acceptance tests of equipment IQ. New versions of operating systems should be reviewed prior to use, and consideration should be given to the impact of new, amended, or removed features on the application. This might lead to formal retesting of the application, particularly where a major upgrade of the operating system has been necessary.

A summary of compliance requirements is as follows:

- Specify version for installation
- IQ—check version installed

GAMP Category 3 Software: Standard Software Packages

These do not normally need extensive verification if the version to be acquired has already been exposed to the marketplace for an extended period. However, new versions are a different matter and should be treated with caution. Effort should concentrate on functionality, critical algorithms and parameters, data integrity (security, accuracy, and reliability), and operational procedures. Change control should be applied stringently since upgrading these applications, while initially easy, can turn out to be painful. User training should emphasize the importance of change control and the compliance integrity of these systems. A supplier audit may be an appropriate defensive measure for critical applications.

A summary of compliance requirements is as follows:

- Specify version for installation
- Specify scope of use
- Develop and approve user procedures
- Develop user training materials
- Review and accept software package documentation
- IQ—check version installed
- Operational qualification (OQ)—verify any data load
- OQ—verify general operation as used
- Performance qualification (PQ)—establish ongoing reliable operation

GAMP Category 4 Software: Custom-Configurable COTS Software Packages

These systems permit users to develop their own applications by configuring or amending predefined software modules. In many cases, the development of additional new application software modules may be needed. Clearly, therefore, each application (of the standard product) is unique.

Particular attention should be paid to any additional or amended code and to the configuration of the standard modules. A source code review of the modified or added code (including any algorithms in the configuration) should be undertaken for critical functionality. In addition, for large, complex, or critical software applications, a supplier audit is essential to determine the level of quality and innate structural integrity of the standard product. The audit must recognize the possibility that the development of the standard product might have involved a prototyping methodology without any eventual users being involved at all! GxP standards require that the development process is controlled and documented. A validation plan should be prepared to document precisely what activities are necessary to validate an application on the basis of the findings of the audit and the complexity of the application.

A summary of compliance requirements is as follows:

- Validation plan
- URS—specify scope of use
- Supplier audit for critical software packages
- Functional specification for configuration in context of software package
- Develop and approve user procedures
- Develop user training materials
- Review and accept software package documentation
- Hardware and software design for bespoke code/macros
- Source code review for critical functionality in custom (bespoke) code
- Specify version of software package for installation
- IQ—check version of software package installed
- OQ—verify any data load
- OQ—verify general operation as used
- OQ—comprehensive user acceptance of configured functions
- PQ—establish ongoing dependable operation
- Validation Report

Testing should cover positive functional testing based on defined user operation (it does what it should do), and risk-focused negative functional testing of all custom software (it does not do what it should not do where the risk assessment suggests a vulnerability) (12).

GAMP Category 5 Software: Custom (Bespoke) Software
For these systems, the full life cycle defined in chapter 4 should be followed for all parts of the system. An audit of the developer is essential to measure their development capability maturity and to examine their quality system. A validation plan should then be prepared to document precisely what activities are necessary on the basis of the insights gleaned in the audit and the complexity of the proposed bespoke system.

A summary of compliance requirements is as follows:

- Validation plan
- URS—full specification
- Supplier audit
- Functional specification
- Develop and approve user procedures
- Develop user training materials
- Hardware and software design—program specifications as necessary
- Extensive source code reviews
- Unit/module testing
- Integration/system testing
- IQ—check installation against specification
- OQ—verify any data load
- OQ—comprehensive user acceptance
- PQ—establish ongoing dependable operation
- Validation report

Testing should cover comprehensive positive functional testing (it does what it should do) and risk-focused negative functional testing of all custom softwares (it does not do what it should not do where the risk assessment suggests a vulnerability) (14).

Optional Extras for GAMP Category 3, 4, and 5 Softwares
Additional requirements may be necessary as appropriate to the project including

- Data migration protocols, records, and reports,
- System installations for development environments, and
- Strategy for phased deployment.

Criticality Impact Analysis

Managers may wish to consider using a *criticality and impact analysis* on functions and components of a computer system to support the planning of individual projects. The criticality of a computer system depends on understanding the criticality of business process it supports. Three categories of computer system impact are defined in relation to the risk posed to product quality, patient safety, data integrity for regulatory submission, distribution, and recall.

- Direct
- Indirect
- None/negligible

The rigor of compliance activities should reflect the impact of the computer system on drug or healthcare product development, manufacturing, or distribution. This concept is illustrated by Figure 5.5 (15). System components are only permitted to exist in three of the four boxes in the matrix—critical components cannot, by definition, have an indirect impact. Chapter 6 discusses the use of GxP assessments to determine criticality. Direct impact systems have one or more critical components and any number of noncritical components. Indirect impact systems cannot have any critical components. Components can themselves be treated as systems and further subdivided into components, and so on. The level of granulation need not be exhaustive; commonsensical approach should prevail. Usually, only one or two stepwise refinements are appropriate.

Criteria for identifying direct and indirect computer systems are given below. It is important to understand that individual computer systems cannot be a higher criticality than the business processes they support.

Direct impact criteria

- Controls or monitors clinical study data
- Manages information used to manufacture or test registration stability lots
- Manages critical manufacturing control parameters
- Manages stock status, location, or shelf life information
- Deals with coding of materials, formulated products, or package components (i.e., labels or label identification)

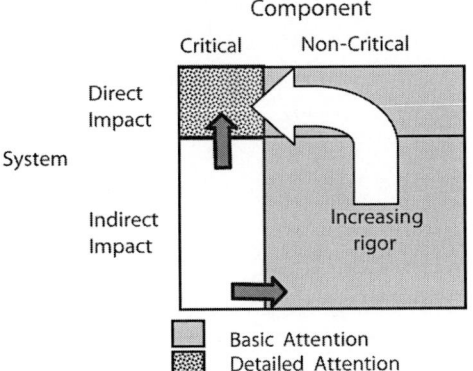

Figure 5.5 Criticality and impact analyses.

- Controls or monitors laboratory testing or other QC checks of data subject to review by a regulatory agency
- Defines materials to be used (i.e., raw materials, packaging components, formulations, etc.)
- Provides data used to support QC product/sample release decisions
- Controls the release and/or distribution of product/samples
- Manages product/samples recall, reconciliation, stock tracing, product history, or product-related customer complaints

Indirect impact criteria

- Provides functionality or has the ability to affect the performance of a direct impact computer system
- Provides data or has the ability to affect the integrity of data used by a direct impact computer system

Examples of direct impact systems include electronic batch record systems, chromatography data systems, MRP II/ERP systems used for batch release. Examples of indirect computer systems include SOP management, training records, document management, and CAPA management.

With a computer system, it should be possible to identify discrete components or segregate aspects of functionality. Critical components/functions should be afforded more validation commensurate with their criticality. Noncritical components/functions do not require the same level of validation.

Critical components/functionality criteria

- Components that are in direct contact with product or with raw material, utility or solvent that itself contacts product
- Controls critical process parameters without independent performance verification
- Monitors critical process parameters without out-of-specification process alarms
- Provides information for product batch records or lot release
- Adjusts or calibrates critical components
- Provides access security
- Controls audit trails content and/or application
- Applies electronic signatures and/or provides the sole record of the signature on a document subject to review by a regulatory agency
- Stores regulated records for retention and/or retrieval

The increasing rigor of compliance with respect to criticality of components is indicated in Table 5.2. The precise requirements will vary and should be governed by a validation plan or change control as appropriate.

Some practitioners have suggested that managing two levels of assurance increases complexity and thus aggravates the likelihood or error. Meanwhile, others maintain that this is the only way to control compliance costs, especially on large or highly integrated computer

Table 5.2 Component Compliance Activities

Critical component	Noncritical component
Specification	Design review
Design	
Source code review	
Development testing	Basic user testing
Testing (installation qualification, operational qualification, performance qualification)	

systems. If criticality impact analyses are used, these can furnish important evidence demonstrating why certain aspects of a computer system cannot affect drug product quality to a regulatory authority. They can also be used to show the level of assurance needed for aspects that do affect drug product quality.

In the future, as more sophisticated science-based risk management approaches become available, these can be applied instead of criticality impact analysis. Care must be taken, however, not to inappropriately apply such new approaches when the basic criticality impact analysis is applicable and sufficient.

Managing Compliance Issues

Potential or actual noncompliances that arise during the course of a project need to be logged and managed. The process used to manage these issues should be defined and referenced in validation (master) plans.

Many pharmaceutical and healthcare companies create a *project compliance issues log* during projects to track compliance issues. The structure of such a log should include (16)

- Project issue number,
- Author and date identified,
- Description,
- Resolution (change control reference or justification for no action), and
- Status (outstanding, in progress, or closed).

Larger projects might consider implementing a risk, actions, issues, decisions (RAID) log rather than confining itself to the simpler project compliance issue log.

A change control process that initiates appropriate corrective action or delivers a documented rationale justifying the acceptance of the defect or characteristic causing the noncompliance without further action should be used. Change controls may prompt revision of SOPs, revision of documents, further training, or other activity. The aim should be to complete open change controls wherever possible within the project. If a change control cannot be completed within the project, then a rationale should be prepared justifying completion of action retrospectively after the project closure.

Validation reports are expected to review the compliance matters raised during the project and itemize any outstanding corrective actions with a corresponding justification of the state of affairs. The issues log should be retained as part of the compliance package, and the responsibility for any outstanding corrective actions should be handed over to the organization managing the ongoing support of the computer system. Periodic reviews should verify that outstanding corrective actions are implemented in a timely manner.

REFERENCES

1. Pharmaceutical Manufacturers Association. Validation concepts for computer systems used in the manufacture of drug products. Pharmaceutical Manufacturers Association Proceedings: concepts and principles for the validation of computer systems used in the manufacture and control of drug products, 1986.
2. Van Vliet H. Software Engineering. 2nd ed. New York: Wiley, 2000.
3. Simmons T. Producing a Quality Plan. Pharmaceutical Automation Updates. West Sussex: Sue Horwood Publishing Limited, 2001.
4. Society of Quality Assurance. Risk assessment/validation priority setting. Computer Validation Initiative Committee, 2001.
5. International Society for Pharmaceutical Engineering. GAMP®5: risk-based approach to compliant GxP computerised systems. Tampa, Florida, 2008. Available at: www.ispe.org.
6. Trill AJ. Regulatory requirements for computer validation, computer systems validation: a practical approach. Management Forum Seminar, London, March 26–27, 1996.
7. International Conference on Harmonisation. Good manufacturing practice guide for active pharmaceutical ingredients. International Conference on Harmonisation Harmonised Tripartite Guideline, November 10, 2000.

8. European Union. EU GMP annex 11 computerised systems. Guide to good manufacturing practice for EU directive 2003/94/EC, community code relating to medicinal products for human use, Vol. 4, 2003.
9. ACDM/PSI. Computer systems validation in clinical research: a practical guide. Version 1.1, December 1998.
10. International Organization for Standardization. ISO 90003: software engineering—guidelines for the application of ISO 9001:2004 to computer software. Geneva, Switzerland, 2004.
11. International Organization for Standardization. ISO 9001: quality management system—requirements, Geneva, Switzerland, 2008.
12. Organisation for Economic Co-operation and Development Environmental Directorate. The application of the principles of GLP to computerised systems. No. 10 OECD Series on Principles of Good Laboratory Practice and Compliance Monitoring, GLP consensus document, environmental monograph No. 116, Paris, 1995.
13. Food and Drug Administration. General principles for software validation, final guidance for industry and FDA staff, Centre for Devices and Radiological Health, 2002.
14. GAMP Forum. Risk assessment for use of automated systems supporting manufacturing process part 1—functional risk, pharmaceutical engineering, May/June, 2003.
15. International Society of Pharmaceutical Engineering. Baseline pharmaceutical engineering guide: qualification & commissioning, 2001.
16. Central Computer and Telecommunications Agency. Managing successful projects with PRINCE2, 1999.

Appendix 5.A   Example of Compliance Determination Statement

Computerized system	Determination reference number: VDS/001

System identification: PLC/002/SMA
System/equipment name   Sterile manufacturing PLC
System name/equipment number 2
Location/department   manufacturing unit A
System/equipment used by: production
System/equipment used for: autoclave control
Justification for whether or not validation is required

(single line strike out inappropriate sentences below)

- The system is used to monitor, control, or supervise a drug manufacturing or packaging process.
- The system manipulates data or produces reports to be used by quality-related decision authorization/approval processes.
- The system is used for batch sentencing or batch records.
- The system manages and stores GxP records.

Acknowledgement of compliance requirements
It is the responsibility of the group(s) using, developing, and supporting the application to notify the QA department of any changes in the use of an application that might impact compliance with GxP regulations.

	Name and title	Signature	Date
Author (including department)			
Approved by user/project manager			
Approved by quality assurance			

Appendix 5.B   Example of Contents for Validation Master Plan

Introduction and Scope
- Author/organization
- Authority
- Purpose and scope
- Contractual status of document

Organizational Structure
- Resource allocation: organizational responsibilities

GxP criticality assessment process
- Define basis of determining criticality
- Justify any prioritization

Compliance strategy
- Description of project life cycle to be adopted (reference to relevant compliance standards)
- Approach to managing suppliers and subcontractors
- A statement to the effect that the computer system will only be authorized for use once satisfactorily validated

Change Control
- Description of change management process to be adopted

Procedures and training
- Identify SOPs to be adopted
- Commitment to training

Document management
- Definition of how documents will be managed and controlled

Timeline and resources
- Target completion date
- Interim milestones as appropriate

References
Appendixes
- Glossary
- Others

Appendix 5.C   Example of Contents for Validation Plan

Introduction
- Author/organization
- Authority
- Purpose
- Relationship with other documents (e.g., validation master plans)
- Contractual status of document

System description
- Define boundaries of system (e.g., hardware, software, operating system, network)
- Constraints and assumptions, exclusions and justifications

Compliance determination
- Rationale behind compliance requirement (may be reference to compliance determination statement)

Project life cycle
- Outline of life cycle being undertaken (reference to compliance standards)
- Approach to compliance for different hardware and software categories (see chap. 4)

Acceptance criteria
- A statement to the effect that the computer system will only be authorized for use once satisfactorily validated
- Description on how project compliance issues will be managed

Role and responsibilities
- Resource allocation: organogram and role descriptions
- CVs (qualifications and experience)

Procedures and training
- Identify SOPs to be adopted
- Training requirements and training records

(continued)

Appendix 5.C   Example of Contents for Validation Plan (*Continued*)

Document review and approvals
- List of documents to be prepared
- Review and approval set in accordance with roles and responsibilities

Supplier and Subcontractor management
- Supplier responsibilities
- Anticipated supplier audits
- Supplier documentation controls

Support program for maintaining operational compliance
- Description on how the compliance status will be maintained

References

Appendixes
- Glossary
- Others

6 | Requirements Capture and Supplier (Vendor) Selection

INTRODUCTION

The capture of requirements and the selection of suppliers entail the following main activities, usually conducted sequentially: user requirements specification (URS), GxP assessment, supplier selection, and supplier audits. Getting this phase right is crucial to ultimate success. The issues that have to be managed can be summarized as follows:

- Suppliers are often expected to work with incomplete or ill-defined user requirements.
  - Users frequently do not spend enough time and effort evaluating and documenting their requirements.
  - Commercial pressures tempt many suppliers to agree a contract but delay working out the details until later
- Suppliers are typically audited late in the user's selection process.
  - There is no strictly defined standard to which suppliers are audited—it is common for different auditing companies to apply various GxP compliance expectations.
  - Suppliers often adopt the most lax standards they believe to be acceptable (i.e., minimal quality assurance).
- Some suppliers have technically sound products but see no commercial benefit in achieving GxP compliance beyond their normal quality practices.
  - There is a widespread lack of understanding of the flexibility that can be applied to aspects of GxP compliance.
  - A supplier may previously have had a bad experience when attempting to achieve GxP compliance.
- A supplier's ability to invest in quality systems supporting GxP compliance may be limited by
  - GxP regulations applying only to a small proportion of their market and
  - Prices being driven down by competitive pressures in their marketplace.

The fundamental challenge is to understand how a cost-effective balance between a supplier's quality system and the user's GxP compliance requirements can be best reached.

USER REQUIREMENTS SPECIFICATION

The business objective to be met by the computer system is expressed in a URS. This must provide a firm foundation for the project. URSs may exist as a single document or as a suite of documents. Supplementary documentation might include business requirements and technical requirements. However, the requirements are collated if they are not precise, concise, and complete; major problems appearing later in the project are almost a certainty!

The URS should not delve into *design* detail; that should be postponed to the functional specification (FS). This may be difficult when specifying bespoke or customized systems, as the design is often anticipated when developing the URS. It may also be tempting to include far too little detail in a URS when a commercial off-the-shelf (COTS) product solution is anticipated.

A multidisciplinary team including production, engineering, and quality assurance staff should draft the URS. End users should be involved as soon as possible and approve the requirements before design and development begin.

Contents

The level of detail contained in the URS varies depending on its relationship with the FS. Information that ought to be considered for inclusion in a URS is provided in Appendix 6.A. Diagrams should be used wherever possible to promote greater understanding and clarity. Spreadsheets are also useful for defining data and clearly showing omitted information.

Some pharmaceutical and healthcare companies give their suppliers a relatively free hand, while others may impose particular equipment and design requirements. A closer examination of any imposed constraints will often expose the fact that preferred equipment and design are not, in fact, critical. Indeed, a measure of choice in the details could be delegated to the supplier and recorded in the design documentation.

The operational requirements section is typically the largest portion of the URS. This information will be presented in the form of textual descriptions, flowcharts, state transition diagrams, or some other similar illustrative form; diagrammatic information is often demoted to appendixes. It should include

- Identification of halt states, error routines, and error recovery,
- Specification of failure modes to protect both the plant and the personnel, and
- Description of any special calculation requirements.

Other important sections in the URS include access security, human interface, and process interface. To be specific,

- Access security should be considered as a means of reducing the risk of unauthorized access, inadvertent modification, or data loss or corruption;
- Security may be achieved through the use of passwords, key switches, or other more sophisticated mechanisms such as those based on biometric features;
- The human interface definition should include screen layouts (or prototypes where these have not been finalized) and other requirements such as configuration pages, alarm pages, mimics, and system responses to data entry errors;
- The minimum information for the process interface includes equipment tag numbers, unit references, descriptions, input/output (I/O) types, and any special treatment, such as segregation requirements; and
- Consideration should be given to the extent of any interface work needed to commission the new system and to identify any changes of functionality in the existing system.

Data may be captured by the computer system through a manual or automated input. User procedures are required for manual data input. Software supporting automated data input such as that used for data acquisition by instrumentation or data migration tools must be compliant with regulatory requirements. Checks should include confirming that any necessary calibration has been conducted and that interfaces are working correctly.

Anticipating Acceptance Testing

Those defining the URS should consider how the requirements might be tested (Fig. 6.1). Alternative phrasing of requirements could considerably clarify the objectives of development testing and user qualification. In response to a poorly written URS, some suppliers have suggested using word searches on "must," "shall," and "will" to determine the minimum set of requirements. All other expressions would be taken to define optional ("nice-to-have") features. It is easy to see how vague expressions of specifications lead to defective acceptance criteria! Test protocols and their relationships to the URS, FS, and other development documentation are discussed in more detail later in this chapter.

Within the URS, mandatory and desirable requirements must be distinguished (Table 6.1). Desired features should then be prioritized in order of their relative importance. The FS should then include a table showing how it complies with the URS. A list of known deviations from the current release of the URS should be highlighted. To make this cross-referencing easier to administer, each URS requirement should be allocated a unique reference number. The GAMP®

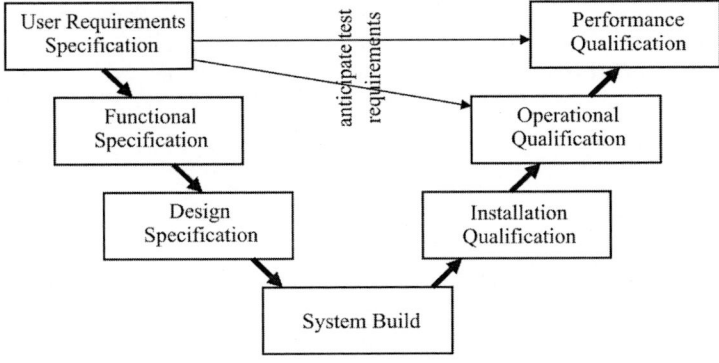

Figure 6.1 User requirements anticipate testing requirements.

Table 6.1 Example of Requirements Numbering in a User Requirements Specification

Specification reference	Requirement	Priority (M/D)	Criticality (C/N)
4.6.1	The XXXX computer system is able to allocate either any or all of the system configurable access rights to each of three user levels as defined in Appendix 6.A.	M	C
4.6.2	Allocation of all configurable access rights to individual users requires onscreen confirmation by system administrator.	D	N

Abbreviations: M, mandate; D, desirable; C, critical; N, noncritical.

Guide recommends that each requirement statement is no longer than 250 words to aid traceability (1). Any assumptions made by the FS relating to such URS deviations should be readily identifiable, so that ensuing misunderstandings can be clarified before the project progresses too far. Ian Johnson, now at AstraZeneca, recalls an instance of an extractor fan fitted on a powder-filling machine for operator protection. The machine was controlled by a programmable logic controller (PLC) that monitored the extract from the fan and displayed an alarm to the operator if the extractor had failed (2). However, the supplier providing the application software had not appreciated the potential danger of powder in the operator's working environment. They had accordingly programmed the powder-filling machine to continue functioning after the alarm of extract failure had been displayed. The URS should have specified the need for an interlock, so that should the operator not notice the alarm and intervene, the powder-filling machine would stop.

Requirements, if appropriate, can be developed, approved, and released incrementally, but care should be taken that interactions and interfaces between software (and hardware) requirements are properly reviewed, analyzed, and controlled (3). While incomplete or ambiguous requirements present an opportunity to proactively manage future clarification of requirements, more often than not, they hinder effective design. Sam Clark, a former FDA investigator, cited an example of a poor requirement.

A novice must be able to use the system in a simple manner and a sophisticated user in a sophisticated manner (4).

It is easy to criticize, but writing a URS is much more difficult than is often believed. The first draft of the document is inevitably unstructured, as authors tend to include information in the order it comes to mind. The creation of a standard format helps, but most authors still customize the layout to reflect their own particular viewpoint. Project managers must accept and plan for rearranging the URS so that a more readable and useful document evolves. Revisions to the URS must be subject to version control so that the project team is kept abreast of updates.

Diligent effort at the early stages will be abundantly rewarded by the avoidance of failures later in the project. The cost of retrospective modifications later in the life cycle can be 10 or 20 times that of instituting the amendment at this stage. Getting it right first time is the cheapest way of doing anything in the long run! Once the requirements have been agreed, they must be documented in a manner understandable by both users and development staff alike. The user may ask a supplier for assistance in producing this document.

Electronic Record/Signature Requirements

Functionality that handles electronic records subject to specific regulatory requirements should be specifically identified in the URS. Individual records requiring ERES controls should also be defined. The following aspects should be outlined for the entire set of electronic records, for a subset, or for individual electronic records (a detailed discussion is provided in chap. 13), as appropriate:

- The ability to discern invalid or altered records.
- The ability to generate accurate and complete copies of records in both human-readable and electronic forms suitable for inspection, review, and copying.
- Secure, computer-generated, time-stamped audit trails to independently record the date and time of operator entries and actions that create, modify, or delete electronic records.
- Retention of previously recorded information when any change is made to records.
- Protection and retention of records and audit trail documentation to enable their accurate and ready retrieval for review and copying throughout the records retention period.
- System access in these cases must be limited to authorized individuals; authority checks must be in place to ensure that only authorized individuals can use the system, electronically sign a record, access the operation or computer system input or output device, alter a record, or perform the operation at hand.
- Where appropriate, operational system checks that enforce only the permitted sequencing of steps and events must be included.
- Use of device checks to determine, as appropriate, the validity of the source of data input or operational instruction should be present.
- During a single, continuous period of controlled system access, the first signing needs to verify the signatory's identity with password, while subsequent signings need to verify only the password. Signings not performed during a single, continuous period of controlled system access must verify both the signatory's identity and his/her password.

Approval and authorization of electronic records should be accomplished using electronic signatures that are unique to individual users and inextricably linked to the electronic records to which they are applied. Where electronic signatures are to be used, each instance of this should be unambiguously specified.

Signed electronic records should contain information associated with the signing that clearly indicates

- The name of the signer,
- The date and time when the signature was executed, and
- The meaning (such as review, approval, responsibility, or authorship) associated with the signature.

These items must be included as part of any human-readable form (e.g., printed) of the electronic record.

Recent Inspection Findings

There were no written specification requirements for the computer system including the software used as the base management database (FDA 483, 2007).

There are currently no approved requirement specifications for XXXX software prior to execution of the installation qualification (IQ), operational qualification (OQ), and performance qualification (PQ) test scripts (FDA 483, 2004).

> Necessary actions have not been predetermined and documented when responding to alarms from the XXXX...(alarm event) is not recorded...(FDA Warning Letter, 2000).
>
> No written requirements or specification for QC laboratory computer software program (FDA 483, 2003).

GXP ASSESSMENTS

GxP assessments provide a useful tool in helping to identify where to focus compliance activities. The assumption here is that GxP processes/functions require a higher level of assurance than non-GxP processes/functions do. Similarly, GxP critical components/devices require a higher level of assurance than non-GxP critical components/devices do. The risk assessment process does not reduce the need for complete specifications, rather it can be used later in the life cycle to improve the usefulness of the testing efforts.

GxP assessments are usually conducted by multidisciplinary teams with expert knowledge of regulatory requirements, the relevant process (development, manufacturing, or distribution), and the computer application.

All GxP functions, processes, components, and devices identified within the GxP assessment should be challenged as part of the design review. Consideration may also be given to occupation health matters, such as the potential effects of the computer system and associated equipment on the personnel who may use or contact the system. GxP functionality includes the use of electronic records and signatures. Hybrid systems must be defined and subject to a verification process to determine whether or not they are robust. It is often useful for processes to be mapped, showing critical points in the process and how various computer system(s) support these critical process points.

Assessments are only as useful as the available information (opinion and documentation) pertaining to the computer system under scrutiny. Such information may be excessive, incomplete, inconsistent, or incorrect. Where it is felt that there is insufficient information to complete the GxP assessment, a preliminary assessment may be conducted and reviewed when further information becomes available during the later phases of the project life cycle.

Identifying GxP Processes and Functions

The decision over whether a process or function affects GxP should be dictated by its operational role and if it is used to maintain GxP master data, as illustrated in Figure 6.2. GxP roles include

- GxP procedural controls,
- GxP decisions (e.g., accept, reject, refer),
- GxP approvals and certifications,
- GxP authorizations, and
- GxP data submissions to regulatory authority.

GxP data maintenance activities include

- Creating GxP master data,
- Modifying GxP master data, and
- Deleting GxP master data.

Examples of GxP processes (functions) include supplier management, procurement, goods receipt, materials management, production control, quality control, batch release, distribution, recall, customer complaints, batch tracking, and compliance management (e.g., SOP management and electronic data archiving).

Examples of non-GxP processes (functions) include capacity scheduling, finance, HR (excluding training), marketing (excluding medical information), purchasing, legal affairs, insurance management, and business reporting.

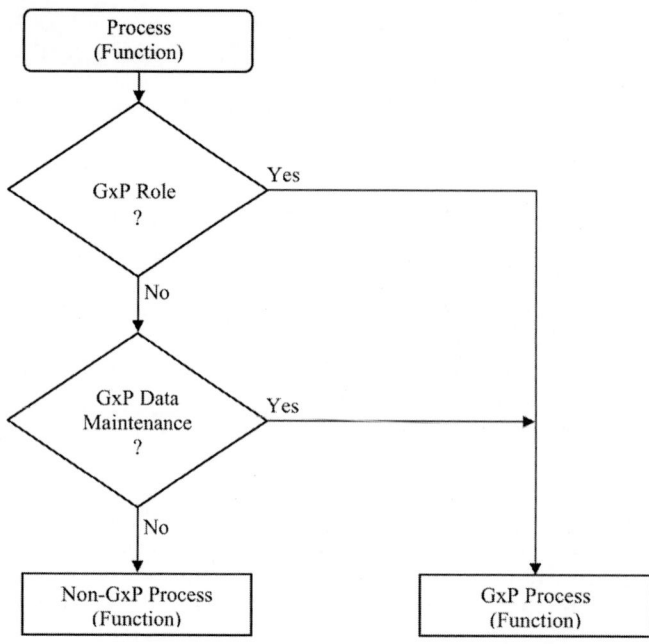

Figure 6.2 GxP assessment decision tree.

Examples of product quality GxP data include study data (e.g., stability trial data, clinical trial data, and patient and animal records/results), regulatory submissions (e.g. stability data and development summary reports), analytical production data (e.g., analytical methods and quality reference data), and compliance management (e.g., indexes to archived documents/records).

Examples of manufacturing GxP data include purchase order information (order number, supplier batch number, supplier quality approval status), bill of materials information (items, quantity, units of measure, conversion factors, work centers, yield factors, critical process parameter), batch information (batch number, batch status, expiry and receipt dates, quantity, potency, conversion factors, and any special instructions or batch record annotations), user security information (name and password), warehousing information (item number, item note, location, type, quality status, shelf life, and retest days), customer order information (order number, customer address, batch number, batch status, expiry date, quantity, potency), distribution information (distributor code number, distributor address, date collected), shipping information (customer order number, customer address, shipping address, shipping notes, dispatch date, goods return note number), secondary batch traceability, inventory control, manufacturing and expiry dates, label control, critical manufacturing process parameters, environmental monitoring, calibration, and maintenance records.

Some systems such as financial management systems will have no impact on GxP unless they contain special functionality that can affect GxP data. Examples would be the use of a zero price to indicate that a product should not be supplied to a particular market/customer. Another example would be financial material reconciliation affecting batch materials usage data. Even if a computer system is not deemed wholly non-GxP, it does not imply that general quality assurance principles are no longer applicable. Good business sense dictates that a quality management approach should always be applied, such as TickIT, IEEE, and SWEBOK.

Identifying Critical Components and Devices

Critical components/devices are usually identified within the architectural design as part of design and development (see chap. 7). Components or devices fulfilling one of the following statements should be considered critical unless there is some form of backup mechanism (backup system or procedural control) for GxP functionality.

- Used to control, monitor, or assess a quality or GxP aspect of production process, including pack integrity

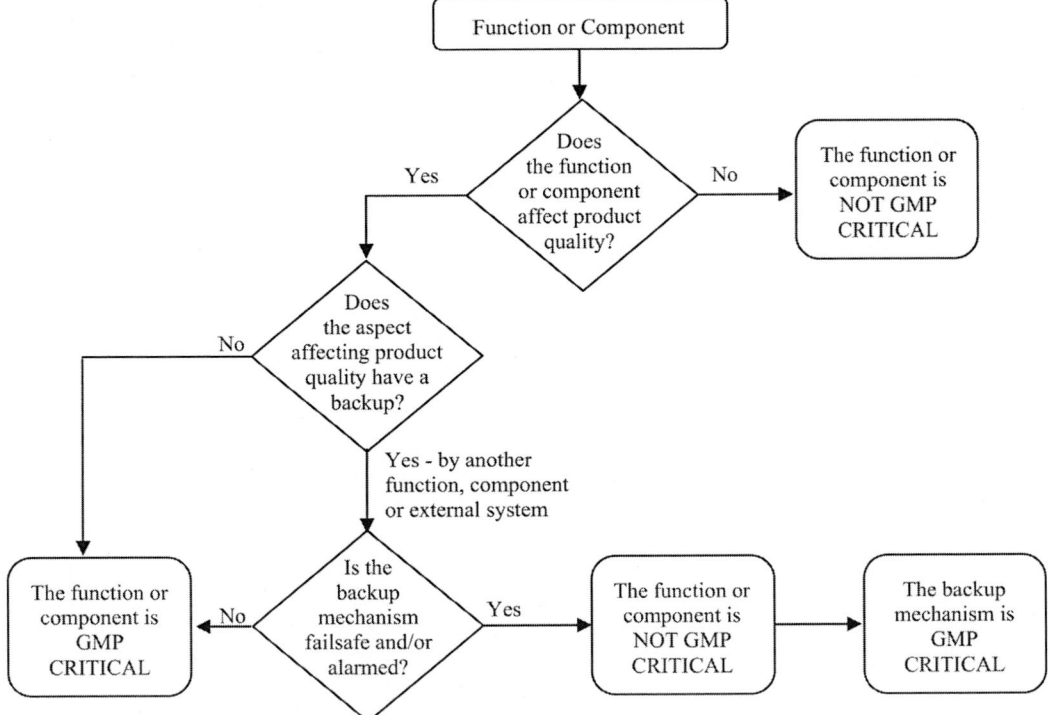

Figure 6.3 Backup mechanism decision tree.

- Used to manage records related to quality or GxP aspect of production process, including pack integrity
- Where a malfunction could result in substandard product quality, pack integrity, or a GxP control being deemed acceptable when in reality it is not
- Used to test or calibrate a critical device/component

Backup systems for GxP functionality include an independent "parallel" or "downstream" component/device to detect any malfunction (e.g., independent monitoring systems). The focus for compliance activities can then shift from the component/device to the backup system, as illustrated in Figure 6.3 (5). Where this implies implementing a backup system that is simpler to validate, this can relieve the burden of dealing with more complex components and devices. For a backup system to be accepted as a validated alternative, it must be able to independently manage key quality assurance functions. Such functions include but are not necessarily limited to

- Provision of independent system access controls,
- Provision of independent functions for alarms (including management and calibration as appropriate of alarm set points, and management of alarm log records),
- Provision of independent data sources for GxP records (e.g., QA monitoring and records for critical parameters, and QA investigations of out-of-specification incidents), and
- Provision of data and functionality for understanding trends over short and long periods.

All these functions must be managed by the backup system, otherwise both the component/device and backup will require validation. Care must also be taken when implementing system changes not to inadvertently undermine the case for directing compliance effort to backup mechanisms.

When conducting a review, care must be taken to address system interfaces. Computer systems receiving GxP data, even if such data is just passing through the machine for processing elsewhere, are likely to require some level of verification since they could potentially compromise the integrity of the GxP data being transmitted.

Recent Inspection Findings

Refer to inspection findings in section "Compliance Determination" in chapter 5.

SUPPLIER SELECTION
Regulatory Expectations

Pharmaceutical and healthcare companies are obliged to determine the suitability of proposed suppliers providing products and services, particularly those providing software. Regulatory authorities neither prohibit nor endorse specific systems despite what some supplier marketing might suggest.

> *...software is a critical component of a computerized system. The user of such software should take all reasonable steps to ensure that it has been produced in accordance with a system of Quality Assurance.* (6)

This requirement is usually met by means of the audit conducted by the pharmaceutical or healthcare company. This examines the quality assurance attributes of the supplier's process, the firm's general capability maturity, and the suitability of their equipment or service suggested for use on the project. Suppliers in this context may be understood to include equipment vendors, service suppliers, or the pharmaceutical or healthcare company's in-house software development department. Regulators hold pharmaceutical and healthcare companies accountable for the use of suppliers whose capability maturity assessment indicates their inability to deliver compliant software that is fit for the purpose (7).

To fulfill regulatory requirements for computer systems, pharmaceutical and healthcare companies are expected to mitigate deficiencies they identify in the quality of their suppliers' software. The MHRA accepts that "Where there is little or no supplier cooperation over validation evidence, it can be difficult to assess the QMS in place at the supplier, let alone have access to source-code and testing records" (8). In such situations, supplier audits are impractical and the industry has often had to rely on functional testing and change control alone. This is an enormous handicap, since functional testing of this or any kind can never furnish an equivalent level of assurance over innate structural integrity that may be derived from an in-depth examination of the supplier's development process. However, the regulatory authorities expect that quality issues form an integral part of the supplier selection process and that wherever possible, quality-minded suppliers should be selected. The implication here is that neglect of quality issues during supplier selection is irresponsible and therefore unacceptable. Technical and quality aspects must be accorded equal weight when determining the fitness for purpose of a product chosen for use in the pharmaceutical and healthcare industry.

Invitation to Tender

Invitations to tender are typically distributed with an accompanying information pack. This should provide contact details for purchasing departments, technical enquiries, etc., and define expected response times. A key document to include within the information pack is the URS. Without a clear understanding of requirements, assumptions, and constraints, the supplier will find it difficult to confidently respond with a clear proposal. In recent years, the regulatory authorities have exerted increasing pressure to improve the content of URS, because of the adverse impact that a poorly drafted URS can have on the ensuing compliance activities. Such an emphasis on investing in a well-prepared URS document has been welcomed by supplier organizations. Other items to be included in the information pack might be copies of relevant company standards (e.g., compliance checklists and a summary of their control and operability philosophy) (Appendix 6.B). Suppliers are normally expected to have their own copies of industry standards and guidelines.

Supplier Proposal

Depending on the nature of the invitation to tender, the supplier's response may amount to little more than a covering letter with accompanying standard literature. This is often the case for COTS products that the supplier believes to be meeting the user's requirements or when a dedicated supplier proposal for a more bespoke or customized solution is to follow. Some suppliers may even draft an initial FS to demonstrate the suitability of the solution they have in mind. FSs are discussed in more detail in chapter 7. Other proposal documentations might include draft project and quality plans.

Proposal Evaluation

The proposals that are received in response to the initial requests need to be evaluated against those requests to highlight any deviations. Preferred supplier(s) are then typically short-listed.

Supplier evaluations should assess a number of factors including

- The capability of the supplier organization within a quality-orientated culture,
- Whether or not a quality management system exists and is applied and maintained,
- The technical competency of staff, as well as the awareness/understanding of pharmaceutical and healthcare industry regulations/practices,
- Whether or not the supplier routinely supplies the pharmaceutical and healthcare industry and is therefore familiar with regulatory expectations, and
- Whether the company is sufficiently financially stable to be able to support the system throughout its operational life.

These factors should be assessed for each supplier being considered, so that the best balance of business fit and compliance can be achieved. An example supplier evaluation matrix is provided in Figure 6.4. Of course, identifying a clear winner in the selection process is seldom clear cut, and the winner rarely meets all the selection criteria. Some pharmaceutical and healthcare companies weigh the importance of various factors and for the outcome compare the sum totals for each supplier. A summary of the supplier selection process applied to larger systems should be retained for possible subsequent inspection (10).

Suppliers have an inherent legal responsibility to all their users that their products and services are fit for purpose (e.g., in the United Kingdom, this responsibility is enshrined in the Sale of Goods Act). Pharmaceutical and healthcare companies will sometimes conduct an audit of their suppliers to assess their quality systems and management. The audit may be part of a supplier selection exercise. Auditors may be available among the pharmaceutical and healthcare company's own staff or engaged and commissioned from specialist audit firms providing these services. Any shortfalls in the supplier's capability uncovered in the audit will have to be mitigated, preferably by the supplier correcting these through process

	Supplier A	Supplier B	Supplier C
Quality Organisation	✓	✓	✓
Quality Systems Exist	✓	☒	✓
Quality System Applied	✓	☒	☒
Quality System Maintenance	✓	☒	✓
Technical Competence	✓	✓	☒
Pharmaceutical Experience	✓	☒	☒
Support Infrastructure	✓	☒	✓
Commercially Robust	✓	✓	☒

Figure 6.4 Example of a supplier evaluation matrix. *Source*: From Ref. 9.

improvement. If not, the pharmaceutical or healthcare companies must initiate action themselves either directly or through third-party support.

Suppliers can benefit enormously if they adopt a positive, enlightened attitude to audits, especially if the auditors are able to direct their attention to process improvements. This is because such improvements will aid the supplier to get their products right first time, thus saving money in the long term. Such has been the experience of the overwhelming majority of firms who have brought their quality system to a level compliant with the GAMP®5 Guide or ISO 9001:2008 (11). It is most important that pharmaceutical and healthcare companies motivate their suppliers in this way as far as possible in a spirit of a long-term business partnership.

The use of qualified auditors is much to be preferred. An example of such qualification would be the formal written examination for ISO 9001:2008 (ISO 90003) and subsequent registration under the auspices of the *International Register of Certificated Auditors*.

Contract of Supply

Contracts of supply should exist for all computer systems acquired from external suppliers. The contract should include terms and conditions to define individual responsibilities, the assignment of responsibilities to third parties, confidentiality, intellectual property rights (IPR), and terms of payment. It is very important that infringements, liabilities, and insurance, and how these are affected by circumstances outside the control of the customer and the supplier are also covered. Contract details should be reviewed during supplier audits as appropriate. EU GMPs require that contract documentation is retained.

Contracts made in relation to customized or bespoke computer systems should make reference to the validation plan, any supplier quality plans, and, where appropriate, the system specification to define the scope of work, goods, and services.

Copies of purchase orders and corresponding supplier dispatch notes should be retained for COTS products. These should specify relevant model/version numbers so that, if need be, supporting information can be traced later with the supplier. Version numbers of technical documents and user manuals provided with the COTS product should also be noted on the purchase order.

A purchasing process that addresses compliance issues is presented in Figure 6.5. This covers supplier selection, award of contract, and delivery. The process should include a determination as to whether or not an audit is required. If the need for an audit is established, the supplier concerned should be contacted and an audit request should be made, explaining the context of the audit.

Recent Inspection Findings

> No documentation available to show that either internal quality audits or an evaluation of the contract software developer had ever been performed (FDA Warning Letter, 2001).

SUPPLIER AUDITS
Applicability

Supplier audits are not appropriate for all computer systems. Chapter 4 has already reviewed supplier audit expectations for different categories of hardware and software. Supplier audits should be undertaken for custom (bespoke) applications and systems' configuration. In addition, supplier audits should be considered for standard software systems and core configurable packages when they are used for GxP critical applications, especially so if the COTS product is highly complex. Figure 6.6 summarizes those situations when supplier audits are appropriate. It assumes that the specific version of the COTS product being considered is in successful use to a sufficiently wide extent. *"In successful use"* here implies that the COTS product is stable and that it is highly unlikely that a significant number of important defects remain to emerge. Of course, by its very nature, this can never be known with anything approaching certainty, and there is no fixed minimum number of successful users. However,

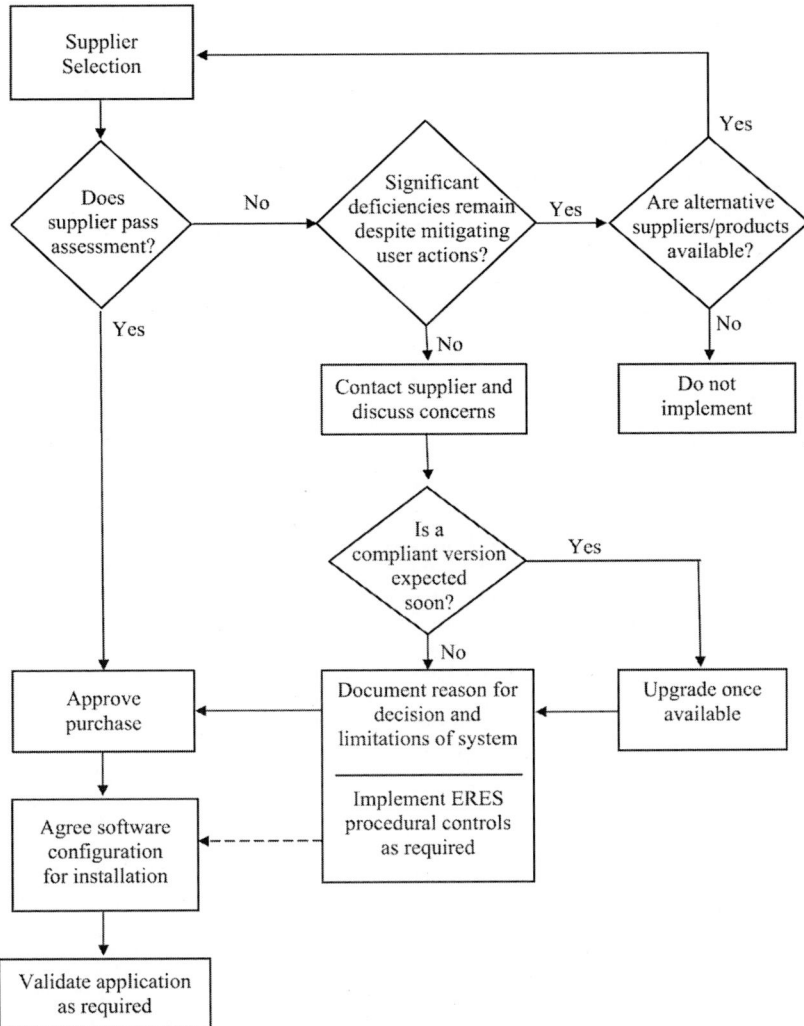

Figure 6.5   Purchasing process addressing compliance issues.

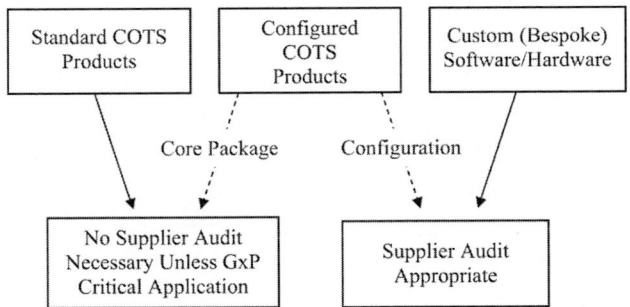

Figure 6.6   Supplier audit determination.

we would expect no less than 100 active applications of exactly the same version (no special adaptations) would be required to be in successful use in this context. The so-called *early adopters* of COTS products, that is, where less than 100 licenses have yet been issued and/or where the product is still under 6 to 12 months old, should treat them as custom (bespoke) computer systems and audit them accordingly.

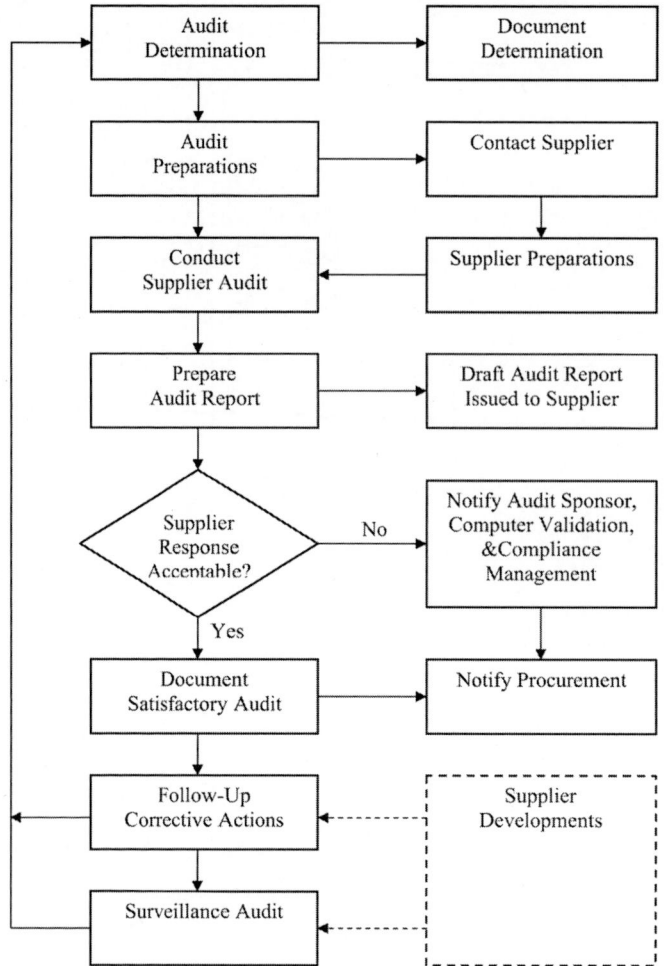

Figure 6.7 Audit process.

Audit Preparation
A typical audit process is outlined in Figure 6.7 and is discussed below.

Supplier Preparations

A supplier may be contacted informally to discuss the most appropriate audit route prior to any formal approach being made to conduct an audit. Any proposal or bid made by a supplier should contain the contact details for the quality managers and/or personnel, so that the auditor can easily liaise with the appropriate representative in the supplier's organization with whom to discuss the next steps. When the auditor contacts the supplier to organize an audit, the supplier should also get the auditor's details (full name, postal address, telephone, fax, e-mail, and even video conference details where appropriate) for future communications.

It is important that, even at this early stage, formal confidentiality agreements should be signed for nondisclosure of proprietary information to unauthorized third parties. Suppliers should consider preparing a standard pro forma that can be completed and faxed/posted to the auditor. The agreement must state whom the signatories represent. Where a pharmaceutical or healthcare company engages the services of a third-party auditor or company, the supplier must ensure that the agreement covers both them and the auditor or his/her company. It should be made abundantly clear that written permission is required from the supplier to authorize the disclosure of the audit findings to the commissioning pharmaceutical or healthcare company, regardless of their favorability, or otherwise to the supplier.

Nonetheless, from a legal standpoint, it would almost certainly be viewed in law that the supplier had conceded this right by allowing the audit to occur in the first instance.

Further guidance for suppliers is summarized in the checklist given in Appendix 6.C. It can be used to monitor progress during audits. It is suggested that suppliers consider taking a photocopy of the checklist and annotate it with user details (proposal number and/or contract number), along with comments as appropriate to the audit. It should then be retained with their user/customer files. By planning audit schedules, conducting preparatory work, and ensuring that the right people are available to present and explain the suppliers' ways of working during the audit, a lot of the stress associated with audits can be relieved. A spirit of resentment felt by suppliers after an audit benefits no one.

ISO 9000 Accredited Suppliers

Many suppliers have quality systems registered as compliant to one of the ISO 9000 quality management standards (11,12). The earlier versions of these standards, based on compliance with procedures, consisted of the following:

- *ISO 9001*: Addresses quality practice for product design, development, production (including inspection and testing), installation, and services
- *ISO 9002*: Addresses quality practice for production (including inspection and testing), installation, and services
- *ISO 9003*: Addresses quality practice for final inspection and testing
- *ISO 90003*: Guidance on applying ISO 9001 to software

Emphasis is now being placed on achieving a documented state of control, with customer satisfaction, management endorsement, and continuous improvement. These were no more than implications in the earlier standards. Hence, we now see the ISO 9000 family with its process emphasis coming into harmony with the *ISPE GAMP®5 Guide*. *The TickIT Guide* version 5 accompanies the ISO standard and may be regarded as a much more detailed guideline for software companies in best practices. In this context, it is a worthy companion to the *GAMP®5 Guide*. Auditors would do suppliers a power of good by commending both to them!

It is important to acknowledge that software can be developed under a quality system registered as compliant with ISO 9001 without necessarily being fully compliant with the guidance given either in ISO 90003 or in The TickIT Guide. It must be remembered that a supplier's certification to ISO 9000 does not necessarily imply a capability to develop GxP compliant software. The auditor should carefully read the ISO 9000 certificate for its qualifying statement and validity duration, for it may not even apply to software development. A system or equipment supplier may subcontract software production and thus not be accredited for this activity. If possible, the auditor should identify such circumstances before the visit and then determine whether the supplier has the necessary safeguards to ensure the quality of subcontracted work. An audit of the subcontractor might then follow as appropriate.

Regardless of whether a supplier is accredited to ISO 9000 or not, the supplier should be aware of how their quality system measures against this standard's requirements. Suppliers should consider preparing a short report mapping their current working practices to the clauses of ISO 9000. This report could then be supplied quickly and easily, if requested, to existing and prospective users. In addition, suppliers should collect reference site material (possibly as sales literature) and user contacts that can be supplied to potential new users. Suppliers may wish to consider collecting testimonials at the end of projects, rather than trying to locate users subsequently who might have moved on by that time.

Postal Audits

Postal audits are usually used as part of the supplier selection short-listing process. They may recommend an audit on a supplier's premises where extra detail is required, but typically such supplier audits are reserved only for the preferred supplier to confirm their acceptability from a compliance perspective.

Howard Garston-Smith, formerly of Pfizer, has published a book on software quality assurance (13). This provides a postal audit checklist reproduced in Appendix 6.D. If the supplier has already prepared an internal ISO 9000 mapping or an internal audit report on how they align to industry standards such as the ISPE GAMP®5 Guide, then this can be offered as an alternative to the auditor's postal checklist. A reduced postal checklist may be agreed, at the very least. Wherever possible, photocopies of actual example documents and test records should be audited. Remember that the pharmaceutical and healthcare companies are themselves being inspected for documentary evidence demonstrating compliance with regulatory requirements.

Auditors should agree response times with suppliers when using the postal checklist. Busy contract periods, illness, holidays, and staff turnover may all delay the responses. Before agreeing to a response time, suppliers should ask to see a copy of the audit checklist so that the amount of work to reply can be gauged.

Visiting Supplier Premises

Audits on supplier premises are used to assess a supplier's quality management system at first hand, with detailed examination of procedures and documentation relating to a product or service. Such audits are expected for suppliers of custom (bespoke) software and systems. Audits are not normally expected for widely distributed standard software or configurable commercial software packages. The case for this reasoning is that the software/system has already been adequately market tested. An audit might still be deemed necessary for configurable commercial packages if the application is particularly complex and/or critical.

Visits to supplier premises are expensive, not only to the supplier concerned, but also to the pharmaceutical or healthcare company. It is important to both parties that they are conducted quickly and efficiently. Reimbursement to suppliers to cover the costs associated with audits is not normally offered or requested. Some suppliers have tried to levy fees, but this has created ill feeling, and supplier firms adopting this approach are generally felt to have blotted their copybook in the marketplace. After all, a quality culture should have already been established in the supplier firm. The audit should be seen by a supplier as an opportunity not just to defend their software development capability maturity but also to sell quality as a distinguishing feature to the product or service on offer. Some suppliers have offered pharmaceutical and healthcare companies the chance to audit collectively in managed groups, perhaps through a user group structure. Suppliers offering this facility are limiting the number of audits to two or three per year and have reduced their annual audit costs by up to 80% (5).

Assembling Audit Teams

Audits are usually conducted by an audit team and led by a qualified, accredited auditor. Accredited auditors have completed a certified development program, have been accredited under an appropriate standard (ISO 9000 in this context), and have conducted a number of qualifying audits. Pharmaceutical and healthcare companies that do not have their own accredited auditors can engage an independent auditor, as we have seen. Names and addresses of accredited auditors can be found in national registers of certified auditors. The International Register of Certificated Auditors is one such register, associated with the *Institute of Quality Assurance* (IQA).

The audit team's size should be kept to a minimum but should adequately represent the following areas of expertise (4):

- Auditing practices
- Computer system engineering
- Computer system quality methods
- Regulatory compliance standards
- Pharmaceutical and healthcare industry compliance practices

Team members should have a preliminary understanding of auditing before they begin an audit. The ISO 10011 auditing standard (2) provides useful material on audit practice, the qualifications of auditors, and the management of audit programs.

Conducting Audits

Audits normally take two days to conduct. One-day audits are possible when the auditor and the firm being audited are well prepared or there is a reduced audit scope. Care must be taken with shorter audits to make sure that the assessment is not unacceptably superficial and lacking the required scope or depth.

A typical audit schedule might consist of

- A general introduction by the supplier giving details of the audit plan, an overview of the supplier's business operations, and quality department organization,
- A presentation by the supplier of the firm's quality management system, perhaps reviewing the supplier's internal ISO 9000 mapping report,
- A presentation by the supplier of the quality process adopted for a particular product or service under scrutiny, perhaps including a review of other user quality expectations with emphasis on pharmaceutical and healthcare industry users,
- An opportunity for the auditor to view actual quality documents and records (example checklist for audits is included as Appendix 6.E), and
- A summary by the auditor of preliminary findings with an opportunity for the supplier to discuss and clarify.

Alternatively, very experienced auditors often have a well-practiced structure or framework by which they conduct audits and draft their reports. This can be of great value where a pharmaceutical or healthcare company decides to audit several potential suppliers for a critical system. In such cases, it usually needs the reports with an identical layout, scope, and depth. This makes comparisons on a level basis easier. Advising the supplier of this in advance is a great help all round.

It is useful if the supplier has nominated a representative to host the audit, ideally someone who has been trained in auditing and therefore can understand and anticipate the auditor's perspective. Training courses are available, for instance, along the lines of industry auditing standards like ISO 10011. Good interpersonal and language skills are often beneficial too, especially where the audit is not conducted in the participants' mother tongue.

The supplier may require a briefing in the proposed audit process objectives and scope. The audit is then conducted, and a report is prepared by the pharmaceutical or healthcare company concerned. The supplier should be given a chance to review and comment on the report so that any factual errors can be corrected, vendor comments can be added, or other necessary clarifications can be made at an early stage. Any corrective actions undertaken by the supplier should be followed up, and so the audit process may loop back by the arrangement of a follow-up visit.

The timing of the audit should be considered here, as some potential members of the team may not be available because of holidays, and their deputies might need to be in place to deal with any unforeseen absences. The logistical challenge of undertaking a supplier audit should not be underestimated. Items to consider are as follows:

- Who is the auditor (by name and function)?
- Will the auditor be assisted by anybody (name and function) or observed by anyone (e.g., trainee or other interested party)?
- Who is the host firm representative (name and function)?
- Who is assisting the host representative (name and function)?
- Where is the audit being conducted; have maps and times been sent out to participants?
- Has hotel accommodation been organized where needed?
- Have presentation slides and materials been prepared, and are copies available for the auditor?
- Will relevant information be brought to the audit by host firm representatives?
- Has a tour of the supplier premises been organized?
- How long is the audit estimated to take? (Do not let the audited organization dictate the duration of the audit; take the time needed to do a proper job.)

It is useful to assign a base camp room for the duration of the audit, a place reserved for the audit team whether or not they are permanently located there. It provides a focus for the audit team, a place for the audit team to leave personal baggage and collect files of information pending review, or leave information behind that has served its purpose.

Subcontractors

The use of subcontractors needs to be clearly understood well before the audit visit, together with the oversight planned by the prime supplier. If the prime supplier does not have adequate control over them, then supplier audits by the pharmaceutical or healthcare company may be required. Key activities and/or documentation under the management and control of subcontractors should be subject to review by prime supplier. The topics listed in the general supplier audit expectations in Appendix 6.D are relevant to subcontractors. Any deficiencies found in a subcontractor's software development capability maturity level must be mitigated by supplementary work to bring it up to an acceptable standard.

Audit Report and Follow-up

The structure of a supplier audit report is suggested in Appendix 6.E and covers the following aspects:

- Introduction and scope
- Summary of audit process
- Review of supplier quality assurance system
- Audit results with list of corrective actions
- Audit outcome (conclusion)

Suppliers should be advised at the outset of an audit that they will be accorded the opportunity to review and correct a draft of the audit report before it is issued, as we have mentioned above. There should also be some agreement as to the timetable expected for the issue of a draft report. It is not proper for a supplier to seek to approve the audit report, since that could amount to an attempt to gag the auditor. By permitting the audit in the first place, the supplier has effectively given the auditor permission to exercise his/her professional impartiality, and this freedom may not subsequently be withdrawn. The contents of the report are the auditor's opinions and should not be influenced, except where they are based on factual misconceptions that must be corrected by the supplier. The auditor has a duty of care to the supplier to permit this. The supplier should retain a copy of the draft audit report with a note of their review comments. A final copy of the approved and issued report must also be furnished to the supplier by the auditor as the basis for future business relationships.

Actions and recommendations should be differentiated. Actions must be completed or progressed to an acceptable state before the computer system can be authorized for use. Open actions should be listed in the project compliance issue log (see chap. 5). Recommendations are just that they do not need to be completed before the computer system is authorized for use. Follow-up audits should track actions and recommendations as appropriate in the context of that audit. The management of actions and recommendations are summarized in Figure 6.8.

The audit report should indicate the significance, importance, and priority to be adopted for the various corrective actions. After being informed of the outcome of the audit, it is the responsibility of the supplier to agree follow-up actions and timescales. If the auditor has been competent enough to engender a positive spirit into the auditor/supplier relationship, the supplier should be defensive at this point but cognizant of weaknesses and motivated to effect improvements. As intimated earlier, here is a marvelous opportunity for audit findings to act as a springboard for an ongoing program of continuous improvement. The ability to demonstrate such an attitude to improvement will always reassure and impress both the user and his/her organization. Consider using the expertise of the auditor to debate options for quality improvement—it is often a free resource within the pharmaceutical or healthcare company, though if external, then consultancy fees might be incurred. Finally, agree how to communicate the completion of agreed follow-up actions with the auditor/user. Not all

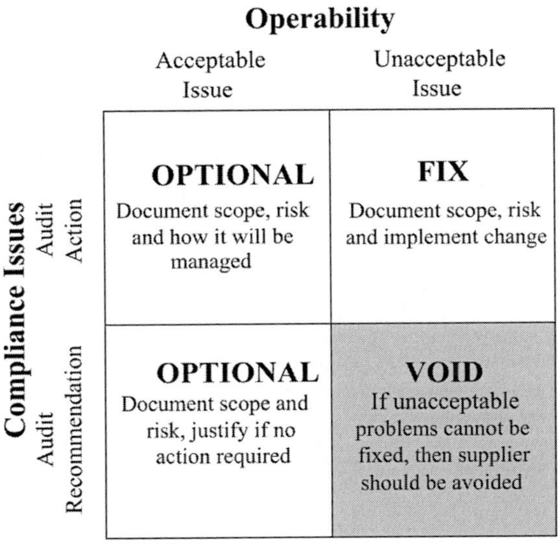

Operability

Figure 6.8   Audit actions and recommendations.

recommendations for action have to be implemented by the supplier, but try to accommodate the auditor's findings.

It is very important that the auditor continues to be willing to act as a mediator between pharmaceutical or healthcare company and supplier after the audit to promote trust, to assist the supplier in any way, and to promote the business partnership. The pharmaceutical or healthcare company may organize a follow-up audit to check whether correction actions have been taken. The number and frequency of recommended follow-up audits should be recorded in the audit report.

Regulatory Access to Audit Reports
There have been many debates over whether or not regulators have the right to scrutinize supplier audit reports. It is the policy of the FDA and other regulatory authorities not to request to inspect such audit reports without due cause [clause 20(c) of Ref. 14 and clause 22 of Ref. 15, respectively]. All that the regulatory inspectors requires is that the pharmaceutical or healthcare company can demonstrate that audits are being performed in accordance with a documented standard procedure. This must include the preparation of written reports and that required corrective actions affecting suppliers and vendors as a result of the audits have been followed up to a satisfactory conclusion. This is usually achieved through presenting audit schedules, audit procedures, validation plans outlining specific audit requests, and validation reports noting progress on audit findings. The regulators do not wish to discourage honest and relevant audit reports being written because these might be scrutinized by them. They regulators realize that such reports are sensitive and confidential and that it is in their interest that pharmaceutical and healthcare companies readily identify supplier and vendor strengths and weaknesses. They want to encourage pharmaceutical and healthcare companies to take effective corrective actions where necessary.

Managing Supplier Deficiencies
Table 6.2 provides guidance for managing the most common and significant problems. Where practical, suppliers should be encouraged to address their deficiencies rather than rely on supplementary user activities. Inclusion of purchase conditions such as withholding final payment until corrective action is completed can be effective for improving supplier practices. However, remember the changes in influence that can be exerted during a project (Fig. 4.4 in chap. 4). Sometimes the supplier has more influence at the end of a project than the customer!

Table 6.2 Examples of Suggested Actions for Managing Supplier Deficiencies

Supplier deficiency	Suggested action
No written process or procedures for development or maintenance activities.	• Do not use supplier for highly critical applications as software quality and support are more likely to be erratic. • Encourage supplier to develop quality management system.
Absent or incomplete design documentation.	• Increase detail of user requirements specification or create design information (if feasible) for critical functionality and test against these specifications to determine if system functions as expected. • Do not use supplier where there are widespread specification gaps for a highly critical system.
Supplier withholds software source code.	• Access agreements should be established to support regulatory inspection. Direct access is not routinely required.
Absent or incomplete testing documentation including lack of source code reviews.	• Supplement with user testing against specification and design documentation, particularly in critical areas of functionality. • Do not use supplier where there are widespread testing gaps for a highly critical system.
Lack of change control process.	• Consider additional testing for patches or upgrades as "fixes" may inadvertently introduce new bugs. • Do not use supplier where review of software patches and upgrades identify widespread undocumented changes.
Software has high number of residual bugs.	• Consider manual process to replace or augment highly critical functionality (e.g., paper trail verification for product release).

Preferred Suppliers

Supplier audits may be conducted as part of individual projects or as part of an overall strategy to select preferred suppliers. Some pharmaceutical and healthcare companies rate the capability maturity level of suppliers within a scoring system. Suppliers might be ranked with a numbering system or with a combination of keywords such as "excellent," "satisfactory," "noncompliant," and "ISO certified." Bernard Anderson of GlaxoSmithKline has suggested the following rankings (16):

- The supplier can be used for any type of software development work.
- The supplier can be used for specific type of software development work.
- The supplier can be used for specific types of software development work subject to *minor* corrective actions being performed.
- The supplier can be used for specific types of software development work subject to *major* corrective actions being performed.
- A documented quality system must be imposed on the supplier in the event a contract is awarded.
- The supplier cannot be used for any regulated system (one might ask just what kind of system could be safely acquired from such a supplier).

Whatever system is used, the ratings must be defined so that users and regulators alike have an unambiguous understanding of their meaning and the suitability of the supplier. Project teams can use the rating system to select suppliers without unnecessarily duplicating audits. Companies using this approach, however, must carefully consider how long a rating remains valid until refreshment of the information is required since a supplier's capability can change, and not always for the better!

User Groups

It may be advantageous to join or establish a user group for a particular computer system to exchange user experiences, disseminate support information, organize collective training, and influence the direction of further development. Later when the systems require replacement, the group might develop a migration strategy. Many suppliers are keen to facilitate user groups as a means of

- Building a shared vision for product developments,
- Gathering direct feedback on system operability and any problems experienced,
- Helping prioritize the correction of known defects,
- Standardize audit requirements, and
- Encourage users to conduct shared supplier audits.

User groups may be confined to staff within a pharmaceutical or healthcare company's own organization, perhaps for reasons of confidentiality or competitiveness. Generally speaking, however, there are greater benefits to be had from the participation of individuals from multiple user companies.

Recent Inspection Findings

No written procedure for vendor audits (FDA 483, 2002).

No approved written procedures for vendor qualification (FDA 483, 2002).

No record of vendor audits performed for suppliers (FDA 483, 2002).

Your firm failed to ensure that the supplier of the XXXX documented all of the required test results to indicate the supplier's quality acceptance of the XXXX delivered to your firm (FDA Warning Letter, 2002).

REFERENCES

1. International Society for Pharmaceutical Engineering. GAMP®5: risk-based approach to compliant GxP computerised systems. Tampa, Florida, 2008. Available at: www.ispe.org.
2. International Organisation for Standardisation. ISO 10011-1 (1990): guidelines for auditing quality systems—part 1: auditing. Geneva, Switzerland, 2000.
3. Food and Drug Administration. General principles for software validation, final guidance for industry and FDA staff. Centre for Devices and Radiological Health, 2002.
4. Grigonis GJ, Wyrick ML. Computer system validation: auditing computer systems for quality, report on behalf of PhRMA computer systems validation committee. Pharm Technol Eur 1994; 18(9):32–39.
5. Wingate GAS. Computer Systems Validation: Quality Assurance, Risk management and Regulatory Compliance for Pharmaceutical and Healthcare Companies. Boca Raton, FL: Interpharm Press, 2003.
6. European Union. EU GMP annex 11 computerised systems, guide to good manufacturing practice for EU directive 2003/94/EC. Community Code Relating to Medicinal Products for Human Use, Vol. 4, 2003.
7. Food and Drugs Administration. Vendor responsibility, compliance policy guides, computerised drug processing 7132a, guide 12. Rockville, MD, 1985.
8. Trill AJ. Computerised systems and GMP—a UK perspective part 3: best practice and topical issues. Pharm Technol Int 1993; 17–30.
9. Reid C. Effective validation strategies. Third Annual Conference on Computer Systems Validation for cGMP Pharmaceuticals. London, March 28–29, 2001.
10. Pharmaceutical Inspection Co-operation Scheme. Good practices for computerised systems in regulated GxP environments. Pharmaceutical Inspection Convention, PI 011-1, Geneva, Switzerland, 2003.
11. International Organization for Standardization. ISO 9001: quality management system—requirements, Geneva, Switzerland, 2008.
12. International Organization for Standardization. ISO 90003: software—guidelines for the application of ISO 9001:2004 to computer software. Geneva, Switzerland, 2004.
13. Garston-Smith H. Software quality assurance—a guide for developers and auditors. Buffalo Grove, IL Interpharm Press, 1997, ISBN 1-57491-049-3.
14. U.S. Code of Federal Regulations title 21, part 820. Good manufacturing practice for medical devices.
15. Food and Drugs Administration. FDA access to results of quality assurance program audits and inspections. Compliance Policy Guide 7151.02. Rockville, MD, 1986.
16. Anderson B. Vendor audits, computer system validation: a practical approach. Management Forum Seminar, London, March 26–27, 1996.
17. Supplier Forum. Guidance notes on supplier audits conducted by customers, 1999. Available at: www.ispe.org/gamp.

Appendix 6.A   Example Contents for User Requirements Specification

Introduction
- Author/organization
- Authority
- Purpose
- Relationship with other documents
- Contractual status of document

System overview
- User perspective of overall function
- Data flow
- Control dependencies
- Operator interface
- Operating environment
- (Reference to separate system overview document if appropriate)
- Assumptions

Operational requirements
- Major functions
- Secondary functions (error handling, monitoring, data integrity checks)
- Start-up and shutdown
- Normal and abnormal operations
- Error and failure reporting
- Recovery and fallback
- Alternative modes of operation
- Electronic records and signature requirements
- Hard copies—print requirements for reports, events and fault logs, preference for printer types and stationary
- Operator control consoles—layout and functionality
- Operator input devices—keyboards and touch-sensitive displays
- Displayed information—layout of mimics, menus, reports, tables, display hierarchies, colors
- Ergonomic factors—operator comfort, and usability of controls and keys
- Critical system timings—throughput, response, and update of data
- Volume of transactions
- Number of simultaneous users
- System security—use of passwords and keys
- Data security—backup and recovery
- Safety of personnel

Design constraints
- Hardware
  - Requirements and environment
  - Standards
  - Interface
  - Tolerance, margins, contingency

- Software
  - Standards and programming languages
  - Interfaces
  - Software packages
  - Databases
  - Operating systems
  - Tolerance, margins, and contingency

- User interface
  - Interface characteristics
  - Environmental constraints
  - Operational constraints

System interfaces

- Content of user displays
- Levels of user access for different user groups
- Digital/analogue inputs/outputs to external equipment
- Serial inputs/outputs to external equipment
- Parallel communications to external equipment
- Network communications to external equipment

Appendix 6.A   Example Contents for User Requirements Specification (*Continued*)

System environment
- Services—electrical power and heat removal
- Environmental conditions—temperature, humidity, noise, and contamination, intrinsic safety in hostile environments

System nonfunctional requirements
- System availability
- Recovery from failure
- Preferred diagnostic methods
- Required level of maintenance support and support period
- Training
- Required documentation
- System hardware maintenance
- Possible enhancements, which may be required in the future

Documentation
- Operator instruction
- Maintenance procedures
- User manuals
- Training materials
- Compliance document package
- Supplier project quality plan

Development issues
- Host development system requirements
- Design methodologies, CASE tools
- Programming languages
- Subsystem testing
- Integration testing
- Configuration control
- Installation considerations
- Support services
- Maintenance requirements
- Expansion capability
- Expected change

Installation considerations
- Conversion/migration instruction from existing system
- Hardware issues including upgrade and maintenance
- Software issues including upgrade and maintenance
- Training

Testing requirements
- Levels of testing to be carried out (unit, module, system, integration, user acceptance, qualification)
- Amount of user versus supplier involvement
- Use of simulation and prototyping

Appendixes
- Glossary
- Others

Appendix 6.B   Supplier Checklist for Receiving Customer Audits

Customer details
- Name
- Proposal and/or contract reference

Preparation work
- Prepare internal ISO 9000 mapping report.
- Prepare internal audit report.
- Prepare/agree confidentiality agreement.
- Train supplier personnel responsible for receiving user audits in ISO 10011 audit process.

Invitation to audit
- Send out and sign confidentiality agreement.
- Agree scope and method of audit.
- Get auditors contact details for communications.
- Agree response time for postal questionnaire.
- Agree date, duration, and participants for full supplier audit.
- Agree supplier review and approval of audit report.

Postal audit
- Complete postal questionnaire, retaining own copy.
- Consider issue of internal ISO 9000 mapping report.
- Consider issue of internal audit report.

Full supplier audit
- Agree audit schedule in advance of audit.
- Facilitate audit logistics.
- Consider issue of internal ISO 9000 mapping report.
- Consider issue of internal audit report.
- Prepare and issue audit presentation slides.
- Take own notes during audit.
- Have summary at closeout of audit.

Audit follow-up
- Review draft audit report, retain your review comments.
- Retain copy of draft and final audit report.
- Agree follow-up actions and timescales.
- Agree how to communicate completion of agreed follow-up actions.

Source: From Ref. 17.

Appendix 6.C   Example Supplier Audit Checklist

- *Audit details*
  What are the name and address of the firm being audited?
  Who is the contact at the firm?
  What are the names and qualifications of the auditors?
  What is the date of the audit?
  Is this an initial or follow-up (surveillance) audit?

- *Business*
  How long has the company existed?
  How is the company organized?
  How many pharmaceutical-related customers does the company have?
  Do they have any customer citations for good work?
  Is the company's pharmaceutical-related business profitable?
  What is the long-term pharmaceutical-related business plan?
  Is the company in any litigation?
  Does the company hold a recognized quality certification?

- *Organization*
  Has a quality control system been established? Is it documented?
  Who is responsible for quality management?
  Is there a quality assurance management structure?
  Is there a project management structure?

Appendix 6.C   Example Supplier Audit Checklist (*Continued*)

Are project work practices documented?
How does project work conform to quality standards?
Has accreditation/registration been achieved? (ISO 9000, specify other)
Is the quality system audited on a regular basis?

- *Employees*
  How many permanent, contract, or temporary people does the company employ?
  How long, on average, do employees stay with the company?
  What is the company's training policy? Are there any records?

- *Planning and requirements*
  Are project and quality plans produced for projects? Who approves them?
  Does planning include a contract review?
  Is there a defined project life cycle? What is it?
  How are project documents controlled?
  How is conformance to customer GxP requirements ensured?

- *Design and development*
  Do design considerations cover reliability, maintainability, and safety?
  Do design considerations cover standardization and interchangeability?
  Are design reviews carried out? Are they minuted?
  Are customers invited to attend design review meetings?
  Are design changes proposed, approved, implemented, and controlled?

- *Coding, configuration, and build*
  Are there guidelines or standards for software programming and hardware assembly?
  How does the company ensure that they conform to current industry requirements?
  Are there records showing projects conforming to company practices?
  What third-party hardware or software is used? Is it supplied by a reputable firm?
  How would changes to third-party products affect the customer's end product?

- *Development testing*
  Are test specifications produced? Are expected results defined?
  Who performs the tests? How is testing organized?
  Are versions of hardware and software inspected?
  Is software "black box" (functional) testing conducted?
  Is software "white box" (structural) testing conducted?
  How rigorous is testing? Are abnormal situations tested?
  How are failed tests documented and corrected?
  Are test results recorded, signed, and dated? Are these records maintained?
  Who signs for overall acceptance of testing?

- *Project completion and software/system release*
  What is the mechanism for deciding that a project is complete?
  Is there a certificate of conformity? Is there a warrantee?
  Are project documents handed over to the customer?
  Are project documents archived?
  Is there an access agreement for regulatory inspections (e.g., ESCROW)?

- Supporting processes
  Is there configuration and version control within projects?
  Does the quality system provide for the prompt detection of failures?
  Are all failures analyzed? Are correction actions promptly taken?
  Are regular internal audits carried out? Are auditing procedures documented?
  Are audits planned? Are corrective actions taken?
  Are audit records stored and available?
  Are responsibilities for document review assigned?
  Are responsibilities for change control assigned?
  Are obsolete documents withdrawn?
  Are changes notified to the customer?
  Are subcontractors audited? How are they managed?
  Are subcontract documentation standards audited?

(*Continued*)

Appendix 6.C   Example Supplier Audit Checklist (*Continued*)

- *Customer relationship*
  Are customers solicited for feedback?
  How are customer responses folded into development plans?
  Are customers kept informed of development plans?
  List other customers provided with a similar service/product that is the subject of this audit.
  Are customers advised of problems found by other users?
  Is revision control maintained for upgrades and patches?
  Who is responsible for ongoing customer support? Is there a support fee?
  What are the response mechanism and timings for customer problems?

Source: From Ref. 1.

Appendix 6.D   Example Postal Supplier Audit Questionnaire

1. Does your company have certification to any recognized internal standards?
2. Do you have a written project management system including the definition of responsibilities within projects?
3. Do you have a separate and independent quality assurance individual or group?
4. Do you have a written quality management system describing the controls on the software engineering process?
5. Is a written software development life cycle methodology in use?
6. Is a written user requirements document a mandatory prerequisite for all projects?
7. Is a written functional specification or design document a mandatory prerequisite for software development?
8. Does a written programming standards guide or coding standards manual exist to define standards for programming?
9. Are structured test plans produced in advance of testing?
10. Are the testing personnel independent of the development personnel?
11. Are test results recorded and retained?
12. Is a written formal change control system for software and documents in operation?
13. Are formal written backup and contingency procedures in place?

Appendix 6.E   Example Contents for Supplier Audit Report

Introduction
- Author/organization and authority
- Purpose
- Relationship with other documents
- Contractual status of document

Scope
- Systems and software covered
- Dedicated audit or shared audit
- Details of single or multiple suppliers covered

Audit process
- Reference procedure to be followed
- Reference checklist used to guide audit (e.g., GAMP)
- Qualifications of audit team
- Identification of lead auditor
- Identification of individuals, with job titles, receiving audit

Quality assurance system
- Describe suppliers quality assurance system.
- Describe supplier's quality assurance organization.
- Reference any independent certification of above noting the supplier's scope of supply defined by the certification (e.g., ISO 9001, ISO 9001-3).

Audit results
- Record of audit observations (good, bad, and ugly)

Appendix 6.E   Example Contents for Supplier Audit Report (*Continued*)

Corrective actions
- Identify critical issues to project implementation.
- Define acceptance criteria for closure.
- Indicate follow-up requirements for project implementation.

Audit outcome
- Use supplier unconditionally.
- Use supplier subject to specified corrective actions.
- Prohibit use of supplier.

References
Appendixes
- Glossary
- Audit team notes (possibly on checklist)
- Examples of documents reviewed (as appropriate)

Others as appropriate

7 | Design and Development

INTRODUCTION

Design and development is the responsibility of the system developer, although system users often take a leading role in the design review. The activities associated with this phase vary between projects but generally follow the established pattern of supplier project/quality plans, functional specification, software and hardware design, and design review. Facilitating requirement traceability is one of the most important activities. Throughout design and development, project managers should beware of extensions to the scope of the user requirement specification (URS) (sometimes referred to as "scope creep"). This is because each modification is likely to lead to a revision on the functional specification and subsequent documents, with associated incurred costs and project delays. If such extensions are still occurring by the time coding has commenced and the supplier's development capability is not up to managing this safely, final software quality may be seriously degraded.

SUPPLIER PROJECT AND QUALITY PLANS

GMP regulatory authorities require work conducted by suppliers for pharmaceutical and healthcare companies to be formally agreed in a contract (1) covering

- GMP requirements,
- Responsibilities of parties involved,
- Inspections of suppliers, and
- Customer agreement of subcontractors.

The contract usually consists of a supplier's proposal, together with a customer's purchase order. Reference may be made to a URS and other documentation. It is advisable to incorporate, or make reference to, the supplier's project and quality plans, which may be separate or combined. These plans share the same purpose as the validation plan and define the supplier's approach to their designated tasks. Without agreement on practices and management, it is unlikely that a project will be completed fully, on time, and within budget. It is not sufficient merely to state that the work must comply with the requirements of various GMP regulatory authorities. Resolving misunderstandings can be a complex and time-consuming task: "Bad contracts can seriously complicate your life" (2)!

The supplier's project and quality plans specify project responsibilities, procedures, and deliverable documentation. The supplier's scope of work usually revolves around design and development, coding, configuration and build, and development testing. The URS and user qualification may also be specified in the plans as requested by a pharmaceutical or healthcare company. The supplier firm should be encouraged to include a statement of its development capability for the service or equipment they are providing. This statement could be based on a firm's ISO 9000 accreditation. Resumes of key staff should be available to support the capability statement.

Pharmaceutical and healthcare companies are expected to review the contractual arrangements before beginning a project. The combined information provided in the supplier audit and supplier project and quality plans gives a good indication of the competence of the supplier and their working relationship with the pharmaceutical or healthcare company. A simple checklist might include the following questions (2):

- Does the customer order reference the final supplier proposal?
- Are the terms and conditions quoted in the customer order or supplier proposal acceptable?
- Are the milestones in the supplier's project plan accepted by all parties?
- Is the program of work in the supplier's quality plan accepted by all parties?
- Are the resource requirements available and used on the contract?
- Are you satisfied that the contract can be delivered in full, on time, and within budget?

If the answer to any of these questions is "no," then the actions and responsibilities to be taken to resolve the deficiencies should be recorded and verified as complete. The contract should not start until any deficiencies are understood and corrected. If a checklist is used, then this should be annotated with answers to the questions and signed as complete by the contract reviewer (who may well be the quality manager) and contract manager (who may well be the project manager). In this simple manner, documentary evidence is produced to demonstrate that the contract for a specific project was reviewed and was satisfactory.

Suppliers should also be encouraged to conduct their own contract reviews. They must be satisfied that they can fulfill the contract and that the pharmaceutical or healthcare company is aware of any necessary support and can provide it.

Contract reviews for computer systems will become increasingly important as pharmaceutical and healthcare facilities become more highly automated and dependent on software. Supplier project and quality plans should, therefore, be adopted as a standard working practice for all projects. A combined project and quality plan will normally be 10 to 15 pages long, but it does depend on the complexity of the project. Remember, however, that the size of a document is not important; the key requirement is quality of content. An example of a content checklist for project/quality plans is given in Appendix 7.A. An organization's commitment to project and quality plans is often a good indicator of their commitment to quality management.

FUNCTIONAL SPECIFICATION

The functional specification is the system-specific response to the URS, specifying a proposed solution to the URS. An external supplier or internal development group within the pharmaceutical or healthcare company typically prepare functional specifications.

Content

Wherever possible, the functional specification should follow the same structure of the URS (Appendix 7.B) and can refer to the URS rather than duplicating information. A requirements traceability matrix (RTM) can be developed, as described later in this chapter. Compliance, omissions, and nonconformances with respect to the URS should all be readily identifiable.

The functional specification should, as far as possible, avoid detailed design and concentrate on defining the operation and user interaction with the computer system. This is generally more difficult than it sounds. In some instances, it is not even applicable because the URS specifically requests particular equipment or a particular design to be used. Similarly, for small projects, it is often more convenient to combine the functional specification and design documents into what is often referred to as a functional design specification, system definition, or system description.

The *GAMP®5 Guide* recommends the following content headings for a functional specification:

- System architecture (scope/overview)
- Functionality (including information flows)
- Data (storage structures and data load)
- Interfaces (users and equipment)

- Nonfunctional attributes (performance)
- Capacities (including expansion capability)

When preparing content for each section heading, it is useful to consider what regulatory authorities such as FDA have indicated they would look for from a system specification (3).

- All inputs that the system will receive
- All functions that the system will perform
- All outputs that the system will produce
- The definition of internal, external, and user interfaces
- What constitutes an error (including alarms and messages), how they are recorded (audible alarms alone are not enough), and how they are to be acknowledged and managed
- All safety requirements, features, and functions that will be implemented
- All ranges, limits, defaults, and specific values that the system will accept
- All performance requirements that the system will meet (e.g., data throughput, reliability, timing, etc.)
- The intended operating environment for the system (e.g., hardware platform, operating system, etc., if this is a design constraint)

GxP functionality and configuration affecting drug product quality should be identified, and acceptable operating limits should be specified (4). Messages for information should be distinguished from alarms generated by unacceptable situations (5).

When considering electronic record and electronic signatures, attention must be given to particular regulatory preferences for functionality to be implemented entirely within the computerized system. Specific electronic record and electronic signature aspects to be covered by the specification include the following:

- Checks should be built in where appropriate for correct data entry and data processing.
- Data should only be entered or amended by persons authorized to do so.
- The identity of person entering GxP data must be recorded by the computerized system or change control process.
- Manual entry of critical process and test data (excluding function and menu selection) must be verified by a second person or electronic means.
- Audit trails should be established either by the computerized system or change control process for the creation and amendment of GxP data.
- Electronic batch sentencing (release) must only be possible by authorized person (a *qualified person* within the European Union).

Many regulatory authorities also expect a high-level diagram to be included with supporting detail as appropriate. An example system overview diagram is given in Figure 7.1.

Architectural Design

Large computer systems often benefit from an architectural design to define the structure and organization of subcomponents comprising the computer system. In essence, it provides the link between the functional specification and the detailed design documentation. The use of diagrams to explain structures should be encouraged. Indeed, a high-level overview diagram of the software is expected by some GMP regulatory authorities (6). The use of COTS products should be clearly identified. An example architectural diagram is shown in Figure 7.2. Architectural designs should only be used for configuration management where the frequency of change is sufficiently limited to make this feasible.

Architectural designs should be included in the RTM, as they will provide the vital linkage between requirements and responding design details. They should also be included within the scope of design reviews.

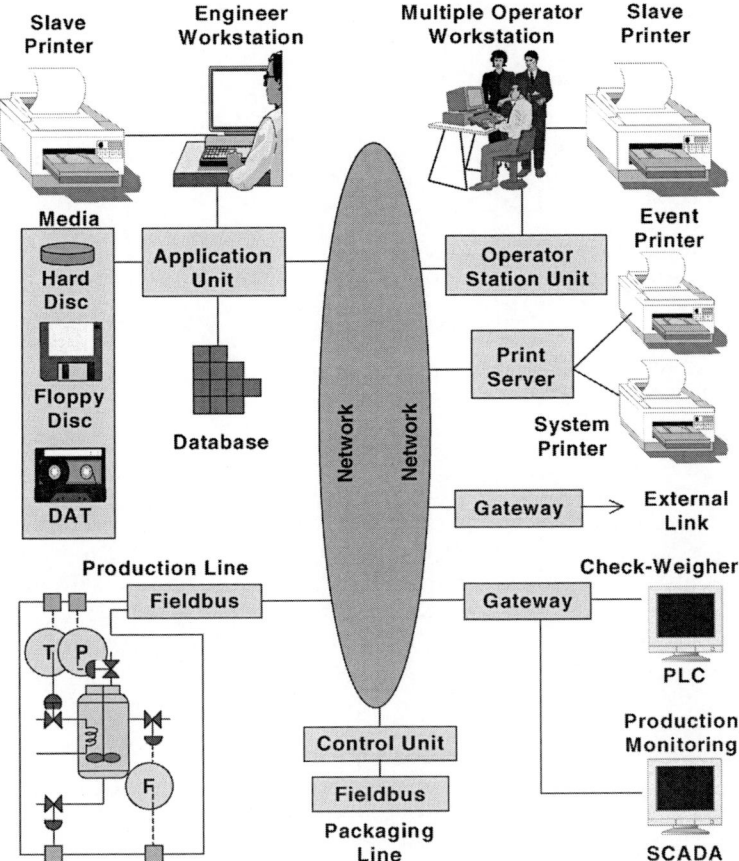

Figure 7.1 System diagram.

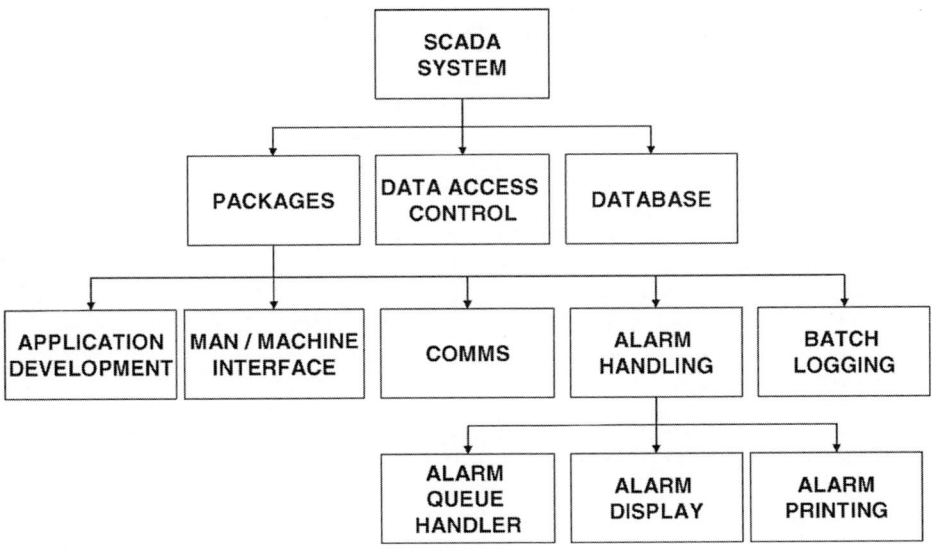

Figure 7.2 Architectural diagram.

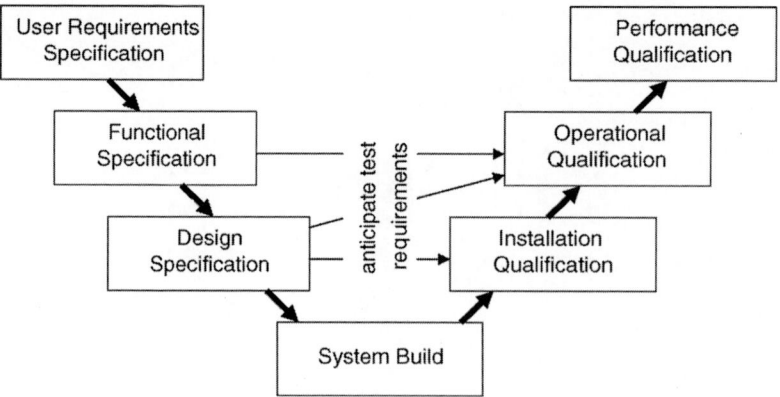

Figure 7.3   Anticipating testing requirements in specifications.

Anticipating Testing

The functional specification should be written in a manner that anticipates functional testing. The testing relationship for functional specifications is illustrated in Figure 7.3. Specifications should be specific, measurable, and achievable, so that testing can clearly demonstrate their acceptance. For instance, a sterilizing cycle time may be specified to be 90 seconds, when a variation of ±5 seconds is quite acceptable. If an acceptable variation of ±5 seconds is not recorded, then a test outcome of 89 seconds will fail.

Dealing with COTS Products

In many circumstances, the supplied system will be based on a standard COTS product and include additional features that are superfluous in the intended context. These features cannot normally be disabled because they are integral to the COTS product. Such redundant features should be included in the functional specification, noted as superfluous and, if possible, rendered inaccessible to users within the implemented computer system. Standard features that support compliance, however, such as audit trails for electronic records, should be used even if not defined within the URS. In such circumstances, it may be necessary to make additional design allowances for the inclusion of these features (e.g., for audit trail functionality, extra storage capacity may be required). Standard documentation for COTS products can be referenced by the functional specification if available for inspection, rather than reproduced. Care must be taken to refer to the correct version of COTS documentation and to keep cross-references up to date following any system upgrades.

Recent Inspection Findings

> An overview with high-level specifications was presented by the firm in support of validation. However, this document was not a controlled record, and it lacked review and approval (FDA 483, 2001).
>
> The system architectural diagrams do not document that the quality unit has approved the diagrams (FDA 483, 2002).
>
> System design documentation including functional and structural design and specifications was not maintained or updated (FDA Warning Letter, 2001).

SOFTWARE AND HARDWARE DESIGN

Software and hardware design may be segregated as two discreet activities, as described here, or combined. In either case, the design describes the implementation of the functional specification, describing it in increasing levels of detail until the components of the design (hardware or software) can be mapped directly to a standard product or implemented as custom (bespoke) elements of the computer system. The increasing levels of detail may be partitioned into different documents for larger systems. It is important to realize that there will be some feedback of information in the design process as the design is refined. The combined contents of the design documentation should address all the items listed in Appendixes 7.C and 7.D at the end of this chapter. They should be cross-referenced to the functional specification to demonstrate how it is being fulfilled. Again, any omissions and assumptions should be unambiguously stated.

The design must ensure that records generated by computer systems conform to their specified content (7). Records covering computer inputs, outputs, and data must be accurate (7–9). The computer system should also include built-in checks of the correctness of data that is entered and processed. Manual entry of critical data should be subjected to a confirmation check within the design that may be either manual or automatic.

Software Design

The software design partitions the functional specification into operational units, referred to as modules. Some modules may be suitable for implementation with COTS software packages, in which case the software packages and any configuration requirements should be defined. Other modules will require custom (bespoke) programming.

Computer system software can be split into four categories, as defined in chapter 4.

- Operating system (including database management software)
- Standard software packages
- Configurable software packages
- Custom (bespoke) programming (including defining user reports)

A list of the software used should be prepared defining the program names, their purpose, version, and the language adopted in their use (not the language in which they were written). The list may exist as a document in its own right or as part of another document. For custom (bespoke) software, version numbers may not be known until design and development has been completed, at which point the list of software should be updated.

Care must be taken when developing application software that uses a COTS software package—here is an example. A hospital used a COTS software package to calculate the dosage of a drug on the basis of a patient's height and weight. A student then developed some front-end application software that allowed dispensary staff to calculate dosages on the basis of imperial units rather than the package's metric units. The application software was regularly modified over a number of years, during which time there was no change control, no supporting documentation produced, no comments embedded in the software code, and no acceptance testing performed. Subsequently, over two years after the application software was written, it was discovered that the conversion between imperial and metric values had been incorrect all along. It is understandable in such circumstances why the suppliers of COTS software packages include a limited warranty clause within their license agreements. A typical one is reproduced below.

Although the software producer has tested the software and reviewed its associated documentation, the software producer makes no warranty or representation, either expressly or implied, with respect to the software or documentation, its quality, performance, merchantability, or fitness for a particular purpose. The licensee assumes the entire risk with regard to the use of the software and documentation.

The design of custom (bespoke) modules should start by defining inputs, functionality, and outputs. Where appropriate, a hierarchy of modules may be described with submodules. Inputs, outputs, and internal module values provide data structures; consideration should be given to the grouping of data items within databases and their access. There may be more than one design document depending on the use of standard software packages and the amount of custom (bespoke) programming. Both used and unused aspects of COTS software should be defined in the document along with self-test and diagnostic facilities.

In many organizations, the software design process remains ad hoc, being based on a programmer's creativity and ingenuity and comprehension of the functionality required. Given the functional specification, usually written in the vernacular or everyday language, an informal design document is prepared on the basis of this. Coding commences, but the design intent gets modified as the system is implemented. Once the software is complete, the design may bare little resemblance to the original software design document. In such circumstances, the design document must be rewritten to fully reflect the final design. The design document must also not be too brief. This is often the case when the design evolves through a prototyping exercise. Programmers who develop prototypes are often very reluctant to fully document their resultant designs. Sam Clark, a former FDA investigator, recalled an inspection where the entire design description for one computer program consisted of the statement *"This program modifies standard errors"* (10). The design description for a category of computer programs in another system was simply defined as *"Those collating data each time a unit is tested or whatever."* The project manager must curb this tendency among programmers, so that a meaningful and complete design is produced.

More methodical design approaches are available from "structured methods." These are sets of notations (a more structured language form, aimed at eliminating the ambiguities of everyday speech) and guidelines that orchestrate a finished software design before coding begins. No particular design methodology is being advocated here. When selecting a suitable design methodology, the skills and experience of the project team applying the methodology must be taken into consideration. Furthermore, the availability of computer-aided software engineering (CASE) tools and integrated projects support environments (IPSEs) supporting the methodology must be considered. It is recommended that development tools, such as CASE and real-time system analysis system design (RTSASD), should be examined to bolster confidence in their capability. The reasons for selecting a particular design methodology should be clearly documented.

Hardware Design

The hardware design describes the equipment constituting the computer system and its configuration in readiness for installation and commissioning. Some GMP regulatory authorities expect to see a high-level schematic diagram of the system's equipment (6). An example of a high-level hardware diagram is shown in Figure 7.4. The equipment will comprise instrumentation, computer-related hardware including communication links and any interface devices, other computer system equipment, and operations staff. The hardware design is likely to consist of a number of documents and make reference to preassembled information from suppliers providing standard models of equipment.

Computer hardware includes all equipments making up the computer system: processing units, user screens, keyboards, printers, and network interfaces. Switch settings and firmware configuration need to be specified. Some computer systems may have elements of computer hardware distributed at a central site or in a central computer room. User screens and keyboard inputs are often placed at distributed locations to be closer to the operator's

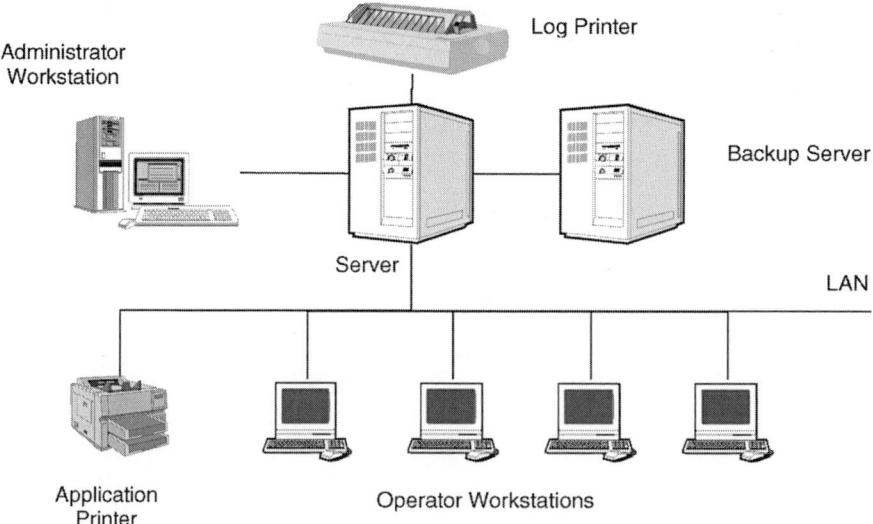

Figure 7.4 Example of high-level hardware diagram.

place of work. Intrinsically safe user screens and sealed keyboards may be required in some classified areas. If the computer system makes use of a fingerprint-sensitive mouse for biometric user identification, then this should be specified. Details of bar coders must also be defined where used.

Instrumentation will include field instruments used in the manufacturing process and other instruments associated with special tasks, such as that of laboratories or monitoring computer room environmental conditions. The accessibility of instruments must be such as to permit their cleaning and maintenance. Siting is also important, and instruments should be installed as close to the point of measurement as possible. The placement of flowmeters in piping dead legs should be avoided. Careful consideration should also be given to the appropriate position of other instruments, such as thermometers and thermocouples, so that they too can fulfill their measurement and control function (5,7). Construction materials that come into direct contact with the pharmaceutical or healthcare production process stream must not contaminate or affect the manufactured product in any way. Instrument lubricants and coolants must not come into contact with in-process product or equipment. A draft calibration schedule may also be prepared.

Computer systems may need an uninterruptible power supply (UPS) or protection from electrical interference such as electromagnetic interference (EMI), electrostatic discharge (ESD), and radio frequency interference (RFI). Sometimes, chart recorders will be needed to monitor the temperature and humidity of the installation site. The accumulation of dust is another danger that can lead to the impairment or breakdown of equipment. Even fire and flood should be considered.

The effectiveness of the software design should not be unduly constrained by hardware limitations. The software may require minimum CPU performance specifications to execute effectively. Memory size and other features of the system (e.g., hard disk, RAM, cache, DAT, and CD-ROM) may also be critical to performance. The clock accuracy also needs defining and taking care to specify the frequency of the power supply (this is 50Hz in European countries but 60Hz in North America), as this could have a fundamental impact on clock performance. Databases should not operate at capacity, and access contentions between multiple processes or users should also be examined. Additional memory and memory management software may be needed to improve the operating speed of any computer system. The transmission limitations of communication links—local area networks (LANs) and wide area networks (WANs)—should not render any real-time processing requirements impossible to achieve.

Minimum performance requirements of hardware (e.g., type of microprocessor) and other key parameters for individual items of equipment (e.g., minimum pages per minute for printers) must be specified, along with any calibration or special maintenance requirements. This will facilitate the identification of suitable replacement and upgrade hardware without the need to revise the hardware design. Designated spare and redundant parts should be included in the hardware design documentation.

Dealing with COTS Products

It may be tempting to assume that COTS software and hardware are inherently fit for purpose since they come from a reputable source and/or are market tested. This is an irresponsible assumption, as in many cases the products have unknown provenance and supplier auditing is frequently impractical. The basic compliance requirements for COTS products should answer the following questions (11):

- What is it (provide title, version, etc., and state why this product is appropriate to fit the purpose of the design)?
- What are the computer system specifications (specify hardware, operating system, drivers, etc., and including version information)?
- What function does the off-the-shelf software/hardware provide?
- How will you ensure that appropriate actions are taken by the end user (specify training, configuration requirements, and steps to disable or exclude the operation of any functionality not required)?
- How do you know that it works (describe any list of known faults)?

- How will the off-the-shelf software/hardware be controlled (this should cover installation, configuration control, storage, and maintenance)?
- How will the maintenance and support of the COTS software/hardware be continued should the original developer/supplier cease to trade or no longer be able to support their products/services or any other reason?

Chapter 15 discusses the use of standard software in more detail and how a design strategy based on exploiting COTS products (as opposed to basing the design strategy on bespoke software) can reduce the compliance effort required by a pharmaceutical or healthcare company.

Recent Inspection Findings

Notably absent documentation for defining the database and version of the database, defining the operating system, defining the location where the database is located or where files are located; defining the security access limiting personnel access to specific fields in the database (FDA 483, 2004).

Firm failed to define and control all customized elements of configurable software package (FDA 483, 2004).

There are no detailed specification documents for any of the computerized process control systems that contain sufficient details on how these systems/software were represented and developed. The only specification documents made available and referred to as the design document were the "system specifications," however, these documents only provide a high-level explanation of what the systems do. They lack sufficient detailed description of specific and complete data structure, data control flow, design bases, procedural design, development standards, and so on to serve as the model for writing code and to support future changes to the code (FDA 483, 2000).

The computer system lacked documentation defining database, operating system, location of files, and security access to database (FDA Warning Letter, 2001).

The computer system uses a purchased custom-configurable software package. The software validation documentation failed to adequately define, update, and control significant elements customized to configure the system for the specific needs of the operations (FDA Warning Letter, 2001).

System layout and wiring were not part of the validation documentation (FDA Warning Letter, 2001).

Documentation regarding layout diagrams was found obsolete (FDA Warning Letter, 2001).

Validation did not address signal lines between detection devices and computer (FDA Warning Letter, 2001).

Diagrams related to system layout, installation, and wiring were not part of the validation documentation (FDA Warning Letter, 2001).

Validation records did not address wiring diagrams (FDA Warning Letter, 2001).

Failure to create and maintain specifications for the software programs (FDA 483, 2001).

DESIGN REVIEW (INCLUDING HAZARD STUDY)

Design reviews are used to confirm that the proposed design fulfills its specification (including compliance requirements) and is suitable for its intended purpose (12). Postponing the search to discover problems and defects until development testing or user qualification is virtually certain to delay the project and to increase the overall expense. This is for the simple reason that getting anything right first time, rather than putting it right retrospectively, is the cheapest and quickest way of accomplishing anything. Perhaps surprisingly, the timing of any such use remains a business risk-driven choice.

An RTM should be considered to verify that the design fully addresses the computer system specification. Additional checks are not necessary if the individual review of each design and development document includes such a check and there have been no omissions or ambiguities in its relationship to other documents.

The design review should include an assessment of threats and controls affecting computer system operability including assessing potential impact between new systems and interfaced existing systems. Three techniques are described in this chapter: hazard analysis and critical process point (HACCP), computer hazard and operability (CHAZOP), and failure mode effect analysis (FMEA). Computer systems failure (complete, partial, or intermittent) may have an adverse effect on drug product quality, pack integrity, or regulatory compliance. The process being controlled by the computer system must be brought into a safe condition following a failure to protect the integrity of the process (13). If risks cannot be managed to an acceptable level, then the computer system cannot be considered fit for purpose and must not be used. The nature of this review often prompts recursive refinements to the design.

The results of the design review must be documented, ideally using report forms, and any corrective actions that arise from the review must be received by consensus and owned by a member of the review team. The actions must be described in detail in the meeting record, with a proposed completion date. The actions must only be signed off and dated as accepted when evidence of their completion is forthcoming. Some actions may require modification in the system design and associated documentation. Other actions may identify specific development testing or user qualification test scenarios. Alternatively, operating procedures may need to be developed or refined. A final report detailing the personnel involved with the review, the topics considered, and the findings of the review is usually written when all of the review actions are completed. The report should be kept under change control.

Ian Johnson of AstraZeneca reports several examples where design weaknesses could all have been revealed in a design review; these concerned a fluid bed dryer, a tablet press, and a water deionizer system (10). The alarms in the fluid bed dryer were not latched by the embedded PLC. Alarms should have remained flagged until acknowledged by an operator pressing the "alarm reset" button. Four alarm conditions occurring in the same fluid bed dryer did not sound a horn or illuminate flash warning lamps, because they were dependent on another status condition being present. This was incorrect. Again, a tablet press had a PLC compression force monitoring system with a compressed air–driven reject mechanism. When the compressed air supply failed, no audible alarm or visible warning was triggered (flashing lamp, audible horn, or a message displayed on an operator terminal). With this particular tablet press, a tablet that should have been rejected but was in fact accepted passed into the acceptance chute. This situation presented a direct threat to product quality. In another case, the PLC-controlled deionizer did not isolate its water supply if the PLC failed. There were no hard-wired alarms or interlocks to alert operations staff and protect the body of deionized water already in the plant, thereby directly threatening drug product quality. A design review would have provided early warnings of these deficiencies before the computer systems were implemented.

When selecting a risk management technique, it is important to appreciate its strengths and weaknesses. Techniques may be more appropriate for use with automated processes, computer systems, or software, as illustrated in Table 7.1. Different techniques may also have a safety or reliability bias.

Once chosen, the risk management technique must be implemented so as not to undermine its effectiveness. Some techniques can be applied at any stage in the project life cycle, while others have a dependency on design information being available. Some techniques are suitable for individuals to implement, while others benefit from a team effort.

Table 7.1 Summary of Hazard Studies

Technique	Suitability	Safety versus reliability bias	Conducted
Computer hazard and operability	System	Both	Team
Failure mode effect analysis	System	Reliability	Individual or team
FMECA	System	Reliability	Individual or team
FTA	Software	Reliability	Individual
Hazard analysis and critical process point	Process	Safety	Team

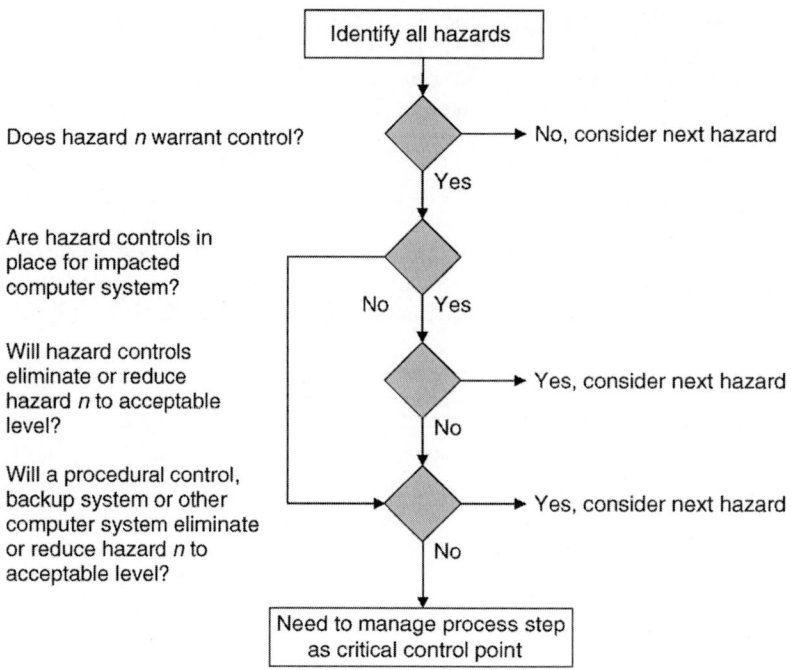

Figure 7.5   Critical control point decision tree. *Source*: From Ref. 16.

Hazard Analysis and Critical Process Point

The application of HACCP is mandated for food and drink GMPs (14), and FDA is exploring its further use for pharmaceutical and healthcare GMPs (15). It provides a process-orientated approach to identifying and reducing known and potential hazards to an acceptable level. The technique is not specifically intended for computer systems, but its principles can be applied with a little modification (Fig. 7.5).

HACCP is best applied by a multidisciplinary team reviewing the design of a computer system. Team members are selected for their particular knowledge of the production process, the computer system, and the software.

Each hazard is assessed in terms of whether it is suitably controlled, either directly by the computer systems affected or by a backup mechanism, or by another computer system. The use of backup mechanisms (backup systems and procedural controls) was discussed earlier in chapter 6. Hazard controls provided by supplementary computer systems are based on the functionality of existing computer systems used elsewhere. More than one hazard control may be used to manage individual hazards. Equally, more than one hazard may be mitigated by an individual hazard control. Hazards not satisfactorily controlled are considered critical control points and will need to be addressed.

A HACCP report outlining the process involved, identified hazards, how individual hazards are controlled, and any resultant recommended actions for hazards that require further controls should be prepared. The report should be reviewed for its accuracy and signed and approved accordingly. Actions should be reviewed for satisfactory closure as part of the validation report. Actions may affect specification and design documentation, and/or testing.

HACCP provides a very basic approach to hazard analysis and control. Computer systems supporting licensed products with medicinal properties should consider more rigorous techniques, such as CHAZOP and FMEA, described below.

Computer Hazard and Operability Study

During the 1960s, ICI developed the hazard and operability (HAZOP) process to identify hazards in chemical plant design. The HAZOP process is now well established and was successfully extended by ICI during the 1980s to include computer control systems.

During the CHAZOP study meeting, the team will systematically go through diagrammatic representations of the system. The experience of the CHAZOP leader will steer the review team through the configuration of the computer system, the software control scheme, the effect of GMP-related functions on the process, and the general operability and security of the computer system. Possible deviations from the design intent are investigated by systematically applying guidewords. Some example guidewords are given below on the basis of CHAZOPs used in the chemical industry (17).

- On, off, interrupt
- As well as, only, other than, part of
- No, not, wrong
- Less, none, more
- Reverse, inverse
- More often, less often
- Early, late, sooner, later, before, after

All guidewords to be used by the CHAZOP review team should be defined and documented. Care must be taken to ensure that all CHAZOP participants understand their precise meaning and whether or not they are suitable in the context in which they are being used. The CHAZOP chairman should remove inappropriate guidewords from the study. At the discretion of the study chairman, new guidewords may be added as appropriate to the computer system being reviewed.

The guideword process can be supplemented by additional topics/questions based on an analysis of previously experienced design deficiencies and operational incidents. For instance, ICI has collated a database of over 350 operational incidents that it uses to refine its CHAZOP study process (18). Some example questions for the CHAZOP study are given in Appendix 7.E. Of particular interest to the study are the effect of partial or catastrophic failures, recovery mechanisms (e.g., rollback and roll-forward), and the general usability of the system (e.g., the need for multiple screens to access data, screen refresh times, meaningful information displays). The list of questions can be expanded with operational and regulatory experience.

CHAZOP is based on a multidisciplinary team reviewing the design of a computer system. Team members are selected for their particular knowledge of the production process, the computer system, and software programs. The CHAZOP meeting is led by a chairman to manage discussions using guidewords and learning from earlier studies. A holistic approach covering hardware failures, software functionality, and human factors (manual dependencies) is required.

A CHAZOP report outlining the process undertaken, the definition of guidewords used, and any resultant recommended actions should be prepared. The CHAZOP report will normally include the completed CHAZOP table. An example of part of a CHAZOP table is shown in Table 7.2. The report should be reviewed for its accuracy and signed and approved accordingly. Actions should be reviewed for satisfactory closure as part of the validation report. Actions may affect specification and design documentation, and/or testing.

The main strength of the CHAZOP process is that it facilitates systematic exploratory thinking. The use of guidewords and deviations prompts the review team to think of hazards that might have otherwise been missed. Recently, both the U.K. Department of Defense (19) and the U.K. Health and Safety Executive (20) have issued a CHAZOP standard. It is important to recognize, however, that the effectiveness of CHAZOP studies is dependent on guidewords and the capture of learning from project postmortems and operational incidents. The relevance of guidewords becomes evident only through practical application. The significance of any learning will be dependent on routine collection and analysis of project postmortems and on the reliable reporting of operational incidents.

Failure Mode Effect Analysis

FMEA provides a hardware-orientated approach to identifying and reducing known and potential hazards to an acceptable level. The FMEA process in Figure 7.6 is based on FDA medical device guidance.

Table 7.2 Example of Computer Hazard and Operability Extract

Operating sequence	Deviations—using guidewords	Possible causes	Consequences	Action required
User log-on	Wrong	User forgot his or her password	Retry, lock out after three unsuccessful attempts. User then has to formally apply to security manager for new password.	None; users trained not to keep passwords secure and not to share them.
		Unauthorized access attempt	Retry, lock out after three unsuccessful attempts.	SOP for periodic review by security manager and escalation to senior management as appropriate.
	As well as	User already logged on at another terminal	If more than one terminal are being used by a user, then likelihood is that at least one has been left unattended and hence is insecure.	Recommend design altered to only allow one active terminal per user at a time.
Enter product code and analytical method	Other than	Invalid product code or analytical method entered	Option given to user to reenter or cancel process initiated and return to main menu.	None.
Enter analytical data	More, less	Invalid numerical data, such as alphabetic characters, zero, or negative values, entered	No check made on range of data entry; may cause data calculation error such as divide by zero elsewhere in system.	Recommend automatic range checks included in design.
Completing user transaction processing	Interrupt	Transaction not completed because of partial system failure or catastrophic system failure	No commitment of data to database and rollback, so all transaction data lost.	None except manual reentry of transaction data.

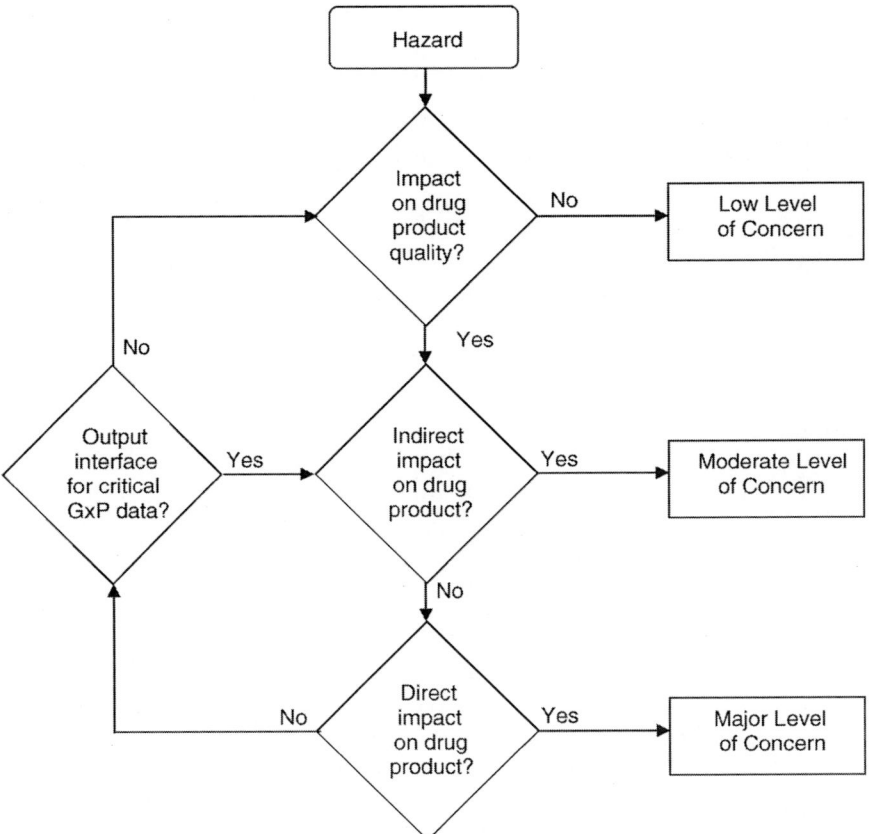

Figure 7.6 Managing levels of concern.

FMEA is best applied by a multidisciplinary team reviewing the design of a computer system. As with CHAZOPs, team members are selected for their particular knowledge of the production process, the computer system, and the software. The team steps through various computer system failures, considering their effect, risk, and how they might be controlled. The outcome of a FMEA is then documented in a template such as the example given in Table 7.3.

More sophisticated FMEAs examine the level of concern over various hazards in terms of GxP criticality (21). Figure 7.6 describes how to determine three levels of concern: low, medium, and high. The decision tree presented solely considers the impact on drug product quality. Some pharmaceutical and healthcare companies may want to include operator safety, business impact, and even the GAMP categories of software affected in their determination of these levels of concern (10).

Once the level of concern is understood, the FMEA team needs to appraise its likelihood. Remote and rare events may be acceptable without further action. The term ALARP is often used for such events; it means reduce risk as low as is reasonably practical. Higher likelihood hazards will demand more attention, as indicated in Figure 7.7. The acceptance of a hazard without control needs to be justified. It is important to note that the FDA does not include "likelihood" in their guidance for medical devices. Instead, they focus on reducing the "level of concern," essentially because any occurrence of a failure in a safety-critical medical device could be catastrophic.

There are many methods of control to eliminate or mitigate identified hazards. The selected control(s) for a hazard should be recorded on the FMEA template. Examples of hazard controls include change in design specification, implementing alarms/warnings/error messages, and instituting a manual process. The most cost-effective hazard control should be selected and described in the FMEA template.

Table 7.3 Example of Failure Mode Effect Analysis Extract

Failure mode	Effect	Hazard	Hazard control	Control verification
Control system power failure	Operator console goes blank.	Facility to collect process data is lost—batch rejected.	Recommend addition of UPS.	IQ check whether UPS supports continued control system operation during power outage.
Control system is defective (short circuit)	Unknown—vendor expects system to freeze and/or operator console to go blank.	Facility to collect process data is lost or corrupted—batch rejected.	None; remote chance event will occur—replace unit on failure; not financially viable to include second redundant control system into architecture.	Not applicable.
Incorrect set point downloaded from control system to product critical instrument	Equipment appears to operate normally.	Unacceptable temperature in manufacturing process not detected.	Recommend addition of independent monitoring system to check process temperate.	Validation of independent monitoring system.
Wire break on control system interface to equipment	Instruments run on last known set point; alarm conditions are not received by control system.	Equipment does not stop. Operator might not notice that monitored process values are unduly static.	Recommend control system design modified to alarm on no instrument signal.	OQ check alarm when control system interface disconnected.
Instrument power failure	Error will not automatically alarm on control system if last received process parameter within acceptable range.	Operator might not notice that monitored process values are unduly static.	Recommend control system design modified to alarm on instrument power failure.	IQ check alarm when instrument power failure.

Abbreviation: UPS, uninterruptible power supply.

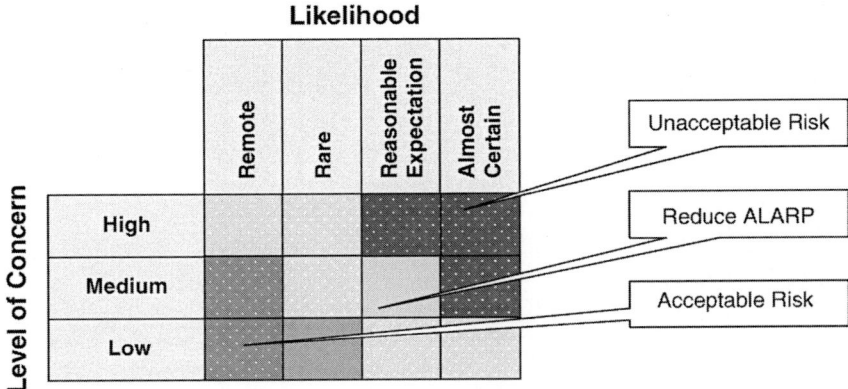

Figure 7.7   Example of decision tree for determining GxP level of concern.

An important factor to consider when examining suitable controls for identified hazards is the time required to restore the full integrity and operability of the computer system if the hazard occurs and whether and/or when to switch over to manual operation. Recovery times are very important for delay-sensitive processes. Control options include

- Hazard mitigation (protective measures),
- Hazard avoidance (inherent safe design), and
- Hazard tolerance (user procedures and training).

Careful consideration should be given over how best to verify the hazard control. Hazard controls that cannot be verified as being in working order will not satisfy regulatory authorities during inspections.

A modified FMEA template is provided in Table 7.4. This calculates relative risk before and after hazard controls are applied (22). Risk is calculated as a function of the likelihood of occurrence, the severity of the hazard, and the probability of detection. Oliver Muth of Pfizer suggests the calculation of a risk priority number (RPN), calculated from multiplying the scores given to likelihood, severity, and detection (10). The *likelihood* of an occurrence is rated on the following scale:

1. Remote, unlikely to occur within two years
2. Code tested, reviewed, and proven to be reliable
3. "Normal code" tested
4. Complex code, interfaces (automated/manual), or many variables from other parts of the application used
5. Complex code, unknown, or test results unknown

The *severity* of the hazard is marked on the following scale:

1. No negative impact
2. Results in minor deviation
3. Deviation from quality profile
4. Causes production or shipment of noncompliant product
5. Could cause injury to a patient using the product of the process

The *probability of detection* is marked on the following scale:

1. 100% automated detection and alarm
2. Sub-100% detection, independent automated detection and alarm
3. Combined manual and automated detection

Table 7.4 Example of Modified Failure Mode Effect Analysis Extract

	Hazard identification							Hazard control						
		Level of concern (GxP)	RPN					Overview description	Control verification (document reference)	Residual RPN				
Description	Cause and consequence		L	S	D	RPN					L	S	D	RPN
Batch failure	Steam temperature not controlled leading to excessive steam temperature	High	2	5	4	40		Add temperature alarm with control limit.	DQ, IQ, OQ		2	5	1	10
Batch time increased	Steam temperature not controlled leading to increased batch time	Low	2	3	4	24		Add temperature alarm with control limit.	DQ, IQ, OQ		2	3	1	5
Batch traceability lost	Corruption of data on save leading to erroneous batch data	High	3	4	4	48		Add checksum check on data save.	DQ, source code review, IQ, OQ		3	4	1	12
Batch traceability lost	Unauthorized data modification leading to erroneous batch data	Medium	4	4	4	64		Put application on network with password security and authorized user access profile.	DQ, source code review, IQ, OQ		1	4	2	8
Batch traceability lost	No backup of data leading to absent batch data	High	2	4	4	32		Put application on network and configure for daily automatic backups.	DQ, IQ, OQ.		1	4	3	12

Abbreviation: RPN, risk priority number.

4. Detected through routine manual quality control checks
5. Not detectable or normally not tested for

It is important to appreciate that risk is a relative concept. An unacceptable high risk might be considered one with a RPN score exceeding 24. All RPN scores below 13 would indicate a low risk (and therefore perhaps acceptable). RPN scores between 12 and 25 might indicate an acceptable risk but one that should be reduced if it is possible to do so within the practical constraints imposed on the project (Fig. 7.6).

Dealing with COTS Products

Configurable and customized COTS software and bespoke (custom) hardware should be subject to a HACCP, CHAZOP, or FMEA as appropriate. It is vital that GxP processes are not compromised, so we must ask the following questions:

- What does the computer system do?
- How might it fail to do what it is supposed to do?
- What causes these failures?
- What is the impact/consequences of these failures?
- How are these hazards to be controlled?

Hazards requiring control can be managed through design modifications (including using alternative COTS products), employing protective measures (e.g., monitoring systems to identify hazard manifestations and take corrective action), or the application of procedural controls.

Nonconfigurable COTS software and standard hardware do not require a HACCP, CHAZOP, or FMEA if operational experience exists to support the view that the COTS products are stable and robust. Operating experience should be considered as suitable when the following criteria are met:

- The intended version COTS product has achieved a sufficient cumulative operating time.
- The intended COTS product has operated in several similar installations (and hence, there has been more chance that hidden errors would be exposed).
- Defects/errors are routinely reported as they occur and are corrected in a timely fashion.
- Meaningful information on reported defects/errors, remedial actions, and status is available.
- No significant modifications have been made, and no errors have been detected over a significant operating time, recommended at least one full year of operation.

The rigor of the analysis of operating experience should be commensurate with the criticality of the computer system. Operating experience should be under conditions similar to the conditions during intended use of the computer system. When the operational time of other versions is included, an analysis of the differences and history of these versions should be made.

Recent Inspection Findings

Risk assessment revealed numerous unanticipated risks that have not been addressed. For example, one such risk is that the computer unit may acquire the wrong patients data (FDA Warning Letter, 2001).

There were three possible causes attributed to this failure in the system risk assessment document, yet there is no implemented strategy to reduce the risk of these failures (FDA Warning Letter, 2001).

ACCELERATED DEVELOPMENT

It is often tempting to implement tools to support design techniques as part of the system/software development processes. Care must be taken, however, not to get carried away with

Table 7.5 Pros and Cons of Prototyping

Advantages	Disadvantages
• User needs are better accommodated. • The resulting system is easier to use. • User problems are detected earlier. • Less effort to realize development of system.	• Limited identification of hazards. • System performance may be worse. • System is harder to maintain. • The prototyping approach requires more experienced team members.

technology! Many tools that promise improved productivity and quality are on the market. However, while these tools may be designed to support particular development techniques/ processes, they are often used out of context. In consequence, the tools are not exploited fully and may even be misused. The promised benefits may not materialize.

Prototyping

It is often difficult to secure and maintain a sufficiently accurate or comprehensive definition of a proposed system's requirements from a prospective user. Prototyping has proved to be a useful approach whereby one or more working models are developed and used as an explanatory model before the real system is implemented. Such working models can be examined by the prospective users to exercise their own understanding of the requirements that must be met. Seeing a working model helps users clarify, explain, refine, and, most importantly, *express* their requirements. Care has to be taken, however, to prevent indiscriminate scope creep, with the prototype taking on a whole life of its own and running away with time and money. Extending the scope beyond requirements to include marginally desirable rather than only essential features ("nice to have"), can become an irresistible temptation to excited prospective users. This must be curbed. Some advantages and disadvantages of prototyping are outlined in Table 7.5.

Typically, prototypes are discarded once the functional requirements and user interface requirements have been fully defined. The final system is then developed through a prospective life cycle of design, coding, configuration and build, development testing, and user qualification. QA need not be involved with the development of prototype software since it is destined to be discarded and does not have to have any significant level of innate quality.

Prototypes, however, do not necessarily have to be discarded. They can be taken forward as the final application. The implications of this approach, however, should be clearly understood before proceeding, as there are great dangers here. Retrospective documentation of prototype systems can cost 25% to 50% more than if documentation was drafted prospectively. Some reengineering is likely to be required, which is expensive and dangerous and rarely delivers the quality that is engineered into a conventional prospective development. Project managers may find themselves under pressure to shortcut the reengineering process, but this will compromise regulatory compliance. If the prototype is being taken forward as the final application, QA involvement from the outset is absolutely essential, otherwise retrospective validation will be required (see chap. 15).

Rapid Application Development

Rapid application development (RAD) emphasizes user involvement, prototyping, software reuse, use of automated tools, and small development teams. The RAD life cycle, sometimes called a *spiral* life cycle, consists of four basic phases.

- Specification
- Design and development
- System build
- Acceptance testing

The specification phase and the design/development phase have much in common and are often merged. They make extensive use of joint application development (JAD) workshops in which developers and prospective users work together to agree the functionality of the planned computer system and develop the prototype. It is vital that key users are present and actively contribute to the JAD workshops by giving an *authoritative* opinion. Otherwise, functional requirements captured may be inaccurate or incomplete, and/or the prioritization of their requirements may be inappropriate.

The system build phase is expected to extensively reuse existing software, rather than to develop new bespoke code. Reused software, however, must be robust and documented, otherwise the benefits of the RAD process will be reduced by the need for a significant reengineering and revalidation exercise.

Successful acceptance testing is used as the basis for authorizing the computer system for installation and use. The focus of the acceptance testing phase is user qualification and user training. It is likely that development testing will be very limited.

Projects implementing RAD impose closely monitored timetables within which activities must be completed. The RAD philosophy is to complete as much of the assigned activities as possible within their designated time constraints. Consequently, aspects of activities must be prioritized so that if necessary, lesser aspects can be sacrificed. For instance, the inclusion of low-priority functional requirements may be sacrificed, or the amount of testing may be limited to enable the design/development and user acceptance activities to be completed within their respective timetables. Obviously, such stringent time management raises questions over the risk of a compromised compliance. Limited testing, for instance, could be seen as failing to meet regulatory requirements. This aspect of RAD must be carefully managed to avoid a costly, retrospective validation.

In conclusion, while RAD projects offer many advantages, they also pose certain risks and thus require careful, competent, realistic management to ensure that they deliver *validated* computer systems. RAD must not be used as an excuse to circumvent necessary life cycle controls and documentation.

Extreme Programming

Extreme programming is a relatively new approach to accelerated development. This aims to deliver software faster than any other widely used approach. It comprises a number of key practices that, it is suggested, must be collectively applied for the approach to work (23). Superficially, some extreme programming practices may appear to contravene conventional principles of good programming practice.

- Systems are planned to go through a series of frequent small incremental developments (i.e., an increment every 1–4 weeks)
- Projects and development have no independent quality oversight. Quality is a collective responsibility.
- Systems are designed to be as simple as possible. Extra complexity is removed upon exposure.
- Programmers write all the code in accordance with rules emphasizing communication through the code, rather than via documentation.
- The inchoate code must pass all its tests before further incremental development can take place. Programmers write their own unit test scripts. Customers write their own user test scripts.
- Systems are tuned rather than developed during final incremental development, leading to authorization for use in the live production environment.

Soundly tested small incremental developments and system releases should bring a high degree of assurance that the computer system is fit for purpose. The lack of formal specification and design documentation, however, means that traditional GxP compliance requirements cannot be satisfied. Instead, design information is documented within the software itself. Testing is conducted against test specifications, rather than predefined design criteria. Thus, it is not a question of compliance activities not actually being undertaken, but rather a case of compliance activities being conducted in a different way.

Another issue with conventional approaches is how quality is managed and controlled. Extreme programming relies on collective responsibility for quality. This does not sit comfortably with current regulatory expectations for *independent* quality oversight, nor, many would allege, does it accord with general experience of human nature! Given the alternative approach to specifications and design described earlier, it may be hard for regulators to accept that extreme programming can delivery high-quality code without some sort of independent project verification. Although not called for by the extreme programming approach, a quality and compliance role may well be appropriate. This may be needed at least until the regulators can understand how extreme programming might be able to work in practice without independent quality oversight.

In the short term, it is unlikely that extreme programming will prove to be acceptable to the regulatory authorities. Nevertheless, it should not be dismissed as a future technique once it has been more widely demonstrated as fulfilling its potential. Regulators appreciate that there are strong cost-benefit drivers throughout industry to find ever faster ways of implementing computer systems. It may well be that tools will emerge that would enable the disciplines of a more conventional, independent measure of oversight to be exerted without slowing down the development process.

REFERENCES

1. European Union. Guide to directive 2003/94/EC. European Commission directive laying down the principles of good manufacturing practice for medicinal products for human use, 2003.
2. International Organization for Standardization. ISO 90003: software engineering—guidelines for the application of ISO 9001:2004 to computer software. Geneva, Switzerland, 2004.
3. Food and Drug Administration. General principles for software validation, final guidance for industry and FDA staff. Centre for Devices and Radiological Health, 2002.
4. Food and Drugs Administration. General principles of process validation. Rockville, MD, Centre for Drug Evaluation and Research, 1987.
5. Kletz T, Chung P, Shen-Orr C. Computer control and human error. Institution of Chemical Engineers, Rugby, UK: 1995.
6. Therapeutic Goods Administration. Australian Code of Good Manufacturing for Therapeutic Goods, Medicinal Products, part 1. Woden, Australia, 1990.
7. U.S. Code of Federal Regulations title 21, part 211, Current Good Manufacturing Practice for Finished Pharmaceuticals.
8. European Union. EU GMP annex 11 computerised systems, guide to good manufacturing practice for EU directive 2003/94/EC, Community Code Relating to Medicinal Products for Human Use, Vol. 4, 2003.
9. Food and Drugs Administration. Input/output checking. compliance policy guides, computerised drug processing, 7132a, guide 7. Rockville, MD, Centre for Drug Evaluation and Research, 1982.
10. Wingate GAS. Computer Systems Validation: Quality Assurance, Risk Management and Regulatory Compliance for Pharmaceutical and Healthcare Companies. Boca Raton, FL: Interpharm Press, 2003.
11. Food and Drug Administration. "Off-the-shelf software use in medical devices."FDA 1252, US Department of Health and Human Services, Centre for Devices and Radiological Health, September 9, 1999.
12. European Union. Annex 15—qualification and validation. Guide to good manufacturing practice for EU directive 2003/94/EC, Community Code Relating to Medicinal Products for Human Use, Vol. 4, 2003.
13. Food and Drugs Administration. Guide to inspection of computerised systems in drug processing (Blue Book), reference materials and training aids for investigators. Rockville, MD, Centre for Drug Evaluation and Research, 1983.
14. Blanchfield JR. Food & Drink—Good Manufacturing Practice: A Guide to its Responsible Management. 4th ed. London: Institute of Food Science and Technology, 1998.
15. Gold Sheet. 2000; 34(5) May.
16. Food and Drugs Administration. Hazard analysis and critical point principles and application guidelines. National Advisory Committee on Microbiological Criteria For Foods, August 1997.
17. Chung P, Broomfield E. Hazard and operability (HAZOP) studies applied to computer-controlled process plants. In: Kletz T, ed. Computer Control and Human Error. Institute of Chemical Engineers, Rugby, UK: 1995.
18. Lucas PR, Wingate GAS. Threats analysis for computer systems, special issue on computer systems validation. Pharm Eng 1996; 16(3).
19. UK Ministry of Defence. Defence Standard 00-58: A guideline for HAZOP studies in systems which include a programmable electronic system, MOD Directorate of Standardisation, Glasgow, UK.
20. Andow P. Guidance on HAZOP Procedures for computer controlled plants. Her Majesty's Stationary Office, 1991.

21. Food and Drugs Administration. Guidance for FDA reviewers and industry: Guidance for the content of pre-market submissions for software contained in medical devices, May 29, 1998.
22. Reid C. Effective validation strategies: put GAMP categories into practice. Business Intelligence Conference on Computer Systems Validation for cGMP in Pharmaceuticals. London, March 28–29, 2001.
23. Beck K. Extreme Programming. Reading, MA: Addison-Wesley, 2000.

Appendix 7.A Example Contents for Supplier Project/Quality Plans

Introduction
- Author/organization
- Authority
- Purpose
- Relationship with other documents
- Contractual status of document

Scope
- Project description

Project plan
- Identify project activities
- GANTT charts
- Milestones and deliverables
- Mechanism to close project
- Risk management

Project organization
- Personnel
- Training
- Nominated contacts
- Mechanism to escalate problems
- Project progress meetings
- Problem escalation process

Quality plan
- Customer quality requirements
  - Customer procedures
  - Independent standards
  - Independent guidelines (e.g., GAMP®5)
- Supplier quality assurance system
- Customer/supplier quality assurance system interface

Deliverables
- Identify types document to be produced through project

References
Appendices
- Glossary
- Others

Appendix 7.B Example Checklist for Functional Specification

System description
- Overview of the process with which the computer system will interact (e.g., laboratory, manufacturing, packaging, labeling, business)
- High-level description of the role of the computer system
- Operational boundaries and interface requirements
- Development tools: CASE tools, computer-aided design (CAD)
- List of constraints including working language affecting computer system

Equipment
- Computer hardware: central processing unit (CPU), random access memory (RAM), memory devices (hard disk, CD-ROM, and so on), storage devices, communication interfaces to equipment and other computer systems, operator terminals

(Continued)

Appendix 7.B   Example Checklist for Functional Specification (*Continued*)

- Computer software: operating system, communication drivers, network controllers, system software, configurable programs, application programs, databases
- Instrumentation: number, type location; model numbers, software versions; tag numbers, data types, valid range
- Controlled elements: valves, heaters, motors and motor starters, relays, solenoids, and so on
- Power requirements with tolerable operating range

Interface

- Human interface
- Number, size, and location of screens (mimic backgrounds and foregrounds)
- Input devices (keyboards, mouse, tracker ball, touch screen, and so on)
- Number, type, and location of printers (alarm, color, report)
- Interface features (information pages, visual/audible alarms, mimics/graphics)
- Security features (levels of access authority, passwords, key switches, biometrics, logging unauthorized access attempts, and so on)
- Process I/O interface
- Operator intervention
- I/O specification for different devices
- Installed spare capacity, including fail-safe redundancy features
- Future expansion with any implications for reduced performance
- Plant external I/O signals (type, format, range, accuracy, timing)
- System-derived inputs (calculated variables: totalizers, analogue-digital converters, interposing relays, frequency, period, validity)
- Communications interface
- Protocols, buffer, cable specifications, line drivers, termination, loading, traffic density, automatic error correction

Process control

- Sequence control: text, flowcharts, and state-transition diagrams
- Continuous control: automatic and control loops, database storage, scan frequencies, complex control schemes (cascade, ratio, predictive)
- Batch control: batch initiation, process decomposition, continuous monitors, recipe handling (timers, volumes, weights, temperatures, pressures, download, and sequence interaction), logging
- Data processing: derived values, data conversions, scaling, frequencies, periods, calculations, algorithms, validity checks, error correction

System attributes

- Alarms: number and priorities, groups, response times, escalation, annunciation, presentation (including color and color changes), conditioning, acknowledgments, viewing, printing and storage capacity, segregation
- Trending: real-time and historical data, histograms, balance sheets, who can configure and view trends, printing and storage capacity
- Events: number, categories, notification, viewing, logging, printing and storage capacity
- Interlocks: hard-wired interlocks, scan rates, suppression, logging, reporting, acknowledgment, computer system handshakes and watchdogs, avoid deadly embrace

Operational environment

- Performance: response times (e.g., screen refresh rates, cycle times, and critical control response times), mean time to failure (MTTF), system remedial action, power failure recovery, start-up, shutdown
- Redundancy: dual operation, segregated I/O on dedicated cards, peripherals, interfaces, internal networks, power supplies, hard disks
- Intrinsic safety: equipment for zoned areas
- Operational safety: fail-safe mechanisms, error handling, database integrity, operator timeouts, fault tolerance, watchdogs, contingency plans, internal and legislative safety compliance
- Operational procedures: operator commands, override control, process monitoring, parameter modification, load scheduling, start-up and shutdown, fault-finding instructions, user manuals and documentation
- Training: formal courses, hands-on training, manuals
- Security: levels, means of access, parameter modification, program access, data security
- Data integrity: archiving, buffer storage, device storage, data recovery, backup, restoration
- Environmental conditions: earthing, filtering, loading, surge protection, temperature, humidity, vibration, electrical interference (electromagnetic interference, electrostatic discharge, radio frequency interference), fire, flood, hygiene
- Maintenance: spares, special handling practices and tools, cleaning, media backups and restoration, service contracts, service instructions, calibration schedules
- Expansion philosophy: tuning variable demand, functional changes, spare capacity, performance, physical size restrictions

Appendix 7.C   Example Software Design Structure

Introduction
Scope
System description
- Module structure
- Interfaces

System data

- Databases
- Files
- Records
- Data types

Module descriptions
- Design
- Functionality
- Interfaces
- Subprogram overview
- Software module/unit data (i.e., databases, files, records, data bases)

Subprogram descriptions
- Operation
- Input/output parameters
- Side effects
- Language
- Programming standards

Interfaces
- Operation
- Timing
- Error handling
- Data transfer

Glossary
References
Appendices

Appendix 7.D   Example Hardware Design Structure

Introduction
Scope (overview)
Design configuration
- Main computer
- Storage devices
- Peripherals
- Interconnections

Input/output

- Digital
- Accuracy
- Analogue
- Isolation
- Pulse
- Range
- Interface cards
- Timing

Environment

- Temperature
- Humidity
- External interfaces
- Physical security
- Radio frequency interference, electromagnetic interference, UV

(Continued)

Appendix 7.D   Example Hardware Design Structure (*Continued*)

Electrical supplies
- Filtering
- Loading
- Earthing
- Uninterruptible power supply

Glossary
References
Appendices

Appendix 7.E   Example Hazard Study Questions

- *Configuration of computer system*
  Is the system implemented according to the intent of its specification?
  Are there sufficient I/Os to enable plant/process to be operated as intended?
  Are all sensors and equipment working correctly as designated?
  What is automatic? What is manual?
  Does the design consider expansion requirements?
  Does the design consider the possibility of performance degradation from probable causes (e.g., disk full)?
  What is the integrity of the power supply?
  Will the system be affected by electrical interference or poor earthing?
  Will the system be affected by ambient temperature or humidity?
  Will the system be affected by dust, contamination, or corrosive materials?
  Have precautions been made for fire, flood, vibration, and shock?
  Have precautions been made for environmental needs and hygiene standards?
  Is the system intrinsically safe?
  To which mode do instruments fail?
  How does the system know if equipment is faulty (instrument, computer, andmanufacturing equipment)?
  Is there any redundancy built in to cover equipment failures?
  What is the system response to and recovery from utility failure?
  How are these facilities controlled?
  Are there hard-wired trips and interlocks? Challenge them.
  How are redundant and standby systems tested?
  Are off-line test or bureau systems used as a source of spares?
  How can the validity of these spares be assured?
  Do suppliers have necessary spares and equipment?
  Can suppliers backup and recreate the present software configuration and restore it?
  Is the equipment still supported by its original company?
  Are parts still available?
  Are other computer systems connected to this system?
  What happens if the connection between systems is lost?
  How are these connections controlled?
  Is the maximum length of communication lines exceeded?
- *Software control scheme*
  Has the software been reviewed, is it backed up, and can it be restored after a fault?
  Is the software documented, with all documents kept in a safe place?
  Will the system recover to a safe state after a power failure?
  Examine all sequence charts.
  Examine summary software flowcharts.
  Examine database structure and content.
  Challenge software functionality for incorrect user input, corrupted data, and an incorrect decision.
  What consequences will follow an erroneous operation?
  How are hazardous situations notified to operations staff?
  Can software recover automatically, and has this been tested?
- *GMP-related functions*
  Who is responsible for directing operations?
  Does supplied documentation adequately cover abnormal plant/process states?
  Are there safe states for operation (start-up, shutdown, holds, normal running, emergency shutdown, maintenance)?
  What happens to the pharmaceutical product in the event of a failure?
  Can the pharmaceutical product be recovered in the event of a failure?
  When is it safe to resume operation?

(*Continued*)

Appendix 7.E   Example Hazard Study Questions (*Continued*)

Is there an operating procedure to cover recovery after partial failure?
Is there an operating procedure to cover recovery after full failure?
Is there a contingency plan?
Challenge all reasonable planned scenarios.
Is the contingency plan periodically reviewed and updated?
Has the contingency plan ever been tested?
Is there a copy of all system parameters and settings?
Can in-process changes to equipment operating parameters be made?
How many product rejects are needed to halt equipment operation?
Is there product reject verification?
How are batch records controlled?
Investigate alarms associated with all failure modes. Are they useful?
What happens when large numbers of alarms are raised together?
Is there a hard copy of valid alarm settings?
Can an interlock be left in an incorrect state?
Identify and challenge event logging and trending.
Are regular (annual?) service reviews conducted?
Are trends in equipment performance and failures monitored?

- *Operation of computer system*
Is the plant/process operating philosophy fully defined and understood?
Are all plant/process operation phases defined (start-up, shutdown, abort, recovery)?
Are operating procedures in place for the computer system?
Are procedures updated and personnel retained after changes to plant/process?
Was appropriate training given to all personnel associated with the system's operation?
Are standards and recommended companies' practices followed?
How does the operator know what equipment is under computer system control at any moment?
Can the operator accidentally change key parameters?
Can the computer system tolerate invalid operator input?
How does the operator know that a production problem has occurred?
How does the operator know what to do (procedures and training)?
Has allowance been made for color-blind operators?
Is the accuracy of data displayed consistent with control needs?
Are standard names, colors, units, symbols, and abbreviations defined?
Are alarms and messages recorded in more than one location in case of a printer failure?
How do we know if any equipment has failed during operation?
Is a fault reporting procedure in place?
Are system reliability and failure rates regularly reviewed for trends?
How frequently are relief instruments used and tested?
Can equipment be operated from more than one location? Could this lead to errors?
Is a service agreement in place?
Are response times defined?
Is equipment maintained and calibrated regularly?
Are maintenance and changes controlled by a procedure?
How are changes between summertime and wintertime controlled?

- *Security*
Do procedures for physical access exist?
Can software be modified without authorization?
What authorization is needed to modify software?
Have access levels been used to define/enable different levels of access?
Are passwords/key locks used to define/enable different levels of access?
Are all changes to software automatically recorded?
Can alarms be suppressed, and readings taken off-line or out of scan?
Are there special security arrangements for critical items?

Source: From Ref. 10.

8 | Coding, Configuration, and Build

INTRODUCTION

Once the computer system has been designed, it can be built. The supplier generally bears all the responsibility for the activities associated with this phase; these cover software programming, source code review, and system assembly. These activities may not involve the pharmaceutical or healthcare company at all, depending on the nature of the relationship with the supplier. In such circumstances, supplier audits may be used to verify that the supplier has the required capability maturity for the task through having suitable controls in place. This, however, will not be possible for commercial off-the-shelf (COTS) software and hardware of unknown pedigree. The acceptability of such products should have been determined as part of the design review.

SOFTWARE PROGRAMMING

Programming is the lowest abstraction for the software development process. Software for the new application may be constructed either by programming in a native programming language like C^{++} or XML[a], line by line, or by assembling together previously programmed software components from software libraries or other sources such as off-the-shelf software. Native language programming might involve the use of assembly language or microcode for real-time or other time-critical operations. Source code is then translated (commonly referred to as compiled) for use on the target computer system.

GMP regulatory authorities expect software to be programmed and maintained under version and change control (1). They also encourage the use of good programming practices (2,3). Such programming practices should be defined and cover the following:

- Software structure (file organization: including comments, headers, and trailers)
- Naming conventions (for folders and directories, file names, functions, and variables)
- Revision convention (configuration/change control, with audit trails as appropriate)
- Code format (style including indentation, labels, and white space)
- Controls on the complexity of code
- Ensuring that there is no redundant or dead code
- Role of compilers (configuration switches and optimization)

The quality of software delivered by the programming work directly determines the effort that will be required to maintain it. Convoluted software often reflects a history of change, perhaps involving the integration of bits and pieces of code from other systems. Such software is hard to follow at the source code level, not only for the author developer, but also, much more importantly, for those who will be obliged to maintain it throughout its useful life. One of the commonest causes of this is poor documentation. Once written, there is not a lot that can be done with such "spaghetti" software. In the short term, it may not be practical to rewrite it, in which case the supporting documentation should be strengthened through a detailed source code review. If the software is expected to have a longer lifetime and especially if portions of the code are intended for future products, then rewriting the code altogether may make the most sense from a cost-effectiveness standpoint. This often turns out to be the case with large business systems such as MRP II/ERP.

---

[a] XML stands for extensible markup language, the universal format for structured documents and data on the World Wide Web.

```
procedure A(var x: w);
begin b(y, n1);
b(x, n2); m(w[x]); y:=x; r(p[x])
end;
```

(A) Unstructured Code Fragment

```
procedure change_window(var nw: window);
        begin border(current_window, no_highlight);
                border(nw, highlight);
                move_cursor(w[n]);
                current_window:=nw;
                resume(process[nw])
        end;
```

(B) Structured Code Fragment

Figure 8.1   Structuring code.

The selected programming language can also have a significant impact on the effectiveness of a piece of software. Data-driven languages are appropriate to data-oriented solutions, while formula-driven languages such as FORTRAN and COBOL tend to lead to algorithm-oriented solutions. A poor choice of a programming language could well impose an inappropriate solution on the task to be accomplished, and the consequential deficiencies will often be deep seated. It is very important, therefore, that the programming language truly complements the software design.

The use of structured indentation and intuitively named functions and variables according to defined conventions aids understanding immediately (see Fig. 8.1). The use of comment, perhaps as a preamble to a subroutine, helps further. Producing code that is more understandable may take some extra effort in the first instance, but in the long run, this is richly repaid in terms of reduced effort to identify and eliminate innate defects during development. Consequently, the testing is accelerated and the ongoing maintenance during operation is less burdensome for users and developers alike. A summary checklist for software production is given in Appendix 8.A.

Programming Approach

The choice of programming practices should be tailored to the particular characteristics of the programming languages in use and be directed toward mitigating their known deficiencies. For instance, the main weaknesses associated with conventional ladder logic include (4)

- Poor facilities for structured or hierarchical program decomposition,
- Limited facilities for software reuse,
- Poor facilities for addressing and manipulating data structures,
- Limited facilities for building complex sequences, and
- Cumbersome facilities for arithmetic operations.

Some programming languages are less forgiving to developers and more prone to errors than others. For instance, error-prone features of programming languages such as BASIC, C^{++}, and RTL/2 include the following (5):

- GOTO, assigned GOTO, and GOSUB
- Floating point numbers
- Pointers
- Parallelism (concurrent processing)
- Recursion
- Interrupts
- Dynamic allocation of memory

Recent research, however, indicates that about 5% of programming code will normally contain logical errors (6) and that these are more likely to arise from the programmer rather than from the language used (7). Good programming standards rigorously enforced by code inspection can greatly diminish this error rate. The best in-house programming standards are those based on well-tried industry standards such as IEC 1131-3 and ISA-S88. There is much help available in the literature and on the Internet in this regard, and there is no excuse nowadays for sloppy and careless programming in the absence of defined standards. An example of good programing standards is given in Appendix 8.B.

Redundant Code ("Dead Code")

Redundant or "dead" code is program logic that cannot possibly execute because the program paths never permit those instructions to be reached. It is not uncommon to find up to 20% of the code being redundant in this way when code from an earlier version is reworked to achieve a slightly different functional purpose. GMP regulatory authorities have expressed concern that redundant code might be unintentionally accessed during system operation, and recommend its removal (8). Redundant code can include the following:

- Superseded code from earlier software versions
- Residual code from system modifications
- Unused features of standard software packages

Source code that has been deliberately commented must not be regarded as dead code since the compiler ignores it and can never therefore become executable instructions.

Wherever possible, redundant code should be removed and the software should be recompiled. This includes where software instructions become dead code because of program modifications. Care, however, must be taken to distinguish rarely used code from redundant code, and not to mistakenly classify the former as the latter! Examples of rarely used code include

- Apparently unused modules in large configurable systems and
- Diagnostic features and test programs that are intended to remain dormant until needed.

There should be no executable code resident in vacant areas of code storage media (sometimes in the form of firmware chips) in computer systems intended to operate in areas subject to electrical disturbances. Transient faults induced by EMI, ESD, and RFI can cause inadvertent jumps to take place to these storage locations. Unused areas of storage should be initialized to contain logic patterns that, if erroneously accessed by the processor, the program will cause the system to revert to a known and predefined safe state. All overlays of reserved storage should be populated with similar logic patterns. A number of hardware and software system level techniques including "error-capturing instructions" and "capability checking" are available that are suitable for implementation within a variety of design constraints (9).

Recent Inspection Findings

Software has been customized, but no design controls were generated (FDA 483, 2006).

The firm has failed to put in place programming standards for the numerous (>100) source code blocks that have been developed and maintained by company personnel (FDA 483, 2001).

Source code blocks contain change control history annotations at the beginning of the code for change history information for each source code program. The firm has failed to ensure that these change history annotations are updated when programming changes have been made (FDA 483, 2001).

The computer system lacked adequate text descriptions of programs (FDA Warning Letter, 2001).

System design documentation including program code was not maintained or updated (FDA Warning Letter, 2001).

SOURCE CODE REVIEW

No one doubts the crucial operational dependence of computer systems on their software, and the importance of professionally developed software is widely appreciated. GMP regulatory authorities hold pharmaceutical and healthcare companies accountable for the "suitability" of computer systems, including software (10), and expect them to take "all reasonable steps to ensure that it [software] has been produced in accordance with a system of Quality Assurance" (11). One GMP regulatory authority is quoted as stating that "there is no room in the pharmaceutical industry for magic boxes" (12).

Comprehensive software testing, in other words, testing that exercises all the pathways through the code, is not a practical proposition except for the very smallest programs. It implies testing every possible logical state that the system can ever assume. Software is often documented using flowcharts that track decision points and processing states. A relatively simple flowchart is given in Figure 8.2. Any path through the software associated with the flowchart is capable of triggering an error and/or failure. And it is not just the pathway that is important—data input and data manipulation will influence whether or not an error and/or failure state is generated. Barry Boehm calculated the number of conditional pathways through the flowchart to be 10^{21} (13). Exception handling for error/failure conditions introduces further complexity, with interrupts creating a "jump" to what would otherwise be a wholly unrelated part of the system. Assuming that one individual test could be defined, executed, and documented each second (a somewhat optimistic assumption in real life), it would take longer than the estimated age of the universe to complete the testing! Indeed, even if batches of a thousand individual tests could be conducted concurrently, the time required to complete overall testing would only be reduced to the estimated age of the universe. It is therefore evident that full functional testing of every pathway is never possible, and much software today is more complex than the example given in Figure 8.2. Other techniques are therefore needed to complement functional testing and measure the quality achieved in software development.

Source code reviews (also known as software inspection) are a proven technique for improving software quality. These are intended to give a degree of assurance of the quality of code along the pathways that can never be functionally tested. We can use these, together with functional testing, to gain an overall measure of software quality (14). It is astonishing that the limitations of functional testing are not widely appreciated. Many software companies and development teams blithely place complete reliance on functional testing as a measurement of quality without realizing the inadequacy of such measures. Quality must be *built into software*—it can *neither* be solely tested in nor can it be *measured by functional testing alone*. Pharmaceutical and healthcare companies must *not* rely on standard license agreements as mitigating the need for effective quality assurance systems, supervision including source code reviews during development, user testing, and supplier audits. Most standard license agreements are nothing more than an abrogation of all responsibility by software developer organizations and can usually be succinctly summarized as *"As is, unsupported, and use at your own risk"* An example checklist for conducting source code reviews is given in Appendix 8.C.

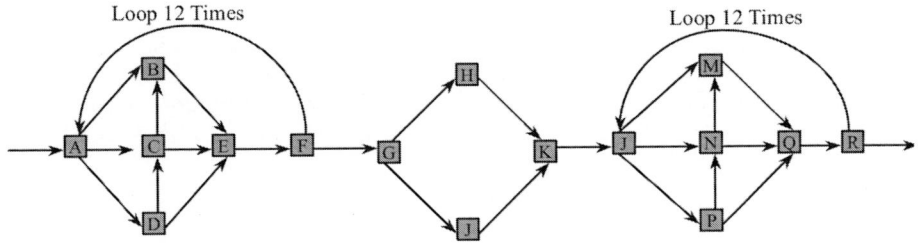

Figure 8.2   Practicalities of comprehensive testing.

Review Criteria

Source code reviews have four basic objectives as follows:

- Exposure of possible coding errors
- Determination of adherence to design specifications, including
  - Affirmation of process sequencing,
  - I/O handling,
  - Formulae and algorithms,
  - Message and alarm handling, and
  - Configuration.
- Determination of adherence to programming practices
  - (e.g., headers, version control, change control)
- Identification of redundant and dead code

International Society for Pharmaceutical Engineering (ISPE)/GAMP has responded to such concerns with a procedure for inspecting software embracing software design, adherence to coding standards, software logic, redundant code, and critical algorithms (5). Source code reviews are particularly useful when verifying calculations that during the systems operation are being updated too quickly to check. Configurations should also be checked against specification.

Redundant bespoke (custom) programming is considered "dead" code and should be removed. The only exception is redundant code strategically introduced to try to protect the commercial confidentiality of proprietary software, usually by confusing disassemblers that might be used by unethical competitor organizations to reverse engineer the product.

COTS software functionality disabled by configuration is not redundant code in the truest sense on the basis that the disabled software is intended to be enabled according to the need of a particular implementation. Examples of where functionality may be disabled without creating redundant code include library software (e.g., printer driver routines or statistical functions), built-in testing software, and embedded diagnostic instructions.

Annotated Software Listing

Listings of software subjected to detailed low-level walkthrough should be annotated with the reviewer's comments. Conventionally, this has entailed handwritten comments on printed software listings, but there is no reason why electronic tools may not be used to support this activity so long as they satisfy compliance requirements (Table 4.4 in chapter 4).

Supplier audits should always include the review of a sample of code to ensure compliance with quality system standards, though such a review will never pretend to assume the rigor of a formal source code review. Where suppliers retain software listings, access agreements should be established (see also chap. 11).

The process of reviewing source code typically consists of the following:

- Check adherence to programming practices (headers, version control, change control)
- Check I/O labels and other cross-reference information
- Check any configuration setup values
- Progressively check functional units for correct operation
- Confirm overall process sequencing

All critical I/O labels, cross-reference information, and configuration should be checked. Formulae and algorithms should be verified against design specification definitions. Where possible, manual confirmation of correct calculations should be undertaken for custom-programmed formulae and algorithms. Message and alarm initiation and subsequent action should be traced to verify correct handling.

As the review progresses, a software listing should be marked up to record the findings made. Any deficiencies in the code should be clearly identified by an annotation. It is just as important to record if no deficiencies are identified. Figures 8.3 and 8.4 provide some examples on how a software listing might be annotated. It should be recognized, however, that the style of annotation would need to be adapted to fit different programming languages and structures.

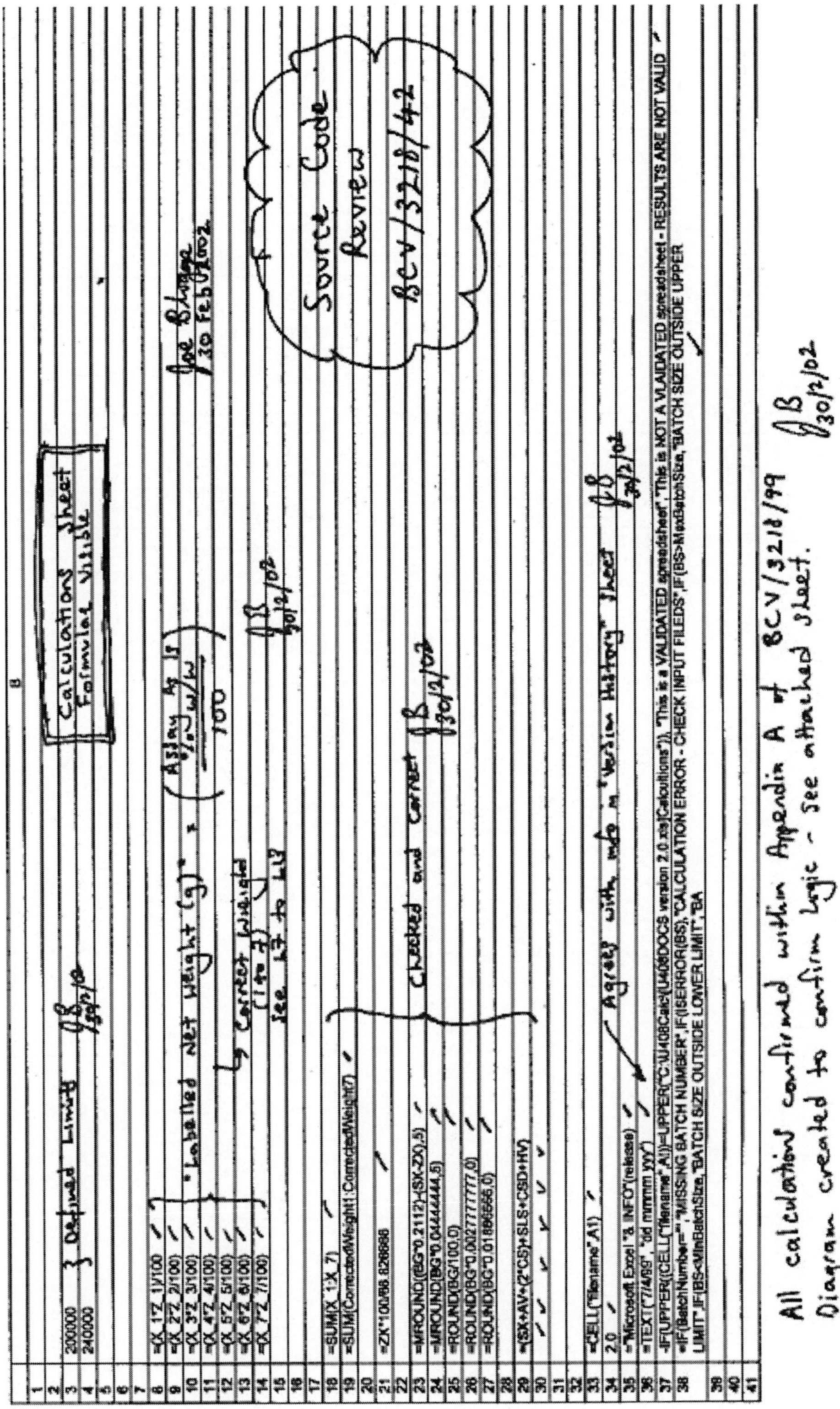

Figure 8.3 Example of annotated software listing.

FC12 - <offline>
"recipe"   Data record transfer management
Name: Family:
Author: PH   Version: 0.1
                         Block version: 2
Time stamp Code: 21/11/2002 14:54:41PM
        Interface: 04/03/2002 07:28:15AM
Lengths (block/logic/data): 00794 00642 00000

Sheet 1 of 2

Module "recipe"
Design Specification.
Ref. A217_FS Section 10.7

Address	Declaration	Name	Type	Initial value	Comment
	in				
	out				
	in_out				
	temp				

none defined AS 28-Nov-2002

Block: FC12   Recipe

Data record transfer can be:
1. TP->PLC, with appropriate parameters SENT to motion controllers after
transfer complete;
2. PLC->TP, with appropriate parameters READ from motion controllers before
data
is passed to TP.

A217_FS Section 11.0
General module description.

Network: 1   Generate datarecord transmit complete oneshot

        A   DB2.DBX   29.2   //bit2, datarecord status, TP's data mailbox
        =   M   59.0   //data record transmit complete oneshot   "dr_tx_complete_os"
        R   DB2.DBX   29.2

Verified one shot data record.
A217_FS Section 11.1

Network: 2   Data record transfer TP to PLC

        AN   DB2.DBX   34.0   //transfer datarecord TP to PLC pb
        ON   M   58.0   //check datarecord transfer not in progr   "dr_tx_idle"   -- Data rec
                                ess   ord transmission idle
        O   Q   124.1   //or running   "connect_motors"   --
        JC   m001   //jump if not required
//start PLC job 70
        L   70
        T   DB2.DBW   10   //plc job number (for TP's interrogation   "hmi_interface".hmi_job_1 --
)
        L   1   //recipe number
        T   DB2.DBW   12   //plc job number parameter 1   "hmi_interface".hmi_job_2 --
        L   DB2.DBW   54   //data record number (that currently on   "hmi_interface".displayed_drecord --
                                display)
        T   DB2.DBW   14   //plc job number parameter 2   "hmi_interface".hmi_job_3 --
//initialise step controller
        L   1
        T   DB127.DBW   90   //TP to PLC datarecord step controller   "general".scratch49   -- data rec
                                                                           ord TP->PLC transfer step controller
m001: NOP   0
        A   DB2.DBX   34.0   //reset the pushbutton request
        R   DB2.DBX   34.0

Verified data record Transfer
A217_FS Section 11.2

Network: 3   Test for datarecord transferred TP to PLC

        AN   M   59.0   //if data record transfer not complete   "dr_tx_complete_os"   --
        ON   DB127.DBX   91.0 //or not at correct step
        JC   m002   //jump
        L   2
        T   DB127.DBW   90   //advance the step controller   "general".scratch49   -- data re
                                                                           cord TP->PLC transfer step controller
        L   0   //reset the PLC job parameters
        T   DB2.DBW   12   "hmi_interface".hmi_job_2 --
        T   DB2.DBW   14   "hmi_interface".hmi_job_3 --
        L   DB2.DBW   54   //on-screen data record   "hmi_interface".displayed_drecord --
        T   DB2.DBW   58   //last sent data record for hmi to disp   "hmi_interface".drecord_in_use -- The
                                lay   last sent data record number
m002: NOP   0

Verified data record test following transfer
A217_FS Section 11.3

Page 1..

SCR REPORT REF. A217_SCR1   A Smith   28-Nov-2002

Figure 8.4   Example of annotated ladder logic listing.

Reporting

The outcome of the source code review will be a report providing an overview of the review,
together with a list of all observations noted and all actions that must be completed. Specific
statements on software structure, programming practice, GMP-related functionality,

information transfer with other portions of the system or other systems, error handling, redundant code, version control, and change control should be made before an overall conclusion on the suitability and maintainability of the software is drawn. A copy of annotated software listings should be retained with the report.

The report may identify some software modifications. How these modifications are to be followed must be clearly defined. A major failing of many reports is the lack of follow-up of outstanding actions. For example, two classes of modification are defined here.

Class A: Software change must be completed, software must be replaced, or supplementary controls must be introduced (e.g., procedural control or additional technical control) before the system can be released.

Class B: Software change does not have to be completed for the system to be released for use. These outstanding changes should be logged in the validation report and subject to periodic review. It is important that these changes do not get overlooked.

It is generally considered that widely distributed COTS software does not need source code review if a reputable developer has produced it under an effective system of quality assurance, and the product requires no more than an application parameter configuration (15). In most computer systems, therefore, source code reviews are limited to custom (bespoke) software and configuration within COTS software products.

Effectiveness

The effectiveness of source code review is often questioned. Inspections conducted by individuals working on their own are of little value, since most of us are incapable of being objectively critical of our own work. The objectivity of others and a willingness to accept criticism are key to any review process. Left to himself or herself, an individual programmer's error detection rates on his or her own code can be as low as 5%. Where a reviewer conducts the inspection in partnership with the software author, however, error detection rates can rise to 30% or more so long as it is not treated as a superficial cursory activity. The time saved in taking corrective action on exposed errors, particularly the structural ones, in advance of testing usually more than justifies the involvement of a colleague.

Examples of real problems identified in source code reviews include the following:

- Version and change control not implemented in so-called "industry standard" PLC and distributed control system (DCS) software
- Functions and procedures in MRP software not having a terminating statement so that the execution erroneously runs into the next routine
- Incorrectly implemented calculations: moving averages in supervisory control and data acquisition (SCADA) systems, material mixing concentrations in DCS systems, flawed shelf life date calculations in laboratory information management systems (LIMS)
- Duplicated error messages because cut and paste functions have been used carelessly
- Interlocks and alarm signal inputs used by PLC software labeled as unused on electrical diagrams

For a PLC or the configuration of a small DCS, the source code reviews will typically require about four days' effort, split between an independent reviewer and the software programmer. The reward of identifying and correcting defects prior to development testing or user qualification has proved time after time to more than compensate for the time and effort required to carry out the review. It really is a most cost-effective way of building quality into software.

Access to Application Code

While rarely invoked, some GxP legislation requires reasonable regulator access to application-specific software, including any source code review records (14,16). For the purpose of regulatory GxP inspections, pharmaceutical and healthcare companies should therefore agree with their suppliers over possible access to application-specific software (say within 24 hours). An example of the wording of such an agreement is given below:

> *[Supplier name] hereby agrees to allow [customer name] or their representative, or a GxP regulatory authority access to view source code listings for [product X] in hard copy and/or electronic format as requested. [Supplier name] also agrees to provide technical assistance when requested to answer any questions raised during any such review. [Supplier name] also agrees to store the original of each version of software supplied to [customer name] until it is replaced plus seven years. In the case of system retirement, the last version shall be stored to retirement plus seven years.*

GxP regulations require that access to the software and relevant associated documentation should be preserved for a number of years after the system or software has been retired. See chapter 12 for more details. Software licenses do not entitle pharmaceutical and healthcare companies to ownership of the software products they have "purchased." All that has been purchased is a license, an official permission or legal right to use it for some period of time under defined conditions of use. Accordingly, some companies have established escrow (third party) accounts with suppliers to retain their access to software, but this is not mandatory. Access agreements directly with the software supplier for the purpose of regulatory inspections are an acceptable alternative. If the software supplier refuses to cooperate, then this poses a dilemma. In such circumstances, it is recommended that pharmaceutical and healthcare companies use other suppliers for future projects (2,17).

Recent Inspection Findings

No procedure for review of source code. No assurance that all lines of code and possibilities in source code are executed at last once (FDA Warning Letter, 2002).

There is no written procedure to describe the source (application) code review process that was performed for the XXXX computer system (FDA 483, 2001).

The firm did not review the software source code that operates the [computer system] to see if it met their user requirements before installation and operation (FDA 483, 2001).

It was confirmed that the [software] listing was not reviewed or approved (FDA 483, 2001).

There was no source (application) code review (FDA 483, 2001).

The firm has failed to perform a comprehensive review of all [softwares] to ensure that appropriate programming standards have been followed (FDA 483, 2001).

The firm failed to document review of source code blocks in change control records (FDA 483, 2001).

SYSTEM ASSEMBLY

Assembly should be conducted in accordance with procedures that recognize regulatory requirements and manufacturer's recommendations. Any risk posed to pharmaceutical or healthcare processes by poor assembly must be minimized. For instance, wiring and earthing practices must be safe.

Assembly should be conducted using preapproved procedures. The quality of assembly work, including software installation, should be monitored. Many organizations deploy visual inspection and diagnostic testing to confirm that the computer system's hardware has been correctly assembled. Some companies tag assembled equipment that has passed such a quality check so that it can be easily identified.

Any assembly problems should be resolved before the system is released for development testing. If necessary, assembly procedures should be revised with any necessary corrections. Packaged computer systems do not need to be disassembled during development testing or user qualification so long as assembled hardware units are sealed.

REFERENCES

1. Food and Drugs Administration. CGMP applicability to hardware and software. Guide 11, compliance policy guides, computerised drug processing, 7132a, guide 11. Rockville, MD, Centre for Drug Evaluation and Research, 1984.
2. Trill AJ. Computerised systems and GMP—a UK perspective, part 1: background, standards and methods; part 2: inspection findings; part: 3 best practices and topical issues. Pharm Technol Int 1993; 5(2):12–26, 5(3):49–63, 5(5):17–30.
3. Food and Drugs Administration. Software development activities. Technical report, reference materials and training aids for investigators. Rockville, MD, Centre for Drug Evaluation and Research, 1987.
4. Lewis RW. Programming Industrial Control Systems Using IEC 1131-3. Control Engineering Series 50. London: Institution of Electrical Engineers, 1995.
5. International Society for Pharmaceutical Engineering. GAMP®5: risk-based approach to compliant GxP computerised systems. Tampa, Florida, 2008. Available at: www.ispe.org.
6. Panko RR. What we know about spreadsheet errors. J End User Comput 1998; 10(2):15–21.
7. Hatton L. Unexpected (and Sometimes Unpleasant) Lessons from Data in Real Software Systems. Safety and Reliability of Software Based Systems. Heidelberg, Germany: Springer-Verlag, 1997.
8. Leveson N. Safeware: System Safety and Computers. Reading, MA: Addison-Wesley, 1995.
9. Wingate GAS. Computer Systems Validation: Quality Assurance, Risk Management and Regulatory Compliance for Pharmaceutical and Healthcare Companies. Boca Raton, FL: Interpharm Press, 2003.
10. Food and Drugs Administration. Vendor responsibility. Guide 12, compliance policy guides, computerised drug processing, 7132a, guide 12. Rockville, MD, Centre for Drug Evaluation and Research, 1985.
11. European Union. EU GMP annex 11 computerised systems, guide to good manufacturing practice for EU directive 2003/94/EC, Community Code Relating to Medicinal Products for Human Use, Vol. 4, 2002.
12. Fry CB. What we see that makes us nervous, guest editorial. Pharm Technol 1992; May/June: 10–11.
13. Boehm B. Some Information Processing Implications of Air Force Missions 1970–1980. Santa Monica: The Rand Corporation, 1970.
14. Food and Drugs Administration. Source code for process control application programs. Compliance policy guides, computerised drug processing, 7132a, guide 15. Rockville, MD, Centre for Drug Evaluation and Research, 1987.
15. Chapman K. A history of validation in the united states: parts 1 and 2—validation of computer-related systems. Pharm Technol 1991; 15:82–96; 16:54–70.
16. U.S. Federal Food, Drug, and Cosmetic Act, section 704, Factory Inspection.
17. Tetzlaff RF. GMP documentation requirements for automated systems: parts 1, 2 and 3. Pharm Technol 1992; 16(3):112–124, 16(4):60–72, 16(5):70–82.
18. ACDM/PSI. Computer systems validation for clinical systems: a practical guide, version 1.1, December 1998.

Appendix 8.A   Checklist for Software Production, Control, and Issue (5)

Software production
- Programming standards
- Command files
- Configuration control
- Change control
- Software structure

Software structure

- Header
- Comments
- Named parameters
- Manageable module size
- No redundancy
- Remove dead development code
- Efficient algorithms

Software headers

- Module/file name
- Constituent source file names
- Module version number
- Project name (and reference code/contract number)
- Customer company and application location
- Brief description of software
- Reference to command file
- Change history

Change history

- Change request number
- New version number
- Date of change
- Author of change
  - Other source files affected

Note: Regulatory authorities treat software as documentation and expect the same standard of control as applied to documentation.

Appendix 8.B   Example of Programming Standards (18)

Naming conventions

Directories	It is recommended that files are stored in an organized folder structure relating to software architecture and/or functions.
Index	An index file should be maintained (e.g., INDEX.TXT) for each directory. The index file should contain a list of all the programs/files in that directory with a short description of their contents/function.
File names	File names should be descriptive and reflect the functions or content of the file. They should only contain alphanumeric characters (possibly plus the underscore character) and should always start with a letter rather than a number.
Extensions	For operating systems that support file extensions, a standard file extension naming convention should be used, e.g.,

filename.DAT—ASCII data file
filename.LOG—SAS log file
filename.SAS—SAS program
filename.SQL—SQL program
filename.TXT—ASCII text file

Variables	Variable names should be intuitive and thereby reflect the contents of the variable. If it is difficult to select a relevant name, then a descriptive label should be used. Names made up of purely numeric characters should be avoided.

Program documentation

All programs and subroutines should include documentation that provides a preamble to the source code in the form of a header comment block. The following information should be included:

Program name	Name of program.
Platform	DOS, UNIX, VAX, Windows, etc.
Version	Version of software (e.g., 6.12 of the SAS package).
Author(s)	Name of the programmer(s) and their affiliation.

(*Continued*)

Appendix 8.B Example of Programming Standards (18) (*Continued*)

Date	Program creation date.
Purpose	A description of what the program does and why it exists.
Parameters	Description of the parameters received from (input) or passed back to (output) the calling program.
Data files	List any data sources for the program (e.g., ASCII files, ORACLE tables, permanent SAS data sets, etc.).
Programs called	List any program calls that may be made external to the program.
Output	List any output files generated by the program.
Assumptions	List any assumptions on which the program relies.
Restrictions	Describe any program restrictions.
Invocation	Describe how the program's execution is initiated.
Change history	This contains change control information for all modifications made to the program. Information should include date of change, name of programmer making modification, an outline description of the modification, and reason for the change. Some of this information needs not be detailed if contained on references change control records.

Program code should be annotated with comments to reinforce the understanding of the code structure and its function. There should be at least one comment per main step, new idea, or use of an algorithm within the program. When a step or algorithm is complex, further comments should be added as appropriate through that section of code. Too much commenting should be avoided as this could hinder rather than aid an understanding of the code.

Program layout

- Each source code statement should appear on a separate line.
- A blank line should be left between each logical section in the source code to aid readability.
- Blocks of source code statements representing nested routines should be indented so that these routines can be more easily identified. For example,

```
IF xxxx THEN DO
        statement;
        statement;
END;
ELSE DO
        statement;
        statement;
END;
```

- All variables should be declared and initialized at the beginning of the program. Default data types should not be used.
- All nonexecutable statements (e.g., variable declarations) should be grouped together in a block preferably at the beginning of the program.
- Complex mathematical expressions should be simplified by separating terms with spaces or by breaking down the complex expression into a number of simpler expressions.
- Conditional branching structures should always bear a default clause to cater for situations outside the programmer's conception. This clause should cause the program to terminate gracefully. In this way, the unexpected termination of the program in an undefined state can be engineered out and avoided.

General practices

- It is good practice to arrange code into small reusable modules. Once such modules have been verified, their reuse should be encouraged to improve quality and remove the need for further verification.
- Possible program input and execution errors should be predicted in advance and handled appropriately in the source code (e.g., division by zero).
- Avoidance of undesirable practices is also important to ensure that the program does not process data in unexpected ways under unexpected conditions. Examples of bad practices to avoid include
  a) commented out code in final versions of programs,
  b) hard-coded data changes in nonconversion programs, and
  c) data processing sequences that vary and are difficult to repeat.

Bad practice examples carry much more weight as a teaching aid than good practice ones.

Output labeling
Output should be labeled with

- The identity of the source program creating it, including version number,
- The date and time generated,
- The identity of the user, and
- The page number and total number of pages.

Appendix 8.C   Example of Checklist for Source Code Reviews (5)

Software reviews
- Review formal issue of software
- Agreed and specified review participants
- Arrange review meeting
- Adequate prereview preparation time
- Conduct review
- Accurate and traceable review minutes
- Systematic coverage of software
  - Software design
  - Adherence to coding standards
  - Software logic
  - Redundant code
  - Critical algorithms
  - Alarms handling
  - Input/output interfaces
  - Data handling
- Agree corrective handling
- Assign corrective actions and completion dates
- Retain original reviewed software
  - Listings
  - Flow diagrams
- Incorporate changes
- Approve changes
- Issue software
- Retain review evidence

Review follow-up
- Ensure successful closure of review
- Escalate if required

Note: Regulatory authorities consider software as a document and expect
it to be treated as such within the quality system supervising its creation.

9 | Development Testing

INTRODUCTION

Development testing is the responsibility of the supplier. It includes establishing the test strategy, conducting unit and integration testing, and conducting system testing in preparation for user qualification. Some organizations refer to system testing as factory acceptance testing. Development testing is based on verifying the computer system's specification and design and development documentation within the practical constraints of being at the supplier's premises. Comprehensive user testing is not usually possible under these circumstances.

Evidence of effective development testing can reduce the amount of subsequent user qualification expected by GxP regulatory authorities. The pharmaceutical and healthcare company will often endeavor to include in their user qualification as many tests as possible from development testing. It should also reduce the time needed to commission the computer system on the pharmaceutical or healthcare company's site, as qualification can focus on confirming an already established operational capability.

The supplier will normally invite the pharmaceutical or healthcare company to observe their own testing as part of a predelivery inspection (PDI). This is particularly important if the pharmaceutical or healthcare company is reducing the planned user qualification based in the expectation of successful development testing. Many pharmaceutical and healthcare companies use predelivery inspection as an opportunity for informal operator training prior to the computer system's arrival on site. If specific training is required for user qualification or the ongoing operation of the computer system, then formal training is needed, and this should be documented in personnel training records.

TESTING STRATEGY

Testing must be carried out according to preapproved test plans and test specifications, and test reports prepared to collate the evidence of testing (i.e., raw data), as illustrated in Figure 9.1. Test reports should be written to conclude each phase of testing and authorize any subsequent phases of testing. Progression from one test phase to another should not occur without satisfactory resolution of any adverse test results.

Test Plans

Testing must include, but not necessarily be limited to, the activities listed under the topics of development testing and user qualification, although the use of these qualification names is not compulsory. Due account must be taken of any test requirements identified by the validation plan, supplier audit, and design review. Testing must not be conducted against an unapproved specification.

Test plans are used to define and justify the extent and approach to testing. Groups or individual test cases are identified together with any interdependencies. Test plans may be embedded within validation plans, combined with test cases (to form what is commonly known as a test specification), or exist as separate documents. Test plans must be reviewed and approved before the testing process they define begins. Test plans and test cases are often referred to as protocols when applied to user qualification. Example content of a test plan is given in Appendix 9.A.

Test Specifications

Test specifications collate a number of individual test cases. The value of preparing effective test cases should not be underestimated. Poor test cases will lead to a weaker measure of product quality than what is possible from the activity, and an inconclusive overall result.

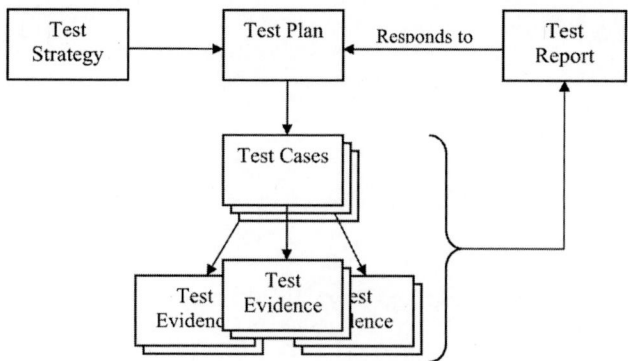

Figure 9.1 Testing philosophy.

These in turn will lead to delays while the uncertainty is considered; problem resolutions are determined and documented, usually with revised test specifications and repeated testing.

The level of detail required in test cases tends to vary considerably. Pharmaceutical or healthcare companies who want to use development testing to justify a reduction in the amount of user qualification should review the test specifications as early as possible. Test instructions down to a keystroke level are not necessary if testers are trained and familiar with the systems being tested. Any assumptions made regarding the capability and training of testers need to be documented in test specifications and supporting training records maintained.

The expected contents of individual test cases are described below (see also Appendix 9.B).

Project title/system name
- Project number and/or system name to be defined in preapproved test specification.
- Major systems should not use duplicate names.

Test reference
- Unique test reference should be defined for each preapproved test case.
- Unique run number should be assigned during testing.
- Default run number should indicate the first test run unless retesting or a particular test requires multiple runs of the test case(s).

Test purpose
- Described a clear objective for each test case in the preapproved test specification.

Reference documents and test prerequisites
- Test case should carry a cross-reference to the part of the system specification that is being tested.
- Any prerequisites such as test equipment, calibration, test data, reference SOPs, user manuals, training, and sequences between different test scripts should be defined in the preapproved test specifications.

Test method
- Define step by step test method.
- Identify data to be input for each step.
- Specify any screen dumps, reports, or observations to be collected as evidence at appropriate steps.
- Define associated acceptance criteria for individual steps as appropriate.
- Test cases must not introduce new system specifications.

Test results
- Register test case deviations in project compliance issue log.
- Cross-reference any project compliance issues in test results.
- Confirm whether acceptance criteria for test method steps are met.

Test outcome and approval
- Define acceptance criteria for an overall successful test outcome.
- Annotate test outcome as appropriate during text execution.
- Insert signature after test execution to assign test outcome.

Table 9.1 Example Test Script

Project title/system name: *UV-visible chromatography system*

Test reference: *CS_TEST_04* Run number: **01**	Test prerequisites: *Test reference CS_TEST_01* *("Log-On") has been* *successfully conducted.*	Reference documents: *User manual CS/01* *functional specification* *CDS_N2_01*

Test purpose: *Verify creation, operation, and reporting of an analytical method that performs spectral analysis of samples*

Test method	Acceptance criteria (expected results)	Actual results:
Step 1: *Put ChemStation into "Advanced" mode. Load test assay method (select "File," select "Load Method," select "test_assay.m" from "\TEST\METHOD" directory on the test server, select "OK").*	Step 1: *None for setup*	Step 1: *Not applicable for setup*
Step 2: *Select "Instrument," select "Setup," select "Spectrophotometer." Enter following parameters: wavelength from "190" to "1100," integration time "0.5," all other values are left as default input.*	Step 2: *None for setup*	Step 2: *Not applicable for setup*
Step 3: *Load "Test Sample CSS05."*	Step 3: *None for setup*	Step 3: *Not applicable for setup*
Step 4: *Select "Run Sample." Print screen dump, initial/date, label and retain as evidence for this test.*	Step 4: *Result identifies sample material as hydrochloric sulfide*	Step 4: *Confirm UV result here*
Step 5: *Select "Close Run," select "Exit."*	Step 5: *None for shutdown*	Step 5: *Not applicable for shutdown*

Test outcome (circle choice): pass/refer/fail		Project compliance issues
Name of tester Signature and date	Name of checker Signature and date	

- Insert signature after test execution to confirm test outcome, noting confirmation as witness or review of test results.
- Name of signer and date of signature must accompany signatures.

Test specifications must be reviewed and approved before the testing they define begins. Test cases can be written in such a way that test results are recorded directly onto an authorized copy of the test specification. Table 9.1 outlines an example test case.

Test Traceability

The requirements traceability matrix (RTM) initially developed for the design review should be extended to track which tests cover which aspects of the specification.

Test Conditions

There are three basic types of testing: coverage testing, error-based testing, and fault-based testing. Tests should aim to expose errors rather than try to prove that they do not exist (we have seen in the previous chapter that proving errors that do not exist is impossible). Testing must not be treated as debugging or snagging.

Coverage-based testing, as its name suggests, is concerned with establishing that all necessary aspects of the computer systems specification and design have been tested. As a general principle, all calls to routines, functions, and procedures should be exercised at least once during testing. In addition, all decision branches should be exercised at least once during testing. The use of a RTM can prove invaluable here, not only as a tool to identify tests, but also to demonstrate afterward what coverage was achieved. Other useful tools include call trees.

Error-based testing focuses on error-prone test scenarios (sometimes referred to as stress testing). It has been suggested that perhaps more than half of the functional tests conducted on a computer system should challenge its operational capabilities. Such testing includes the following:

- Boundary values (*guidewords: minimum, zero, maximum*)
  Many problems arise when the design fails to take account of processing boundaries, such as data entry, maximum storage requirements, and maximum variables scanned at highest scan frequency.
- Invalid arguments (*guidewords: alphanumeric, integer, decimal*)
  Includes operator data entry, acknowledgements, state changes, open circuit instruments, instruments out of range, and instruments off-scan.
- Special values (*guidewords: null entry, function keys*)
  Includes totally unexpected operator input and checking for undocumented function key short cuts.
- Calculation accuracy (*guidewords: precision, exceptions*)
  Includes precision to a number of decimal places, underflow and overflow, division by zero, and other calculation exceptions.
- Performance (*guidewords: sequence, timing, volume of data*)
  Includes execution of algorithms, task scheduling, system load, performance of simultaneous operations, data throughput, I/O scanning, and data refresh.
- Security and access (*guidewords: user categories, passwords*)
  Includes access controls for normal and privileged users, multiuser locking, and other security requirements.
- Error handling and recovery (*guidewords: messages, alarms*)
  Includes software, hardware, and communication failure. Logging facilities are also included.

Fault-based testing focuses on the ability of tests to detect faults. This approach may artificially seed a number of faults in the software and then require the overall testing régime to reveal at least 95% of them. Seeding must be conducted without reference to existing test specifications. Test practitioners do not commonly adopt fault-based testing, although it provides a useful measure on how effective testing has been conducted.

Test Execution and Test Evidence

Independence in testing is essential. No one can be relied on to be wholly objective about his or her own work, and this is especially true in the highly creative activity of software development. Personnel who designed or developed the computer system under test should not conduct testing.

The collection of test evidence should concentrate on the main object of each test case. There must be sufficient evidence (e.g., screen prints, printed reports, or written observations) to allow an independent reviewer to reach the same conclusion of the test outcome (i.e., whether or not the acceptance criteria were met). Written observations must provide evidence to support the test objective and its conclusion. It is not necessary to collect evidence for every step of the test method. Neither, in general, is it necessary to collect test evidence to demonstrate correct data entry or command keystrokes. Setup configuration should be defined in the test specification rather than being treated as test evidence. Files used to support testing need to be archived.

Test evidence may be collated separately or attached to test cases. The GAMP®5 Guide provides templates for collecting test evidence separately (1). Table 9.1 provides an example of test case that must be approved prior to testing but can then be used to directly record testing. Whichever approach is used, a cross-reference should be made to and from separate test evidence and the test case it supports.

All raw data collected as test evidence should be initialed and dated. Observations made as test evidence should be documented as they occur with timings in addition to dates when appropriate. Supporting hard copy printouts, screen dumps, logs, photographs, certificates, charts, annotated drawings and listings, and reference documents must be identified with the tester's initials and dated at the time the evidence was produced. The use of ticks, crosses,

"OK," or other abbreviations to indicate that actual results satisfied expected results should be avoided unless their meaning is specifically defined in the context of testing. It is better to faithfully record the actual results obtained.

Test Outcome
The outcome of each test is compared against acceptance criteria to ascertain whether the result fulfills the criteria without deviation. The concluding test outcomes are documented and approved as a "pass," "refer," or "fail."

- *Pass*: Signifies that the test result meets the acceptance criteria as detailed in the test script in full, without deviation from them in any way.
- *Refer*: Signifies that the test result is ambiguous in that a deviation has occurred but the test still potentially fulfills the intent of the acceptance criteria. An example here might be a typographical test case error not affecting the integrity of testing. All referred test outcomes need to be registered in the Project compliance issue log. Referred test outcomes must be either resolved as "pass" or "fail" before an overall conclusion to testing can be drawn. There is no need to formally raise a deviation for inconsequential nonconformances so long as this justification is documented.
- *Fail*: Signifies that the test result does not fulfill the acceptance criteria.

Independent Checks
Test outcomes recorded by testers need independent verification for validated applications. There are two main ways to manage the checking of test evidence, and the meaning inferred from check signatures varies accordingly.

- *Witness test results* as they occur. This requires personnel independent to the tester to monitor testing as it progresses and check and countersign test results/outcomes as each test is completed. This approach need only be used to document critical observations where physical test evidence such as printouts cannot be obtained (e.g., audible alarm sounded).
- *Review test results* after testing is complete. This is often cheaper and hence the preferred method. It requires that sufficient evidence be collected to support the test results/outcomes for each test case.

Independent checks should clearly document a review of corroborating evidence. It is this review that will give credence to the independent check if it were ever challenged during a regulatory inspection. Simply stating a pass or fail test outcome without any test evidence is unlikely to satisfy regulatory inspection.

Test Failures
All test failures must be documented, reviewed, and analyzed to identify the origin of the failure. The person approving the test results must consider the consequences of failure on the significance of the test results already obtained. Single or multiple tests may be abandoned. If the analysis of a test failure results in an amendment to the test case, controlling specification, or software, then the relevant documentation must be amended and approved. Further testing requirements must be agreed in accordance with the relevant change control procedure. Retest of single, multiple, or all tests may be required. Deviations from the test case acceptance criteria, where there is no risk to GxP or safety, may be accepted with the approval of the user and QA. Such concessions must be recorded and justified in the test report. Managing deviations and the use of project compliance issue logs are discussed further in chapters 3 and 5.

Test Reporting
The results of testing should be summarized in a test report that states

- System identification (program, version configuration),
- Identification of test specifications,

- Resolution to referred test outcomes, with justification as appropriate,
- The actions taken to resolve test failures, with justification as appropriate, and
- Overall determination on whether testing satisfies acceptance criteria.

The test report must not exclude any test conducted including those repeated for failed tests. The test report may be combined in a single document with test results. A successful overall testing outcome authorizes the computer system for use. Test reports are not necessarily prepared by QA; however, they should be approved by QA.

Managing Changes During Testing

It is likely that changes to the system will be required during testing to correct inherent software defects exposed by test failures. It is important for the developer to manage such changes under careful change control. Supplier and user organizations should not apply undue pressure on developers to make and release changes too quickly such that change control might be compromised. After all, it is very tempting for developers under the pressure of unexpected project delays to carelessly correct one defect and in the process create another in an apparently unconnected function. Thus, when changes are made, the design must be carefully considered, and the requirements for the regressions testing the system must be derived directly from this understanding. Then, and only then, can regression testing demonstrate that the change has not inadvertently created defects in other parts of the system.

Test Environment

Test environments can be quite complex depending on the size of the application and the need to provide configuration management of version upgrades. Small applications such as spreadsheets may be developed, tested, and released from a single environment. Larger applications generally warrant a segregated, if not separate, test environment.

For very large applications, there are typically three working environments, as illustrated in Figure 9.2: development, test, and holding. Software development for new and modified code is conducted in a dedicated development environment. When the software is ready for testing, it is moved to the testing environment for unit, integration, and system testing. The testing environment may be a different physical installation or a segregated area in the development environment. Either way, strict configuration management must be observed. Only when testing has been successfully completed, can the software be moved into the holding area as master source code. The holding area needs to be a highly protected separate environment in which access is restricted to those with authority to release approved software versions. If testing has been unsuccessful, the software is returned to the development environment for revision before subsequently coming back to the holding area for repeated testing until a successful outcome is achieved and the software is ready for release.

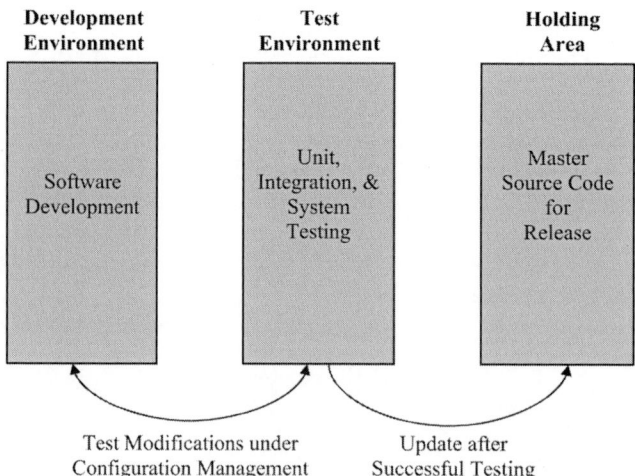

Figure 9.2   Commercial test environment.

Recent Inspection Findings

> Test inputs are not always documented ... Expected results are not always defined. Two comparisons done did not state whether or not the results were acceptable. The procedure states that the application "validates" if computer and manual results "are the same." There is no definition of "same" with acceptable variation specified. Unused XXXXXX printouts were routinely discarded with no explanation (FDA Warning Letter, 2000).

> Test results often consist of check marks only. The inspection found that data in numerous records were altered, erased, not recorded, recorded in pencil, or covered in white-out material. Therefore, there is not a complete record of all data secured in the course of each test (FDA Warning Letter, 2000).

> Test results were found reported in pencil on uncontrolled pages. Test documents included multiple of sections of test forms, which were crossed out without initials, dates, or explanation. The procedure calls for the same individual who writes/revises the software program to validate the program. Test results lacked indication of review or approval. The test report generated from these activities lacked a document control number (FDA 483, 2000).

> Firm failed to ensure that the supplier of the XXXX documented all the required test results to indicate the supplier's quality acceptance of the XXXX manufactured and delivered to your firm (FDA Warning Letter, 2002).

UNIT AND INTEGRATION TESTING

Unit testing (also known as *module testing*) is often done concurrently with coding and configuration as program components are completed. Unit testing should be extensive but not necessarily exhaustive, the aim being to develop a high degree of confidence in the essential functionality of modules.

Unit tests must be accompanied by integration testing. Integration testing exercises the interfaces between components and typically ensures that subsystems that have been separately developed to work together correctly. Testing should ensure a high coverage of internal control flow paths, error handling, and recovery procedures, paths that are difficult to test in the context of functional (or "black box") testing, as we have seen in the previous chapter.

Structural (White Box) Testing

Together, unit and integration testing are often referred to as structural (or "white box") testing. Tests exercise the components and subsystems of the design in isolation, using known inputs to generate actual outputs that are then compared with expected outputs (Fig. 9.3). Coverage-, error-, and fault-based testing should be applied, as described earlier.

It is important that the pharmaceutical and healthcare company has confidence in the structural testing as well as in the functional testing of the computer system. Both complement one another and together provide the measure of quality of the overall system. Records of unit testing and integration testing (including test specifications and results) should be kept by the supplier and retained for inspection, if requested, by the pharmaceutical or healthcare company. Any test harnesses, emulations, and simulations used during testing must be specified, and assurance in their capability must be demonstrated.

It is recommended that about 80% of the development testing effort is focused on unit testing and integration testing to establish the inherent structural correctness of the computer system. The remaining testing effort is applied to system testing.

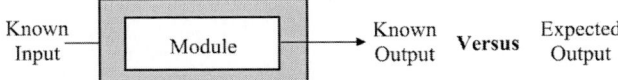

Figure 9.3 Structural "white box" testing.

Acceptance Testing of Commercial Off-the-Shelf Products

The system developer should pay careful attention to the use of commercial off-the-shelf (COTS) products and associated necessary acceptance testing. COTS products that are successfully used by a wide user base for many years will require less testing than those with a limited user base or that are new to the market. The following acceptance testing recommendations are made for COTS products (2):

- Test that the functions performed by the COTS software or hardware meet all specified requirements.
- The interfaces through which the user or other software invokes COTS functionality should be thoroughly tested.
- Test that all functions that are not required and remain unused cannot be invoked or adversely affect the required functions, for example, through erroneous inputs, interruptions, and misuse.
- Verify that all functions that are not required, remain unused, and are not access protected do have procedural controls in place.
- All errors discovered and traced to a COTS product during testing must be reported to the vendor, and the design review must be revisited as necessary.

Any open source software (OSS) used will also require testing. The level of testing should be commensurate with the criticality of functionality provided. Ideally, fault-based testing can be conducted so that some indication of innate quality (albeit a very weak measure) can be derived. Fault-based testing will not always be feasible for OSS depending on access to its source code and the availability of supplementary design-related information such as user manuals. Alas, the very nature of OSS means a supplier audit, which is what is really needed in these circumstances and is not possible. Consequently, the use of OSS (especially for critical functionality) is not recommended, but sometimes, the lack of any viable alternative makes its adoption unavoidable.

Recent Inspection Findings

Ian Johnson (3) recalls the instance of a PLC-controlled granulator that failed when challenged by operators deliberately entering inappropriate values for control parameters. Entry of zero for the run duration or the stopping torque would cause the device to run indefinitely. Entry of zero revolutions per minute for the motor speed did not disable the motor, as it should have done. Unfortunately, no memory was available to implement any warning messages, or to provide some entry editing function or reject an invalid value. As the granulator was entirely dependent on the PLC, the whole system was abandoned.

SYSTEM TESTING

System testing is conducted by the supplier to verify that the computer system's intended and defined functionality has been achieved. Such functional testing is often referred to as "black box" testing because it does not focus on the internal workings of a system (components and subsystems), but rather, the focus is on the complete system as a single entity (Fig. 9.4).

System testing by suppliers of COTS products is sometimes called *alpha testing* and is used as the basis for releasing a product to market. Some suppliers will also invoke *beta testing*,

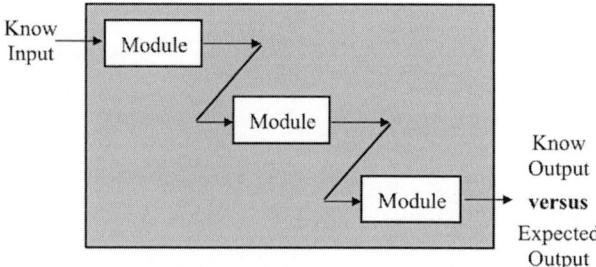

Figure 9.4 Functional "black box" testing.

whereby a selected band of trusted users are invited to evaluate COTS products before their general release. This is done in the full knowledge that inherent defects may well emerge and the trusted users may run that risk. In this way, the supplier can verify the robustness of their products in the privacy of a smaller group of partners before making any necessary revisions prior to public exposure of the product in the wider market.

Functional (Black Box) Testing

Functional testing is based on testing the system from a user's perspective, that is, without knowledge of the internal architecture and structure of the system. Inventory checks are made by visual inspection, while functionality is verified by running the computer system. Test scenarios should include

- Checking hardware components against equipment list,
- Checking switch settings (e.g., interface card addressing),
- Checking whether any equipment calibration is calibrated as required,
- Checking bespoke and COTS software versions loaded against configuration management plan,
- Exercising inbuilt software diagnostic checks,
- Verifying system operation against design intent,
- Challenge testing against operating ranges (e.g., data entry and performance),
- Challenge testing security and access,
- Verifying start-up and shutdown routines,
- Verifying data backup and recovery routines,
- Verifying that communication interfaces are operating, and
- Verifying alarm and event status handling.

Interface functionality is often tested using simulation utilities to avoid the inconvenience of setting up associated equipment and instrumentation, with the added burden of any calibration required. The use of simulators may entail additional verification, as discussed in regard to software tools in chapter 4.

Tests not conducted as part of the system testing must be included in user qualification. Safety functions should include functional testing to ensure that safety devices operate as intended in normal operating conditions and include exploring the consequences of a component failure and the effect this will have on the system. Calibration records must be kept to support user qualification as required (4,5).

Automated testing tools can be used to great effect during system testing and are discussed in more detail in chapter 4. The use of any automatic testing tools must be agreed with the pharmaceutical or healthcare company, preferably in the supplier project/quality plan.

Upgrade Compatibility

The upgrade path for superseded versions of the computer application also needs to be verified. Users expecting to upgrade existing applications should not experience problems. Upgrade tests should not be limited to functional testing but should also exercise data structures. An informed regression testing strategy needs to be employed.

Recent Inspections Findings

Common testing problems observed by GMP regulatory authorities include the following (3):

- Poor choice of test cases
- Failure to define the intent of tests
- Failure to document the test results

The success of development testing should be based on identifying and correcting deficiencies rather than merely looking at the initial pass rate. After all, it is far more important to detect deficiencies now than being deluded into believing that the system is fully functional, only to be embarrassed later on during user qualification or when the system is found to be noncompliant during live operation.

PREDELIVERY INSPECTION

PDIs are used to verify the system build against detailed hardware and software design, coding and configuration programming standards, hardware assembly practices and any relevant regulatory requirements, or industry guidance relating to these areas. Practitioners may be more familiar with the phrase "midway audit" in conjunction with GCP computer systems (6).

Many pharmaceutical and healthcare companies find PDIs useful to help prompt their suppliers to ask for help and clarifications during design and development. Often suppliers have multiple concurrent projects, in which case work on individual projects tends to slip behind schedule and become rushed toward the end of the designated project timetable. Individual projects may need to be brought back on schedule, and if so, the pharmaceutical or healthcare company may be able to help by extending timescales, providing additional resources, or clarifying requirements.

PDIs are based on visual assessments and are distinct from physical testing described earlier in this chapter. PDI typically covers observation/verification of the following (7):

- Drawings and layout diagrams
- Adoption of good programming practice
- Assembly checks as appropriate
- User interface functionality
- Unit, module, and integration test records

PDIs need not be a single event. In some situations, the PDI may be best conducted in parts, examining various elements of a system as they are completed. The scheduling and scope of PDIs should be carefully considered to maximize their benefit.

It should be recognized that there will be situations, especially on smaller projects, where the cost of attending the PDI may outweigh the benefits and risks in terms of schedule. In these cases, the inspection can be postponed until delivery on site; this is a business cost-benefit decision. An example where single PDI might be appropriate on a large project is instructive. This might be where a project team is sequentially rolling out a number of similar applications, and a PDI on the first application may be all that is needed depending on the differences between the similar applications. PDIs are also not appropriate for COTS products, because, by definition, they are already released to market, and so development and testing are complete.

Not many pharmaceutical and healthcare companies currently conduct PDIs, although the concept has been identified as good practice for some time. This is because PDIs are often hard to justify, especially when project budgets are tight; they are often considered as only desirable, not essential. Experience has shown, however, that they have proved very useful and effective, giving early warning of potential problems and helping to build a partnership with the suppliers. It is important to avoid situations where the supplier wants to release a system for delivery (for cash flow reasons), while the pharmaceutical or healthcare company is equally keen to accept delivery (and get on with the project). It is recommended that projects do not wait until the user qualification stage to fix known problems that are more easily corrected before installation of the computer system at the pharmaceutical or healthcare company's site.

REFERENCES

1. International Society for Pharmaceutical Engineering. GAMP®5: risk-based approach to compliant GxP computerised systems. Tampa, Florida, 2008. Available at: www.ispe.org.
2. Jones C, Bloomfield RE, Froome PKD, et al. Methods for assessing the safety integrity of safety-related software of uncertain pedigree (SOUP). UK Health and Safety Executive, contract research report 337/2001, 2001.
3. Wingate GAS. Computer systems validation: quality assurance, risk management and regulatory compliance for pharmaceutical and healthcare companies. Boca Raton, FL: Interpharm Press, 2003.
4. European Union. EU guide to good manufacturing practice for EU directive 2003/94/EC, Community Code Relating to Medicinal Products for Human Use, Vol. 4, 2003.
5. U.S. Code of Federal Regulations title 21, part 211, Current good manufacturing practice for finished pharmaceuticals.
6. Stokes T. Validating computer systems, part 4. Appl Clin Trials 2001; 10(2).
7. International Society of Pharmaceutical Engineering. Baseline pharmaceutical engineering guide: qualification & commissioning, 2001.

Appendix 9.A   Example Test Plan

Introduction
Scope (overview)
Test plan

- Specific areas not tested
- Test procedure explanation
- Action in the event of failure
- Logical grouping of tests
- How to record test results

Test requirements

- Personnel
- Hardware
- Software
- Test harness
- Test data sets
- Referenced documents

Test procedure

- Unique test reference
- Cross-reference to specification
- Step-by-step method
- Expected results (acceptance criteria)

Test results

- Raw data
- Retention of results
- Method of accepting completed tests

Glossary
References
Appendices

Source: From Ref. 1.

Appendix 9.B   Example Test Structure

Unique reference
Objective

- Single sentence

Resources requirements

- Specific to tests

Step-by-step procedure

- Repeatable procedure
- No unrecorded prerequisite requirements
  - Information
  - Experience

Acceptance criteria

- Smart
  - Specific
  - Measurable
  - Achievable
  - Realistic
  - Timed

Testing requirements

- Personnel
- Hardware
- Software
- Test harness
- Test data sets
- Referenced documents

Bottom line test result

- Pass/fail categories

Observations

- Additional information
- Acceptance concession

Source: From Ref. 1.

10 | User Qualification and Authorization to Use

INTRODUCTION

The purpose of the user qualification stage is to verify the operability of a computer system. Authorization to use the computer system after user qualification has been successfully completed is documented through a validation report.

Use of the term "qualification" is not universally accepted. User qualification may also be known as user acceptance testing or user verification. As the FDA has conceded, there is no consensus on the use of testing terminology, especially for user site testing (1). For the purposes of this book, the term "qualification" is used to embrace any user testing that is conducted outside the developer's controlled environment.

User qualification differs from development testing in that it is performed using the installed system as it is intended to be used at the user site. User qualification is typically conducted under the supervision of the user organization. Development testing, meanwhile, does not require any user involvement, and indeed, commercial off-the-shelf (COTS) systems users are, in general, seldom consulted. Care must be taken when planning computer systems not to unnecessarily duplicate development testing during user qualification. Computer systems testing can be combined with process/equipment/facility testing where this makes sense.

QUALIFICATION

Qualification is the responsibility of the pharmaceutical and healthcare company, although suppliers often assist. This phase consists of four sequential activities, as illustrated in Figure 10.1: site preparation, installation qualification (IQ), operational qualification (OQ), and performance qualification (PQ). IQ, OQ, and PQ should be applied to computer systems as indicated by key regulatory guidance (2–4).

The relationship between qualification and system specifications is indicated in Figures 10.2 and 10.3. Site preparation ensures that the setup requirements for the computer system are complete; IQ verifies the installation, configuration, and calibration of delivered equipment to the software and hardware design; OQ verifies the operational capability to the system specification; and PQ verifies the robust and dependable operation of the computer system. The inclusion or exclusion of tests between these qualification activities is usually based on convenience.

Test Documentation

Qualification should follow the same principles that were outlined for the computer system's development testing and discussed in chapter 9. Test specifications (also known as qualification protocols) must be written, reviewed, and approved before testing begins. It is especially important that the qualification meets the so-called *S.M.A.R.T.* criteria (5).

Specific: Test objectives address documented requirements.
Measurable: Test acceptance criteria are objective, not subjective.
Achievable: Test acceptance criteria are realistic.
Recorded: Test outcome evidence is signed off and, where available, raw data is attached.
Traceable: Test records, including subsequent actions, can be traced to defined system functional requirements (it does what it is supposed to do).

Many consultancy firms offer pharmaceutical and healthcare companies access to their standard qualification protocols for a fee. However, such test specifications should be adapted to reflect the specific build configuration of the system being tested.

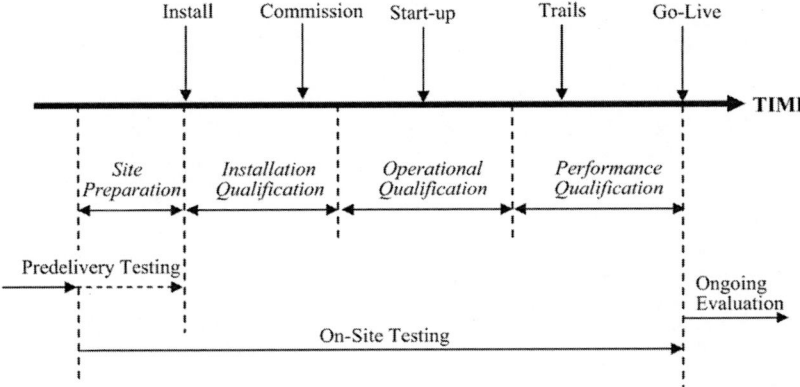

Figure 10.1 Qualification time line.

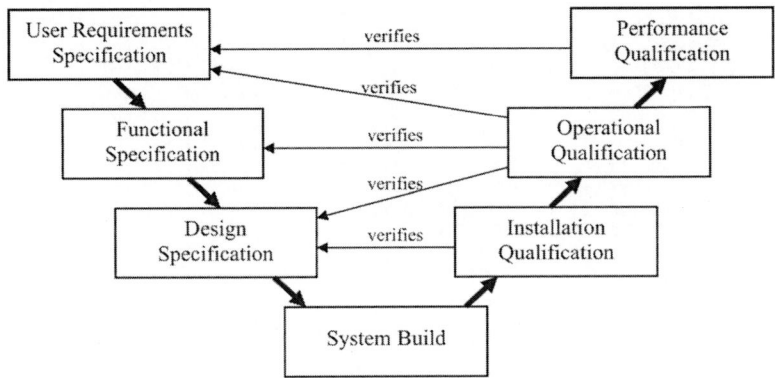

Figure 10.2 Verifying system specifications.

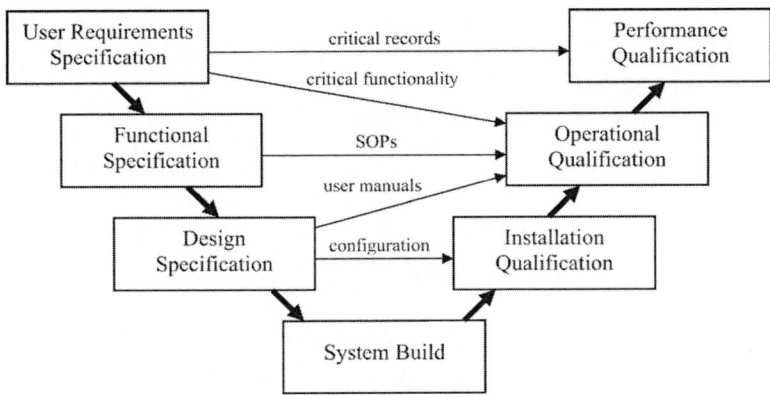

Figure 10.3 Supporting test requirements.

Test specifications can in theory be written during system development. In practice, however, while they may be drafted during development, they often need details confirmed with information that is only available after system development is complete.

User qualification can begin once test specifications have been approved. Figure 10.4 outlines the test management process. For a specific function to be tested, it is necessary to

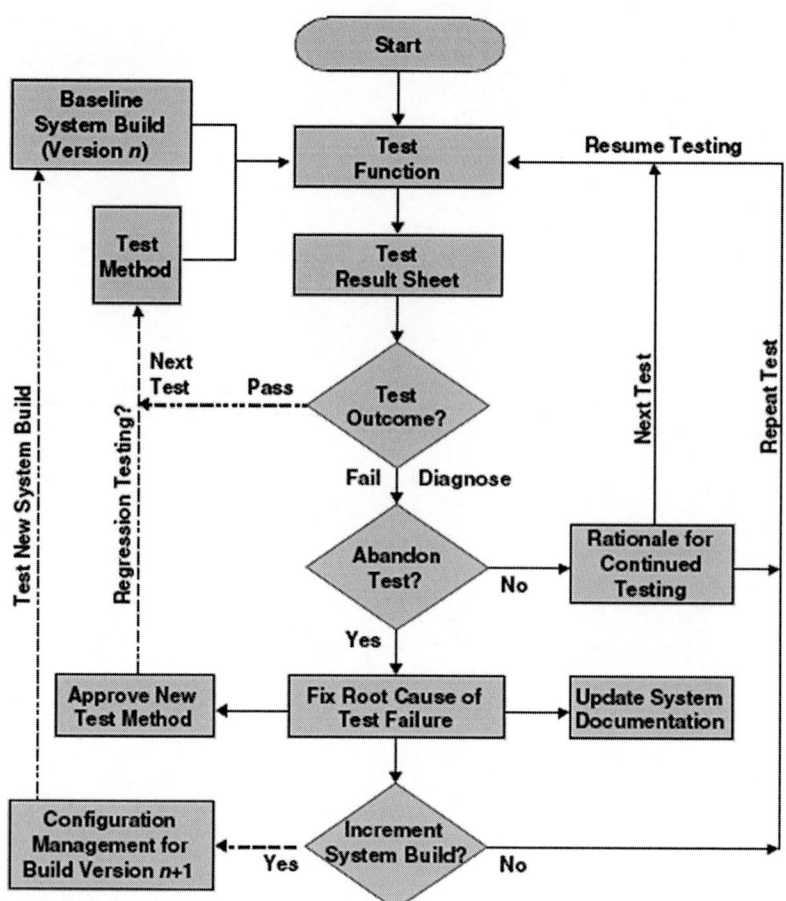

Figure 10.4   Test management.

have a test method and a known system build configuration. Test results should be recorded for all test methods executed. The outcome of the test should satisfy predefined acceptance criteria, in which case the testing may proceed to the next test. All test failures must be recorded, and their cause must be diagnosed beyond doubt. It may be necessary to abandon that test, but this does not necessarily mean that the overall testing activity has to cease there and then. Where testing continues after the failure of an individual test, a rationale should be prepared and approved to record the justification for this decision to proceed. Examples where there may be a clear justification to proceed included a limited hardware failure or isolated software failures, or even a software failure with a limited impact. In some instances, the software itself may be defect free in that respect, but the apparent failure is actually due to an incorrect test execution process, in which case the test may be repeated. In other instances, the individual test may be abandoned, but the overall testing continues with the next logical test. Where tests are repeated, for whatever reason, the original test results as well as the retest results should be retained. The most important factor throughout is *never* to ignore a test failure that *could* point to a fundamental design flaw. This is to deceive oneself, and such an action is bound to end in tears. Such failures must be explored to allay suspicion before much other testing ensues.

Test failures will normally require a root cause fix. Some tests might fail on a cosmetic technicality, such as an incidental typographic error. In this situation, the necessary amendments can be marked up on an existing copy of the test method while taking care not to obscure the original text. The reason for making the amendment and the person effecting it, together with the time the amendment was made, should be clearly identified. Other tests might trigger a failure because a test method is clearly in error. In these situations, it may be acceptable to annotate a fresh clean copy of the test method and rerun the test. Again, the reason for making the amendment and the person effecting it, together with the time the amendment was made, should be clearly identified.

Hopefully, most tests will uncover technical system deficiencies rather than test method inaccuracies. Technical deficiencies should be corrected, and system documentation should be updated to reflect any change made. It may be appropriate to increment the system build version under configuration management. New test methods may need to be prepared to test any changes made. If a new system build is created, then overall testing should be reviewed to determine where a comprehensive retest is required or whether relevant regression testing will be sufficient.

A test report should be prepared to complete each qualification activity (IQ, OQ, and PQ), summarizing the outcome of testing. Any failed tests, retests, and concessions to accept software despite tests on it having that failed must be discussed. Not every test has to be passed without reservation to allow the next qualification activity to begin so long as any permission to proceed is justified in the reports and corrective actions to resolve any problems are initiated. Each report will typically conclude with a statement authorizing progression to the next qualification activity.

Design reviews should be revisited as appropriate to consider errors discovered during qualification. All errors identified in a COTS product should be reported to the supplier, and a response should be sought. If no satisfactory response is forthcoming, the seriousness of the failure should be assessed and the ensuing decision, with any mitigating further actions, should be recorded.

The requirements traceability matrix (RTM) should be updated with details of test specifications and test reports. It should be possible to track a user requirement through functional specification, design, system build, development testing, and user qualification.

Stress Testing

Testing must include worst-case scenarios, sometimes referred to as stress testing. The Pharmaceutical Manufacturers' Association (United States) has promoted the model illustrated in Figure 10.5 to explain the boundaries that should be exercised. It is not sufficient just to test a computer system within its anticipated normal operating range. Instead, testing should verify correct operation across a proven acceptable range. This range should exceed the control range. Processing outside the control range will cause an alarm or error to be generated. It is important that the system does not fail when it should be alarm or error handling. Testing to the point of physical failure (destructive testing) is not required and should indeed be avoided. If such severe testing is required, then it should generally be conducted using simulation techniques.

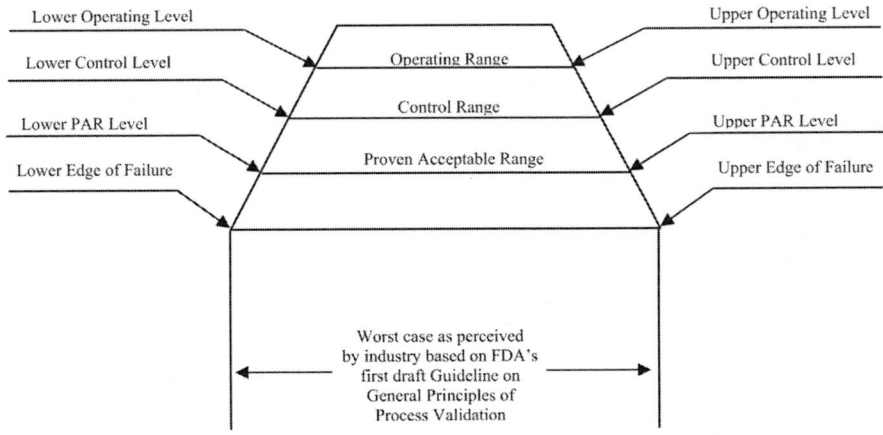

Figure 10.5 PMA stress testing model.

Test Environment

It is becoming common to have separate development, QA, and live environments within which different levels of testing can be conducted. Development and QA environments are what is termed *off-line*, that is, independent of the day-to-day operating processes. The live environment, however, is in contrast *operational*. The aim is to progress testing through each environment such that

- Development testing takes place in the off-line development environment,
- User acceptance testing occurs off-line in the QA environment, and
- Online user acceptance takes place in the live environment.

The management of development testing and controls for associated test environments are discussed in chapter 9.

It is vital that the QA and live environments are equivalent, so that test results between the two can be regarded as equivalent. Without such equivalence, there is no assurance that a satisfactory test outcome in one environment will be replicated in the other environment. The QA environment should therefore be subjected to IQ, demonstrating that this is, from a testing standpoint, equivalent to the intended live environment. Transport mechanisms used to move or replicate the application from one environment to another should be verified.

OQ is normally conducted in the controlled off-line QA environment. Alternatively, OQ may be conducted with the final system installed in situ prior to its release for use in the live environment. Unlike OQ, PQ must *always* be conducted in the live environment.

It is vital that the QA environment is maintained under strict configuration management. There should be no software development in the QA environment. Software should be prepared in the development environment and then, when completed, transported to the QA environment. Source code reviews should be conducted in the QA environment. If this approach is taken, then strict configuration management and change control within the development environment are not required. This should facilitate faster software development.

Testing operations are rarely as sequential between the various test environments as the illustration in Figure 10.6 might imply. It is quite normal for testing to iterate backward through the test environments when tests fail to deliver the expected results or when testing is conducted on an incremental enhancement to an existing system. In particular, the live environment may be used to provide "snapshot" dynamic data for the QA environment rather than to laboriously load dummy dynamic data. Similarly, the configuration for the development environment may take for its basis the IQ from the QA environment, which is equivalent to the live environment.

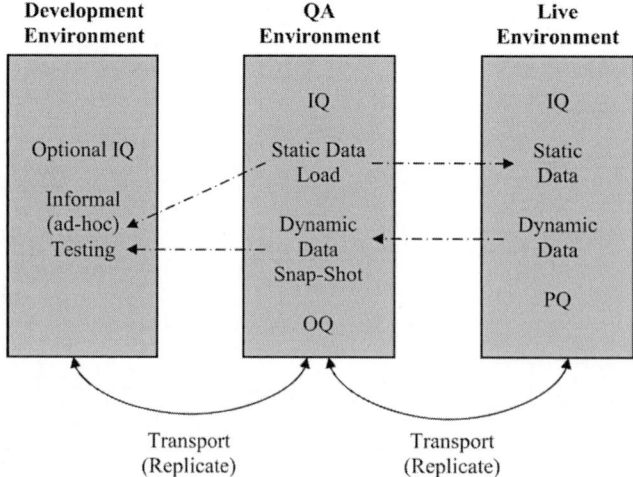

Figure 10.6   Test environments.

Table 10.1 Changing Focus of Testing Through Project Life Cycle

	Development testing		User qualification		
	COTS vendor	System integrator	Installation qualification	Operational qualification	Performance qualification
Test scope	Whole product (hardware and software)	Customization associated with COTS products	Hardware platform, software install, data load, interfaces to other systems	Complete integrated system as it is intended to be used	Complete system in operational environment
Focus	Release certification of product as fit for purpose	Any COTS product configuration, new bespoke (custom) hardware and software	Installation of computer system for use in working environment	GxP critical processes	GxP data and records, operational performance
Test strategy	Comprehensive (white box) testing of product	Extensive (black box) testing of application, including stress testing	Check completeness, confirm that interfaces work	Check user functionality, challenge testing on process level	Confirm user functionality in operational environment

Abbreviation: COTS, commercial off-the-shelf.

Training should be conducted whenever possible within the QA environments. It is likely that training will involve setting up case study situations with supporting dummy records. If the live operating environment is used for training, then care must be taken to restore any records added, modified, or deleted as a result of the training course exercises. Such data manipulation for training purposes, however, is not without risk of human error and the possible impact that it could have in the live environment.

Leverage Development Testing

The scope and depth of user qualification can be reduced if reliance can be placed on the adequacy of the supplier's development testing. Commercially available software that has been successfully tested by its supplier does not require the same level of user testing by the pharmaceutical or healthcare company. Supplier audit and predelivery inspection can be used to provide confidence and evidence in taking this approach.

Table 10.1 shows how the focus of the testing changes as development testing and user qualification progress. Inadequate development testing means that additional user qualification will be expected in compensation. For example, a lack of structural (white box) testing during system development would require more rigorous user testing later on. Structural testing may not be possible, especially for COTS products, so, extensive functional (black box) testing should be considered with significant stress testing. Consideration should be given to the effects of transportation (e.g., system vibration or disassembly of equipment for transit) and any necessary additional user testing required confirming the continued integrity of the system on arrival at its destination.

Parallel Operation

Computer systems replacing manual ways of working should be at least as effective as the older manual process. If they are not, then they should not be authorized for use. It is for this reason that some regulations call for manual ways of working to be run in parallel with the replacement computer system until the hoped-for improved effectiveness is demonstrated. In practice, a back-out strategy for the replacement new computer system is usually developed with procedures as necessary, so that if testing demonstrates that the transition will not be successful, the status quo ante can be restored. Operations can return to the original ways of working, be they manual or automated. It always makes good business sense to have a contingency plan.

Running the legacy system, manual or automated, in parallel with the new system for the period of the process PQ is often not a practical option. In such circumstances, processes, such as additional data checks and report verification, should be temporarily operated in parallel with the computer system until the completion of PQ.

Beta Testing

As indicated earlier, some pharmaceutical and healthcare companies agree to conduct beta testing for suppliers. Beta testing involves customers taking delivery of a system prior to its general release who then use it in its intended operating environment and report any problems experienced back to the supplier. The advantage to the customer is early access to a system or application. The disadvantage to the customer is that there may be yet unknown high impact defects. Beta systems can therefore not be considered as "standard" or fully tested, as we explained earlier. More information on standard systems can be found in chapter 7. Pharmaceutical and healthcare companies must never use beta-ware as part of a validated computer system.

Recent Inspection Findings

No qualification undertaken (IQ, OQ, PQ) (FDA 483, 2006).

The validation manager could not identify which test scripts were intended to demonstrate IQ, OQ, or PQ (FDA 483, 2004).

IQ, OQ, PQ not performed (FDA Warning Letter, 2002).

Failure to exercise appropriate controls over and to routinely calibrate, inspect, or check automatic, mechanical, or electronic equipment used in the manufacturing, processing, and packaging of a drug product according to a written program designed to assure proper performance (21 CFR 211.68) in that the IQ, OQ, or PQ performed for the redacted was not performed (FDA Warning Letter, 2002).

Completed IQ/OQ/PQ data not available for XXXX computer system server (FDA 483, 2002).

No documentation detailing IQ, OQ, and PQ of XXXX system (FDA 483, 2001).

Failure to perform/maintain computer validation in that there was no validation protocol to show how the system was tested and what were the expected outcomes; there was no documentation to identify the operator performing each significant step, date completed, whether expected outcomes were met, and management review (FDA Warning Letter, 2000).

The XXXX form that documents approval to migrate the program to the production environment was not signed off by quality control (FDA 483, 2002).

The firm failed to define or describe the use of the various development, test, and production environments (FDA 483, 2001).

The test report generated from these activities was not approved by the quality unit (FDA 483, 2000).

Firm lacks records of qualification of off-the-shelf software program used to manage regulated information (FDA 483, 2006).

SITE READINESS ACTIVITIES

The physical site of the computer system should be prepared. Some organizations treat such site preparation as part of commissioning.

Site Preparation

The suitability of the operating environment for the computer system to be deployed (6) needs checking against that defined in the system's specification. The physical location should be compliant with any original vendor or system integrator's recommendations. The placement of the computer system, including the building of any special rooms or housing, associated wiring, and power supply voltages, must be confirmed as adequate and in-line with preapproved engineering line diagrams (ELDs).

Instrumentation must be accessible to facilitate operations and be covered by maintenance and calibration schedules (7). Loop checks should be made for instrumentation. Inputs and outputs must be checked to provide strong assurance of accuracy.

Environmental requirements outlined in the hardware design, such as temperature, humidity, vibration, dust, electromagnetic interference (EMI), radio frequency interference (RFI), and electrostatic discharge (ESD), should also be checked in comparison with their acceptable bounds. Once these checks are complete, in situ qualification of the computer system can begin.

Commissioning

The physical installation of a computer system, often known as commissioning, should be conducted according to preapproved procedures. Commissioning records should document fulfillment of any relevant vendor/supplier installation recommendations. Commissioning activities will include

- Interface card addressing checks,
- Field wiring checks (loop testing),
- Input/output continuity testing, and
- Calibration and tuning of instrumentation.

Computer hardware will require electrical earths and signal earths for intrinsic and nonintrinsic safety to be achieved. Wiring diagrams should be available as appropriate to site-specific installations.

Commissioning often involves an element of "snagging" to address any unforeseen issues and fix any installation errors. It should be possible to repeat installation instructions, if this is a more appropriate corrective action. Verification of the installation is documented through a process of IQ.

Calibration

Instrumentation should have its predelivery calibration verified and any remaining calibration set. Calibration should be conducted with at least two known values.

The following advice is based on the ICH Good Manufacturing Guide for Active Pharmaceutical Ingredients (4):

- Control, weighing, measuring, monitoring, and test equipment and instrumentation that is critical for assuring the quality of pharmaceutical and healthcare products should be calibrated according to written procedures and an established schedule.
- Calibrations should be performed using standards traceable to certified standards if these exist.
- Records of these calibrations should be maintained.
- The current calibration status of critical equipment/instrumentation should be known and verifiable.
- Equipment/instruments that do not meet calibration criteria should not be used.
- Deviations from approved standards of calibration on critical equipment/instruments should be investigated. This is to determine if these deviations effect the quality of the pharmaceutical or healthcare products manufactured using this equipment since the last successful calibration.

The GAMP® Good Practice Guide for Calibration Management (8) further suggests the following:

- A calibration master list for instruments should be established.
- All instrumentation should be assigned and tagged with a unique number.
- The calibration method should be defined in approved procedures.
- Calibration measuring standards should be more accurate than the required accuracy of the equipment being calibrated.

- Each measuring standard should be traceable to a nationally or internationally recognized standard where one exists.
- Electronic systems used to manage calibration should fulfill appropriate electronic record/signature requirements.
- There should be documentary evidence that all personnel involved in the calibration process ate trained and competent.

The contents for a calibration master list are suggested below (8).

- Asset
- TAG
- Device description, manufacturer, and serial number
- Device range (must satisfy process requirements)
- Device accuracy (must satisfy process requirements)
- Process range required
- Process accuracy required
- Calibration range required (to satisfy process requirements)
- Calibration frequency (e.g., 6 months, 12 months)
- Device criticality (process critical, product critical, or noncritical)

Calibration certificates should be prepared where they are not provided by third parties and retained as a regulatory record. Many pharmaceutical and healthcare companies are installing computer systems to manage calibration master lists, and calibration certificates that support resource scheduling. Such computer systems should be validated. An example calibration certificate is shown in Figure 10.7.

Self-calibrating features should not be relied on to the exclusion of any external performance check. The frequency of periodic checks on self-calibrating features will depend on how often it is used, its scale, criticality, and tolerance. Typically, annual checks should be conducted on self-calibrating features.

Recent Inspection Findings

Failure to assure computer equipment is routinely calibrated, inspected, or checked according to a written program design to assure proper performance (FDA Warning Letter, 2000).

Inadequate standard operating procedure (SOP) for review and evaluation of calibration reports from outside contractors (FDA 483, 2001).

No procedure for corrective and preventative action when equipment outside calibration range (FDA 483, 2001).

Procedures for calibration of various instruments lacked some or all of the following information: persons responsible for the calibration, specifications or limits, action taken if a test fails, and a periodic review by management (FDA Warning Letter, 2001).

No QA program for calibration and maintenance of the XXXX system (FDA 483, 2002).

There is no documentation that equipment calibration was performed when scheduled in your firm's procedures (FDA Warning Letter, 2001).

Your procedures for calibration are incomplete, for instance, no predetermined acceptance criteria (FDA Warning Letter, 2002).

Failure to maintain calibration checks and inspections (FDA Warning Letter, 2002).

DATA LOAD

The reliance that can be placed in a computer system is fundamentally determined by the integrity of the data it processes. It must be recognized that data accuracy is absolutely vital in the business context. However well an application works, it will be fundamentally

Calibration Test Sheet	Electronic Temperature Transmitter
Department	Complies with Procedure Number:
Service	Temperature Element Serial Number:
	Temperature Transmitter Serial Number:
Location/Use	Control Loop/Tag Number:
	Instrument Range:
	Critical Device ☐ Noncritical Device ☐

Electronic Temperature Transmitter

Manufacturer:	Type and Model:
Process Range _____to_____	Device Accuracy _____ ± ˚C
Calibrated Range _____to_____	Specified Process Accuracy _____ ± ˚C

Calibration

Standard RTD Serial Number	Standard RTD Temperature	Signal Output (mA)	Temp. Output Equiv. (˚C)	Error (˚C)	Pass/Fail

Post Adjustment Calibration

Standard RTD Serial Number	Standard RTD Temperature	Signal Output (mA)	Temp. Output Equiv. (˚C)	Error (˚C)	Pass/Fail

Test Equipment Details

Equipment	Manufacturer	Model Number	Serial Number	Certificate Number
Digital Multimeter				
Standard Reference				
Standard RTD				
Standard RTD				
Standard RTD				

Conclusion

The combination of the above Test Equipment is able to calibrate a device to an accuracy of ....˚C

Comments/Observations:

Test Performed and Recorded by:	Name:	Signature:	Date:
Checked by:	Name:	Signature:	Date:

Figure 10.7   Example calibration certificate.

undermined if the data it processes is dubious. Data load is a key task that must be adequately managed to satisfy business and regulatory needs. Loading of data can be broken down into five basic steps: data sourcing, data mapping, and then data collection, data entry, and data verification.

Data Sourcing

Data sourcing consists of defining, in existing systems or documentation, the master reference (prime source) for the data entities required to support the new system. In some instances, data may need to be created because it does not already exist electronically.

A top-level definition of static data as fixed, and dynamic data as subject to changes is not necessarily as clear as it sounds. Most data actually change in practice, but it is the *frequency* of change that is important when considering what are static and dynamic data. It should be possible to check static data against a master reference to verify that it is correct. No such check can typically be done for dynamic data, because by its nature it changes frequently, so a check can only be made against its last known value. Examples of static dynamic data include recipes and supplier details. Examples of dynamic GxP data include date of manufacture, batch number, notification of deviation, planned change, analytical results, and batch release.

Data Mapping

Data mapping is the process of identifying and documenting, for every field being populated in the new system, where the data is to be found in existing systems (or documents). The mapping of each field will be classified as follows:

Simple: There is an obvious legacy field equivalent, or lack of equivalent, to the new system field.

Complex: There is information in the legacy environment but, before it is suitable for entry into the new system, the field length or format needs to be changed. Perhaps, the field needs to be transformed, several fields need to be combined, a field in the legacy system needs to be split to feed several fields in the new system, or there may be a combination of all or some of these.

Data mapping should consider any electronic record implications such as maintaining audit trails during data migration. Electronic record requirements are discussed in more detail in chapter 13.

Data Collection

The method of data collection is affected by the approach taken to loading data into the new system (i.e., electronic or manual). The criteria used to decide whether to load manually or electronically include

- Whether a standard program exists for the data transfer of the particular business object in the new system,
- The availability of the data in electronic form,
- The number of records that need to be transferred,
- The feasibility within the constraints of the project (e.g., time, available resource with the appropriate skill sets), and
- Expected error rates.

Data Entry

Data entry needs to be verified as accurate against master references (system sources and/or documents). Data from different sources may need to be aggregated during migration, or perhaps some reformatting might be required (e.g., field lengths). The manipulations need to be verified as having been conducted correctly. The creation of backup copies of the original data should be regularly scheduled following defined procedures, to provide a fallback position in the event of problems.

Data Verification

Data verification should immediately follow the data entry and precede any further processing of the data. Where this is not possible, checking must be conducted as soon as possible and a risk assessment must be performed to address the potential consequences of erroneous data input.

Checks are required for transcription errors. Manual data entry errors cannot be assumed to be negligible. If spreadsheets are used as a medium to transfer data, then error rates typically in the range 20% to 40% should be expected. Regulations expect an additional check on the accuracy for quality critical data entered into a system manually (4,6,7). Additional checks on the accuracy of quality critical data can be done by an identified second person or by the system itself.

It may be possible to justify a sample check for particular categories of data where a business case can be made to justify the omission of checks on all records. It is important to recognize that sampling is not appropriate for quality critical data because it could impact product quality and patient/consumer safety.

Recent Inspection Findings

No auditing procedures to ensure electronic data are traceable to original data contained in paper records (FDA 483, 2006)

Input data validation methods not always defined; validation not conducted after XXXX data was migrated to new server (FDA 483, 2002)

No written instruction for entry of raw data into database (FDA 483, 2005)

No established written procedures for manually transcribing raw GMP data to Word document table and onward to Excel spreadsheet (FDA 483, 2004)

No established written procedures for compiling data summary from database to report (FDA 483, 2004)

No record that data transcription is reviewed for accuracy by a second person (FDA 483, 2005)

Further inspection findings can be found for data maintenance in chapter 11.

INSTALLATION QUALIFICATION

IQ provides documented verification that a computer system has been installed according to written and preapproved specifications (9).

The integration of the computer system (hardware, software, and instrumentation) must be confirmed in readiness for the subsequent OQ activity. Some practitioners have referred to this as the *testing of static attributes* of the computer system. The importance of completing the IQ before commencing the OQ can be illustrated by a recent incident in which a pharmaceutical company had over 35% of the instrumentation for a multiproduct plant but did not have available calibration certificates. There were various reasons for this, but none were recorded! Some instruments were no longer used, some had extended recalibration periods, and some had been undergoing calibration for several weeks. The net effect was that the computer system under qualification was clearly not in a controlled state suitable for the OQ, and in consequence, it was not ready for use.

Scope of Testing

IQ should focus on the installation of the hardware platform, software installation, loading of data, and setting up the interfaces to other systems. This will include the following:

- Inventory checks
- Operational environment checks
- Diagnostics checks
- Documentation availability

IQ testing should embrace the test environments as discussed earlier in this chapter. Appendixes 10.A and 10.B provide checklists that may be used in the development of an IQ protocol.

Inventory Checks

FDA and other regulatory authorities require that all major items of equipment be uniquely identified. All the components of the system specified in hardware design should be present and correct including printers, visual display units (VDUs), and touch screens, keyboards, and computer cards. Minimum performance requirements for hardware components should be met or exceeded. The identifying serial numbers and model numbers of all the major items must be recorded. The question as to whether units of equipment need to be dismantled to check their component details is often raised. If a unit is sealed in such a way that dismantling would invalidate the manufacturer's equipment warranty, then disassembly should not be attempted. It is not required in these circumstances. The IQ should simply check the unique identity of the sealed unit. Processing boards that are clip fastened into slots in a rack should have their serial numbers recorded, along with their slot position within the rack. It is worth checking with suppliers in advance of delivery whether their equipment does in fact have unique identifiers.

The correct versions of software must be installed, and appropriate backup copies must be made. The correct versions of firmware must also be checked for their presence. This may include a physical inspection of an electronically programmable read only memory (EPROM) to read its label. The configuration of databases and the content of any library information should also be checked. The last three generations of backup should be retained. The storage medium for the software must be labeled with the software reference name and version. Facilities should exist to store the backup files in a separate and secure place (7). Fireproof cabinets or rooms should be used wherever possible.

Operational Environment Checks

Operating environment checks should include those on power supplies, ambient temperature and humidity, vibration and dust levels, EMI, RFI, and ESD as relevant to the needs of the computer system. This list of operational environment requirements is by no means exhaustive and may be extended or even reduced depending on what is known about the system. EMI and RFI might be tested with the localized use of mobile or cell telephones, walkie-talkie communications receivers/transmitters, arc welding equipment, and electronic drills. The aim is to test the vulnerability of the computer system to interference in situations that must be considered normal working conditions.

Diagnostics Checks

Diagnostic checks are normally conducted as a part of the IQ. Such checks include those of the built-in system configuration, conducting system loading tests, and checking timer accuracy. Software drivers, such as communication protocols, will also require testing.

Documentation Availability

All documentation furnished by the supplier should be available. User manuals, as-built drawings, instrument calibration records, and procedures for operation and maintenance (including calibration schedules) of the system should all be checked to verify that they are suitable. Supplier documentation should be reviewed for accuracy in its specifications of the various versions of software used, and approved as fit for purpose. It is recommended that checks are made to verify that contingency plans, SOPs, and any service level agreements (SLA) are also in place. Any specific competencies supposed to be acquired before the IQ/OQ/PQ through training should also have been achieved—these records should be checked.

Recent Inspection Findings

The IQ protocol stipulated that all required software be installed, but the protocol did not state what software was required (FDA 483, 2002).

Software used "out of the box" without deviation report or investigation into configuration error (FDA 483, 2002).

Headquarters has failed, despite deviations and problem reports, to establish adequate control of software configuration settings, IQ, and validation (FDA 483, 2002).

OPERATIONAL QUALIFICATION

OQ provides documented verification that a computer system operates according to written and preapproved specifications throughout all its specified operating ranges (9).

OQ should only commence after the successful completion of the IQ. In short, it comprises user acceptance testing, for it is necessary to demonstrate that the computer system operates in accordance with the functional (design) specification. Individual tests should reference appropriate functional specifications. Testing should be designed to demonstrate that operations will function as specified under normal operating conditions and, where appropriate, under realistic stress conditions.

An OQ summary report should be issued on completion of OQ activities. Simpler computerized systems may combine the IQ and OQ stages of user qualification into a single activity and document this accordingly. More complex computerized systems may be divided into subsystems, and these then may be subjected to separate OQ. These exercises should then be complemented by a collective OQ demonstrating that the fully integrated system functions as intended.

Scope of Testing

OQ should focus on GxP critical processes. It should

- Confirm that user functionality works, including hazard controls,
- Verify that disabled functionality cannot be accessed,
- Check the execution of decision branches and sequences,
- Check important calculations and algorithms,
- Check security controls—system access and user authority checks,
- Check alarm and message handling—all important error messages designed into the system should be checked so that they appear as intended under their relevant error conditions (it may be wholly impractical to check *all* the error messages),
- Confirm the creation and maintenance of audit trails for electronic records, and
- Confirm the integrity of electronic signatures including, where appropriate, the use of biometrics.

Additional tests demanded or recommended as a result of the findings of the supplier audit, software inspection, or design review activities should also be included. Appendixes 10.A and 10.C provide checklists that can aid in the development of an OQ protocol.

Test Reduction

The OQ may be based on a repetition of a chosen sample of the development testing tests to reduce the amount of OQ testing conducted (6). As discussed earlier, this is only permissible where extensive development testing has been successfully conducted (i.e., without significant defects emerging), and recorded. The suitability of such documentation must be reviewed and approved by QA for this purpose. The test sample for OQ must include, but not be limited to, tests originally conducted as emulations and simulations. Simulation and emulation specifically for qualification should be avoided (5). If the repeated tests of the chosen sample do not meet their acceptance criteria (i.e., if fresh system defects emerge), then the causes of such failures must be thoroughly investigated, and an extended sample of tests must be repeated if confidence in the new system is not to be fatally undermined. The advantage in this approach is that commissioning time on the pharmaceutical and healthcare company's site is reduced, and the system can become fully operational sooner provided all is well. It might be argued that repeating the supplier's development testing does not contribute to an increasing level of assurance of the fitness for purpose of the system. However, practical experience suggests that crucial deficiencies are often discovered in systems even at this late stage in the life cycle. This is very worrying, for obvious reasons—it implies that much of the preceding effort to confirm the innate quality of the system has missed its target. Here are just a few examples of such late-stage failures.

- Backup copies of the application software did not work.
- A computer system froze when too many concurrent messages were generated.

- The operator of a control system would never become aware of concurrent alarm messages as the graphic pages bearing them had banners that only permitted the display of the latest-generated alarm.
- When "on" and "off" buttons were pressed simultaneously, the computerized system initiated an equipment operation.
- Computer software was able to trigger the controlled equipment into operation despite the fact that the hard-wired fail-safe lockout device had been activated.

Verifying SOPs

Operations personnel must be able to use all operating procedures *before* the computer system is used for live use. User SOPs can be used to confirm system functionality. Any competencies required to conduct these tests, including training on user SOPs, should be given and recorded before testing begins.

System Release

Computerized systems are often released into the live environment following completion of OQ. An interim validation report or an alternative document such as a system release note should be prepared, reviewed, and approved to authorize the use of the system. The interim report should address all aspects of the validation plan up to and including the OQ. Several draft validation reports of this kind may be required to phase the roll out of components of the overall system or where a phased roll out is planned to multiple sites.

Recent Inspection Findings

Deficient OQ of computer system in the documents do not have place to record name or initials of performer performing test, rather, they only document that of the person who observed the test being performed (FDA 483, 2005).

Deficient OQ of computer system in the secure data storage, backup, and retrieval functionality tests not performed (FDA 483, 2005).

No testing of the [computer] system after installation at the operating site. Operating sites are part of the overall system and lack of their qualification means that the system validation is incomplete (FDA 483, 2000).

There was no assurance that complete functional testing has been performed (FDA Warning Letter, 2001).

Inadequate qualification in that no power failure simulations were performed as required by the firm's protocol (FDA 483, 2002).

There was no testing of error conditions such as division by zero, inappropriate negative values, values outside acceptable ranges, etc. Testing of special values (input of zero or null) and testing of invalid inputs are not documented. Validation did not test outside limits. The procedure does not call for error condition testing.

Alarm system is unable to store more than XX transgressions, and these transgressions are not recorded (FDA Warning Letter, 2000).

There is no secondary review of alarm events, and any corrective actions are not recorded (FDA Warning Letter, 2000).

Did not verify that computer system would correctly calculate amounts (FDA 483, 2006).

Control program algorithms not described and tested to establish their correct functioning (FDA 483, 2006).

Testing has not included test cases to assess the password security system (FDA 483, 2001).

PERFORMANCE QUALIFICATION

Verifying whether or not a computer system is fit for its intended purpose often means designing tests that are directly related to the manufacture of drug products. PQ, therefore, provides documented verification that a computer system functions correctly, according to written and preapproved specifications, while operating in its specified operating environment (9).

PQ should only commence after the successful completion of the OQ stage. It comprises product performance and/or process PQ. At this stage, the pharmaceutical and healthcare company must demonstrate that the completed installation ("as-built") of the computer system at the site is operating in accordance with the intent of the URS. PQ is sometimes also referred to as a part of process validation, where the computer system supports a production process. A fundamental condition within PQ is that changes may not be made to the computer system during testing. If the need for change emerges as a result of test failures, PQ must be repeated in its entirety. The underlying principle here is that the change may have disrupted system stability and reproducibility.

Scope of Testing

PQ should focus on GxP data and records, and operational performance. It must prove that

- GxP records are correct and
- Automated processes are reproducible.

The degree of testing will also be influenced by the amount of OQ testing already conducted. Appendixes 10.A and 10.D provide checklists that can be used to assist the development of a PQ protocol.

Product Performance Qualification

Product PQ is a quality control activity that aims to verify the correct generation of GxP records. A matrix approach might be required to cover the practical range of acceptable variations. Some examples of product PQ tests are

- Creation of batch reports (start-up, sequencing, and closeout of consecutive batch processes),
- Data/analysis checks of custom user reports,
- Structure and content checks for label variants, and
- Checks of presentation details on product packaging variants.

Batch reports for PQ include batch records (e.g., those relating to key manufacturing steps such as media fills, cleaning), product release (i.e., sentencing), packaging (including labeling), and product batch distribution records (for batch tracking and recall). The PQ for multiproduct applications should cover necessary variants. The PQ exercise should test the system's operation in handling a minimum of three production batches, or five for biological applications. The number of consecutive batches required is, however, not fixed and will depend on the process being validated.

The content and format of batch records must be defined within the system specification. Automated batch records must provide an accurate reproduction of master data (7) and deliver a level of assurance equivalent to a double manual check, bearing in mind that manual checks can identify and record unexpected observations (7,10). Computer systems releasing batches must be designed to demand an authorization for each batch, and the identity of responsible person giving this must be recorded against the batches (6,7,11). All batch records require quality control inspection and approval prior to release and distribution of the product (7). The identity of operators entering or confirming data should be recorded. Authority to change data and the reasons for such changes should be recorded in an audit trail. Similar requirements apply to labeling and packaging.

Process Performance Qualification

Process PQ is a quality assurance activity that aims to verify that the automated process is reproducible. Process PQ is sometimes referred to as *postimplementation review* and is based on performance monitoring rather than testing. Examples of some process PQ topics are as follows:

- Demonstrating that the correct functionality of the system is not disrupted during acceptable daily, calendar, and seasonal operating environment variations

(e.g., variations in power supply, temperature, humidity, vibration, dust, EMI, RFI, and ESD)
- Demonstrating that an acceptable level of service continuity is achieved (e.g., availability, failure on demand, and reliability)
- Demonstrating the effectiveness of SOPs and training courses
- Demonstrating that the users are being adequately supported (e.g., through a reduction in the rate of enquiries received from them, with a decreasing number of outstanding responses/resolutions to their questions)

Variations in temperature and humidity might be monitored over a period of time using a portable chart recorder as part of the PQ. Vulnerabilities to ESD, vibration, and dust are more difficult to measure. All that may be possible in this context is to periodically review whether these have affected live operations in any way. If this is the case, then it should be clearly stated and the causes should be followed up as part of the ongoing support program for maintaining operational compliance.

Service organizations should set up processes to collect and analyze operational performance data. Examples of performance charts are shown in chapter 11. Performance metrics to be tracked and acceptable service levels to be met should be specified in SLAs. Performance charts might include monitoring the training and help desk activity as indicted in the bullet points above.

Recent Inspection Findings

PQ not completed for system software. Activities not completed include review of discrepancies and deviations, and review of audit trail log (FDA 483, 2005).

Testing was not conducted to insure that each system configured could handle high sampling rates. Validation of the system did not include critical system tests such as volume, stress, performance, boundary, and compatibility (FDA Warning Letter, 2000).

Further related inspection findings can be found in section "Performance Monitoring" in chapter 11.

AUTHORIZATION TO USE

Pharmaceutical and healthcare products should not be released to the market where the processes and equipment used to manufacture them have not been properly validated. This includes necessary validation of computer systems. Annex 11 of the *European Guide to GMP* imposes specific rules regarding the validation of computerized systems (6) when these are used for recording certification and batch release (12). The only possible exception to this rule should be when all of the following criteria are met:

- The pharmaceutical medicines and healthcare products (e.g., medical devices) concerned are for life-threatening diseases or situations.
- There is no equivalent pharmaceutical or healthcare product available in the marketplace.
- The supply of available treatments or medicines has fallen to a critically low level.

In such extreme situations, justifications for releasing pharmaceutical and healthcare products to market under these most exceptional conditions must be fully documented by responsible personnel, approved by their senior management, and agreed in advance with relevant regulatory authorities.

Validation Report

Validation reports are prepared in response to validation plans. Their purpose is to provide to management a review of the success of the validation exercise and any concessions made

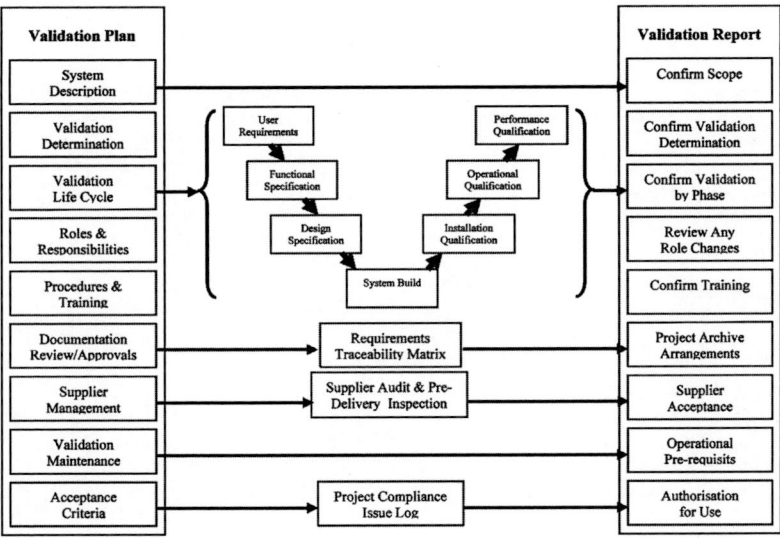

Figure 10.8 Relationship between validation plans and reports.

during it. The objective of the report is to seek management endorsement of the completion and acceptance of the validation conducted. Validation reports may also document failed validation and instruct design modifications and further testing. FDA and other regulatory authorities may request a translation if the original document has been drafted in a language other than English, so that their inspectors can scrutinize the document themselves during an inspection.

Validation reports should be prepared by the person instructed and authorized by management to do so in the validation plan or in another relevant procedure. Where this is not the case, then the authority under which the report is written should be stated.

It is recommended that validation reports follow the same structure as its corresponding validation plans, so that the two documents can be read side by side and the course of the compliance activities can be followed step by step. Figure 10.8 illustrates this relationship.

The *GAMP®5 Guide* suggests that the validation report should include the following information (9):

- Reference to the controlling specification for the phase
- Confirmation that all tests or verification were executed and witnessed (if applicable) by suitably qualified and authorized personnel; this includes all supplier factory testing and site acceptance testing
- Details of any supporting resources involved—names, job titles, and qualifications
- Locale and environment for any testing
- Confirmation of the dates over which the phases occurred, with explanations of delays and actions taken to resolve them
- Confirmation that all tests and activities were subjected to regular project team and QA reviews, with reference to supporting evidence

Each phase should have a clear unambiguous statement drawing a conclusion on the success or otherwise of the activities undertaken. Details of test outcomes, test certificates, documentation, etc., should be included as appropriate to support the conclusion. Test environment(s) should be described in outline, and any test prerequisites should be discussed in case they restrict, or even undermine, the value of activities undertaken. If the validation report is read in this way, then the overall validation conclusion to the completed project should come as no surprise provided each phase has satisfied its predetermined acceptance criteria.

Issue No.	Author and Date Identified	Description	Resolution	Justification	Status
98	E. Thomas 10 Oct 2003	IQ Test Failure – wrong version of application software loaded	No Action Annotate correction to test record and accept test result against original version observed	Test script had typo – correct version of software was loaded actually correctly as required	Closed
99	S. Pattison 22 Oct 2003	OQ Test Failure – Standard reports would not print when requested	Change Control Reference 37 Printer set-up error. Re-configure printer and retest	Not Applicable	Closed
100	B. Neilston 22 Oct 2003	OQ Test Failure – configured function does not save updated records	No Action Software error identified and confirmed by vendor	Function is not used in this application, no impact elsewhere	Closed

Figure 10.9   Example part of a project compliance issues log.

Validation reports should reference the project compliance issue log (or equivalent) and identify and justify each and every issue not resolved by a corrective action during the project. All deviations and interventions to the pharmaceutical and healthcare company's validation plan or supplier's project/quality plan must be recorded, their impact on validation must be assessed, and their root cause must be investigated. Resolutions to deviations must be commensurate with the criticality of the deviation. Further discussion on deviation management can be found in chapter 3.

Figure 10.9 provides an example of part of a project compliance issue log, which can be used within a validation report. The table provides details of the variance, why it occurred, and how it was resolved. It also furnishes a written justification for situations where a corrective action is not possible or appropriate. Similarly, suppliers may supply a report summarizing their own verification work, which can also be referenced by the validation report.

The validation report authorizing use of the computer system should not be issued until all operation and maintenance requirements, including document management, calibration, maintenance, change control, security, and so on have been put in place. It is essential that the validation status of the system does not become compromised. Revalidation will be required if operational compliance controls are not being implemented. The costs of revalidation can be in excess of five times that of ensuring operational compliance in the first place. Management must ensure that their organization's investment in validation is not effectively jettisoned.

QA must approve validation reports. For European pharmaceutical and healthcare companies, this might be the qualified person (13).

Validation Summary Report

Validation summary reports are usually prepared to accompany validation master plans, although this is not necessarily always the case. They provide an executive summary of the validation report and need to be approved by QA. Details of deviations should not be included; the report simply provides a walk through the succession of project stages, identifying key deliverables. The GAMP®5 Guide suggests the following contents (9):

- Provide the mapping of the activities performed against those expected in the validation (master) plan
- Provide a summary of the compliance activities undertaken
- Provide reference to evidence that these activities are in compliance with the stated requirements

System Name	Electronic Batch Record System
Controlling Specification Reference	EBRS/FS/03
Validation Plan Reference	EBRS/VP/02

FINAL SYSTEM VALIDATION APPROVAL

The signatories below have reviewed the validation package for the *[name of the supplier (vendor), and name of system]* computer system. The review included the assessment of the phase reports listed below, including details of the execution of approved test scripts, test phase conclusions based on test phase acceptance criteria, and resolution of items listed in issues log. The determined validated status is derived as a culmination of this review process.

Key Validation Package Documentation	Document Reference	Acceptance Criteria Satisfied (Yes/No)
Supplier Audit	EBRS/SA/01	Yes
Design Review	EBRS/DR/02	Yes
Source Code Review	EBRS/SCR/01	Yes
Predelivery Inspection	EBRS/PDI/01	Yes
Installation Qualification — Peripherals	EBRS/IQ1/01	Yes
Installation Qualification — QA Test Environment	EBRS/IQ2/01	Yes
Installation Qualification — Production Environment	EBRS/IQ3/01	Yes
Operational Qualification — User Functionality	EBRS/OQ1/03	Yes
Operational Qualification — Interfaces	EBRS/OQ2/02	Yes
Operational Qualification — Security	EBRS/OQ3/01	Yes
Performance Qualification	EBRS/PQ/01	Yes
Project Issues Log	EBRS/PIL/12	Yes
Validation Report	EBRS/VR/01	Yes

VALIDATION STATUS DECLARATION

In consequence, we determine that the *[name of system]* has been validated in accordance with requirements of its Validation Plan, and we authorize its use by suitably trained and qualified personnel. We affirm that this system must be maintained in order to preserve its validated status.

APPROVAL DATE

[must be entered after approval signatories below have been added, but prior to first date of use]

Each individual signing below approves the validation status of the [name of system] computer system.

Name	Job Title	Signature	Date
[System Owner/User]			
[Quality and Compliance]			

Figure 10.10   Example format of a validation certificate.

- Confirm that the project documentation has been reviewed and approved as required by appropriate personnel
- Confirm that training has been provided and documented as planned
- Confirm that documentation has been created to show that all the compliance records have been securely stored
- Specify the approach to maintaining operational compliance when the system is in use
- Confirm that all project compliance issued that were logged during the project have been resolved satisfactorily
- Report that the project has been successfully completed

It is sometimes necessary to modify the original intent of a computer system or compliance strategy to some degree to achieve an acceptable outcome. The validation summary report should highlight and justify such changes of direction. As for validation reports, validation summary reports should be made available to the FDA in English.

Validation Certificate

The concept of a validation summary report can be taken a stage further in the form of a validation certificate. Such certificates consist of a one-page summary statement defining any constraints on the use of the computer system. An example of validation certificate is shown in Figure 10.10. Validation certificates are sometimes displayed alongside the computer system itself, where the system is a single discrete item. Certificates for distributed systems do not normally make sense, since there are too many potential points of use alongside which to display such a certificate. Compliance determination statements (described earlier in chap. 5) can be presented to an inspector with reciprocal validation certification as the very highest level of evidence of validation. If validation certificates are produced then they should be approved by QA.

Recent Inspection Findings

Validation records fail to support conclusions in validation report (FDA 483, 2005).

Validation was accepted even though there were unresolved issues (FDA 483, 2005).

Validation report was not approved until two years after validation activity was completed (FDA 483, 2004).

Computer enhancement was identified as needed to correct labeling deviations, but this enhancement was still not implemented over one year later (FDA 483, 2002).

Failure to perform/maintain computer validation in that there was no documentation to show if the validation was reviewed prior to software implementation (FDA Warning Letter, 2000).

The inspection reports that the documents reviewed did not define the system as being validated but was a qualification document (FDA Warning Letter, 2001).

The firm has failed to generate validation summary reports for the overall program throughout its software life cycle (FDA 483, 2001).

The validation summary should include items such as how the system is tested, expected outcomes, whether outcomes were met, worst-case scenarios, etc. (FDA Warning Letter, 2000).

Password master list was made globally available in validation report (FDA 483, 2002).

No validation report was written following execution of validation protocol (FDA 483, 2002).

Validation report was approved, although deviations were not adequately investigated (FDA 483, 2002).

REFERENCES

1. Food and Drug Administration. General Principles for Software Validation, Final Guidance for Industry and FDA Staff. Centre for Devices and Radiological Health, 2002.
2. European Union. EU GMP Annex 15—Qualification and Validation, Guide to Good Manufacturing Practice for EU Directive 2003/94/EC, Community Code Relating to Medicinal Products for Human Use, Vol. 4, 2003.
3. Food and Drug Administration. Glossary of Computerized System and Software Development Terminology. Rockville, MD. 1995.
4. ICH. Good Manufacturing Practice Guide for Active Pharmaceutical Ingredients, ICH Harmonised Tripartite Guideline, November 10, 2000.
5. Wingate GAS. Computer Systems Validation: Quality Assurance, Risk Management and Regulatory Compliance for Pharmaceutical and Healthcare Companies. Boca Raton, FL: Interpharm Press, 2003.
6. European Union. EU GMP Annex 11 Computerised Systems, Guide to Good Manufacturing Practice for EU Directive 2003/94/EC, Community Code Relating to Medicinal Products for Human Use, Vol. 4, 2003.

7. U.S. Code of Federal Regulations Title 21, Part 211. Current Good Manufacturing Practice for Finished Pharmaceuticals.
8. International Society for Pharmaceutical Engineering. GAMP Good Practice Guide—Calibration Management. Tampa, Florida, 2002. Available at: www.ispe.org.
9. International Society for Pharmaceutical Engineering. GAMP®5: risk-based approach to compliant GxP computerised systems. Tampa, Florida, 2008. Available at: www.ispe.org.
10. Therapeutic Goods Administration. Australian Code of Good Manufacturing for Therapeutic Goods, Medicinal Products, Part 1. Woden, Australia, 1990.
11. Food and Drug Administration. Identification of "persons" on batch production and control records, compliance policy guides, computerized drug processing, 7132a, guide 8. Rockville, MD, Centre for Drug Evaluation and Research, 1982.
12. European Union. EU GMP Annex 16—Certification by a Qualified Person and Batch Release, Guide to Good Manufacturing Practice for EU Directive 2003/94/EC, Community Code Relating to Medicinal Products for Human Use, Vol. 4, 2003.
13. Article 12 of EU Directive 75/319/EEC and Article 29 of EU Directive 81/851/EEC.

Appendix 10.A

Example Qualification Protocol Structure
Introduction
Test plan
- Specific areas that have not been tested, with justification for this
- Test procedure explanation
- Action in event of failure
- Logical grouping of tests
- How to record test results

Test requirements
- Personnel
- Hardware
- Software (including configuration)
- Test harness
- Test data sets
- Referenced documents

Test prerequisites
- Relevant documents must be available.
- Test system must be defined.
- Critical instruments must be calibrated.

Testing philosophy
- Witness and tester must be agreed by customer.
- Test results must be countersigned by both witness and tester.

Test procedure format
- Unique test references
- Controlling specification reference (cross-reference)
- Title of test
- Prerequisites
- Test description
- Acceptance criteria
- Data to be recorded
- Further actions

Test procedure execution
- Endorse the outcome as pass or fail.
- Attach raw data.
- Report unexpected incidents and noncompliances.
- Failed tests may be completed or abandoned.
- A change or a repair may trigger a fresh set of tests to verify the patch.

(Continued)

Appendix 10.A   (*Continued*)

Test results file
- Test progress section
- Passed test section
- Failed test section
- Test incident section
- Review report section
- Working copies of test scripts
- Test result sheets and raw data

Test evidence
- Raw data
- Retention of test results
- Method of accepting completion of tests

Glossary
References

Source: From Ref. 9.

Appendix 10.B

Example Installation Qualification Contents
Scope
- Visual check on hardware.
- Power-up and power-down.
- Inventory of software installed (with versions).
- Verify configuration parameters.
- System diagnostic testing.
- Verify acceptable operating environment (e.g., power supply, electromagnetic interference, radio frequency interference).
- Computer clock accuracy testing.
- Check that all the standard operating procedures are in place.
- Check that the documentation has been produced and is available, including the user manuals and calibration certificates.
- Confirm that training has been conducted.

Appendix 10.C

Example Operational Qualification Contents
Scope
- Start-up and shutdown of application.
- Confirm user functionality (trace the test results back to the user requirements).
  - Correct execution of decision branches and sequences.
  - Correct display and report of information.
  - Challenge user functionality with invalid inputs.
  - Verify process control limits work as intended.
- Verify deselected or disabled functionality cannot be accessed or reenabled.
- Check application-specific calculations and algorithms.
- Check security controls—system access and user authority.
- Check alarm and message handling—all error messages.
- Verify that trips and interlocks work as intended.
- Check creation and maintenance of audit trails for electronic records.
- Verify integrity of electronic signatures.
- Ensure that backup, media storage arrangements, and restore processes exist and have been tested.
- Ensure that archive, retention, and retrieval processes exist.
- Check for existence of business continuity plans, including recovery after a catastrophe.
- Verify battery backup and UPS cutin upon a power failure.

Appendix 10.D

Example Performance Qualification Contents
Scope of product performance qualification
- Check batch reports.
  - ➢ Production records against plant logbooks for inconsistencies
- Check data accuracy and analysis for custom user reports.
  - ➢ Cycle counting
  - ➢ Period ending cycles
  - ➢ Inventory reconciliation
  - ➢ Release processes
- Check label variants.
  - ➢ Structure
  - ➢ Content
- Check product packaging variants.
  - ➢ Presentation details

Scope of process performance qualification
- Operability during daily, calendar, and seasonal operating variations.
  - ➢ Environmental (e.g., variations in power supply, temperature, humidity, vibration, dust, electromagnetic interference, radio frequency interference, and electrostatic discharge)
  - ➢ Peak user loading
- Acceptable level of service continuity is maintained.
  - ➢ System availability (planned and unplanned downtime)
  - ➢ Access denial on demand
  - ➢ Security breach attempts
  - ➢ Data performance (e.g., network, database, disk)
- Effectiveness of standard operating procedures and training.
  - ➢ Suitability of standard operating procedures (Be concerned if an avalanche of change requests has appeared!)
  - ➢ Competency assessment scores for recipients of training
- User support.
  - ➢ Reduction in number of enquiries received from users
  - ➢ Number of outstanding responses/resolutions to user enquiries decreasing
  - ➢ Monitor upheld change requests

Appendix 10.E

Example Contents for a Validation Report
Introduction
- Author/organization
- Authority
- Purpose
- Relationship with other documents (e.g., validation plans)
- Contractual status of document

System description
- Confirmation of the identification of the system scope and boundaries (e.g., hardware, software, operating system, network)
- Confirm constraints and assumptions, exclusions and justifications

Compliance determination
- Confirm rationale behind compliance requirement (may be reference to compliance determination statement)
- Confirm rationale updated as necessary to address any changes in system scope

Project life cycle
- Confirm completion of life cycle phase by phrase
  - Identification of specification documentation.
  - Summary of key findings and corrective actions from the design review.
  - Summary of key findings and corrective actions from the source code review.
  - Summary of test results including any test failures with corrective actions from test reports. Summary should cover installation qualification, operational qualification, and performance qualification.
  - Confirmation that all operation and maintenance prerequisites are in place.
- Review project compliance log and satisfactory resolution of items

(Continued)

Appendix 10.E   (*Continued*)

Role and responsibilities
- Review any role changes
- Provide additional CVs (qualifications and experience) as appropriate

Procedures and training
- Confirm that training in standard operating procedures is delivered
- Confirm that training records are updated

Document review and approvals
- Lists all documentation produced that should be readily available for inspection
- Identify RTM where developed
- Confirm project document archive arrangements

Supplier and subcontractor management
- Summary of key findings and corrective actions from any supplier audit reports
- Summary of key findings and corrective actions from any predelivery inspections

Support program for maintaining operational compliances
- Description of how the compliance will be maintained

Conclusion
- A clear statement that the validation plan has been successfully executed with a review of any outstanding actions or restrictions on use of system. All deviations from the validation plan must be justified or resolved.

References
Appendixes
- Glossary
- Others

11 | Operation and Maintenance

INTRODUCTION

In practice, the operation and maintenance of computer systems can be far more demanding than system development. Over the lifetime of a computer system, more money and effort are put into operation and maintenance than that on the original project implementation, and good maintenance can substantially extend the useful life of what are more and more expensive assets. Consequently, the operation and maintenance of computer systems should be a high-profile role. Pharmaceutical and healthcare companies that ignore this are more likely to be forced to replace systems earlier than they need to because their systems have degraded faster as a result of change than they needed too. Degrading system documentation and functionality will also affect the ongoing level of compliance.

This chapter reviews key operation and maintenance activities from a quality and compliance perspective.

- Performance monitoring
- Repair and preventative maintenance
- Upgrades, bug fixes, and patches
- Data maintenance
- Backup and restoration
- Archive and retrieval
- Business continuity planning
- Security
- Contracts and service level agreements (SLAs)
- User procedures
- Periodic review and revalidation

Reliable operation does not indicate that a computer system is compliant, although such evidence can be used to support validation. Regulatory authorities uncovering operational issues concerning a computer system during an inspection are likely to follow up with a detailed inspection of system validation. Such inspections are often referred to as "for cause" and are discussed in more detail in chapter 14.

PERFORMANCE MONITORING

The performance of computer systems should be monitored to establish evidence that they deliver service levels required. The intent is also to anticipate any performance problems and initiate corrective action as appropriate. Performance monitoring can be seen as an extension to process performance qualification. A key step is the identification of appropriate performance parameters to monitor.

Performance Parameters

Depending on the risks associated with an application, the type of computer systems, and the operating environment, the following system conditions might be checked:

Servers/Workstations/PCs

- CPU utilization
- Cache memory utilization
- Disk capacity utilization
- Interactive response time

- Number of transactions per time unit
- Average job waiting time
- Print queue times
- I/O load
- System alarm/error messages
- Condition/readiness of business continuity measures
- Trip count for uninterruptable power supplies (UPSs)

Network

- Availability of components (e.g., server and routers)
- Network loading (e.g., number of collisions)

Applications

- Monitoring application error/alarm messages
- Response times

Procedures should exist describing monitoring activities, data collection, and analysis. Operational observations are typically recorded in logbooks with the time and date, comment, and signature of the person making the observation. Some logbooks also have entries noting any corrective action (perhaps the reference to a change request) against the observation. Key performance metrics may also be plotted to give a visual presentation of performance (Fig. 11.1).

Statistical analysis such as statistical process control (SPC) may be used to derive performance parameters as well as track and trend for alert/alarm conditions. Automated monitoring tools may be available to assist in the collection of relevant data. A record of any such tools used should be maintained, and any validation requirements should be considered.

Status Notification

The notification requirements of out-of-specification results will vary depending on the criticality of the deviation. Some deviations may need immediate attention such as alerts identifying the loss of availability of I/O cards or peripheral devices. Other observations such as the above recommended disk utilization will gather information to be used by periodic reviews. All parameter deviations should be diagnosed, and any corrective action should be progressed through change control.

The mechanism employed to notify the status of monitored parameters should be carefully considered. The timeliness of communication should be commensurate with the degree of GxP risk particular parameters pose. All deviations on GxP parameters affecting product quality must be reported to QA. Example notification mechanisms include

- Audible or visual alarms,
- Message on the system console,
- Printed lists or logs,
- Pager message to system operators,
- E-mail to system operator,
- E-mail to external services, and
- Periodic review.

Procedures and controls must be established to ensure that status notification is appropriately handled. For instance, distribution details must be maintained to ensure that e-mails are received by the right people. Validation of specific notification mechanisms may be appropriate.

Monitoring Plan

A monitoring plan should be developed to identify parameters to be monitored and specify the warning limits and frequency of observation. The time intervals and warning limits for

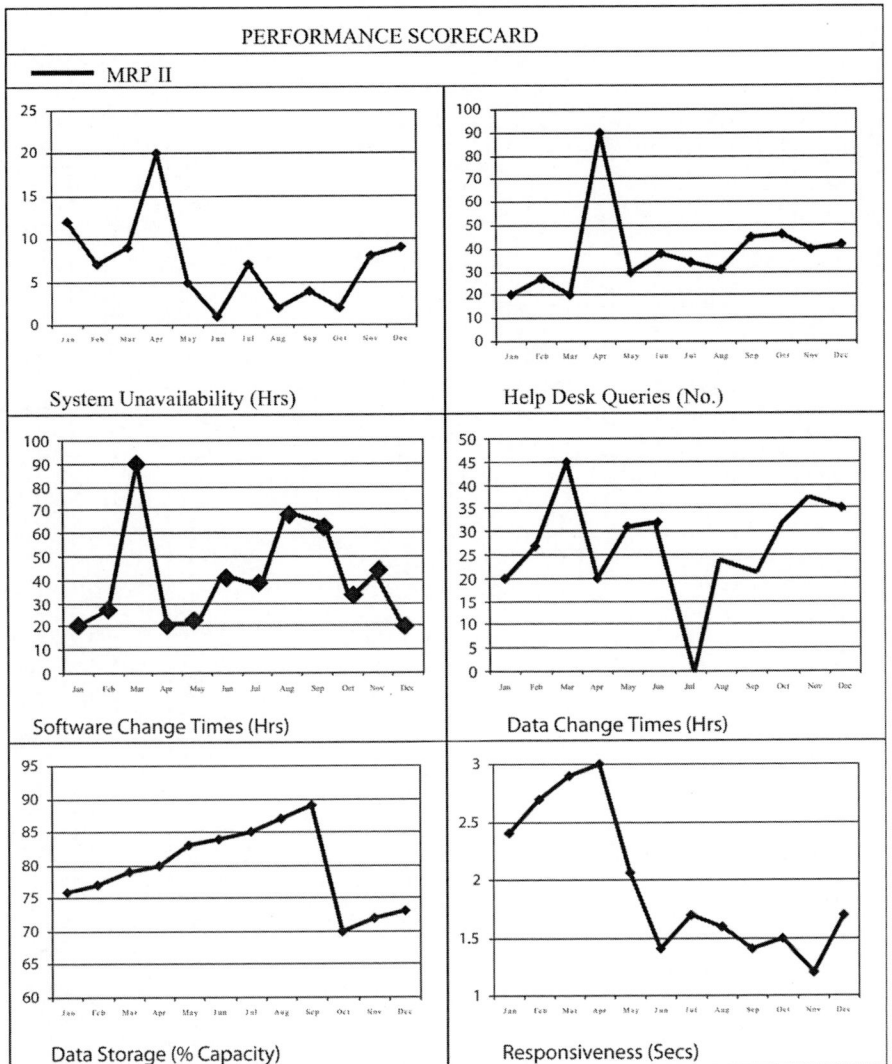

Figure 11.1 Example performance scorecard.

monitored performance parameters must be adequate to take corrective timely action where appropriate. Regulatory expectations will be invoked when certain phrases are used to describe monitoring intervals. Frequent typically indicates hourly or daily. Regular typically indicates weekly or monthly. Periodic typically indicates quarterly, annually, or biannually.

Some firms use reliability-centered maintenance (RCM) as part of their preventative maintenance strategy.

The GAMP®5 Guide (1) suggests a tabular format for monitoring plans (Table 11.1). The structure of the table includes identification of the monitored parameter with warning limit, frequency of observation, monitoring tool, notification mechanism, when and where results are documented, and the retention period for these results. Monitoring records should be maintained and retained for appropriate predefined retention periods in a safe and secure location.

Recent Inspection Findings

Routine checking of computer system is not performed according to a written program designed to assure proper performance (FDA 483, 2007).

Table 11.1 Example Monitoring Plan for Server-Based LIMS

Monitored parameter	Warning limit	Frequency of observation	Monitoring tool	Notification mechanism	Where monitoring records are documented	Retention period
CPU utilization	Average over 25% in 24-hr period	Every 10 min	System procedure	System console	File with 24-hr CPU statistics	6 mo
Disk filling grade	Over 90%	Hourly	System procedures	E-mail to system operator	E-mail directory	30 days
System error message	Error count increased by severe system error (defined in the tool)	Every second	Tool "CheckSys"	Message to operator pager with error number	According to standard operating procedure "Problem Management"	According to appropriate GxP regulations
Critical batch jobs	If batch job is lost	Every 10 min	System procedure	E-mail to system operator; automatic restart of batch jobs	E-mail directory	30 days

- All monitor jobs
- Fullbackup.com
- Dircheck.com
- Check print_queues.com
- Stop_database.com
- LIMS

Monitored parameter	Warning limit	Frequency of observation	Monitoring tool	Notification mechanism	Where monitoring records are documented	Retention period
Critical processes	If process is not running	Every minute	Tool "CheckSys"	E-mail to system operator	E-mail directory	30 days

- LIMS
- Pathworks
- Oracle
- Perfect disk
- UCX
- DECnet
- Security audit

Not all critical alarm reports describe the investigation, provide an assignable cause for the alarm, or describe the corrective actions performed, conclusions, and final recommendations (FDA 483, 2001).

No corrective/preventative action taken to prevent software errors due to the buildup or temporary files in computers used to control [computer system] (FDA 483, 2001).

No controls or corrective action after frequent XXXX software errors causing computer lockup (FDA 483, 2001).

Personnel will receive their XXXX via an e-mail that has been sent from an e-mail distribution list. The firm has failed to implement controls to document that these distribution lists are maintained updated with the current approved list of users (FDA 483, 2001).

Further related inspection findings can be found at the end of the chapter in the section dealing with deviation management.

REPAIR AND PREVENTATIVE MAINTENANCE

Routine repair and maintenance activities should be embodied in approved standard operating procedures (SOPs). Instrumentation, computer hardware elements, and communication network components should all be covered. The following areas should be addressed:

- Scheduling maintenance
- Scheduling calibration
- Recommended spares holding
- Documentation

Scheduling

The frequency of maintenance should be defined in these SOPs and, unless otherwise justified, comply with the OEM's recommendations. Maintenance frequencies may be determined by recalibration requirements and reliability-centered preventative maintenance calculations. Advice can be sought from supplier organizations, but it should not be solely relied on because it is highly unlikely that they do not fully understand the precise nature of the pharmaceutical or healthcare application. Justifications for periodic inspection intervals should be recorded, remembering that they can be modified in the light of operational experience. Any change to recalibrations periods or preventive maintenance intervals, however, must be controlled.

Repair and maintenance operations should not present any hazard to the pharmaceutical or healthcare product (2). Defective elements of computer systems (including instrumentation and analytical laboratory equipment) should, if possible, be removed from their place of use (production area or laboratory bench) or at least be clearly labeled as defective. It is unlikely, unfortunately, that the precise time of failure will be known. This often leaves operations staff with a dilemma of what to do with the drug products that might or might not have been made when the computer system was defective. Indeed, was there an initial partial failure and a period of degraded operation before any fault was recognized? No specific guidance can be given except to consider the merits of each situation case by case and ensure that a quality check is performed on product batches made during the period when the computer system is suspected of malfunction or failure. It is best to play safe when considering the number of product batches that are potentially substandard and assume worst-case scenarios.

Calibration

Calibrated equipment should be labeled at the time of each calibration with the date of the calibration and next calibration due date. This label facilitates a visual inspection of equipment to check whether it is approaching its next calibration date or is overdue. The label should also include as a minimum the initials of the engineer who conducted the calibration. Some companies also include space for a full signature and printed name, but this should not prove necessary if initials are legible and can be traced to the appropriate engineer. Where labels are used, however, care must be taken to apply them to a clean dry area so that they do not fall off. Labels should be considered as an aide memoir, with the master record being kept elsewhere (perhaps handwritten in a plant logbook or a calibration certificate in an engineering management system) in case the labels become detached. Calibration procedures must be agreed on and, wherever appropriate, refer to national calibration standards.

The GAMP® Good Practice Guide for Calibration Management (3) adds the following regulatory expectations:

- Each instrument should have a permanent master history record.
- All instrumentation should be assigned and tagged with a unique number.
- The calibration method should be defined in approved procedures.
- Calibration frequency and process limits should be defined for each instrument.
- There should be a means of readily determining the calibration status of instrumentation.
- Calibration records should be maintained.
- Calibration measuring standards should be more accurate than the required accuracy of the equipment being calibrated.
- Each measuring standard should be traceable to a nationally or internationally recognized standard where one exists.
- All instruments used should be fit for purpose.
- There should be documentary evidence that all personnel involved in the calibration process are trained and competent.
- A documented change management process should be established.
- Electronic systems used to manage calibration should fulfill appropriate electronic record/signature requirements.

A nonconformance investigation should be conducted when a product quality critical instrument is found out of calibration or fails a recalibration. The investigation process should include the following steps (3):

- Previous calibration labels/tags should be removed where applicable.
- An "out-of-calibration" label should be attached to the instrument.
- The failure of the instrument should be logged, and this information should be made readily available.
- A nonconformance report should be raised for the failed instrument before any adjustments are made.
- The action to repair, adjust, or replace the instrument should be followed by a complete calibration.
- The QA department should be informed to investigate the potential need for return or recall of manufactured/packaged product.
- The nonconformance report should be completed, approved, filed, and retrievable for future reference.

Spares Holding

A review should be conducted on the ready availability of spare parts. The availability of some spare parts may be restricted. Special arrangements should be considered if alternative ways of working are not possible while a computer system awaits repair. There may be a link here to business continuity planning discussed later in this chapter.

Spare parts should be stored in accordance with manufacturer recommendations. Model numbers should be clearly identified on spare parts. Version numbers for spare parts containing software or firmware should also be recorded so that the correct part is retrieved when required.

Care should be taken when considering the use of equivalent parts for superseded items. The assumption that the change is "like for like" is not always valid. A medical device company operating in the United Kingdom, for instance, once bought replacement CPU boards for their legacy analytical computer systems. The original boards had a 50-Hz clock, but the replacements came from the United States with a 60-Hz clock. Unfortunately, it was a time-critical application, and the problem was only discovered after a computer system had been repaired and put back into operation. Another medical device company in the United States recalled a workstation associated with a medical system because a so-called equivalent video display unit reversed the left/right perspective of medical image data. This image reversal could potentially have lead to erroneous medical diagnosis. Not all "like-for-like" changes are as dangerous as these examples, but they do illustrate the point not to assume that there will be no impact of change, hence, the recommendation that evidence of equivalence needs be collected (e.g., supplier documentation or supplementary user qualification) and retained.

Documentation

Maintenance and repair documentation may be requested during an inspection by a GMP regulator. Documentation for maintenance activities must include a description of the operations performed, who conducted the maintenance and when, and the results confirming that the maintenance work was completed satisfactorily. Calibration certificates should be retained. Repair records, meanwhile, should include a description of the problem, corrective action taken, acceptance testing criteria, and the results confirming that the repair work has restored the computer system to an operational state. Repair logbooks can be used to record nonroutine repair and maintenance work.

Records should be kept regardless of whether or not the work was conducted by a contractor service supplier. If such engineering support is provided by an external agency using their own procedures, then those procedures must be subject to approval by the pharmaceutical or healthcare company before they are used. Repair logbooks should note visits form external staff, recording their names, the date, and the summary work conducted so that additional information held by the supplier can be traced in future if necessary. It is important that service arrangements defining when suppliers are used by pharmaceutical or healthcare companies to conduct maintenance and repair work are formally agreed. Such agreements are often embedded in contracts called SLAs. The GAMP®5 Guide promotes the development of a maintenance plan to define roles and responsibilities (1). It is unacceptable to the GMP regulatory authorities not to have documentary evidence demonstrating management control of these activities.

Recent Inspection Findings

No software or hardware maintenance is routinely performed on the GMP computer systems (FDA 483, 2008).

No written procedures for maintenance of QC computer system (FDA 483, 2003).

No documented maintenance procedures (FDA 483, 2002).

Failure to perform/maintain computer validation in that there was no documentation to show if problems were experienced during the process and how they were solved (FDA Warning Letter, 2000).

No calibration was performed prior to [system] use (FDA Warning Letter, 2000).

Your firm does not have a quality assurance program in place to calibrate and maintain equipment according to manufacturer's specifications (FDA Warning Letter, 2000).

Calibration records not displayed on or near equipment and not readily available (FDA 483, 2001).

UPGRADES, BUG FIXES, AND PATCHES

This section concentrates on software upgrades, bug fixes, and patches. It is important to appreciate some basic practicalities of what happens in real life when considering compliance activities.

Why Upgrade?

When upgrading software, it is prudent to establish why the upgrade is necessary. Practitioners usually cite one or more of the reasons below.

- Vendors do not support earlier version.
- Upgrading to establish common operating environment between new and existing systems.
- Are you hoping that the upgrade will fix bugs in the existing product you have already bought?
- Are you wanting to use new features promoted as part of the upgrade?
- Do you really need the new features offered as part of the upgrade?
- How many known bugs are associated with these new features?

User licenses can give suppliers the right to withdraw support for their products as soon an upgrade becomes commercially available. This effectively forces users to upgrade immediately. The latest PIC/S computer validation guidance recommends that unsupported computer systems should be withdrawn from service (4).

Most suppliers will support their respective hardware and software for at least the three latest versions. If an entirely new product supersedes an existing product, then there is usually some period of grace to migrate to the new product. Some suppliers, however, have deliberately built in discontinuity into their product upgrades. This aspect should be carefully considered. Upgrading software may also necessitate upgrading hardware, disk size, and processor. Equally, upgrades to hardware may require a supporting upgrade to software.

To maintain a common operating environment, the existing systems need to be upgraded. The networked computer systems in many organizations, for instance, are moving toward the use of a standardized desktop configuration. It can be very difficult to run two or more versions of the same software product across the network.

If the case for an upgrade is based on a new feature, then check when will the new feature be delivered. Quite often, the scope of a new release is cut back to meet shipping dates. Remember too that new features will have their own bugs. Try and use market-tested software. Do not feel the urge to upgrade to be at the leading edge unless there is a compelling business case. Pharmaceutical and healthcare companies should consider waiting until the software has developed some kind of track record. A typical waiting period might be six months for a widely used piece of software. Where a pharmaceutical or healthcare company consciously decides to be an early adopter, additional development testing and user qualification are likely to be required to establish confidence in the software.

Bug Fixes and Patches

Software firms knowingly release their products with residual bugs. Remember that it is not practical to test every single aspect of the software's design, refer to chapter 11. Patches to large software products like MRP II may contain many hundreds of bug fixes. Such large patches should not come as a surprise, remember that average, commercial programs have about 14 to 17 bugs with various degrees of severity per thousand lines of software. MRP II/ERP products can have many millions of lines of code.

Programmers typically rely more on actual program code than on documentation when trying to understand how software works to implement a change. It is easy to miss potential impacts of changes on seemingly unrelated areas of software when relying on the personal knowledge and understanding of individuals rather than approved design documents. Programmers also often take the opportunity when making a change to make further modifications that are not specifically authorized or defined in advance. Not too surprisingly, up to one in five bug fixes in complex software can lead to the introduction of a further new bug, the so-called software death cycle.

The adoption of good practices such as those defined by GxP validation should improve software quality. Original document sets need to be reviewed after a number of changes have been implemented to see if a new baseline set of documents needs to be generated. The URS typically only requires updating when a significant change is made if the URS is written at a suitable high level, as discussed in chapter 6. Bug fixes and patches should not normally require a URS update. Functional specifications may need updating to reflect changes imposed by patches, but they do not normally require modification to address bug fixes. Design documents, meanwhile, may be the only documentation affected by bug fixes.

Pharmaceutical and healthcare companies should evaluate whether or not to immediately take a patch when it first becomes available. Patches should only be taken if they support the bug fixes needed. Unless there is a driving operational requirement to apply the patch, it is recommended to wait and evaluate the experience of other firms applying the patch just in case the patch includes new bugs that make the situation worse rather than better. It may also be more effective to implement a number of patches together rather than individually.

Major upgrades may be required to implement specific bug fixes. Upgrades tend to be feature focussed, not quality focussed, in an attempt to attract new users. If a specific bug fix is required and it is critical to many customers, then there is a good chance that it will have been addressed. Suppliers typically prioritize bugs, especially for large applications, in an attempt to fix all critical bugs for a new release.

Installation and Validation

When a major upgrade is being planned, it is worthwhile considering bringing forward the next scheduled periodic review to determine whether any revalidation can be combined with the upgrade effort. Revalidation is discussed in more detail later in this chapter. Patches and bug fixes, meanwhile, are typically managed on the basis of a change control and an installation qualification (IQ). In either case, the scope of change needs to be understood prior to installation and validation. Supplier release notes should be consulted.

Some operational qualification (OQ) activity may be required to verify the upgrade—confirming that old and new functionalities are available and work. In addition to directly testing the change, sufficient regression testing should be conducted to demonstrate that the portions of the system not involved in the change were not adversely impacted. Existing OQ test scripts may be suitable for reuse with the savings that they bring. The amount of OQ testing will depend on the complexity and criticality of the computer system and the supplier's own release management of the new version. If the supplier has conducted rigorous testing, then the pharmaceutical and healthcare company's OQ can be limited to a selection of functional tests confirming key operations. Do not assume, however, that supplier activities have been conducted without evidence, perhaps from an earlier supplier audit.

Before installing an upgrade, patch, or bug fix, a backout strategy should be defined with approved procedures as appropriate. If the installation is in trouble, users will be keen to return to the original computer system while the upgrade, patch, or bug fix is reevaluated. It is often not practical to roll back and reinstall the original hardware or software once an upgrade has been conducted even when the upgrade brings severe problems. The cost of an organization rolling back to an original installation often far outweighs the money back for the purchase price of the upgrade. The message is clear in regard to implementing upgrades: do not implement automatically, look before you leap.

Upgrade Considerations

When deciding whether or not to upgrade, it is important to take account of the following issues:

- New version functionality should be downward compatible with the previous version(s).
- New versions should be able to process data migrated from the previous version(s).

Suppliers usually make sure that their products are backward compatible so that legacy systems can be seamlessly replaced by new systems. Suppliers typically develop their upgrades for use on the same hardware platform. Full compatibility, however, is more than this. The new product must have all the functionality that the old product had. New functions can be added, but previous functions must not be removed. For example, newer versions of word processing software typically can read formatted text documents written on older versions.

With every software upgrade, either of the application or of an operating system, the validity of previously recorded data files should also be checked. This can be achieved by comparing the data derived from a legacy system with the data derived from the system upgrade.

Operating System Upgrades

Figure 11.2 shows a flowchart of compliance considerations for operating system upgrades. Operating systems are considered to include software defined by GAMP category 1 software. Examples include OS/2, MPE, Microsoft Windows, Fix DMACS, and system software such as database management products like Oracle.

The preferred solution is to maintain a version of operating system software certified as suitable by the supplier of the application using the operating system. It is recognized that this is not always possible. For instance, the application supplier might go out of business or be strategically positioning to migrate users to a successor product.

Applications whose supplier has certified the operating system upgrade will require minimum testing. Suppliers should be encouraged to certify operating system upgrades if certification is not already available. It may be in the interests of the pharmaceutical manufacturer to share the costs of certification if this reduced overall upgrade costs and improves regulatory compliance. It is unlikely that the pharmaceutical manufacturer will have sufficient operating system and supplier application knowledge to conduct a design-level impact assessment. It is far better for the application supplier to do this.

If a criticality assessment is conducted and determines that the operating system upgrade has a significant effect of application operability (i.e., critical change), then the upgrade should not be implemented. A noncritical upgrade can be accepted if additional testing based on a risk

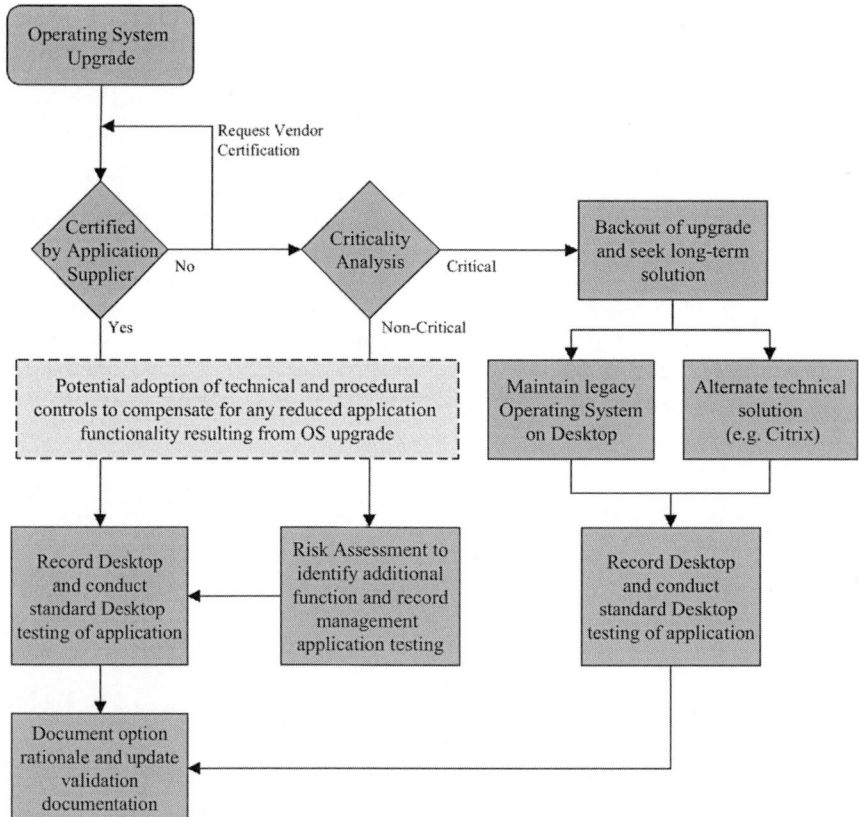

Figure 11.2 Operating system upgrades.

assessment is conducted. Full application operability may no longer be available with the upgrade, in which case, technical and procedural controls can be introduced to compensate for it. It is assumed that any additional operating system functionality will not be available unless the end application is modified to exploit this. In either situation, the desktop build with any configuration must be recorded and tested. Standard desktop tests will include

- Application launch test,
- Desktop conflict testing to determine whether resident applications interfere with each other, and
- Application regression testing prompted by operating system release notes.

Extra tests identified by the risk assessment will involve checking critical application functionality, management of regulated records and signatures, and any newly invoked technical and procedural controls.

The decision to maintain a legacy operating system on the application desktop should not be taken lightly. The legacy operating system will be needed until the user application is replaced. This may be many years away, and the operating system may fall out of support during this period. Careful consideration should be given to business continuity issues if problems are discovered later involving the software and there is no formal support agreement in place.

Beta Software

Many software vendors distribute early versions of their software, called beta versions, usually for free to interested customers. This software is still under test and must not be used to support regulated pharmaceutical and healthcare operations. Users of beta software are supposed to help the software vendor by reporting bugs they discover. The software vendor makes no promises to fix user-discovered bugs before final release of the product concerned. For the likes of Microsoft, it has been suggested that 90% of the bugs reported against beta software are already known by the vendor. Cynics have suggested that beta testing is a marketing ploy to make potential customers think of themselves as stakeholders in the success of the new product release. Fifteen percent of software firms do no formal testing, relying, instead, entirely on beta testing, before releasing their products to market.

Emergency Changes

Exceptional circumstances may require changes to be made very rapidly. In an emergency situation, there is but one thing that matters: getting the system up and running as soon as possible. Because of time constraints at the time when the emergency change is made, it may be necessary to complete and approve documentation retrospectively and therefore proceed while accepting a degree of risk. It is important to consider that for some critical systems that could impact public health, it may be better to stop operations and go through a formal prospective change control process rather than accepting the risk associated with an emergency change.

If emergency changes are allowed to occur, the process must be defined in an approved procedure. The use of this procedure should be monitored to ensure that it is not abused by being deployed for nonemergency changes.

Figure 11.3 depicts the so-called emergency change scenario in which changes are made to software: it is recompiled and deployed into use before associated documentation (detailed design and, where appropriate, functional specifications) is updated. It is unlikely that testing will be conducted to preapproved test specifications; rather, release is based on a documented conclusion to informal testing. Backout plans will be needed until formal testing is successfully completed just in case the emergency change needs to be withdrawn.

It is important to note that the structure of the software can degrade quickly when such quick fixes are made because of the danger that followup specification and testing will not catch up and properly control and document the change. If emergency changes are not managed properly, future maintenance becomes more and more difficult. Document maintenance activities should be planned to repair any structural degradation incurred.

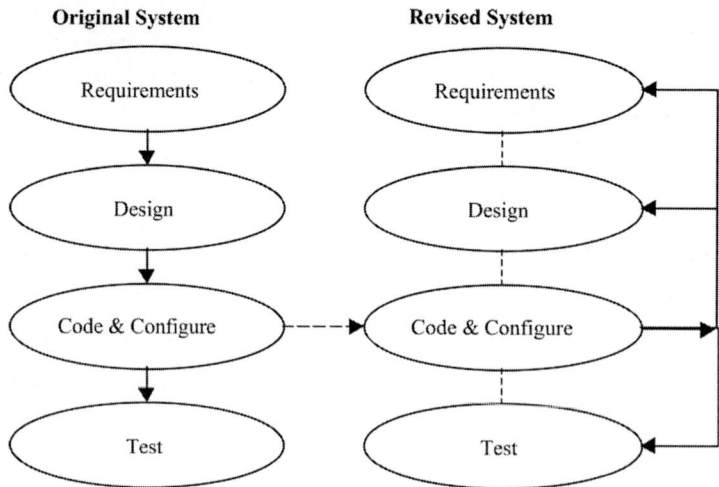

Original System   **Revised System**

Figure 11.3   Emergency change process.

Availability of Software and Reference Documentation

All custom (bespoke) software source codes must be available for regulatory inspection [e.g., OECD recommendation (5)]. Relevant COTS product reference documentation should also be available for inspection, recognizing that propriety COTS source code is usually only available at the supplier's premises, and access may not be available for regulatory inspection.

Copies of retained software must be stored in safe and secure areas, protected within fire safes. Where access to software is restricted, formal access agreements should be established, for example, ESCROW accounts. Responsibility for the maintenance of the copied software and keeping reference documentation up to date and its duration of storage must be agreed.

Prioritizing Changes

The risk assessment process presented in chapter 7 can be used to assist scheduling change requests. Without such an approach, prioritizing changes can become a cumbersome activity and, in extreme circumstances, use vital resource that would be better focused on implementing change. Care must be taken when applying the risk assessment process, however, because the data associated with a change could alter whether or not its associated function is critical. For instance, using an active ingredient without an associated batch number is more significant than using a pencil without an associated batch number.

Recent Inspection Findings

No documentation that software changes and bug fixes in product software did not affect other components in the system and did not require a full revalidation of the entire software (FDA 483, 2007).

Multiple software fixes were incorporated from version to version, but no performance validation studies were performed to verify acceptability (FDA 483, 2004).

The program was not controlled by revision numbers to discriminate one revision from another (FDA Warning Letter, 2001).

There were no written SOPs for hardware and software change control and software revision control (FDA 483, 2001).

Although the firm has in place change control for program code changes, the quality unit has failed to put in place procedures to ensure that the system design control documentation XXXX is updated as appropriate when program code changes have been made. Design control documentation has not been updated since the initial release *(three years ago)* (FDA 483, 2002).

Software used "out of the box" without deviation report or investigation into
configuration error (FDA 483, 2002).

Documentation regarding system and software changes not fully available (FDA 483, 2004).

DATA MAINTENANCE

Data maintenance is required throughout the data life cycle (Fig. 11.4). Data may be captured
by a manual or automatic input. User procedures are required for manual data input, and their
effectiveness should be audited. Software supporting automated data input such as that used
for data acquisition by instrumentation or data migration tools requires verification. Checks
should include confirming whether any necessary calibration has been conducted and if
interfaces are working correctly.

It is important to appreciate that some data may be transient and will never be stored to
durable media, while other transient data may be processed to derive data before being stored.
Both transient and stored data must be protected from unauthorized, inadvertent, or malicious
modification. It is expected that a register of authorized users, identification codes, and scope
of authority of individuals to input or change data is maintained. Some computer systems
"lock down" data, denying all write access. Security management is discussed in more detail
elsewhere in this chapter.

Data Protection

It is important to appreciate that some data may be transient and will never be stored to
durable media, while other transient data may be processed to derive data before being stored.
It is important to consider the "commit and roll back" capability of the system. Some data may
be lost during unforeseen system outage if this capability is not available. The likelihood and
criticality of such data loss are application specific and should be subject to a risk assessment.
Additional controls and design features will be needed to compensate and mitigate any
unacceptable risks.

Both transient and stored data must be protected from unauthorized, inadvertent, or
malicious modification. It is expected that a register of authorized users, identification codes,
and scope of authority of individuals to input or change data is maintained. Some computer
systems "lock down" data, denying all write access. Security arrangement is discussed in more
detail elsewhere in this chapter.

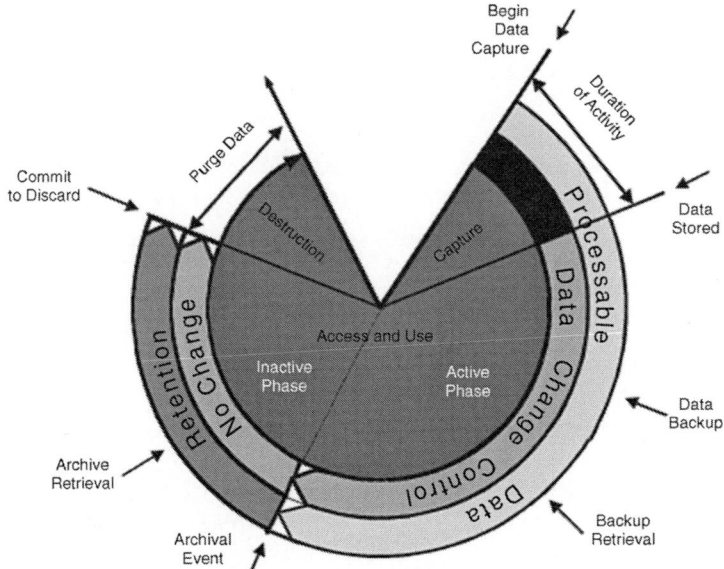

Figure 11.4 Data life cycle. *Source*: From Ref. 6.

Managing Data Changes

Authorized changes to stored data must be managed under change control. Data changes should be approved before they are implemented, and data entry should be checked to confirm accuracy. Some regulatory authorities require a second verifying check for critical data entry and changes. Examples of data requiring such a second check include manufacturing formula and laboratory data. The second check may be conducted by an authorized person with logged name and identification, with timestamp, via a computer keyboard. For other computer systems featuring direct data capture linked to databases and intelligent peripherals (e.g., in a dispensary), the second check may be part of the validated computer system functionality (4). Built-in checks might include boundary checks that data is within valid range or authority checks to verify that the person making the change has specific authority to do so for the data item concerned.

Audit trail information supporting change control records should be maintained with or as part of their respective change control records. This information should include who made the data change, nature of change, and date/time the change was made. Audit trail information may be maintained in paper or electronic or hybrid form. Whatever medium is chosen, audit trail information must be preserved in conjunction with its corresponding data. Security arrangements should be equivalent to those protecting master data. Audit trails should be available in human readable form for purpose of inspection.

Retention of Raw Data

Raw data must be retained in its original format for a period of time as defined by cGMP requirements. Regulatory authorities do accept nonelectronic copies provided they are exact copies of original data records. For instance, GMP and GLP predicate rules state that [clause 180(d) of Ref. 7 and clause 195(g) of Ref. 8] "records required by this part may be retained either as original records or as true copies such as photocopies, microfilm, microfiche, or other accurate reproductions of the original records."

If the original records are transcribed to create a copy, then the original records need to be kept. This is because of the human error associated with transcription. However, if the original was scanned or copied by means such as microfiche/microfilm, then the original could be destroyed so long as such reproductions are a "true and accurate copy" (9). In addition, microfiche/microfilm images must note whether original records have been altered in any way, and if so, then a copy of the original record must also be available. The disposal of original records must be conducted in accordance with defined procedures and authorized by local management. These procedures should be approved by QA. Archiving is discussed in more detail later in this chapter.

Recent Inspection Findings

No written procedure describing how raw data is transcribed into database worksheets. No record that data transcription is reviewed for accuracy by a second person (FDA 483, 2005).

Computer system was never challenged to determine if raw data could be deleted (FDA 483, 2003).

No SOP describing practice of deleting files from engineering and warehouse computer systems (FDA 483, 2004).

No records of review of a software fix and correction of electronic records (FDA 483, 2003).

Data retrieval function was not evaluated/tested/challenged, and as a result, the firm was unable to retrieve a large amount of GMP data (FDA 483, 2004).

No assurance of the integrity of electronic data recording because changes to security codes and permissions made without explanation or justification (FDA 483, 2005).

Data is written to a computer hard drive, stored for two weeks, and then overwritten (FDA 483, 2008).

The equipment's computer used for filling operations, which retain equipment errors that occur during filling operations, lacked the capacity to retain electronic data. After every 15th filling operation, the information was overwritten because of the storage capacity of the equipment's hard drive (FDA Warning Letter, 2001).

No control over changes operators can make to processing data (FDA 483, 2002).

Firm failed to maintain all laboratory original data even though this option was available (FDA 483, 2001).

When the capacity of the floppy disk is filled, the original data is not retained as a permanent record. Rather, the data on the floppy disk is overwritten and/or deleted (FDA 483, 2001).

No record to document that the quality unit reviews process operation data in computer system's data historian (FDA 483, 2001).

No procedure detailing file management for files stored/retrieved from network server (FDA 483, 2001).

No procedure governing XXXX data file management for file stored on server (FDA 483, 2001).

Raw data was not properly recorded or reviewed; changes in raw data were not initialed or dated (FDA Warning Letter, 2000).

Corrections to raw data were noted to be obscured with white correction fluid or improperly voided (no initials, date, reason, or explanation of change) (FDA Warning Letter, 2000).

Raw data was lost (FDA Warning Letter, 2000).

There is no evidence that the server has proper security to protect and secure the data (FDA 483, 2005).

BACKUPS AND RESTORATION

GxP regulations require pharmaceutical and healthcare companies to maintain backups of software programs including configuration, data input, and operational data in accordance with defined procedures. Backups provide a means of recovering computer systems and restoring GxP records from loss, corruption, physical damage, and unauthorized change. Periodic backups may also be required to avoid memory shortages and degraded performance. Without backups and a restoration capability, most companies cannot recover from a major disaster regardless of other preparations they have made. Restoration processes need to be verified as part of validation.

Strategy

Options for backup and restoration are summarized in Table 11.2. Pros and cons must be balanced to meet the company requirements. More than one strategy for backup and restoration may be deployed as appropriate. The strategic approach to be adopted should include consideration of the following topics:

- Common policies/procedures/systems will facilitate a consistent backup/restore approach to different applications and infrastructure, which can help simplify managing recovery.
- Standardized desktop configuration should reduce the variability to be managed during recovery.
- Adopting a thin client computing architecture concentrates recovery processes on a few key servers, thus reducing overall workload and numbers of personnel involved.
- WORM media (write once, read many) offers high security and integrity for backups.

Scheduling

The scheduling requirements for different computer systems will vary, and the needs of individual systems must be assessed. Many organizations perform backups at intervals of between one and sixty days, although the frequency will vary depending on the criticality of

Table 11.2 Backup and Restoration Options

Strategy	Description	Pros	Cons	Cost
Traditional backup to tape	Manual process of copying data from hard disk to tape and transporting to secure facility	Simple-to-implement technology, multiple price point devices/software are available.	Manual transportation and storage prone to risk and error. Potentially long lead time to restoration. Not always practical given available "windows" of processing time.	Low
Backup to electronic tape vault	Copying data from disk to a remote tape system via a WAN link	Data is accessible in shorter time frame, services become standardized, WAN link process falls, and exposure to risk/errors in manual methods is reduced.	WAN links can introduce latency into backup process; depending on vault provider, storage may be difficult to restore; data restoration times potentially lengthy.	Medium to high
Disk monitoring	Copying data written to one disk or array of disks to a second disk or array of disks via a WAN link	Instantaneous restoration of access to data possible (depending on WAN link availability and synchronicity of primary and mirrored arrays).	WAN links can introduce latency into production system operations; some mirroring systems reduce production system performance; logic errors may be replicated from original to mirrored data sets.	High

the computer system, rate of change affecting the computer system, and the longevity of the associated storage media. A register of backup activity for each computer system must be kept. It is strongly recommended that backup activities are automated through networked storage devices.

Procedure

A procedure should be established for conducting backups and restoration. The procedure should cover

- Type of backup: full or incremental,
- Frequency of backup (daily, weekly, monthly depending on the computer system concerned),
- Number of separate backup copies (usually two, one stored remotely),
- Labeling of storage media with backup reference,
- Storage location for backups (local, and remote if critical),
- Number of backup generations retained,
- Documentation (electronic or paper) to be retained to provide a history of the backups and restorations for the live system,
- Recycling of storage media for reuse, and
- Retention of installation media for COTS software.

It is generally recommended that three backup copies are kept, one for each of the last three backups. This system is sometimes referred to as grandfather-father-son backups. Each backup should be verified before it is stored in a secure location (10), preferably a fireproof safe. Environmental controls in storage area should be carefully considered to avoid unnecessary degradation of backup media as a consequence of excessive heat, cold, and humidity.

Any change to the backup procedure and any necessary reciprocal modification to the restoration procedures made must be carefully considered. There have been several instances where incorrect backup procedures have not been tested an, subsequently, backups could not be restored.

Media

The appropriate backup media can vary; examples include diskettes, cartridge tapes, removable disk cartridges, or remote-networked host computers. The retention responsibilities for backups are the same as that for other documentation and records. Stored backups should be checked for accessibility, durability, and accuracy at a frequency appropriate for the storage medium (2,11). Beware of wear-out of media when purposely overwritten for reuse. Different media have different life spans depending on retention period and number of read/write accesses.

Recent Inspection Findings

No SOP is being followed for reviewing and approving files to check whether they are backed up properly (FDA 483, 2005).

There were no written SOPs for backup (FDA 2001).

There is no established written procedure that describes the steps taken to back up the XXXX disks to ensure data recovery in the event of disk loss or file corruption (FDA 483, 2002).

Firm procedures did not specify the frequency of backing up raw data files (FDA 483, 2002).

Data cannot be backed up because of a malfunctioning floppy drive (FDA 483, 2003).

Files for the computer had not been backed up or archived since *[date]* (FDA 483, 2004).

There is no evidence that the electronic raw data SOP is being used to assure that files are reviewed, approved, and backed up appropriately (FDA 483, 2005).

Data cannot be backed up because of malfunctioning drive (FDA 483, 2003).

Raw data held in computer system is not backed up (FDA 483, 2004).

ARCHIVING AND RETRIEVAL

Archive should not be confused with taking backups. Backups of data and software can be loaded to return the computer system back to a known operational state. Backups are usually taken on a daily or weekly basis, and backup copies are retained for a number of months. In contrast, archive records need to be accessible for a number of years perhaps by people who were not involved in any way with their generation.

Archiving Requirements

GxP data, records, and documentation including computer validation should all be archived. An example schedule of computer validation documents requiring archive is given below (see also appendixes in chap. 13 identifying other regulated records).

- Access management record
- Change control documents
- Data migration records
- Decommissioning records
- Design specifications (including functional specifications)
- Design review documents
- Incident management records
- Periodic reviews
- Programming standards
- Qualification documents (i.e., IQ, OQ, PQ)
- Source code and source code review
- Supplier audit reports
- Test evidence
- User documentation (e.g., SOPS and training records)
- User requirements specification
- Compliance determinations, plans, and reports

Internal audit reports from self-inspections monitoring a pharmaceutical or healthcare company's compliance with its own quality management system do not have to be retained once corrective actions have been completed so long as evidence of those corrective actions is kept (e.g., change control records). Supplier audit reports and periodic reviews are not internal audits and should, meanwhile, be retained.

The integrity of archived records is dependent on the validation of the systems from which they were taken and the validation of systems used for archiving and retention of those records. Chapters 12 and 13 discuss special requirements for regulated electronic records/ signatures and long-term archiving solutions, respectively. SOPs for archiving and retrieval of software and data must be specified, tested, and approved before the computer system is approved for use.

Retention Periods

Retention periods for data, records, and documentation are the same regardless of the medium (electronic or paper) (11). R&D records should be generally archived for 30 years, although in specific circumstances, longer periods may be appropriate.

The retention time for validation documentation relating to a drug product's manufacture is at least one year after the product's expiry date. The retention time for validation documentation relating to a drug product exempted from expiry dates varies depending on whether it is supplied to the United States or to Europe. For the United States, it is at least three years after the last batch has been distributed (11), while for Europe, documents must be retained for at least five years from its certification (2). The United Kingdom's IQA Pharmaceutical Quality Group suggests that all documentation be retained for a period of at least five years from the last date of supply (12). An effective solution for many organizations has been to store their documents for a period of seven years after the date of drug product manufacture supported by the computer system. The aim should be to discard and purge the data, records, and documents when they are no longer needed to support the operation of the computer system.

One exception to the seven-year rule of thumb is source code. Source code need only be retained for three years so long as change control and source code review documentation is retained longer to support any relevant investigations. It is generally not feasible to establish a legacy computer environment greater than three years old because superseded hardware and software mean that the current system is incapable and/or no longer under support agreements/warrantee for such setups.

Storage Requirements

Archives, like backups, should be stored at a separate and secure location (2). Critical documentation, records, and data should be kept in a fireproof safe. In some cases, it is acceptable to print copies of electronic records for archiving, but advice should be sought from regulatory authorities. Clinical trial data are often stored on microfiche or other electronic mediums. It should not be possible to alter such electronic copies such that they could by interpreted as master records (13).

Temperature and humidity may have a bigger impact than that in the case of backups because of the extended duration of storage. The storage environment should be periodically evaluated to confirm that stable storage conditions exist. Environment data should be recorded and maintained. Some firms use automated monitoring systems for this purpose.

Retained media are likely to require at least one refresh during their retention period. Different media have different life spans, and the manufacturer's recommended refresh intervals vary. CD ROMs, for instance, typically have a 10-year life span and a 5-year refresh recommendation. DAT usage should not exceed 20 times for read/write operations and are typically considered to have a five-year life span without copy. Tapes, meanwhile, may be accessed perhaps up to a hundred times but require retensioning. It is recommended to take a new copy of a tape every 12 months to retain it. The process of data migration is discussed in chapter 10. Data migration will be required not only as part of normal media management but also when media becomes obsolete during the retention period. Long-term preservation issues for archives are discussed in chapter 12.

Retrieval Requirements

Archive information required by regulators, including those stored electronically, must be accessible at their site of use during an authorized inspection. It should be possible to give inspectors, if requested, a true paper copy (accurate and complete) of master documentation regardless of whether the original's medium was magnetic, electronic, optical, or paper within 24 hours of the request. Longer retrieval periods of up to 48 hours may be agreed for information that is stored remotely from the site being inspected. True copies must be legible and properly registered as copies. Where large volumes of information are archived, the use of manual or automated supporting indexes is recommended to ease retrieval. Software applications, scripts, or queries used for manipulating or extracting data should be validated and maintained for the duration of the retention period.

It is vital that retained records are not compromised. Unlike backups that by their nature are routinely superseded by newer copies, archives are irreplaceable historical records. The content and meaning of archived information must not be inadvertently or maliciously changed. Consequently, access to retained records should be read-only. After each use, the storage media should be given an integrity test to verify that it has not been corrupted or damaged. Logs of archive access should record media retrieved and returned, and the success of subsequent integrity testing should be documented.

Recent Inspection Findings

There were no written SOPs for archival (FDA 483, 2001).

BUSINESS CONTINUITY PLANNING

Business continuity plans define how significant unplanned disruption to business operations (sometimes referred to as disasters) can be managed to enable the system recovery and business to resume. Disruptions may occur as a result of loss of data or outage of all or part of the computer system's functionality. The range of circumstances causing disruption can range from accidental deletion of a single data file to the loss of an entire data center from, for instance, fire.

Business continuity plans are sometimes referred to as disaster recovery plans or contingency plans. There are two basic scenarios.

- Suspend business operations until the computer system is restored.
- Use alternative means to continue business operations until the computer system is restored.

Suspending business operations may entail scrapping work in progress or continuing work in progress to completion using alternative means. It may be possible to use alternative means to support business operations for some time before final suspension awaiting restoration of the original computer system. The duration for which alternative means can be supported will depend on the overhead to operate them including the effort to retrospectively enter interim operational data into the original computer system to bring it up to date.

Procedures and Plans

Procedures and plans supporting business continuity (disaster recovery plans and contingency plans) must be specified, tested, and approved before the system is approved for use. Topics for consideration should include catastrophic hardware and software failures, fire/flood/lightning strikes, and security breaches. Procedures need to address (10)

- Specification of the minimum replacement hardware and software requirements and their source,
- Specification of the time frame within which the replacement system should be in production, based on business considerations,
- Implementation of the replacement system,
- Steps to revalidate the system to the required standard, and
- Steps to restore the data so that process activities may be resumed as soon as possible.

The procedures and plans employed should be retested periodically, and all relevant personnel should be aware of their existence. A copy of the procedures should be maintained off-site.

The regulators are interested in business continuity as a means of securing the supply of drug products to the user community. The GxP requirement for business continuity plans covering computer systems is defined in EU GMP annex 11 (the FDA has similar requirements).

> There should be available adequate alternative arrangements for systems, which need to be operated in the event of a breakdown. The time to bring the alternative arrangements into use should be related to the possible urgency of the need to use them. For example, information required to effect a recall must be available at short notice. The procedures to be followed if the system breaks down should be defined and validated. Any failures and remedial actions taken should be recorded. (Clauses 15 and 16, EU GMP Annex 11)

There are seven basic tasks to be completed for business continuity planning.

- Identify assets and/or business functions that are vital to the support of critical business functions.
- Assess interdependencies between critical computer systems/applications.
- Identify vulnerable points of failure and make changes to reduce or mitigate them.
- Select recovery strategy to meet appropriate time frames for restoration.
- Develop business continuity plan.
- Prepare procedural instructions and conduct training.
- Verify business continuity plan through verification exercise.

Major threats are identified in Table 11.3 with suggested controls to support continuity of business operations. Leading disaster scenarios in one survey were system malfunction (44%),

Table 11.3   Threats and Controls for Business Continuity Planning

Threats	Controls
• Water damage (e.g., leaky pipes and floods)	Water detection to provide early warning of leaks and other water hazards (e.g., condensation)
Fire/heat damage (e.g., arson, equipment overheating, lightning strikes)	Detection of preignition gases, smoke, and other indicators of impending fire to enable proactive response that will ensure health and safety of personnel and prevent loss of data and equipment to fire
	Suppression of fires (e.g., sprinkler systems, gaseous extinguishing systems, using noncombustible materials in facility, restricting storage of combustible consumables such as paper)
	Use of fireproof cases, cabinets, and safes
Power failure	Continuity of electrical power in the presence of an electrical outage (e.g., use of an uninterruptable power supply—uninterruptable power supply) or surge (e.g., electrical conditioning)
Network failure	Network backup and restoration facilities at local and intersite level. Restoration of communications external to company
System malfunction (software, hardware, human error)	Detection of contamination levels (dust, food and drink, production materials) that can accumulate in equipment and lead to system malfunction
	Monitoring hours worked by individuals and/or mundane nature of work that might result in loss of concentration and hence introduction of human errors (data errors and user operation errors)
Malicious/accidental damage (e.g., hackers)	Logical firewalls and user access systems requiring combination of physical and logical password elements
	Physical security of corporate computing, data centers, and telecommunications facilities
Other factors (forced evacuation for environmental hazards, aircraft crashes)	Provision of and training in evacuation procedures and safe areas

human error (32%), software malfunction, computer viruses (7%), and natural disasters (3%) (14). Plan for general disaster scenarios, it is too easy to get bogged down trying to identify every conceivable catastrophic situation. It is also important to remember that threats are relative. Water extinguishers to suppress a fire, for instance, should not be treated as bringing a new threat of water damage.

Verification is not normally possible through comprehensive testing. Some companies may claim that they can test element of a system in isolation, accepting the disruption this often involves. Testing disaster scenarios, however, are, by their nature, catastrophic and not to be knowingly invoked. Simulation provides a much more practical approach. Simulation exercises are based on rehearsals whereby teams walk through what they would do in a disaster scenario using procedures and, possibly, some support systems. Simulations can prove useful training events. The approach to verifying business continuity planning will depend on the particular opportunities and constraints affecting a company.

Redundant Systems and Commercial Hot Sites

In the event of a disaster, dedicated redundant systems at a separate locality that must be far enough distant not to have been affected by the disaster are brought on line. Users are either relocated to the backup facility or provided remote access to the backup system via some sort of preestablished network connection. User applications typically have a target time for restoration of redundant systems and commercial hot sites of between 1 to 2 and 7.5 hours, respectively.

Besides being the most reliable method of recovery, with minimal business disruption, redundancy also tends to be the most expensive. A commercial hot site, for this reason, is often a more acceptable alternative from a cost perspective provided a slightly longer recovery window is acceptable to the business.

Service Bureaus

Some companies elect to back up systems against failure by contracting with a service bureau for emergency recovery. Essentially, it is an insurance policy whereby the pharmaceutical or healthcare company leases a standby system. User terminals and printers are installed in the client offices with network connection to the service bureau that may be at the service supplier's premises or a mobile facility that is driven onto site. User applications typically have a target time to restoration within 24 hours. The problem with commercial mobile facilities is that their service providers often require up to 48 hours to guarantee deployment.

This approach to business continuity planning requires

- The interdependency between critical and noncritical applications to be understood so that when the service bureau is invoked, it can operate independently or so that other critical cosystems are also restored and
- The most recent application versions to be restored with current data.

This solution can be very complex where there are several applications involved, as each application typically requires its own service bureau. Many companies are not considering the use of Internet and Intranet linking to support restoration.

Backup Agreement

This approach involves a site being provided with a backup by a partner organization. This does not mandate a redundant system but more often mandates utilization of spare computing capacity at the partner organization. User applications typically have a target time to restoration within 24 hours. Practical problems include maintaining current system versions on partner organizations and finding mutually convenient time to test the backup facility. Maintaining the partnership can be complex. Another issue is how to ensure that the partner's computer systems are not themselves brought into the disaster scenario by placing too higher demand on their computer systems when the backup is invoked.

Cold Sites

Cold sites involve preparing an alternate backup system. Company-owned cold sites have the drawback of being expensive to outfit. Such an investment can however be used for off-site storage and training when not activated. An alternative is to employ a commercial cold site that might be shared between a number of client companies. As with service bureaus, cold sites may have mobile facilities that are driven to a client's site. The risk with cold sites is that because they are shared, it is possible that they may not be available if a disaster has already hit one of the sharing parties. User applications typically have a target time to restoration of between 24 and 72 hours. Longer than 72 hours typically means that the business has come to a complete stop.

Manual Ways of Working

Define manual ways of working for application during system outage. Remember that on restoration, some reprocessing of data input to the original or backup system (catch up) will be required and this must be planned for. Manual records made during the outage, even once input into the restored system, must be retained.

Software Licenses and Support Contracts

Loss of software support for aging versions of business critical systems can create significant business continuity and regulatory risks. Pharmaceutical and healthcare companies should provide a definitive statement on how they will maintain critical systems where support has historically been provided by third parties but that support is no longer available or set to expire. Measures need to be established to prevent adverse impact to product quality and product data and to ensure business continuity during any system outage.

The U.S. Uniform Computer Information Transaction Act gives vendors the power to deactivate software without a court order so long as this is defined in a license agreement (1). Users are to be given 15 days' notice of any turnoff. This raises several key compliance concerns.

- Notification of software license termination. What if warnings of software termination for whatever reason go astray? The vendor may not hold the company's current address, the company's name may have changed through merger or divestment, the employee who signed the agreement may have left the company, or the employee who signed the agreement may be absent from work for holiday, birth of a child, or sickness.
- Business continuity. While the loss of a word processing package will be generally irritating, the loss of a server might be critical if it led to the outage of a network. The effects of disabling software may not be limited to the target company and extend through supply chains. The ability to turn off software will not be limited by national boundaries. Key suppliers (or equipment, drug ingredients, and services) may not be able to function and fulfill their commitments to pharmaceutical and healthcare companies. Distribution and wholesale of drug products, often outsourced, may itself be halted because of disabled software, which could affect the availability of vital products to patients. Joint ventures, partnerships, and intercompany may also be in jeopardy.
- Consequential loss. Questions have been raised if the turnoff of software led to the corruption or loss of GMP data. Pharmaceutical and healthcare companies will be forced to assign significant resources on checking licensing agreements of COTS products.
- Unauthorized disabling of software. Another concern is that disabling codes for potential use by the vendor could also be used by hackers.

Design features to disable software are not new. In the 1990s, a chemical manufacturer suffered the loss of an MRP system when they failed unwittingly to renew a support contact over the New Year period. The software was automatically disabled in mid-January, with severe business impact. The software vendor had not escalated the license issue when there was no reply to a renewal request sent a few months earlier.

FDA has indicated that such features may compromise management of electronic records and electronic signatures and have indicated that software products with such features should not be used in validated systems (15). Unfortunately, suppliers may insist on the right to use such features or charge a higher price to compensate for its absence. Pharmaceutical and healthcare companies should

- Know the terms for software,
- Write procedures, if necessary, to ensure that record integrity is maintained in case the software stops functioning,
- Assess how automatic restraints impact compliance and validation, and
- Make sure that the above issues are considered when purchasing software.

Inspection Findings

There were no written SOPs for disaster recovery (FDA 2001).

Following flood damage in September 1999 to your facility and equipment, you or your employees failed to evaluate the raw data storage conditions or implement any procedures or changes to existing procedures to alleviate future damages (FDA Warning Letter, 2000).

SECURITY

Hardware, software, and data (local and remote) should be protected against loss, corruption, and unauthorized access (10). Physical security is required to prevent unauthorized physical access by internal and external personnel to computer system hardware. Logical security is required to prevent unauthorized access to software applications and data. The network and/or application software should provide access control.

Management

SOPs for managing security access (including adding and removing authorized users, virus management, and physical security measures) must be specified, tested, and approved before the system is approved for use. Topics to be covered include the following:

- Issue unique user ID codes to individual users.
- Passwords should be eight characters long (16).
- Do not share personal passwords or record them.
- Do not store information in areas that can be accessed by unauthorized persons.
- Do not download from the Internet.
- Applications are protected from viruses: virus check all floppy disks, CDs, hard disk drives, and other media from internal and external sources.
- Do not disable virus checks.
- Do not forward unofficial messages containing virus warning (may be a hoax and unnecessarily increase traffic or may further propagate a real virus).
- E-mail over Internet is not secure without public key infrastructure (PKI).
- Do not send messages from someone else's account without authorized delegation and management controls.
- Do not buy, download, or install software through unauthorized channels.
- Do not make unauthorized copies of software or data.
- Amendments to electronic records should be clearly identified and not obscure original record.
- Use of electronic signatures is controlled.
- Electronic links used to transfer data are secure.
- Take backups of software and data.

Passwords should be securely issued to their users, ensuring that the users concerned have been authorized to access the computer systems for which the passwords are being granted. Merely issuing a user ID and sending an e-mail to the user with the password enclosed are insufficient. It is very difficult to guarantee that unauthorized staff might have access to the e-mail or the user's account. The identity of the user should be authenticated before a password is issued. Some pharmaceutical and healthcare companies do this by verbally communicating passwords in two halves, one half to the user's line manager and the other half to the user, respectively. Neither party can use a portion of the password to gain access to a system without knowledge of the other party's portion of the password. In the process proposed, the line manager authenticates the user as authorized for the computer system concerned before giving the user the other half of the password they need.

Once users have been granted access of a computer system, it is common practice to prompt them to renew their passwords every few months (e.g., expire every 90 days for networked users). There is no formal regulatory requirement, however, to change passwords that are still secure. Many users struggle to remember passwords that change frequently, often reverting to writing the passwords down or using passwords that can be easily memorized such as family names and vehicle license plate numbers. Some pharmaceutical and healthcare companies are looking at random alphanumeric passwords with longer expiry periods to improve overall security (6). Such passwords, by their nature, are virtually impossible to guess but also harder to remember. The issue of remembering passwords is compounded when users have access to a number of computer systems each nominally having individual passwords. It is very tempting to set all systems to share user IDs and associated passwords, in which case validation of the controlling mechanism needs careful validation.

User Access (Profiles)

The rules and responsibilities for assigning access rights should be specified in procedures approved by QA. Access rights need to be documented and reviewed regularly to ensure that they are appropriate. All users need to receive appropriate training to their user access privileges. Default user access should be no access. Users with changing authority levels should have their access rights modified to accurately reflect their new role. Meanwhile, access rights for those who are no longer authorized to use a system should be immediately removed. Screen locks should be used to prevent unauthorized access from unattended user terminals.

Computer Viruses

The vulnerability of computer services to computer viruses is not easily managed. Besides deploying antivirus software, the only other defense is stopping unauthorized software and data by loading them on computer systems and building firewalls around networked applications. This is a prospective approach that assumes that existing computer services are free from computer viruses. However, this approach cannot entirely remove the threat of computer from computer services. The source of authorized software and data may itself be unknowingly infected with a computer virus. Novel viruses can also break through network firewalls. It is therefore prudent to check software and data related to their computer services that are used within an organization.

The management of computer viruses is primarily based on prevention:

- Strict control of access to computer services
- Policies forbidding the use of unauthorized software
- Vigilant use of recommended antivirus software to detect infections

Procedures should be established covering

- Stand-alone computer systems including laptops
- Client workstations
- Network servers providing file services to PC workstations
- Compact disks (CDs)

- Digital video disks (DVDs)
- Other removable storage media

Virus checking should be performed on all computer systems and removable storage media if

- They originate from an external organization (including but not limited to universities or other educational establishments, research establishments, training organization, external business partners),
- Their origin is unknown (including but not limited to unsolicited receipts),
- They have been used with other computer systems or removable storage media of unknown status (including but not limited to being sent off-site for repair, maintenance, or upgrade),
- They are received for demonstration, training, or testing purposes,
- They belong to a representative or employee of an external organization and are to be used in conjunction with in situ computer equipment, and
- They were last on an external system for business, educational, training, or private purposes (including but is not limited to software acquired electronically from external networks or the Internet).

Regular virus-checking arrangements (sweeping) should be defined with service providers. Local instructions will be needed for users to carry out the necessary checks. It is important to understand that virus-checking software only checks for known viruses. Updates to the antivirus software must be applied when available. The application of multiple antivirus software utilities may be recommended to offer higher combined detection coverage of viruses. Only vetted and approved antivirus software utilities should be used.

Detected computer virus should be reported so that the virus is removed and the integrity of the computer system is restored. If a virus is found or suspected, then the following steps should be taken:

- No application must be run on the affected computer system. Any error or warning messages displayed must be recorded along with details of any unusual symptoms exhibited by the computer system.
- Local support staff must use their judgment as to whether or not it is safe to save data and/or exit any currently executing application in a controlled manner. Where it is determined that this is not safe to do, the machine must be powered down immediately.
- Every effort must be made to find the source of the virus. The virus must be identified, and instructions must be sought from the antivirus software documentation or elsewhere on how to remove it. Unresolved virus infections must also be noted.
- After investigation, infected removable storage media should be destroyed, but if important data is needed, the virus must be removed under the supervision of the IT support contact. Systems that may have come into contact with the diskette must be checked immediately.
- Computers must be rebooted by using clean, write-protected system diagnosis disks. This will ensure that a true analysis of the computers is performed without any viruses being resident in memory. All local hard drives must be scanned. If the virus has been supplied from an external source, then that source should be noted. If no virus is detected, then this should be recorded.
- Any servers that may have come into contact with the virus must also be checked immediately. Any computer system that has come into indirect contact with the infected computer system via removable storage media must also be checked.
- All deleted data files and software must be restored from backups or the original installation media. Local computer drives should be checked after restoration to verify that they are still clear of any computer viruses.

- Crisis management will be required where a computer virus has manifested itself causing a computer system malfunction. Senior management should be kept informed of the incident and corrective actions being undertaken, and the wider user community should be warned of the incident to reenforce vigilance.
- Deploying antivirus software without validation may be a necessity to control virus attacks or avoid anticipated attacks. Virus attacks may pose a more significant risk to GxP data than lack of validation.

An example of virus incident form is shown in Figure 11.5.

VIRUS INCIDENT FORM			
Notifying Person	Name & Function of person initiating this form		Date:
System Name:		Serial/Asset No.	
Company/Department:		Site/Location:	
System Type:	E.g. Server, Desktop, Portable, Other (please specify)		
Operating System:	E.g. DOS/Windows, Windows 95, Windows NT, Other (please specify)		
VIRUS DETECTION & REMOVAL			
Name and/or Description of Virus			
Detection Method			Time/Date:
Symptoms of any Malfunction Observed			Time/Date:
Removal Method			Time/Date:
Verify Clean and Approve for use:	Signature of IT Service Engineer		Date:
VIRUS INVESTIGATION & FOLLOW UP ACTIONS			
Suspected Source of Infection			Time/Date:
Potential other systems affected and Corrective action			
Any necessary validation complete:	Signature of QA/Validation Representative		Date:
Closure of Incident:	Signature of Security Manager		Date:

Figure 11.5   Example virus incident form.

Recent Inspection Findings

The computer system used to monitor and control manufacturing equipment lacked appropriate controls to ensure that only authorized personnel had access to the system (FDA Warning Letter, 2001).

There is no written procedure to describe the process that is used to assign, maintain passwords, and access levels to the control system (FDA 483, 2001).

Access and access level control are not documented (FDA 483, 2004).

No written procedures describing the responsibilities of the system administrator (FDA 483, 2004).

There was no procedure to remove people who no longer should have access to the computer or recall or periodically change passwords (FDA 483, 2004).

[*System*] password function was disabled (FDA 483, 2002).

Once the computer system is accessed by one employee, it is left open and can be used by other employees without knowledge or consent of original employee (FDA 483, 2004).

No SOP describing the process of assigning access levels (FDA 483, 2004).

There is lack of assurance of control because too many individuals (five of nine users) were given system administrator level access (FDA 483, 2005).

No procedure is being followed to assure that the server has proper security to protect integrity of data (FDA 483, 2005).

No SOP for network, server, and workstation security including password, levels of access, and lockout procedures (FDA 483, 2006).

Software failed to demonstrate adequate security in that analysts have the ability to overwrite original data, and there are no individual user names and passwords limiting access to the system (FDA 483, 2007).

Systems are not secured or protected with access passwords (FDA 483, 2007).

Six people share the same password to access the production control system, and they can make modifications to production steps, alarm points, and user identification numbers without any traceability to specific person making the changes (FDA 483, 2008).

There were no written SOPs for virus detection (FDA 483, 2001).

There was no validation data to demonstrate that an authorized user of the corporate WAN did not have access to analytical data on the laboratory's LAN (FDA 483, 2001).

The client/server password system failed to adequately ensure system and data integrity in that passwords never expired and could consist of four characters (FDA 483, 2001).

You failed to have adequate security controls for your XXXX systems because your system, once accessed by one employee, is left open and available for other personnel to gain access to the original employee's analytical test results (FDA 483, 2002).

There was no established written procedure that addressed the access code for the software development room and notification of team members of the changes (FDA 483, 2002).

Users could grant authority to themselves or any other person high-level access within the application (FDA 483, 2001).

The firm failed to produce an approved list of personnel currently authorized to use the [*computer system*] (FDA 483, 2001).

System security has not been defined (FDA 483, 2001).

No security system to prevent unauthorized changes to computer database (FDA 483, 2001).

CONTRACTS AND SERVICE LEVEL AGREEMENTS

Contracts should be established with all suppliers. For standard items of equipment and software, this can take the form of a purchase order. For support services, it is common practice for users of computer systems to establish a SLA with their suppliers.

SLAs should unambiguously define the system being supported, the services to be provided, and any performance measures on that service (1). Examples of services that might be provided include

- Developing and/or installing software upgrades, bug fixes, and patches,
- System management and administration,
- Support for underlying IT infrastructure,
- Use of any particular software tools, and
- Routine testing and calibration.

Other relevant information normally held in appendix or schedule to the SLA includes user and supplier contact details, definition of fixed costs, charge-out rates, and penalty payments as appropriate. Contractual terms and conditions might also be included if not managed as a separate document. Escalation management processes should be documented and understood.

Service providers should have formal procedures in place to manage their work. They can, however, agree to use customer procedures if this is more appropriate. Pharmaceutical and healthcare companies should reserve the right to audit use of whatever governing procedures are being used. Service providers should be audited just like other suppliers (chap. 6). This is especially important for system development, IT infrastructure, and maintenance activities. Audit reports should be retained, and any audit points should be followed up as required.

Service levels should be periodically reviewed, and summary reports should be prepared. Performance measures should be established with target minimum service levels. Responsibilities for collecting data to support performance measures should also be agreed along with any calculations to be used to derive performance levels. Trending topic areas may provide a useful indicator regarding emerging issues. Consideration should be given to who will receive SLA reports and how often such reports are required. As a minimum, such reports should be reviewed when considering contract renewal.

USER PROCEDURES

Experience suggests that human error accounts for up to one-fifth of system malfunctions (17). This emphasizes the importance of accurate and practical user procedures accompanied by suitable training.

User procedures for operating and maintaining the computer systems, control system, or laboratory system must be specified, approved, and, where possible, tested before the systems are approved for use (18). User procedures can make good use of role activity diagrams (RAD) to help readers understand the specific responsibilities associated with different roles. An example of RAD is shown in Figure 4.5 in chapter 4.

The FDA's current interpretation of 21 CFR Part 11 requires that SOPs maintained electronically must have electronic audit trails irrespective of whether they are approved and distributed using a paper-based system. The mere fact that the electronic original will be used as the basis of the updated version makes them electronic records. The only exclusion is where the documents are completely retyped upon every update.

Procedures should be put in place to pick up possible system errors as well as human error or misuse. It is important to track trends and demonstrate proactive management of issues. Statistical analysis should be applied to data gathered.

Recent Inspection Findings

No documentation regarding procedures for use of the system (FDA 483, 2004)
No written procedures describing system operation (FDA 483, 2004)
No written instructions describing which software manual versions are to be used (FDA 483, 2004)

PERIODIC REVIEW

Computer systems, as critical items of equipment, should be periodically reviewed to confirm that their validated status has been sustained (19). Validation reports concluding the implementation of a computer system should identify when the first periodic review is expected. The selected interval for periodic review needs to be justified. Many companies conduct periodic reviews every 12 months for their most critical systems. Less critical systems do not generally warrant such regular review. It is recommended that intervals between periodic reviews do not exceed three years to reduce the risk of undetected deviations.

It may be possible to collectively review a number of less critical systems by the product they support (e.g., through annual product reviews) or by the physical area in which they reside (e.g., laboratory, manufacturing line). Sometimes periodic reviews combine process validation and computer validation. If either of these approaches is taken, then the coverage (list of systems) must be defined for the review.

The following criteria can be used when evaluating suitable intervals between periodic reviews and the scope of review:

- Nature of use—potential impact on the quality of drug and healthcare products
- Character of system—size and complexity of the computer system, and how easily unauthorized changes can be made
- Extent of design changes—cumulative effect of changes to the computer system (including software upgrades) made since the last (re)validation exercise
- System performance including any system failures—any problems experienced with the system's operation (e.g., user help desk enquiries, system availability, access control, data accuracy)
- Changes to regulations—effect of changes made to regulatory and/or company requirements since last (re)validation exercise

Organizations often establish a review panel to conduct periodic reviews. Before the panel meets, the chairman should estimate the scope of the review and the time needed to undertake the review and determine the size and composition of the review panel. The level of review should be based on a documented risk assessment. Members on the review panel should include operations staff and management, system support staff, and quality assurance. User communities of networked applications should be represented by their departmental functions.

The review panel meeting should only take a few hours if all the necessary information for the periodic review is collated before the meeting. Table 11.4 identifies some topics for consideration in the periodic review. The review meeting must be recorded either by minutes or as a formal report. It will normally begin by reviewing progress on actions assigned at the last meeting and close by assigning a new list of actions that should be assigned to individuals with target dates for completion.

A particularly important decision to make during a periodic review is whether or not revalidation is required. At a certain point in time, maintaining an old system becomes too ineffective for the expense incurred. There are no predefined metrics to base this decision on, but certain characteristics signal system/software degradation.

- Frequent system failures (partial or catastrophic).
- Significant growth in size of software modules/subroutines (possible emergence of complex system structure and spaghetti code).
- Excessive and increasing maintenance effort (possible difficulty in retaining maintenance personnel—key knowledge being lost).
- Documentation does not adequately reflect actual system (need to refer to supplementary change controls to understand system).
- Cumulative effect of change (impact of system functionality changes on patient/consumer safety, impact of data/record changes on regulatory submissions).
- Over three years since last validation/revalidation.

The greater the number of such characteristics, the greater the scale of potential reengineering required. In fact, it may get to a stage where it is more cost-effective to entirely replace the system. Pharmaceutical and healthcare companies are encouraged to collect their own metrics to make this decision process more objective. Typically, such decisions are very subjective, and care is taken to make sure that the decision is not unduly influenced by dominant personalities rather than real needs.

Occupational Health

Consideration must be given to the potential effects of the computer system and associated equipment on the personnel who may use or come into contact with the system. Typically,

Table 11.4 Example Periodic Review Topics

Topic	Comments
Performance	Check critical process performance parameters and whether any problems are potentially due to supporting computer system.
Procedures and training	Check that training records are current.
	Examine the need for refresher and induction courses for new employees (permanent and temporary staff, consultants, and contractors).
	Standard operating procedures should be reviewed on a biennial basis and hence do not require retraining within that time unless something has changed.
Change control	Have the change control procedures been correctly adopted? Is the cumulative effect of change understood? Have company or regulatory computer validation standards been changed?
	Does the URS adequately describe the current use of the computer system?
	Check what has changed with computer system instrumentation, computer hardware, and computer software. Do design documents reflect these changes?
	Check whether any unauthorized changes have been made. Conduct spot checks to compare running systems with documentation.
	Check requirements traceability to verify whether installation qualification/operational qualification/PQ testing covers the system as used.
	Review the criticality of any outstanding change requests and how long they have been outstanding.
Calibration and maintenance	Check software copyrights and licenses. Some software applications cease to function upon expiry of a license.
	Check maintenance and calibration schedules.
	Exercise uninterruptable power supply batteries and check ongoing records monitoring the operating environment (e.g., humidity and temperature).
Security	Review physical access arrangements and any attempted breaches.
	Review accuracy of lists of active users. Review user access profiles for access rights that are no longer required. Review unauthorized access attempts.
Data protection	Check lockdown of user access to alter data. Check audit trail of any data maintenance activities.
Backups	Verify whether backups and archive copies are being taken and can be restored.
Business continuity	Review any service level agreements to check that details are correct, still appropriate, and that the supplier is aware of his obligations.
	Walk through contingency and disaster recovery plans to check whether they are still applicable.

these risks are associated with the interfacing to visual display units (VDUs) and environmental conditions.

Recent Inspection Findings

While the individual changes have been reviewed during the change control process, a comprehensive review of all the collective changes has not be performed to assure that the original IQ/OQ remains valid and to assure that the [computer system] does not require requalification or revalidation (FDA 483, 2001).

While the individual changes have been reviewed during the change control process, a comprehensive review of all the collective changes has not been performed to assure that the XXXX does not require requalification or revalidation (FDA 483, 2001).

No controls or corrective action after frequent HPLC software errors caused computer lockup (FDA 483, 2001).

On XXXX, a laptop computer was swabbed and tested for detection of XXXX. There is no documentation of whether and when this item was decontaminated and whether and when it was used in the XXXX and subsequently in the XXXX facility (FDA 483, 2002).

Automated analytical equipment was left in service even though system software reliability had been questioned due to frequent malfunctions that had impeded quality control procedures (FDA 483, 2002).

No root cause was determined to initiate necessary corrective action for multiple software errors (FDA 483, 2005).

Other related inspection findings can be found at the end of chapter 3 in section "Managing Deviations."

REVALIDATION

Computer systems undergo change even to sustain their original design intent. Operating systems and software packages will require upgrading as vendors withdraw support for older products. New technology may prompt hardware changes to the computer system and supporting computer network infrastructure. Unless documentation is completely revised to embed changes, the document will have to read in conjunction with change control records. As, progressively, more changes are made, it will become harder and harder to accurately understand the current system as a whole. This will make the rigor of future change control harder because the impact of proposed changes on the existing system will be harder to evaluate. Hence, the value of validation will tend to decline until the computer system validation and associated documentation are rebaselined by a revalidation exercise.

If a periodic review identifies the need to reestablish or test the confidence in the validated status, the computer system should be revalidated. Equally, if significant changes have been made or if regulatory requirements have altered, it may be deemed prudent to revalidate a computer system. In practice, the attention of operational staff to quality procedures and records often wanes unless they are carefully coached or monitored (see also section "Ensuring a State of Inspection Readiness" in chap. 14). As the period between successive revalidations increases, so does the likely amount of revalidation work required (Fig. 11.6). Intervals of between three and five years between revalidations are typically appropriate.

Revalidation does not necessarily imply a full repeat of the validation life cycle; partial requalification is acceptable when justified. An analysis of changes implemented can be used to help determine how much revalidation is needed. Were there changes evenly spread throughout the system (sporadic) or were there focal points? Computer systems with modular architectures may allow revalidation to be segregated to particular functional elements.

The testing strategy should ensure that all critical functions are subject to comprehensive retesting regardless as to whether they have changed or not (Fig. 11.7). GxP assessments discussed in chapter 6 can help identify what is critical functionality. Comprehensive testing should also be conducted on non-GxP critical areas of the system functionality that have changed since original validation. All other used functionalities need only representative

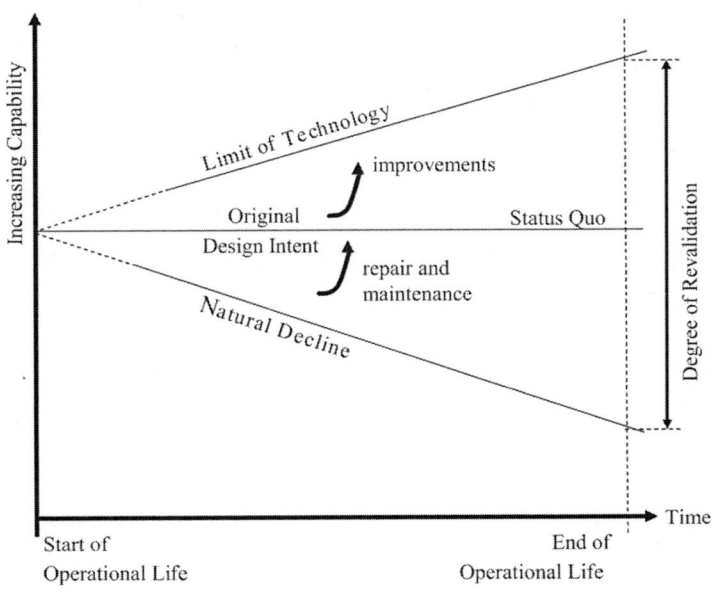

Figure 11.6 Degrading validation.

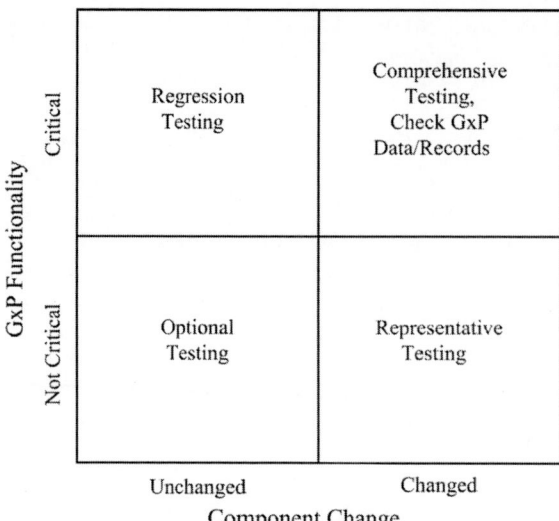

Figure 11.7   Focus of revalidation testing.

testing. Additional checks for GxP data over and above routine data maintenance should also be considered.

Revalidation may be synchronized to coincide with computer system upgrades in a bid to make the most effective use of resources. Such strategies should be defined and approved in advance.

Revalidation can often be conducted without restricting release of the drug products whose manufacturer is supported by the computer system. Authorized quality assurance personnel must approve release of drug products during revalidation. In Europe, this should be a qualified person.

Recent Inspection Findings

There were no written SOPs for revalidation (FDA 483, 2001).

REFERENCES

1. International Society for Pharmaceutical Engineering. GAMP®5: Risk-Based Approach to Compliant GxP Computerised Systems. Tampa, Florida, 2008. Available at: www.ispe.org.
2. European Union. EU GMP Annex 11 Computerised Systems. Guide to Good Manufacturing Practice for EU Directive 2003/94/EC, Community Code Relating to Medicinal Products for Human Use, Vol. 4, 2003.
3. International Society for Pharmaceutical Engineering. GAMP® Good Practice Guide: Calibration Management. Tampa, Florida, 2001. Available at: www.ispe.org.
4. Pharmaceutical Inspection Co-operation Scheme. Good Practices for Computerised Systems in Regulated GxP Environments. Pharmaceutical Inspection Convention, PI 011-1, Geneva, 2003.
5. OECD. GLP Consensus Document: The Application of GLP Principles to Computerised Systems, 1995.
6. Gold Sheet. Electronic records/signatures requirements. F-D-C Reports Inc., October 2000.
7. U.S. Code of Federal Regulations Title 21, Part 211. Current Good Manufacturing Practice for Finished Pharmaceuticals.
8. U.S. Code of Federal Regulations Title 21: Part 58. Good Laboratory Practice for Nonclinical Laboratory Studies.
9. Food and Drug Administration. Use of microfiche and/or microfilm for method of records retention. Compliance Policy Guide 7150.13, 1989.
10. ACDM/PSI. Computer Systems Validation in Clinical Research: a Practical Guide, Version 1.1, December 1998.
11. U.S. Code of Federal Regulations Title 21, Part 211. Current Good Manufacturing Practice for Finished Pharmaceuticals.
12. UK IQA. Pharmaceutical Supplier Code of Practice Covering the Manufacturing of Pharmaceutical Raw Material, Active Ingredients and Excipients, document reference no. P00020, issue 2, Institute of Quality Assurance, Pharmaceutical Quality Group, 1994.

13. Food and Drug Administration. A memo on current good manufacturing practice issue on human use pharmaceuticals. Human Drug Notes, 1995; 3(3).
14. Toigo J.W. "Disaster Recovery Planning." London: Prentice Hall, 2000.
15. FDANews.com. Devices & diagnostics letter. March 2001, 28(9).
16. ISO 17799 on Information Security Management.
17. Wingate GAS. Computer Systems Validation: Quality Assurance, Risk Management and Regulatory Compliance for Pharmaceutical and Healthcare Companies. Boca Raton, FL: Interpharm Press, 2003.
18. International Conference on Harmonisation of Technical Requirements for Registration of Pharmaceuticals for Human Use. Good Manufacturing Practice Guide for Active Pharmaceutical Ingredients. ICH Harmonised Tripartite Guideline, November 10, 2000.
19. U.S. Code of Federal Regulations Title 21, Part 210. Current Good Manufacturing Practice in Manufacturing, Processing, Packaging, or Holding of Drugs; Part 211. Current Good Manufacturing Practice for Finished Pharmaceuticals.

12 | Phaseout and Withdrawal

INTRODUCTION

The end of the operational life of a computer system needs to be managed. This chapter discusses the implications of phasing out computer systems as a result of site closures, divestments, and acquisitions. Various system management and record management options are discussed. Key steps for all these situations include

- Retirement of the legacy system,
- Archiving of electronic records and documentation,
- Migration to a replacement system where appropriate, and
- Final decommissioning.

SITE CLOSURES, DIVESTMENTS, AND ACQUISITIONS

Disentangling computer systems as part of site closures, divestments, and acquisitions is becoming more complex as systems become more integrated. A decade ago, systems could be switched off with little consequence. Nowadays, record retention, data integrity, and security access requirements for GxP information mean that the management of computer systems needs careful planning.

Site Closures

There are no additional or reduced regulatory requirements for closing sites. Computer systems should be maintained in a state of compliance up until the very last day of their operational life. GxP records must be archived and stored for the required retention periods. Archived records should be readily retrievable to support critical quality operations like recall, customer complaints, and batch investigation. Computer systems should then be decommissioned, as discussed later in this chapter. Some computer systems may be disassembled and sent for installation at other sites as part of a program of drug product transfers.

Site Divestments

Divested sites can typically expect a regulatory inspection within the first year after sale. Regulatory authorities will typically be interested in how operational aspects of the business were managed through the divestment process.

There are two pivotal transition dates during site divestments. First, there is the date of sale/purchase for the geographic site with computer systems in situ as a going concern, and second, there is the date at which the inventory of work in progress is handed over as part of the ledger of assets. Disentanglement of computer systems must take account of data responsibilities as well as operational dependencies between site systems and the other retained systems in the divesting organization.

Systems Management

Compliance issues affecting system management during divestment can be summarized into three topics.

- Validation of the computer systems
- Operation and maintenance controls
- Inspection support and dependencies

New owners of legacy computer systems are dependent on the validation conducted before they took over responsibility for the systems. Due diligence exercises are usually conducted by the new owner before taking possession, followed by a supplier audit on the divesting organization's support organization. Replacement systems introduced by the new owner should, of course, be validated to the new owner's standards. This will include any data migration of records from legacy systems to new systems. Table 12.1 presents various system management options.

Typically, organizations that are divesting sites will want to sever all dependencies with the divested site other than links that may be required for an ongoing business relationship.

Table 12.1 System Management Options

Option 1	Option 2	Option 3
Retain computer systems and operate applications on behalf of divested site.	Transfer computer systems as is for divested site to operate applications.	Sever computer systems and require divested site to migrate to new system.
Advantages (largely to new owner)	*Advantages*	*Advantages (largely to former owner)*
• Continuity in business operations, no process changes. • Best option in terms of lowest immediate cost.	• Continuity in use of computer system. • New owner empowered to make own changes. • Less disruption and potential cost compared to option 3.	• Intellectual property protected. • Divesting organization does not become external service provider. • No inspection liability for divesting organization.
Disadvantages (largely to former owner)	*Disadvantages*	*Disadvantages (largely to new owner)*
• New owner locked into divesting organization for ongoing operation and maintenance support. New owner requested changes managed in context of divesting organization environment and priorities. • Potential confidentiality issues concerning shared processes/data. • Divesting organization could be included during regulatory inspection of new owner because of dependency on original integrity of production data.	• New owner may still require divesting organization's network and shared servers ("open system"), and hence extra controls may be required. • Divesting organization could still be inspected as a result of new owner regulatory inspection if computer systems have cross-reference dependencies on divesting organization documentation. Significant less risk of inspection than option 1.	• Discontinuity in use of legacy systems (may also be advantage to new owner). • Divestment of site may be delayed in order to bring new system into operation (may also be disadvantage to divesting organization). • Probably most disruptive and expensive option.
Implementation activities	*Implementation activities*	*Implementation activities*
• New owner conduct supplier audit on divesting organization as external service provider. • Formal contract of supply required. • Service level agreement established for maintenance and inspection support.	• New owner conducts due diligence on divesting organization's validation. Controlled copy of all relevant documentation made available to site, marked "copy of original." • Local procedures should be made autonomous by new owner.	• Agree on replacement system. • Conduct data migration from legacy computer systems. • New owner validates new systems in accordance with new owner standards.

This will reduce the regulatory dependency between the divesting organization and the new owner and the inspection vulnerability that it brings. For instance, a divested site may continue, for some period, to use the divesting organization's networks and MRP II system. An inspection on these computer systems at the divested site could result in regulatory corrective actions not only at the site but also across the divesting organization, even though the inspection was not directly on the new owner's organization. Some divesting organizations set a threshold of six to twelve months' support from the date for sale after which the new owner is expected to be self-sufficient. The new owner will be keen to preserve operational continuity including the transition to any new system through this period. Limited resources in the divesting organizations may mean that they cannot afford to divert operation staff to support the ongoing business of the sold site for any longer period.

The operation and maintenance of regulated computer systems have already been discussed in chapter 11. The new owner should ensure that whoever is supporting their computer systems (divesting organization, third party, or internal support group) is effectively managing the following requirements:

- Performance monitoring
- Repair and preventative maintenance
- Upgrades, bug fixes, and patches
- Data maintenance
- Backup and restoration
- Archive and retrieval
- Business continuity planning
- Security
- Contracts and service level agreements (SLAs)
- User procedures
- Periodic review and revalidation

The new owner should ensure that operation and maintenance procedures are clearly marked as approved and operated by them when they take over responsibility for supporting the legacy systems.

Both the new owner and the divesting organization may have particular sensitivities around inspection readiness. Regulatory observations on the new owner could imply corrective action for the divesting organization. Equally, the new owner will, at least for a period, be dependent on inspection support by the divesting organization for existing systems until they become sufficiently familiar with them. A transitional support agreement is typically built into the sale/purchase contract, possibly as a SLA. Both the divesting organization and new owner are usually keen for the new owner to become independent of the divesting company as soon as reasonably possible. Transitional arrangements, both technical and inspection support, typically last for less than a year.

Records Management

Compliance issues affecting records management during divestment can be summarized into four topics.

- Records retention
- Records retrieval
- Access controls
- Data integrity

Records retention affects both the divesting organization and the new owner. Figure 12.1 illustrates the various record management scenarios that might exist. The divesting organization is accountable for historical product data within required retention periods. Examples of GxP records include batch records and supporting information such as analytical results. Contracts should specify any transition period after the date of the site sale during

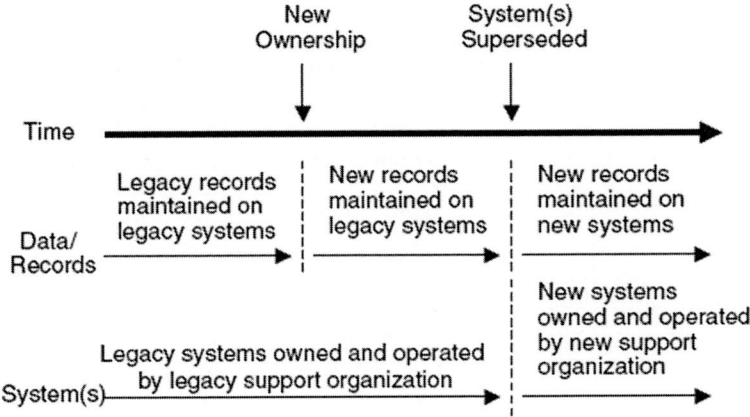

Figure 12.1   Divestment timeline.

which work in progress is completed and subsequently owned by the divesting organization. A complete copy of relevant product inventory information should be taken by the divesting organization. Operational data, meanwhile, typically becomes the responsibility of the new owner from the date of sale/purchase. Examples of GxP records would include calibration records, change control records, etc. Contracts should specify that the new owner will maintain historical records for a defined period and, where necessary, provide copies in support of a batch investigation by the divesting organization.

Just as for closing sites, GxP records should be maintained on systems that facilitate timely access in support of critical quality operations like recall, customer complaints, and batch investigation. The divesting organization will need to establish suitable record retention and retrieval systems. Alternatively, the divesting organization could ask the new owner to retain GxP record and provide a retrieval service where it has been agreed that the new owner will maintain legacy data. In this scenario, the new owner becomes a service provider, and formal contracts with SLAs should be agreed and audited by the divesting organization. The regulations require ready access to records and documentation; there are no requirements prohibiting this being the new owner on behalf of the divesting organization.

Access controls are needed to restrict change to authorized users and protect information from authorized modification (inadvertent or malicious). Computer applications managing records may be under the control of the divesting organization or the new owner.

Change control and audit trails are key aspects requiring management to assure data integrity and detect corruption. Reference should be made to chapter 11 where data maintenance is discussed. In addition, there may be electronic record management requirements, and more information regarding regulatory expectations in this regard can be found in chapter 13.

Site Acquisitions

It should be recognized that what is a phaseout for one organization may be a phase-in for another organization. Divestments are also acquisitions depending on which side of the fence an organization is sitting on. The new owner needs to consider both system management and records management requirements, as already indicated. Once new owner computer systems are installed, data migration will be needed from the legacy systems that can then be decommissioned. Migration and decommissioning are discussed further later in this chapter.

RETIREMENT

Computer system retirement describes the process of taking a system out of operation, that is, the system is not in active or routine use. This decision may be taken for a number of reasons including because the application is redundant, the computer technology is obsolete, it cannot comply with new regulatory requirements, or perhaps a replacement system with added functionality is planned. It is important to understand that retirement does not necessarily indicate the end of a computer system's life. A computer system may be brought out of retirement if required unless it has been fully decommissioned (scrapped).

When a GxP computer system is retired, the request is often made and implemented through a change control process. A retirement plan should be formulated to address the steps needed to retire the system, identify what (if any) new system will replace the current system, timelines for the retirement process, and the individuals responsible for the retirement process. The rationale for retiring the system must be documented.

The process of transferring records from the current to the new system should be an element of the project plan and must be qualified. Measures should also be in place to ensure that archived records on the retired system can still be accessed and read.

Once retirement is complete, a retirement report should be prepared in response to the retirement plan. After this, a decision as to whether or not to switch off and decommission the system can be taken (A checklist for system requirement is provided in Appendix 12.A).

Electronic Records Management

An electronic records management framework should be formulated and deployed. Steps within the framework might include the following (1):

- Determine and document which records need to be retained.
- Maintain a system for tracking the locations where electronic records are stored (hard drives on mainframes and personal computers, CDs, DVDs, and other media). This system is required to enable timely retrieval of electronic data.
- Ensure that the storage media can be read, maintaining mechanical tools such as microfiche readers and logical tools such as record indexes as required.
- Provide for off-site storage of the records needed for disaster recovery.
- Ensure that contracts with consultants, services providers, and other third parties require compliance with the company's record policies and permit periodic audits.
- Document policies and procedures for creating, storing, destroying, and indexing different types of information. Disposition should cover evidence that a record was destroyed, when it was destroyed, who destroyed it, and how it was destroyed.
- Ensure that similar records are treated similarly, whether paper or electronic.
- Follow authorized procedures in purging electronic records.
- Develop a procedure to suspend the disposition of records if a lawsuit is filed or is imminent.
- Document that policies and procedures have been followed in retaining and disposing of electronic records.
- Educate employees and other personnel authorized to use the company's advanced technologies about the company's records retention policy.
- Conduct periodic audits to ensure compliance with the company's records retention policy.
- Identify persons responsible for compliance with records programs.
- Provide review of the framework to adapt to changing technology, evolving company directions, and emerging judicial and regulatory trends.

A regular review of data stored in the archive is essential, not only as indicated earlier to detect any degradation of the storage media, but also to determine if the archive technology or record is becoming redundant. Periodic assessments will be needed to decide whether or not to maintain the archive electronic records. It may be decided only to maintain critical records such as those involved with batch records, batch sentencing, and recall, over longer periods of time. Once the retention period is over, a follow-on decision will need to be taken as whether to retain the electronic records for a further period or whether to destroy them. The minimum retention times for some example electronic records are indicated in chapter 12.

No electronically stored data should be destroyed without management authorization and relevant documentation. Other data held in support of computerized systems, such as source code and development, validation, operation, maintenance, and monitoring records, should be held for at least as long as the records associated with these systems are required to be retained (e.g., section 9 of Ref. 2).

Long-Term Preservation of Archive Records

FDA has clearly stated in an industry guide and conferences that compliance extends beyond the retirement of a computer system. For example (3),

> Recognising that computer products may be discontinued or surplanted by newer (possibly incompatible) systems, it is nonetheless vital that sponsors retain the ability to retrieve and review the data recorded by the older systems. This may be achieved by maintaining support for the older systems or transcribing data to the newer systems.

Long-term storage presents its own special challenges. The FDA expectations are summarized below (3).

- All versions of application software, and software development tools involved in processing of data or records should be available as long as data or records associated with these versions are required to be retained.
- Any data retrieval software, script, or query logic used for the purpose of manipulating querying, or extracting data for report generating purposes should be documented and maintained for the life of the report.
- [Pharmaceutical and healthcare companies] may retain these themselves or may contract vendors to retain the ability to run (but not necessarily support) the software.
- Although FDA expects [pharmaceutical and healthcare companies] or vendors to retain the ability to run older versions of software, the agency acknowledges that, in some cases, it will be difficult for [pharmaceutical and healthcare companies] and vendors to run older computerized systems.

The content of an electronic record must therefore be maintained in a form that is readable after the system used to create it is obsolete. For instance, a document originally stored today in Microsoft Word 7 format might need to be retained for regulatory reasons for 30 years when Microsoft Word 7 will no longer be available. This issue is compounded as Microsoft Word, for instance, has links to other applications that may be used to generate and maintain inserted content (e.g., PowerPoint diagrams) in the electronic record. It is insufficient just to store the text, as the record should appear to retrievers in its original format. Furthermore, the file formats may be dependent on systems software (operating systems, databases, compilers, etc.) and hardware. Potentially, software and hardware will need to be archived, but the practicality of this must be questioned.

A strategy must be put in place to migrate electronic records to new types of media as and when they are introduced. Media reliability is a potential problem but is fairly well understood. For instance, DAT and CD-ROMs have a notional operational life of 5 and 10 years, respectively, if they are not copied and kept in good storage conditions. It is more likely that the media technology will become obsolete within the electronic records operational lifetime. Media technology is currently being superseded every five years. The content of old media archive will need to be copied to new media archive to prevent any loss. It is wise not to rely on a single archive copy just in case the operational life of an archive copy degrades earlier than expected.

Wherever possible, employ standard data formats for archive copies to assist in any recovery process when original equipment to read specialist data formats may not be available. Industry standards are not widely used at present, with products often specifically implementing new functions and standards as a means of retaining existing customers and attracting new ones. Portability seems a long way off.

Pharmaceutical and healthcare companies need to keep appropriate computer systems that are capable of reading electronic records for as long as those records must be retained. Maintaining a legacy computer system just to read old records can be expensive, especially when this strategy might still require transfer to a new system or format at a later date when maintenance becomes impractical.

Where system obsolescence forces a need to transfer electronic documentation from one system format to another, the process must be recorded step by step and its integrity must be verified (1). An exact copy must be verified prior to any destruction of the original media. The obsolete system could alternatively be maintained as a legacy system, an approach that can be expensive and one that might still require transfer to a new system or format at a later date when maintenance will become impractical.

If the existing system is not validated, then the integrity of the data within the system cannot be relied on. Data cannot simply be transferred to a new electronic repository without data verification.

Metadata Considerations
Archive records needs to be accessible for a number of years, perhaps by people who were not involved in any way with its generation. For this reason, other related information needs to be stored alongside the original information, and this is usually referred to as metadata. Not only does metadata help to provide a context, which makes the information easier to retrieve, it can also be used to store vital regulatory evidence such as audit trails. Metadata is discussed further in chapter 13.

Retention Considerations
It is important to remember that the quality of the original record and its electronic capture can undermine its integrity. Image capture techniques may reproduce an original record very accurately, but if the original has insufficient dots per inch for clear reading, then the reproduction may not be useable. Electronic records are often not nearly as rugged and durable as their paper counterparts. The following factors may affect their life expectancy (1):

- Quality of storage medium
- Number of times the medium is viewed
- Care in handling
- Storage temperature and humidity level
- Cleanliness of storage environment
- Quality of the recorder used to write to the media

Business continuity plans should prompt the development of alternative media storage for critical records (e.g., paper or fiche) to enable the retention of access to these records in the event of a system failure or access to critical records once the system has been switched off.

Archiving Options
The long-term archive of electronic records would seem to be fraught with difficulties. Options for a way forward that would allow the original system and software to be decommissioned include the following:

- Maintain records on legacy system (time capsule).
- Emulate old software on new hardware/software.
- Migrate electronic records to new system.
- Store data in an industry standard format.
- Take a printed paper copy, microfilm, or microfiche.

An assessment must be performed and documented to determine the most appropriate method for preserving archives. Selection of the appropriate method must be considered within the context of the size, complexity, scope, and business impact of the system to be decommissioned. The method chosen must be documented using the appropriate change control form.

Maintain Legacy Computerized System
Retaining the legacy computer system as a "time capsule" provides the best method to achieve regulatory compliance in that original software and configuration functionality is maintained (4).

However, it is unlikely that the hardware and software will be supported by the supplier for the extended period that some record retention periods require. Any inability to maintain legacy systems will increase the likelihood that the retrieval may be unsuccessful. Therefore, it is recommended that this method is not relied on for periods of a few years beyond the supplier support for that system.

The key steps are as follows:

- Back up the entire system for contingency protection in case of failure.
- Reduce user access to "read-only" operation in relation to required electronic record, amend SOPs accordingly, and verify.
- Maintain the ability to restore the application, data, and operating environment on a vendor-supported hardware environment.
- Operate system only when needed.
- Ensure that integrity of electronically signed records is demonstrable.
- Verify record retrieval relevant to GxP processes.

Emulation of Old Software on New Hardware/Software
Suppliers sometimes provide this facility as part of an upgrade or replacement product. This option, if available, is a useful alternative to migrating records to entirely new computerized archive system. The integrity of the emulation facility must be verified.

The key steps are as follows:

- Back up the entire system for contingency protection in case of failure.
- Ensure that search and sort query reporting facilities are available or developed.
- Ensure that integrity of electronically signed records is demonstrable.
- Verify emulation created including record retrieval relevant to GxP processes.

Migrate Electronic Records to a New System
Electronic records are copied, possibly reprocessed, to make them accessible by a new computerized archive system. This can be a large and complex task but has the advantage that the new system is specifically designed for the purpose. This method, however, should not be used where the integrity of the original records being migrated can be disputed unless data accuracy checks are implemented. Data load requirements are discussed in chapter 10.

The key steps are as follows:

- Back up the entire system for contingency protection in case of failure.
- "Mirror" legacy data architectures within new system/database(s).
- Verify data migration including any support programs used.
- Ensure that search and sort query reporting facilities are available or developed.
- Ensure that integrity of electronically signed records is demonstrable.
- Validate new system created including record retrieval relevant to GxP processes.

Store Data in an Industry Standard Format
This approach works well with simple data configurations (e.g., small self-contained data tables). Because industry standard formats are used, the risk of technical obsolescence is reduced, and consequently, the likelihood of archive migration is minimized. Examples might include RTF rather than Microsoft Word 7 formats. Electronic records can also be stored as images (e.g., PDF format), although this increases storage volume requirements significantly. This method should not be used where there is a loss of data processing capability (e.g., search and sort cannot be run, and spreadsheet formulas are lost when the records are converted).

The key steps are as follows:

- Capture any necessary "metadata" in converted electronic records.
- Verify data migration including any programs used to generate output to archive media.

- Ensure that search and sort query reporting facilities are available or developed.
- Ensure that integrity of electronically signed records is demonstrable.
- Verify record retrieval relevant to GxP processes.

Take a Printed Paper Copy, Microfilm, or Microfiche
This sounds simple but may not be practical because the volume printing can be enormous. Printing may also be complicated where electronic records are made up of distributed data that requires electronic queries to retrieve it. These data structures are usually by far the most efficient storage mechanism for the electronic records. Printing can multiply the scale of archive task by a factor of one hundred. When large volumes of information are archived in this way, it is often pertinent to build a companion index to aid search and retrieval. A simple computer system can typically be developed to do this. Any programs or tools used to generate records suitable for archiving on paper, microfilm, or microfiche must be verified.
     The key steps are as follows:

- Capture any necessary metadata in converted electronic records.
- Verify data migration including any programs used to generate output to archive media.
- Ensure that search and sort query reporting facilities are available or developed.
- Ensure that integrity of electronically signed records is demonstrable.
- Verify record retrieval relevant to GxP processes.

     Regulatory authorities do accept printed copies provided prints are exact copies of original records (see discussion in chaps. 13 and 14). It is not necessary to reprocess archived information to prove the integrity of historical records, rather, it is expected that archived information can be used as constructive evidence to support the accuracy of historical records.

REPLACEMENT SYSTEMS
Companies should set and review a migration strategy that addresses both near-term and long-term corporate needs for individual computer systems. When migrating from manual to computerized systems of upgrading computer technology, the following implications should be considered:

- Configuration flexibility and capacity for expansion
- Financial cost
- Installation impact on operations
- Integration capability
- Performance improvement
- Personnel requirements
- Technology risk
- Compliance requirements
- Vendor capability

     Computerized systems being replaced should be run in parallel for a time with the new system to verify the migration (5,6). It is unacceptable, however, to rely on parallel operating as the sole basis of validation (7). The replacement system must be validated in its own right.
     Computer systems employed should meet or exceed the compliance requirements for the manual functions they replace (8). The new computer system must be at least as reliable as the computer system it replaces. Pharmaceutical and healthcare regulations do not mandate parallel operation of manual systems being replaced by computerized systems. If it is decided to have a period of parallel operation, then it should be run with the purpose of demonstrating that the computerized systems are at least as good as the old manual system, and only then can the manual system be decommissioned.
     Practitioners should not necessarily run the system in parallel until there are no "bugs;" the real question is whether the bugs can be managed. Parallel operation, of course, may not

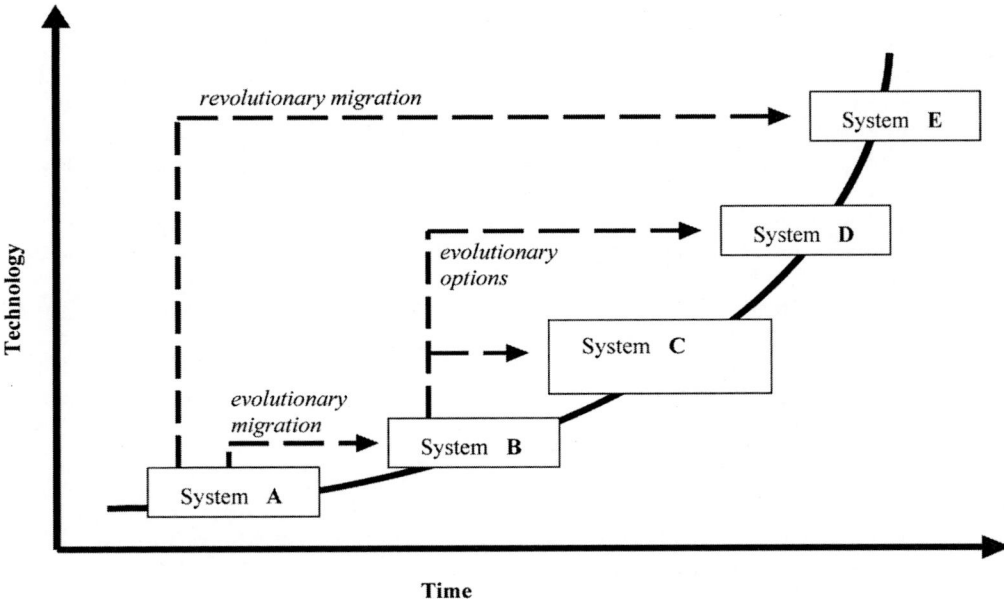

Figure 12.2 Migration routes.

always be possible or desired. The personnel requirements to run two systems together may be considered too high, or perhaps would require two production facilities.

Migration Strategy

Once a computer system has been implemented, the pharmaceutical and healthcare company must appreciate that computer technology is continually advancing. The next generation of microprocessor technology and software (half the price of double the functionality) has been arriving on average every two or three years, and there seems to be no reason to suspect that this trend will not continue. The next generation may consist of an upgrade to the computer system or its replacement. The various migration options are shown in Figure 12.2 (9). Not every option to upgrade may be taken, but care must be taken not to slip unknowingly into obsolescence when older versions are no longer supported by their suppliers. There may alternatively be reasons for ceasing all updates and establishing a legacy system.

Regular upgrades following an evolutionary migration are associated with low technology risks, but the combined compliance effort for every upgrade can be considerable. One aspect of computer systems that can be overlooked is the upgrading of hardware components (such as printers, monitors, instruments) and system software (such as operating systems and standard packages). In particular, system software is continually being upgraded, and while upward compatibility may be claimed on the initial release of an upgrade, confidence without supporting evidence should be limited. In this situation, it is recommended that installation of the upgrade is delayed until the new release is market tested.

Major step changes in technology across several generation upgrades (a revolutionary approach) will reduce the overall compliance effort, but the technology risk can be high. Examples of step changes include the cutover of large systems, such as MRP II/ERP, where parallel operation may not be practical because of the large volume of data and user interaction. To reduce the risk, larger systems are usually implemented in stages with phased cutovers for main functional elements. Within the MRP II/ERP systems, cutovers might include financials, customer services, and manufacturing.

Legacy Systems

Updated software and hardware are usually installed only if they include bug fixes or extend functionality, or if support for the old version is being removed. New versions of products, however, do not always bring operational benefits. Early adopters may find bugs yet undiscovered by the supplier. Equally, new product versions may actually degrade overall system performance (e.g., the original system memory is insufficient for the new data processing requirements). In these circumstances, it is advisable to retain and operate the original system, wait a period (perhaps six months) for a favorable track record to be established by other industry practitioners with the updated products, and only then install the revisions.

It may be possible to upgrade some systems because suppliers are no longer supporting their software or hardware products. A decline in the number of users of a product may lead a supplier to question the financial viability of their continued support of the product. Pharmaceutical and healthcare companies must discuss this topic with their suppliers so that a suitable compliance strategy can be planned.

Legacy systems are quite acceptable provided the original system has been validated to current GMP requirements and its compliance status is being maintained. Compliance activities will include the following:

- Establishing version and change control
- Collating documentary evidence that the software and hardware provided by a supplier have been developed and maintained under a quality assurance regime supporting regulatory compliance
- Reviewing documentation and preparing any supplementary information required to make the documentation complete
- Investigating the supply chain of any second-hand software and hardware used to maintain the system to establish whether it came from the original supplier and that it has not suffered any damage
- Testing critical features with additional tests to supplement, where necessary, supplier testing

If validation is not practical, pharmaceutical and healthcare companies should consider selecting and replacing legacy software and hardware with equivalent products or replacing the entire computer system. This may involve using alternative suppliers. New software or hardware in a legacy system will require verification to confirm that its functionality operates as required and that it does not affect what remains of the original system.

DECOMISSIONING

Computerized systems are generally decommissioned when they have become technologically obsolete, they have become too unreliable, or the process they are controlling has become obsolete. Decommissioning may also take place after an adverse regulatory inspection demands their replacement. The computerized system may, nevertheless, still be needed at a later date to support a new or rejuvenated process. The compliance requirements of decommissioning must be carefully considered. There are compliance issues if its documentation is needed in relation to a future recall of a drug product or if the system is used again in the future. Documentation may also be required if, for any reason, there is a regulatory investigation affecting the system.

Decommissioning will normally be based on an established shutdown procedure. There may, however, be special decommissioning operations that have not been used before on the live system. Operations management must ensure that decommissioning hazards are identified and that procedures are defined to avoid any accidents.

When decommissioning is complete, a short report should be composed to pass on any learning points. Only when this report has been issued and any archiving has been completed, can operations managers relinquish their responsibility for the system.

Validation costs for the new system could be halved if it is similar to the original application. If there is any possibility of the system being used again, then it should be dismantled and tagged, carefully packaged and labeled, and stored in a secure location. Documentary evidence of regulatory compliance must be archived. Retention periods for archived records are discussed earlier in this chapter. System specifications, verification protocols, user manuals, and maintenance procedures could prove very useful if the system is reused.

REFERENCES

1. Kahn RA, Vaiden KL. If the Slate is Wiped Clean – Spoliation: What It Can Mean for Your Case, Business Law Today. American Bar Association Publication, May/June 1999.
2. OECD, . GLP Consensus Document: The Application of the Principles of GLP to Computerised Systems. Environment Monograph No. 116, Environment Directorate, Paris, 1995.
3. Food and Drug Administration. Computerised Systems Used in Clinical Trials, Guidance to Industry, April 1999.
4. PDA. Good Practice and Compliance for Electronic Records and Signatures: Part 1—Good Electronic Record Management (GERM), published by ISPE & PDA, 2002. Available at: www.ispe.org.
5. Australian Code of Good Manufacturing for Therapeutical Goods. Medicinal Products—Part 1, section 9. Woden, Australia: Therapeutic Goods Administration, 1990.
6. European Union. EU GMP Annex 11, Computerised Systems, Guide to Good Manufacturing Practice for EU Directive 2003/94/EC, Community Code Relating to Medicinal Products for Human Use, Vol. 4, 2003.
7. Tetzlaff RF. GMP documentation requirements for automated systems: parts 1, 2 and 3. Pharm Technol 1992; 16(3): 112–124, 16(4):60–72, 16(5):70–82.
8. UK Department of Health. Good Laboratory Practice: the Application of GLP to Computer Systems. London, United Kingdom Compliance Programme, Department of Health, 1989.
9. Salazar JM, Gopal C, Mlodozeniec A. Computer migration and validation: a vendor's perspective. Pharm Technol June 1991.

Appendix 12.A

Example Retirement Checklist

This checklist provides the activities, concerns, and issues that may need to be addressed when a system is retired.

- Determine and document rationale for retiring system.
- Determine the impact of system retirement on other systems or users.
- The records retention requirements for the specified records will determine whether or not the records must be archived in a format that will allow for subsequent inspection of the records.
- If the system is being replaced by another system, retrieve archived records for loading into the replacement system and verify the migration.
- Develop the retirement schedule for system.
- Communicate the retirement schedule to client community.
- Document client community approval for retirement.
- Determine what system-related documentation should be archived (e.g., source code, life cycle documentation, user and technical manuals, security, and system change control logs, etc.).
- Document final disposition of system hardware and software.
- Retire any system specific SOPs.
- Determine appropriate storage medium for archived materials (e.g., ASCII format files, printed records stored to microfiche, etc.).
- Remove access to the system.
- Clean up any system logical/symbols/menu references.
- Delete the software and associated files from the system.
- Notify all affected personnel to discontinue regular system support activities (such as regular backups, preventive maintenance, etc.).

13 | Electronic Records and Electronic Signatures

INTRODUCTION

Many countries have now introduced regulations governing the use of electronic records, and the legal equivalence of electronic signatures to handwritten signatures. The basic requirements are based on established GxP expectations. Interpretation of the electronic record and signature regulations, and appropriate methods for achieving compliance has been subject to much debate and discussion in the industry. This chapter discusses the practicalities of compliance with U.S., EU, and Japanese regulations on electronic records/signatures and other principal international regulatory requirements and expectations. Topics covered include

- Practical definition of what constitutes an electronic record,
- Audit trails for creation, modification, and deletion of electronic records,
- Operational checks to verify authorized users,
- Logical and physical security measure for access control,
- Training for use of electronic records and electronic signatures,
- Legal admissibility of electronic signatures,
- Validation of procedural and technical controls,
- Impact of applying a risk-based approach, and
- Implications for electronic batch records.

ELECTRONIC RECORDS

Electronic records are defined here as records used in submissions for product licensing and those used for GxP decision/review processes. Electronic records may be automatically created by computer systems, created by direct manual entry of information to a computer system, or transcribed into a computer system from an original source sometimes after the original information was recorded. Electronic records include records created by scanning paper records. Appendix 13.A helps identify examples of GxP electronic records. Financial, legal data protection, and other non-GxP records are not covered here.

FDA looks to predicate regulations to identify records that when stored electronically will require electronic record controls (1). The predicate regulations, alas, were developed on the whole without this use in mind. Consequently, there are few explicit electronic record/ signature references, and consideration should also be given to expectations relating to general references to data and documents. Not surprisingly, there remains significant ambiguity in what exactly, on a practical level, is to be considered within the scope of definition of an electronic record (e.g., whether status flags, configuration parameters, and software programs are considered as electronic records). In response, the FDA has suggested that risk assessments be conducted to identify predicate rule records that may impact pharmaceutical or healthcare product quality and safety and hence require special management to preserve data integrity (1). Other regulatory authorities expect pharmaceutical and healthcare companies to make their own determination on the basis of published GxP regulations and guides on what are critical records in their computer systems and to apply electronic record controls accordingly (2).

Regardless of terminology, the process of identifying most important records is basically the same. Risk assessment and criticality are inextricably linked. International Society for Pharmaceutical Engineering (ISPE) has distinguished high- and lower-impact records with a view to the risk posed to patient and consumer health (1). Examples of high-impact records include product quality disposition decisions, batch records, laboratory test results, and clinical trial results. Examples of low-impact records include training, computer setup, and

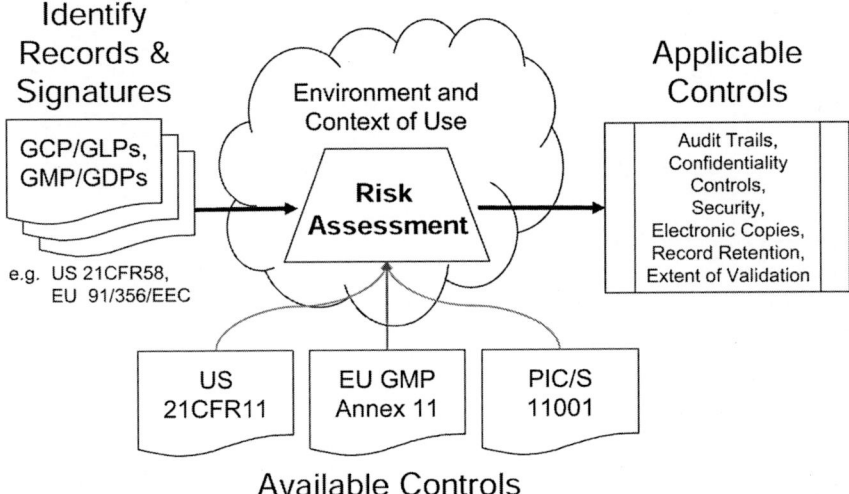

Figure 13.1   Electronic record risk management.

configuration parameters. The premise is to identify primary records protecting patient/ consumer health.

ISPE/GAMP has published guidance to help distinguish critical records and appropriate controls (3). Figure 13.1 outlines the basic concept being promoted. The process can be used to identify all records requiring specific management and control. The level of control should be commensurate with the importance of the record. Computer system validation is all that is necessary for low-risk records. Particular technical controls will tend to be needed to address high-risk records.

The risk assessment process can be conducted by examining record types to see if they are GxP or non-GxP, and then applying criticality checks, likelihood, and probability of detection criteria, as illustrated in Figure 13.2. The most critical records should be linked to direct patient/consumer impact. GxP noncompliance and broken license conditions are severe in their own right but not as critical as patient/consumer health in this analysis. Its likelihood will be influenced by the degree of human error in how the record is input and used. The probability of detection needs to take into account the probability of the impacted record being used. Once failure modes are understood, the appropriate design controls can be introduced. These should be documented and validated as part of the computer system life cycle discussed earlier in this book.

The FDA excuses electronic records from 21 CFR Part 11 where they are printed, and it is the printed copy that is used rather than the electronic version (1). The electronic record in these circumstances is considered "incidental." FDA will, however, challenge how such printed copies are used in practice by the organization to determine whether there is still a dependency on the electronic version. It is recommended that pharmaceutical and healthcare companies document their use of electronic and printed copies within SOPs. Printed copies must not be taken in an effort to sidestep regulatory requirements.

Record Life Cycle

A data flow analysis should be conducted to identify the creation and maintenance of electronic records. The life cycle of a record is shown in Figure 13.3.

Electronic records are created when their component raw data is processed and stored as a compiled record to a durable media. From this point on, electronic records may require audit trails to be maintained, as discussed later. Intermediate data is not considered an electronic record. Examples of electronic data used to compile electronic records include calculations used to determine sample potency range, individual temperature readings from an autoclave used to plot a temperature profile, individual points used to plot a peak in a chromatogram,

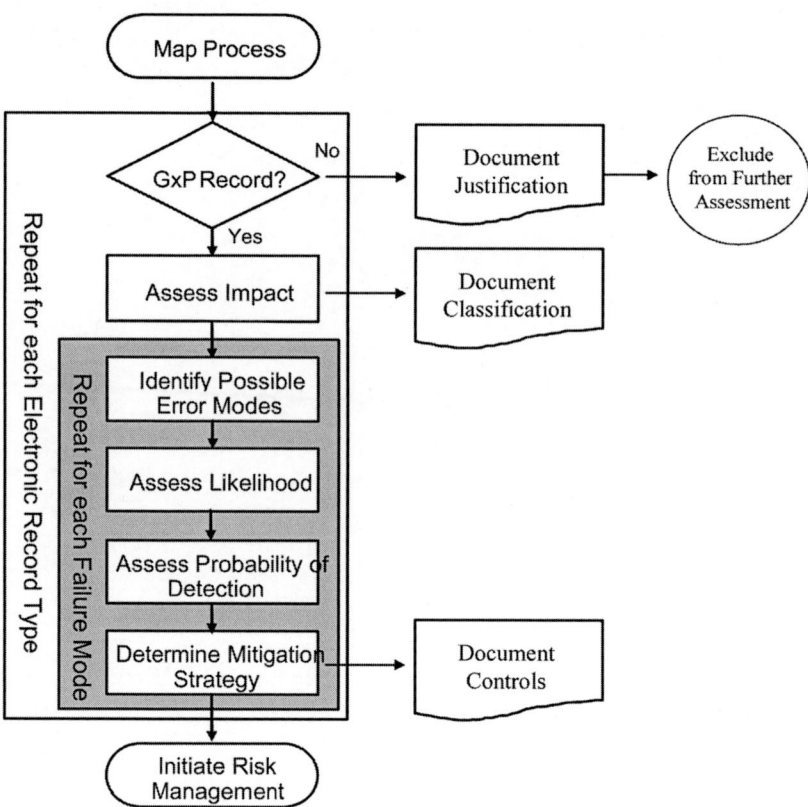

Figure 13.2   Electronic record risk assessment process.

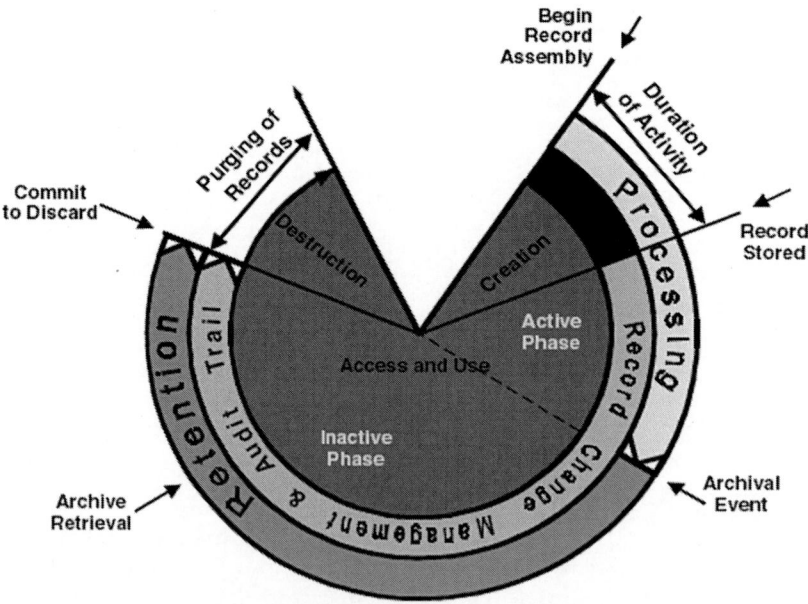

Figure 13.3   Electronic record life cycle. *Source*: From Ref. 5.

and configuration/control parameters used for equipment setup. Electronic raw data must be protected from alteration, periodically backed up and retained in a secure environment, and not deleted without necessary archiving. Data maintenance requirements are discussed in chapter 11.

FDA advises caution when specifying systems to create default raw data (4). A risk assessment should be conducted to determine if this is a sensible approach. Default data may not be appropriate in all circumstances, and inappropriate data capture may be deemed falsification of an electronic record.

It is important to appreciate that some data may be transient and will never be stored to durable media, while other transient data may be processed to derive data before being stored. Systems that only handle transient data are excluded from 21 CFR part 11. These are systems that acquire data, temporarily store it in files, which have no user access but as part of normal work flow pass that data on to a printer or another system before the process task is complete and the data is purged. Electronic buffers (including temporary files) cannot be considered transient data if user modifications to committed data are permitted. Battery backups for long-term retention of temporary storage invalidates the definition of transient data as to do situations where multiple cycles of so-called transient data are stored before being purged.

Another important consideration in modern computer systems is whether electronic records physically exist as single entities or whether they are split into connected elements that may exist in different physical locations. Computer and data architectures must support construction (and if need be reconstruction) of electronic records from their constituent parts.

Audit Trails

Audit trails log who created, modified or deleted the record and when ("time stamp"). Audit trails should explicitly identify either who or what created, modified and deleted the electronic record or allow that information to be unambiguously determined. FDA has suggested that predicate regulations may be used to determine whether or not audit trails on specific records are warranted (1). In practice risk assessments often provide a more practical way to determine when audit trails are appropriate (4).

Electronic audit trails are recommended for the most critical electronic records. An example audit trail is shown in Figure 13.4. This example does not imply any preferred format but rather has been included here to help demonstrate the principle of construction.

Hybrid audit trails electronically logging "last changed by" with date and link to related paper-based change records are acceptable for critical records so long as previous versions of the record are maintained. It may be possible in some cases to fulfill the audit trail requirements with a transaction database log. Some database designs require the user to execute a "commit record" step, while others commit the data as soon as the next field is tabbed to. In cases where a conscious decision to commit the record is required, data entered should not be defined as an electronic record until this action is taken, thus potentially simplifying the audit trail. In cases where there is no "commit" step, the audit trail should start as soon as each data item is entered.

Entirely paper-based change records alone should be sufficient for noncritical electronic records. Basic data maintenance controls described earlier in chapter 11 apply. Write-protected files with MS-DOS creation dates may provide sufficient control in some circumstances.

Audit trails must be available for the duration of their record's retention period and protected from any form of alteration. It should be possible to establish the current value and all previous values of a record by using the audit trail. Normal working practices (procedural and in-built computer controls) should prevent audit trail content being altered without definitive authorization by a second documented supporting party. Audit trails need to be available with their electronic records in human-readable form for purpose of inspection.

Time Stamps

Time stamps have three basic components: date, clock time, and time zone. The use of dates must be defined to avoid any misinterpretation (e.g., is 02/03/04 understood as February 3, 2004 or March 2, 2004). System clocks should be set to required levels of accuracy

Figure 13.4 Example audit trail.

File reference	Name	Time	Date	Record name	Data value	Unit	Action
Bx5 ProdX	Jim Smith	12:45:17	July 13, 1999	Temperature1	55	°C	Modify
Bx23 Prod Z	Rita Davies	12:40:03	July 13, 1999	Pressure1	17	bar	Create
Bx23 Prod Z	Rita Davies	09:32:45	July 13, 1999	Weight3	2362	g	Create
Bx23 Prod Z	Fred Jones	11:15:21	July 12, 1999	Weight3	Deleted	g	Delete
Bx23 Prod Z	Fred Jones	11:10:06	July 12, 1999	Weight3	2632	g	Modify
Bx23 Prod Z	Fred Jones	11:01:43	July 12, 1999	Weight3	2630	g	Create
Bx23 Prod Z	Jim Smith	10:13:42	July 12, 1999	Weight2	1750	g	Create

(e.g., hours and minutes). Time zones should be specified except where they can be unambiguously determined.

The application of timestamps should be periodically reviewed. Checks should be made to verify authorized clock changes, such as the change between summertime and wintertime, have been correctly implemented. Checks should be also made for unauthorized modification of system clocks and drift. Networked computer systems can be used to synchronize clocks. Procedural controls should be established to prevent unauthorized system clock changes in the absence of technical means.

Metadata

The FDA has in the past promoted the ability to reprocess electronic records, that is, to retrospectively process necessary raw data again using the same or equivalent conditions to "prove" the integrity of original records. Such processing requires metadata, data about data. Audit trail information is insufficient to reprocess electronic records. Details of the software originally used to create and maintain the records are also required to reprocess records together with hardware platform dependencies.

FDA has now reconsidered and now only requires the meaning and content of electronic records to be preserved (1). This is typically achieved through appropriate validation of supporting computer systems and applying audit trails where necessary to individual electronic records. Metadata will normally be managed through computer validation rather than as part of the electronic record as required previously by FDA. This is consistent with other regulatory authorities, which only expect constructive evidence to support the accuracy of electronic records.

Regulatory Inspection of Electronic Records

During the course of an inspection, it must be possible to provide the inspector to view regulated records either in electronic form or in paper form (human-readable form). Analogous to today's paper-based environment, companies must be able to make requested data available within a reasonable period (typically a few hours for on-line data, and between 24 to 48 hours for archived data). This is achieved by displaying the data on screen or by printing it out.

Databases are usually more able to meet the individual requirements of inspectors than is currently the case with paper-based filing systems. However, because the systems used can only be operated in accordance with their specifications, it cannot be assumed that they will be able to answer every conceivable query. For each individual case, it must, therefore, always be clarified with the inspector or agency how the data can best be collected for the purposed of the inspection—on the basis of what is technically feasible. This also applies to formats and media used to send data in electronic form to regulatory authorities.

If it is not possible to evaluate the requested electronic record without the corresponding application, then the inspector or agency should be consulted. In such circumstances direct access to computer terminals should only be given to trained personnel in accordance with established SOPs. The inspector does not necessarily have to access information directly. The inspector may agree to witness a trained user interrogate a company's computer systems on his behalf.

Copies of electronic records sent to regulatory authorities must be true and accurate. The copying process must preserve the content and meaning of the electronic record. Computer systems should be designed to facilitate the provision of copies of electronic records in common portable formats such as PDF or ASCII. Standard automated conversion or export facilities should be used in preference to proprietary methods. The ability to search, sort and trend records should be sustained as far as practically possible.

Record Maintenance

The World Health Organization GMPs suggest that electronic records should be stored and protected by backup transfer on magnetic tape, microfilm, paper printouts, or other means (6). There is no obligation to maintain electronic master copies of records where accurate nonelectronic copies exist. FDA has recently announced a similar position with the proviso that GxP processes do not refer back to the electronic version of the record (7). If GxP processes refer back to electronic records then FDA consider any disposition to paper or other nonelectronic media as incidental and consequently expect the electronic records to be maintained in electronic form. When printing an electronic record that will be retained for GxP purposes remember to authenticate it either through validation or with a dated handwritten signature applied directly to the print.

Retention periods for electronic records should be same as equivalent paper records. During the retention period stored records must be readily available. This applies to records stored on electronic and nonelectronic media. Electronic records like their paper record counter-parts should be purged at the end of their retention period. Procedures for disposal should be defined and require management authorization for final destruction of records. Some organizations keep a log of purged records for a further retention period so that they can demonstrate management and control of the purging process. Issues that need to be managed for long-term archiving for electronic records are discussed further in chapter 12.

E-mail messages, including attachments, should not be used as electronic records unless the e-mail system is validated as fit for this purpose. Validation requirements for e-mail include verifying integrity, authenticity and confidentiality through appropriate use of protocols, encryption and public key infrastructure. Individual e-mail messages can be managed as electronic raw data, prints taken with dated signatures annotated and an electronic master copy maintained.

Software Programs and Configuration

Compiled software including firmware is not considered an electronic record under the scope of regulations like 21 CFR part 11. Instead software source code and configuration are considered analogous with Standard Operating Procedures (7). GERM recommends that source code listings be retained and the software managed under change control (5). Where software listings are not available for commercial off-the-shelf (COTS) products then the version number should be recorded and any user specified operational parameters (setup) documented.

Recent Inspection Findings

The Quality Unit has failed to define and control electronic records (FDA 483, 2004).

Audit trails not maintained for raw data files (FDA 483, 2002).

Files could be deleted without an audit trail (FDA 483, 2004).

There is no actual information available as to the actual time for critical control steps (FDA 483, 2006).

Test results generated on HPLC could be deleted or overwritten without traceability (FDA 483, 2003).

LIMS test results can be edited and flagged as invalid with reason for result being invalid in the comments field, but the text comments are not included in hard copy when test results are reported for approval (FDA 483, 2003).

There is no record to determine the identity of the individual who reprocessed the data (FDA 483, 2004).

Inadequate control and protection of GMP laboratory data. All users log on at level that permits them to modify dates and times, turn on or off or modify audit trails, delete files without leaving record of deletion (FDA 483, 2006).

Audit trail software not properly configured to capture deletion, copying and renaming of files (FDA 483, 2006).

The electronic records are not protected in that no password system is in place, software programs do not secure files from accidental alteration and data loss, and data can be modified or deleted (FDA 483, 2006).

The software did not secure data from alteration, inadvertent erasures, or losses (FDA 483, 2007).

The software did not provide a working audit trail to track changes made to data (FDA 483, 2007).

Software not capable of performing audit of computer generated data from original raw data (FDA 483, 2007).

The software program used to control process equipment during production allows engineering access by several individuals, and there is no traceability for changes these individuals make (FDA 483, 2008).

ELECTRONIC SIGNATURES

The purpose of an electronic signature in a computer application is to enable an individual to authorize an electronic record (e.g., author, review, approve, comment, etc.). Appendix 13.B helps identify examples.

Electronic signatures can be based on nonbiometrics, biometrics or digital technology. An example of a nonbiometrics signature is the use of the traditional user identification (user ID) and password combination. Examples of biometrics signatures are fingerprints, hand geometry and retinal scans. Digital signatures can be based on cryptographic user keys.

The application of electronic signatures may be explicit or inferred by predicate regulations. For example, within U.S. CFR 211 (cGMP for finished pharmaceutical products) master production and control records are required to have the full handwritten signature of the person preparing the record and the signature of an independent checker, and signatures of persons performing and checking laboratory tests are required. It is important to appreciate, however, that most predicate rules were not written in anticipation of electronic signature requirements and not too surprisingly they do not comprehensively identify all expected signings. The same regulation, for instance, does not specifically identify recall, investigation, or out of specification records as requiring signature. Care must be taken not to rely too heavily on predicate rules.

The Preamble to the U.S. Government's 21 CFR Part 11 Rule on Electronic Records and Electronic Signatures says that the FDA will treat initials as signatures. EU and other international regulations are not this specific. In many instances initials are used as a convenient paper-based means of providing identification rather than as a legally binding short-form of a handwritten signature. If companies are to treat initials as identification rather than signature then they need to review current practice and convert where initials are effectively used as signatures into full handwritten signatures.

It is recommended that a work flow analysis is conducted to identify checkpoints appropriate for electronic signature. Not all existing handwritten signing or initialing need to be transposed as electronic signatures. In many instances signatures and initials have been implemented to facilitate identification of an individual rather than any legal signing (8). Consequently, the availability of audit trail information identifying individuals can remove historical instances of handwritten signatures and initials. A good example of this is the use of initials for nonsignificant activities recorded on batch records. Only significant or critical activities formally require signature. Nonsignificant entries on batch records only require the identification of an individual where relevant. Electronic signatures on electronic batch records are therefore not needed for all signatures and initials found on their equivalent paper records. Caution is in order as FDA has indicated that all signatures performed electronically, whether or not they are required by predicate rules, must comply with part 11. Therefore it is advisable to limit electronic signings to those required.

Admissibility

Regulatory authorities such as FDA, MHRA, MHLW and TGA expect electronic signatures to be legally binding electronic equivalent to handwritten signatures (1,2,9,10). FDA goes further and requires firms to notify them, in writing, of the use of electronic signatures as an equivalent to handwritten signatures. A standard format letter is provided for this purpose in a docket on the FDA Web site www.fda.gov.

Individuals that apply electronic signatures to electronic records are accountable and responsible for actions initiated under their electronic signatures. Electronic signatures should be declared within the pharmaceutical and healthcare company's organization to be the legally binding equivalent of the person's handwritten signature, or initials. Users should be trained to appreciate this equivalence. The consequences of falsifying data or signatures must be made clear.

- Employees should be disciplined for failure to follow company procedures regarding the use and/or administration of electronic record and/or electronic signatures.
- Employees should be considered for dismissal if they have deliberately falsified electronic records or electronic signatures.

User acknowledgement that they understand the significance of electronic signings should be documented. This can be done as part of the user request for system access.

Signature Attributes

Electronic signatures must be uniquely associated to one person, and must not be reassigned to another person. Before authorizing the assignment of an electronic signature, the company must identify the individual in question. If a person leaves the company, the signature is not transferable.

The signature application process must, by appropriate technical (computer controlled) and procedural means, ensure as a minimum that signature creation

- Can only be applied by its rightful owner,
- Cannot, with reasonable assurance, be derived, and the signature is protected against forgery using currently available technology,
- Can be reliably protected by the legitimate signatory against the use of others, and
- Can be linked to the data to which it relates in such a manner that any subsequent change of the data is detectable.

In addition signature creation must not alter the record being signed or prevent such records from being presented to the signatory prior to the signature process. Electronic signatures should be verified at the point of signing to ensure with reasonable certainty that the signature is authentic. Detected discrepancies must be alerted. The signature verification process itself must allow the contents of signed records to be reliably established and any security relevant changes to be detected.

Electronically signed records must contain the following information and this information must be visible each time the record is viewed or printed out:

- Name of the signatory
- Date and time of the signature
- Reason for signature (e.g., review or release)

E-mail messages should not be used to authorize GxP activities or approve GxP documentation unless the e-mail system is validated and individual e-mails comply with electronic record requirements.

Linking a Signature to an Electronic Record

Electronic signatures need to be unequivocally linked to their respective electronic records, and in such a way that they cannot be removed as the preamble to part 11 says by "ordinary means" (e.g., cut and paste). With electronically signed records, the link can be ensured by, for example, a unique relationship within a database or by an additional check using hash algorithms (the hash value of the record is signed)[a]. This unequivocal linking may present something of a technical challenge, but has been eloquently achieved in some applications designed to capture and embed handwritten signatures to documents, for example, PenOp and Entrust.

Identification Codes and Passwords

Administration of electronic signatures based on the combination of user ID and password must be designed in such a way that the misuse of an electronic signature requires the cooperation of at least two people (e.g., divulging of one's password to a colleague). Only the owner of the signature must know the combination, which typically means only the owner knows their secret password.

User Identification

The unique identifier could be a personal identifier. It does not need to be secret. Old tried and trusted technologies such as a log on, entered from the keyboard, or more effectively from a card reader or bar code are satisfactory but these are being superseded by newer ones which are on the way.

Passwords

The secrecy of the password is paramount for the integrity of the nonbiometric signature to be guaranteed. Thus a policy must be in place making this clear, and rigidly enforced. It is usual for the deliberate sharing of password to be a dismissable offense. Should such action be necessary, it should be publicized within the organization as a mechanism for ensuring the importance of the policy?

Secret passwords need to be sensibly constructed and maintained. They should be memorized and changed at regular intervals. These requirements are often seen as mutually exclusive! Frequent changes mitigate against remembering the password while never changing or "flip-flopping,", that is, changing between two at the prescribed intervals, risks their accidental exposure.

Guidelines need to be developed to manage this situation and should include the following:

- A minimum number of characters for passwords
- Mixed alpha and numeric characters
- Avoiding obvious combinations like one's car registration number or dog's name
- Not incrementally changing a character, so that it is possible to work out the current password from the key (starting combination) and the date.
- Setting suitable intervals between prompted password changes based on a risk assessment (4).

It was not uncommon in the past for passwords to be legally shared between teams of staff working together. This is acceptable practice so long as users are restricted to read-only access. Shared codes and passwords must not however be used where unique identification of an individual is required such as electronic signatures.

Operating procedures must be defined that specify the action to be taken if passwords, identification cards or the like are lost or compromised in any way. Staff occasionally forget their passwords or an attempt at intrusion is made. The software governing access should react to multiple attempts to gain access using an invalid password (say three) by locking out the individual and send an alarm to a responsible person to investigate, take appropriate action, and record the outcome. Some organizations require passwords to be changed every three months but there is no regulatory expectation the force password changes at particular

---

[a] A hash algorithm is a basic technique in asymmetric cryptography; it is an irreversible mathematical function that yields a certain value when used with a data file, for example, used with a document it always yields the same value but it is impossible to calculate the document from the hash value.

intervals. Indeed it could be argued that changing passwords too frequently will encourage staff to write then down because they will be unable to remember them.

It must be ensured that the unauthorized use of a user ID/password combination for an electronic signature is detected by the system and that the company's relevant authorities are notified immediately. It must always be ensured that the system design does not permit such misuse—this must be verified as part of the validation.

A suitable escalation procedure should be in place that enables, for example, a typing error when entering a password to be handled differently from an attempt to deliberately falsify a signature. If an authorized user incorrectly types in a password after three attempts, the system blocks the user from using this function and logs the incident. If a user attempts to sign in to an area for which they have no authorization, the system also logs this in a file. The appropriate specified authorities, such as the administrator or system owner, are notified immediately (e.g., by automatic e-mail).

Old identifiers should be removed when staff leave and must not be reissued, at least for a number of years (not less than 10) or there will be potential for repeating identifier password combinations and confusing audit trails. These passwords or ID cards must be immediately deactivated.

Hybrid Solutions

Hybrid solutions are systems that use handwritten signatures on printouts of electronic records as the means of approving those electronic records. All such handwritten signatures should be dated. The use of initials in lieu of full signatures is acceptable so long as this equivalence is stated in SOPs and accepted by the signer.

Handwritten signature must be linked to the associated electronic record. Including the unique file name and the date/time it was printed on the printout can help facilitate this. The paper and electronic copies of the record can be compared, if needed, to verify that they have the same content. The meaning of the signature should also be clear, either by labeling or annotating printouts with wording such as "Approved by." Labeling of printouts may be accomplished as part of the printing process, by a manual application of a stamp, or writing directly on the paper.

Digitized copies of handwritten signatures (e.g., bitmap images) are not in themselves electronic signatures, they are simply handwritten signatures recorded electronically. Use of uncontrolled bitmaps or other facsimiles of a signature, would not comply with the electronic signatures requirement, and may mislead viewers of the document into thinking that a valid signature had been given, when this may not be the case.

FDA has until recently only considered hybrid solutions as an interim measure until new computer systems can be implemented which fully comply with all 21 CFR Part 11 requirements. This position has now changed (1) and FDA in line with other regulatory authorities will allow the use of a hybrid solution as part of a final system. In either case robust procedures must be implemented for hybrid solutions to ensure electronic records are contemporaneous with printed copies.

Recent Inspection Findings

> Your written responses dated XXXX and YYYY stated that you would formalize the policy regarding electronic data and signatures and notify the FDA. You have not provided this documentation. This response is inadequate (FDA Warning Letter, 2000).
>
> You failed to certify to the FDA that the electronic signatures are legally binding (FDA Warning Letter, 2001).
>
> With regards to your responses concerning the use of electronic records and signatures, we find your reply inadequate. 21 CFR 11.100 requires that prior to the time of use, firms must certify to the Agency that the electronic signatures in their system, used on or after August 20, 1997, are intended to be the legally binding equivalent of traditional Handwritten signatures (FDA Warning Letter, 2001).
>
> No written procedures that would hold individuals accountable for actions taken under their electronic signatures. It is vital that employees accord their electronic

signatures the same legal weight and solemnity as their traditional handwritten signatures. Absent such written and unambiguous policies, employees may be adapt to make mistakes, under the erroneous assumption that they will be held to a lower level of accountability than they might otherwise expect when they execute traditional handwritten signatures (FDA Warning Letter, 2002).

The firm's assessment of the computerized systems such as XXXXX (inventory control system) and XXXXX (LIMS System) found them to be noncompliant with 21 CFR Part 11 requirements. For example, the firm indicated that XXXXX exhibited deficiencies in the area of "Signature/Record Linking" (FDA 483, 2001).

The electronic record requires electronic signatures, for which there is no timestamp on the record. (FDA Warning Letter, 2001).

Electronic documents are not electronically signed and there is no signed hard copy record (FDA Warning Letter, 2000).

The computers used to [. . .] were not part 11 compliant. They were not scheduled to be brought into compliance until an unspecified time during. . . . Remediation solutions and project plans had not yet been developed (FDA 483, 2004).

Use of electronic signatures does not meet the requirements of 21 CFR Part 11 (FDA 483, 2004).

OPERATING CONTROLS
Device Checks

Appropriate measures must be taken to ensure the validity of the sources for data and/or commands. Validation of the automatic interfaces, or a check of the input medium in the case of manual inputs, is performed as part of system validation. For example, if several sets of scales are connected to a network, only calibrated scales with the correct weighing range may be accessed. Similarly, for example, it should only be possible to use radio scanners assigned to a particular dispensary for weighing raw materials. In addition personal identification devices (e.g., company identity badges or ID cards that are used in conjunction with a password) should expire after a period and only be issued to authorized users. On expiry such devices should need formal renewal.

The use of devices should be fail-safe. Do not assume, however, fail-safe operation without thorough checking. A large pharmaceutical manufacturing site in the United States once found, for instance, that Visa credit cards could be used to gain access through their site-specific card-swipe system (11).

The security devices such as strip or bar code readers need to be tested prior to their first use and at regular intervals thereafter. Device checks can be incorporated into routine internal audit procedures. Many of these checks and procedures may already be in place as part of "Good IT Practice" to protect commercial confidentiality of information. A thorough review of IT security procedures and practices is nevertheless recommended to ensure compliance with electronic record/signature regulatory requirements.

Second Person Checks

Many regulatory authorities require a second person to check input of quality critical data into computer systems. The second check is not strictly necessary if the computer itself checks data the validity of data entered (e.g., format checks, limit checks). The nature of any automated check should be commensurate with the criticality of the data involved at its susceptibility to error during manual data entry (e.g., checksum used for batch record numbers). The U.S. GMPs for finished pharmaceutical products (21 CFR Part 211) were updated in 2008 to reflect that second person checks can be replaced by automated verification so long as one person still checks operations are properly performed in the first instance (12). Other regulatory authorities take a similar position.

Sequence Checks

The observance of critical sequences must be assured. System function checks should be implemented to verify steps that need to be performed in a particular order. For example, in the process "Input Data→Check Data→Release Data," the system must no permit step 2 to be performed before step 1, and step 3 must not be performed before step 2. Similarly when a document is first created, the system should automatically check whether another document with the same file name exists. If the file name is already in use on the system, the system needs to force the user to change it. After confirming the acceptability of a file name the document can be stored.

Continuous Sessions System Access

There is a potential for unauthorized access when a user terminal is temporarily vacated with the application open but the risk should not be exaggerated (11). The expectation is that this risk will be managed by locking out the access device after a defined period of inactivity. Care should be taken to define mechanisms with practical inactivity intervals because too shorter intervals will pose excessive inconvenience and too long intervals will pose higher security risks. The security situation needs to be seen in the context of the total security system from the perimeter fence to the seat in front of a terminal in a manufacturing suite or a dedicated office. Access around sites is often controlled and restricted and frequently for all but the most sensitive of tasks, others trained and authorized to carry out the same tasks will be around in the same area. These factors all mitigate against the need to have a very short lockout time.

Inactivity timeout period should be set in proportion to the risk of unauthorized access to the GxP applications concerned.

- A default inactivity timeout period of 40 minutes might be appropriate for most GxP applications when procedures require and systems allow users to manually lock the user interface (e.g., desktop, window) to the application or logoff the application when they leave it unattended.
- An inactivity timeout period of between 5 and 10 minutes might be more appropriate for GxP applications when procedural controls (i.e., manual lockout or logoff) are not practical.
- An inactivity timeout period is not required when an application confirms the identity of the user for all key inputs (e.g., authorizations and approvals) through a user identification and password check, or application of electronic signature.

If a user identifier or password fails for what ever reason then the application must require that both parts of the identifier/password combination are reentered and checked.

Special consideration needs to be given where applications of different risk of unauthorized access are on the same desktop. In this situation if and the desktop is the only mechanism to manage inactivity timeout (i.e., versus being controlled by the application server) then the inactivity timeout period should be set to meet the requirements of the most critical application. For short periods of inactivity FDA have suggested the use of a screen saver might be appropriate to prevent data entry until a correct password has been entered to reactivate the session (4).

Computer systems should be designed to limit the number of login attempts and record unauthorized login attempts (4). If a terminal were left open inadvertently and another person (authorized or not) entered the secret part of her/his password combination, the application should reject it as being incompatible with the identifier entered earlier. The application must then demand that both parts of the identifier/password combination are reentered and checked.

The use of electronic signatures during continuous sessions and timeouts also require management. The very first instance of the signature requires full input of the signature (user identification and password) unless the same user identification and the same password were entered at login in which case only the password is required. An exception here are initial start-up passwords, which must be changed when first used. All subsequent signatures only require input of the password provided that the person that initially logged continues to use the system without interruption.

Open and Closed Systems

Computing environments can be classified as open and closed. Computer system whose access is controlled by authorized individuals are referred to as closed system. This also applies to systems with modem access if a secured form of dial-in is used. Authorized individuals may be staff from any department within the organization who are responsible for GMP-relevant data, including internal or external personnel who are responsible for system maintenance.

Open systems refer to computer set-ups in an environment where a specific person who is responsible for the stored data does not control system access. A good example of an open system is the Internet. Specialist controls are required such as encryption and digital signature standards like Public Key Infrastructure (PKI) to provide necessary assurance in electronic records and electronic signatures.

Recent Inspection Findings

No safeguards to prevent unauthorized use of electronic signatures that are based on identification codes/passwords when an employee who has logged onto a terminal leaves the terminal without logging off. This is serious because another employee or individual could impersonate the individual who has already been logged on, and thereby easily falsify a record. The resulting batch production record, for instance, would not be an accurate and reliable indication of the lot's history. Moreover, in such an environment it would be fairly easy for the genuine logged on employee to disavow a signature as false, and thereby seek to avoid responsibility for actions under his/her signature (in the basis that is fairly easy for someone else to apply his/her electronic signature) (FDA Warning Letter, 1999).

Failure to establish and implement adequate computer security to assure data integrity in that during this inspection it was observed that an employee was found to have utilized another person's computer access to enter data into the XXXX computerized record system [21 CFR 211.68(b)]. Review 21 CFR part 11 for regulations pertaining to the utilization of electronic records and signatures, and security controls pertaining to both (FDA Warning Letter, 2001).

EXPECTED GOOD PRACTICE

Regulatory authorities such as FDA and MHRA have basic good practice expectations associated with the management and control of electronic records and electronic signatures. For instance, Annex 11 on Computerized Systems of the Guide to the EU GMP Directive 2003/94/EC (13) outlines the following expectations:

- Validation of systems to ensure accuracy, reliability, consistent intended performance, and the ability to detect invalid or altered electronic records.
- Backup of electronic records, their audit trails, and related documentation must be retained for a period at least as long as that required for the subject electronic records and must be available for review and copying by regulatory agencies.
- Determination that personnel (including external suppliers), who develop, maintain, or use electronic record/electronic signature systems have documented education, training, and experience to perform their assigned tasks.
- Security measures employed should be documented and approved.
- The release of batches of finished pharmaceuticals using a computer system for sale or supply regulated by European Union should allow only for a qualified person to release the batches and should clearly identify and record the person releasing the batches.
- Adequate alternative arrangements need to be available in the event of a computer system breakdown to maintain access to electronic records for business continuity purposes. The time to bring the alternative arrangements into use should be related to the possible urgency to use them (e.g., access to electronic records to effect a recall must be available at short notice).

These expectations logically extend to GCP, GDP, and GLP applications. MHRA are currently awaiting confirmation of legal status with use of electronic signatures to GCP and GLP applications.

Validation

GxP regulations require pharmaceutical and healthcare companies to maintain a system of documentation, and this includes any computer systems supporting the management and control of electronic records. Take, for example, EU GMP Article 9 of Ref. 13. Article 9.1 requires that documents be clear, legible, up to date, and retained for the appropriate period and Article 9.2 goes on to anticipate electronic records, the main requirement here being that supporting computer systems are validated. FDA also states a validation requirement for electronic records and electronic signatures (1).

It must be possible to demonstrate that the computer system is able to store the electronic record for the required time, that the data is made readily available in legible form and that the electronic record is protected against loss or damage. Both technical and procedural controls (Appendix 13.C) should be specified and verified including audit trail functionality and the successful application of electronic signatures to records. The identity and version of software and hardware must be recorded (4).

Backups and Archives

EU Directive 2003/94/EC sets out the legal requirements for electronic records within the context of GMP documentation. There is no requirement to maintain electronic copies of records in preference to other media such as microfiche or paper.

Electronic records (including associated electronic signatures and audit trails) must be accessible in a readable form for the duration of the retention period. The retention period depends on the time periods prescribed. Consequently appropriate backup and archive procedure should be established.

Backup copies of electronic records should be maintained on independent storage media either as electronic files within another computer system or on a dedicated storage medium (e. g., CD, DVD). Backup copies should be contemporaneous with the original source information (4). FDA recommends that backup and recovery logs are maintained to facilitate an assessment of the nature and scope of any data loss resulting from system failure (4).

Archiving should also utilize independent storage media. Appropriate measures must be taken to ensure data availability and integrity. In particular, it must be checked whether a different medium and/or data format is necessary for the archiving period. It is not necessary to be able to reprocess records from raw data (4). It must be possible, however, to reconstruct records from their constituent parts if this is how they are stored by within a computer or data architecture. The actual application software, operating system, hardware, and procedures used to manipulate data and records need not be retained so longs as records are stored in such a way as to facilitate basic search and sort queries. See also chapter 12.

Training

Training records should be maintained that demonstrate that individuals, as appropriate, have sufficient education, training, and experience to develop, use, and maintain computer systems that support electronic records and electronic signatures. Training can be limited to the specific operations a user is required to perform (4). It is expected that training should be conducted by qualified individuals on a continuing basis, as needed, to ensure users are familiar the system and any changes in functionality that may occur from time to time and the system is modified and upgraded. Refer to chapter 3 for more guidance on training.

Security

Suitable mechanisms must be put in place to control system access. *ISO 17799 Information Security Management* is a good practice standard and is often quoted by European regulators making observations concerning information security management. It has general commercial applicability and is used outside the pharmaceutical and healthcare industry. It recognizes the existence of regulatory requirements in certain industry sectors. As a standard user organizations can be certified and audited by an independent assessor. While such independent certification is not accepted by the regulator in lieu of their own inspections, it does provide clear evidence that an organization is committed to and has achieved basic good practice.

ISO 17799 includes implementation guidance including for instance risk management. Relevant topic areas in ISO 17799 include

- Security policy/organization,
- Personnel security,
- Physical/environmental security,
- Communications and operations,
- Access control, and
- System development and maintenance.

The standard attempts to encourage a security culture and shares many of the expectations of part 11. For instance, to improve personnel security ISO 17799 recommends definition of security in job responsibilities, personnel screening, training and awareness, and incident reporting. Access controls recommended by ISO 17799 also match part 11, for example, user registration, user ID and password management, definition of user responsibilities, user authentication, and monitoring system access for unauthorized access attempts. ISO 17799 further recommends that procedures be established with the aim of preventing the exposure of information through covert channels (e.g., computer virus, worms, or other harmful software).

Procedures and controls should be put in place to prevent unauthorized access and potential alteration of electronic records via external software applications that try to sidestep security features (4). Depending on the outcome of a risk assessment it may also be appropriate to defend against authorized attempts by external software to browse, query and report data extracts (4).

At any point in time it should be possible to provide a list of authorized users with job titles (roles) and access privileges (4).

In summary, electronic records must be protected against loss, damage, and unauthorized alteration. The level of security control and associated validation will be commensurate with risk posed to electronic records. Further guidance on security practice can be found in chapter 11.

Business Continuity Planning

Plans should be established to protect electronic records throughout their retention period. Plans should also aim to preserve timely retrieval of electronic records for business and regulatory scrutiny purposes. ISO 17799 prompts the following:

- Are projections of future capacity requirements made to ensure that adequate processing power and storage are available?
- Is there a managed process in place for developing and maintaining business continuity throughout the organization?
- Has a risk assessment been carried out to identify possible interruptions to business processes, that is, equipment failure, fire and flood?
- Have plans been developed to maintain or restore business operations in the required time scales following interruption to, failure of, critical business processes?
- Has a single framework of business continuity plans been maintained to ensure that all plans are consistent, and to identify priorities for testing and maintenance?
- Are business continuity plans tested regularly to ensure that they are up to date and effective?

Further guidance on business continuity can be found in chapter 11.

Recent Inspection Findings

Master production records are generated from a computer as electronic records without any apparent controls to assure authenticity and integrity (FDA Warning Letter, 2001).

In the event of that there is an equipment alarm or process utility alarm the computer system does not retain the alarm information as a permanent electronic record (FDA 483, 2002).

There is no documentation to establish that the system by which these [electronic] records were produced has been properly validated (FDA Warning Letter, 2001).

The firm did not validate software for electronic records and electronic signatures (FDA Warning Letter, 2000).

Your firm failed to validate the electronic documentation system [*and associated electronic records and signatures*] prior to implementation (FDA Warning Letter, 2000).

With regards to your responses concerning the use of electronic records and signatures, we find your reply inadequate. 21 CFR 11.10 requires these systems to be validated and to employ procedures and controls designed to ensure authenticity, integrity, and where appropriate, the confidentiality of electronic records. This part also required that adequate controls exist to ensure the distribution of, access to and use of documentation for system operation and maintenance. Your system must also guarantee that only authorized individuals can access the system. Please be aware of these requirements if you decide in the future to institute the use of electronic signatures/records (FDA Warning Letter, 2001).

The firm has not fully implemented procedures for control of all documents for their electronic records and electronic signatures (FDA Warning Letter, 2000).

Several laboratory instruments (including HPLCs and GCs) were considered non-compliant because of limited security of saved analytical methods (FDA 483, 2001).

The firm's assessment of the computerized systems such as XXXXX (inventory control system) and XXXXX (LIMS System) found them to be noncompliant with 21 CFR part 11 requirements. For example, the firm indicated that XXXXX exhibited deficiencies in the area of security (FDA 483, 2001).

Review of your XXXX files reveals they have not been properly validated, access to your system has not been limited, as well as other significant deficiencies (FDA Warning Letter, 2001).

Our investigator noted that the laboratory is using an electronic record system for processing and storage of data from the XXXX and HPLC instruments that is not set up to control the security and data integrity in that the system is not password controlled, there is no systematic backup provision, and there is no audit trail of the system capabilities. The system does not appear to be designed and controlled in compliance with the requirements of 21 CFR Part 11, Electronic Records (FDA Warning Letter, 2002).

IMPLICATIONS FOR EXISTING SYSTEMS

While compliance with electronic record/signature regulatory requirements is not without its challenges for new systems, they are small by comparison with those involved in bringing legacy systems into compliance.

Regulatory Expectations

Regulatory authorities expect electronic record/signature requirements to be addressed although some leniency may be given to older legacy systems. Shared regulatory expectations include the following:

- Drawing up a timetable indicating how and when compliance with electronic record/ signature requirements will be achieved in a company
- Creating an inventory of GMP-relevant computer systems
- Evaluating individual computer systems regarding their compliance, and creating a plan of what is to happen to these systems (e.g., whether they will be replaced by compliant systems or upgrades)

However, the FDA and other regulatory authorities expect more than planning to take place however. Meaningful progress is expected. Prioritization is accepted as it is widely recognized that it will take some time for all computer systems to come into full compliance. In the transition period procedural controls are expected to be put in place to compensate for any technical deficiencies.

Common Practical Issues

ISPE/GAMP identified the following common technical issues affecting practical compliance (11), and they are still very relevant today:

- *Password expiry*: How to manage when systems do not facilitate automatic periodic change. Also, how to make passwords are not repeated or take forms that are easily guessed (e.g., car registration number, street names, family names).
- *Retention of data*: Which data is required for retention and can any data be discarded. May be practical issues on volumes of data that need to be retained and how this can be managed. It must be practical to search data to find items of interest; otherwise why retain?
- *Audit trails*: How to approach systems that do not facilitate electronic audit trails.
- *User profiles*: In complex computer systems it is not always practical to have individual user profiles as the management of many thousand unique variants is too difficult. The role of all-powerful super users needs to be defined and controlled.
- *Timeouts*: What is an appropriate timeout period? Some systems do not facilitate timeouts when the user screen is not actively used. What is acceptable solution?
- *Virus management*: Viruses are a major threat to modern computer systems; problems with full compliance to part 11 should not prevent an organization deploying virus management tools.
- *Electronic signatures*: When should these be used (for instance, at the point where a record is authorized/approved or when the information is captured in a regulated document such as a batch record)?
- *Hybrid systems*: What constitutes a practical hybrid solution? How do we ensure paper and electronic records are contemporaneous?

If such practical issues cannot be overcome through application of risk management and procedural controls then replacement may be needed for remediation.

Management Approach

ISPE/GAMP suggest the following key management steps (14):

1. Agree the objective with senior management, gaining their support and approval. This is not a trivial task and may require the approval of significant resources.
2. Compile a list of systems, assign system owners, and identify those that need to be brought into compliance. Communicate the objective, including the support of management, to everyone involved.
3. Meanwhile, an *agreed* interpretation of electronic record/signature requirements for your organization must be developed. This is (politically) the most difficult step and is best done with a small team of informed individuals led by a senior technical manager. Adequate time for debate is necessary to allow all team members to be able to justify the decisions to others when challenged later.
4. Form a team to assess the level of compliance for every legacy GxP system against the agreed interpretation. This is most easily done using a checklist assessment, perhaps based on Appendix 13.C, and should be done together with the system owner.
5. Evaluate the strategic options for each system and agree the actions. There are five basic strategies:

   - Stop the activity (*this is unlikely to apply in many cases*).
   - Retire the system and return to paper (*there are still a few activities which were computerized by an enthusiastic amateur and which add complexity for little or no benefit*).
   - Develop an interim solution [*putting manual procedures in place as an extra layer of control to prop up the computerized system*].

- Upgrade the computerized system.
- Replace the computerized system *(here migration, record retention and retrieval become serious issues).*

    This is the most difficult technical step since sufficient knowledge of the application software to be able to make realistic estimates of the effort involved in updating versus replacement may take some time. The last three options are the most realistic and the latter the most expensive, involving as it does, specialist programming in often superseded languages for an application with a limited life.

6. Develop a master plan. It is sensible to include a prioritization step in assessing which systems should be replaced/upgraded first, a decision that should again involve the system owner. Factors affecting prioritization include the following:
   - The GxP criticality of the system
   - The extent of noncompliance (large, medium, small)
   - The age of the system or software and when its operational life is expected to end

Master Plans

The scope of master plans need not be limited to particular regulatory authorities or regulatory requirements such as 21 CFR part 11. Many pharmaceutical and healthcare companies have developed a more generic organizational plan to collectively address the various electronic record/signature requirements of those regulatory authorities that inspect their operations.

    Master plans should be reviewed and maintained on a regular basis, as business conditions may dictate changes to the actions originally agreed. Showing progress against this agreed plan is a vital part of being able to demonstrate progress toward compliance for legacy systems. Example progress charts are presented in Figure 13.5.

    Arguing with regulatory authorities that computerized systems cannot be rescued in terms of electronic record/signature compliance, or that there was no point in archiving data from a nonvalidated system is not a defensible position. As a bare minimum, interim measures will be expected to have been implemented until a "final solution" is implemented. Appendix 13.C outlines the use of procedural and technical controls applicable to both pharmaceutical and healthcare companies and their suppliers. The application of interim measures is discussed further in chapter 15.

Recent Inspection Findings

    We strongly encourage you to perform a thorough and complete evaluation of all your electronic records in accordance with 21 CFR part 11 as well as guidance generated by the FDA to assure conformance to our requirements. Do not limit your evaluation solely to the examples cited above (FDA Warning Letter, 2001).

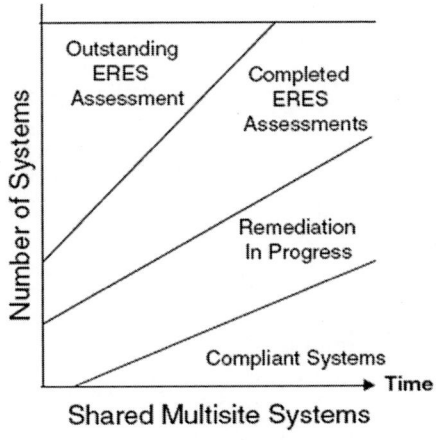

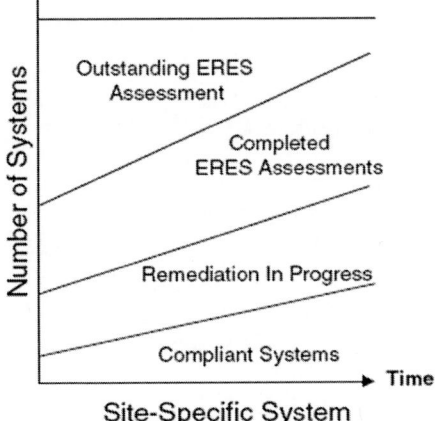

Figure 13.5 Example progress charts.

In addition, we further request details regarding steps your firm is taking to bring your electronic cGMP records into conformance with the requirements of 21 CFR part 11; Electronic Records; Electronic Signatures. ... please outline your firm's global corrective action plan, including timeframes for correction, to address this Part 11 issue (FDA Warning Letter, 2000).

There was no indication during the inspection that the XXXX system [*and associated electronic records and signatures*] was being validated. In fact there was no evidence that a concurrent manual system was in place (FDA Warning Letter, 2001).

No assessments of requirements, or procedures for such assessments, to assure integrity of electronic records (FDA 483, 2005).

IMPLICATIONS FOR NEW SYSTEMS

Electronic record and/or electronic signature requirements must be specified and taken into account during any selection process for all new computerized systems. Relevant third-party suppliers of bespoke systems should have requirements contractually defined.

Pharmaceutical and healthcare companies are expected to work with key individual suppliers and wider industry user groups to help suppliers develop electronic record/ signature-compliant COTS products and that such products will be migrated into use as they become commercially available. A checklist assessment is suggested for COTS products to evaluate their level of compliance, perhaps based on Appendix 13.C. Current versions of COTS products need not necessarily be specifically customized to provide full electronic record and electronic signature functionality; the development risk with bespoke development must be balanced with the complexity and criticality of any change. It should be possible to compensate for the lack of key software functionality by adding user procedural controls.

Open Source software must be fully evaluated by the user organization to assess relevant electronic record/signature functionality since there is no supplier accountable for functionality definition, product development, or maintenance (8). Caution should be exercised since it is very difficult to truly demonstrate the trustworthiness of such software in the absence of life cycle development and support documentation.

Hazard Study

The PDA have recommended what is essentially a hazard study process to reveal where record integrity may be compromised (8). This following checklist has been developed for use with both new and existing systems.

1. Layout the basic work flow of computer applications and conduct a data analysis to identify electronic record creation and maintenance (include identification of supporting raw data).

2. Answer such questions as the following:
   - Where do the records go?
   - Who uses them, internal and external to the company?
   - How are they used?

3. Identify critical steps along the work flow where the integrity of records may be compromised through use or transmission.
   - Incomplete records
   - Duplicate records
   - Communications corruption
   - Transmission gaps/ chain of custody issues
   - Opportunities for record corruption

4. Identify levels of control that exist or will be needed for these records.
   - Identify how records are secured, backed up, and archived.
   - Identify how records are restored to active systems from backup.
   - Examine disaster recovery and security requirements.

5. Determine the extent of validation of the computing environment.

Application of this checklist can be incorporated into the hazard study process discussed in chapter 7.

Handling Electronic Batch Records

Batch records must provide a full history of the manufacturing-related operations constituting the production, final review and release of a batch (also known in some counties as a "lot") of medicinal product. Approval of a batch signifies that it has been manufactured in accordance with GMP and that the final product satisfies its licensed specification. In Europe approval of API and finished pharmaceutical products can only be given by a qualified person (15).

Conventionally it has been necessary to provide a single consolidated full printed copy of the batch record for both final review and approval and subsequent inspection by regulatory authorities. Nowadays integrated computer systems with an over-arching data architecture mean that different data components of electronic batch records will exist in multiple computer systems. There is no need for each computer system to send its part of the final batch record to a single computer system (sometimes known as a historian) responsible for final physical collation and storage so long as the component parts can be virtually compiled using computer logic. This macro approach reflects what happens at a micro level within a computer system where data is logically organized within a coordinated data architecture but will may be physically distributed throughout the computer's memory. The important thing is that the batch record can be collated and presented in a human-readable form for review or inspection if needed.

Review and inspection in human-readable form does not necessarily mean that a paper copy is required. Our comfort with print can constrain the use of suitable alternative arrangements. It should be acceptable to rely on validated electronic review of batch records using computer screens, especially if the entire batch record is held electronically. Many systems now facilitate this approach.

The ability to review an entire batch record electronically facilitates another functional enhancement, namely "review by exception" (RBE). On the basis that manufacturing-related operations are validated it is feasible to limit batch review to critical process exceptions because everything else is in control and within manufacturing and product specifications. RBE reduces and potentially eliminates need for reviewing acceptable data and trends (16). GMP decisions such as release, quarantine, and reject can be made from reviewing manufacturing batch exceptions. If the responsible person for batch release needs to follow up on an exception, or is just curious, then it must be possible to retrieve the full batch record for review.

It needs to be emphasized that RBE should not be entered into lightly. Not all regulatory authorities currently accept RBE as described in relation to the use of computer systems for batch disposition. It is strongly recommended, therefore, that RBE is only applied to stable manufacturing processes that are well understood with defined critical quality attributes (CQA) and critical process parameters (CPP). Such processes should produce products that rarely have an out of specification result.

INSPECTION ANALYSIS

Pharmaceutical and healthcare companies should review their computer systems in regard to common regulatory observations so that mitigating action can be taken. An analysis of FDA inspection findings referring to electronic records and electronic signatures is given in Figure 13.6. This analysis is based on a review of 17 Warning Letters and "483" form observations issued by FDA that reference 21 CFR part 11 since it became effective in August 1997. It is interesting to note that only one of these occurred after 2003. A full list of computer related regulatory inspection findings reviewed in this book can be found in chapter 14.

The most common observation made by FDA concerns the lack of, or incomplete, audit trails. This is often associated with the incorrect identification of electronic records. Specifically, the Warning Letters referred to chromatography data systems (CDS), electronic document management systems (EDMS), databases, batch records, change records and device history records.

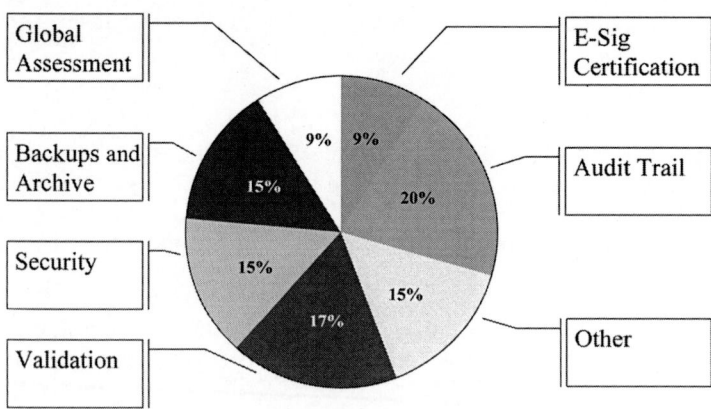

Figure 13.6 Part 11 Warning Letters observation analysis.

The lack of validation or incomplete validation was the next most common observation. The need for prospective validation of electronic record/signature capability during computer system implementation is stressed in two of the six Warning Letters making an observation on validation. The computer systems concerned were computer-aided drawing (CAD), process control systems, record keeping systems, and EDMS.

The next most cited group of observations concerned backup and archive. Systematic backups are required to defined schedules. Backups and archives must be maintained for the duration of the record retention requirements and records readily retrievable. The Warning Letters making these observations referred to CDS, Spreadsheets, electronic drawings and the implied use of a CAD application, complaint files, and Device History Records.

The next most cited group of observations concerned backup and archive. Systematic backups are required to defined schedules. Backups and archives must be maintained for the duration of the record retention requirements and records readily retrievable. The Warning Letters making these observations referred to CDS, Spreadsheets, electronic drawings and the implied use of a CAD application, complaint files, and Device History Records.

Security as a topic is referred to the same number of times as backup and archive. Security issues raised stress the need to limit access to computer systems to protect records, and in one instance deficient password controls are mentioned. Computer systems referred to include CDS, CAD, record keeping systems, and spreadsheets.

Failure to submit certification to FDA that the use of electronic signatures in an pharmaceutical and healthcare company's organization has the same legal standing as handwritten signatures accounts for just under one in ten Warning Letter observations. This is a simple observation to correct with the issue of a single letter of declaration given to FDA as described earlier in this chapter.

Three of the Warning Letters referred to a wider organizational review of electronic record/ signature requirements beyond the scope of the particular computer systems that were the focus of the original inspection. Pharmaceutical and healthcare companies should ensure they have a compliance plan that covers the whole part of their organization subject to 21 CFR part 11.

The remaining Warning Letter observations covered a variety of topics that only appeared once or twice as an observation and did not group naturally with the analysis above. These observations concerned human-readable copies of electronic records for electronic drawings and complaint files, taking hard copies of electronic change control records, continuous session controls in relation to integrity of batch records recording operator actions and detecting invalid records.

CURRENT REGULATORY DEVELOPMENTS

At this time this book was written two major regulatory guidance documents are being prepared: a revision to U.S. CFR part 11 on electronic records and electronic signatures, and a revision to EU GMPs impacting use of electronic records and signatures. Both of these initiatives may take some time before final regulatory requirements are issued.

U.S. CFR part 11 is being updated to reflect the principles already established in "scope and application" guidance (1). Such changes are consistent with the approach to electronic records and signatures discussed in this chapter.

European proposals distributed for industry comment regarding EU GMP chapter 4 (concerning records and documents) and annex 11 [concerning electronic records and signatures (17)] are potentially much more controversial (18). These changes include the following:

- Extending the scope of currently what would currently be considered electronic records (e.g., all raw data generated will need to be retained and subjected to electronic record controls)
- Applying controls without making reference to the acceptability of a taking a risk-based approaches (e.g., challenge testing data limits, and detailed controls expected for databases)
- Applying controls that are not practical with current technology and would likely inhibit use of electronic records and signatures (e.g., audit trails must be immutable rather than protected by "normal" means, printouts of records must include all metadata, and linking electronic signatures to all metadata)

Concerted industry feedback will hopefully lead to a revision that is less prescriptive and promotes a risk-based approach. EFPIA and ISPE/GAMP suggest that companies should be able to define their approach to proportionate to the level of risk. The application of procedural controls to compensate for practical technical limitations should also be allowed. If these principles are accepted, and they should be given EMEA agreement on ICH Q9, then the EU GMP revisions will also be consistent with the approach to electronic records and signatures discussed in this chapter. More importantly, European requirements would be broadly aligned with current thinking behind U.S. CFR Part 11.

REFERENCES

1. Food and Drug Administration. 21 CFR Part 11. Electronic Records; Electronic Signatures—Scope and Application, Guidance for Industry. Rockville, MD, 2003.
2. Pharmaceutical Inspection Co-operation Scheme. Good Practices for Computerised Systems in Regulated GxP Environments, Pharmaceutical Inspection Convention, PI 011-1. Geneva, 2003.
3. GAMP Forum. Risk assessment for use of automated systems supporting manufacturing processes Part 2—risks to records, pharmaceutical engineering, 2003.
4. Food and Drug Administration. Computerised Systems Used in Clinical Investigations: Guidance for Industry. Rockville, MD, 2007.
5. PDA. Good Practice and Compliance for Electronic Records and Signatures: Part 1—Good Electronic Record Management (GERM). ISPE and PDA, 2002. Available at: www.ispe.org.
6. World Health Organization. WHO Expert Committee on Specifications for Pharmaceutical Preparations, thirty-second WHO Technical Report. Geneva, 2000.
7. Compliance Policy Guide. Computerised Drug Processing, 7132a: Source Code for Process Control Application Programs (Guide 15). Rockville, MD: Food and Drug Administration, 1987.
8. PDA. Good Practice and Compliance for Electronic Records and Signatures: Part 3—Models for System Implementation and Evolution. ISPE and PDA, 2003. Available at: www.ispe.org.
9. Directive 1999/93/EC of the European Parliament and of the Council of December 13, 1999 on a Community Framework for Electronic Signatures. Official Journal of the European Communities, January 19, 2000.
10. Japanese Pharmaceutical Manufacturers Association. Guideline for the Application of ERES in Production Control and Quality Control for Human Drug Manufacturing, 2002.
11. Selby D. Practical implications of electronic signatures and records. In: Wingate G, ed. Validating Corporate Computer Systems: Good IT Practice for Pharmaceutical Manufacturers. Buffalo Grove, IL: Interpharm Press, 2000.
12. U.S. Code of Federal Regulations. Current Good Manufacturing Practice for Finished Pharmaceuticals, CFR Title 21, Part 211. Washington: U.S. Government Printing Office, 2008.
13. European Union, EU Guide to Good Manufacturing Practice for EU Directive 2003/94/EC, Community Code Relating to Medicinal Products for Human Use, Vol. 4, 2003.

14. International Society for Pharmaceutical Engineering. GAMP® Good Practice Guide: Risk-Based Approach to Electronic Records and Signatures. Tampa, Florida, 2005. Available at: www.ispe.org.

15. European Union Directive 2001/83/EC. Qualified Persons, Article 5.1, 2001.

16. ISPE. GAMP®5: Risk-Based Approach to Compliant GxP Computerised Systems. Tampa, Florida: International Society for Pharmaceutical Engineering, 2008. Available at: www.ispe.org.

17. European Commission. Draft Annex 11—Computerised Systems (EU Guideline to Good Manufacturing Practices for Medicinal Products for Human and Veterinary Use), Public Consultation Document. Brussels, April 2008.

18. McDowall RD. Major changes proposed for European Union good manufacturing practice (GMP) annex 11, Scientific Computing, 2008.

Appendix 13.A   Example Electronic Records

Electronic records can be identified by searching regulatory requirements for the key words "record" and "document." This appendix is based on U.S. Code of Federal Regulations and EU directives, and is not intended to be exhaustive. A more definitive list has been published by ISPE/GAMP.

Summary of references in GCP

- Consent documents (informed and institutional review board)
- GCP protocols and amendments
- Clinical investigation and changes
- Financial disclosure forms and reports
- Investigator statement
- New drug application forms and submission statements
- Clinical study data and ownership statements
- Investigational drug shipment and disposition

Summary of references in GLP

- Equipment maintenance and calibration records
- GLP protocols and amendments
- QA audit records
- Standard operating procedures
- Final study reports and QA statements
- Training records
- Job descriptions

Summary of references in GMP

- Equipment cleaning maintenance records
- Master production and control records
  - Components specifications
  - Drug product containers and closures specifications
  - In-process materials
  - Packaging material
  - Labeling specifications
  - Drug products specifications
  - Procedures and specifications

- Batch production and control records, including
  - Products from contractors
  - Production records
  - Packaging records
  - Laboratory tests results (QC records)
  - Reprocessing of batches
- Biological sterilization
- Laboratory tests
- Out of specification investigations
- Customer complaints
- Standard operating procedures
- Training records
- Job descriptions

Appendix 13.A   Example Electronic Records (*Continued*)

Summary of references in GDP

- Distribution and shipment records
- Adverse event reports
- Recall records
- Customer complaint records
- Standard operating procedures

Source: From Ref. 13.

Appendix 13.B   Example Electronic Signatures

The regulated use of signatures can be determined by searching regulatory requirements for the key words "signature," "initial," "approval/approved," "authorization/authorized," and "certify." This appendix is based on U.S. Code of Federal Regulations and EU directives, and is not intended to be exhaustive. A more definitive list has been published by ISPE/GAMP.

Summary of references in GCP

- Consent documents (informed and institutional review board)
- GCP protocols and amendments
- Clinical investigation and changes
- Financial disclosure forms and reports
- Investigator statement
- New drug application forms and submission statements
- Clinical study data ownership statements

Summary of references in GLP

- GLP protocols and amendments
- Exact transcripts of raw data and changes to raw data.
- QA audit records
- Authorization for animal treatments
- Changes to, and deviations from, standard operating procedures
- Final study reports and QA statements

Summary of references in GMP

- Major/critical equipment cleaning, maintenance and use
- Master production control and batch production control records
  - Components
  - Drug product containers
  - Closures
  - In-process materials
  - Packaging material
  - Labeling
  - Drug products
  - Procedures and specifications
  - Products from contractors
  - Final batch production record
- Laboratory tests
- Out-of-specification investigations
- Significant steps in production (e.g., dispensary and weighing)
- In-process controls
- Formal checks, where appropriate
- Deviations and unusual event records
- Rejection of batches
- Reprocessing of batches
- Recovery of batches
- Standard operating procedures

Appendix 13.B Example Electronic Signatures (*Continued*)

Summary of references in GDP

- Distribution and shipment records
- Adverse event reports
- Return records
- Recall records
- Customer complaint records
- Standard operating procedures

Source: From Ref. 13.

Appendix 13.C Procedural and Technological Controls for 21 CFR Part 11

Clause	type of Control	responsibility	notes
11.10	Procedural	Pharmaceutical manufacturer	This clause specifies a number of specific controls. The pharmaceutical organization will need to demonstrate a system of self-inspection audits to demonstrate compliance with the procedures and controls listed below.
11.10 (a)	Procedural	Pharmaceutical manufacturer	ER/ES systems need to be validated.
	Technological	Supplier	ER/ES system should be able to identify changes to electronic records to detect invalid or altered records. In practice, this means having an adequate audit trail, which can be searched for information, e.g., to determine whether any changes have been made without the appropriate authorizations.
11.10 (b)	Technological	Supplier	ER/ES systems should allow electronic data to be accessed in human-readable form.
	Technological	Supplier	ER/ES systems need ability to export data and any supporting regulatory information (e.g., audit trails, configuration information relating to identification and status of users and equipment).
11.10 (c)	Procedural	Pharmaceutical manufacturer	Pharmaceutical organizations should specify retention periods and responsibilities for ensuring data is retained securely for those periods.
	Procedural	Pharmaceutical manufacturer	Pharmaceutical organization needs a defined, proven and secure backup and recovery process for electronic data.
	Technological	Supplier	ER/ES Systems should be able to maintain electronic data over periods of many years regardless of upgrades to the software and operating environment.
11.10 (d)	Procedural	Pharmaceutical manufacturer	Pharmaceutical organization needs procedures defining how access is limited to authorized individuals. Managing super-user account should be given special consideration.
	Technological	Supplier	ER/ES Systems should restrict access in accordance with preconfigured rules that can be maintained. Any changes to the rules should be recorded.

Appendix 13.C   Procedural and Technological Controls for 21 CFR Part 11 (*Continued*)

Clause type of responsibility notes
Control

Control			
11.10 (e)	Procedural	Pharmaceutical manufacturer	Pharmaceutical organization needs procedure to maintain the audit trail (see 11.10 (c) above).
	Technological	Supplier	ER/ES systems should be capable of recording all electronic record create, update, and delete operations. Data to be recorded must include as a minimum: time and date, unambiguous description of event, and identity of operator. This record should be secure from subsequent unauthorized alteration.
11.10 (f)	Technological	Pharmaceutical manufacturer and supplier	Where operations are required in a predefined order, e.g., in batch manufacture, the ER/ES system should enforce that ordering through the system's design.
11.10 (g)	Procedural	Pharmaceutical manufacturer	Pharmaceutical organization needs procedures defining how the operations are to be performed and that staff have been trained in their use.
	Technological	Supplier	ER/ES Systems should restrict use of system functions and features in accordance with preconfigured rules that can be maintained. Any changes to the rules should be recorded.
11.10 (h)	Technological	Pharmaceutical manufacturer and supplier	Where pharmaceutical organization requires that certain devices act as sources of data or commands, the ER/ES system should enforce the requirement.
11.10 (i)	Procedural	Pharmaceutical manufacturer	Pharmaceutical organization's staff who use electronic record/electronic signature systems must have the education, training, and experience to perform their assigned tasks.
		Supplier	Supplier requires procedure to demonstrate that persons who develop and maintain electronic record/electronic signature systems have the education, training, and experience to perform their assigned tasks.
11.10 (j)	Procedural	Pharmaceutical manufacturer	Policy needed to describe the significance of electronic signatures, and the consequences of falsifying them, both for the pharmaceutical organization and the individual.
11.10 (k)	Procedural	Pharmaceutical manufacturer	Pharmaceutical organization needs procedures covering distribution of, access to, and use of operational and maintenance documentation once the system is in operational use.
	Procedural	Pharmaceutical manufacturer	Pharmaceutical organization must ensure adequate change control procedures for operational and maintenance documentation.
	Technological	Supplier	Where systems documentation is in electronic form, an electronic audit trail should be maintained, in accordance with 11.10 (e) above.
11.30	Not covered by this table		Requirements for open systems.
11.50	Technological	Supplier	ER/ES Systems must ensure signed electronic records contain information associated with the signing that clearly indicates all of the following: 1. The printed name of the signer 2. The date and time when the signature was executed 3. The meaning (such as review, approval, responsibility, or authorship) associated with the signature

Appendix 13.C Procedural and Technological Controls for 21 CFR Part 11 (*Continued*)

Clause Control	type of	responsibility	notes
			These items are subject to the same controls as other electronic records The information can be stored within the electronic record or in logically associated records, but must always be shown whenever the record is displayed/printed.
11.70	Technological	Supplier	ER/ES systems must provide a method for linking electronic signatures, where used, to their respective electronic records, in a way that prevents the signature from being removed, copied or changed to falsify that or any other record.
11.100 (a)	Procedural	Pharmaceutical manufacturer	Pharmaceutical organization must ensure uniqueness of electronic signature, and that they are not reused or reallocated.
	Technological	Supplier	ER/ES System should enforce uniqueness, prevent reallocation of electronic signature, and prevent deletion of information relating to the electronic signature once it has been used.
11.100 (b)	Procedural	Pharmaceutical manufacturer	Pharmaceutical organization needs to verify the identity of individuals being granted access to ER/ES system.
11.100 (c)	Not applicable		Not a functional requirement for electronic records and signatures.
11.200 (a) (1)	Technological	Supplier	ER/ES systems providing nonbiometric electronic signatures need at least two distinct components.
11.200 (a) (1)	Procedural	Pharmaceutical manufacturer	Pharmaceutical organization needs to establish how it will ensure that both components of electronic signature are entered if session has not been continuous (this can be through system design, or operating procedure if necessary).
	Technological	Supplier	ER/ES system should enforce that both components are entered at least at the first signing, and following a break in the session.
11.200 (a) (2)	Procedural	Pharmaceutical manufacturer	Pharmaceutical organization must ensure staff who only use their own electronic signature, not anyone else's even on their behalf, as that would be falsification [see also 11.10 (j)].
11.200 (a) (3)	Procedural	Pharmaceutical manufacturer	Pharmaceutical organization needs procedure that users do not divulge their electronic signature (e.g., passwords).
	Technological	Supplier	ER/ES System should not provide any ordinary means of accessing electronic signature information.
11.200 (b)	Not covered by this table		Biometrics requirements.
11.300 (a)	Procedural	Pharmaceutical manufacturer	System users must be identifiable through unique combination of user identification and user password.
	Technological	Supplier	Passwords must not be disclosed should be regularly changed.
11.300 (b)	Procedural	Pharmaceutical manufacturer	Pharmaceutical organization needs procedures to cover: removal of obsolete users; changing of profiles as user roles change; periodic checking of identification codes and passwords for inconsistencies with current users; periodic changing of passwords.

Appendix 13.C Procedural and Technological Controls for 21 CFR Part 11 (*Continued*)

Clause type of responsibility notes
Control

	Technological	Supplier	System should force passwords to be periodically changed and also enable id/password combinations to be rendered inactive without losing the record of their historical use.
11.300 (c)	Procedural	Pharmaceutical manufacturer	Pharmaceutical organization needs procedure for management of lost passwords.
11.300 (d)	Procedural	Pharmaceutical manufacturer	Pharmaceutical organization needs procedure to describe how response to attempted or actual unauthorized access is managed.
	Technological	Supplier	System should provide notification of attempted unauthorized access and should take preventative measures (e.g., lock a terminal after three failed attempts, retain card).
11.300 (e)	Procedural	Pharmaceutical manufacturer	Pharmaceutical organization should define how any devices or tokens that carry user/id or password information are periodically tested and renewed.

Note: Pharmaceutical manufacturer is the organization that is going to use ER/ES system in regulated environment, Supplier of ER/ES System (this could of course be a separate internal function of the pharmaceutical organization, such as the Information Systems department).

14 | Regulatory Inspections

INTRODUCTION

Regulatory inspections are conducted before a new drug or device can be approved, to verify production method and technology changes, and periodically verify every two or three years that GxP practices are being maintained. Inspections are used to determine if processes are adequately validated with documentary evidence that provides a high degree of assurance that a specific process will consistently produce a product meeting its predetermined specifications and quality characteristics (1).

This chapter discusses what to expect during inspections, how inspectors approach their work, and how to manage the process of receiving an inspection. Specifically, inspections by the U.K. Medicines and Healthcare Products Regulatory Agency (MHRA), and U.S. Food and Drugs Administration (FDA) are explored. Preinspection questionnaires and inspection checklists used by the regulatory authorities are attached as appendixes to this chapter (Appendix 14.A–C).

INSPECTION AUTHORITY

The inspection authority of the FDA, MHRA and other regulatory authorities is broadly the same although specifics vary. Taking FDA as an example, the Agency has legal authority to gain access to all regulated companies' facilities including vehicles that carry regulated products. This remit covers the use of equipment, computer systems, and personnel with production, warehouses, packaging, and distribution facilities. The FDA has the authority to inspect records, files, papers, processes, controls and facilities bearing on whether prescription drugs are adulterated, misbranded, or in some other way violate GxP regulations. No distinction is made between active pharmaceutical ingredients (APIs) and finished pharmaceuticals, and failure of either to comply with cGMP constitutes a failure to comply with the requirements of the Federal Food, Drug, and Cosmetics Act. It is policy not to examine internal audit and supplier audit reports without due cause because they do not want the company to compromise the detail in these reports on the premise they might be inspected. The FDA, however, are not allowed access to financial data and information, sales data (other than shipping and distribution), pricing information, personnel records (except training records and CVs), and research data (other than for product being inspected). While this distinction in theory is quite clear, sometimes it is hard in practice to split items of GxP and non-GxP information that may exist together in a single record.

INSPECTION PRACTICE

The FDA is sometimes quoted as saying "*In God we trust, everyone else needs documentation.*" This phrase neatly captures a strong and common theme to GxP inspections conducted by the various national regulatory authorities around the world. Computer validation requires the documentary evidence that a system was developed, operates and is maintained in accordance with predefined acceptance criteria, that is, demonstrably fit for purpose. The FDA is primarily looking for evidence of bad practice and fraud. This stringent approach was reinforced by the "Generic Drug Scandal" in the late 1980s when the FDA uncovered instances of fraud by pharmaceutical companies. Other regulatory authorities such as the MHRA have much more of a "partnership" approach. Both approaches have their merits.

Approach to Organizational Capability

The emphasis of inspections is moving away from particular products toward general operational capability. This move was first evident in the quality systems inspection technique (QSIT) adopted by FDA for medical device inspections in January 2000. Companies are considered "out of control" if any one of the main quality management controls inspected is found noncompliant with regulatory requirements (2).

- Complaint handling
- Corrective and preventative action
- Management oversight
- Production and in-process controls (including design)

The success of the inspection technique led to the development of the Systems Based Approach for full and abbreviated inspections of pharmaceutical and healthcare companies. Full inspections are conducted for the initial inspection of a facility, or where a facility has a history of poor compliance, significant changes have taken place, or for any other cause deemed appropriate. Abbreviated inspections are applicable when a pharmaceutical or healthcare company has a record or GMP compliance, with no significant recall, or product defect or alert incidents, or with little change in scope or processes comprising the manufacturing operations of the firm within the last two years. Both full and abbreviated inspection will satisfy biennial inspection requirements.

Full inspections will cover all, and abbreviated inspections at least two, of the following:

- *Quality system* (including status of required computer validation, change control, and training/qualification of QA staff)
- *Facilities and equipment systems* (including equipment IQ/OQ, computer system validation (CSV), security, calibration and maintenance, and change control)
- *Materials system* (including validation and security of computerized or automated processes, change control, and training/qualification of personnel)
- *Production system* (including contemporaneous and complete batch production documentation, validation and security of computerized or automated processes, change control, and training/qualification of personnel)
- *Packaging and labeling system* (including validation and security of computerized processes, change control, and training/qualification of personnel)
- *Laboratory control system* (including calibration and maintenance programs, quality and retention of raw data, validation and security of computerized or automated processes, system suitability checks, change control, training/qualification of personnel)

These focal points should be rotated in successive abbreviated inspections. The frequency of abbreviated inspections will be based on the pharmaceutical or healthcare company's specific operation, history of previous coverage, and other priorities determined by the FDA. The manufacturing operations of some firms may be limited and an abbreviated inspection may itself comprise inspection of the entire firm, for example, contract laboratory, in which case abbreviated inspections are synonymous with full inspections.

The FDA district office managing an inspection is responsible for determining the depth of coverage given to each pharmaceutical and healthcare company, and whether an computer inspection expert is required, to assess the state of compliance.

For a pharmaceutical or healthcare company to be considered in a state of control, there should be no "objectionable" deviations identified in any one focal point covered during an inspection. Whether or not a Warning Letter is issued will depend on the seriousness and frequency of the problems found. It should be possible to determine from a FDA 483 whether or not a Warning Letter is likely based on the following guidance:

- Quality system
  - Pattern or failure of QA personnel to review/approve procedures/documentation
  - Pattern of failure of QA personnel to assure compliance with SOPs

- Facilities and equipment
  - Pattern of failure to qualify equipment including computers
  - Pattern of failure to establish/follow change control process

- Materials system
  - Lack of validation of computerized processes
  - Pattern of failure to establish/follow change control process

- Production system
  - Lack of validation of computerized processes
  - Pattern of failure to establish/follow change control process

- Packaging and labeling
  - Lack of validation of computerized processes
  - Pattern of failure to establish/follow change control process

- Laboratory control system
  - Lack of validation of computerized and/or automated processes
  - Pattern of failure to establish/follow change control process
  - Pattern of failure to retain raw data

Full inspections may be recommended as a consequence of an adverse abbreviated inspection. The issuance of a Warning Letter or undertaking of other significant regulatory action will normally warrant a full inspection to verify remedial actions as satisfactorily complete and thereby close out immediate FDA concerns. Failure to satisfy regulatory authorities such as FDA can result in heavy fines (see chap. 1) and restrictions on future product approvals and marketing licenses.

An important aspect to this new approach is the expectation that pharmaceutical and healthcare companies will implement any corrective actions identified as the result of a site inspection across the whole of their operations. Effective coordination of corrective actions is vital for large multi-national organizations. FDA and MHRA already share information and have the ability to readily trend data and track repeated offenses on particular topics across multiple sites in a firm's organization. We should expect other regulatory authorities to follow suite.

Approach to Individual Computer Systems

Most regulators follow a top-down approach similar to the four-level review process described by the FDA (3):

Level 1: Recognize how the computer system interacts with operations.
Level 2: Evaluate the quality procedures used by companies to control their operations.
Level 3: Examine documentation in the compliance package supporting and computer system.
Level 4: Review software source code as appropriate.

The first review level is necessary to confirm the inspector's understanding of the criticality and history of computer systems and to set the inspection priorities. This will involve discussions with the pharmaceutical and healthcare company's senior technical management and a tour of the facility. A list of known software and data errors may be requested. The most significant inspection observations are likely to be those that can link the computer system to an actual or potential drug or healthcare product quality problem.

The second review level should identify poorly defined or missing procedures within the pharmaceutical and healthcare company's quality system. This will affect the expectations of the third review level and the scope and detail of compliance documentation. The inspector will want to understand the respective roles and relationships between departments and with suppliers.

The third review level examines the document sets for particular computer systems identified in the first review level. Validation plans and validation reports are typically amongst the first documents to be inspected. If the review of a computer system is not superficial then the main life cycle documents identified in chapter 4 may be inspected. The inspector is likely to ask to see evidence of system specification, design, test, qualification, supplier management, data maintenance, change control, training and security. Sometimes inspectors will ask for supplementary information to be sent onward to them if they are seeking clarification of an issue.

The fourth review level is usually only invoked by specially trained inspectors for software configurations and customizations, but may be extended to standard software packages where deficiencies are identified.

Throughout the review process where customary or reasonable compliance evidence is lacking or incomplete, inspection scrutiny may be increased. Conversely, if the preliminary review of the compliance evidence does not raise apparent or suspect problems, the scrutiny may be reduced. Once identified inspectors will pursue weak spots such as lack of documentation, or inconsistencies. They will examine employee performance for common errors (training or ways of working at fault). The inspector will establish the degree of any compliance gap between company practice, company procedures and regulatory requirements. It is worth presenting information to an inspector in a form which readily understandable and meets their expectations. Use industry terminology where ever possible.

Mutual Recognition Agreements

The concept behind MRA is that one regulatory authority will accept the findings of another authority with the confident in the rigor of the inspection process and hence negate the reason to conduct their own inspection of the same pharmaceutical or healthcare company. This is all good theory, but requires harmonized inspection standards, practices, reporting and training.

Regulatory inspections conducted under the MRA have already begun although progress on individual agreements is often a start/stop affair as various issues are worked through. Initial pilots are almost always based on inviting an inspector from one authority may participate as an observer in an inspection by the other authority. Budget constraints are being imposed by most national governments on their respective regulatory authorities however and it is not likely to be long before MRA inspections become a regular occurrence. In the interim, it is reasonable to expect inspection findings being shared between different regulatory authorities. FDA inspection findings are available to the MHRA anyway under the U.S. Freedom of Information Act. A reciprocal arrangement, other than the MRA, does not exist to give FDA open access to MHRA inspection findings.

INSPECTION PROCESS

Receiving an Inspection Request

When a request to conduct an inspection is received the pharmaceutical or healthcare company's senior management should be immediately notified. Notice of an inspection may be received by a number of people in a pharmaceutical or healthcare company so it is important that a procedure exists describing who the request is to passed onto and how. Usually, the focal point is the Head of Quality.

Having received an inspection request the Head of Quality will appoint an inspection response team manager. The inspection response team manager should contact the regulatory authority concerned to confirm the date, time, duration, site and topic of the inspection. It is not unknown for inspectors to arrive at the wrong site, or to try and inspect systems or product that are not located at the site proposed for inspection. The inspector may request advance information and/or documentation. The response to these requests must be carefully considered as information may be interpreted out of context by the inspector.

At this stage the pharmaceutical or healthcare company may wish to consider asking the inspector to sign a confidentiality agreement. During the inspection proprietary information must be respected.

Preparing for an Inspection

A SOP should be prepared to describe how inspections are to be managed from the notification of an inspection, through its completion. Such procedures are usually applicable to multiple sites within a pharmaceutical and healthcare company's organization, ensuring inspectors are treated in the same fashion no matter which site they inspect. Advice on how to handle

Table 14.1 Inspection Response Team Roles and Responsibilities

Roles	Responsibilities
Team manager	• Manages inspection response team • Acts as company's direct interface with inspector when organizing logistics for inspection
Inspection coordinator	• Manages control room • Coordinates scribe and runner
Host and deputy	• A senior manager • Represent site management • Welcome inspector and establish commitment of company to support inspection and its outcome
Senior quality representative	• Own inspection process • Agree company position on inspection topics • Agree response to inspection findings • Provide knowledge of how computer systems are used in support of GxP
Regulatory affairs representative	• Provide knowledge of regulatory submissions with direct and indirect reference to use of computer systems
Technical representative	• Provide technical backup on deployment and maintenance of computer systems (IT, process control and laboratory applications)
Operations representative	• Provide knowledge of how computer systems are used
Supporting quality representative	• Support inspection of compliance documentation from retrieval of appropriate documents to walking through validation conducted
Scribe/secretary	• Keeps minutes of inspector's comments and observations • Keeps a record of documents request and provided to inspector
Escort	• Accompanies the inspector during the inspection at all times
Runner	• Brings and removes documents requested by the inspector

Notes: Typically, the host for an inspection is the site QA manager. It is usually polite for the site director to attend opening and closing meetings.

inspection scenarios (good and bad) and particular inspectors should be captured in training materials rather than the SOP.

The structure and membership of the inspection response team should be agreed be in accordance with predefined internal guidelines. inspection response teams are usually established at a site level. The inspection response team manager should not have to negotiate release of key personnel. Table 14.1 suggests inspection response team roles and responsibilities. One individual may fulfill more than one role, but careful consideration should be given to whether certain mixes of roles actually conflict. Named deputies should be recorded in case primary nominations are not available for whatever reason.

Key preparation steps for an inspection include the following:

(a) Prepare personnel to receive audit, possibly include training in how to interface to inspectors for those who are unfamiliar with inspection requirements. Notify site to inspection so that general preparations can be put in place. A site briefing may be appropriate.

(b) Obtain room/office for the inspector that is isolated from employees: the inspection room. In parallel, allocate a room or office as the inspection response team's control room. The inspection room should not be too close to the control room.

(c) Identify what information and resources may be needed during the inspection. What was reviewed and outcome of previous inspections and what corrective actions are closed, in progress and not started? Review problem logs and change control records. Consider if there are any topics the company would like to take the opportunity to brief the inspector with.

(d) Gather documentation for key computer systems together in the control room. Arrange files into a logical accessible order. Typical documentation to get ready should include the following:
- Organizational charts
- Training records
- Validation master plans
- Change control records
- Problem logs
- System requirements and overviews
- Development methodology
- Validation plans and reports
- Testing records

(e) Perform a quick walk-through of key computer systems, and user workstations in the facility at their point of use. Consider conducting a mock inspection. Pull the records from archives (can information be retrieved in a timely manner?). Review documentation for obvious errors—a fresh pair of eyes! Identify potential problem areas and have answers prepared. Final computer validation reports should be available in English for the FDA.

The preparation for inspections should include a risk assessment based on the drug product being processed, the production process involved, and the technology mix including the use of computer systems, and a review of the company's internal audit and regulatory inspection history.

Hospitality

Hospitality must not be perceived as influencing the inspection. Regulators are typically required to pay their own accommodation costs and usually have a fixed daily allowance. Suggest suitable local hotels that fit their pocket. Hotel reservations can be made on their behalf but check they are comfortable with the arrangements. The pharmaceutical and healthcare company should also consider local transport requirements from the airport or train station to the site, and daily commuting to and from the hotel. If the inspector is making his/her own way to the site under inspection, then reserved car parking would be courteous.

Only company representatives hosting the inspector should stay at the hotel to avoid accidental discussions being overheard—it is not unknown for inspectors to overhear conversations in the hotel bar! Company administration staff should check that no company employees or suppliers are booked into the hotel for the duration of the visit. Make sure there are not too many company representatives acting as host at any one time as it gives the opportunity for the investigator to play one representative off another. It also makes for a more congenial atmosphere.

The pharmaceutical and healthcare company should consider establishing a policy whereby personnel are to decline to comment to inspectors queries outside company premises. Indeed personnel should be required to notify site security who will mobilize an official company response to off-site queries. Only nonwork issues should be discussed out of work, otherwise personnel to say run of conversation is inappropriate to discuss/chat about—and should if necessary walk away.

Arrival of the Inspector(s)

Site security should be briefed on the expectation of an inspection. First impressions count so security need to be courteous, and the site needs to be generally tidy and in a state of good repair.

Upon arrival, the inspector should present himself/herself to the site reception or gatehouse. The nominated Host will usually go the meet the inspector and take him/her to the designated inspection room. Once on site, an Escort and Scribe should accompany the inspector at all times. The Scribe will record all remarks, observations, questions, and responses made by both the inspector and company staff. If other authorities arrive with the inspector, note who they are and why they are there. This information should be relayed to the Control Room.

When at the designated inspection room try to agree to use it as a base for the inspector. Confirm the purpose and scope of the inspection. How long will the inspection last? Is this a routine or "for cause" inspection? What documentation would they like to see? Who would they like to speak with during the inspection? Do they have any other requirements? Create an agenda for the inspection with the inspector. Inspectors will not always have a predefined agenda and an agreed plan will help the inspector structure the inspection as well as helping the host organize logistics to make the inspection as efficient as possible. Request daily wrap up meetings during the inspection and final closure meeting.

Conducting the Inspection

Performance of company personnel on the day(s) of the inspection counts, presenters and supporters. It is not over until the regulatory is traveling back home.

Do not assume anything, always repeat inspector questions and ask for clarification if required. Inspectors may ask open-ended questions or make nonspecific request. This may be because they are unsure themselves of exactly what they want and are just fishing. As with most things a sense of balance should pervade. Do not question every request in detail as this will almost always annoy the inspector. Only address the specific point being raised by an inspector when answering questions—do not elaborate. Do not explain you answer unless specifically requested to do so. Let the inspector follow through their process. You might be trying to help but you could end up confusing the situation. Beware of informal "off-the-record" questions because everything is on the record. Do not get "friendly" with the inspector. Further, do not be tempted to speak when the inspector is quiet. Silence is generally good, not bad. Inspectors may employ long gaps between questions to encourage loose talk. Do not argue with inspection observations. Instead prepare evidence to present to the inspector to address his/her concern.

Inspectors will typically assume everything is GxP critical unless justified with rationale, and even then they are likely to spot check and challenge such justifications. Types of inspection questions related to computer systems include the following (4):

- Quality management and system development methodology
- Use of tools and standards
- Use of supplier (roles and responsibilities)
- Document control (draft, review, approve, superseded, withdraw)
- Change control
- Access controls (passwords and user log-ons)
- Data sources and entry/capture including contemporaneous transcription
- Data processing
- Data archiving, storage and retrieval
- Information security management (including virus checking)
- Internet links
- Remote access
- Electronic records and audit trails
- Signatures and status control
- IT infrastructure (including network firewalls)
- E-mail transactions/interactions
- Configuration and version control
- User training

There will be uninitiated questions, inquisitive questions, skeptical questions, adversarial questions, and long pregnant pauses from the inspector. Personnel should be instructed to only state only what they know to be true, and not to guess or speculate. Personnel should be firm and sincere when answering questions. Having said this they must not become adversarial. If they do not know then they should let the inspector know that this and that they will get back to him or her to follow up on their request. It is perfectly acceptable to admit you do not know, but make sure the question is not left unanswered. Open issues should be noted by the Scribe and logged by the inspection response team. Follow-up responses should be discussed with the inspection response team and positioned accordingly before the inspector is given the answer or information.

Above all there should be a consistent in approach by personnel to the inspector. There should be a clear objective of thoroughness and clarity—of trying to do the right thing and not shirking responsibility. Be sensitive to responsibilities and demeanor of the inspector—he may be just having a bad day! Make best out of deficiencies, concentrating on positive aspects, what has been done to put situation right and what is planned. Avoid the use of jargon and should not use undefined terms during the inspection. It is also important that personnel are briefed and sensitive to possible national language differences, for example, "warm feeling" means in control in the United Kingdom—means out of control in the United States.

The inspector may ask for documentation that is outside their inspection authority. Do not provide such documents without due consideration. Inspectors will have some reason for the request, if you are unsure on the validity of the request gently explore this with the inspector. Be careful not to refuse documentation by citing strict interpretation of the regulations, be cooperative where this is possible. Consider if the inspector's line of inquiry without documentation is alternative proof available. For instance share audit schedules rather than audit reports as proof of auditing.

Make a list and copies of all documents provided to the inspector during the inspection. Mark documentation given to the inspector as appropriate (confidential, restricted, uncontrolled, controlled, etc.).

Only provide documentation specifically requested. Provide copies of requested documents as per company SOP in a timely manner. Lengthy lag times in responding will make the inspector suspicious that there is a problem. Some questions are appropriate to answer quickly such as SOPs, some require slower response such as technical detail. Inspections of computer systems are predicated on the assumption that pharmaceutical and healthcare companies have effective record retention and retrieval systems (5). Significant problems may arise during inspections where these systems are inefficient or ineffective (6).

Pharmaceutical and healthcare companies should have a company policy that no cameras, videos, or recording devices can be used without prior written permission. This policy should apply to inspectors too. If pushed by an inspector, the company should take and process photographs and send a copy to the regulatory authority concerned.

Do not employ any delay tactics. On the contrary facilitate a swift inspection and let the inspector go home—that is what both parties are really after. We do not want a repeat visit.

The majority of inspectors working for regulator agencies do not have specialist knowledge of computer systems and technology. Should the assigned regulatory inspectors responsible for an inspection be particularly anxious about the compliance of computer systems, then advice and assistance can be requested from a specialist inspector within the agency. Remember that if the discussion of an issue is getting bogged down in technical detail then it might be useful to position a common sense type of explanation. This approach is after all what many inspectors will use to determine there is a potential problem in the first place.

Demonstrations may prove useful to an inspector by facilitating a less time consuming overview of functionality. Obviously the demonstration should reflect how the system is used in real life. The availability and suitability of demonstrations (including simulations) should be carefully planned. Demonstration software needs to be verified in its own right.

Keep control of the inspection by leading the inspector as much as possible through the agreed agenda and processes being audited. Remain calm and cordial at all times. Do not let company staff argue amongst themselves in front of the inspector and make sure those put in front of an inspector do not have an axe to grind. Do not make hasty commitments, some inspectors make lots of suggestions and this might just be an indication that they do not understand fully how company manages issues.

Sometimes inspectors will ask for supplementary information to be sent to them once their site visit is finished. Such documentation must be controlled in the same fashion as documents given to the inspector during the inspection. Remember to agree timings of delivery of any documentation. Timings should not be agreed that cannot be achieved. The inspector will generally be understanding of reasonable time constraints.

Daily Washup with Inspector

While inspectors are under no obligation to conduct daily washup meetings they can be very useful to the inspectors and the inspected. The meetings can provide a useful means of getting/giving early feedback on the good and the no so good from the inspectors perspective. Washups offer in particular pharmaceutical and healthcare companies two main benefits.

- The opportunity to provide requested information to the inspector that could not be supplied earlier and thereby possibly close what might otherwise be left open issues
- The opportunity to clarify outstanding questions/issues that are not satisfactorily closed so that closure can be planned

The attendance at daily washup meetings should be limited to the host, senior members of the inspection response team, and a scribe. A separate site washup can be held afterward with the full inspection response team and other invitees as appropriate. The daily washup should be used as the beginning of preparations for the next day's inspection.

Do not volunteer "war stories" about fixing the system. You may think this will impress the inspector but it would not because the inspector will be worried about the project being out of control. A good project is one that is well managed so that there are not situations warranting heroic action!

After the Inspection

The inspection response team will normally conduct an internal debriefing immediately after the inspector has left the site. A more formal inspection report should be written soon afterward. The inspection report will summarize the inspection and include an index of all documentation provided to the inspector. In addition, the inspection report will capture the corrective actions that the pharmaceutical or healthcare company will share with the regulatory authority to close any adverse observations made by the inspector. There may also be other lessons that will be acted on that will not be openly shared with the regulatory authority.

It is important that the inspection findings be presented to senior management in an honest, direct, and timely fashion. It may be many weeks, even months, before the inspector officially presents his inspection findings back to the pharmaceutical or healthcare company. This is too long to wait to keep senior management informed of the implications of the inspection.

Inspection Findings

The inspector will normally write back to the pharmaceutical or healthcare company after the inspection to confirm significant findings (positive and negative). The letter can take many weeks to arrive. The FDA takes a slightly different approach. A citation of noncompliance, known as a "483," will be issued by the FDA at the close of the on-site inspection with a pharmaceutical or healthcare company. An opportunity to clarify issues is given before the close of the inspection and the formal issue of the citation.

Observations concerning the compliance of computer systems might be logged as specific items or incorporated within the text of the system's associated equipment/process. FDA will consider the lack of computer validation as a significant inspection findings and log it as a 483 noncompliance citation. MHRA meanwhile may take a more lenient view depending on the criticality of the system on GxP operations. The lack of a written detailed description of individual computer systems (kept up to date with controls over changes), its functions, security and interactions (EU GMP annex 11.4); a lack of evidence for the quality assurance of the software developed process (EU GMP annex 11.5), coupled with a lack of adequate validation evidence to support the use of GxP related computer systems may very well be either critical or major deficiency. Ranking will depend on the inspector's risk assessment.

Figure 14.1   Potential impact of lack of regulatory compliance.

The escalation of regulatory action is illustrated in Figure 14.1. Decisions on whether or not noncompliance merits pursuit of regulatory action will be based on a case-by-case evaluation. The general criteria for regulatory action are the same for most regulatory authorities.

- Nature and extent of deviations
- Effect on product quality and data integrity
- Adequacy and timeliness of planned corrective measures
- General compliance history

Regulatory citations for computer compliance by FDA should reference the applicable predicate regulations. Enforcement by MHRA and other European regulatory authorities is through annex 11 on computerized systems in EU GMP directive. They too will generally refer to the governing GMP requirement when citing computer system noncompliance.

Inspection records are typically confidential to the inspected company and the regulatory authority. In the United States, 483 (listing findings) and FDA establishment inspection reports (EIR) are available to the public in accordance with the U.S. Freedom of Information Act. Other countries also have Freedom of Information laws that make records available although sensitive information will be redacted before records are released into the public domain.

Global Commitments

Care must be taken when making commitments to regulatory authorities not to inadvertently imply a global commitment to universal corrective action across an entire organization. While pharmaceutical and healthcare companies have an obligation to share learning across their organizations this is not the same as making a formal commitment to specific corrective actions. Most noncompliances will be location specific to an individual site or facility. Only systemic issues should be considered for global commitments. Indeed regulatory authorities will expect global commitments for such issues. Global commitments should be made in a timely fashion as part of a proactive recognition and management of an issue. Regulatory authorities will generally take further regulatory censure if they feel like they are having to persuade an organization to make a global commitment. This could mean issuing a Consent Decree for instance.

Poor Excuses

Many excuses have been given to GxP regulatory authorities when inspections have found computer systems to be out of compliance. Sam Clark, a former FDA investigator now working for Kempers-Masterson, listed some of the excuses offered to him when he was inspecting computer systems (6). Clark listed these excuses under two categories. Firstly, some responses were from pharmaceutical companies who simply did not validate and computer system that was inspected. The excuses offered included the following:

- We do not have the resources.
- We have used the system for years.
- We do not have anyone who can do that.
- It was done, just not documented.
- We got the system from a reputable supplier.

None of these excuses could be accepted. Pharmaceutical and healthcare companies should not release drug products whose manufacturing practice is not validated and maintained under operational compliance.

Secondly, other excuses were presented by pharmaceutical companies for incomplete, inconsistent, or missing documentary evidence supporting validation.

- We do not need written procedures-all our people are professionals.
- We have excellent training programs.
- We have done it this way for years and have not had any problems.
- It was done, just not documented.

Without documentation, there is no physical evidence that validation took place, regardless of whether it was sufficient. Hence, the saying "If it ain't written, it ain't done." GxP regulatory authorities may well believe on a personal level that a pharmaceutical or healthcare company did conduct suitable validation but they are unable to accept a computer system is compliant without documentary evidence. It is imperative that pharmaceutical and healthcare companies collate documentation supporting their validation as evidence to be presented to GxP regulators on inspection.

ISO 9000 and Validation

Questions are often raised concerning the acceptability of ISO 9000 accreditation of pharmaceutical and healthcare companies and their suppliers in lieu of validation. GxP regulators do not accept this position. ISO 9000 and other software development processes do provide foundation for validation, but they do not replace the specific needs of GxP validation. This perspective is supported by recent research, which suggests that ISO 9000 and other software development processes in general help improve bad practices rather than improve good practices (6). Theses who are familiar with ISO 9000 will also know that the annual follow-up audits supporting an organization's ongoing certification by an accredited body almost always uncover problems with management procedures and their application, even though some of these audits are very brief-perhaps only a day long. Holding an ISO 9000 certificate does not guarantee high quality work; it is just an indicator of capability. GxP regulators would seem to be tight in their cautious attitude toward ISO 9000.

ENSURING A STATE OF INSPECTION READINESS

No matter how well a pharmaceutical or healthcare company believes it conducts validation, it will count for nothing unless during an inspection the regulator understands what has been done and can easily find his way around supporting documentation. Pharmaceutical and healthcare companies need to demonstrate they understand their responsibilities and are actively controlling compliance. To this extent a key feature in any compliance exercise is inspection readiness.

For many organizations inspection readiness consists of a frenzied few weeks preparation getting ready before a regulatory inspection. However for organizations that may receive inspections without prior notice or who are regularly inspected several times a year by various regulatory authorities this is not an efficient means of ensuring inspection readiness. A significant change in mindset is required to improve inherent inspection readiness (Fig. 14.2).

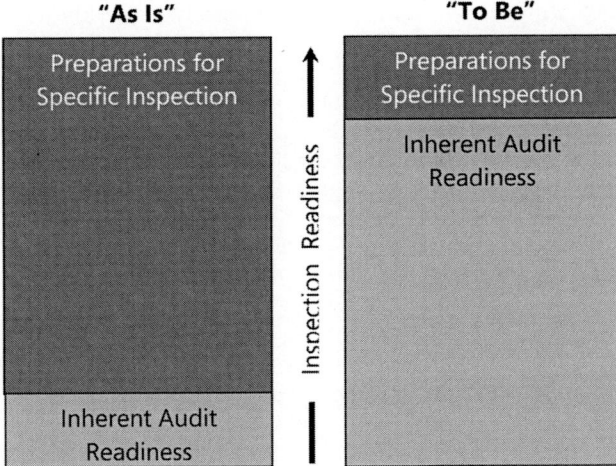

"As Is"

"To Be"

Inspection Readiness

Figure 14.2   Improving inspection readiness.

A three tiered approach may be appropriate to handle different levels of inspection readiness.

Tier 1: Highly regulated sites subject to regular inspection by regulatory authorities that routinely challenge computer systems compliance against published regulatory requirements

Tier 2: Regulated sites subject to inspection by regulatory authorities who occasionally challenge computer systems compliance against published regulatory requirements

Tier 3: Other regulated sites subject to inspection by regulatory authorities who have no published regulatory requirements for computer systems but who might challenge basic management and control practices

Maintaining a state of inspection readiness will be most relevant for tier 1 and tier 2 sites with supporting activities prioritized to support tier 1 sites. Conventional preparation specific to individual inspections will be more appropriate for tier 3 sites.

Inventory of Systems

An inventory of systems and knowledge of which ones are GMP critical must be maintained and available for inspections. A MHRA preinspection checklist has this as one of its opening topics. The availability of otherwise of this information is a clear indicator of whether management is in control of their computer systems. The use of an inventory need not be limited to inspection readiness; it could also be used for determining supplier audits and periodic reviews, etc. Many pharmaceutical and healthcare companies use a spreadsheet or database to maintain this data. Where a site's inventory is managed between a number of such applications (perhaps one per laboratory, one for process control systems, one for IT systems) care must be taken that duplicate entries are avoided and equally some systems are missed and not listed anywhere. It should be borne in mind that where spreadsheets and databases are used to manage an inventory then should satisfy regulatory requirements just like any other GxP computer application.

System/Project Overviews

Management overviews should be available for systems and projects giving a succinct summary of the scope of the system, essentially drawing boundaries and identifying functionality and use of the system/application concerned. Top level functional diagrams and physical layout diagrams are highly recommended. It is also worthwhile considering developing some system maps showing various links between systems, dealing with both manual and automatic interfaces. Care must be taken to keep system maps up to date as new systems are introduced, old systems are decommissioned, and as the use and interfaces of some systems are modified to meet evolving user demands. Regulators are often interested in system interfaces, manual and electronic, and the compliance status of connected systems. As a rule of thumb, all systems providing GxP information (data, records, documents, instructions, authorizations, or approvals) to a validated computer systems should themselves be validated together with the interface.

Some regulators have requested guidance be given by pharmaceutical and healthcare companies on what is of particular relevance in terms of GxP functionality within their corporate computer systems. Such GxP assessments often fit neatly in the system overview. The reason for this request by regulators is to help them concentrate on key aspects of the system during an inspection without getting bogged down in aspects of the system, which are not of a prime concern. It is easy for a regulator who is unfamiliar with a corporate computer system to get lost in its extensive and complex functionality (information overload). Needless to say, any GxP assessment information presented to a regulator must be understood and carefully justified.

Validation Plans/Reports and Reviews

It is likely that during a GxP inspection a regulator will ask whether or not a particular system(s) has been validated. This line of investigation may stop with a yes/no response from the pharmaceutical or healthcare company. The line of investigation may however lead to a follow-up request to see the validation plan and validation report for a system described as validated. Many of the computer systems used today have been in use over many years, and the regulator may also ask for any evidence of any validation reviews. These documents are, not too surprisingly, vital in demonstrating GxP compliance. It is not very clever to let a regulator discover a system in use with a validation plan but an incomplete or nonexistent validation report. Equally if the system has been used for many years it is more than reasonable to expect a recent validation review. Validation plans, reports, and reviews should be checked to make sure they exist, are approved, and meet current regulatory expectations. In some instances pharmaceutical and healthcare companies when considering this point may put in place a review program to check the items discussed above are complete and in place.

Documentation

It is vital to be able to easily locate documentation. Compliance documentation that exists but cannot be retrieved as required during an inspection is worthless—it might as well have not been prepared in the first place. To this end an index to documentation should be maintained. All documentation supporting compliance should be available at site during inspections.

A procedure should be developed describing how to handle requests by regulators for documentation. Where requested, access to master (or copies of) documents (including raw data such as test evidence) should be provided within reasonable time scales, normally 24 to 48 hours depending on circumstances. Canadian Health Products and Food Branch Inspectorate for instance require records to be accessible within 48 hours (7). FDA has similar requirements for off-site paper based archives (8,9). Service level agreements between central support functions and sites should define the service levels for access to documentation.

Controlled copies of centrally held validation plans and associated validation reports should be issued to sites in advance of any regulatory inspection. Access to electronic copies of centrally held protocols and reports can be facilitated during regulatory inspections to avoid unnecessary delays waiting for paper master copies to arrive. Such access can be facilitated through e-mail or a shared system directory. In such circumstances it should be clearly stated to the regulator that these electronic copies may not adhere to regulatory electronic record/signature requirements but are being provided to assist the inspector in advance of hard copies being delivered to site.

Sometimes pharmaceutical and healthcare companies elect to store documents on microfilm or microfiche. The requirements to allow this have been discussed in chapter 13. However in addition it is advisable to provide microfilm and microfiche readers so that inspectors can view records during inspections (10).

Presentations

In practice, computer systems are not perfect, and projects implementing applications will typically raise many management issues—that is life in the real world! The validation of any system/application will present its own special problems and solutions. Rationales need to be prepared and documented to demonstrate how problems and solutions have been managed. It is important to present a system/application in a positive light. Knowing how to effectively position problems and solutions will dramatically enhance the overall perception of compliance. The aim must be not to mislead an inspector, rather just to present issues in the vein of a glass half full rather than a glass half-empty. If all reasonable endeavors have been taken by a pharmaceutical or healthcare company to validate a system/application then this should normally be sufficient to satisfy an inspector, remembering that reasonable endeavors might include replacement where an original system/application cannot be validated to meet current regulatory expectations.

It is useful to prepare a small presentation of each system that may be subject to an inspection that can be offered during an inspection. Remember, however, that some inspectors will not want an introductory briefing. Presentations should consist of perhaps four or five slides and certainly less than a dozen. The presentation slides should not be too detailed but provide a broad picture describing a system/application and facilitate discussion. It is worthwhile letting the legal department look over the slides because they may be a danger of too high a level of information being interpreted as misleading if the detail of a system/application is examined. There is a careful balance to be struck between too much information and concise clarity. The slides should be in a suitable state to provide the inspector with a copy if requested.

Internal Audit Program

An internal audit program should be established if it does not already exist to cover the use of computerized systems. A schedule of audits should be planned placing priority on key topics subject to inspection such as data center, laboratories and manufacturing lines. It is useful to create a set of metrics to benchmark audit outcomes and monitor progress against audit actions. The audit should only mandate corrective actions where company policy, procedures, or regulatory requirements are not fulfilled. The audit can also be used to make recommendations for sharing examples of best practice with other sites, or adopting best practice from other sites. Recommendations should not be included in audit metrics.

Mock Inspections

A mock inspection program should be developed if one does not already exist. Mock inspections should be as realistic as possible. Mock inspections on computer systems may be conducted as part of a more wide ranging mock inspection or as a topic of a mock inspection in its own right. An inspection checklist is provided in Appendix 14.D to help structure a mock inspection. Appendix 14.E provides a supplementary checklist for inspecting electronic records and electronic signatures.

The opportunity should be taken to actively coach personnel receiving the mock inspection, clearly identifying areas for improvement. If necessary be prepared to withdraw individuals from the front line of a potential inspection if they are not readily capable of fulfilling this role. Sometimes doing yet more training will not be enough. It is important to accept that not everybody is suitable for putting in front of an inspector.

Trained Personnel

Last but by no means least, the availability and use of trained presentation personnel during inspections is key. Those who present to an inspector should be permanent employees otherwise there may be an impression of dependence on quality from temporary staff whose loyalty and long-term commitment to a pharmaceutical or healthcare company could be questioned. Presenters need to be knowledgeable about systems/projects they are asked to front. They need to understand compliance strategy, and appreciate why certain project decisions were taken. The position papers, slide packs, validation plans/reports/reviews should all help in this respect as long as the individuals concerned have enough time to study and digest the information they contain before they present them to the inspector.

Individuals can feel quite exposed when they are informed they may be required to participate in an inspection, especially if they are likely to be asked to answer an inspector's questions. Individuals will benefit from training in this regard, and senior management can have confidence in how company members will interact with an inspector.

Presenters should be educated in what to expect in the way of inspection protocols and regulatory practice. This aspect of training is likely to be tailored to the individual regulatory authorities as for instance the FDA have a very different approach to many EU national regulatory authorities such as the MHRA. Those who front during an inspection need to be aware of these differences. Mutual recognition agreements should also be understood as information presented to one regulator in one context could be shared with another regulator out of context. Fronting an inspection can be a complex affair!

Training courses should be considered for the following:

- How to respond to inspector's questions
- How to escort/host an inspector
- How to provide copies of documentation to the inspector
- How to conduct yourself in front of an inspector
- How to report inspection findings to senior management

Training must cover what to say and what not to say. How to react when asked a question. How questions might be asked or phrased by an inspector. How to ask for clarification if requests are unclear. The aim is to remove any unnecessary fear.

Knowledge Management

Pharmaceutical and healthcare companies often rely on the personal knowledge and skills of individuals without formally managing this knowledge as a key corporate asset. Projects often do not employ suitable measures to safeguard and retain knowledge and skills particular to discrete project phases. Project documentation can become difficult to understand if it is overtaken by numerous change control records. For large systems documentation may become so complex in terms of number of documents or in terms of location of storage that it becomes very difficult to retrieve in a timely manner. Change control records may become fragmented and give insufficient information to retrospectively understand change. Old and new computer system documentation may not be reconcilable if audit trails are not clearly maintained during changes to terminology or development methodologies. Furthermore, changes made over time may also inadvertently move system functionality away from its original intent.

The release of permanent staff from projects back into the business and their subsequent interdepartmental movements makes their return to support inspections difficult and unreliable. Inspection readiness can be further frustrated by key staff taking on external positions or leaving the business for other reasons, for example, voluntary redundancy. Many projects have a high dependence on contracted resource, turnover of such staff can be high. Once staff are dispersed there may be an irretrievable loss of knowledge.

Succession plans need to be established, and proper handovers when staff leave.

Refresher training should be considered for support staff. The reasons and benefits of historical changes in system functionality, terminology, and development methodologies must be documented in an easy to access and readily understandable way. An understanding of technological issues throughout the life of the system must be retained. Any outsourcing must clearly define user compliance accountabilities and mutual user/supplier responsibilities.

PROVIDING ELECTRONIC INFORMATION DURING AN INSPECTION

Regulatory authorities such as FDA and MHRA may make requests to access electronic copies of documentation and records. The FDA, for instance, has a legal right to access such information electronically under the part 11 (Electronic Records and Electronic Signatures) regulation. It is important to distinguish the difference between electronic documents/reports, electronic copies of desktop applications such as spreadsheets and databases, and electronic records that might be held on distributed/relational databases. The first two are relatively easy to extract as an entity to give to the inspector/investigator. The latter is much more difficult.

Provision of Electronic Documents and Reports
The provision of electronic copies of documents/reports should be defined in procedures describing the general manual approach taken. Many inspectors may find authorized paper copies of documents/reports more useful as they are often easier to read than electronic text.

Provision of Electronic Copies of Desktop Applications
The provision of electronic copies of desktop applications such as spreadsheets and simple databases should be defined in procedures describing the general manual approach taken. Many inspectors will be able to execute these applications on their own computer systems. Because of this authorized copies of relevant operating procedures and associated validation should normally be provided with the copy of the desktop application.

Provision of Electronic Records
The provision of electronic copies of records held on distributed/relational databases will need technical support to extract the right information to meet the regulators needs without the regulator having to have sophisticated and expensive computer technology to read the information in a meaningful way. It is unlikely the inspector/investigator will have the technical capability to read such information (e.g., they do not have their own SAP system to load data onto for investigation). For this reason provision of electronic records from distributed/relational databases in not typically useful to the inspector/investigator and alternative ways of providing relevant information should be explored. A high level procedure should describe the general process.

Direct Access to Electronic Information by Regulators
Direct access to electronic documentation and records should not be offered to the inspector/investigator. If direct access is requested by the inspector/investigator then the legal department should be informed. The inspector/investigator is not an employee of the company and would have to be properly approved, involving authorization, suitable training, and competency to have access. Such access could also violate the security (e.g., "closed system" status) of the companies computer systems. Similarly, inspectors/investigates should not be permitted to connect their own computer systems to pharmaceutical or healthcare company's systems. Again, such access could violate the security (e.g., closed system status) of the companies computer systems. Instead, an inspector who asks to see electronic documentation or electronic records can watch an authorized user query a system and make a printout.

Use of Computer Systems by Regulators
Operational use of a computer system should not be offered to the inspector/investigator. Inspectors/investigators do not have the right to use company computer systems by themselves to access electronic information. Inspectors/investigates can watch an authorized user access a computer system but they must not themselves directly use the computer system. If direct access is requested by the inspector/investigator then the legal department should be informed. The inspector/investigator is not an employee of the company and would have to be properly approved, involving authorization, suitable training, and competency to have access. Such access could also violate the security (e.g., closed system status) of the companies computer systems.

Electronic Copies of Information
During inspections an inspector/investigator may request to see archived document, or documents not held on the site under inspection. As discussed earlier in this chapter, pharmaceutical and healthcare companies should provide information in a timely manner and allow the inspection to flow naturally in accordance with the expectations of the inspector/inspector. If the physical transport of original documentation is not fast enough, then fax copies could be presented with the agreement of the inspector/investigator. The time to fax large documents may not make this approach practical. In this situation it may be that just the main body of documents are faxed without appendixes and attachments. If this is still too slow then again with the agreement of the inspector/investigator electronic copies

might be retrieved directly from company databases, or sent to the site under inspection by e-mail, and printed locally. In this latter situation the inspector/investigator must understand the printed copies are being presented to aid the inspection by removing delays. These printed documents are not claimed to compliant with electronic record/signature requirements. This approach must only be taken upon a specific request/authorization from the inspector/investigator.

Where copies of electronic documentation and records are provided to an inspector/investigator to take away they should be provided on read-only media (preferably write-once, read-only). The same issuance process should be followed as per paper documentation (e.g., signed hand-over of copy to inspector/investigator, additional exact duplicate copy made on the same media provided to the inspector/investigator for retention by site). Ad-hoc electronic reports from computer systems specifically requested during inspections do not have to be verified (11).

INSPECTION ANALYSIS

A summary of FDA Warning Letters referring to computer systems compliance issues that were issued between 2000 to 2009 are listed in Appendix 14.F. There are 98 entries relating to computer systems and 93 for medical devices (a total of 191 observations). The distribution of observations relating to computer systems by type is analyzed in Figure 14.3.

A life cycle analysis of the FDA Warning Letters listed in Appendix 14.F is presented in Figure 14.4. Twenty-eight percent of the observations relate to a lack of validation planning or deficient risk management. Another 20% of the observations related to system specification and supplier selection, design and development, system build, and development testing. This figure may seem low given that just under 28% of observations related to user qualification. It could be argued that many user qualification observations could be avoided if there were better system development. The medical device regulatory authorities have already recognized this trend from their inspection analysis and are putting more emphasis on system development during their inspections. Pharmaceutical and healthcare companies should also expect increasing regulatory focus on system development.

Operation and maintenance accounted for almost 29% of the FDA Warning Letters observations related to computer systems. The majority of these are for data integrity and system security. Many observations are examples of bad practice and highlight the need for ongoing operational compliance after initial validation. The final 4% of observations were related to phaseout and withdrawal.

The distribution of observations for computer systems and medical devices is remarkably similar (Appendix 14.F). The only basic variation being for medical devices there is slightly more emphasis on system specification and supplier selection, design and development, system build, and development testing compared with computer systems.

The potential causes of compliance failure are as follows (12):

- Inadequate documentation of plans.
- Inadequate definition of what constitutes the computer system.
- Inadequate definition of the expected results.
- Inadequate specification of the software (e.g., user requirements, functional specification).

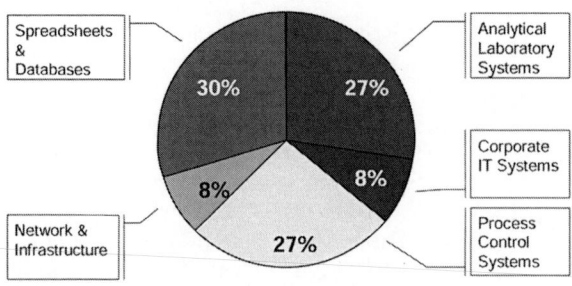

Figure 14.3 FDA observations by type of computer system.

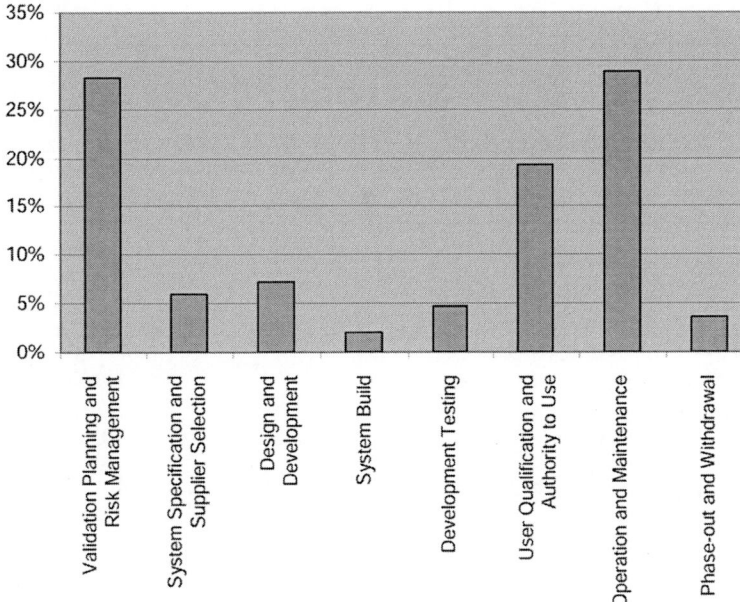

Figure 14.4 FDA observations by life cycle phase.

- Software does not meet specification.
- The source code for the software is not available.
- Inadequate specification of the computer hardware and operating environment for which the system is designed to work.
- The computer hardware or operating environment differs from the specification.
- The way the system should be used is not defined.
- Inadequate consideration given to centralized IT infrastructure, for example, network management, procedures and responsibilities.
- The intended use of the system is clearly defined, but users are not aware of it, or do not adhere to it.
- The system has been inadequately tested, or the testing has been inadequately documented.
- Documented standard procedures for the development, maintenance, operation (including security) or use of the system are inadequate.
- Documented procedures for disaster recovery are inadequate.
- System developers or other personnel involved with system implementation and use are not properly qualified, trained or competent.
- Documentary evidence to demonstrate qualification, training and competence level of personnel involved with the system is not available.
- Documentation for all or part of the validation does not exist, or cannot be located.
- Evidence of review and approval of compliance documentation by qualified staff is not available.
- Inadequate change control over any element of the system (i.e., hardware, software, procedures, people).

REFERENCES

1. Food and Drug Administration. General Principles of Process Validation. Center for Drug Evaluation and Research: Rockville, MD, 1987.
2. Food and Drug Inspection Monitor. "FDA to do system-based audits of drug companies," Vol. 5, No. 10. Washington Information Source Co., October 2000.
3. The Gold Sheet. July 1996, 30(7).

4. MCA. Computerised Systems and GMP—Current Issues, Top Ten GMP Inspection Issues. Royal Garden Hotel, London, September 24, 2002.
5. Food and Drug Administration. Compliance Program Guideline Manual, 7346.832, Pre-approval Inspection Investigations. Rockville, MD, 1992.
6. Wingate GAS. Computer Systems Validation: Quality Assurance, Risk Management and Regulatory Compliance for Pharmaceutical and Healthcare Companies. Boca Raton, FL: Interpharm Press, 2003.
7. Canadian Health Products and Food Branch Inspectorate. Good Manufacturing Practice Guideline, 2001.
8. U.S. Code of Federal Regulations Title 21: Part 203. Prescription Drug Marketing.
9. U.S. Code of Federal Regulations Title 21: Part 205. Guidance for State Licensing of Wholesale Prescription Drug Distributors.
10. Food and Drug Administration. Use of microfiche and/or microfilm for method of records retention. Compliance Policy Guide 7150.13, 1989.
11. Food and Drug Administration. Investigations Operations Manual. Office of Regulatory Affairs, May 2002.
12. ACDM/PSI. "Computer Systems Validation in Clinical Research: a Practical Guide." Association of Clinical Data Management (ACDM) and Statisticians in the Pharmaceutical Industry (PSI), version 1.1, December 1998.
13. Pharmaceutical Inspection Co-operation Scheme. Good Practices for Computerised Systems in Regulated GxP Environments, Pharmaceutical Inspection Convention, PI 011-1. Geneva, 2003.
14. Regina Brown. Inspecting a laboratory computerized system. GAMP Americas Meeting. Philadelphia, March 22, 2001.
15. FDA. Guideline to Inspections of Computerised Systems Used in Food Processing Industry, October 1998.
16. Janis Halvorsen. Georgia GMP Conference. Gold Sheet, November 2000.

Appendix 14.A Preinspection Questionnaire

The following information is sometimes requested by regulatory authorities prior to an inspection.

1. Details of the organization and management of IT and other computer services (from business IT systems to process control) on site.
2. State corporate policy on procurement of hardware, software and systems for use in GxP areas.
3. IT/computer services standards and SOPs? (Attach list.)
4. Provide a list of all GxP related computerized systems on site by name and application for business, management, information and automation (equipment and process control) levels. Indicate the totality of the inventory of computerized systems and indicate links with other sites/networks, etc.
5. For the systems identified as GxP related, has the company identified the critical systems, interfaces, sub-systems, modules and/or programs that are relevant to GxP, product quality and safety? If so, please cross-refer to lists provided for question "4" above.
6. What documentation generally exists to provide up-to-date descriptions of the systems and to show physical arrangements, data flows, interactions with other systems and life cycle and validation records? Comment as to whether all of these systems have been fully documented and validated.
7. Comment on the qualifications and training aspects of personnel engaged in design, coding, testing, validation, installation and operation of computerized systems? (specifications, job descriptions, training logs).
8. What is the firm's approach to assessing suppliers of hardware, software and systems?
9. How does the firm determine whether purchased or "in-house" software has been produced in accordance with a system of quality assurance?
10. What project management standards and procedures are in place for the development of applications and validation work? (List key titles and reference numbers.)
11. What approach is taken to the validation and documentation of older systems where original records are inadequate? (Summarize and list systems undergoing retrospective documentation and justify the continued use of these systems.)
12. Has the firm determined whether GxP critical systems conform with electronic data processing needs, accuracy and controls (including retrieval of archived records) for quality records as required by 91/356/EEC article 9 and EU GMPs 4.9 (inter alia)?

Source: From Ref. 13.

Appendix 14.B GLP Inspection Checklist

- The main focus is on quality of drug products and integrity of associated data.
- The integrity of the data and how it is maintained gives her an overall judgment of product quality. Therefore, procedures should be in place to assure the integrity of all processes and data. During an inspection, missing data is a cue that something is amiss and causes her to search further in this area.
- Key questions during an inspection:
  - who has access to the data?
  - how is access controlled?
  - what operations are permitted (read, write, edit, and delete)?
  - how can you demonstrate that what is reported is the same as that stored?
  - have you evidence that backup and restore of data has been tested and can be demonstrated?
- A company policy and guideline on CSV is expected. Documentation, SOPs are reviewed; diagrams, flowcharts on the systems are requested.
- All systems should be validated and calibrated before implementation.
- Change control of the system is reviewed, if it is lacking, this is viewed as a QA oversight.
- Audit trails must exist and restrictions on delete functions are required.
- Passwords on the system must be controlled and changed periodically.
- If electronic signatures are used, procedures must be in place for how they are used and maintained.
- Part 11—the adequacy and timeliness of planned corrective measures. The company is expected to have a reasonable timetable and must be able to demonstrate progress and see the corrective actions that have been executed.
- Lot systems and data are typically reviewed.
- For a chromatograph system, stability tests results were examined to compare the paper record with the electronic record.
- All raw data should be retained.
- Security measures must be in place, especially for HPLC systems.
- Quality Control personnel must know everything about the system, the validation, training records, etc. For example, an individual may be asked about data that resides on a system and then asked to retrieve the archived data in question, this is to ensure the individual knows what they are talking about.
- GMP training is required for all people involved in the manufacturing process.
- Overall the functional, operational and security features are investigated.

Source: From Ref. 14.

Appendix 14.C GMP Inspection Checklist

- Determine the critical control points (base investigation on FMEA or other hazard analysis technique). Examples would be
  - Pasteurization,
  - Sterilization,
  - pH control,
  - Temperature control,
  - Cycle timing,
  - Record keeping, and
  - Control of microbiological growth.
- For those critical control points controlled by computerized systems determine if failure of the computerized system may cause drug adulteration.
- Identify computerized system components including the following:

  Hardware inventory
  - Input devices
  - Output devices
  - Signal converters
  - Central processing unit
  - Distribution system
  - Peripheral devices

  Software inventory
  - Inventory of files (program and data)
  - Documentation:
  - Manuals
  - Operating procedures

(Continued)

Appendix 14.C GMP Inspection Checklist (*Continued*)

Hardware
Obtain a simplified drawing of the computerized system (covering major computer components, interface, and associated system/equipment). For computer hardware, determine the manufacturer, make, and model number.

Software
For all critical softwares, determine
- Name,
- Function,
- Inputs,
- Outputs,
- Set points,
- Edits,
- Input manipulation of date, and
- Program overrides.

Version control
- Who developed software (standard, configured, customized, bespoke).
- Software security to prevent unauthorized changes.
- Computerized systems input/outputs are checked.

Obtain simplified drawing of overall functionality of collective software within computerized systems.
Data
- What data is stored and where?
- Is data distributed over a network—how is it controlled?
- How is compliance to electronic record regulations achieved?
- How is data integrity verified?

Personnel
- Type (developer, user, owner)
- Training records

- Observe the system as it operates to determine if
  - Critical processing limits are met,
  - Records are accurate,
  - Input is accurate (sensor or manual input),
  - Time keeping is accurate, and
  - Personnel are trained in systems operations and functions.

- Determine if the operator or management can override computer functions. How is this controlled?
- How does the system handle deviations from set or expected results?

Check all alarms, calculations, algorithms and messages
Alarms
- Types (visual, audible, etc.)
- Functions
- Records

Messages
- Types (mandate action?)
- Functions
- Records
- Determine the validation steps used to insure that the computerized system is functioning as designed.
  - Was the computerized system validated upon installation?
    - Under worst-case conditions?
    - Minimum of three test runs?
  - Are there procedures for routine maintenance?
    - User manual
    - Vender supplied manual
    - Third-party support manual
    - Management manual

  - Does the equipment in place meet the original specifications?
  - Is validation of the computerized system documented?
  - How often is system

Appendix 14.C GMP Inspection Checklist (*Continued*)

- Maintenance performed,
- Calibrated, and
- Revalidated?
  - Check scope and records of any service level agreements.
- Are there procedures for revalidation? How often is revalidation conducted?
- Are system components located in a hostile environment that may effect their operation? (ESD, RFI, EMI, humidity, dust, water, power fluctuations). Are system components reasonably accessible for maintenance purposes?
- Determine if the computerized system can be operated manually. How is this controlled?
- Automated CIP (cleaning in place).
  - How is automated verified?
  - Documentation of CIP steps.
- Automated SIP (sterilization in place).
  - How is automated sterilization verified?
  - Documentation of SIP steps.
- Shutdown procedures
  - Does firm use battery backup system?
  - Is computer program retained in control system?
  - What is the procedure in event power is lost to computer control system?
  - Have backup and restore procedure been tested?
- Is there a documented system for making changes to the computerized system?
  - Is there more than one change control system (hardware, software, infrastructure, networks)? For each challenge, consider the following:
  - The reason for the change
  - The date of the change
  - The changes made to the system
  - Who made the changes
    - How do they interface Challenge change history, verify audit trail?
- What are the auditors impressions of
  - Presentation of validation,
  - State of documentation,
  - State of compliance,
  - Maintaining validation, and
  - Requirements for revalidation.

Source: From Ref. 15.

Appendix 14.D Mock Inspection Checklist

The following checklist is intended to test whether a site/facility is ready for a regulatory inspection. The site/facility should be able to present an informative and convincing response during a mock inspection to all of these questions without the need to defer answering for more than a day.

1. Please provide list of critical computer systems used at your site/facility.
2. Please provide evidence of governance controls.
3. For your three most critical computer systems please provide the following:
   - Validation report confirming validated status
   - Explanation of original validation
   - Explanation of site and central validation responsibilities for initial validation and ongoing maintenance
   - Explanation of major changes since implementation and a list of all changes within past year
   - Explanation of critical system functionality, demonstrate testing
   - Explanation of approach to ensure accuracy of data migration from old to new systems
   - Explanation of how site staff are trained on use of the system
   - Explanation of any support/maintenance that is outsourced
4. For networked computer systems please provide the following:
   - Network diagram showing flow of information between different computer systems
   - List of interfaces and an explanation of interface controls
   - Specification of supporting network infrastructure
   - Explanation of approach to verification of hardware platforms

(*Continued*)

Appendix 14.D Mock Inspection Checklist (*Continued*)

5. Please explain approach to computer security.
   - Administration of user-ids/passwords
   - Control of electronic signatures
   - Virus prevention/detection activities
   - Protection from outside threats, e.g., hackers
6. Please provide name and contact details for potential detailed inspection of central team.

Appendix 14.E Electronic Record/Signature Inspection Checklist

- Review firms' record-keeping requirements.
- Predicate record-keeping requirements even if not electronic.
- Determine if the firm has procedures for providing electronic and paper copies of records.
- What is the overall security of the electronic record-keeping system?
  – Can records be altered without a trace?
  – Do systems by design fail to record noncompliant information?
  – Are password systems robust (sticky notes, same as user names, easily guessed strings)?
  – Is access restricted to the system? Normally or when station is unattended?
  – What are the procedures in the event passwords or tokens are compromised?
- Documentation of the following:
  – Functional requirements specifications
  – Design specifications high level and detailed
  – Code documented and commented
  – Testing plans, documented test results
  – Review of all documentation by knowledgeable people
  – Release criteria and maintenance plan
  – Validation plans, procedures, and report
- Does the firm know their own deficiencies and have specific corrective action plans?
- Can the firm document progress toward achieving their corrective action plans?
- Does the firm "maintain a validated state"? Is validation documentation current and readily available?
- Has the firm trained IT and technical personnel on FDA regulations?
- Have administrative controls been put into effect?

Source: From Ref. 16.

Appendix 14.F Recent FDA Warning Letters

Company name	Life cycle								Application							Number of paragraphs in Warning Letter dealing with different computer issues
	Validation planning and risk management	System specification and supplier selection	Design and development	System build	Development testing	User qualification and authority to use	Operation and maintenance	Phaseout and withdrawal	Analytical laboratory system	Process control or monitoring system	Spreadsheet and database applications	Corporate computer system	Computer network infrastructure and services	Medical device	Electronic records and electronic signatures	
02/99 Hydro Medical Sciences Inc.	X					X	X		X							1
04/99 Fairbanks Memorial Hospital						X	X				X					1
04/99 General Electric Company					X	X	X							X		9
04/99 Florida Blood Services	X		X	X		X	X		X					X		4
05/99 Glenwood LLC				X		X	X									2
05/99 Picker International Inc.			X	X	X	X	X							X		12
06/99 Cypress Bioscience Inc.						X	X				X					1
06/99 Solvay Pharmaceuticals B.V.							X			X		X				6
07/99 Gensia Sicor Pharmaceuticals Inc.						X	X		X							2
08/99 Drager Medizintechnik GmbH	X						X				X					1
08/99 Linweld Inc.					X	X	X					X	X	X	X	11
10/99 Synthes						X	X							X	X	5
11/99 Apheresis Technologies Inc.							X				X				X	1
12/99 Hoffman-LaRoche	X		X				X		X	X						3
02/00 Abbott Chemistry Analyser	X					X	X							X		1
02/00 ERBE Electrosurgical Systems	X						X							X	X	1

(Continued)

Appendix 14.F Recent FDA Warning Letters (*Continued*)

Company name	Life cycle								Application							Number of paragraphs in Warning Letter dealing with different computer issues
	Validation planning and risk management	System specification and supplier selection	Design and development	System build	Development testing	User qualification and authority to use	Operation and maintenance	Phaseout and withdrawal	Analytical laboratory system	Process control or monitoring system	Spreadsheet and database applications	Corporate computer system	Computer network infrastructure and services	Medical device	Electronic records and electronic signatures	
02/00 Avidcare Monitoring Systems	X	X					X							X		1
03/00 Schein Pharmaceutical Inc.			X		X	X			X				X			4
03/00 Johnson Matthey					X	X	X		X							1
04/00 Harper Hospital							X		X							4
05/00 Intersurgical Limited	X					X	X		X					X		2
05/00 Schering Laboratories	X						X								X	1
06/00 Poly Implants Sa						X	X							X		1
06/00 Medical Industrial Equipment Limited							X							X	X	4
06/00 Sani-Pure Food Laboratories						X	X		X							4
06/00 A&L Laboratories			X			X					X					2
06/00 Jiangsu Hengrui Medicine						X	X		X							2
07/00 Integrity Pharmaceuticals Corporation						X	X		X (system not specified)							2
07/00 Rhodia Inc.							X		X							1
08/00 Baxter Healthcare Corporation						X			X	X					X	3
09/00 Leiner Health Products	X												X		X	2
10/00 Spolana a.s.	X					X	X		X							6
10/00 Contract Pharmacal Corporation						X	X			X						1
11/00 Alcon Laboratories Inc.		X	X		X	X			X							2

Date	Company										Count
11/00	SOL Pharmaceuticals Limited			X				X		X	3
11/00	Sybron Chemicals	X		X				X	X	X	1
12/00	Societa Italiana Medicinali					X		X		X	1
12/00	Chemrich Holdings		X	X		X	X	X		X	1
01/01	Pharmacia Corporation (Sterile)		X	X		X	X	X	X	X	11
01/01	Pharmacia Corporation (API)				X	X		X		X	9
01/01	Aventis Behring				X		X			X	1
01/01	Biological Research Solutions					X		X		X	2
01/01	Allergy Laboratories	X			X	X		X		X	1
01/01	DSM N.V.			X	X	X		X	X	X	3
03/01	Eli Lily and Company				X		X	X		X	1
03/01	Zeus Scientific Inc.	X		X		X		X		X	1
04/01	Stough Enterprises	X	X		X				X	X	2
04/01	Cardiomedics Inc.	X					X			X	2
04/01	Neurocontrol Corporation			X	X		X	X		X	1
05/01	Zenith Goldline Pharmaceuticals				X		X	X		X	1
06/01	Meridian Bioscience	X		X	X	X	X	X	X	X	2
07/01	Cardinal Health	X					X			X	5
07/01	SeQual Technologies Inc.	X					X	X		X	1
07/01	Esolyte Inc.			X	X	X	X	X		X	1
07/01	Kaken Pharmaceuticals				X		X	X		X	2
07/01	EP MedSysems	X		X	X	X	X	X		X	1
07/01	Aventis Bio Services		X	X			X	X		X	1
07/01	American Blood Resources Association		X					X		X	4
08/01	Paradigm Medical Industries		X	X		X	X	X	X	X	2
08/01	Farouk Systems Inc.	X	X			X		X		X	1
08/01	Medical Instruments Technology	X	X	X		X		X	X	X	3
09/01	Pharmakon Labs	X		X		X	X			X	1

(Continued)

Appendix 14.F Recent FDA Warning Letters (Continued)

Column groups: "Life cycle" covers columns *Validation planning and risk management* through *Phaseout and withdrawal*; "Application" covers columns *Analytical laboratory system* through *Electronic records and electronic signatures*.

Company name	Validation planning and risk management	System specification and supplier selection	Design and development	System build	Development testing	User qualification and authority to use	Operation and maintenance	Phaseout and withdrawal	Analytical laboratory system	Process control or monitoring system	Spreadsheet and database applications	Corporate computer system	Computer network infrastructure and services	Medical device	Electronic records and electronic signatures	Number of paragraphs in Warning Letter dealing with different computer issues
09/01 Utah Medical Products	X													X	X	1
09/01 Braun Medical	X													X	X	1
09/01 Christ Hospital	X						X			X						2
09/01 Cleveland Medical Devices	X					X	X							X		3
09/01 Dentsply International Inc.					X									X		1
09/01 Total Medical Info. Mgt Systems		X	X				X				X			X		1
10/01 Northeast General Pharma.							X		X							1
10/01 Michigan Instruments							X				X			X	X	1
10/01 Bunnel Inc.	X										X			X		1
10/01 Luneau	X						X							X	X	2
10/01 Neil Laboratories						X			X							2
10/01 Sorenson Development, Inc.	X						X			X	X			X	X	1
12/01 Natural Technology Inc.						X				X						1
12/01 Medical Device Services	X										X	X				3
12/01 Cardinal Enterprises				X		X			X						X	2
01/02 Sysmex Corporation										X				X		1
01/02 Pharmaceutical Distribution System						X				X						1
02/02 GOJO Industries							X			X				X		
04/02 American Dental Technologies						X	X							X		1
04/02 A-Vox Systems Inc.					X									X		1

Date / Company	1	2	3	4	5	6	7	8	9	No.
07/02 Earlham College	X			X	X	X		X	X	1
10/02 Abbot Laboratories	X		X	X	X					1
11/02 Braun Medical Limited	X			X	X		X	X		1
01/03 Shepeard Blood Centre				X						4
02/03 Merits Health Products	X		X	X	X		X			1
02/03 MAK-SYSTEM	X	X		X				X		1
03/03 Advanced Radiation Measurement					X			X		2
05/03 ConMed Corporation	X	X		X	X					1
06/03 Roche Diagnostics	X		X		X	X				2
06/03 MagneVu	X		X	X						2
07/03 Mercy (Blood) Medical Center	X	X		X	X			X		2
07/03 South Texas Blood Centre		X		X		X				1
07/03 Mercury Medical Centre	X		X		X			X		1
08/03 Image Analysis Inc.	X			X	X					1
09/03 Laborde Diagnostics	X		X	X	X	X				1
09/03 Con-Cise Lens				X	X	X				1
09/03 Consolidated Machine Corporation	X			X	X	X				1
10/03 Nutra Med Inc.	X				X		X	X		1
10/03 Medi-Stat	X	X		X						1
10/03 Tri-State Analytical Lab LLC	X	X			X			X		1
12/03 Eldon Biologicals	X		X	X	X	X				1
03/04 Ocuserve Instruments Inc.	X	X		X						2
03/04 Michigan Instruments Inc.	X	X		X	X					4
03/04 MedRx Inc.	X	X	X	X						1
04/04 Cordis Corporation	X	X		X						1
04/04 Pilgrim's Pride Corporation	X		X			X				1
04/04 Tri-Med laboratories Inc.	X					X		X		1
04/04 Surgilight Inc.		X		X		X				1

(Continued)

Appendix 14.F Recent FDA Warning Letters (*Continued*)

Company name	Life cycle								Application						Electronic records and electronic signatures	Number of paragraphs in Warning Letter dealing with different computer issues
	Validation planning and risk management	System specification and supplier selection	Design and development	System build	Development testing	User qualification and authority to use	Operation and maintenance	Phaseout and withdrawal	Analytical laboratory system	Process control or monitoring system	Spreadsheet and database applications	Corporate computer system	Computer network infrastructure and services	Medical device		
05/04 Nipro Diabetes Systems	X	X	X	X	X	X								X		2
05/04 Virotek LLC	X									X				X		1
07/04 Aap Impantate	X									X						1
07/04 Genetere Inc.															X	1
08/04 SinTea Biotech	X		X													1
08/04 Bigmar-Bioren SA						X				X				X		1
09/04 Roche Diagnostic Corporation		X											X			1
09/04 Guardian Drug Company						X					X					1
09/04 General Medical Company						X				X						1
10/04 Tecan	X				X	X	X						X	X		2
10/04 Laser Therapeutics Inc.															X	1
10/04 Computerized Radiation Scanners	X		X		X	X								X		1
11/04 Clarity Inc.	X									X						1
11/04 Ortek AG	X									X						1
12/04 Precision Piece Parts	X									X						1
12/04 Advanced Sterilization Products	X										X					1
12/04 Advanced Sterilization Products		X				X	X							X		1
02/05 Byron Company Inc.	X									X						1
02/05 Electronic Data Systems											X					3
03/05 Baxter Healthcare							X							X		1

Company												Count
03/05 Tosoh Bioscience	X						X			X		1
04/05 PIE Medical Inc.	X						X			X		1
04/05 Inoveon Corporation	X									X		1
04/05 Rosenthal Eye				X						X	X	1
05/05 Electrologic of America Inc.										X		1
05/05 Biolmagene	X				X		X			X		1
05/05 Philips Medizin Systeme Gmbh	X						X			X	X	4
06/05 Millenium Technologies	X					X						1
06/05 Arrow International Inc.	X					X						1
06/05 ConMed Corporation							X			X		1
07/05 Hitachi Medical Systems Inc.							X			X		1
08/05 Tepnel Diagnostics Inc.	X						X			X		1
08/05 Estill Medical Technologies Inc.	X						X			X		1
09/05 Nephron Pharmaceuticals Corporation						X	X					1
10/05 Biolase Technology Inc.		X								X		1
11/05 Visual Technologies Network	X						X			X		1
12/05 Guidant Corporation	X					X	X			X	X	1
01.06 National Genetics Institute					X	X	X					3
01/06 Gambro Dasco	X			X			X			X		8
04/06 Agile Radiological Technologies	X						X			X		1
04/06 Agile Radiological Technologies	X				X							1
04/06 Neurotone Systems Inc.	X						X			X		1
05/06 Neil Laboratories Inc.			X			X	X					1
07/06 Concord Laboratories			X		X	X	X			X		3
08/07 Sonotech Inc.	X						X			X		1

(Continued)

Appendix 14.F Recent FDA Warning Letters (Continued)

Company name	Life cycle								Application						Electronic records and electronic signatures	Number of paragraphs in Warning Letter dealing with different computer issues
	Validation planning and risk management	System specification and supplier selection	Design and development	System build	Development testing	User qualification and authority to use	Operation and maintenance	Phaseout and withdrawal	Analytical laboratory system	Process control or monitoring system	Spreadsheet and database applications	Corporate computer system	Computer network infrastructure and services	Medical device		
09/06 Vapotherm	X												X			1
09/06 Pointe Scientific Inc.	X									X	X					2
09/06 Terumo Cardiovascular Systems	X						X	X						X		1
10/06 Patterson Technology	X	X	X			X	X							X		7
11/06 EP Medisystems	X					X								X		1
12/06 Perkins Electronic Company	X					X	X							X		4
12/06 Welch Allyn Company	X										X					1
02/07 Actavis Totowa LLC									X						X	1
04/07 Seryx Inc.	X															1
04/07 Newport Medical Instruments	X						X							X		1
04/07 Medico Laboratories Inc.							X		X					X	X	2
04/07 G&B Electronic Design Limited	X										X					1
04/07 MRL	X	X									X					1
04/07 BTI Filtration														X		1
04/07 Abbott Laboratories						X	X	X		X				X		1
05/07 Rhytec Inc.	X									X						1
05/07 Medical Wire & Equipment Limited	X															2
05/07 Defibtech LLC			X	X	X		X							X		1
07/07 Avazzia Inc.	X		X	X	X	X	X							X		2
08/07 Liener Health Products LLC									X						X	1
09/07 Pulse Biomedical Inc.		X													X	2

Date / Company	1	2	3	4	5	6	7	8	9	10	11	12	Total
09/07 International Biophysics Corporation	X											X	1
10/07 Dialysis Dimensions	X											X	1
10/07 Chiu Technical Corporation	X					X						X	1
10/07 Arrow International Inc.	X				X				X	X		X	1
01/08 Omron Healthcare Inc.	X								X	X			1
02/08 Spinal Inc.	X	X										X	1
02/08 North West Medical Physics	X					X	X					X	3
02/08 Siemens Medical Solutions	X					X						X	3
02/08 Medical Actions Industries					X	X	X				X	X	1
03/08 Baxa Corporation						X	X					X	1
04/08 Xoran Technologies	X				X	X			X			X	1
04/08 iScreen					X	X						X	3
04/08 Merck & Company Inc.					X	X			X			X	1
04/08 Innovations for Access						X						X	1
04/08 Polymer Technologies Systems						X						X	1
05/08 Philips Medical Systems	X		X		X	X	X					X	3
05/08 MedX Corporation	X		X	X	X	X						X	4
05/08 Heartsine Technologies	X			X		X						X	1
05/08 National Biological Corporation	X					X						X	2
05/08 Safer Sleep LLC	X											X	1
08/08 GE Healthcare	X					X	X	X				X	6
08/08 Sandoz								X					1
09/08 Mavidon Medical Products	X											X	1
09/08 Fall Prevention Technologies	X				X	X			X			X	2
10/08 Steris Isomedix Inc.	X								X			X	1

(Continued)

Appendix 14.F Recent FDA Warning Letters (*Continued*)

Company name	Life cycle								Application							Number of paragraphs in Warning Letter dealing with different computer issues
	Validation planning and risk management	System specification and supplier selection	Design and development	System build	Development testing	User qualification and authority to use	Operation and maintenance	Phaseout and withdrawal	Analytical laboratory system	Process control or monitoring system	Spreadsheet and database applications	Corporate computer system	Computer network infrastructure and services	Medical device	Electronic records and electronic signatures	
10/08 Oncology Tech LLP	X													X		1
10/08 Larson Medical Products Inc.			X		X	X								X		1
10/08 Encore Medical LP						X	X	X		X				X		1
02/09 Genzyme Corporation						X	X				X					1
05/09 Lupin Limited							X				X	X				3
Subtotal medical devices	*55*	*13*	*17*	*7*	*11*	*26*	*49*	*11*						*93*		*93*
Subtotal computer systems	*40*	*7*	*7*	*9*	*5*	*39*	*49*	*1*	*36*	*35*	*40*	*12*	*10*		*29*	*101*
Grand total	*95*	*20*	*24*	*16*	*16*	*65*	*98*	*12*	*36*	*35*	*40*	*12*	*10*	*93*	*29*	*194*

Note: 'X' denotes nature of citation.

15 | Compliance Strategies

INTRODUCTION

This chapter takes a look at various compliance strategies that can be adopted around organizational roles, outsourcing, standardizing computer applications and software reuse, segregating GxP aspects of integrated systems, retrospective validation of legacy systems, and use of statistical techniques to support compliance activities.

ORGANIZATIONAL STRUCTURES

Questions often arise regarding the relationship of internal versus external suppliers, especially within large pharmaceutical and healthcare organizations, and the corresponding role of quality and compliance. Expectations for these organizational structures are discussed below. It is important to appreciate that regardless of the role of internal and external suppliers it is the end user that regulatory authorities hold accountable for ensuring computer systems are compliant and fit for purpose (1).

Quality and Compliance Roles

Regulatory authorities require pharmaceutical and healthcare companies to have a quality organization (sometimes referred to as a quality unit). The role of the quality organization covers both operational quality (individual project/system support) and Compliance Oversight (corporate governance of management practices). Table 15.1 compares R&D (GCP/GLP), manufacturing and distribution (GDP/GMP), and medical device regulatory clauses relating to quality organization responsibilities for computer compliance.

Operational quality and compliance oversight groups can exist as separate groups, or as a single group, depending on the size and structure of the organization. Some controls on who does what however do need to be specifically managed. For instance, a quality professional providing direct project/system support on one system should not be allowed to audit that same system because this would compromise the auditor's independence. One pharmaceutical company describes this way of working as "QC at home, QA away."

GDP/GMP Quality Unit

The quality unit must be independent of those parts of the organization responsible for testing (2) and production (3) and has a critical role in overseeing the whole qualification and validation process (4). It is expected to

- Be involved in all quality matters (5),
- Review and approve all appropriate quality related records and documentation (5), and
- Ensure timely notification of compliance issues to management (2,5).

The main responsibilities of the quality unit should not be delegated (5). FDA believes that such accountability will result in more consistent and reliable compliance (6).

GCP/GLP Quality Unit

The British Association for Research Quality Assurance (BARQA) has interpreted international GCP/GLP regulations and expects the GCP/GLP quality unit to (11)

- Conduct GCP/GLP awareness training, validation training and change control training,
- Review and approve validation and change control procedures,

Table 15.1 Quality and Compliance Organizational Roles

Regulatory responsibilities (reference numbers)

	Organizational roles	
	Operational quality	Compliance oversight
GCP/GLP (1–3,7–9) (FDA refer to quality assurance unit)	• Ensure that all data is reliable and processed correctly.	• Set policy. • Responsible for procedures applicable to the QA unit for in-house and purchased systems. • Compliance auditing. • Compliance monitoring (review and inspection).
GDP/GMP (3–6) (FDA refer to quality control unit)	• Ensure that validations are carried out. • Oversee whole qualification and validation process. • Review and approve validation protocols and validation reports. • Review changes that potentially affect product quality. • Determine if and when revalidation is warranted.	• Set policy. • Oversight of validation procedures. • Compliance auditing. • Make sure that internal audits (self-inspections) are conducted. • Review effectiveness of QA systems. • Conduct GDP/GMP training.
Medical devices (10)	• Establish quality plans.	• Set policy. • Establish quality system procedures. • Conduct quality audits. • Review performance of quality system.

- Review quality plans and key validation documents [i.e., validation plan, requirements, test plan, test results, acceptance, record retention (archiving), change control],
- Advise projects on software development,
- Review changes (individually or part of periodic review process), and
- Conduct system audits (including system development, software, operation and use).

Concept of Internal Supplier

Internal suppliers exist as distinct entities with a pharmaceutical or healthcare organization and are separate from end user organization. Common examples of internal suppliers include corporate IT departments, corporate engineering functions and central laboratory support groups. The basic role of the quality unit remains unchanged although end user responsibility may be delegated in part to quality personnel within the internal supplier organization. Ultimate accountability remains with the end user organization's quality unit who should have line management outside the internal supplier organization to demonstrate its independence.

It is important to recognize that compliance standards are the same for internal suppliers as end users. Internal suppliers may believe they can abdicate responsibility for compliance – making this the responsibility of the end user. This is a serious misjudgment. End user validation is typically highly dependent on compliant work by the internal supplier organization. Regulatory authorities could ask to inspect internal supplier organizations when they realize this dependency. This almost never happens with an external supplier.

Any change in the so-called internal supplier organization or associated ways of working must be carefully managed. Care must be taken not to inadvertently create a discontinuity in support or system documentation. Years later, for example, tracing the documentation between two different quality management systems can be quite difficult to explain in a credible manner. Transitions between organization structures and quality management systems are fertile ground for noncompliance.

Central Development and Support Groups

In an effort to exploit standardization many organizations have established central groups to develop and support common systems. The objective is to establish consistent, effective and efficient business processes and to minimize development, support, and compliance costs. As such, site adaptation of applications is strongly discouraged if not forbidden. Examples of situations where central development and support groups make sense include the following:

- Multiple locations served by a single shared implementation of an application (e.g., MRP II)
- Multiple locations sharing the design for their own implementation of a common application [e.g., LIMS, distribution systems, and common distributed control system (DCS)]
- Multiple locations deploying their own specific implementation of a single version of a preferred product [e.g., DCS, chromatography data system (CDS), and near infrared (NIR) instruments]

Central development organization for a particular system may be separate or combined with its central support organization. However, if a central development organization exists without a reciprocal central support organization or an acting custodian (e.g., lead site), common systems tend to diverge and overall management control is lost. Both central development and central support groups should have quality unit support.

Pharmaceutical and healthcare companies tend to have an ebb and flow in regard to centralized and decentralized organizations. This is often reflected in the harmonization or disparity in compliance practices adopted between sites or geographic regions of a company. The cyclic nature of the organizational changes must be managed to minimize the impact on consistent compliance standards and practices. Centralized departments must not loose touch with the hands-on experience of operating site. Decentralized departments must ensure suitable support network is established with focal points for maintaining a common vision and approach. Both extremes of organizational structure are hard to maximize efficiencies; it gets easier in the middle ground.

External Supplier Responsibilities

Goods and services, including software-based systems (12), must correspond to their description and be of a merchantable quality (fit for purpose) (13–15). Both GxP regulatory requirements and commercial contract law share the objective of computer systems being "fit for purpose" and this should be achieved through good professional practice.

Although GxP requirements hold the pharmaceutical and healthcare companies directly accountable for all aspects of computer compliance, in contract law if the supplier knows the customer's application intent (regardless of the product's common usage), the goods or services must be fit for that intended purpose. This does not mean pharmaceutical or healthcare companies can defer regulatory observations of noncompliance and the liability for corrective actions directly on to their suppliers. Rather it open up the possibility for pharmaceutical and healthcare companies, after receiving a noncompliance observation from a regulatory authority, to take their supplier separately to court if under commercial contract law it is felt the supplier's actions were responsible for the regulatory deficiencies.

Duty and Standard of Care
Duty of care is based on avoiding reasonably foreseeable adverse consequences. The failure of "duty of care" implies negligence. It has been successfully applied to deficiencies in

- Design,
- Construction,
- Inspection,
- User instructions, and
- Data security.

and hence covers some basic attributes of GxP. In addition, there is the general expectation of safe operation (16–18). Data security, which includes access security, is mandated in many counties by laws protecting individuals and organizations from the misuse of information (19).

Within the United Kingdom, a "standard of care" is imposed on the equipment producer who is liable to compensate the pharmaceutical or healthcare company for personal loss but not for corporate damage (20). The concept of standard of care is very similar to that of duty of care. Prosecution for negligence of care must usually be brought within some limited period from the date of supply. In the United Kingdom, this period is three years from loss or awareness of loss and cannot be brought after 10 years from the date of original supply. Other legislation may strengthen the regulation affecting some aspects of supply, such as supply chains (21).

Breach of Contract
A successful suite for damages (breach of contract) must satisfy a "reasonableness" test demonstrating negligence. Exclusion clauses are usually implemented as a defense against damages. However, within the United Kingdom, their use is limited, and negligence can never be the subject of such a clause (22). Indemnities can be used to pass damage responsibility to a third party who supplied the source of system noncompliance. There may, of course, be a join responsibility between the primary subject of the breach of contract and the third party, in which case responsibility may be shared. In this way (depending on supply roles), a combination of system integrator, equipment hardware supplier, and software supplier may be held accountable for breach of contract between the end customer and the primary supplier. Examples of accountability include software installed on an inappropriate hardware plat form, system inappropriately implemented as customer solution, or operational instructions not followed during maintenance servicing. Indemnity is strictly controlled through common law; there must be no doubt of accountability. Secondary contracts limiting the liability of the initial contract are legally permissible, but are unlikely to pass the reasonableness test. The EU directive on unfair terms in consumer contracts (93/113/EEC) interprets all ambiguous contract clauses in favor of the end customer, which in the case of compliance is likely to be the pharmaceutical or healthcare company (23).

Associated with the breach of contract is "misrepresentation." This is a misleading understanding given outside the contract, but that is integral to the contract agreement. Such an understanding might involve the qualification and experience of personnel implementing the contract. Pharmaceutical and healthcare companies or their suppliers may be allowed to rescind the contract, but only to recover costs where misrepresentation is fraudulent or negligent (24).

Legal Defensive Positions
The overwhelming majority of contracts are brought to a successful close. If fulfillment of the contract is disputed, however, the prosecuted party has four basic defenses:

1. Presentation of an ISO 9000–accredited quality system, adherence to established company and industry practices, and use of competent personnel.
2. Demonstrating the likelihood that the adverse consequence was introduced by the user subsequent to delivery by the supplier. Evidence of predelivery inspection and testing is required.

3. Presentation of a "development risk" whereby the custom (bespoke) nature of an application is presented as a source of acceptable risk. This argument, however, is usually self-defeating because in such circumstances it is "reasonable" to apply more rigorous development practices.
4. The goods or services supplied conform to the customer's formal requirements and it was these requirements that were deficient. In practice, user requirements are rarely precise enough to begin debating this defense.

These defenses highlight the importance of mutual respect and partnership within a working supplier-customer relationship. It is in both parties' interest to ensure contracts are fair and rigorous.

Liability of Personnel
Employees can, in theory, be sued for breach of contract if they are shown not to have taken reasonable care in their duties. In practice, this rarely happens because of the limited recoverable resources from the individual. Instead, the employee is subject to disciplinary action and the possibility of dismissal.

Negligent work by an employee under employer management or established employer practice is the responsibility of the employer. It is the employer's responsibility to demonstrate their management and practices were not negligent to defend against this position. Company directors representing the employer may be accountable for the employee's negligence if they have a duty covering the negligence and there is "gross negligence." Proving gross negligence in the absence of unambiguous evidence is extremely difficult.

Contractors under a "contract for services," like employees, can be sued for breach of contract where they are shown not to have taken reasonable care in their duties. In practice, however, because of their limited recoverable resources, it is far more likely that they will be dismissed. The position of contractors as "independent" for the purpose of prosecution for negligence is complex. Independence implies that the contractor worked outside employer management and employer practice. This is rarely the case, and contractors are treated by the law as employees.

Regulatory Authority Responsibilities
GxP regulatory authorities also have a duty of care to the pharmaceutical and healthcare companies inspected, but what constitutes their duties is not precisely defined. Few cases have been successfully brought against GxP regulators.

OUTSOURCING
Outsourcing can be a very attractive means to reduce the cost of ownership associated with computer systems. With added pressures on pharmaceutical and healthcare companies to reduce headcount, the transfer of personnel to the outsourcing company a part of the "deal" can be an added benefit. Outsourcing however should not be gone into lightly. The pharmaceutical and healthcare company will become entirely dependent on the outsourcing company for the computer systems included. Poor levels of service often have a direct impact on the operation of the pharmaceutical or healthcare company. Breaking away from one outsource company back to the pharmaceutical or healthcare company or to another outsourcing company can be a very painful experience.

Regulatory Requirements
Pharmaceutical and healthcare companies are accountable to the GxP regulatory authorities for the actions undertaken by outsourcing company. FDA regulations, for example, simply require that personnel have the appropriate combination of education, training and experience

to perform their assigned tasks (25). It is further expected that training in current good manufacturing practice shall be conducted by qualified individuals on a continuing basis and with sufficient frequency to assure that employees remain familiar with the GxP requirements applicable to them. European Union regulations meanwhile discuss extensively the roles of the contract giver and contract acceptor. Due diligence on behalf of the pharmaceutical or healthcare company is expected not only on the technical ability of the outsourcing company to perform the desired job but also that any outsourcing company meets the regulatory compliance requirements (26). A supplier audit as presented in chapter 6 should therefore be conducted. This principle is consistent with the expectations regarding system suppliers discussed earlier in this book. If the outsourcing company operates in a way that results in regulatory noncompliance, then the contracting pharmaceutical or healthcare company will have a regulatory compliance issue as well. It is the responsibility of the pharmaceutical and healthcare company to find suitable business partners.

Planning and Supervision

Good contract management is vital for successful outsourcing. The following checklist is based on material from David Begg Associates (27):

- Prepare a written "statement of requirements" for the outsource company to tender against (make sure no misunderstandings before work starts). Identify what needs to be done to minimize cost and ensure that the necessary information and expertise remains in-house. Develop an exit strategy just in case outsourcing relationship irrevocably breaks down.
- Agree a "contract" defining mutual responsibilities. Specify standards, processes and procedures to be used.
- Review and maintain a "service level agreement" (SLA). Provide ongoing compliance oversight of activities being outsourced.

QA should be involved at the outset in helping to define compliance requirements. Clear responsibility needs to be given to particular QA departments to ensure ongoing provision of resource for review and audit activities. There also needs to be a clear escalation process for the QA function to progress any compliance issues identified.

Statement of Requirements

Specific and comprehensive requirements for the scope of work and impact/dependencies with other IT activities need to be defined. This might include a separate SLA. The roles and responsibilities of both client and supplier need to be included.

- Project/quality plans specifying policies, procedures, deliverables and milestones
- Specification of expected service levels with alert and escalation processes
- Design constraints and standards to be adopted
- Security and data protection requirements
- Change control processes to be adopted
- Record retention requirements
- Training and education requirements
- Transition arrangement including knowledge transfer for take-up of service
- Principles by which outsourcing partnership might be dissolved without risk to service levels

Selection of an outsource partner to support a GxP application must ensure proper due diligence. This will normally involve a conducting a supplier audit. Corrective actions to address any adverse observations should ideally be completed before a formal contract is signed. However, if this is not possible, then corrective action plans should be included in the contract arrangements.

Contracts

There is no standard outsourcing contract to cover every outsourcing opportunity, however, it is important to cover a checklist of major clauses.

- General provisions: For example, definitions and governing laws and regulatory requirements. The outsource partner should be given formal permission to use client software. There may be a need to arrange software licenses for the outsource partner.
- Service particulars: Based on the statement of requirements.
- Change control: Definitive clause specifying that both client and supplier will comply with a defined change control procedure.
- Access security: Specification of policies and standards to be adopted. Procedures and working practices would be detailed in an accompanying schedule.
- Confidentiality and data protection: Specification of nondisclosure of sensitive information.
- Record retention: Definitive expectations for regulated records and signatures.
- Intellectual property: To cover software developed by the client, outsource partner, and any third party.
- Managing poor performance: Aim is to address faults without delay and encourage good performance. Care must be taken to avoid defensive language. A *force majeure* clause should be included to cover instances where failure to meet contractual commitments is beyond the reasonable control of either party.
- Dispute resolution: Principles and process need to be clearly defined and include any special criteria used to determine fault on quality and compliance issues since many of these will be attributable to failings of both client and outsource partner.
- Audit rights: Client rights to audit supplier records, procedures, processes and records to confirm quality of service.
- Contract management: With consideration for extension or termination.

Financial details, warrantees, liabilities and indemnities will also be covered. Contracts are likely to have a number of schedules (appendixes) giving more detail on certain clauses.

Service Level Agreements

Initial SLAs will normally be a contained in a schedule to the main contract. The main topics covered by SLAs are listed below.

- Description of project/service: Defined scope with any exclusions clearly identified.
- Quality management: Standards, compliance responsibilities, audit arrangements.
- Risk management: Early identification, reporting and escalation of issues and risks that may develop into business or regulatory threats.
- Inspection readiness: Ensure that inspection readiness and inspection support for GxP systems are not degraded (e.g., timely provision of system documentation and system knowledge to client site under inspection, and ability to handle direct regulatory inspection of outsource partner).
- Interfaces: How constraints and dependencies between project/service and other IT activities are handled.
- Periodic review: Frequency and type of progress review meetings (link to quality management).

It is important that SLAs are maintained separately to the contract as a living document. There should be regular reviews to ensure they meet business and regulatory compliance requirements. Changes may be made at any time through the agreed change control process.

Organizational Capability

The outsourcing company should have a designated quality manager and quality management system. The effective use of the QMS should be demonstrable, as too the capabilities of the quality manager. Outsourcing companies may need to consider recruiting suitable qualified personnel. Additional training may be required to fulfill regulatory expectations. Some pharmaceutical and healthcare companies transfer members of their organization to the outsourcing company either as a secondment or to be directly employed by the outsourcing company. It is very important that the outsourcing company's organization, structure and culture support GxP principles.

Training, documentation, and change management, together with configuration management, self-inspection and managing deviations (as discussed in chap. 3) are vital supporting compliance practices. It is important to demonstrate unequivocally that they work well to ensure any potential regulatory inspection. Poor practices will totally undermine a regulatory authority's confidence that the computer systems they are inspecting are being effectively and compliantly managed.

McDowall identified documentation practices and change management as particular topics that an outsourcing IT organization may not fully appreciate pharmaceutical and healthcare regulatory expectations (28). It is not just an issue of having SOPs or working instructions but also following them and having documentary evidence that the procedures are being followed. Software engineers are frequently not trained on GxP documentation practices. The use of pencils instead of pen; the use of typewriter correction fluid instead of marking a single strike-out and writing alongside the right information (initialed and dated) for corrections; and the use of post-it notes and regulatory information written on scraps of paper is commonplace in many IT departments. The documentation of changes is also often poor. Documentation may be incomplete, not detailed enough, missing review and approvals, or lacking rigor of change specification and testing. This situation is exasperated with many outsource organization have a high staff turnover so that even it software engineers are trained they replaced by new staff who do not appreciate GxP requirements. Some pharmaceutical companies routinely see a "sawtooth" characteristic in quality metrics as experienced outsource staff are replaced by new inexperienced staff.

The pharmaceutical and healthcare company together with the outsourcing company should anticipate possible regulatory inspection. Consideration should be given as to whether the outsourcing company is inspection ready and would know how to handle an inspection or inspection request. Regulatory inspections and knowledge management are discussed in chapter 14.

Transition

A process of disentanglement usually has to be undertaken to transfer systems to the outsourcing company. Compliance issues can be divided into the following categories:

Systems Management
The operation and maintenance of regulated computer systems has already been discussed in chapter 11. The outsourcing company should effectively be managing the following requirements:

- Performance monitoring
- Repair and preventative maintenance
- Upgrades, bug fixes, and patches
- Data maintenance
- Backup and restoration
- Archive and retrieval
- Business continuity planning
- Security
- Contracts and SLAs
- User procedures
- Periodic review and revalidation

Records Management
Compliance issues affecting the management records held on the outsourced computer systems can be summarized as follows:

- Records retention
- Records retrieval
- Access controls
- Data integrity

Contracts should specify that the outsourcing company will maintain historical records for a retention period defined by the pharmaceutical and healthcare company. Means to ensure timely record retrieval also need to be established. Record retrieval will be required to support, amongst other activities,

- Audits from the pharmaceutical and healthcare company,
- Inspections by regulatory authorities, and
- Critical quality operations like recall, customer complaints, and batch investigation.

The administration of access controls is usually passed to the outsourcing company. Access must be restricted to authorized users. Users may be from both the pharmaceutical and healthcare company and the outsourcing company. Access controls must protect information from authorized modification (inadvertent or malicious). Security in general is discussed in chapter 11.

Data maintenance practices to assure data integrity and detect corruption should be instituted if they are not already established. Change control and audit trails are key aspects requiring management. Reference should be made to chapter 11 where data maintenance is discussed in more detail. In addition there may be electronic record management requirements and more information regarding regulatory expectations in this regard can be found in chapter 13.

Ongoing Oversight

It is important to agree at the outset management and controls concerning security, confidentiality, intellectual property, documentation ownership, and compliance oversight. These topics should be included in legal contracts defining the outsourcing service to be provided.

The pharmaceutical and healthcare company's QA staff should retain ongoing involvement in the following key compliance activities:

- Approve the outsourcing company's quality plans so that compliance requirements are visible and understood from the outset.
- Review work at regular agreed intervals.
- Audit the work against agreed plan and standards.
- Manage modifications through change control (ensure appropriate level of participation from pharmaceutical, healthcare, and outsourcing companies).
- Ensure outsourcing company completes and properly organizes all compliance documentation.
- Conduct periodic compliance reviews as part of any contract renewal process.
- Keep the outsourcing company up to date with regulatory developments and compliance expectations (possibly conduct tailored training programs).
- Monitor knowledge retention in the outsourcing company and in the pharmaceutical/ healthcare company's organization concerning the use and compliance of relevant computer systems.
- Define and use problem escalation and resolution processes as appropriate, do not let compliance issues remain unresolved.

Pharmaceutical and healthcare companies should not assume that the outsourcing company will conduct particular activities unless it defined in service agreements. At least one major pharmaceutical company has fallen foul of this principle, resulting in their "world-class" outsourcing company not doing some "good practice" configuration management and documentation for system modification because these activities were not specified as required in their contract. In the end the pharmaceutical company had to replace the computer system concerned because retrospective validation was deemed as too expensive.

STANDARDIZING COMPUTER APPLICATIONS

Standardized computer applications are defined here as those using common software across a number of installations (e.g., use of COTS products and/or shared use of custom applications across multiple sites). Corporate computer system strategies of many pharmaceutical and healthcare companies are now based on the use of standard software because of the advantages on offer.

- *Standard release documentation*: The specification and testing documentation is shared amongst many installations, so its unit cost per application should be less than that for custom (bespoke) software.
- *Wide user base*: A large user community implies that if there are any problems they will be discovered quickly and rectified.
- *Less effort to validate*: Leverage on central development so that less supplementary work is required by end users.

Approach to Validation

The approach to standardized software should follow a variant of the V-Model called the X-model (Fig. 15.1). Assuming that the standardized software has been developed under a suitable quality management then end user qualification can be abridged from the full custom (bespoke) software life cycle.

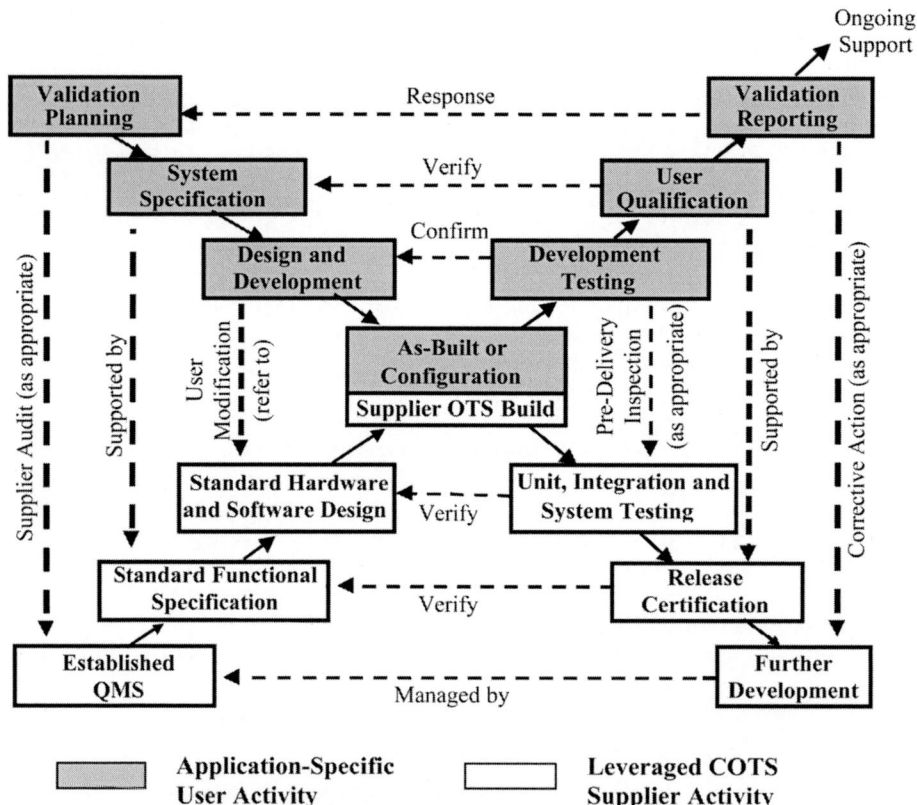

Figure 15.1 X-model life cycle for standardized software.

Getting the right balance between end user qualification, system development, and development testing is vital. User qualification should concentrate on the end application and therefore include the following (29):

- System specification (refer to but do not repeat standard software documentation)
- Configuration details including any macros used to build the application
- Definition and testing of any customization including custom (bespoke) developments
- Verification of critical algorithms, alarms and parameters
- Integrity, accuracy and reliability of static and dynamic data
- Operating procedures are complete and practical
- System access and security

The relationship between user qualification and development of the standard application must be clearly understood and described in an application's validation plan. Users should review and accept standardized application release documentation. Supplied documentation must match the version of the standard software being implemented.

Figure 15.2 and Table 15.2 suggests the general split in documentation between a user and supplier documentation for COTS software. Access agreements should be established that support regulatory inspection of any software and documents not released to the user.

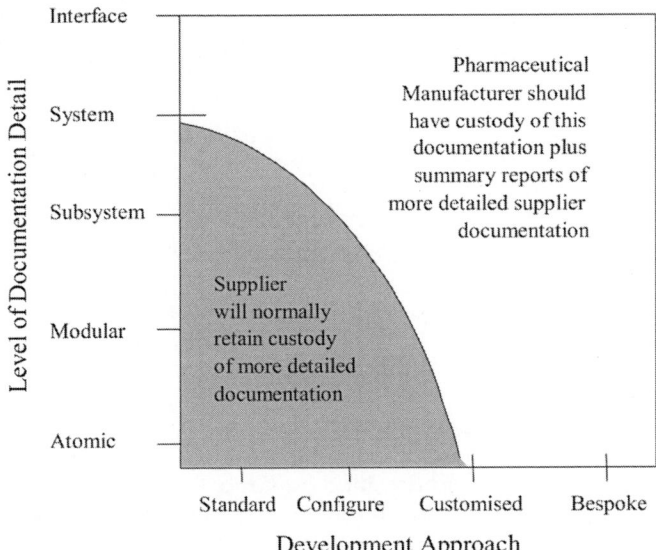

Figure 15.2 Custody of documentation.

Table 15.2 Documentation for Standard Software

Standardized application release documents	User compliance documents
Quality plan	Validation plan
Product specification	User requirements specification
Product design	Functional specification
Program specifications	Configuration details
Source code review	Design review
Development testing	Installation qualification
Product release certification	Operational qualification
Change control	Performance qualification
Product development plans	Validation report
Service level agreements/warrantees	Change control

Managing User Modifications

It is important to understand that users are often tempted to modify standardized applications and thereby undermine the standard status. There are basically four types of modification that need to be managed.

- *Configuration*: Setting process parameters and process paths. This modification does not impinge of the standard software status.
- *Customization*: Rewriting portions of standardized application code to meet specific user requirements. This modification makes the standardized application non-standard. Detailed specifications and structural (white box) testing will be required for the modifications and other aspects of remaining system functionality altered by the change.
- *Custom (bespoke) element developments*: Writing extra software to complement the standardized application. These modifications may impinge on standard software status, but can be compensated by overall functional (black box) testing. Custom code must itself be fully validated, including structural (white box) testing.
- *Upgrade versions*: Caution is needed when implementing new versions or bug fixes to standardized applications. Release documentation should confirm continued quality of software. If serious doubts exist over software quality then common sense should prevail and the software should be treated as customized or entirely custom (bespoke) and compliance expectations set accordingly.

If the standard status of software has been compromised, then the following steps should be taken to recover the situation:

- *Review and document concerns*: Do not hide or ignore issues. Quality and compliance are, after all, really about good business sense; if there is a problem, then fix it in the most appropriate way.
- *Determine and document action plan*: Identify supplementary work that can be undertaken to compensate for any concerns. This may be achieved through a risk assessment process.
- *Raise concerns with supplier*: A supplier audit should be considered for external suppliers, possibly positioned as free consultancy on pharmaceutical and healthcare requirements. Be realistic on corrective action planning. Prioritize where effort needs to be placed.
- *Work with supplier*: Possibly offer ongoing free consultancy. For critical applications it may be worth considering the placement of one of the customer's quality engineers in the supplier organization to help the supplier understand and address issues.
- *More user acceptance testing (qualification)*: Increment rigor of user testing commensurate with application to improve confidence in software.
- *Replace application*: Finding an alternative source of supply may be necessary as the only practical solution to longer-term compliance. Pharmaceutical and healthcare companies should not disregard this option out of hand.

Software Reuse

Pharmaceutical and healthcare companies and suppliers are faced with the task of balancing increased programming efficiency offered by reuse and the potential hazards reuse may incur. It has been suggested that the reuse of small amounts of software can actually introduce more problems than writing the whole application from scratch because the new software must fit around the reused software. To reap the dividend of reuse, it has been recommended that at least 70% of a program must consist of reused software components of proven functionality (30). Furthermore, it must be understood that, while reused software may be configured, any customization will negate its proven component functionality and the software must be considered as custom (bespoke). Caution is also required when considering reuse of software of unknown pedigree, or open source software. Without an audit trail to its original development, such software cannot be treated as standard software and should be subject to the more rigorous compliance requirements of custom (bespoke) software.

Examination of some tableting PLC software revealed the original code to have been written in Spanish, with subsequent functional revisions in German and English before a final modification for a French application. It is important to realize with software such as this that older portions of the software may not have been developed to current compliance requirements, and features from earlier versions that are no longer needed may still remain. This situation occurs quite regularly to suppliers who are asked by pharmaceutical and healthcare companies to provide standard software with a few additional features. Pharmaceutical and healthcare companies should be aware that such developments increase the compliance requirements because the software can no longer be considered "standard."

A special case of reuse involves the portability of software across a range of operating platforms. Standard programming languages, communication protocols, and application environments should be significantly reducing the modifications required to adapt software for different computers and operating systems. Practitioners sometimes use the term *open systems* to describe standard software capable of running on a variety of system architectures. As noted above, however, it is important to distinguish between customized and configured software when considering the compliance implications of reuse. Practitioners should not underestimate the problems they may experience with portability.

SEGREGATING INTEGRATED SYSTEMS

Use of integrated applications increases the complexity of the overall "system" that in turn impacts the complexity of the compliance activities required. In some cases, it is difficult to conclusively demonstrate that functions not requiring validation do not affect functions that do need validation. This situation often leads to increases in the scope of validation to include functions, which taken separately on their own merits, would not be considered as requiring validation.

Isolating GxP Functionality for Validation

A strategy for segregating integrated systems into those requiring validation and those that do not is considered here. This strategy can be extended to segregating distinct modules in large computer systems such as MRP II systems. A clear definition of system/module boundaries is required. This often prompts additional compliance activities for automated and manual interfaces.

Individual computer systems should be validated when they are

- Creating, modifying, or deleting GxP master data,
- Used for GxP processes and functions, or
- Providing GxP data to other systems for use in GxP processes and functions.

Interfaces should be validated when GxP data is being output from or input to those computer systems identified using the above criteria.

The identification of GxP processes and functions has already been discussed in chapter 6 as part of GxP Assessments. Compliance Determinations Statements should be prepared for each system to document the rationale for where validation is and is not deemed necessary. Validation would then be conducted for those systems and modules that require it as described in chapter 4.

It may be appropriate in some circumstances to implement and validate independent monitoring systems for critical GxP processes rather than validate the primary system. Chapter 5 provides guidance on identifying critical components and devices where this approach is appropriate.

Validation is not required for individual systems that have no GxP functionality. However, the following controls are expected across the integrated systems to protect the integrity of the validated systems.

- Contemporaneous management of GxP data replicated in multiple systems.
- The integrated architecture of systems is robust against individual system failures.

Change control during operation and maintenance should verify the Compliance Determination Statement is not affected by modifications to individual systems. The use of individual systems often changes over time and at some point it is possible that a non-GxP system may be used in a GxP context. It is important not to inadvertently undermine the compliance rationale for the overall integrated system.

Separating Computer Network Infrastructure

Separating the compliance activities for applications and the computer network infrastructure should reduce potential duplication of testing of common infrastructure shared by multiple applications. GxP applications should be validated as outlined in chapters 5 to 10. Testing multi-site applications can be based on a comprehensive test at a single site of shared functionality across multiple sites. In addition separate tests may be needed to test site-specific functionality. OQ testing should include at least one test to verify operability from each user site.

Computer network infrastructure should be verified to support validated applications. Bristol Meyer Squibb has adopted a three-level model to assist the verification of their computer network infrastructure (31). This approach is summarized in Table 15.3. Layer 1 comprises computers that provide shared resources such as servers, hosts, mainframes, and mini computers. Layer 2 is the network infrastructure (e.g., hubs, routers, and switches). Layer 3 comprises the user desktop environment (i.e., workstations, personal computers and laptops).

Functional specification(s) should be developed for the host machine, its operating system, and utilities. The scope will include the use of any servers. Design documentation should cover the actual configuration and setup of the computing hardware and associated equipment.

IQ needs to cover both hardware and software aspects. Hardware installation of the host computer should be documented with the installation method. Components added to standard hardware should also be recorded (e.g., memory, NIC card, and hard drives). Operating system details together with any patches and upgrades must be documented. For larger system particular use of modules, utilities, or library functions should also be recorded so that the software environment is defined.

OQ should include backup and recovery, data archive and retrieval, security, system administration procedure verification, start-up and shutdown, UPS continuity, communications loss and recovery, and systems redundancy challenges such as mirrored drives, secondary systems, fail-safe systems.

PQ of the network should cover loading tests as appropriate to verify network performance. Such testing is not always appropriate as PQ and may be included instead as part of ongoing performance monitoring.

A final summary report should be prepared for the computer network infrastructure to summarize the results of the verification exercise. A case study on computer network architecture is provided in the second part of this book.

Desktop Controls

Desktops are a fundamental component of the IT infrastructure that provide access to company assets (e.g., business applications, data, internet, printers, etc.) over a network

Table 15.3 Infrastructure Compliance Documentation

Compliance documents	Layer 1	Layer 2	Layer 3
Functional specification	Y	Y	Y
Design documentation	Y	Y	N
Installation qualification	Y	Y	Y
Operational qualification	Y	N	Y
Performance qualification	N	Y	N
Summary report	Y	Y	N

connection. Desktop hardware can take the form of stationary personal computers, portable personal computers (laptops), radio frequency terminals (RFT), personal digital assistants (PDA), and mobile phones. GxP regulations expect the implementation of controls to mitigate risks threatening reliability, security, and confidentiality.

Verification of systems and processes associated with maintaining the integrity of the desktop environment should be based on the identification, assessment and mitigation of known hazards. Desktop hazards that should be assessed include:

Virus infections can cause data changes and disturb application functionality. Mitigated through use of virus protection software (virus definition files must be kept current), isolating desktop/network (implementing network firewalls, restricting email and internet access), and minimizing data held on the client.

Spyware infections can breach security/confidentiality and disturb application functionality. Mitigated through use of spyware protection software, isolating desktop/network (implementing network firewalls, restricting email and internet access), and automatically monitoring software installed on clients.

Unauthorized user changes can cause data changes and disturb application functionality. Mitigated through the introduction of policies on appropriate usage of computerized systems, implementation of policies to restrict the installation of software, locking down clients (software configuration and physical hardware controls) and limiting privileges to install software to particular system administrators, automatically monitoring client configuration, and auditing clients to identify whether any unauthorized changes have been made.

Radio frequency interference can disturb application functionality and impact operation of associated equipment. Mitigated through testing system as part of original validation, restricting usage of radio devices near computerized systems, shielding equipment from interference, and periodically checking capability of equipment to resist interference.

Electromagnetic interference can disturb application functionality and impact operation of associated equipment. Mitigated through testing system as part of original validation, restricting usage of heavy-duty electrical equipment near computerized systems, shielding equipment from interference, and periodically checking electromagnetic capability of equipment to resist interference.

Incorrect date and time when client used as calendar or clock can invalidate electronic records created or managed by clients. Mitigated through locking down software configuration to prevent user authority to change system clock, use of time service from server clock rather than rely on client clock, and recording system clock changes in protected event log. Auditing or periodic checking as controls are not recommended because of issue of limited sample check and lag to detect issues that may be compromising regulated electronic records.

Client application conflicts can disturb application functionality and compromise electronic records. Mitigated through processes and standards to mange the build of new clients and their subsequent upgrade.

Failure/loss of client can lead to the loss of data and electronic records. Mitigated by frequent synchronization of clients to servers to replicate local data, minimize amount and sensitivity of data held on client, establish business continuity plans and procedures including potential use of rollback facilities.

Unauthorized remote access can lead to breach in security and confidentiality. Mitigated through use of hard/soft token to authenticate users, use of Virtual Private Networks (VPN), and implementing network firewalls.

There are three basic desktop configurations and these can be categorized into two basic setup scenarios (Thin Client and Thick Client). Thin Clients process data on the server using software installed on the server. Thick Clients process data locally on the client using software installed on the client. These two scenarios are useful because they can be used to differentiate between different levels of risks associated with desktop hazards.

- Computer applications operate exclusively on a server (referred to as *Thin Client* applications)[a].
- Computer applications are distributed between a client and service (referred to as Thick Client applications).
- Computer applications that operate exclusively on the desktop (considered equivalent to Thick Client application).

Three levels of desktop build of increasing rigor are suggested to deal with the different vulnerabilities of Thin and Thick Clients.

Unrestricted build for applications outside the scope of GxP regulations. No special mitigation control instituted for GxP purposes.

Restricted build for GxP applications on Thin Clients and less critical applications subject to GxP regulations on Thick Clients. Users not able to change system clock or calendar.

Controlled build for critical applications on Thick Clients subject to GxP regulations. Users not able to change system clock or calendar. Electronic audit trails for electronic records implemented rather than relying more of change control records. More rigorous conflict testing between GxP applications.

Centralized management will be needed to govern standard desktop builds. All client configurations (physical and logical) must be specified and maintained. Conflict testing is required between different applications on desktop builds. The deployment of client applications must be controlled. The installation of builds on new clients and upgrades must be managed. Virus protection must be maintained. The integrity of client builds should be periodically reviewed and include the effectiveness of security arrangements. An example risk map for handling desktops is shown in Figure 15.3.

VALIDATING LEGACY SYSTEMS

It is now widely understood and accepted that computer systems should be validated prospectively. It is not acceptable to implement a computer system and attempt to validate it after it has been installed for use. This said, where a system has had a change in use to bring it with scope of an existing validation related regulation, or new validation related regulations have been introduced such as electronic records and signatures to include the computer system within their scope, then retrospective validation is acceptable.

Validating existing systems, however, can be more than five times more expensive than if that same system had been validated when it was new. Practitioners should, therefore, consider whether it is cheaper to implement a replacement system rather than conduct retrospective validation.

Setting Priorities

It is often necessary to prioritize projects when addressing a significant backlog. Priorities for validating different computer systems should be set according to a defined strategy. Some projects may be given a higher priority because a regulatory inspection that is likely to include the system is imminent, or there is outstanding noncompliance from a previous regulatory inspection, or the computer system is supporting a process subject to a new drug regulatory submission. Equally, a lower priority may be given to computer system systems that are soon to be replaced. Some pharmaceutical manufacturers, for instance, decided when prioritizing not to validate those systems due for replaced within a year. If such a stance is taken it is important that the system is replaced within the stated time frame. It is all too easy to delay the

---

[a]Prior to personal computers, all computer systems operated on a server and the user interface to the computer system was primarily through a display terminal. When personal computers became available, software was designed to emulate the simple functions of a terminal and these were commonly referred to as "Green Screen" or by name of the emulation software (IRMA, Reflections, Rumba). After development of Client Server technology, these types of components on desktops became know as "Thin Client."

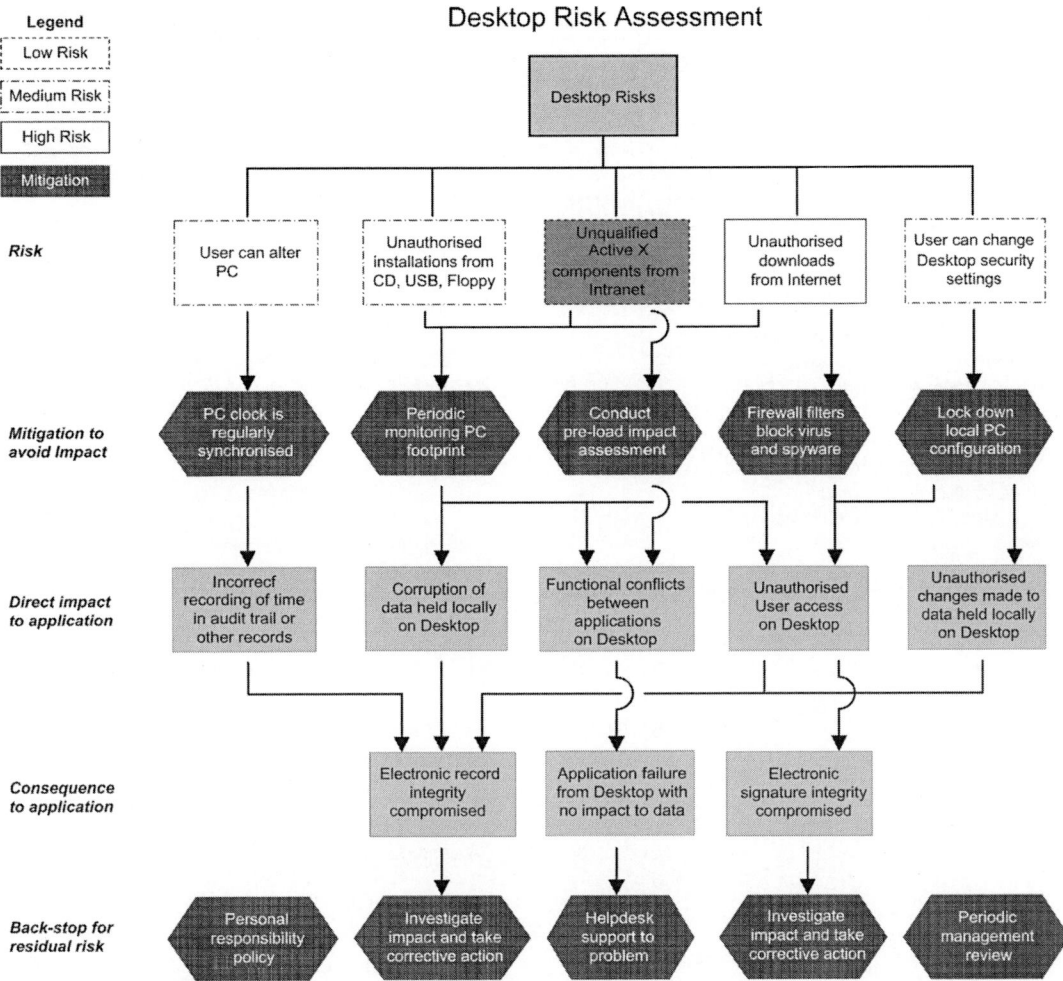

Figure 15.3 Example desktop risk map.

replacement of a system so that it is permanently to be replaced within the year—such situations are not acceptable to the GxP regulatory authorities.

The first step of determining an order of work is to define levels of risk and system characteristics that affect risk. Individual computer systems can then be classified against the set criteria and a weighted risk factor calculated. The state of existing compliance is then calculated and subtracted from the weighted risk factor to given a compliance gap. The compliance gaps can then be compared between systems to order work.

Three levels of risk are suggested here (low, medium, and high) although some pharmaceutical and healthcare companies may like to consider five levels of risk to match the system integrity levels defined by IEC/ISO 61508 for safety critical systems. Each system should be rated against a number of weighted risk factors to determine an overall level of risk. Seven example risk factors are considered in Table 15.4.

- System development
- Security practice
- Performance history
- Support service
- Visibility of use
- Regulatory exposure
- Remaining life

Table 15.4 Example Risk Factors and Weightings

Risk factors	Low risk (score 1)	Medium risk (score 2)	High risk (score 3)
System development (weighting × 1)			
Standard software	Commodity, off the shelf, application	Used in complex or critical application	Not applicable
Configuration	Not applicable	Only parameters set, no custom code	Custom macros or customization
Customization	Not applicable	Not applicable	Customized software
Custom application	Not applicable	Not applicable	Custom software
Support service (weighting × 2)			
Supplier capability	QMS and SLA	QMS or SLA	No QMS or SLA
Staff turnover	<3%	4–8%	>8%
Dependency on contractors	<30% of staff	30–50% of staff	>50% of staff
Spare parts	Spares and/or alternate system available on site.	Only available off site <24 hr (unless cannibalize?).	Only available off site >24 hr; no alternative supply
Data/software backups	Regular backups	Infrequent backups	No routine backups
Performance history (weighting × 1)			
Downtime	<1 hr (or one occurrence)/yr	<1–8 hr (or 1–5 occurrences)/yr	>8 hr (or >5 occurrences)/yr
User changes and system upgrades	None within last year	<3 user changes and <1 system upgrade in last year	>3 user changes and/or >1 system upgrade in last year
		None planned	Some planned
Security practice (weighting ×1)			
Physical access	Restricted by physical barrier (e.g., locked room)	Restricted by location only (e.g., panel key, removed keyboard)	No restrictions
Logical access	Different levels of password access for users and system administrator	System protected by single level of password access	No password protection in use
Virus management	Automatic	User dependent	No management
Extent of use (weighting × 3)			
Size	<10 users	10–25 users	>25 users
	<500 records	500–5000 records	>5000 records
	<20 key critical control points	20–60 key critical control points	>60 key critical control points
	<100 I/O	100–500 I/O	>500 I/O
Replication	One-of-a-kind application on site	Multiple systems on site used in same or similar manner	Application running multiple sites in same division of company
Regulatory exposure (weighting × 4)			
Inspection history	Not covered by or no comments from last inspection	Observations from last inspection	Critical observations from last inspection
Submission	Not applicable	No new submissions, general inspections still expected	Preapproval inspection/ expected <1 yr
GxP criticality	No/negligible impact	Indirect impact	Direct impact
Remaining life (weighting ×2)			
Expected remaining operational life	Planned withdrawal within 2 yr	Anticipated life approximately 3–5 yr	No planned replacement

Abbreviation: SLA, service level agreement.

Multiplying the score for each row in Table 15.4 with its corresponding weighting and taking the sum across all the rows yields a total which can be used determine the level of risk. A total score of between 21 and 35 is considered a LOW risk, a score of between 36 and 49 is considered a MEDIUM risk, and a score of between 50 and 63 is considered a HIGH risk.

A worksheet should be developed to log the risk assessment. It must be stressed that Table 15.4 is only given as an example. Pharmaceutical and healthcare companies should give careful consideration to which risk factors and weights are best suited to their business.

The state of compliance for each computer system can be determined from examining its associated documentation. The examination is not intended to be a detailed review. Rather it should be a rough-cut evaluation delivering a quick result. Locating and retrieving what documentation exists is likely to be a much more time consuming task than the examination of the documentation itself. Documentation should be marked according to a scale such as: 1: does not exist; 2: exists but needs work to fulfill current regulatory requirements; and 3: exists and adequate to fulfill current regulatory requirements. Document names will vary between systems so for guidance generic document types are suggested in chapter 3. Again, worksheets should be developed to log the document examination. The sum of marks given for the generic document types give the state of compliance.

The compliance gap is calculated by subtracting the "state of compliance" score from the maximum possible "risk assessment" score for that systems level of risk. The maximum possible risk assessment scores for LOW-, MEDIUM-, and HIGH-risk systems are 35, 49, and 63, respectively. To avoid negative scores the state of compliance assessment should be designed so that its maximum score is equal or less than the maximum possible risk assessment score for a LOW risk system. The compliance gap score can be included in the system inventory. The priority attached to projects should be based on tackling the systems with the highest compliance gap scores first.

Completion of validation across a number of legacy computer systems, whether by remediation or replacement of individual systems, should be achieved within two to three years from the outset of the overall program of work. Status reports should be periodically prepared to demonstrate progress.

Hazard Control

When prioritizing projects is important to consider critical dependencies on particular computer systems. Hazards must be controlled. A stepwise approach to hazard control is given below.

- Assess each computer system to determine whether or not it can influence the strength, identity, security, and purity, or otherwise, quality of drug product must be assessed. The assessment should be conducted in accordance with a defined process and the outcome of each assessment recorded.
- Precisely, how a computer system impacts drug product attributes should be documented. Those computer systems that impact drug product attributes require validated. The decision to validate or not to validate shall be approved by an authorized person as part of the compliance determination.
- All computer systems should be considered in scope unless reliance can be placed on an independent downstream system. A downstream system may be a manual system, a further computer system, or a nonsoftware based item of equipment.
- Downstream systems based on computer systems must be validated. Downstream systems based on manual ways of working and/or nonsoftware based items of equipment should be periodically challenged at suitable intervals during its operational life.
- If the downstream system is a checking device, and is not a separate computer system (i.e., it forms part of functionality of the computer system under review), then the whole system including the checking device should be validated. A regime of sampling the output of the computer system will not be accepted as a downstream quality check.
- A remedial action plan is required where a compliance gap is determined against a computer system's compliance requirement. For significant compliance gaps system replacement may be more cost-effective than revalidation.
- Once critical computer systems are compliant with regulatory expectations then remaining computer systems should be validated.

Hazard control can help focus effort and thereby rapidly establish significant GxP improvements. This is likely to be especially important where skilled resource and/or available time to bring systems into compliance is limited.

Interim Measures

Interim measures are additional controls applied in relation to computer functionality that support critical quality related activities. They are implemented where compliance gaps are considered to exist, to provide added assurance of control and to justify the continued use of a computer system. Interim measures are used to supplement or replace defined computer functionality. Examples of interim measures include the following:

- Independent manual procedures used in parallel to support computer system functionality
- Comparison of data sampled from specific functions with independently derived data
- Independent computer systems to monitor critical quality related activities
- Independent downstream computer systems to detect quality failures
- Combination of the above

The type of interim measure implemented should be appropriate to the computer functionality being addressed. Computer functionality being addressed should be mapped so that appropriate interim measures can be identified. The mapping should include both a workflow analysis and a data flow analysis. Controls that are already in place may provide the basis for the interim measures. Critical activities that should be given particular consideration for interim measures include the following:

- Stages in the operational process where status change occurs such as approval of a raw material or intermediate product
- Critical processing activities that are reliant on computer systems such as dispensing
- Label information and printing
- Product quality related specifications held by or used by computer systems
- Approval of product to release to the market
- Access points where GxP data can be modified or deleted

Interim measures do not eliminate the need for full corrective actions; they do not resolve actual computer system compliance issues. Full corrective solutions must still be planned and implemented to bring computer systems into compliance. If interim measures are implemented this activity must be properly planned and form part of an overall plan to install permanent corrective solutions. Interim measures should be kept as simple as possible.

Validation Activities

The following checklist is based on work by the German APV for practitioners validating existing computer systems that were not, or were only partially, developed in accordance with compliance requirements (32). Some practitioners prefer to use the term retrospective evaluation to highlight that the exercise is founded on the principle of a compliance gap analysis and consequential remedial actions. It is important to realize that the validation of existing systems takes more effort than prospective validation and rarely meets achieves the same standard.

- Stop or justify the continued use of the computer system.
- Freeze the computer system to stop any changes during subsequent validation.
- Conduct a compliance gap analysis on the GxP-relevant components and functions of the system with reference to the past operational experience. Assess the completeness of documentation, outstanding internal audit observations, and outstanding regulatory commitments.
- Prepare a validation plan.
- Create/revise the documentation describing the computer system.

- Conduct a design review.
- Inspect critical application software, conduct an IQ, conduct an OQ with emphasis on GxP component and functions of the system, and conduct a PQ.
- Prepare a validation report.
- Release the computer system for use, if necessary implementing system modifications and/or additional organizational measures under change control.

The general approach to legacy systems is the same as validating new systems (refer to chap. 4). The content and structure of validation plans should fulfill the recommendations outlined in chapter 5. Validation plans usually have an additional section giving a brief history of the system from its original procurement, through any developments, to the current system configuration. The validation plan should indicate the new and existing documentation that will be used to support validation of the computer system. Some prospective activities may not be possible to conduct, however, such as supplier audits if the supplier is no longer trading, source code reviews if there is no access to source code and relevant design documentation, and Development Testing if detailed design information is not available. If original design and development documentation is missing and/or the change history is missing or incomplete but there is evidence to demonstrate ongoing reliable operation then the computer system can be treated like OSS (see relevant comments in chap. 9).

Some pharmaceutical and healthcare companies combine the intent of URS and functional specification into a document called a System Specification for legacy systems. The System Specification will include a statement to the effect that the document represents not only a description of the system in use, but also that this description fulfills user requirements for the system. Although the original design intent of the computer system may have changed it may not be necessary to totally rewrite existing specification documents. Instead it may be possible to write a short frontispiece to existing documents, defining the changes and their impact on the original design.

Supplier audits should be conducted where practical for custom (bespoke) and critical applications. Emphasis will be placed on the level of support available from the supplier. Remember that the supplier may be a function within the pharmaceutical or healthcare company's organization. In such instances, the supplier audit becomes an internal audit and document search.

Software and hardware design documentation may have to be reverse engineered, both at module and system level. The GAMP® Special Interest Group on Legacy Systems recommends reverse engineering only for custom (bespoke) software elements, COTS software at this level only needs configuration to be defined (32). Software logic flows should be described and flowcharts developed as appropriate. All algorithms need to be defined. Hardware configuration items should be listed.

A design review should be conducted before testing begins. This will normally involve developing a Requirements Traceability Matrix (RTM). If no detailed design information is available then cross-references should be made between the newly prepared System Specification, available operator manuals, and user procedures. Source code reviews will be expected for custom (bespoke) software under the control of the pharmaceutical and healthcare company and redundant code identified should be removed.

Development testing by definition for an existing system should have already been conducted although original test records may be incomplete, insufficient or missing. Test protocols should be reviewed to ensure that they reflect the current operating environment. Some pharmaceutical and healthcare companies take the opportunity to supplement their user qualification will additional tests to unit, system, and integration tests that might otherwise be conducted as a separate activity.

User qualification should comprise of IQ, OQ and PQ. The IQ effectively baselines the system for OQ and can be conducted while the system is making pharmaceutical and healthcare grade products. The OQ should cover all functional aspects now defined in the System Specification. Some OQ testing such as safety-related test and disaster recovery tests may have to be delayed until a planned facility shutdown takes place. Some facilities may not have a planned shutdown for more than a year, in which case consideration should be given to especially planning one for the validation project. The final phase, PQ, can use, but must not rely solely on, historical evidence of dependable operation. Product PQ should be conducted over larger number of samples than product PQ for new systems to increase confidence that the existing systems as used are working correctly. For instance, it has been suggested that the product PQ should review at least 30 batches of manufactured drug products.

Procedures and user manuals may be outdated with users relying on typed or hand-written instructions to supplement or replace old manuals. Procedures for operating the computer system should be reviewed and updated as necessary to reflect the current use of the system. Training records should be current and reflect training in these updated procedures. Access rights should be checked as appropriate and authorized. Role specifications may need to be updated. Business continuity plans should also be reviewed and amendments made as required.

Finally, a validation report should be written in reply to the validation plan. Internal and third party SLAs may need to be established to ensure validation is maintained. Arrangements for effective change control and configuration management must be put in place.

Recent Inspection Findings

Retrospective validation may be conducted for a well-established process used without significant changes to [drug product] quality because of changes in raw materials, equipment, systems, facilities or the production process. This validation approach may be used where:

1.   Critical quality attributes and critical process parameters have been identified;
2.   Appropriate in-process acceptance criteria and controls have been established;
3.   There have not been significant process/product failures attributable to causes other than operator error or equipment failures unrelated to equipment suitability; and
4.   Impurity profiles have been established for the existing [drug product]

Once an existing process has been validated retrospectively, and the process needs to be revalidated because of changes that may effect the quality of a [drug product], the validation should be done prospectively, or in certain limited cases, concurrently. Most important, these changes should be controlled by a formal change control system that evaluates the potential impact of proposed changes on the quality of the [drug product]. Scientific judgment should determine what additional testing and validation studies should be conducted to justify a change in a validated process (FDA Warning Letter, 2000).

It could be difficult to retrospectively validate a computer system if there were changes and revisions that were not documented and the cumulative affects of many revisions had not been assessed. Lack of sufficient system documentation would make it impossible to perform meaningful retrospective validation. FDA concludes that the XXX and YYY systems lack adequate validation and therefore are unacceptable for use in the production of drug products. Please indicate whether you can perform a retrospective validation of XXX and YYY systems or rely in the interim on manual operations, which use source documentation until the new validated computer systems are functional (FDA Warning Letter, 2001).

Manual verification of calculations and inventory checking with the existing computer software that has been found to be problematic is not an adequate reason for lack of validation. Existing computer software should be validated or replaced (FDA Warning Letter, 2001).

We continue to find proposed timeline to complete validation of the XXXX system to be unacceptable. The XXXX system should not be in use unless they have been completely validated to current standards (FDA Warning Letter, 2002).

Software "bug" that could result in erroneous release not scheduled for correction...Headquarters has allowed a workaround for a software problem to be in place for eight years (FDA 483, 2002).

There is no documented program for remediation (FDA 483, 2004).

STATISTICAL TECHNIQUES
When statistical sampling is used it is recommended that professional statistical support is used rather than relying on ad-hoc advice. It is vital that statistical techniques are used appropriately.

Approach to Projects
Statistical sampling can be considered as part of a testing strategy for projects implementing/deploying multiple systems that are the same or very similar (i.e., within an acceptable delta). All systems must be tested but full testing is only required for the selected sample. A similar approach, sometimes referred to as matrix validation, is used in the context of validating manufacturing equipment and processes.

The determination of the sample size must be documented. An important aspect to consider in applying statistical sampling is the need to predefine the acceptability of "similar systems." If the systems and their operational environment are exactly identical then a sample size of one may be sufficient. If the systems are not identical, then consideration needs to be given to what is an acceptable delta for the differences between those similar systems. Some of the deltas that one can consider may include the differences in software (operating systems, 3rd party tools, application program) version, patches, and fixes, as well as the deltas in hardware and equipment, instrument or other peripheral that are the components of the system. Great care must be taken in justifying an acceptable delta. Computer systems should be considered separate applications and validated accordingly when there is significant variation.

Approach to Data
Data checking can be a resource intensive process. Statistical sampling can provide a viable method to reduce the effort, resources, and time required to check data while retaining a high degree of assurance that the required level of data accuracy is being maintained.

Data can be classified into different types, each type with a different level of acceptable accuracy. Three basic classifications are described here by way of example.

- Critical data (includes GMP data) is required to be 100% accurate. This can only be established by a 100% check, preferably independently by two persons, to minimize the likelihood of mistakes, for example, due to fatigue and other random errors.
- Significant data (if this is to be distinguished from critical data) is required to have a predetermined acceptable accuracy (e.g., has at most a 5% error rate). This can be established by a randomly drawn sample so long as a small risk is accepted that even though the sample strongly indicates that the error rate is below the predetermined acceptable level, in fact the "true" error rate is above the predetermined acceptable level. This is an inevitable consequence of using a sample. The only alternative is a 100% check as above.
- Other data (can be divided up into further subcategories) is required to have a predetermined acceptable accuracy (e.g., have at most a 25% error rate). This can be established as above for significant data by a randomly drawn sample so long as a small risk is accepted.

The objective of statistical sampling is to establish likely values for the true error rate in the population of data being considered. If the true error rate was known, the probabilities of given numbers of errors in samples could be obtained mathematically using standard statistical distributions. Statistical inference allows the reverse process, from an observed error rate in a sample likely and possible true error rates can be inferred. Likely data population error rates are defined by the 99% single upper confidence limit and possible data population error rates by the 99.9% single upper confidence limit on the sample error rate.

Large populations of data (in excess of 5000 items) can be regarded as infinite and thus a binomial approximation to the hypergeometric distribution can be applied. It is assumed errors occur randomly throughout the data population. If data within the population has been obtained from different sources in different ways, however, there may be an expectation that error rates for these sub-populations may differ. If this is the case then the data population should be split into "strata" and analyzed separately. Note for populations less than 5000 items it is recommended to check all items rather than take a sample.

The *likely error rate*, as stated earlier, is defined as all values less that the 99% single upper confidence limit on the population error rate, that is,

$$100 * \left\{ p + 2.3263 * \sqrt{\left[\frac{p(1-p)}{N} \right]} \right\}$$

If extra assurance is required, the *possible error rates* are defined by the 99.9% single upper confidence limit on the population error rate, that is,

$$100 * \left\{ p + 3.0902 * \sqrt{\left[\frac{p(1-p)}{N} \right]} \right\}$$

where p is the observed proportion of errors in the sample and N is the sample size.

Tables in Appendix 15.A are provided to support the statistical analysis. Extra tables can be easily developed to support other error rates and smaller data populations if need be. To determine the required sample size from the tables, follow the steps below.

1. Select the target error rate (5% or 25% for the tables provided).
2. Select the observed error rate that is believed likely to become true and use that (rounding up as necessary) to chose a column in the table. Rounding up will give a sample size large than is strictly required but eases use of the table.
3. Identify the smallest sample size that the chosen column gives a likely error rate less than the target error rate (e.g., for an observed error rate of 3.5% is applicable to the table for error rates not exceeding 5%, and yields a sample size of 1050).
4. Obtain a random sample of this size and measure the error rate. Note that the sample must be (effectively) random to avoid potential bias from unknown or ignored influences on the data population.
5. If the observed error rate in the sample is equal to or less than that the predefined acceptable level then no further action is required. Nevertheless it is recommended that the opportunity be taken to correct any errors found and investigate any commonalties between the errors to identify any root cause that might affect the rest of the data population.
6. If the observed error rate is greater than the predefined acceptable level then repeat step 3 using the observed error rate. Note that part of the required sample has already been taken. In the example given in step 3 if the observed error rate is 4% then a further sample of 1050 is required.

The tables with *likely error rates* will normally be used unless a very cautious approach is being taken in which case the *possible error rates* should be used.

Recent Inspection Findings

Failure to establish and maintain procedures to ensure that sampling methods are adequate for their intended use and are based on a valid statistical rationale (FDA Warning Letter, 2000).

No documentation to support statistical techniques used (FDA 483, 2002).

No written description or written instruction of how analysis is performed on statistical software (FDA 483, 2005).

Deficient validation of statistical software system in that not all raw data checked with active data files for accuracy, no review of data discrepancies, no audit trail, PQ acceptance criteria not met, deviations between files not resolved (FDA 483, 2005).

Program used to calculate linear regression has not been validated (FDA 483, 2003).

REFERENCES

1. Food and Drug Administration. Computerised Drug Processing; Vendor Responsibility, Compliance Policy Guide 7132a.12, Sec. 425.200, ORA. Rockville, MD, 1987.
2. U.S. Code of Federal Regulations Title 21: Part 58. Good Laboratory Practice for Nonclinical Laboratory Studies.
3. European Union. EU Guide to Good Manufacturing Practice for EU Directive 2003/94/EC, Community Code Relating to Medicinal Products for Human Use, Vol. 4, 2003.
4. PIC/S Recommendations for Validation Master Plan and Installation/Operational Qualification, 2001.
5. International Conference on Harmonisation of Technical Requirements for Registration of Pharmaceuticals for Human Use. Good Manufacturing Practice Guide for Active Pharmaceutical Ingredients, q7 document, 2000. Available at: www.ich.org.
6. U.S. Code of Federal Regulations Title 21: Part 211. Current Good Manufacturing for Finished Pharmaceuticals, Federal Register, 2006.
7. International Conference on Harmonisation of Technical Requirements for Registration of Pharmaceuticals for Human Use. Guideline for Good Clinical Practice, ICH Harmonised Tripartite Guideline, 1996.
8. Organisation for Economic Co-operation and Development. Principles of Good Laboratory Practice to Computerised Systems. Paris, 1995.
9. United Kingdom Department of Health. The Application of GLP to Computer Systems. The Principles of Good Laboratory Practice, United Kingdom Compliance Programme, London, 1995.
10. U.S. Code of Federal Regulations Title 21, Part 820. Good Manufacturing Practice for Medical Devices.
11. BARQA. Regulatory Compliance and Computer Systems. Conference Procedings, 1997.
12. Lloyd IJ, Simpson MJ. Computer Risks and Some Legal Consequences, Safety and Reliability of Software Based System. New York: Springer-Verlag, 1997.
13. United Kingdom Sales of Goods Act, 1974.
14. United Kingdom Supply of Goods and Services Act, 1982.
15. United States Food, Drugs and Cosmetics Act.
16. United Kingdom Supply of Machinery Regulations, 1992.
17. United Kingdom Health and Safety at Work Act, 1974.
18. United Kingdom Environmental Protection Act, 1990.
19. United Kingdom Data Protection Act, 1984.
20. United Kingdom Consumer Protection Act, 1987.
21. United Kingdom Product Safety Regulations, 1994.
22. United Kingdom Unfair Contract of Terms Act, 1977.
23. Unfair Terms in Consumer Contracts, EU directive 93113/EEC, 1993.
24. United Kingdom Misrepresentation Act, 1967.
25. Current Good Manufacturing Practices for Finished Pharmaceutical Products 21 CFR 211.25(a).
26. European Union Food Manufacturing Practice for Pharmaceuticals, Medicines Controls Agency, 1997.
27. David Begg Associates. Computers and automated systems quality and compliance. York, June 24–27, 2002.
28. McDowall RD. Regulatory compliance considerations when outsourcing (part 1 and part 2). Eur Pharm Rev 2002.
29. International Society for Pharmaceutical Engineering. GAMP®5: Risk-Based Approach to Compliant GxP Computerised Systems. Tampa, Florida, 2008. Available at: www.ispe.org.
30. Hatton L. Unexpected (and Sometimes Unpleasant) Lessons from Data in Real Software Systems, Safety and Reliability of Software Based Systems. New York: Springer-Verlag, 1997.
31. Williams Y, Torres J. Documentation of infrastructure qualification and system validation. IVT Conference on Network Infrastructure Qualification & Systems Validation. Philadelphia, October 8–9, 2002.
32. International Society for Pharmaceutical Engineering. GAMP® Good Practice Guide: Legacy Systems. Tampa, Florida, 2003. Available at: www.ispe.org.

Appendix 15.A Error Rate Tables

Table 15.A.1 Likely "True" Error Rates (%) for Observed Error Rates (%) in Samples of Given Sizes with Target Errors Rate of at Most 5%

Sample size	Observed error rate in sample					
	1%	2%	3%	3.5%	4%	4.5%
350	2.24	3.74	5.12	5.79	6.44	7.08
700	1.87	3.23	4.50	5.12	5.72	6.32
1050	1.71	3.01	4.22	4.82	5.41	5.99
1400	1.62	2.87	4.06	4.64	5.22	5.79
2100	1.51	2.71	3.87	4.43	4.99	5.55
2800	1.44	2.62	3.75	4.31	4.86	5.41
3500	1.39	2.55	3.67	4.22	4.77	5.32
7000	1.28	2.39	3.47	4.01	4.54	5.08
10500	1.23	2.32	3.39	3.92	4.44	4.97
14000	1.20	2.28	3.34	3.86	4.39	4.91
17500	1.17	2.25	3.30	3.82	4.34	4.86

Note: True error rate is defined at 99% single upper confidence limit.

Table 15.A.2 Likely "True" Error Rates (%) for Observed Error Rates (%) in Samples of Given Sizes with Target Errors Rate of at Most 25%

Sample size	Observed error rate in sample					
	4%	8%	12%	16%	20%	24%
10	18.42	27.96	35.91	42.97	49.43	55.42
25	13.12	20.62	27.12	33.06	38.61	43.87
50	10.45	16.93	22.69	28.06	33.16	38.05
100	8.56	14.31	19.56	24.53	29.31	33.94
200	7.22	12.46	17.35	22.03	26.58	31.03
300	6.63	11.64	16.36	20.92	25.37	29.74
350	6.44	11.37	16.04	20.56	24.97	29.31
700	5.72	10.39	14.86	19.22	23.52	27.76
1050	5.41	9.95	14.33	18.63	22.87	27.07
2100	4.99	9.38	13.65	17.86	22.03	26.17
7000	4.54	8.75	12.90	17.02	21.11	25.19
14000	4.39	8.53	12.64	16.72	20.79	24.84

Note: True error rate is defined at 99% single upper confidence limit.

Table 15.A.3 Possible "True" Error Rates (%) for Observed Error Rates (%) in Samples of Given Sizes with Target Errors Rate of at Most 5%

Sample size	Observed error rate in sample					
	1%	2%	3%	3.5%	4%	4.5%
350	2.64	4.31	5.82	6.54	7.24	7.92
700	2.16	3.64	4.99	5.65	6.29	6.92
1050	1.95	3.34	4.63	5.25	5.87	6.48
1400	1.82	3.16	4.41	5.02	5.62	6.21
2100	1.67	2.94	4.15	4.74	5.32	5.90
2800	1.58	2.82	4.00	4.57	5.14	5.71
3500	1.52	2.73	3.89	4.46	5.02	5.58
7000	1.37	2.52	3.63	4.18	4.72	5.27
10500	1.30	2.42	3.51	4.05	4.59	5.13
14000	1.26	2.37	3.45	3.98	4.51	5.04
17500	1.23	2.33	3.40	3.93	4.46	4.98

Note: Possible error rate is defined at 99.9% single upper confidence limit.

Table 15.A.4 Possible "True" Error Rates (%) for Observed Error Rates (%) in Samples of Given Sizes with Target Errors Rate of at Most 25%

Sample size	Observed error rate in sample					
	4%	8%	12%	16%	20%	24%
10	23.15	34.51	43.76	51.83	59.09	65.73
25	16.11	24.77	32.08	38.66	44.72	50.40
50	12.56	19.86	26.20	32.02	37.48	42.66
100	10.06	16.38	22.04	27.33	32.36	37.20
200	8.28	13.93	19.10	24.01	28.74	33.33
300	7.50	12.84	17.80	22.54	27.14	31.62
350	7.24	12.48	17.37	22.06	26.61	31.05
700	6.29	11.17	15.80	20.28	24.67	28.99
1050	5.87	10.59	15.10	19.50	23.81	28.07
2100	5.32	9.83	14.19	18.47	22.70	26.88
7000	4.72	9.00	13.20	17.35	21.48	25.58
14000	4.51	8.71	12.85	16.96	21.04	25.12

Note: Possible error rate is defined at 99.9% single upper confidence limit.

16 | Capabilities, Measures, and Performance

INTRODUCTION

The ability to perform cost-effective compliance is dependent on the organizations understanding of requirements and their compliance capability. This chapter applies the established capability maturity model (CMM) to computer compliance. Examples of compliance metrics and measures are examined to draw lessons. The metrics cover prospective validation as well as operation and maintenance of computer systems. Lean manufacturing and six sigma are promoted as tools that organizations can use to streamline and improve the performance of their compliance processes.

COMPLIANCE CAPABILITY

Experience has shown the significant advantages for suppliers of computer systems (internally within pharmaceutical and healthcare company organizations, systems integrators, and equipment vendors) who improve their compliance capability. In particular, the risk of noncompliance is reduced, and the process of achieving compliance becomes more cost effective and time efficient.

A framework recognizing the symbiosis between both process and product was first proposed by the Software Engineering Institute (SEI) at Carnegie Mellon University and called the CMM. CMM is based on five evolutionary levels of capability from ad hoc, chaotic processes to mature, disciplined processes (1).

> Level 1: The quality process is characterized as ad hoc and occasionally even chaotic. Few processes are defined, and success depends on individual efforts and heroics.
> Level 2: Basic project management processes are established to track cost, schedule, and functionality. The necessary process discipline is in place to repeat earlier successes on projects with similar applications.
> Level 3: The quality processes for both management and engineering activities is documented, standardized, and integrated into a standard quality process for the organization. All projects use an approved, tailored version of the organization's standard quality process for developing and maintaining systems.
> Level 4: Detailed measures of the quality process and product quality are collected. Both the quality process and products are quantitatively understood and controlled.
> Level 5: Continuous process improvement is enabled by quantitative feedback from the process and from piloting innovative ideas and technologies.

An assessment of an organization against CMM can be used to generate a profile from which an organization can identify necessary initiatives to support an improvement in their quality assurance capability. In the same manner, the profile can help organizations prevent management oversight of crucial activities supporting a level of capability. Walter Royce suggests that only 25% of companies can be considered level 3 or above (2), that is, at a competency level similar to ISO 9000.

An adaptation of SEI's CMM for application to computer systems compliance is shown in Figure 16.1. This is not a definitive adaptation and is based on the principles of CMM. An additional sixth level (0) has been added to deal with organizations that have not yet embarked on any compliance capability.

CAPABILITY APPRAISALS

Pharmaceutical and healthcare companies should consider placing themselves within the framework. Organizations will often have a capability profile that includes elements of

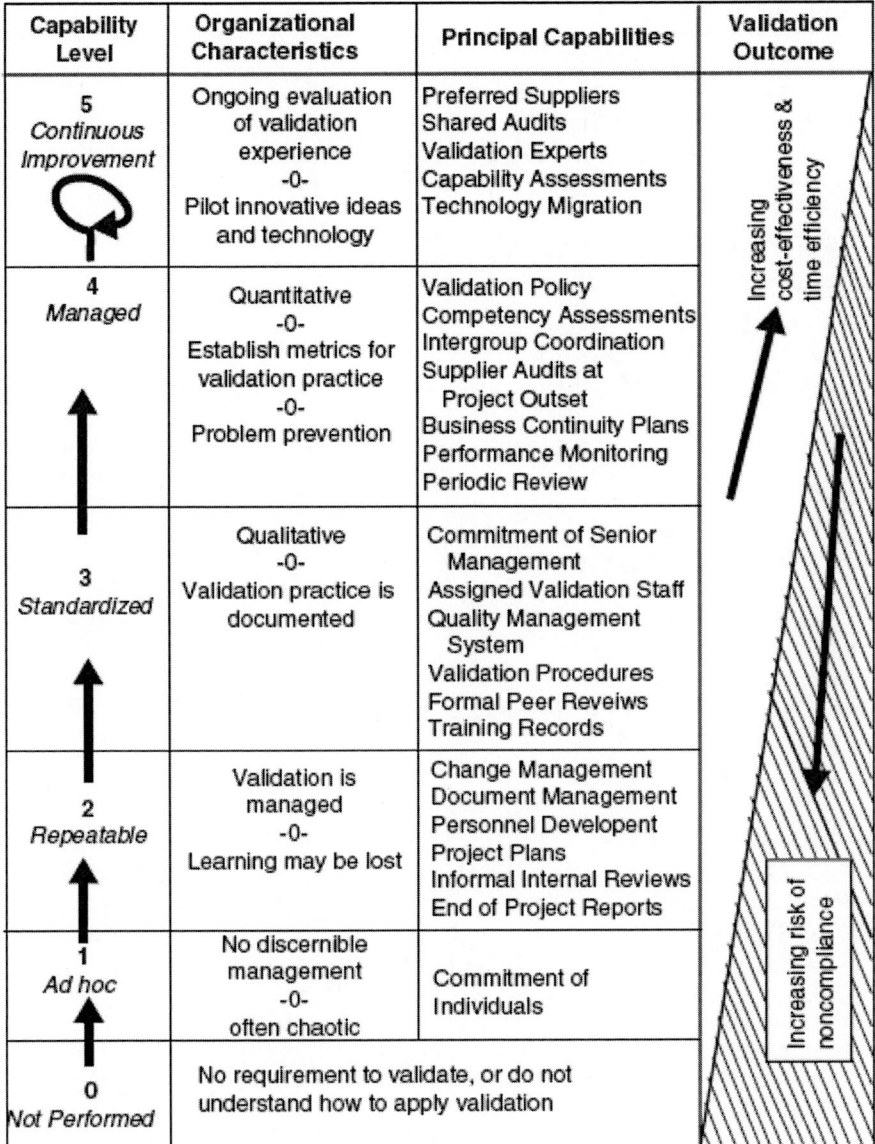

Capability Level	Organizational Characteristics	Principal Capabilities	Validation Outcome
5 *Continuous Improvement*	Ongoing evaluation of validation experience -0- Pilot innovative ideas and technology	Preferred Suppliers Shared Audits Validation Experts Capability Assessments Technology Migration	Increasing cost-effectiveness & time efficiency
4 *Managed*	Quantitative -0- Establish metrics for validation practice -0- Problem prevention	Validation Policy Competency Assessments Intergroup Coordination Supplier Audits at Project Outset Business Continuity Plans Performance Monitoring Periodic Review	
3 *Standardized*	Qualitative -0- Validation practice is documented	Commitment of Senior Management Assigned Validation Staff Quality Management System Validation Procedures Formal Peer Reveiws Training Records	
2 *Repeatable*	Validation is managed -0- Learning may be lost	Change Management Document Management Personnel Develoment Project Plans Informal Internal Reviews End of Project Reports	
1 *Ad hoc*	No discernible management -0- often chaotic	Commitment of Individuals	Increasing risk of noncompliance
0 *Not Performed*	No requirement to validate, or do not understand how to apply validation		

Figure 16.1   Computer compliance capability model.

capabilities from several levels. The assessed level of capability will be that with which an organization is entirely compliant.

A sample questionnaire is given in Appendix 16.A to help evaluate which level of compliance capability an organization fits (3). The best way to conduct an appraisal is by an unannounced surprise audit. Mature organizations should by their nature be inspection ready. When conducting an appraisal, specific examples should be documented to demonstrate a capability and a note taken as to whether the capability is readily observable in more than one context. During any appraisal care must be taken to assess the true organization capability rather than massaging the assessment outcome. If the whole audit process takes longer than 5 to 10 man-days effort inclusive of auditor and auditee then it probably indicated that evidence is not readily available.

CAPABILITY CHARACTERISTICS

The characteristics associated which each level of compliance capability can be summarized as follows:

Level 1: Unpredictable performance in terms of costs, schedule and quality. Possibly less that 20% of issues raised will be resolved. There are often difficult team interactions, mainly because there are no defined processes to assist implementation of computer systems.

Level 2: Repeatable performance from project to project in terms of costs, schedule and quality, but no performance improvements. Typically less that 60% of issues raised will be resolved. There are likely to be some difficult team interactions, but basically the team will support each other.

Level 3: Better performance on successive projects in terms of costs, schedule and quality. Less that 25% of issues raised remain unresolved. Team members mutually support each other.

Level 4: Significant performance benefits in successive projects in terms of costs, schedule and quality. Less that 10% of issues raised remain unresolved. Computer implementations are trustworthy and consistently delivered with full functionality, in budget, and to schedule. Highly cooperative team interactions.

Level 5: Continually improving performance benefits on successive projects in terms of costs, schedule and quality. 100% of issued raised are resolved. Teams are cohesive and seamless. Level 5 organization is typically specialized in niche expertise.

In the capability framework presented, it has proven quite difficult to align compliance activities between level 3 and level 4. The divide suggested is based on an established project-by-project capability of level 3 compared with the ongoing inherent organizational capability of level 4. In the framework, level 4 equates to a fully compliant GMP regime.

CAPABILITY ASSESSMENT OUTCOMES

Level 1 and level 2 assessment outcomes usually denote pharmaceutical and healthcare companies whose senior managements are still not wholly committed to sustainably achieving compliance and rely on their subordinates without the practical support they could offer. Computer compliance is often characterized by firefighting.

Pharmaceutical and healthcare companies typically like to think of themselves at level 3. A compliance capability below level 3 will almost certainly be regarded by GMP regulatory authorities as insufficient for GMP. Noncompliance may not be identified on an initial or limited inspection. Pharmaceutical and healthcare companies must not become complacent and should prepare for further and detailed inspections. The regulator's position with individual cases of GMP noncompliance will vary with the severity of the deficiencies they find. Generally, however, they will give the pharmaceutical or healthcare company a period of time to take corrective actions before they take the matter further.

Less than 1% of organizations have a level 5 capability. Level 5 signifies the opportunity for pharmaceutical and healthcare companies and their suppliers to reap the reward of tangible benefits discussed earlier in chapter 1. Principal capabilities associated with level 5 might include selecting preferred suppliers, conducting joint Supplier Audits with other organizations and sharing audit reports, developing in-house compliance experts, conducting internal capability assessments to identify improvement opportunities, and planning for technology migration to exploit any new innovations.

SUPPLIER CAPABILITY ASSESSMENTS

A slightly different situation exists with suppliers of computer systems to the pharmaceutical and healthcare industry. Suppliers generally understand the benefits of a quality approach, but unless the pharmaceutical and healthcare industry forms a significant proportion of their sales, then it is unlikely they really understand how their quality approach relates to the requirements of regulatory compliance, despite what the supplier's salesmen might say. For this reason, while a supplier might be level 3 or 4 on the CMM, the same supplier is likely to be level 2 on the compliance capability framework. It is very important for pharmaceutical and healthcare companies to undertake a Supplier Audit to determine the actual capability of their suppliers, and, in particular, to assess the competence of personnel assigned to their project.

BENEFITS OF IMPROVING CAPABILITY

Annual returns on original investment of an enhanced quality assurance capability for computer systems should be in excess of four fold. Stepping up one level in the capability framework should reduce costs by 20% to 25% (4). It typically takes an organization about two years of concerted effort to go up a level of capability. This is because capability is linked with culture. It is relatively easy to establish policies and procedures; it is much harder to build a complementing inherent quality culture. For instance QA groups are often perceived a striving for 100% perfection on computer compliance to mitigate all risk of noncompliance ("zero tolerance"). Consequently development groups often push back on QA, sometimes to the extent of compromising basic quality assurance practices ("dumbing down"). This is because they loose sight of "fit for purpose" in the pharmaceutical and healthcare industry means not only that the system works and fulfills industry standards, but also that the computer system satisfies regulatory requirements. A developing organizational capability must break down these barriers and foster a collaborative working environment.

PROJECT VALIDATION METRICS

Pharmaceutical and healthcare companies have the opportunity to use validation to reduce the cost of ownership for the computer systems they use. The cost of validation to a project represents an investment that will be more than returned in lower maintenance costs, which can anecdotally be reduced by between 50% and 80%. With maintenance perhaps being responsible for half the lifetime cost this gives a return on investment of one to three years.

Figure 16.2 collates some project validation metrics (3) to help practitioners understand compliance, and the allowance that should be made during project planning. The metrics will help challenge project planning where resource requirements seem excessive or too low to be credible. Current best practice using lean sigma and a risk-based approach suggest validation should cost no more than between 5 and 10% project cost.

When reviewing Figure 16.2, remember that there was no standard definition between the sources identified as what exactly constituted validation. The percentage of overall project cost is a rough indicator and not surprisingly increase as the complexity and customization of systems increase. The criticality of the system will impact cost too. It has been suggested that validation of critical systems can account for in excess of 50% overall project costs (5).

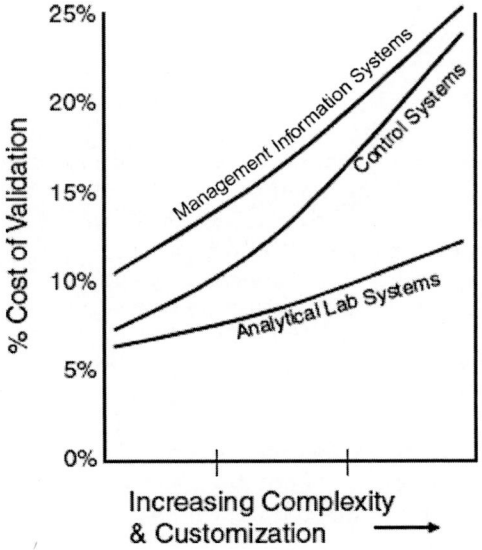

Figure 16.2 Historical validation cost as percentage of project cost.

Table 16.1 Comparing the Cost of Validating COTS Software

Type of application	Relative effort to validate
Custom (bespoke) application	100%
Configured COTS application	55%
COTS application without configuration	25%

Analytical laboratory systems in Figure 16.2 include analytical instruments with coupled laptops or personal computers (e.g., HPLC, GC, LC), and chromatography data systems. The metrics provided here assume that analytical laboratory systems are based on commercial off-the-shelf (COTS) products. LIMS are specifically excluded here and included in the management category of computer systems since they predominantly provide an information management function. Validation costs should be low because of the standard nature of the applications and only increase slightly with the relative size and complexity of the applications.

The control systems referred to in Figure 16.2 cover programmable logic controllers (PLCs) as the simplest example, supervisory control and data acquisition (SCADA) systems, and distributed control systems (DCS). These systems are typically based COTS products but have extensive configuration. As the systems get larger there also tend to be a growing number of customized interfaces to sub-systems and control instruments. Control systems may have many hundreds, even thousands, of associated instruments. This leads to a larger increase in validation costs compared with analytical laboratory systems relative to the growing size and complexity of the overall application.

Management Information Systems in Figure 16.2 include simple MRP, LIMS, MRP II, and integrated ERP systems. The relative increase in validation costs compared with growing size and complexity is relatively linear but with a greater rate of increase than analytical laboratory systems. This is because these applications typically involve extensive configuration. Customization is not such a significant factor, system functionality being provided by plug-in modules provided by the supplier of the core product or a certified product partner.

It is important to understand too that customization will further increase validation costs (e.g., custom PLC applications will likely incur a 10% validation costs as rather than the 5% indicated in Figure 16.2 which assumed a configured COTS application). Table 16.1 compares the relative increased costs associated with custom applications, configured COTS applications, and standard COTS products that do not require configuration. The cost of validation decreases the more standard software can be leveraged. The exploitation of standard software is explored further in chapter 15.

DESIGN AND DEVELOPMENT METRICS

An analysis of over 130 computer projects of various sizes (summarized in Table 16.2) emphasizes the benefits of conducting design and development reviews. It is suggested that the combination of effective design reviews and source code reviews should reduce overall project costs by about 10% compared with projects not implementing these reviews (this saving is achieved by detecting errors before testing). Too often, however, such reviews are ill defined and ineffective. Without effective design reviews, the design effort may be doubled because of the need to clarify ambiguous specifications or correct errors during coding and testing. Similarly, ineffective or missing source code reviews typically incur up to an additional 25% coding effort during testing to correct errors. A detailed analysis of validation costs by Murtagh emphasizes how source code reviews take much less effort to perform when conducted by the software developer's organization rather than the user because the developer's organization is more familiar with the type of software and better understand the design intent (6). Indeed, this principle holds true for all those aspects of compliance that can be supported by a supplier organization. It is not uncommon to find about two-thirds of a pharmaceutical and healthcare company's QA department input to a project revolving around resolving supplier related quality and compliance issues.

Table 16.2   Project Metrics

Life cycle phase	Typical project effort				Typical error detection capability	Normalized effort to fix error
	Analytical laboratory systems	Control systems	Management systems	Web-based systems		
System specification and selection						1
Design and development	40%	35%	25%	55%	20–45%[a]	2–5
Coding, configuration, and build	20%	25%	40%	15%	5–30%[b]	3–10
Development testing and user qualification	40%	40%	35%	30%	50–75%	5–30

[a]Design review.
[b]Source code review.

TESTING METRICS

As previously stated in chapters 9 and 10 testing should be designed to detect errors in the developed computer system. If the testing process itself is not robust, however, then it too will induce errors and rework. The testing conducted on 85 computer systems used across primary and secondary pharmaceutical manufacturing in several companies is analyzed here to examine test failures and how they were managed to closure.

Test failures were attributed to a number of causes as illustrated in Figure 16.3. Operator error executing the test case accounted for 1% of test failures. These tests were repeated once the error was understood. Incorrect set-up also accounted for 1% of test failures. These tests too were repeated with the correct set-up once the error was understood. Clarity problems with the test method and acceptance criteria accounted for 40% of test failures. Only the remaining 58% of test failures did what they should have done, detected system errors. That is, 42% of test failures processing was avoidable if a more robust test process had been adopted. Of the errors identified 37% were classed as significant, 63% errors were classified as not significant. Resolution of these errors impacted specification and design documents.

Major amendments to documentation was required to address 18% of the errors identified, the rest only required minor document change. Remember not all changes are limited to a single document. All document changes, however, need to go through change control that in practical terms means rework and delays.

The follow-up actions to test failures are analyzed in Figure 16.4. The vast majority of test failures (78%) were accepted as cosmetic with no further action. The test case required revision and reissue so that the test could be repeated for 11% of the test failures. For further 10% of test failures the test case was deemed acceptable but incorrectly executed. These tests could be rerun without modification once the tester understood where the test was misapplied. Finally, 1% of tests prompted hardware repair and a repeat test. The data collected highlight the need to train test staff to execute tests right first time, but then also to quickly recognize when a test

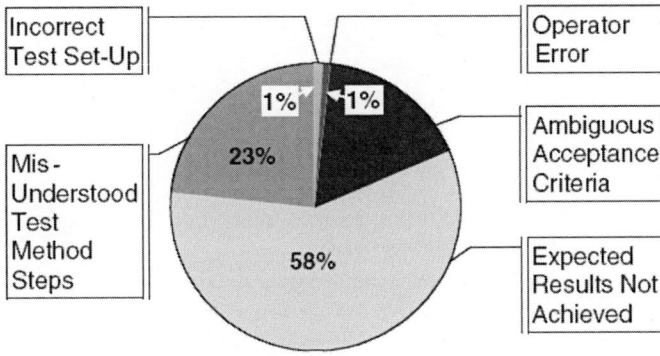

Figure 16.3   Test failure analysis.

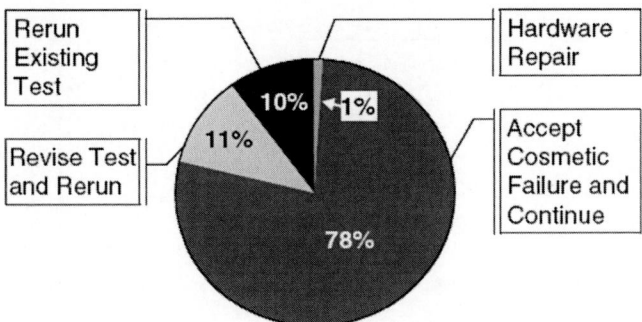

Figure 16.4   Test failure action.

failure is cosmetic so that testing can progress without undue interruption to overall test execution.

USER QUALIFICATION METRICS

The division of effort put into user qualification is shown in Figure 16.5. About one third of the total effort is used to prepare test cases. It is important test cases are clear and cover all the requirements of the computer system. Test execution and collation of testing evidence including preparing test reports accounts for just over a half of the user qualification effort. User qualification, however, often uncovers issues with specification and design documentation. Correcting specification and design deficiencies typically accounts for about 15% of the effort put into user qualification. Corrective activity higher than this indicates poor development. Corrections to specifications and design documentation must not be ignored as it undermines compliance.

The effort required to conduct an installation qualification broadly increases in a linear fashion with the size of a computer system. The effort required to conduct an operational qualification, meanwhile, tends to increase exponentially compared with the complexity of the computer system. The effort to conduct a performance qualification, like IQ, tends to increase in a linear fashion compared with the size of a computer system.

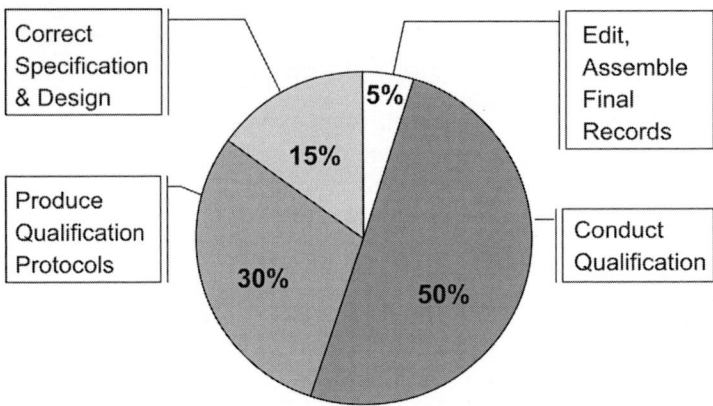

Figure 16.5   Split in qualification effort.

UNDERSTANDING CONTRIBUTORY FACTORS

Specific factors that contribute to the overall increased effort on computer validation projects include more comprehensive procedures and training, higher level of detail in documentation, increased testing, and more rigorous document control (Fig. 16.6).

Additional documentation and testing are the primary factors that make validation more expensive than conventional quality practices. Extra "quality assurance" approvals add effort and sometimes the perception of bottlenecks to the validation process. Additional "quality and

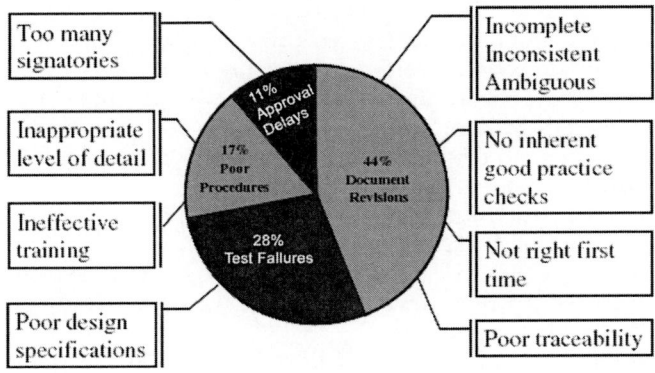

Figure 16.6 Factors adding effort to validation.

compliance" approval signatories are often a result of various departments not agreeing responsibilities and duplicating effort rather than being a regulatory requirement. The regulatory requirement for approvals is minimal. The same basic principle of over-engineering can contribute to the additional procedural controls often associated with validation. Controls do need to be robust, but complexity is often added as a result of departmental politics and matrix organizational responsibilities rather than regulatory requirements.

Another important factor to appreciate is the impact of late change during computer system projects. It is generally understood that during computer system implementation late changes can have a very high impact compared with making modifications early. The relative impact of change during a project and operation of small and large computer systems is summarized in Figure 16.7. Collected data from 130 projects supports these observations (7).

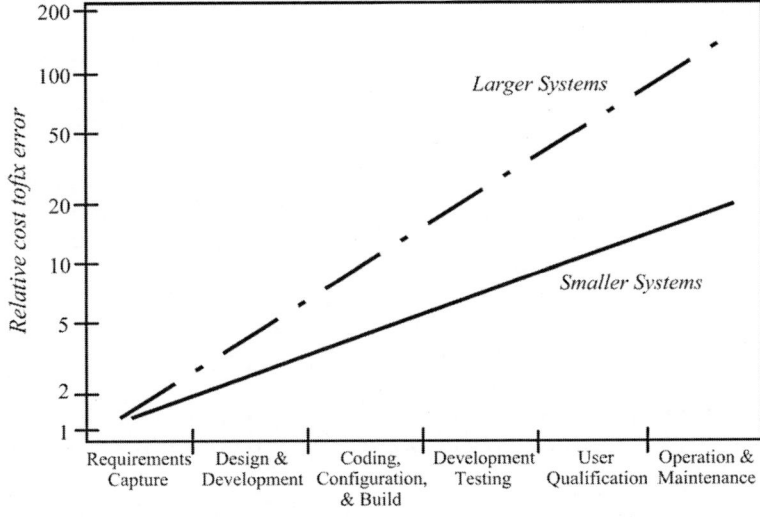

Figure 16.7 Relative cost of change.

RULES OF THUMB

- Typically projects afford about 40%-20%-40% of effort expended to (a) system specification, design and development, (b) coding, configuration and build, (c) development testing and user qualification.

- The combination of effective design reviews and source code reviews should reduce overall project costs by about 10% compared with projects not implementing these reviews. (This saving is achieved by detecting errors before testing.)
- Increased project effort spent on system specification, design and development should more than pay for itself in project pull-through.
- Typically, 50% of design effort is expended during coding and testing either to clarify ambiguous specifications or correct errors.
- Typically, 20% of coding effort is expended during testing to correct errors.
- Typically, 75% of errors are associated with 25% of the software.
- System testing typically only exercises 55% of errors without tracing tests to system requirements. With traceability to system requirements up to 80% of errors may be exercised during system testing.
- Experience suggests that more than 10% of defects remain undetected at the point when the system is authorized for use.

OPERATION AND MAINTENANCE METRICS

Users should anticipate that there will be some operational issues with computer systems. Most computer systems are brought into use has residual defects, known and unknown. Although suppliers will have appraised the criticality of defects and prioritize efforts to resolve them, they will not know how their software is going to be used and without knowledge of the context it is not possible to make a definitive risk assessment. The acceptability of some threshold of residual defects is subjective and often falls to a sales and marketing decision (8). Suppliers will often position that defects will be fixed later as a patch or as part of the next major software upgrade. Consequently in reality some defects may be left with no commitment to fix them.

The confidence that can be placed in a computer system will be based on the amount of testing conducted. More testing should at a very basic level increase the chance of detecting defects (8). Care must be taken not to assume better testing is necessarily based on higher coverage of predefine specifications. Risk management requires more intensive testing of most critical functionality, and less testing elsewhere.

The reliability of the computer system during operation can be defined as the number of failures as a function of the number of times the system is used. It has been suggested that it takes about two years to discover all the defects in programs with less than five defects per thousand lines of code, and greater than eight years to discover all the defects in programs with more than 30 defects per thousand lines of code (9). It has also been suggested that the rate of discover fits a bell curve (10).

We should really be interested in dependability (trustworthiness) and not just reliability. By its nature critical functionality must work correctly otherwise there will be a serious failure, perhaps even catastrophic failure. The dependability of a computer system will depend on the

- Success of removing defects prior to use (with focus on most critical defects),
- Rate at which defects subsequently discovered during operation are resolved, and
- Rate at which new defects are introduced by system changes.

In response, software maintenance is not limited to the correction of errors. Maintenance activities cover corrective maintenance, adaptive maintenance, perfective maintenance, and preventative maintenance.

- *Corrective maintenance* deals with the repair of errors.
- *Adaptive maintenance* deals with adapting software to changes in the operating environment such as new hardware or the next release of an operating system. Adaptive maintenance does not lead to changes in system functionality.
- *Perfective maintenance* mainly deals with accommodating new or changed user requirements. It concerns functional elements of the computer system. Perfective maintenance also includes activities to increase the system's performance or to enhance the user interface.

- *Preventative maintenance* concerns activities aimed at increasing the system's maintainability, such as updating documentation, adding comments, and improving the modular structure (architecture) of the computer system.

It is worthwhile noting that the IEEE combines adaptive and perfective maintenance activities under the title of adaptive maintenance. Data has been published that suggests that half the maintenance effort involves correcting errors, and half involves modifying the user to meet changing user needs including dealing with upgrades (11). In reality the amount of effort directed at the latter will depend entirely on the organizations investment strategy and architecture philosophy. For this reason, it is only possible to make meaningful metric observations on those maintenance activities focused on correcting errors.

CORRECTIVE MAINTENANCE METRICS

The annual corrective maintenance costs from approximately 250 computer applications is surprisingly consistent (12). Not surprising the maintenance effort decreased for older applications on the basis that an increasing proportion of errors are corrected over time. Annual corrective maintenance costs would seem as a rule of thumb to decrease by about one-sixth year on year, see Figure 16.8. The initial corrective maintenance costs were more dependent on the size of the application rather than initial error rate. This is because fewer but bigger errors tend to be the addressed in the early years of operation. These maintenance figure assume there is no other user driven enhancements, or system platform upgrades, etc.

When an error is to be corrected the time to implement a change can vary enormously depending on the nature and scope of the change. The change control process should not unduly waylay changes. Ineffective change control can delay changes by many weeks or months not because of the complexity of analyzing the proposed change and assessing its wider impact on the existing computer system but because an inability to process the paperwork in a timely fashion. The performance of the change control process can typically be greatly improved by

- Instituting a rapid initial appraisal of change requests to filter out rejected changes,
- Ensuring that the change management process does not have bottlenecks, and
- Automating the change control process with electronic review and approvals.

Research has also been conducted into the so-called software death cycle. It has been suggested that in some cases up to one in three changes introduce a new error. A more typical metric might be is one in five changes. There is a strong dependency on specification and design documentation. Poor documentation encourages maintenance staff to hack a solution,

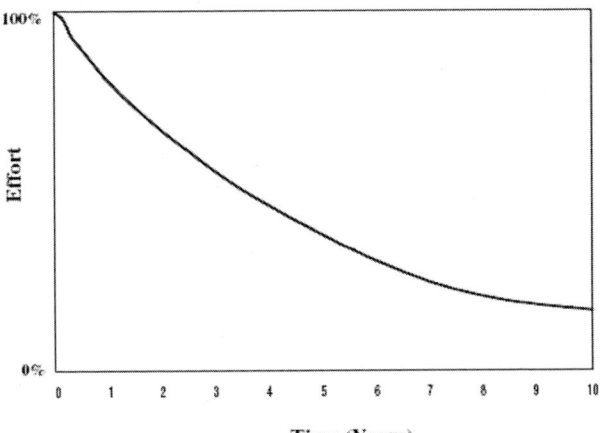

Figure 16.8 Corrective maintenance cost.

relying on their personal knowledge of the particular application to avoid introducing new errors. Trying to avoid an appropriate level of detail in specification and design documentation during projects is a false economy.

DEPENDABILITY METRICS

Operational dependability is a vital element of GMP compliance. An insight into the operational problems experienced with computer systems is given by an analysis of a consultancy firm's database of over 350 computer system malfunctions experienced by a number of international chemical and pharmaceutical manufacturing companies in the 1990s (3). The results are presented in Figure 16.9.

- Poor application design and programming errors accounted for 29% of malfunctions, indicating the importance of a supplier's project capability. Some of these problems were due to poor change control of the installed computer system by the pharmaceutical company and the lack of documentation provided by servicing engineers. It is all to easy for operations staff to make changed on quiet shifts and forget to records what they did; it then comes as a great surprise to the responsible managers that the documentation describing their system is our of date.
- The importance of conducting Supplier Audits for COTS software is highlighted by the 18% of malfunctions attributed to standard software.
- The importance of training system operators is demonstrated by the 20% of malfunctions attributed to human error. Companies must ensure training is given with approved SOPs before operators are required to use the computer system.

Unfortunately, the remaining malfunctions could not be diagnosed because a simple reboot of the software resumed normal operation and subsequent investigation could not identify any reasons for the malfunction.

The extra cost associated with initial validation also affects operation and maintenance. It has been suggested conventional quality effort for operation and maintenance processes may be doubled. However, if validation has been successful case study evidence suggests the overall cost of operation and maintenance may be reduced by up to 75%.

A selection of operational lessons gathered from a book considering management issues for systems dependability are listed below (13). While there are undoubtedly others lessons relevant to pharmaceutical and healthcare systems, these would seem to convey the key points of learning.

- Management should be commensurate with the criticality of the system.
- Ensure competency of operations staff as individuals and teams.
- Control access to the system, including keys and passwords.
- Control the use of system overrides.
- Communicate learning from incidents.
- Ensure that essential records are kept and maintained.
- Monitor changes and maintenance to the system.

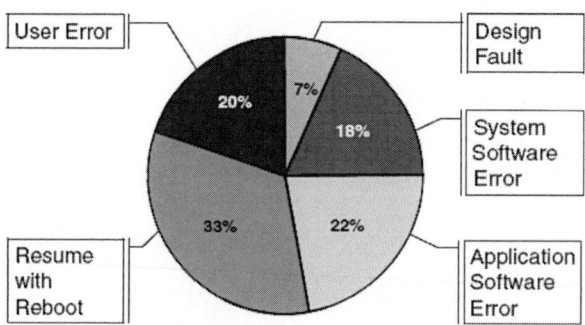

Figure 16.9   Malfunction diagnosis.

- Ensure that manufacturer's recommended operating instructions are followed.
- Ensure that appropriate national and international standards are adopted.
- Audit and follow-up outstanding issues with suppliers and subcontractors.
- Ensure that contingency plans are practical.
- Maintain a positive attitude amongst operations staff.
- Regularly audit systems to verify their specifications are still current and that they perform as intended.

It is evident that organizations must be ever vigilant of their GxP computer systems and continually develop their compliance capability.

RULES OF THUMB

- Maintenance costs often exceed original project costs.
- Corrective maintenance costs typically reduce by about one-sixth year on year.
- Typically, annual support costs average about 10% of original project cost charged annually, index linked to rate of inflation.
- As many as one in five changes introduce a new defect.
- Effective validation should reduce maintenance costs by about 75%.

PROCESS IMPROVEMENT

Many pharmaceutical and healthcare companies are now considering process improvements for their compliance practices. Two main approaches are typically adopted on the basis of the established process improvement methodology's commonly known as lean manufacturing and six sigma. Lean manufacturing is aimed at removing redundant steps and wait time from processes. Six sigma is aimed at reducing process variability. Both lean manufacturing and six sigma look at actual working practices rather than what is supposed to be happening. Together lean manufacturing and six sigma (sometimes combined and referred to as lean sigma) offer powerful tools to improve business efficiency including computer compliance. Case studies have shown how effort reductions of 50% to 72% are possible through applying lean sigma (14,15). A lean sigma approach to compliance improvement is shown in Figure 16.10.

LEAN PRACTICES

Ways of working that have not been subjected to a focused performance review typically offer fertile ground for improvement. The basic approach to leaning compliance practices can be summarized in the following five key steps:

Define the Problem/Opportunity

- What are you trying to characterize?
- What are the scoping boundaries?

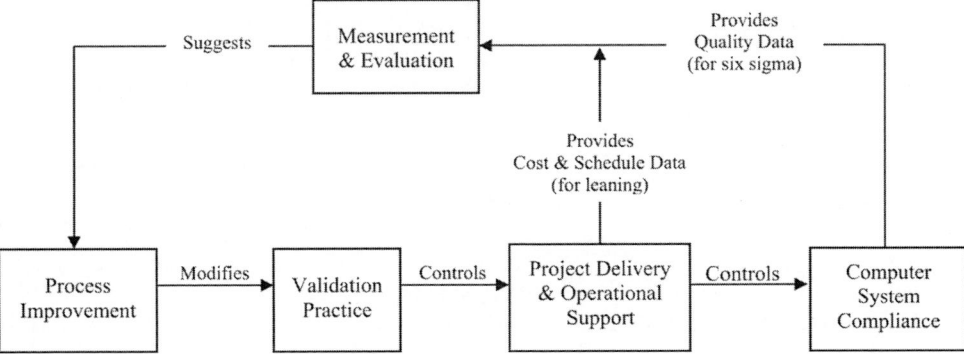

Figure 16.10 Compliance process improvement.

- What is the business case for validating?
- Who/what are the process suppliers, inputs, outputs and customers?
- What process metrics are appropriate?

Baseline Current Way of Working

- What is my baseline?
- How should I collect data to baseline performance?
- What are the key equipment, process, and product parameters?
- How capable is the current process against what my customers require?
- How capable is the current process against what my suppliers require?
- What are the failure modes?

Analyze Opportunities

- What is the current process flow?
- What sources of variation are relevant?
- Cause and effect: what effects my key equipment, process and product parameters?
- How can the process be systematically optimized?

Make Improvements

- What solutions help verify or improve the process?
- What are the costs, benefits, and risks associated with each solution?
- Do pilot runs confirm hypotheses?
- How best to implement improvements?

Realize Benefits

- Validate and document revised process.
- How to monitor revised process to preserve gains and maintain control?

The "fish bone" diagram, as illustrated in Figure 16.11, can be used to structure the identification of numerous opportunities for removing waste from current ways of working. Each will then have top be quantified and opportunities prioritized for implementation. There are seven basic types of waste.

- Overproduction: Developing optional software features that are not critical or mandated, preparing unnecessary reports, unnecessary duplication of information between documents
- Waiting: Staff unavailable when needed (meetings, reviews, and approvals), processing corrective actions monthly rather than straight away, and delays to critical path
- Transportation: Physical movement of people and documentation
- Inventory: Too many documents, too many people, poor organization
- Extra processing: Conducting activities that are not necessary (e.g., too many signatories), maintaining documents that do not need to be kept current, rework to correct defects
- Motion: Sequential activities that could be conducted in parallel, inability of staff to resolve issues referred to them without handoff to someone else
- Defects: Data and document errors, miscommunication

Collecting data to analyze how personnel spend their time can provide very useful baseline information. Figure 16.12 shows an activity analysis for QA staff at two different sites. In this example less than half the time available is spent actually preparing, reviewing, and approving compliance documents. There would appear to be a lot of wasted time in fruitless meetings and chasing documents round their distribution for review and approval. Why is this? One reason might be that documents are being prematurely released before they are ready to hit critical path target dates in the project plan. Another reason might be that

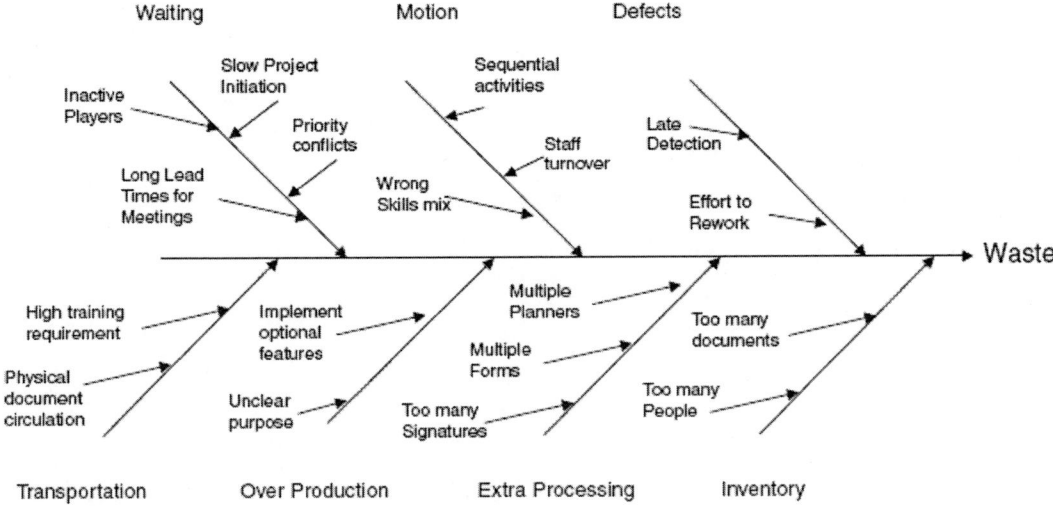

Figure 16.11 Fish bone diagram identifying waste.

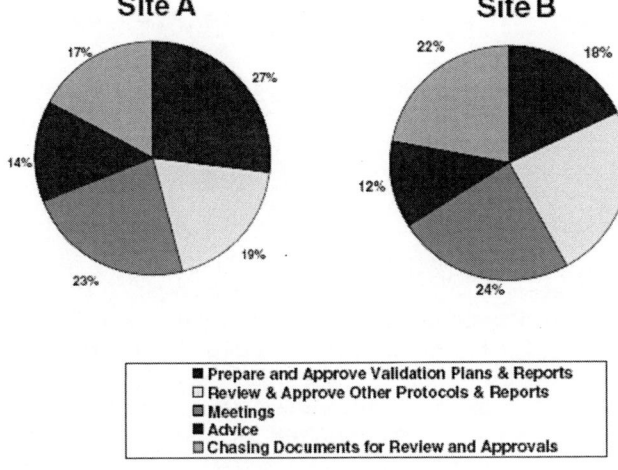

Figure 16.12 Example validation staff activity analysis.

documents go through many revisions each with half a dozen or more signatories thus creating project delays. In theory there should be no need for revisions if the document is right first time. The need for large numbers of signatories must also be challenged. Further investigation is required and corrective action taken.

SIX SIGMA PRACTICES

The six sigma process can be used to benchmark the capability of a compliance practices and hence indicate the significance of any opportunity for improvement. Average capability is characterized by a three sigma performance. Six sigma indicates a world-class performance. Beyond six sigma is not considered cost effective (16).

Some opportunities identified by pharmaceutical and healthcare companies in their software engineering processes include the following:

- Project startup time
- Size of certain key documents

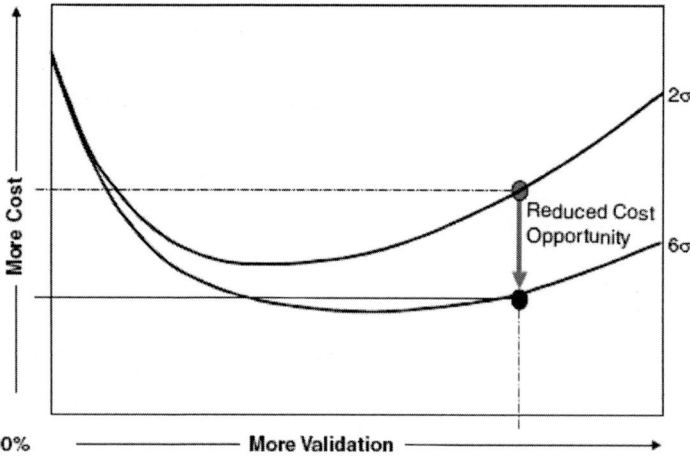

Figure 16.13 Six sigma improvement opportunity.

- Number of signatories on individual documents
- Document review and approval cycle times
- Clarity of requirements (checks on ambiguous words)
- Amount of evidence collected during testing
- Testing time for similar systems

Figure 16.13 presents a cost versus compliance curve and is based on the compliance strategy discussion in chapter 2. This graph takes Figure 2.1 from chapter 2 a step further to illustrate the six sigma opportunity that pharmaceutical and healthcare manufacturers have to improve their level of compliance and reduce costs at the same time. The large dot on the two sigma plot is meant to represent point B from Figure 2.1, that is, the commonsensical approach to ensuring sufficient, but not too much, compliance effort is conducted to fulfill regulatory requirements and avoid major noncompliance. The same point is marked on the six sigma plot to illustrate how more capable processes can reduce compliance costs.

Consider an example such as executing test cases. A project may run many hundreds of test cases. Defects observed could be based on all issues relating to ambiguous test instructions and acceptance criteria. Test cases should not have residual problems. They should have been reviewed beforehand. Appendix 16.B.1 can be used to approximate the six sigma capability for the process. Appendix 16.B.3 is then used to indicate the cost of quality as a percentage of the cost of ownership. The cost of quality includes the cost of failure (scrap, rework), cost of appraisals (self-inspections, regulatory inspections, and supplier audits), and cost of prevention (procedures, validation planning, and training).

To demonstrate how six sigma capability can be calculated using Appendix 16.B.1 lets assume we have 120 test cases of which 15 have ambiguities that are not discovered until test execution. The yield of correct test cases is therefore 0.87. Two critical-to-quality (CTQ) characteristics have been discussed above in relation to case studies (ambiguous test instructions and ambiguous acceptance criteria), that is, N = 2. Assuming the CTQ characteristics are evenly split then the defect rate per CTQ characteristic is $((1 - 0.87)/2) = 0.065$, and consequently the defects per million opportunities (DPMO) is 65,000. This equates to an approximate six sigma value of 3.0 using Appendix 16.B.2. Now examine Table 16.B.3 that indicates subjecting overall compliance to the same sigma level will result in the cost of quality to about 25% to 40% of the cost of ownership of the computer system. This is similar to some of the anecdotal examples given for cost of quality in chapter 1. There would seem to be plenty of opportunity for improvement.

Another example might be to improve the review and approval process. Some pharmaceutical companies have successfully reduced cycle times by an order or magnitude. The breakthrough is usually made when the team looking at process improvement analyze real

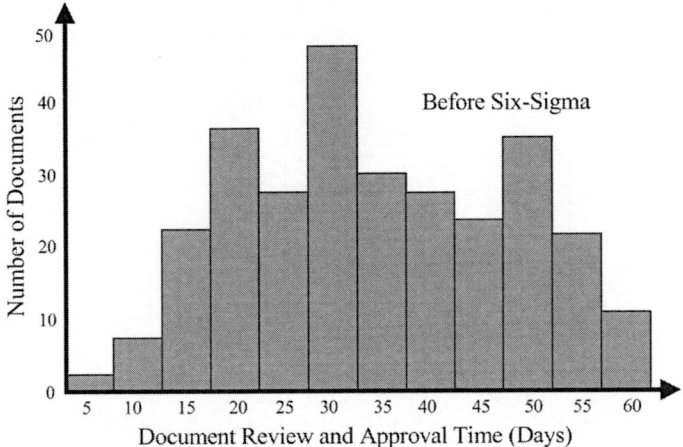

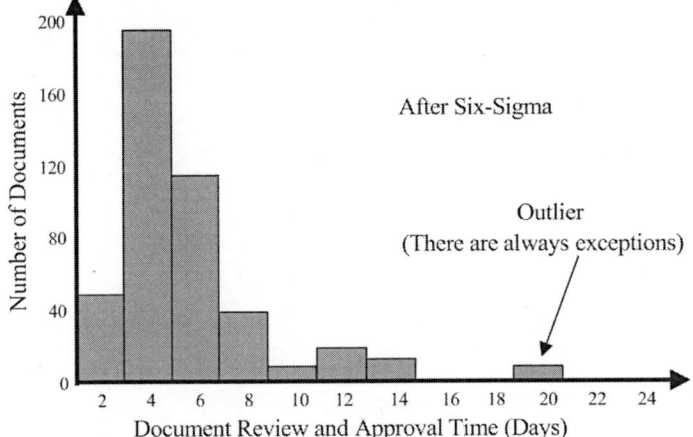

Figure 16.14 Example document cycle time distributions

data on how their processes are operating and see what is actually happening in practice. A realistic cycle time can be targeted for improvement should be based on actual current practice. In this case, as a result of six sigma improvements, project managers should be able to better anticipate and schedule document review and approvals.

Graphically plotting activities is a useful way to illustrate the starting situation and the impact of any process improvement. Common plots include bar charts, cats whiskers, control charts, and scatter-grams. Figure 16.14 provides an example of how a cycle time for the general review and approval of documents might be drawn. The review and approval cycle times seem excessive. Project critical paths are likely to bottleneck in these circumstances. Unnecessary complexity and effort is probably being added to projects to manage late approvals. Figure 16.15 shows how a control chart can be used to measure existing practice for particular document types and how target improvements might be set.

The improvement to be made could be to institute a weekly approval meeting for documents. Documents need to be circulated in advance of the meeting (a minimum advance circulation time should be set). Attendees at the meeting must review documents before the meeting. Any revisions to documents need to be agreed in the meeting and changes made directly to documents so that documents can be concluded (signed) at the meeting. Nominated attendees must assign deputies authorized to approve documents on their behalf when they cannot attend meetings themselves. This process requires a lot of self-discipline. Of course, just letting project managers know that document review and approval cycle times are being monitored may be enough in itself to prompt improvements.

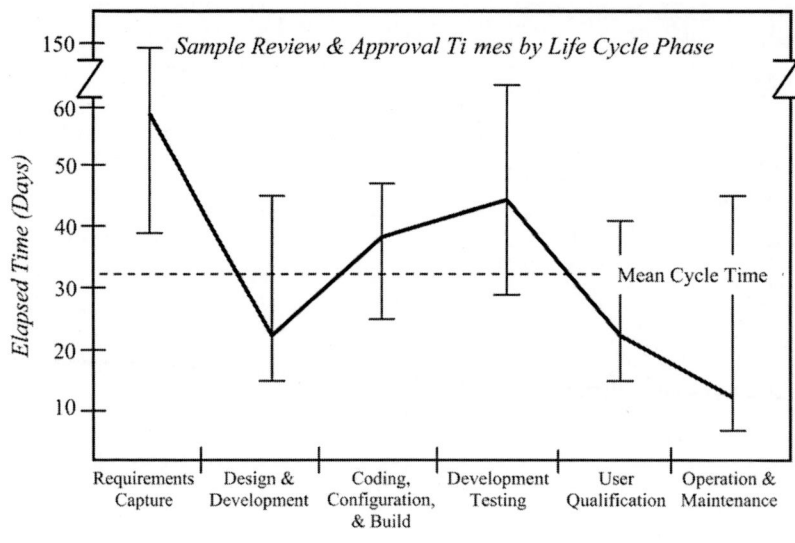

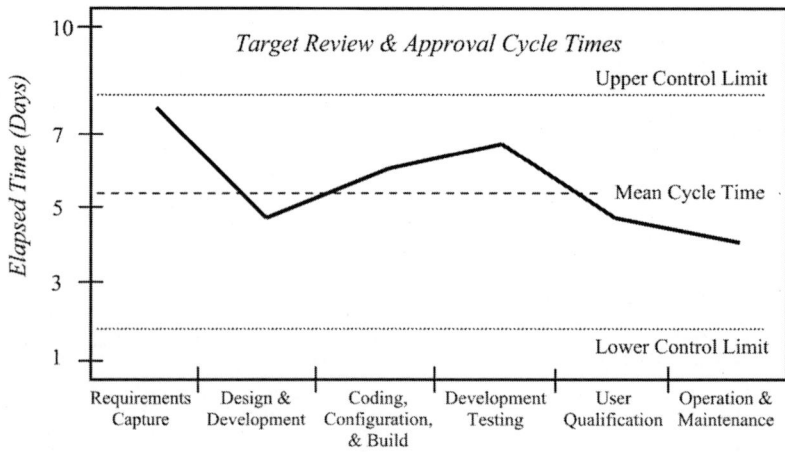

Figure 16.15   Example six sigma control charts.

REFERENCES

1. Paulk MC. The evolution of SEI's capability maturity model for software. Software Process: Improv and Pract 1995; 3–15.
2. Royce W. Software Project Management: A Unified Framework. Reading, UK: Addison-Wesley, 1998.
3. Wingate GAS. Computer Systems Validation: Quality Assurance, Risk Management and Regulatory Compliance for Pharmaceutical and Healthcare Companies. Boca Raton, FL: Interpharm Press, 2003.
4. Herremans P. Manage system development capabilities. Institute of Validation Technology Conference on Computer & Software Validation. London, February 21–22, 2002.
5. Somerville I. Software Engineering. 8th ed. Reading, UK: Addison-Wesley, 2006.
6. Murtagh R. Identifying Improvements to current industry practice for the validation of automated tablet compression machines. MSc dissertation, University of Manchester Institute of Science and Technology. Manchester, 2002.
7. Grady RB. Practical Software Metrics for Project Management and Process Improvement. Englewood Cliffs, NJ: Hewlett-Packard Professional Books, 1992.
8. Hamlet D, Maybee J. The Engineering of Software. Reading, UK: Addison-Wesley, 2001.
9. Jones C. Software quality: a survey of the state of the art. Software Productivity Research, 2003.
10. Stein RT. The Computer System Risk Management and Validation Life Cycle. Paton Press, 2006.
11. Vliet HV. Software Engineering: Principles and Practice. 2nd ed. New York: Wiley Press, 2000.

12. Maxwell KD. Applied Statistics for Software Managers. Software Quality Institute Series. Saddle River, NJ: Prentice Hall, 2002.
13. Redmill F, Dale C. Life Cycle Management for Dependability. New York: Springer-Verlag, 1997.
14. Ryan TM. Applying lean-sigma methodology to optimise the software validation lifecycle strategy in combination medical device manufacturing. MSc dissertation, Dublin Institute of Technology. Dublin, Ireland, 2009.
15. Wingate GAS. GAMP®5 Drivers, Objectives and Benefits, Launch of GAMP®5. ISPE U.S. Conference on Manufacturing Excellence. Florida, February 25–28, 2008.
16. Harry M, Schroeder R. Six Sigma. Garden City, NY: Doubleday Press, 2000.

Appendix 16.A   Compliance Capability Questionnaire

Level 2 questions
- Does the project follow a formally documented project planning process?
- Are estimates of cost and scheduling (including intermediate milestones) documented for use in planning and tracking project progress?
- Do project plans identify work packages and responsibilities for their delivery?
- Do all affected group and individuals agree their responsibilities?
- Are adequate resources and time provided for project planning?
- Does the project manager review planning both on a periodic and event-driven basis?
- Is the actual project performance (e.g., cost and schedule) compared with original plans and are corrective actions taken when they differ?
- Do all affected group and individuals agree to any change in their responsibilities?
- Is someone on the project specifically tasked with tracking and reporting progress?
- Are measurements used to determine the status of activities and deliverable on the project?
- Are project tracking activities and results periodically reviewed with senior management?
- When changes occur are the necessary amendments made to project plans?
- Do projects follow project and quality management policy requirements?
- Are project team members trained in the procedures they are expected to use?
- Is progress on project deliverables subjected to periodic review?
- Is a documented procedure used for selecting suppliers and subcontractors?
- Are changes to subcontractors notified to the pharmaceutical or healthcare company?
- Are periodic technical interchanges held with subcontractors?
- Are performance issues followed up with suppliers and subcontractors?
- Does the project manager review supplier and subcontractor performance on both a periodic and event-driven basis?
- Is a defined quality management system used on projects?
- Do quality plans identify quality assurance activities and deliverables?
- Are internal audit results provided to affected parties?
- Are software quality assurance issues not resolved by the project addressed by senior management?
- Are adequate resources and time provided for quality assurance activities?
- Are measurements used to determine the cost and schedule status of quality assurance activities?
- Are quality assurance activities reviewed with senior management on a periodic basis?

Level 3 questions
- Does the organization follow a written policy for both the development and maintenance of computer systems?
- Does the organization have a documented and maintained quality management system?
- Does the organization collect, review and make available performance data related to the use of the quality management system?
- Do users of the quality management system receive adequate training?
- Is the review and maintenance of the quality management system planned, monitored and audited?
- Is there a training policy?
- Are training requirements planned covering both management and technical skills?
- Are adequate resources put into training?
- Are measurements used to determine the quality of training?
- Is training reviewed with senior management on a periodic basis?
- Are projects planned in accordance with the quality management system?
- Are project activities and deliverables reviewed and audited by quality assurance personnel?
- Is consistency maintained across different projects?
- Is there a written policy that guides the establishment and management of multi-disciplined teams?
- Do internal groups work together in collaboration?
- Are inter group issues identified, tracked and resolved?

(Continued)

Appendix 16.A Compliance Capability Questionnaire (*Continued*)

- Are measurements used to determine the status of inter group coordination activities?
- Are inter group relationships reviewed with senior management on a periodic and event-driven basis?
- Is effort focused on critical aspects of the computer system development and maintenance?
- Is the change control process defined, proceduralized and robust?
- Do personnel understand and received training to enable them to discharge their change control responsibilities?
- Are the volume and nature of changes measured and monitored?
- Is there a mechanism for verifying that the originator of a change request is satisfied by the change implementation?

Level 4 questions
- Is there a written policy for quantitatively measuring management of development and maintenance of computer systems?
- Is there a defined quantitative measurement process?
- Is the management performance of development and maintenance of computer systems controlled quantitatively?
- Are adequate resource provided for quantitative measurement process activities?
- Are quantitative measurements reviewed with senior management on a periodic and event-driven basis?
- Are documented correlations made between historical management and actuals?
- Are historical management and actuals used to improve planning on current projects?
- Do projects use measurable and prioritized goals for managing quality?
- Are measurements used to determine the status of activities for managing quality (e.g., the cost of poor quality)?
- Are the activities for managing quality planned in advance for projects?
- Are the activities performed for quality management reviewed with senior management on a periodic basis?
- Is return on investment evaluated, monitored and reported to senior management?

Level 5 questions
- Are defect prevention activities planned?
- Is there a formal process to identify common cause defects?
- Once identified, are common causes of defects prioritized and systematically eliminated?
- Is training given in defect prevention?
- Are defect prevention activities subject to quality review and audit?
- Does the organization follow a defined process to management technology changes?
- Are new technologies evaluated to determine their effect on quality and productivity?
- Do senior management sponsor the introduction of new technology?
- Do people throughout the organization participate in process improvement initiatives?
- Are improvements continually made to process management?
- Are process improvement initiatives reviewed with senior management on a periodic basis?

Source: From Ref. 3.

Appendix 16.B
Six Sigma Tool Box

Table 16.B.1 Six Sigma Capability

Step	Action	Equation
1	Select a process.	Not applicable
2	How many times was the process run?	Not applicable
3	How many process runs did not exhibit defects?	Not applicable
4	Calculate yield.	Step 3/step 2
5	Calculate the defect rate from step 4.	1 − step 4
6	Determine the number of things that could potentially cause the observed defects.	N = number of CTQ characteristics
7	Calculate the defect rate per CTQ characteristic.	Step 5/step 6
8	Calculate the DPMO.	Step 7 × 1,000,000
9	Convert the DPMO into a six sigma value using Table 16.B.2.	Not applicable
10	Draw conclusions using Table 16.B.3.	Not applicable

Abbreviations: DPMO, defects per million opportunities; CTQ, critical-to-quality.

Table 16.B.2 Six Sigma Conversion Table

Sigma value	DPMO	Sigma value	DPMO
0.0	933,193	3.2	44,565
0.2	903,199	3.4	28,717
0.4	864,334	3.6	17,865
0.6	815,940	3.8	10,724
0.8	758,036	4.0	6,210
1.0	691,462	4.2	3,467
1.2	617,911	4.4	1,866
1.4	539,828	4.6	968
1.6	460,172	4.8	483
1.8	382,088	5.0	233
2.0	308,537	5.2	108
2.2	241,964	5.4	48
2.4	184,060	5.6	21
2.6	135,666	5.8	9
2.8	96,800	6.0	3
3.0	66,807		

Note: This table includes 1.5 sigma shift.
Abbreviation: DPMO, defects per million opportunities.

Table 16.B.C Cost of Quality

Sigma level	Defects per million opportunities	Cost of quality
2	308,537 (uncompetive)	>40% of cost of ownership
3	66,807	25–40% of cost of ownership
4	6,210 (industry average)	15–25% of cost of ownership
5	233	5–15% of cost of ownership
6	3.4 (world class)	<1% of cost of ownership

17 | Practical Troubleshooting

Rarely does everything go exactly to plan. Practitioners should be prepared to deal with unforeseen issues that potentially compromise the regulatory compliance being sought. This chapter shows how risk-based thinking might be employed to resolve common issues. However, in presenting this offering we need to emphasize that the individual circumstances may change what is appropriate and acceptable. As consultants would say "It depends"! Practitioners need to use their regulatory knowledge and judgment to deem what is suitable or not. The needs of the pharmaceutical and healthcare company will vary based on many different factors.

RISK MANAGEMENT (SIGNIFICANCE OF ISSUES)

Effective troubleshooting requires an understanding of the significance of a deviation to planned activities in the context of patient safety and regulatory compliance; otherwise practitioners will tend to default to conservative risk averse response. Primary factors to consider are

- Likelihood of operational issues based on inherent susceptibility of system and track record of any operational issues caused by software bugs and hardware breakdowns
- Ability to manage and control operational issues based on usability of specifications, verification documentation, and change control records, and whether or not there are any independent checks that would detect erroneous operation and trigger mitigation
- Process being automated and the impact of unaddressed erroneous operation on product quality
- Implications of substandard product on patient safety (e.g., sterile injectables generally have greater consequence than topical creams and ointments)

The most serious issues, referred to here as critical issues, are associated with situations capable of resulting in death or temporary/permanent disability, there is a direct violation of a regulatory licence, or there is fraud or falsification of regulated records. Affected healthcare and drug products must be reviewed and potentially withdrawn from use. Supply of product should stop until a corrective action is completed.

Major issues are those that are not critical, but are judged to significantly impact the quality of the healthcare or drug product concerned, or may result in a product that does not consistently meet the requirements of its regulatory licence. Supply of product can proceed so long as a corrective action plan is immediately put in place and executed. Systemic major issues seen putting undue risk on patient/consumer safety should be elevated to be treated as critical issues.

Minor issues are those that are considered to breach defined standards but do not meet the criteria of a critical or a major issue. Minor issues have a relatively low probability of affecting the quality or usability of the healthcare or drug product concerned. Supply of product can proceed so long as a corrective action plan is being progressed. Multiple minor issues may be collectively treated as a serious issue.

The reality is that most issues that occur impacting computer will be minor in nature. Critical, major and minor issues all need resolving but the level of detail and hence amount of work should be less for noncritical issues. Some hints and tips on how to approach typical issues are given below. This must not be taken as definitive guidance, and personal judgment will be required in all cases to ensure a common sense approach is taken.

PLANNING AND PROJECT INITIATION
Compliance Determination Statement
Potential issues include compliance determination not documented; there are errors in the determination, and determination did not consider an applicable regulation such as electronic records and signatures.

The determination of a compliance requirement is not itself a regulatory requirement, although it is hard to proceed in a meaningful way if some kind of assessment is not conducted. As such issues impacting compliance determinations should be seen as a minor issue.

If a compliance determination was effectively documented in a validation plan then no further action is needed. Neither is a compliance determination needed if the computer system already satisfies all applicable regulatory requirements. However, it will probably be necessary to assess which regulations are applicable and hence it will only take a little more effort to document the outcome of the assessment as a compliance determination statement.

Validation Plan
The lack of a Validation Plan does not necessarily imply a significant compliance gap exists. Such issues are typically deemed minor in nature unless critical functionality is impacted in which case they could be treated as a critical issue.

For noncritical applications, a validation plan may not be required if there is a suitable specification that has been fully tested. However, for critical applications that potentially impact drug product safety and patient safety, a validation plan will typically be expected to scope remedial actions. The extent of validation should be commensurate with risk. Critical applications should be validated with more rigor than noncritical applications.

User Requirement Specifications
Common issues include specification missing or never generated, specification does not cover all functions that the system performs (perhaps it is out of date following modification), and specification does not include sufficient detail to allow objective testing. Retrospective preparation of a User Requirements Specification is typically challenged as non-value added activity. However, with such a document it is not possible to demonstrate that all system functionality has been tested, and this can be very serious if critical functionality is not verified.

Unless system requirements are documented elsewhere a user requirement specification should be prepared. Critical functionality should be described in detail. A high-level summary of requirements will be suitable for other functionality.

Supplier Assessment
Common issues include no supplier assessment performed, supplier assessment performed for different version of system/software to that currently being used, and assessment performed but for different product from the same supplier. A capable supplier is a key factor to assuring quality of work conducted and hence robust compliance. Ongoing support will also be crucial if defects are found that need timely remediation. The importance of this increases for critical applications that can impact drug safety and hence patient safety.

A retrospective supplier assessment should be conducted for critical applications using sources such as local performance history, Internet research, or even telephone interviews with the supplier. A supplier audit should only be expected if the computer system has significant design defects or operational issues, and there is a dependency on routine supplier support. If the audit reveals insufficient specification and testing then additional user qualification may be required. If the supplier has gone out of business then the practicality of the situation means use what information is available. In such circumstances supplier assessments may be limited to local performance history and Internet research. Regulatory authorities will expect sources of alternative third-party support to be evaluated and if unsuccessful then plans for system replacement should be made. For noncritical applications supplier audits are not necessary.

SPECIFICATION, DESIGN, AND DEVELOPMENT
Design Specification (including Functional Specification)

Potential issues include missing documentation or where design specifications exist they do not unambiguously trace back to user requirements. The significance of this is similar to user requirement specification issues—without a robust document it is not possible to demonstrate that all system functionality has been tested, and this can be very serious if critical functionality is not verified.

Retrospective functional specifications should be prepared to a level of detail commensurate with criticality of functionality. There is no need to duplicate functionality that can be referenced in user manuals.

Re-engineering of design documentation is only required for custom (bespoke) elements of critical applications where source code is available, otherwise just record configuration and identify any custom (bespoke) software. The configuration of commercial off-the-shelf (COTS) software packages should be recorded for all applications. More detail should be considered if there are many years of service left, there are frequent modifications or operational incidents. It is not necessary to retrospectively develop design documentation to a level from which source code can be written.

Where there is a lack or incomplete traceability of design specification to user requirements then traceability should be reestablished. Which tool to use is a subjective decision. Many practitioners prefer to develop a requirements traceability matrix (RTM), while others prefer to state direct links within specification and design documents. Both approaches are equally compliant if implemented properly. If the computer system has COTS software then cross-references can be made to supplier documentation (e.g., user manuals and technical reference material) without the need to duplicate content.

Design Review

The lack of a design review or a deficient design review is not typically considered a significant issue. Nevertheless, if a design review has not been conducted for a critical application or does not meet current standards then it is recommended that it is conducted once to confirm robust design and identify any currently unknown problems. The design review will also confirm traceability of user requirements through to design. It is not necessary to retrospectively conduct a design review for noncritical applications unless there is a history of operational problems. The outcome of the design review can be documented in a separate document or as part of the validation report.

Programming Standards and Source Code Review

Common issues include no or inadequate programming standards, no or only cursory source code review performed, and significant/numerous software modifications made since last source code review. There are few examples of regulatory findings on such issues that are generally not treated as serious deficiencies unless critical functionality is implicated.

Programming standards should be prepared for user written software in critical applications that is likely to undergo further modification. Source code reviews should then be conducted when future software modifications are made. Retrospective source code reviews are not generally expected unless there is a history of operational problems impacting functionality attributed to custom (bespoke) code. If the software was developed by an external supplier who is no longer in business then no further action is required other than documenting this constraint as part of the supplier assessment. Corrective actions identified in source code reviews that have never been finished should be followed up and completed. Source code reviews conducted online by viewing software on computer screens do not need repeating but the original mode of review should be stated in the validation report.

TESTING
Development Testing

Testing is a key activity and deficiencies can pose a critical risk. So if it is discovered that no development testing has been conducted, or that it was not documented or only superficially documented, then this should be treated very seriously.

In such circumstances for critical applications, additional testing should be conducted to ensure that design specifications are fully tested. If the supplier assessment determined poor testing practice prevailed during system development then the handling of test failures, execution of tests without pre-approved test cases, missing approval signatories, etc., should be reviewed. It may be that some tests need repeating or corrective actions completed (the focus should be on critical functionality). If the supplier refuses to complete corrective actions and escalation has failed to resolve the issues then this should be reported in the validation report and additional user qualification considered to compensate. For noncritical applications, missing tests can be covered by functional testing within the operational qualification.

Data Loading

Incorrect or missing data can be catastrophic. As the saying goes "garbage in, garbage out." It is crucial that any GxP data deficiencies are promptly addressed. The significance of the issue will be commensurate with the risk passed to drug quality and patient safety.

For critical applications where there are known data errors, a sampling plan should be created and executed to determine overall data accuracy and the nature of any remedial action needed. It may be that a further larger sample is required, or indeed a 100% data check is required. All erroneous data identified should be corrected. In extreme circumstances, a data reload may be the most appropriate course of action. For noncritical applications it may be acceptable to continue to use the computer system without a data review if it can be justified as acceptable to correct data errors as they are discovered during routine operation.

Data migration records may also be missing or deficient. This is less serious than erroneous data. A retrospective account of the data migration, as it was done originally, should be prepared, and any corrective actions that need to be taken are identified. Automated data migration tools should be verified before use. If it was not verified for the original data load and it is not being used again as part of a remedial action, then there is no need to retrospectively verify its operation. The use and verification tools should be included in the retrospective account of data migration.

Installation Qualification

A common problem is that although an original IQ was performed and documented, subsequent upgrades to the system have occurred and installation records have not been maintained. Other issues include not having any installation documentation in the first place or perhaps only having the installation instructions or an installation certificate. Such deficiencies are usually treated as minor issues.

For critical applications, the installation process should be repeated and results properly recorded. An inventory of hardware and software should be created and setup checks completed. It is quite acceptable to make use of installation instructions in user manuals. If configuration management records show conflicting details then these must be resolved as part of the installation qualification. For noncritical applications the generation of an inventory of hardware and software is sufficient. Original installation records may still be useful even if they were not approved at the time they were used so long as they are reviewed and found acceptable.

Operational Qualification

Common issues include vague test cases, do not reference pre-defined specifications, and cannot be reconciled with user requirements or design documents to verify all required functions were tested. Deficient testing of critical functionality is very serious unless there is a downstream system that verifies operation. There are numerous examples of regulatory findings for deficient functional testing.

For critical applications the OQ must demonstrate all functionality works as intended. Comprehensive test records are expected. There should be a clear pass/fail outcome for each test. Testing should focus on critical functionality if the system has been used for many years without operational problems, and it is based on a COTS software package with little or no customization. For noncritical applications only critical functionality needs testing.

Performance Qualification

The lack of performance qualification prior to the computer system being brought into unrestricted use is typically not a significant issue because the system will have already been in use for some period of time. Because of this such situations are rarely treated as more than a minor issue.

For critical applications, the operational history should be reviewed and any performance issues must be resolved. For noncritical applications the operational history still needs to be reviewed but it is likely that continued operation can be justified without further actions even if there have been performance issues.

It is worth noting that if calibration and system suitability testing has been routinely conducted during system operation then PQ requirements may have already been satisfied. Furthermore, if the computer system replaced a manual process then it is not worth worrying about how to satisfy the EU requirement to demonstrate the computer system works as least, as well as the manual process—it is too late!

REPORTING AND AUTHORIZATION TO USE
System Release

Typical issues include the lack of documented authorization (dated signatures) for the actual use of the computer system in its operating environment, or that formal authorization was given after the computer system was put into use. Such deficiencies are typically treated as a minor issue unless there has been an actual operational incident impacting product quality or patient safety in which case the authorizing the use of that computer system use will become a critical issue.

For critical and noncritical applications work out when the computer system was first used and record that date with an explanation of how the date was determined. Work area log books can be a useful source of information to bolster personal memories. If possible refer to any installation records to determine the version originally used. Then retrospective justify original use of the system and document this in an up-version to the validation report. Change controls for system upgrades can be used to position subsequent authorizations.

OPERATION AND MAINTENANCE
Supplier Support

The loss of hardware or software support for older systems is a recurring problem. This can pose a significant risk where the computer systems concerned are used in the supply of medically critical drug products. Regulatory authorities will expect systems to be brought out of use if there are serious issues impacting public health; however for medically critical products this may not be an option if there are no alternative drugs available.

For critical applications with the potential to directly impact consumer/patient safety it is important the computer system is still within warrantee and service level agreements so that any future issues can be rapidly resolved. Therefore, the onus is on reestablishing some form of interim support agreements from the original vendor or new third party until a replacement system can be implemented. For noncritical applications, there is not an overwhelming case to replace older systems no longer supported by their suppliers if the business is happy to take the financial consequences if they breakdown.

Change Control

A lack of proper change control is another serious issue impacting computer systems. The significance of the problem posed will be related to the number and scope of changes already made without sufficient control. Depending on the role of the computer system consumer/ patient safety could have been compromised.

For both critical and noncritical applications appropriate change control should be immediately established and a review of changes already implemented conducted to determine if revalidation is required.

REGULATORY EXPERIENCE

Malcolm Olver who is an MHRA inspector with over 20 years experience of computer systems compliance has offered the following advice to practitioners (1). To appropriately appraise the significance of issues it is vital to have relevant knowledge of the automated processes and the healthcare and drug products impacted. Original risk assessments may have been valid at the time but situations change and the assessments can be invalidated as events overtake them. It is important to *"see the whole picture"* and take a holistic viewpoint on current risks posed combined with the need to satisfy GMP regulations. If GxP principles are fulfilled then remedial actions should be acceptable. It is where GxP principles are compromised that regulatory authorities will deem unacceptable shortcuts have been taken and implement sanctions on healthcare and pharmaceutical companies.

Care must be taken not to justify poor practices in the mistaken belief they are appropriate or that the regulatory authority is making unreasonable demands. Standards typically evolve over time and firms must stay alert to industry developments, and not just keep abreast of the letter of regulations. To keep vigilant and maintain management support may take a shift in an organization's culture. A quality and compliance mindset couples with the pursuit of continuous improvement must be evident to realistically avoid future compliance issues.

The bottom line is that if at any point a computer system is found to jeopardizing product quality or patient safety then consideration should be given to taking it out of use. The only reason possibly to keep it in use is where the drug product impacted is medically critical to keep patients alive, supplies are low, and there are no alternative drug products available. In such situations regulatory authorities for the markets involved should be immediately consulted and mutual agreement made on how to proceed. For other computer system compliance deficiencies healthcare and pharmaceutical companies will be expected to take appropriate and timely remedial action commensurate to the risk posed.

REFERENCE
1. Malcolm Olver. 'Risk Based Approaches to ASTM 2500—A Regulators View', ISPE European Conference on Product and Process Quality, Manchester, England, 2008.

18 | Concluding Remarks

Compliance practices for computer systems have been presented in line with current expectations of GxP regulatory authorities and industry practice. This chapter now concludes the first part of this book by reviewing some fundamental concepts and industry trends that will affect how we validate in the 21st century. Specific guidance for different types of computer system can be found in the case studies presented in the second part of this book. The authors of these case studies are themselves experienced practitioners and they have been encouraged to focus their papers on the key issues impacting their respective subject matter.

COMPUTING ENVIRONMENT

Today's computing environment is illustrated in Figure 18.1. Four tiers of computer systems are considered:

1. Measurement and control—providing data acquisition and actuator directives
2. Process management—providing process monitoring and control
3. Operations management—providing management control across multiple processes
4. Enterprise management—providing management control across multiple processes at multiple locations

The computer systems used within each of these levels tend to have common characteristics that will influence their validation. For instance, measurement and control systems are generally configurable commercial off-the-shelf (COTS) instruments. Examples include control instrumentation, analytical instrumentation, and medical devices.

Process management systems include real-time control systems, spreadsheets, and databases. They often include bespoke programming although they may be based on commercial off-the-shelf products. Examples of process management systems include programmable logic controllers (PLC), supervisory control and data acquisition (SCADA), industrial PC systems, distributed control systems (DCS), chromatography data systems (CDS), and blood processing systems. Measurement and control systems and process management systems are often integrated together. Process Analytical Technology (PAT) is a special case implementing noninstrusive process instrumentation.

Operations management systems include manufacturing execution systems (MES), laboratory information management systems (LIMS), building management systems (BMS), and engineering management systems (EMS). These systems are not necessarily dedicated to a particular plant or laboratory and may be shared, for instance, between several manufacturing units on a site. While operations management systems may have response-time dependencies, these are generally much less stringent than the real-time operations required by process and measurement and control systems. IT infrastructure plays a very important role is supporting these and enterprise systems.

Enterprise management systems include manufacturing resource planning (MRP II) systems, commercial Marketing and supply applications, and electronic document management systems (EDMS) and are typically based on large IT packages. These packages often claim to be configurable, although such configuration may be extensive and almost akin to bespoke programming. Enterprise management systems are often implemented to coordinate operations across multiple sites, possibly in different countries.

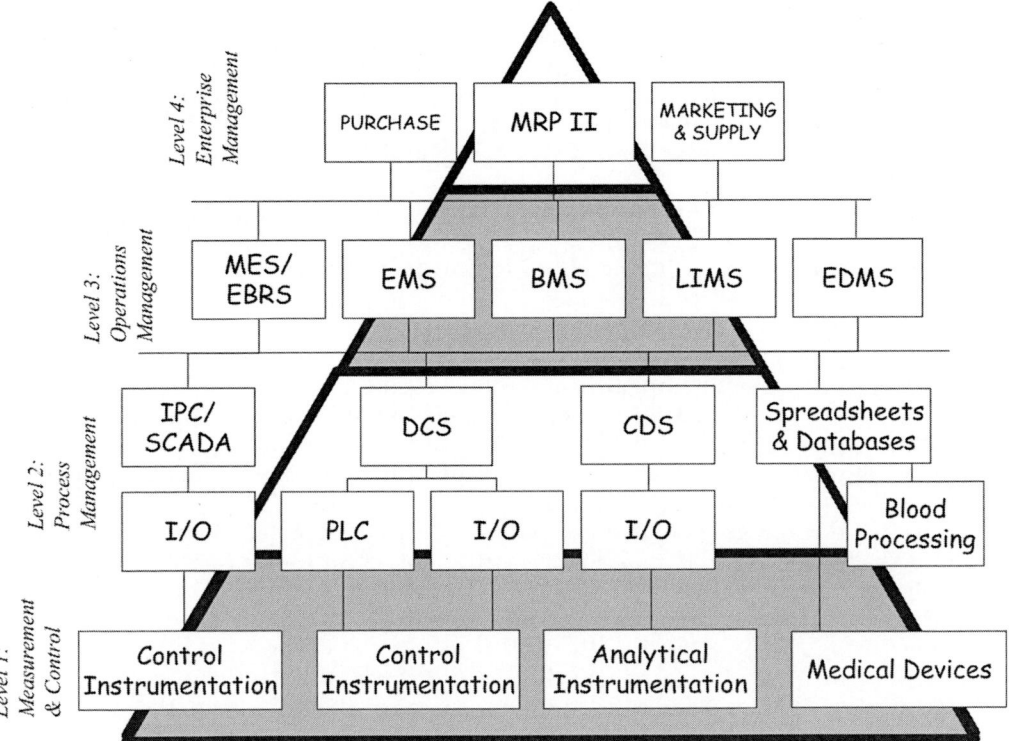

Figure 18.1 Computer system hierarchy.

THE BUSINESS CASE

A key influence in today's business environment is the return on investment for computer systems and the cost of ownership for what are often expensive assets. Automation strategies need to exploit the following benefits that computer systems offer:

- More flexibility
- Higher efficiency
- Faster operations
- Improved consistency
- Less human error
- Real-time performance data

All this counts for naught however, if either the system does not operate correctly in support of ensuring pharmaceutical and healthcare products are effective, safe and comply with regulations. The cost of noncompliance can be huge as discussed in chapter 1. The challenge is to determine how much compliance is enough. Excessive compliance may increase confidence in regulatory compliance but it is expensive and will not necessarily bring any further assurance in the process being automated.

The focus should be on establishing high process capability. That in turn will ensure computer systems are robust, reliable and secure. Compliance must not unduly hinder exploitation of new technology. Quality departments in many pharmaceutical and healthcare manufacturing organizations have a reputation for slowing down or even preventing the implementation of new technology. Compliance must not be seen as a constraint and should not be managed as such. Pragmatic solutions must be found to enable the implementation of new technology whilst ensuring compliance.

GOLDEN RULES REMAIN UNCHANGED

Many light-hearted publications have published lists of golden rules for compliance practitioners. None of these lists, however, is carved in stone, and there is some flexibility in deciding what should be included in the way of guidance. The checklist given below was first published in *Validating Automated Manufacturing and Laboratory Applications* (1), and essentially remains unchanged. It should help practitioners concentrate on 10 key compliance issues.

- Plan and monitor compliance—adopt a proactive project management style.
- Use competent personnel and train where necessary.
- Document compliance including collating raw data as supporting evidence —ensure everything is reviewed and approved.
- Implement a regime of change control covering projects and operational use of the computer system.
- Specify procedures for compliance and follow them.
- Develop system specifications with testing in mind and test using preapproved qualification protocols.
- Use the approval of summary validation reports to authorize the use of a computer system.
- Operate and maintain validated computer system in a state of control.
- Periodically review the compliance status of computer systems and initiate revalidation where necessary.
- Archive compliance evidence for future retrieval.

Remember that the GxP requirements for computer data are the same as for regulated records and documentation. Similarly, computer systems should satisfy the same basic expectations held for regulated equipment.

INDUSTRY CONSENSUS

A common industry approach to computer compliance has emerged providing the basis for the international acceptance of compliance work by various GxP regulatory authorities. The basis for consensus has been the various editions of the GAMP®5 Guide (2). The benefits of a widely adopted industry framework for computer compliance have been extolled by Tony Trill of the MHRA and include (3)

- Stabilizing standards and their interpretation at a realistic level
- Linking compliance to existing standards (e.g., ISO 9000)
- Reducing the costs of compliance
- Shortening the time required for compliance
- Ensuring appropriate documentation is produced for compliance projects
- Improving the quality and reliability of delivered systems

This all makes good business sense and should eliminate the need for corrective work because of misunderstood compliance requirements. Similar harmonization would appear to be occurring on the topic of electronic records and electronic signatures. Mutual recognition agreements (MRA) between various national regulatory authorities such as FDA, MHRA, TGA, and MHLW and the work of the International Conference for Harmonisation (ICH) are helping to formally consolidate harmonized computer compliance requirements.

VALUE FOR MONEY

So, how much compliance is insufficient, enough, or too much? There is no panacea, but a key step is concentrating effort where it is needed. Emphasis should be placed on assuring the critical aspects of the system (4). There are often just a few critical functions and/or components that affect product quality, supported by a collection of measurements and manual interactions. The nature of any manual interactions will have a significant impact on the amount of compliance required. The collection of product data is another important factor in assessing the degree of control taken by a computer system.

Even so the cost of compliance can appear high if the total cost of ownership for a computer system is not taken into account. This book has looked at compliance metrics and methods for identifying and implementing performance improvements. No organization can afford to ignore improvement opportunities in the current economic climate. Having said this it is important not to be over zealous such that satisfactory GxP compliance is compromised. The cost of noncompliance (e.g., nonissue, delay, or revoking of manufacturing licence) can be significantly more expensive than, what in hindsight, might appear to be marginal cost savings.

RISK MANAGEMENT
The FDA has recently highlighted the importance of risk management as part of 21st century compliance (5). Other regulatory authorities such as MHRA share this perspective. Without risk management computer compliance costs can quickly become prohibitive. Taking the highest level of compliance for all aspects of a computer system will not necessarily lead to discernible increased patient/consumer safety.

ISPE have published two important concept papers for functional risk management and electronic record/signature risk management (6,7). These papers marked the start of a new era is applying risk management as an integral part of computer system compliance. Pivotal GAMP® guides from the ISPE have subsequently been published (2,8).

This book has positioned the role of risk management throughout the computer system life cycle and in the controls required for electronic records and electronic signatures. Central to effective risk management is the need for practitioners to understand the impact that computer system functionality has on the healthcare and drug products ultimately being supplied. Design for risk management has also been discussed in terms of using backup systems, independent monitoring systems, and segregating regulated and nonregulated aspects for compliance within systems.

KEY ROLE OF SUPPLIERS
There has been an increasing use of COTS software packages because of improved availability and the advantages offered over custom (bespoke) software. The advantages and disadvantages of COTS software packages and custom (bespoke) software are summarized in Table 18.1. Computer systems are also being integrated into ever more complex and

Table 18.1 Comparing COTS Software Packages and Custom (Bespoke) Applications

Approach	Advantages	Disadvantages
Custom (Bespoke) Software	Complete change freedom	Expensive, unpredictable developments
	Smaller, often simpler implementations	Typically late delivery, over budget and reduced functionality
	Often better performance	Single platform dependency
	Control of development and enhancement	Often immature and fragile with an undefined maintenance model
	Clear understanding of user requirements	Drain on expert resources
COTS software packages	Predictable license costs	Up front license fees
	Broadly used, mature technology	Multiple-supplier incompatibilities, integration is not always trivial
	Available now	Frequent upgrades
	Dedicated support organization	Dependency on supplier, limited warrantees
	Hardware/software independence	Run-time efficiency sacrifices
	Rich in functionality	Functionality constraints and/or unnecessary features that consume extra resources

Abbreviation: COTS, commercial off-the-shelf software.

Table 18.2 Benefits of Cooperation Between User Organizations and Suppliers

User organization	Supplier
Meet user needs	Satisfy customer
Be easier to set up	Be handed over sooner
Be in production sooner	Be paid sooner
Break down less often	Fewer warranty visits
Be easier to repair	Shorter warranty visits
Be easier to further develop	Be easier to modify and/or upgrade
Be used more effectively	Good reference sites for new customers
Cheaper overall	Cheaper overall
Preferred supplier	Repeat business

Source: From Ref. 3.

sophisticated network architectures and it becomes harder to segregate GxP from non-GxP elements. A holistic approach is required so that key aspects are not inadvertently missed.

The key role suppliers have been widely acknowledged. To avoid duplicating tasks in their entirety or in part, pharmaceutical and healthcare companies and suppliers should work together in partnership. They must be able to work efficiently as a combined team and, as such, must be able to communicate effectively to streamline compliance activities. The mutual benefits of cooperation between customers and suppliers are outlined in Table 18.2.

ORGANIZATIONAL CHANGE

Many pharmaceutical and healthcare companies are now operating in a tougher economic environment than they have ever experienced before. New blockbuster drugs are not replenishing the drugs coming off patient. There is a general push by governments to reduce the cost of medicines. And more recently the "credit crunch" has hit the world economy. The response by many organizations is a wholesale review of their business operations and restructuring of their organizations.

At such times it is vital that senior management must give and hold to a clear vision for compliance. Compliance principles need to be incorporated into the culture of an organization. Those tasked to champion the compliance cause must believe in its intrinsic value to the business. Without sustained management backing for computer compliance, skills will remain in the domain of consultants instead of being instilled as just another competency of their permanent employees. Corporate memory of why compliance is required, what is needed, and how best to achieve compliance can be quickly lost at a time of downsizing and outsourcing. Going back about 20 years, it was this lack of organizational competency that was at the heart of many high-profile computer compliance issues and high cost of remediation.

THE FINAL ANALYSIS

Computer systems compliance has been a topical issue for over two decades. The importance of computer systems compliance is set to continue with regulatory interest in the role of new technology and ever more pervasive use of computer systems to support the development, manufacture, and distribution of pharmaceutical and healthcare products. In the final analysis, pharmaceutical and healthcare companies have no choice but to validate and sustain the compliance of their computer systems, otherwise their license to market a drug or healthcare product may be withheld or withdrawn.

The cost of compliance can be excessive if not properly managed. Common sense and good judgment are vital. Over the years new ways of working have emerged and industry practice has been steady improving in terms of effectiveness and efficiency. That said, the pursuit of cost-effective compliance is set to accelerate with the tougher economic climate impacting the world economy. This book does not pretend to present the definitive approach to computer compliance but hopefully it provides a robust grounding in terms of principles and their practical application that will be relevant for many years to come.

REFERENCES

1. Wingate GAS. Validating Automated Manufacturing and Laboratory Applications: Putting Principles into Practice. Buffalo Grove, IL: Interpharm Press, 1997.
2. ISPE. GAMP®5: Risk-Based Approach to Compliant GxP Computerised Systems, International Society for Pharmaceutical Engineering, Tampa, Florida. Available at: www.ispe.org, 2008.
3. Wingate GAS. Computer Systems Validation: Quality Assurance, Risk Management and Regulatory Compliance for Pharmaceutical and Healthcare Companies. Boca Raton, FL: Interpharm Press, 2003.
4. ASTM. E2500-07 Standard Guide for Specification, Design and Verification of Pharmaceutical and Biopharmaceutical manufacturing Systems and Equipment. American Society for Testing and Materials, 2007.
5. FDA. Pharmaceutical CGMPs for the 21st Century: A Risk Based Approach". Available at: www.fda.gov/oc/guideline/gmp.html, 2002.
6. GAMP Forum. Risk Assessment for Use of Automated Systems Supporting Manufacturing Processes. Pharmaceutical Engineering, 2003.
7. GAMP Forum. Risk Assessment for Electronic Records and Electronic Signatures. Pharmaceutical Engineering, 2003.
8. ISPE. GAMP® Good Practice Guide: Risk-Based Approach to Electronic Records and Signatures. International Society for Pharmaceutical Engineering, Tampa, Florida. Available at: www.ispe.org, 2005.

19 | Case Study 1: Computerized Analytical Laboratory Systems

Ludwig Huber
Labcompliance

INTRODUCTION

Computers are widely used in analytical laboratories for instrument control, data evaluation, and data management, and are subject to all validation and verification activities. Verification and validation activities to assess computer systems used in analytical laboratories cover all life cycle phases, from user requirements specifications, design, development, and manufacturing to installation and operation. In this study, users of computer systems will find guidelines on

- How to define user requirements and functional specifications,
- Which type of documented evidence the vendor should provide to prove that the system was developed according to recognized standards,
- How to validate and document software developed in the user's own laboratory,
- How to proceed in cases where vendors do not give evidence of development and verification,
- How to qualify a computer system at installation and for operation,
- How to evaluate computer systems in analytical laboratories retrospectively, and
- How to ensure an ongoing performance control during routine analysis.

This case study does not cover verification activities during development at the vendor's site. This is described elsewhere (1).

A computer system, as used in an analytical laboratory, consists of computer hardware, peripherals, and software to perform a task. It can perform different tasks.

- Instrument control, data acquisition, and data evaluation
- Laboratory information management
- Archiving of electronic records

Figure 19.1 shows an example of a complex computer system that includes computerized analytical instruments for the collection and evaluation of data, servers for data review and centralized archiving, and a laboratory management system.

On a computer used in analytical laboratories, we generally find three different software categories.

1. System software, such as operating software (Windows 2000, Windows XP, Windows VISTA or UNIX®), drivers, and file management, supplied by software vendors. These are supplied with the computer in a machine-executable format that cannot be modified by the user and is not unique to any one user's system. The correct function of this software is verified whenever an application runs under the system software.
2. Standard application software, for example, COTS chromatography software, generally supplied by an instrument vendor. The correct function should be verified during and at the end of development. The user must perform acceptance testing prior to use.
3. User-specific application software, written by the user or by a third party for a specific user to meet the specific functional needs in the user's laboratory. Examples are macros to customize a system for specific user needs. This software must be validated prior to and during routine use.

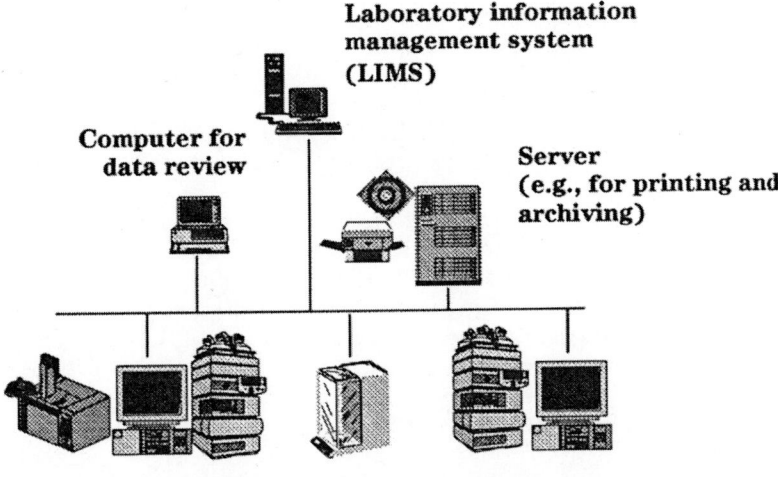

Figure 19.1 Computerized laboratory.

OVERVIEW OF VALIDATION STEPS ASSOCIATED WITH COMPUTER SYSTEMS

Figure 19.2 illustrates the validation steps in an analytical laboratory from a user's point of view. For a specific project, validation activities should follow a ten-step validation plan.

1. The user requirements are set. These describe the analysis problem and include instrument performance requirements for a specific analysis task.
2. From the user requirements, the type of analytical equipment, and computer system, the functions and functional specifications are derived.
3. The user then selects a standard instrument and appropriate options.

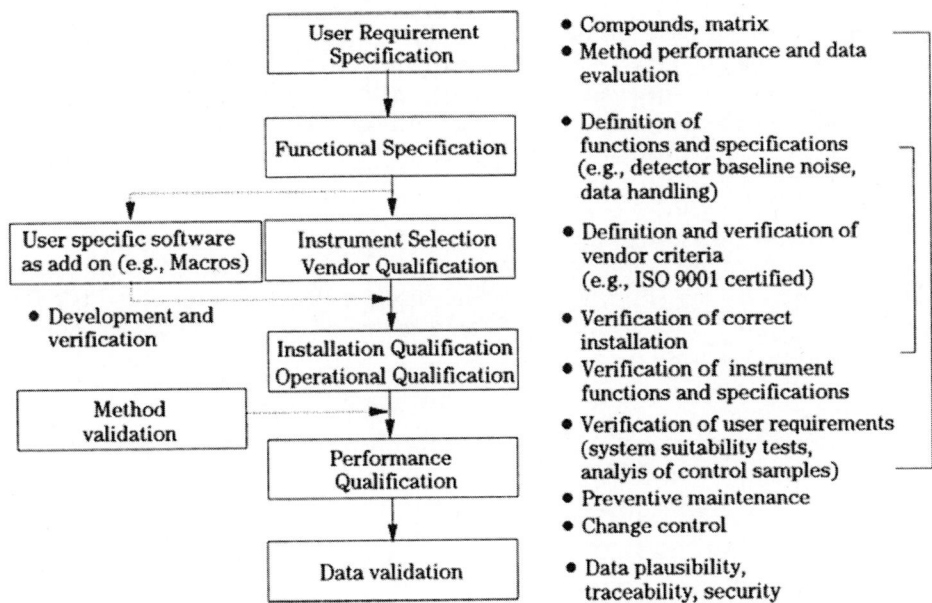

Figure 19.2 Validation life cycle.

4. A vendor should be selected who develops hardware and software equipment in accordance with a quality assurance system, for example, ISO 9000:2008. (*Steps 2 to 4 are sometimes called the design qualification. The user should verify that the design of the computer system meets the requirements and that the vendor meets the vendor qualification criteria.*)

5. If the standard software supplied by the vendor does not cover all of the user's requirements, user-specific software is developed as an add-on macro by the user, the vendor, or a third party.

6. The modules are installed and put together as a system. Correct installation and operation should be verified against functional specifications as defined by the user, a process that is called installation qualification (IQ) and operational qualification (OQ).

7. The proper functioning of analytical methods should be verified on the new system. This covers testing of significant method characteristics, for example, limit of detection, limit of quantification, selectivity, and linearity. If the method has not been validated or if its scope did not cover the new instrument, the method should be newly validated or revalidated.

8. The performance of the complete system should be validated against the user requirements specifications. The system combines the instrument hardware, computer hardware and software, and the analytical method. In chromatography, it also includes a column and reference standards for calibration. This validation, usually referred to as system suitability testing, tests a system against documented performance specifications for the specific analytical method. Analytical systems should be tested for suitability prior to and during routine use, practically on a day-to-day basis.

9. When analyzing samples, the data should be validated. The validation process includes documentation and checks for data plausibility, data integrity, traceability, and security.

10. A complete audit trail that allows the final result to be traced back to the raw data should be in place. According to Food and Drug Administration (FDA) regulation 21 CFR Part 11 on electronic records and signatures, this audit trail must be computer generated and independent of the operator.

Table 19.1 lists in chronological order the steps that a user of computerized analytical equipment can follow for the entire validation process.

The type and degree of validation of a computerized analytical system depend on its complexity. For example, the functions of a simple, computer-controlled system, with little or no flexibility regarding data input or evaluation, can be verified by executing holistic tests (2) and by comparing the test results with anticipated results. On the other hand, a more complicated computerized system with online databases and extensive flexible data evaluation requires complex validation.

ASSESSMENT OF THE NEED FOR VALIDATION AND THE VALIDATION PLAN

Before validation can begin, management should assess whether the system requires formal validation. Criteria to be considered are whether the system will be used in a regulated or quality standard environment and how critical the data generated by the system are. Examples of such environments are as follows:

- Good Laboratory Practice (GLP)
- Good Manufacturing Practice (GMP)
- Good Clinical Practices (GCP)
- ISO 17025

When the decision has been made about the need for a system validation, sufficient resources should be allocated. For larger projects, the recommendation is to form a validation team consisting of quality assurance personnel and technical experts. All validation activities at the user's site should follow a validation plan (3–5).

Table 19.1 Recommended Validation Steps with Examples

Step	Explanation	Examples
Define user requirements	Criteria: compounds, matrix, detection limit, precision, selectivity, accuracy, concentration range, qualitative or quantitative, throughput, regulatory day requirements, e.g., 21 CFR Part 11 (6)	Analysis of phenoxy acid herbicides in drinking water, detection limit: 0.01 µg/L qualitative and quantitative information, electronic archiving of data, 30 samples/day. Method and equipment: solid phase extraction and HPLC/diode array detection. A computer and associated software control the instrument and acquire and evaluate data.
Define functional specifications	Define intended equipment hardware, software, and system functions and operational limits; define regulatory requirements	HPLC: binary gradient, flow rate range and system 0.2 to 2 mL/min, diode array detector with 10-mm path length, baseline noise limits $\leq 4 \times 10^{-5}$ AU, computer for integration, quantification, peak purity check, interactive and automated spectral library search, qualitative and quantitative report, system suitability testing and archiving of method parameters with raw data file, limited and authorized access to the system and data through user ID and passwords, electronic audit trail, electronic signatures; signatures must be bound to electronic records.
Select and qualify vendor-purchased equipment	Develop criteria for vendor selection and check if criteria are met	Quality system, availability of validation documentation, local support response time, reputation and experience; information through documentation available from the vendor.
Develop user-specific software; qualify modules and systems prior to routine use	Macro, spreadsheet calculations, installation qualification, operational qualification	Customized reporting, statistical evaluation. Check if shipment complies with purchase order; test equipment (e.g., test the precision of amounts and retention times); verify correct software installation.
Validate methods (optional, if the methods are not already validated)	Specify validation parameters and acceptance limits; define and execute validation experiments	Limit detection, limit of quantification, selectivity, linearity, precision, accuracy, ruggedness.
Assure ongoing performance	Develop and implement schedules and procedures for periodic maintenance, for calibration, and for initial and ongoing performance qualification; develop and implement procedures for error detection, recording, and handling; develop procedures for change control	Calibration of balances, calibration of ultraviolet grating for wavelength accuracy, exchange of lamps, system suitability testing, analysis of quality control samples and evaluation of results using control charts. ROM check at system boot-up, automated shutdown of pump if leak is detected in an autosampler, or authorization of changes to user-written change control software.
Assure validity, security, integrity, and traceability of data	Develop and implement security-relevant procedures	Limited system access through user-specific passwords, data file integrity through checksum or other routines.

Abbreviation: HPLC, high performance liquid chromatography.

USER REQUIREMENTS AND FUNCTIONAL SPECIFICATIONS

Requirements specifications define how the system will be used. For an analytical system, these can include the following:

- The type of compounds and matrix
- The expected limits for detection and quantitation
- The expected precision and accuracy

- The number of samples in a given time frame and mode of operation: manual or automated
- The type of information: qualitative and/or quantitative
- The type of computer and IT environment for instrument control, data acquisition, and data evaluation, for example, LAN-based networked data system
- The type of information calculated from original data, printed in the report, and stored on electronic media
- Archiving of data

From the requirements specifications, the user can derive the instrument type and its minimal functional specifications. For example, if an instrument is scheduled to run overnight, the number of samples should be specified so that the system can inject automatically. The UV/visible detector's baseline noise specification can be determined from the specified detection and quantitation limit of an high performance liquid chromatography (HPLC) analysis. The required data evaluation will determine the demands on the evaluation software.

The next step is either to select an existing system for the analysis task or to purchase a new system. When a new system is purchased, it is often purchased not just for a specific analysis but also for use in general applications in the laboratory. In this case, the requirements specifications should include a representative mix of the anticipated applications, and the functional specification should be set such that the instrument can handle all of the requirements. Next, the user should look for an instrument on the market that best meets these requirements. If the selected system does not provide all of the functions—for example, regarding software—the user can decide whether to develop these himself/herself or ask the vendor or a third party to do so.

RESPONSIBILITIES OF VENDORS AND USERS

New software and computerized systems in analytical laboratories are usually purchased from a vendor. Such COTS software must be validated as the FDA part 11 validation draft guidance (7) states, "Commercial software used in electronic record keeping systems subject to 21 CFR Part 11 needs to be validated, just as programs written by end users need to be 'validated'." A frequently asked question is who is responsible for the validation of such a system: the vendor or the user? The Organisation for Economic Co-operation and Development (OECD) states clearly in consensus paper number 5, "It is the responsibility of the user to ensure that the software program has been validated" (3). This is also the practice of the U.S. FDA, as specified by Ron Tetzlaf, a former FDA investigator, "The responsibility for demonstrating that systems have been validated lies with the user" (4). However, it is obvious that product quality cannot be solely achieved by testing in a user's laboratory. This must be incorporated during design and development. Therefore, the OECD makes a further statement in a new consensus paper: "There should be adequate documentation that each system was developed in a controlled manner and preferably to recognized quality and technical standards (e.g., ISO 9000:2008)." Furman et al. (2) from the U.S. FDA also make it clear: "All equipment manufacturers should have validated their software before releasing it."

- The vendor is responsible for assuring that the system is developed, tested, and supported according to proper development and change control procedures.
- The user is responsible for the entire validation. One part of this overall validation process is to obtain documented evidence about the proper development according to documented standards. To have this assurance, a vendor assessment program should be developed that may include formal written procedures for the selection, evaluation, and qualification of vendors.

One criterion for vendor qualification should be whether the vendor has a documented quality assurance program that follows recognized quality standards, for example, ISO 9000:2008. This registration is usually sufficient for laboratories that must comply with ISO 9000:2008 or with a laboratory accreditation standard such as ISO 17025. However, it is the author's experience that regulatory agencies do not always accept such third-party evaluation or successful registration according to ISO 9000:2008 or the equivalent as proof of a vendor's qualification. They expect vendors to have other proofs of qualifications such as documented familiarization of their staff with software development practices, GLPs and cGMPs, procedures for software backup, archiving and periodic integrity checks, a software tracking and response system, or references from internal or external users of the system. Frequently, additional documents are requested, such as development validation certificates and an assurance by the vendor of accessibility to the source code by regulatory agencies.

VALIDATION OF NEWLY PURCHASED SYSTEMS WITHOUT EVIDENCE OF VALIDATION FROM THE VENDOR

If a vendor is not able or willing to provide documented evidence of validation, the user should consider selecting another vendor. This recommendation is easy to follow if there are a number of competitors for the same or similar products. If this is not the case, for example, when special software for an emerging technique is required, a user may purchase the software anyway but should evaluate the vendor and perform more thorough testing and keep more detailed documentation. For example, for an evaluation of the vendor, checklists can be sent to the vendor, with questions on the following topics:

- The company (history, size, financial status, member of employees)
- The organization (quality assurance department)
- Certifications (ISO 9000:2008)
- Hardware and software development (quality and technical standards)
- Testing and verification of life cycle phases (reviews of requirements specifications, design, code inspections, test traceability matrix from test cases back to requirement specifications, whether there are release criteria for the phases)
- Source code (guaranteed accessibility to regulatory agencies, where stored, escrow account)
- The product (how many are sold to the target industry)
- Security (how unauthorized changes are prevented and whether disaster plans exist)
- Support (response, language, phone, on-site, modem, support plan existing)
- Handling of failures and enhancement requests (formal procedures)
- Change control (who initiates and authorizes changes, version identification, and revision history)
- People qualification (training programs)
- User documentation (what is archived and for how long)
- Customer training (topics, frequency, next location)
- Equipment hardware (how specifications are verified)

A detailed checklist for vendor qualification has been developed by the author. If the answers to the checklist are not satisfactory, a direct audit should be considered.

If no documentation on validation during development can be obtained from the vendor, the user should evaluate the system retrospectively. For example, the 21 CFR Part 11 draft validation guidance (7) states, "When the end user cannot directly review the program source code or development documentation, more extensive functional testing might be warranted than when such documentation is available to the user." Detailed test cases with well-characterized test data sets and known results should be developed to evaluate the correct functioning of all software programs. In the case of completely computerized analytical systems, the analysis results with reference standards or quality control samples should be compared with known results from the standard or sample. Besides this holistic test, it is recommended to verify the performance of each individual subsystem. Such checks should include the proper response of the equipment to inputs from the computer. The tests should also check the system's error-handling capabilities. The system should recognize and display any wrong entries—for example, entries that are out of the system's operating range. Another simple test would be to check how the program responds when alphanumeric data are entered into fields that are designed to accept numerical values. In addition, the mathematical formulas should be verified with alternative methods of calculation.

VALIDATION OF USER CONTRIBUTED SOFTWARE (FOR EXAMPLE, MACROS)

Application software developed by the user should be fully validated and documented by the user. Such software may be a stand-alone software package (e.g., for statistical data evaluation), or it may be an extension to purchased standard software (e.g., a macro to enhance functionality). The development and validation of such software should follow a documented procedure, and the source code should be available. The effort involved in validation depends very much on the size and complexity of the program. The development of large programs should follow the software development life cycle and can take several months or years. Validation can take several weeks, and the documentation will be extensive. On the other hand, the validation of small programs can be done in a few hours, and the documentation may be only a few pages. The development, validation, and documentation of such small programs require the following steps at minimum:

1. Describe the problem, how the problem is solved currently, and how the newly developed program will solve it
2. Identify responsibilities for development, test, and approvals
3. Describe the task and the system requirements (hardware, system software, standard software)
4. Describe the program in terms of the functions it will perform
5. Document formulas and algorithms used within the code
6. Write and document the code in such a way that it can be understood by other people whose knowledge and experience are similar to the programmer's. Print the code.
7. Develop test cases and data sets with known inputs and outputs. Include test cases with normal data across the operating range and at the boundary, and some unusual cases with incorrect inputs. The results should be calculated by the new program and also by using alternative methods. The development of an automated test procedure that can be executed as often as possible is recommended. Test procedures and results should be documented, reviewed, and signed off.
8. Develop user documentation with information on how to install, test, and operate the program.
9. Describe and implement procedures for data backup and security routines for limited access to authorized people.
10. Develop a procedure to authorize, test, document, and approve any changes to the software and documentation.

For combined systems, vendor-updated software revisions may be critical, especially if the updated version supplied by the vendor will have an effect on the interface between the vendor's and the user's softwares (e.g., if the meaning of a macro command has been changed). The user should obtain information from the vendor on how the updated version may affect the interface. The user should also test his or her software after it has been integrated into the vendor's updated standard software. More details are found elsewhere about standard operating procedures (SOPs) for developing and validating simple as well as complex applications software developed in the user's laboratory (8).

PREINSTALLATION

Before the instrument arrives at the user's laboratory, serious thought must be given to its location and space requirements.

- A full understanding of the new equipment must be obtained from the vendor well in advance: required bench or floor space and environmental conditions, such as humidity and temperature, and, in some cases, utility needs, such as electricity and compressed gases for gas chromatographs.
- Care should be taken that all of the environmental conditions and electrical grounding are within the limits as specified by the vendor and that correct cables are used.

- If environmental conditions may have an influence on the validity of test results, the laboratory should have facilities to monitor and record these conditions, either continuously or at regular intervals.
- Any special safety precautions should be considered (e.g., for radioactivity measurement devices), and the location should also be checked for any devices generating electromagnetic fields nearby.

INSTALLATION

Once the instrument arrives, the following should be considered:

- The shipment should be checked by the user for completeness.
- It should be confirmed that the equipment ordered is what was in fact received. Besides the equipment hardware, other items should be checked (e.g., correct cables, other accessories, and documentation).
- The documentation should be checked for completeness (operating manuals, maintenance instructions, SOPs for testing, safety, and validation certificates).
- For more complex instrumentation, wiring diagrams should be generated, if not obtained from the vendor.
- An electrical test of all modules and systems should follow.
- The impact of electrical devices close to the computer system should be considered and evaluated if a need arises. For example, when small voltages are sent between sensors and integrators or computers, electromagnetic energy emitted by poorly shielded nearby fluorescent lamps or by motors can interfere with the transmitted data.
- When complex software is installed on a computer, the correctness and completeness of the installed program and data files should be verified. Vendors can assist this process by supplying installation references files and automated validated verification procedures. In this case, the integrity of each file is verified by comparing the cross-redundancy check (CRC) of the installed file with the checksum of the original file recorded on the installation master. Modified or corrupt files have different checksums and are thus detected by the verification program. Verification reports include a list of missing, changed, and identical files.

The installation should end with the generation and sign-off of the installation report—in pharmaceutical manufacturing referred to as IQ document. The hardware and software should be well documented with model, serial, and revision numbers. For larger laboratories with lots of equipment, this should preferably be a computer database. Entries for each instrument should include the following:

- In-house identification number
- Name of the item of equipment
- The manufacturer's name, address, and phone number for service call; service contract number if available
- Serial number and firmware revision number of equipment
- Software with product and revision number
- Data received
- Date placed in service
- Current location
- Size, weight
- Condition when received (e.g., new, used, reconditioned)
- List of authorized users and responsible person

It is recommended to make copies of all important documentations; one copy should be placed close to the instrument; the other should be kept in a safe place. An identification sticker should be put on the instrument with information about the instrument's serial number and the company's asset number.

LOGBOOK

An electronic or bound paper logbook should be prepared for each instrument in which operators and service technicians record all equipment-related activities in chronological order. Information in the logbook can include the following:

- Logbook identification (number, valid time range)
- Instrument identification (manufacturer, model name/number, serial number, firmware revision, date received, service contract)
- Column entry fields for dates, times, and events (e.g., initial installation and calibration, updates, column changes, errors, repairs, performance tests, quality control checks, cleaning, and maintenance, plus fields for the name and signature of the technician making the entry)

OPERATIONAL QUALIFICATION AND ACCEPTANCE TESTING

After the installation of hardware and software, the hardware should be calibrated where required. An operational test should follow a process that is referred to in pharmaceutical manufacturing as OQ. The goal is to demonstrate that the equipment's hardware and software operate "as intended" in the user's environment.

For a computer system in an analytical laboratory, OQ can mean, for example, verifying correct communication between the computer and other hardware. As part of the product documentation, vendors should provide operating procedures for the tests, limits for acceptance criteria, and recommendations in case these criteria cannot be met. The documentation should also include algorithms for critical calculations and procedures on how to verify the algorithms in a user's environment. If the user finds the tests recommended by the vendor inappropriate or insufficient, the user can design and perform other or additional tests.

Chemical standards used for instrument calibration or qualification tests should be traceable to national standards.

The documentation of testing should include the following:

- The description and unique identification of equipment.
- Test items.
- Acceptance criteria.
- Summary of results.
- The data.
- Names and signatures of persons who performed the tests.
- The instrument should be labeled with the calibration and qualification status, indicating the dates of the last and next calibration and OQ.

QUALIFICATION OF SOFTWARE

The correct functioning of software loaded on a computer system should be checked in the user's laboratory under typical operating conditions and under high load conditions. As FDA's 21 CFR Part 11 industry guidance states, the type and extent of testing should be based on the risk the computer system has on the product quality and data integrity: We recommend that you base your approach on a justified and documented risk assessment and a determination of the potential of the system to affect product quality and safety, and record integrity (7) This is in line with recent references for computer validation such as GAMP®5 (9).

During the equipment hardware test, as described in the previous section, many software functions are also executed, such as instrument control, data acquisition, peak integration, quantitation, file storage and retrieval, and printing. Therefore, after successful completion of hardware tests, it can also be assumed that the software operates as intended. There are two situations where software verification independent of the equipment hardware may be necessary.

1. If not all critical software functions are executed during the hardware verification
2. If a verification of the software functions should be done without a need for equipment testing

Generate Master Data

1. Generate master chromatogram.
2. Develop method for evaluation.
3. Generate master result.
4. Save results electronically and on paper.

Verification

1. Select master chromatogram and method.
2. Run test (manually or automatically).
3. Compare new results with master data.
4. Report verification results.

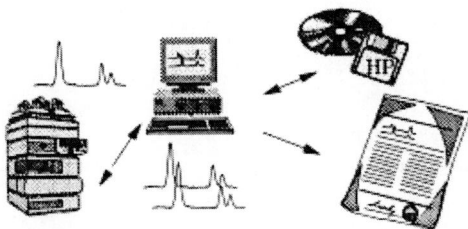

Test items

- Data transfer
- Data acquisition
- Integration
- Quantification
- Storage
- Retrieval

Figure 19.3 Verification process of chromatographic software.

This is the case after a change on the computer system—for example, if a new operating system has been installed or if new hardware, such as CD-ROMs or a hard disk, has been installed on the computer system.

The following paragraphs describe a procedure for the verification of important chromatography software functions without injecting a sample (Fig. 19.3). The concept has been described in detail elsewhere (1). It is very generic and can also be used to test and verify the correct functions of other software packages.

Well-characterized test chromatograms derived from standards or real samples are stored on disk as a master file. Chromatograms may be supplied by the vendor as part of the software package or can be recorded by the user. This master data file goes through normal data evaluation from integration to report generation. Results are stored on the hard disk. The same results should always be obtained when using the same data file and the same method for testing purposes.

Preferably, tests and the documentation of results should be done automatically, always using the same set of test files. In this way, users are encouraged to perform the tests more frequently, and user-specific errors are eliminated. In some cases, vendors provide test files and automated test routines for verification of a computer system's performance in the user's laboratory. Needless to say, the correct functioning of this software should also be verified. If such a software is not available, the execution of the tests and the verification of actual results with prerecorded results can be done manually.

Successful execution of such a procedure ensures that

- Executed program and data files are loaded correctly on the hard disk,
- The current computer hardware is compatible with the software,
- The current versions of the operating system and user interface software are compatible with the application software, and
- Data are correctly transferred between the equipment and the computer (if this feature is supported by the system).

In addition to typical functions required by the application, other functions required by regulations and internal company policies should be tested. These include the following:

- Limited and authorized access to the system and data. This can be achieved by trying to enter the system with correct and incorrect combinations of passwords and user IDs.
- Electronic audit trail. Check if the audit trail records events as specified in the functional requirement specification document.
- Electronic signatures. Check if a signature includes the full name of the person who signed, date and time, and a meaning.

When data are transferred between computers through a network, the accuracy of data transfer should be verified. This can be achieved by comparing printouts before and after transfer or comparing harsh factors before and after data transfer. More detailed information on the qualification and testing of networks using hash factors can be found in Ref. 10.

ROUTINE MAINTENANCE AND ONGOING PERFORMANCE QUALIFICATION

When the installation is complete and the equipment and the computer system are proven to operate well, the computerized system is put on routine analysis. Procedures should exist that show that "it will continue to do what it purports to do."

Each laboratory should have a quality assurance program that is well understood and used by individuals, as well as by laboratory organizations, to prevent, detect, and correct problems. The purpose is to ensure that the results have a high probability of being of acceptable quality. Ongoing activities may include the following:

- Preventive instrument maintenance
- Performance verification and calibration
- System suitability testing
- Analysis of blanks and quality control samples
- Ensuring system security

PREVENTATIVE MAINTENANCE

Operating procedures for maintenance should be in place for every system component that requires periodic calibration and/or preventive maintenance. The idea is to replace critical maintenance parts before they have a negative effect on the quality of analytical data.

- Critical parts should be listed and should be available at the user's site.
- The procedure should describe what should be done, when it should be done, and what the qualification of the engineer performing the tasks should be.
- System components should be labeled with the date of the last and next maintenance.
- All maintenance activities should be documented in the instrument's logbook.
- Suppliers of equipment should provide a list of recommended maintenance activities and procedures (SOPs) on how to perform the maintenance.

All maintenance activities should be recorded in a maintenance logbook. To make this more convenient, modern equipment includes electronic maintenance logbooks where the user enters the type of maintenance, and the equipment records this activity together with the date and time.

CALIBRATION

Operating devices may become miscalibrated after a while (e.g., the temperature accuracy of a gas chromatography (GC) column oven or the wavelength accuracy of a UV/visible detector's optical unit). This can have an impact on the performance of an instrument. Therefore, a calibration program should be in place to recalibrate critical items of an instrument.

- All calibrations should follow documented procedures, and the results should be recorded in the instrument's logbook.
- The system components should be labeled with the date of the last and next calibration.
- The label on the instrument should include the initials of the test engineer; the form should include his/her printed name and full signature.

SYSTEM SUITABILITY AND QUALITY CONTROL SAMPLE ANALYSIS FOR ONGOING PERFORMANCE QUALIFICATION

The analysis of standards or quality control samples with the construction of quality control charts (Fig. 19.4) has been suggested as a way to incorporate quality checks on results as they

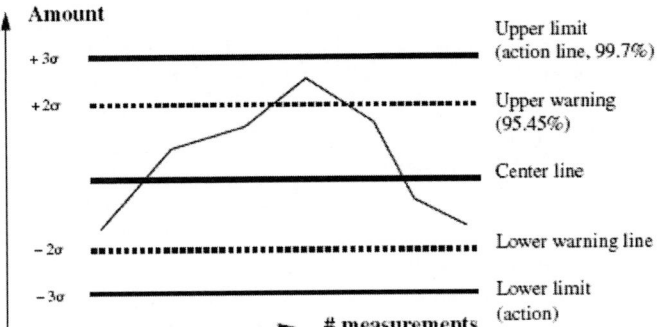

Figure 19.4 Quality control chart with warning lines and control lines.

are being generated. Such tests can then flag values that may be erroneous for any of the following reasons:

- Reagents are contaminated
- GC carrier gas is impure
- HPLC mobile phase is contaminated
- Instrument characteristics have changed over time

For an accurate quality check, quality control samples are interspersed among actual samples at intervals determined by the total number of samples and the precision and reproducibility of the method. The control sample frequency will depend mainly on the known stability of the measurement process—a stable process requiring only occasional monitoring.

Control samples should have a high degree of similarity to the actual samples analyzed; otherwise, one cannot draw reliable conclusions on the measurement system's performance. Control samples must be so homogeneous and stable that individual increments measured at various times will have less variability than the measurement process itself. Quality control samples are prepared by adding known amounts of analytes to blank specimens. They can be purchased as certified reference material (CRM) or may be prepared in-house. In the latter case, sufficient quantities should be prepared to allow the same samples to be used over a longer period of time. Their stability over time should be proven, and their accuracy should be verified, preferably through interlaboratory tests or by other analysis methods.

The most widely used procedure for the ongoing performance control of equipment through quality control samples involves the construction of control charts for these samples. These are plots of multiple data points versus the number of measurements from the same samples using the same processes. Measured concentrations of a single measurement or the average of multiple measurements are plotted on the vertical axis, and the sequence number of the measurement is plotted on the horizontal axis. Control charts provide graphics tool to demonstrate statistical control, monitor a measurement process, diagnose measurement problems, and document measurement uncertainty. The most commonly used control charts are X-charts and R-charts as developed by Shewart. X-charts consist of a central line representing either the known concentration or the mean of 10 to 20 earlier determinations of the analyte in control material. The standard deviation has been determined during method validation and is used to calculate the control lines in the control chart. Control limits define the bounds of virtually all values produced by a system in statistical control.

Control charts often have a centerline and two control lines with two pairs of limits: a warning line at $m \pm 2s$ and an action line at $m \pm 3s$. Statistics predict that 94.45% and 99.7% of the data will fall within the areas enclosed by the $\pm 2s$ and $\pm 3s$ limits. The centerline is either the mean or the true value. In the ideal case, where unbiased methods are being used, the centerline would be the true value. This would apply, for example, to precision control charts for standard solutions.

When the process is under statistical control, the day-to-day results are normally distributed about the centerline. A result outside the warning line indicates that something is wrong. Such a result need not be rejected, but documented procedures should be in place for suitable action. Instruments and sampling procedures should be checked for errors. Two successive values of the quality control sample falling outside the action line indicate that the process is no longer under statistical control. In this case, the results should be rejected, and the process should be investigated for its unusual behavior. Further analyses should be suspended until the problem is resolved.

CHANGE CONTROL

Software has one distinct advantage over hardware: it does not change its performance characteristics over time. Theoretically, there should be no need to revalidate software as long as the hardware and environmental conditions do not change. However, almost 100% of all softwares written will be changed following their release for use. There are three reasons for a software change.

- To correct errors
- To adapt software to changes in its operating environment
- To enhance the software (e.g., to add functionality)

In addition to software changes, there may be hardware changes to a computer system. The processor may be upgraded, the system may get additional disk space, or new memory chips may be installed. All software and hardware changes to a computer system may influence the performance and correct functioning of the system and may need a reverification or revalidation. A control system should exist that describes what should be revalidated after a change to the system. Procedures should be available for changes to a system purchased from a vendor as well as for software developed in-house.

Software developed and validated by the user should also be revalidated by the user. Principally, software redevelopment and testing should follow the same procedure as for newly developed software. Procedures should include information on who authorizes changes, who executes the changes, and who can finally release the changed version. Compared with new software, the amount of testing can be reduced through intensive reuse of previously developed test files and test procedures.

If computer systems are upgraded with new operating systems or when new hardware is added, the user should thoroughly document the upgrade and perform acceptance testing as for a new system. The use of existing test files and automated procedures can make this process very efficient.

If the user purchases a software upgrade, the supplier should supply documentation with a description of change, a statement that the upgrade has been validated by the supplier during development. The supplier should provide information on

- System requirements for the upgraded version,
- How to install the upgrade,
- Impact on macros written by the user, and
- Recommendation on what to test in the user environment.

The installation should be documented, and the user should perform an acceptance testing before the system is authorized for use.

REFERENCES

1. Huber L. Validation of Computerized Analytical and Networked Systems. Buffalo Grove: Interpharm Press, 2002.
2. Furman WB, Layloff TP, and Tetzlaf RF. Validation of computerized liquid chromatographic systems. J AOAC Intern 1994; 77(5):1314–1318.
3. Organisation for Economic Co-operation and Development. Compliance of Laboratory Suppliers with GLP Principles. Series on Principles of Good Laboratory Practice and Compliance Monitoring, No. 5, GLP Consensus Document, Environment Monograph No. 49, Paris, 1992.

4. Tetzlaf RF. GMP Documentation Requirements for automated systems: Part III, FDA inspections of computerized laboratory systems. Pharm Technol 1992; 71–82.
5. Organisation for Economic Co-operation and Development. The Application of the Principles of GLP to Computerised Systems. Series on Principles of Good Laboratory Practice and Compliance Monitoring, No. 10, GLP consensus document, environment monograph No. 116, Paris, 1995.
6. Code of Federal Regulations, Title 21, Food and Drugs, Part 11, Electronic Records; Electronic Signatures, Final Rule, Fed Register 1997; 62(54):13429–13466.
7. FDA. Guidance for Industry, 21 CFR Part 11, Electronic Records; Electronic Signatures, Scope and Applications, 2003.
8. Huber L. Macro and spreadsheet quality package, Labcompliance, 2008. Available at: www.labcompliance.com/books/macros.
9. ISPE. GAMP®5 Good Automated Manufacturing Practice, a Risk-based Approach for Compliant GxP Computerized Systems (version 5), 2008. Available at: www.ispe.org.
10. Huber L. Network Quality Package, Labcompliance, 2008. Available at: www.labcompliance.com/books/network.

| **Case Study 2: Chromatography Data Systems**

Bob McDowall

McDowall Consulting

INTRODUCTION

Chromatography is an analytical technique used in virtually all areas of the pharmaceutical and biotechnology industries to detect or measure compounds during the course of product development and manufacture. It can be used for the measurement of active ingredients, raw materials, impurities and determining the stability of active substances in final preparations. The chromatograms from these analytical methods produced are generated, displayed, integrated and results calculated by a software application called a chromatography data system (CDS).

This chapter presents some approaches to prospectively and retrospectively validating client server networked CDS based on case studies; in addition the business benefits that can be exploited from the implementation of electronic signatures when upgrading an existing system or implementing a new one are presented.

OPERATIONAL CAPABILITY

Figure 20.1 shows the typical work flow performed by a CDS, which is summarized below, and further details can be found in the books by McDowall (1) and Dyson (2).

Method Files

The start of the data acquisition operation of a chromatography data system is to build a method file. This tells the data system how to acquire data and process and interpret the results. A method file should control the following:

- The data sampling rate of the analogue to digital (A/D) converter (3) or direct data acquisition from a detector
- When to start and stop the integration of the chromatogram
- Whether peak areas or heights should be used
- Retention time windows and identification of the analytes and internal standard
- Allocation of the method to calculate the analyte amount or concentration

The parameters used to control the chromatograph can be found in the method file or a related instrument file associated with a specific method.

A name, number or a mixture of both should identify individual method files within the system. In addition, the system should be able to provide facilities for version control of method files to ensure that control is maintained over the method for the lifetime of its use. Part of the control function must be access control to identify the individuals who can create, modify or delete analytical methods. If a method has been modified then copies of the modifications must be stored with the data processed by that method. This is to provide an audit trail for the data and results produced by a version of a method. However, when developing methods, flexibility with method files is essential and a default method should be available to acquire data and then feedback to a normal method.

Naming Conventions

When a laboratory uses a client server CDS there will be a requirement to devise and implement naming conventions for method, sequence and all data files within the data system. Typically a database is the most effective way of organizing and managing these files and to aid efficient archiving and unambiguous identification of these files. Any naming convention system must aid users, quality assurance, and regulatory inspectors.

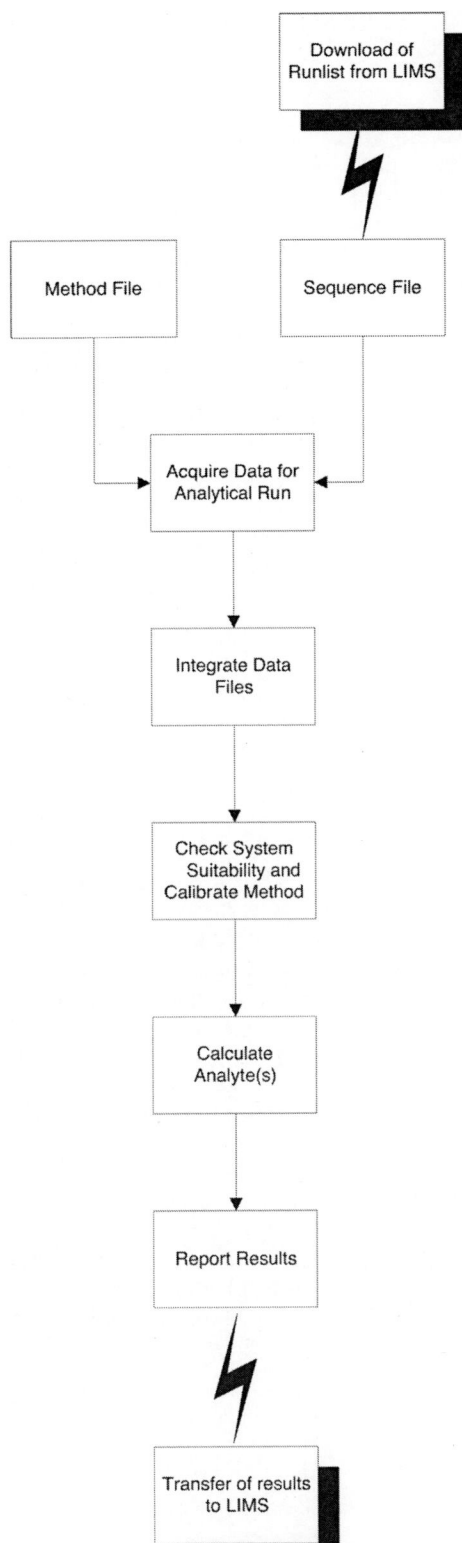

Figure 20.1 Work Flow for a typical chromatography data system.

A naming convention should be based on the work flow undertaken by a laboratory. This is to allow efficient archiving of data but just as importantly, the efficient retrieval of data. Some ideas might be the following:

- Organize the data around drug products or development projects, as this is how the work is structured and how project teams are organized as this will help store data to aid 21 CFR 11 compliance for ready retrieval of electronic records.
- Major subdivisions of each project should be based around the type of work done, for example, method development, method validation, preformulation, etc.

Sequence File

The sequence file is the run list or order that the samples, standards, quality control samples and blanks will be injected into the chromatograph; this is essential as it puts in context to the individual data files. Each sequence file or each injection must be linked with a method file to process the resulting data. For laboratories with large numbers of samples for a single method, the sequence file will usually be linked with a single method. Smaller laboratories may need the flexibility to link the sequence file with several methods during the course of a single analytical run for maximal use of equipment resources.

Each sample to be analyzed should be identified in the sequence file usually as one of the following types:

- Unknown
- Calibration standard
- Quality control sample
- Blank (either matrix or reagent)

Depending on the data system involved, at least the first three options should be available to a user.

Interpretation of Chromatographic Data

After the method file and the sequence file have been set up the analytical run is started and data are collected. A data file containing the detector reading values will be obtained for each chromatographic run and sample injected. It is important from scientific and regulatory considerations that the data files must not be capable of alteration and to this end each one contains a checksum to detect alteration.

Moreover, they must not be overwritten either if the same sample information is assigned to an assay or if the disk becomes full. This is an area for consideration when validating the chromatography data system; as it is important to know what happens to data files, especially in a regulated environment.

The data system will interpret each data file, identifying the individual peaks and fitting the peak baselines according to the parameters defined in the CDS method as shown in Figure 20.2. The data systems should have the ability to identify whether the peak baselines have been automatically or manually interpreted. This is a useful feature for compliance with part 11 to indicate the number of times a chromatogram has been interpreted.

Most data systems should be able to provide a real-time plot, so that the analyst can review the chromatograms as the analytical run progresses. In addition, the plotting options of a data system should include

- Fitted baselines,
- Peak start/stop ticks,
- Named components,
- Retention times,
- Timed events, for example, integration start/stop,
- Runtime windows and user-defined plotting windows, and
- Baseline subtract.

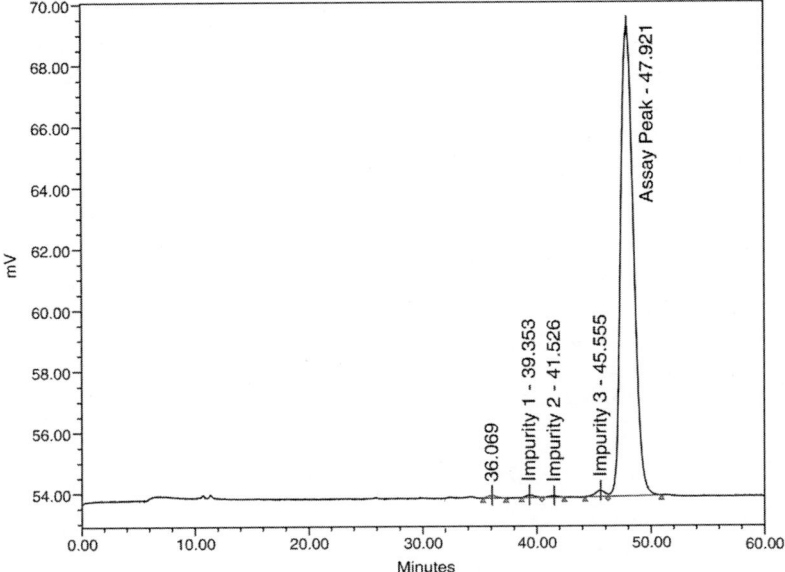

Figure 20.2 A typical chromatogram of an active substance separation from impurities/degradation products.

Each of these options should be enabled or disabled by a user.

An overlay function should be available to enable comparisons between results and samples. This will be used to compare and contrast chromatograms from the same run sequence as well as chromatograms from different sources. The maximum number of overlays will vary from data system to data system but a minimum of six to eight is reasonable and practicable. More overlays may be technically possible, but the amount of useful information obtained may be limited. Overlays that can be offset by an amount determined by the user are useful to highlight certain peak information. Ideally, the overlay screen should have hidden lines removed and be able to be printed.

Calibration

Calibration is an area where care has to be exercised by a user as often the calibration methods lack statistical rigor. Within a pharmaceutical analysis laboratory, the number of calibration model options that can be successfully used is usually limited to

- Bracketed standards at one concentration or amount for bulk drug or finished product assays,
- Response function for all analytes,
- Average by amount for bulk drug and finished products, and
- Multilevel or linear regression for related substances and degradation products.

Within each calibration type, the data system must be able to cope, sufficiently flexibility, with variations in numbers of standards used in a sequence and types of standard bracketing. The incorporation of a blank standard into the calibration curve should always be an option.

Each plot of an analyte in a multi-level or linear regression calibration model must contain an identifier for that calibration line and the analyte to be determined. The calibration curve should show all calibrating standards run in any particular assay. In assays containing more than one analyte it will be necessary to interpret all the calibration graphs before the calculation of results. Again, this is an area that is poor for data system as many only offer one line fitting method for all analytes in the run resulting in compromises.

User-Defined Analytical Run Information

The system should be capable of collating user-defined parameters (e.g., height, area, ratios, concentrations, etc.) for selected analytes from a sequence of runs. After collation system defined and/or user-defined calculations (custom calculations) will be carried out on the data generated. The type of calculations required should include mean, standard deviation, analysis of variance and possibly significance testing.

Reports and Collation of Results

Ideally, the report following an individual chromatogram should contain both elements that are user definable and those, which are standard; this should enable the laboratory to configure a report. At the end of the analytical run, a user-defined summary report containing information such as sample ID, area or height, baseline and calculated analyte concentration should be created. This report can either be printed out or transferred to a laboratory information management system (LIMS) for further analysis and interpretation.

Instrument Control

The primary interaction of the CDS with analytical instrumentation is with the output from the detector; however, there are other considerations such as instrument control. These can vary from system to system and the following options are available:

- Contact closures for the control of chromatographic valves or associated equipment during analysis is usually available for other supplier's equipment.
- Instrument control by the data system so that set up of a chromatograph can be achieved from a single workstation.
- Remote monitoring of the chromatography system output including the instrument conditions.
- The ability to list the items of equipment (pump, detector, etc.) used for a particular analysis is a function to help to automate the administrative records associated with an analysis and help meet GMP compliance.

ARCHITECTURE OF A NETWORKED CDS

A typical networked chromatography data system will consist of several hardware components, as shown in Figure 20.3.

- Chromatograph: this is the instrument that performs the analytical separation and can be a high performance liquid chromatograph (HPLC), gas chromatograph (GC) or a capillary electrophoresis (CE) instrument.
- Combined data capture and instrument control via a data server in the laboratory or data capture only via an A/D converter from the instrument detector to the CDS. The A/D unit converts the continuous detector analogue signal to a number of discrete digital data readings. Typically the data server and the A/D unit will have buffering capability if the network is temporarily unavailable to prevent data loss.
- Network: transport medium for moving the data from the instrument to a server for secure data storage.
- Server for running the CDS the application and data storage. This can be a single or multiple servers especially if terminal emulation is used to run the application.
- Workstation (client): for operating the CDS setting up an instrument, checking that the separation is working correctly and interpreting the resultant chromatograms after the run is finished and reporting the results.

KEY REGULATORY REQUIREMENTS AND ISSUES WITH A CDS

Before discussing how to validate a chromatography data system, it is important to understand the regulatory requirements and their interpretation. The responsibility for the validation rests with the system or business process owner but from experience most do not understand fully

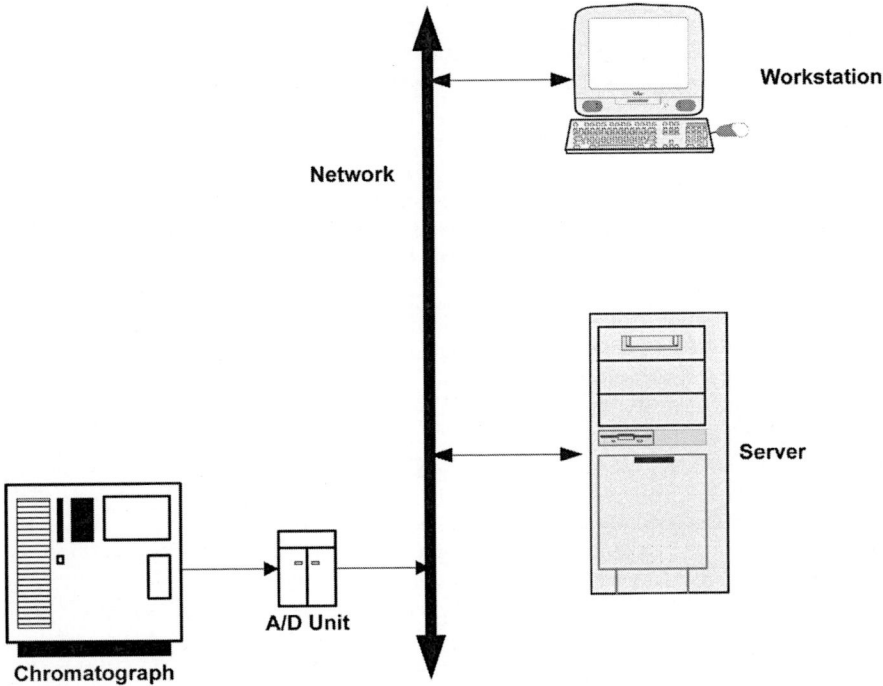

Figure 20.3 Schematic diagram of a typical networked chromatography data system.

the regulations they work under or the risk mitigation strategies that need to be undertaken when validating a CDS.

The regulations and guidelines have a view on what is expected during the implementation and release of a CDS as well as what is expected when the system is operational and when it is retired. In general, the emphasis is concerned with generating the evidence to demonstrate that the computerized system is accurate when validated and continues to be so when it is operational and that there is sufficient evidence of management awareness and control. The evidence of an activity usually means that it must be documented, although the format of documentation (paper or electronic) is not stated.

The definition of performance qualification (PQ) is documented verification that the computer-related system performs it functions in accordance with the computerized system specification while operating in its normal operating environment (4).

The major point to make is that the laboratory must test the CDS as they use it and not how the supplier has tested it (i.e., in the laboratory's operating environment, using the laboratory's analytical methods, specifications and capacities and using the laboratory's networks).

REGULATORY INSPECTIONS
A variety of pertinent FDA inspection observations impacting CDS are shown below.

- *The CDS was not validated to ensure the system produced accurate and precise data.*
- *There was no documentation to show that the system's ability to handle overload situations in an orderly fashion.*
- *There was no assurance of the program's behavior when working at its limit. Functional testing that includes volume and stress testing was not conducted to demonstrate the system's behavior.*
- *Confidential and unique user log-ins and passwords were not assigned to each analyst to ensure data authenticity and integrity. Each workstation had a single log-in name and password, which was shared with all users.*

- *There were no automatic computer generated time-stamped audit trails to ensure authenticity and integrity of analytical data that was acquired and processed with the CDS. Analyst's transactions were not documented to show whether the analytical data were modified, copied or deleted.*
- *There was no documented evidence that the CDS was adequately configured and performed as intended.*
- *The firm did not have a system administrator that was responsible for system configuration and control of access to configuration tools that can modify or delete electronic records. System administrator permissions and rights were given to some QC analysts who were also responsible for analyzing samples.*
- *There was no control over how analysts interacted with analytical data on the system.*
- *The universal log-in and password system gives users rights and permissions to edit, modify and delete data files. The system was not configured to deny analysts rights to directories and users did not have read / write access to analytical data on the system. Users could not only modify their records but all records on the server. There was no written documentation that established what limits and rights the IT groups assigned QC laboratory users.*
- *There was no documented evidence to show that the firm periodically restored analytical data from its tape backup medium to ensure that data files could be reconstructed and were not corrupted. IT personnel did not know how to reconstruct the graphic data on workstations and referred us to analysts in the laboratory to perform system administrator tasks.*
- *There was no documentation to show that analytical data on the chromatography network could not be altered or modified by authorized users of the corporate network. The networks are connected by a router, which enables data packets to move between networks. The chromatography network did not have capabilities for tracking and controlling the integrity of each sample throughout its retention period. There were no protocols that explained the logical security procedures in place to prevent unlimited and unauthorized access to chromatographic data files.*

Concord Laboratories Warning Letter

Concord Laboratories received a Warning Letter in July 2006 (5), and the key observations were the following:

- *Laboratory records fail to include the initials or signature of the person who performs each test §211.194(a)(7). Specifically, laboratory analysis records for analyses performed on HPLC XX and YY do not indicate which analyst performed the injections.*
- *Failure to maintain complete records of any modification of an established method employed in testing [21 CFR §211.194(b)]. Specifically, the records of laboratory methods stored in the computer system do not include the identity of the person initiating method changes.*
- *Appropriate controls are not exercised over computers or related systems to assure that changes in analytical methods or other control records are instituted only by authorized personnel [21 CFR §211.68(b)]. Specifically, the observations were as follows:*
    - a. *Laboratory managers (QC and R&D) gained access to the computer system through a common password. Analysts were not required to use individual passwords; they operated the system following the log-in by the laboratory managers.*
    - b. *Because of the common password and lack of varying security levels, any analyst or manager has access to and can modify any HPLC analytical method or record. Furthermore, review of audit trails is not required.*

In essence, sharing user identities has caused all of these citations for the company.

Key Inspection Learning Points

Some of the key learning points from these inspections and FDA Warning Letters that we need to remember for the validation of any CDS are as follows:

- The CDS must be validated and the scope of work includes documenting any customization or configuration of the system.

- Include in the PQ testing, capacity tests for stress and overload conditions to comply with §211.63 ("adequate size"). The nature and extent of these capacity tests will vary depending on the architecture of the individual CDS system and also how an individual laboratory uses it.
- Effective preservation of electronic records is vital to passing any inspection: have a procedure, follow it and have documented evidence that it works. Use redundant hardware such as redundant array of inexpensive disks (RAID) and uninterruptible power supplies (UPS) as a first line of defense against electronic record loss.
- Change control is vital and the process must include the IT department and the network.
- Security must be enabled, documented and tested. Ensure that user identities are never shared so that any action in the system can be attributed to an individual.

A CDS has also been at the center of the Able Laboratories fraud case from 2005 where data was falsified by laboratory management and staff to pass products that otherwise would have been rejected (6,7).

EXPLOITING THE BENEFITS OF ELECTRONIC SIGNATURES
Mapping and Understanding the Current Process
The first task when considering implementing electronic signatures is to map the current process. This is relatively quick and the current laboratory high level process is shown in Figure 20.4. We can see that there are parallel electronic and paper activities when chromatographic analysis is undertaken. For example, when a chromatograph is set up, a paper record (Lab Book) needs to be updated and checked. When results are calculated the report and chromatograms printed out and the Lab Book updated and checked again.

It is important to analyze the current process.

- What are the process metrics? For example,
  - How many samples are analyzed, and
  - What are the turnaround times?

- Once this information has been obtained, analyze the turnaround times and find out the reasons for fast and slow turnaround.

Answers to these questions will give the information needed to start to improve the process and make it more effective and efficient.

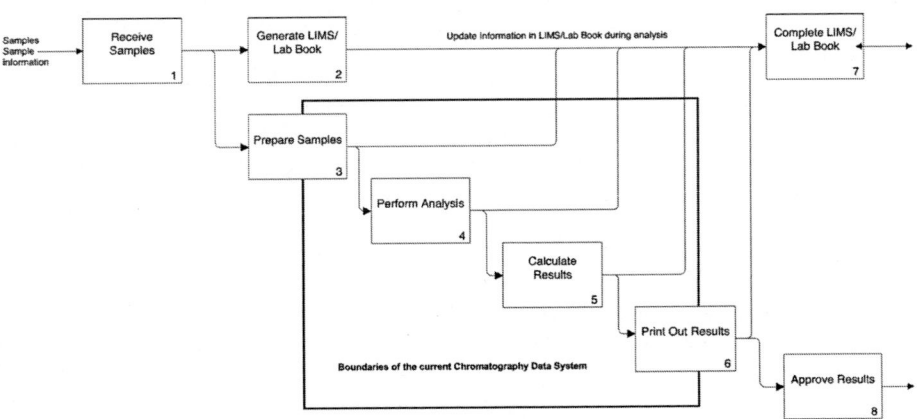

Figure 20.4 The current process highlighting the boundaries of the current version of the chromatography data system.

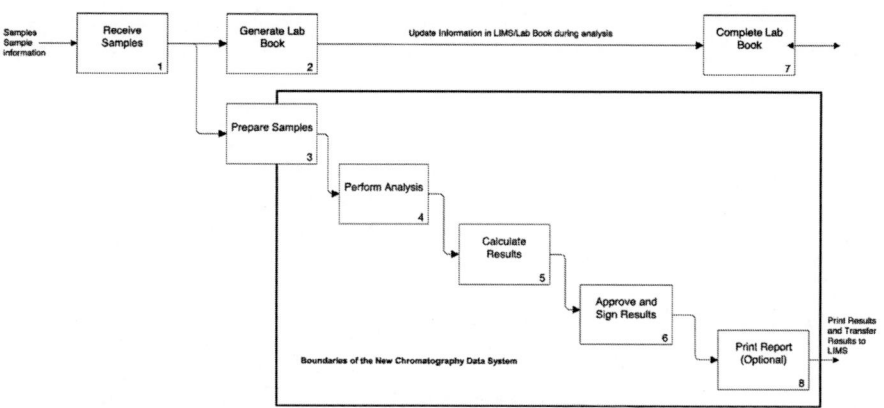

Figure 20.5 The redesigned process highlighting the extended boundaries of the new version of the chromatography data system.

The boundaries of the current version of the chromatography data system are also shown in Figure 20.4. In the current system the approval of results occurs outside of the chromatography data system on paper.

Optimizing the Work Flow to Use Electronic Signatures

Knowing the problems and improvement ideas from the analysis of the current ways of working, a new process can be designed to exploit the use of electronic signatures. It is important at this stage to ensure that the new process is compliant with 21 CFR 11 and any predicate rule requirements and that the new version of the CDS can support the new process as well. For example, where in the process will signatures be used and where will identifications of actions be sufficient?

In the example, the redesigned process is shown in Figure 20.5; the main differences are the following:

- Elimination of the need to update the Lab Book for chromatographic analysis. This is a quick win that is estimated to save about 0.3 to 2.6 full-time equivalents or person years (FTE). This is independent of implementing electronic signatures in the CDS.
- Expanding the scope of the CDS. In effect the approval of electronic records and calculated results takes place in the CDS and the printout is an option.
- Using the CDS to carry out all calculations rather than use a calculator or spreadsheet, this streamlines the whole process for calculating, reviewing and approving results.

The benefits of the process redesign when the CDS is linked to the LIMS would be in the region of 6 to 12 FTE. This is a surprising benefit but enables more capacity to be generated with the current laboratory resources. This is against a one off cost of about two FTE for the process redesign, linking the system to the LIMS and validation of the CDS and the data link to LIMS (8).

LIFE CYCLE APPROACH TO VALIDATION

Chromatography Data System Life Cycle

In GAMP®5 a series of life cycle models applicable to the various software categories are presented (9). However, as a CDS is a relatively simple configurable product and the custom calculations can be controlled effectively via a SOP, a simplified life cycle model is appropriate than the one presented in GAMP®5 for category 4 software and is presented in Figure 20.6. The left-hand side of the V represents the specification stages of the application, the bottom is the system build, and the right-hand side is the configuration, verification, and testing stages of the life cycle.

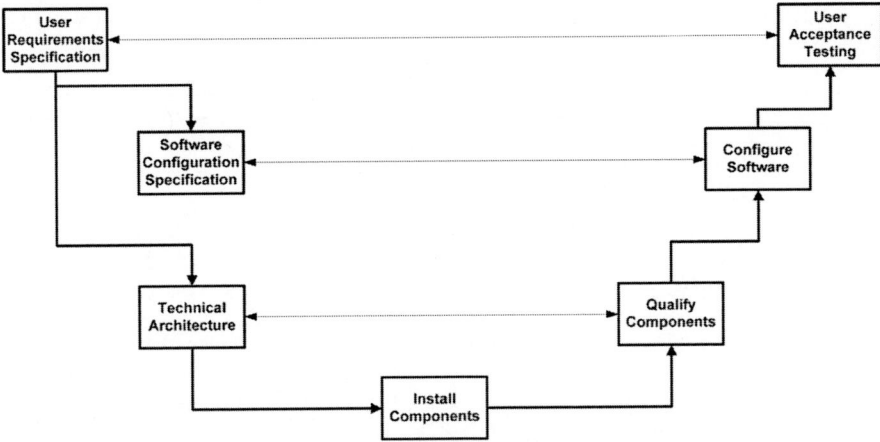

Figure 20.6 Simplified life cycle for a chromatography data system. *Source*: Adapted from Ref. 10.

The documents that could be produced for a CDS validation are presented in Table 20.1 and the key ones are discussed in more detail in the next section. Taken together all of these documents will provide the validation package to support the contention that the chromatography data system is fit for purpose. Note that this is a suggested minimum list and may be extended, depending on the size and extent of the system.

Table 20.1 Typical Documentation for a CDS Validation

Document name	Outline function in validation
Validation plan	• Documents the intent of the validation effort throughout the whole life cycle • Defines documentation for validation package • Defines roles and responsibilities of parties involved
Project plan	• Outlines all tasks in the project • Allocates responsibilities for tasks to individuals or functional units • Several versions as progress is updated
User requirements specification	• Defines the functions that the CDS will undertake • Defines the scope, boundary and interfaces of the system • Defines the scope of tests for system evaluation and qualification
System selection report	• Outlines the systems evaluated either on paper or in-house • Summarizes experience of evaluation testing • Outlines criteria for selecting chosen system
Risk analysis and traceability matrix	• Prioritizes system requirements: mandatory and desirable • Classifies requirements as either critical or noncritical • Traces testable requirements to specific PQ test scripts
Supplier audit report and supplier quality certificates	• Defines the quality of the software from suppliers perspective (certificates) • Confirms that quality procedures matches practice (audit report) • Confirms overall quality of the system before purchase
Purchase order	• From supplier quotation selects software and peripherals to be ordered • Delivery note used to confirm actual delivery against purchase order • Defines the initial configuration items of the CDS
Configuration specification	• Defines the configuration of the system policies • User types and access privileges • Default entries into the audit trail defined
Technical architecture	• IT platform(s) defined, e.g., terminal servers, database server together with resilience features • Operating systems and service packs • Operating environments: production, validation, etc.

(Continued)

Table 20.1   Typical Documentation for a CDS Validation (*Continued*)

Document name	Outline function in validation
Installation qualification	• Installation of the components of the system by the supplier after approval • Testing of individual components • Documentation of the work carried out
Operational qualification	• Testing of the installed system • Use of an approved suppliers protocol or test scripts • Documentation of the work carried out
User acceptance test (e.g., PQ) test plan	• Defines user testing on the system against the URS functions • Highlights features to test and those not to test • Outlines the assumptions, exclusions and limitations of approach
PQ test scripts	• Confirmation of software configuration • Test script written to cover key functions defined in test plan • Scripts used to collect evidence and observations as testing is carried out • Documents any changes to test procedure and if test passed or failed
Written procedures	• Procedures defined for users and system administrators including definition and validation of custom calculations, account management and definition of logical security • Procedures written for IT related functions • Practice must match the procedure
User training material	• Initial material used to train super users and all users available • Refresher or advanced training documented • Training records updated accordingly
Validation summary report	• Summarizes the whole life cycle of the CDS • Discusses any deviations from validation plan and quality issues found • Management authorization to use the system

Abbreviations: CDS, chromatography data system; PQ, performance qualification.

KEY VALIDATION DOCUMENTS FOR A CDS

The main validation documents will be presented in this section, typically in the order in which they are written and used in the system development life cycle, however there are differences that will depend on individual circumstances.

Specifying the Chromatography Data System Requirements

User Requirements Specification (URS): Defining the System Functions

The first document in the validation is usually the URS as this can influence the validation strategy outlined in the validation plan. From Figure 20.6 that the system requirements are related to the tests carried out in the user acceptance tests (PQ). Therefore, it is important to define the requirements for the basic functions of the CDS, the adequate size, 21 CFR 11 requirements and consistent intended performance in the URS. Remember that the URS provides a laboratory with the predefined specifications to validate the CDS; without this document the system cannot be validated (11).

It is important to realize that the URS is a living document and must be updated as the system changes and evolves; for example, an URS should be written to select a system, it will then be reviewed and updated to reflect the selected CDS and version that will be validated and the functions specific to the laboratory where it will be installed. The reason for this is that the selected system may have more or less features that contained in the original URS, therefore the URS must be updated to reflect the system to be validated.

The main elements in an URS should include the following major areas; each requirement must be individually numbered and written so that it can be tested as noted later in this section (1).

- Overall system requirements such as number of users, locations where the system will be used and the instruments connected to the system; whether terminal emulation be used

- Compliance requirements from the predicate rule and 21 CFR 11 such as open or closed system definition, security and access configuration of the software application including user types, requirements for data integrity, time and date stamp requirements, electronic signature requirements
- Data system functions defined using the work flow outlined in Figure 20.1; but ensure that capacity requirements are defined such as maximum number of samples to be run, custom calculations and reports for the initial implementation and roll-out, etc.
- IT support requirements such as backup and recovery, off line archive, and restore
- Interface requirements such as whether the CDS will be a stand-alone system or it will interface with a LIMS, and if so, how

System Specification Issues to Consider
Therefore, the first stage in the considerations for validating a CDS is to define all functions in a URS; for example, some or all of the following requirements will be included in the document (1):

- Data capture rates across all chromatographic techniques connected to the CDS. For example, conventional chromatography with a run time in the order of 20 minutes a data capture rate of 1 Hz is usually adequate. However, for capillary GC, 10 to 20 Hz may be appropriate and for CE a higher rate may be required depending on the overall migration time and the analyte peak shape.
- Several chromatographs may be linked into a data server or an A/D unit in the laboratory. Consider if crosstalk (the interference from one channel to another) could be an issue if the A/D chip is multiplexed across two or more channels and / or total sampling or buffering capacity of the data server.
- Has the maximum number of injections for an analytical run been defined? This is a critical component, if 100 vials are routinely injected in a run, the system cannot be tested with a run of only 10 samples as a user has not demonstrated adequate size. The specification must match the use of the system including replicate injections.
- Some data systems will be configured to collect data from diode array detectors (DAD). If this is required, especially to analyze product, then the data collection and analysis will need to be checked as part of the adequate size as some data files can be in the Mb range. The file delete option should not be enabled to protect the electronic records generated.
- Virtually, all client server CDS systems will have a buffering capacity within the data servers (if acquiring digital data from chromatographs via network interfaces). Therefore, so part of the adequate size requirements must be the ability to capture and buffer data if the network is unavailable, followed by the successful transfer of data to the server when the network connection is reestablished.
- How many users will there be on the system at the same time and will the system still perform its functions reliably? This number may be lower than the number of concurrent licensed users but it is still important to define this in the URS and test during the PQ. If the system becomes unreliable or unstable as the number of users increases then the system owner cannot state that the system has adequate size or can perform as intended.

These are some of the considerations for each installation of a CDS. Once installed in a laboratory environment the CDS becomes unique. The network location, server support, operating systems, software patches and laboratory configuration make each application different even if in just the smallest regard and testing needs to confirm the CDS works under each specific operating environment.

Documenting the System Requirements for Traceability
Although not mentioned in the regulations specifically, traceability of system requirements to the testing phase is important for any system including a CDS, therefore the way that system requirements are presented and managed is important.

Table 20.2 How System Requirements for CDS Capacity Can Be Documented

Requirement number	Data system feature specification	Priority M/D
3.3.01	The CDS has the capacity to support 10 concurrent users from an expected user base of 40 users.	M
3.3.02	The CDS has the capacity to support concurrently 10 data acquisition channels from an expected 25 total number of channels.	M
3.3.03	The CDS has the capacity to support concurrently 10 digital acquisition channels from an expected 25 total number of channels.	D
3.3.04	The CDS has the capacity to control concurrently 10 instruments from an expected 20 total number of connected instruments.	M
3.3.05	The CDS has the capacity to simultaneously support all concurrent users, acquisition and instrument connects while performing all operations such as reprocessing and reporting without loss of performance (maximum response time is <10 sec from sending the request).	M
3.3.06	The CDS has the capacity to hold 20 GB of data live on the system.	D

Abbreviation: CDS, chromatography data system.

It is all very well the regulations stating that a user must define their requirements in a URS, what does this mean in practice? Table 20.2 illustrates one way that capacity requirements can be documented; each requirement; note that each requirement meets the following criteria:

- Uniquely numbered.
- Written so that it can be tested, if required, in the PQ.
- Prioritized as either mandatory (M = essential for system functionality) or desirable (D = nice to have but the system could be used without it). This prioritization can be used in risk analysis of the functions and also for tracing the requirements through the rest of the life cycle as will be discussed in a later section.

Each requirement must be written so that it can be tested if required; according to IEEE standard 1233 (12) a well-defined requirement must address capability, condition, and constraint. Remember as shown in Figure 20.6 that the URS functions are related to the tests carried out in the qualification phase of the life cycle. If the requirements are not specified, how can they be tested? Further discussion on CDS user requirements can be found in (1).

Review of the URS

Ideally, an independent group of users (persons not involved in writing the document) should evaluate the URS and challenge each requirement and any interfacing requirements for the chromatographs or any other computer applications. If any missing requirements or inconsistencies can be found at this stage they are easy and inexpensive to correct. Therefore, the extra work in ensuring that the system requirement specification is correct are time and resources well spent; problems that can be rectified at this stage are far cheaper to solve than those identified later in the life cycle. When the system requirements specification is complete, the outline selection tests can be generated that can be used to select a potential system and reused later in the life cycle during the PQ testing.

Validation Plan

The name for this document varies so much from laboratory to laboratory: validation plan, master validation plan or validation master plan or even quality plan. Regardless of what it is called in an organization it should cover what steps will be taken to demonstrate the quality and compliance of the CDS in the laboratory. Ideally, it should be written as early in the process as possible to define the overall steps that are required and the documents to be produced from each. See chapter 5 for more details.

System Selection

The purchase of a new CDS system should be a formal selection process to see if an application matches the main requirements of the URS. The outline tests can be used to screen and select the system; an in-house test can be an option if there is sufficient time and resources to do this. A selection report would be the outcome of this phase of the work and would form part of the supporting evidence for the CDS validation.

Supplier Audit

The majority of the system development life cycle for a commercial CDS will be undertaken by a third party: the supplier; this is shown in Figure 20.6 as all of the operations under the horizontal line. The European Union GMP annex 11 on computerized systems states (13):

> *The software is a critical component of a computerized system. The user of such software should take all reasonable steps to ensure that it has been produced in accordance with a system of quality assurance.*

Many CDS suppliers are certified to ISO 9000 of some description and offer a certificate that the system conforms to their quality processes. This is fine but remember that there is no requirement for product quality in ISO 9000 and product warranties do not guarantee that the CDS is either fit for purpose or error free. If the system is critical to GxP operations it is better to consider a supplier audit (1). Supplier assessments including the role of audits are discussed in more detail in chapter 6.

Requirements Traceability and Risk Assessment

The next stage in the process is to carry out a risk assessment of each function determined on if the function is business and/or regulatory risk critical (C) or not (N). Table 20.2 for URS now has two additional columns added to the as shown in Table 20.3. This approach allows priority and risk to be assessed together.

Only those functions that are classified as both Mandatory and Critical are tested in the qualification phase of the validation (1,14). Therefore in Table 20.3 functions 3.3.03 and 3.3.06 are not considered for testing, as they do not meet the criteria. Of the remaining four requirements these all constitute capacity requirements that can be combined together and tested under a single capacity test script, which, in this example, is called test script 05 (TS05). In this way, requirements are prioritized and classified for risk and the most critical one can be traced to the PQ test script.

Table 20.3 Part of a Combined Risk and Analysis and Traceability Matrix for a CDS

Requirement number	Data system feature specification	Priority M/D	Risk N/C	Test
3.3.01	The CDS has the capacity to support 10 concurrent users from an expected user base of 40 users.	M	C	TS05
3.3.02	The CDS has the capacity to support concurrently 10 data acquisition channels from an expected 25 total number of channels.	M	C	TS05
3.3.03	The CDS has the capacity to support concurrently 10 digital acquisition channels from an expected 25 total number of channels.	D	N	–
3.3.04	The CDS has the capacity to control concurrently 10 instruments from an expected 20 total number of connected instruments.	M	C	TS05
3.3.05	The CDS has the capacity to simultaneously support all concurrent users, acquisition and instrument connects while performing all operations such as reprocessing and reporting without loss of performance (maximum response time is <10 sec from sending the request).	M	C	TS05
3.3.06	The CDS has the capacity to hold 20 GB of data live on the system.	D	N	–

Abbreviation: CDS, chromatography data system.

Technical Architecture

This document is typically written by the IT department taking into consideration the recommendations of the supplier in terms of server sizing (minimum processor power, memory, and disk sizing, etc.) and corporate standards. A technical architecture will document the servers and their operating systems that will constitute the system, for example,

- Database server,
- Application server,
- Terminal emulation for the application, for example, Citrix, and
- Operating system and service packs installed on each server.

A diagram is a very useful way of illustrating how the components come together to constitute the overall system and helps to collate the individual server specifications.

Configuration Specification

The way that the CDS application will be configured must be documented and one way is in a configuration specification. The content of this document will encompass

- Definition of user types and their access privileges,
- Configuration of system policies: functions turned on and any settings, for example, password length, use of electronic signatures, audit trail configuration,
- Definition of context sensitive default entries for the audit trail, and
- Locations of equipment and identification of which data server each chromatograph is connected to in the laboratory.

Installation Qualification and Operational Qualification

Installation Qualification

Establish an initial configuration baseline by taking an inventory of the whole system including hardware, software and documentation. For networked CDS systems the installation qualification (IQ) should cover the following:

- Server (for data storage) installation by the IT department, server supplier or manufacturer
- Installation of the A/D units or data servers to the corporate LAN
- Processing or data review workstations either the IT department or contractors working on their behalf (typically with an operating system configured to corporate requirements)
- Network connection of the workstations to the corporate LAN
- Installation of the CDS application software for data processing on the workstations
- Connection of the chromatographs to the A/D units or data collection servers

This work is typically supported by suppliers, system administrators from the laboratory, and IT department depending on the complexity of the configuration of the CDS. Planning is essential; retrospective documentation of any phase of this work is far more costly and time consuming.

Operational Qualification

The operational qualification (OQ) is carried out after the IQ and is intended to demonstrate that the application works the way the supplier says it will. Most suppliers will supply OQ scripts. These of necessity will only cover a subset of functions and will not be a substitute for the user acceptance tests or PQ tests.

What should be in an OQ? Here this depends on a supplier and the marketing approach to this "value-added" package. The purpose of an OQ is to show that the software and system works the way that the suppliers state it should. The amount of OQ testing can be relatively small, as the supplier has carried out the bulk of the work at their development site. The main focus of OQ should be to test application specific customization. Where there is a lot of laboratory customization of the application, for example, chromatographic spectral library for specific user compounds, then a supplier's OQ package is of less or little help here.

Assess IQ and OQ Documentation
The regulations require that before execution the test protocols have to be approved by the QC/QA unit and also that whatever is written in them needs to be scientifically sound [clause 160 (15)]. Here is an example of a Warning Letter sent by FDA to Spolana (16).

> Furthermore, calibration data and results provided by an outside contractor were not checked, reviewed and approved by a responsible Q.C. or Q.A. official.

Never accept documentation from a supplier without evaluating it and approving it. Check not only coverage of testing but also that test results are quantified (i.e., have supporting evidence) rather than solely relying on qualified (e.g., pass/fail) terms. Quantified results allow for subsequent review and independent evaluation of the test results. Further, ensure personnel involved with testing are trained appropriately by checking documented evidence of such training such as certificates is current at the time that the work was carried out.

Configuring the Application
Following the installation and qualifying the CDS application software, it should be configured according to the configuration specification.

- Defining user types and the access privileges for each one
- Setting up user accounts and allocating each one a user type
- Turning on or off the system policies and inputting the defined settings from the configuration specification
- Entering the default entries for the audit trail

The set up of the system should be documented against the configuration specification. This has two advantages: the first is the documentation for the validation document suite and the second is for disaster recovery purposes

User Acceptance Testing or Performance Qualification
The PQ stages of the overall qualification of the system can be considered as the user acceptance testing, undertaken by trained users and based on the way that the system is used in a particular laboratory. Therefore, a CDS cannot be considered validated simply because another laboratory has validated the same software: the operations of two laboratories may differ markedly even within the same organization. The functions to be tested in the PQ must be based on the requirements defined in the URS and with the numbering of individual requirements can be traced back to the system requirements via the traceability matrix (17,18).

The number of PQ test scripts needed for a CDS typically falls in the range of 15 to 30 to provide adequate coverage for the important functions documented in the URS depending on the complexity of the system.

Testing system functionality should consider the following (1,19):

- Security and access control
- Data acquisition from the different types of chromatograph interfaced to the system
- Crosstalk of A/D converters where used and dependent on the design of the A/D unit
- Calibration methods used within the laboratory: whether they are mathematically correct
- Analyte calculation
- System suitability test parameters
- Reporting data
- Sample continuity
- Unavailability of the network: buffering of the data server
- Remote processing over the network

- Data acquisition and data processing using a DAD and/or dual wavelength detector
- Creation and management of DAD spectral libraries
- Custom calculations implemented within the system
- Macros used to perform functions automatically
- System capacity tests, for example, analyzing the largest expected number of samples in a batch, number of users on the system
- Interfaces between the CDS and other software applications, for example, LIMS

Testing should also consider any electronic record/signature requirements (e.g., 21 CFR Part 11) and other regulatory requirements.

- Preservation of electronic records, for example, backup and recovery; archive and retrieve
- Data file integrity
- System security and access control including between departments or remote sites
- Audit trail functions
- Date and time stamps
- Electronic signatures
- Identifying altered and invalid records

Note that all aspects of the system that need to be tested must be defined in the URS (1).

TESTING CONSIDERATIONS
Analytical Run Capacity
First consider an analytical run and the capacity test considerations that will need to be evaluated versus your user requirements. The maximum number of vials to be injected in a single run should be defined in the URS. Testing should include standards, samples, quality control and blank reagents that are to be used as part of normal working procedures. A test should be designed to run the maximum samples including replicate injections.

Unavailability of the Network
There will be times when the network is unavailable and data will be buffered in the A/D unit or data server. It is important to ensure that this function works during the PQ. The worst-case example for the buffering will be defined in the URS and will be the number of injections with the longest run time. The run should be started, then the network is disconnected and the data accumulated in the A/D unit or data server until the end of the run when the network is reconnected and the buffered data are transferred to the server. There should be no loss of data integrity including time and date stamps in any of the buffered and transferred files if this test is to pass.

System Capacity
The capacity of the system needs to be tested in a way that reflects the way the system will be used and there are several approaches to take depending on the architecture of the CDS. For example, if there is a 30 user license then one of the simplest ways of assessing the capacity is to run all chromatographs simultaneously, however this will only test the data acquisition and transfer to the sever via the network. As the A/D units, buffer acquired data until transferred to the server this test will also implicitly evaluate the transfer with the network traffic at the time of the test. However, one of the main causes of performance degradation will be integration of data and this must also be included as part of any test of system capacity.

Logical Security and Access Control
While logical security appears at first glance to be a very mundane subject, the inclusion of this topic as a test is very important for regulatory reasons as it is explicitly stated in 21 CFR 11 and the predicate rules. Also when explored in more depth it provides a good example in the design of a test case.

The test design should consist of the following basic components as a minimum:

1. A test where the incorrect account fails to gain access to the system
2. A single test where the correct account and password gain access to the system
3. A test where the correct account but minor modifications of the password fail to gain access to the software
4. A test where the account is locked and there is an alert to the system administrator

The important considerations in this test design are as follows:

- Successful test cases are not just those that are designed to pass but also are designed to fail. Good test case design is a key success factor in the quality of validation efforts. Of the test cases above most are designed to fail to demonstrate the effectiveness of the logical security of the system.
- The test relies on good practices to ensure that users change or are forced to change their passwords on a regular basis and that these are of reasonable length (minimum six to eight characters).

Other test case designs are defined below.

- Boundary test: The entry of valid data within the known range of a field, for example, a pH value would only have acceptable values within 0 to 14.
- Stress test: Entering data outside of designed limits, for example, a pH value of 15.
- Predicted output: Knowing the function of the module to be tested, a known input should have a predicted output.
- Consistent operation: Important tests of major functions should have repetition built into them to demonstrate that the operation of the system is reproducible.
- Common problems: Both on the operational and support aspects of the computer system should be part of any validation plan, for example, backup works, incorrect data inputs can be corrected in a compliant way with corresponding audit trail entries. The predictability of the system under these tests must generate confidence in the CDS operations (trustworthiness and reliability of electronic records and electronic signatures) and the IT support.

The format of the document and more detail of PQ testing see the book by McDowall (1).

Personnel and Training Records

All personnel involved with the selection, installation, operation and use of a CDS should have training records to demonstrate that they are suitably qualified to carry out their functions and maintain them. It is especially important to have training records and curricula vitae of installers and operators of a system as this is a particularly weak area and a system can generate an observation for noncompliance. Major suppliers of CDS will usually provide certificates of training for their engineers but also the IT Department staff responsible for the network, backup, etc., require evidence of their education, experience and training.

The types of personnel involved that could be involved in a validation are as follows:

- Suppliers staff: They were responsible for the installation and initial testing of the data system software, left copies of their training certificates listing the products they were trained to work on. These were checked to confirm they were current and covered the relevant products and then included in the validation package.
- System managers: Training in the use of the system and administration tasks were provided by the supplier and documented in the validation package.
- Users: They were either analytical chemists or technicians whom had their initial training by the supplier staff to use the data system and this was documented in their training records.

- Consultants: Any consultants involved in aiding a validation effort must provide a curriculum vitae (resume) and a written summary of skills to include in the validation package for the system as required by the regulations (U.S. GMP §211.25).
- IT staff: Training records and job descriptions outlining the combination of education, training and skills that each member has.

Training records for CDS users are usually updated at the launch of a system but can lapse as a system becomes mature. To demonstrate operational control, training records need to be updated regularly especially after software changes to the system. Error fixes do not usually require additional training, however major enhancement or upgrade should trigger the consideration of additional training. The prudent laboratory would document the decision and the reasons not to offer additional training in this event.

To get the best out of the investment in a CDS, periodic retraining, refresher training or even advanced training courses could be very useful for large or complex ones. Again this additional training should be documented.

Service Level Agreement

In the case of outsourcing the support for the hardware platforms and network that run the chromatography data system software to the internal IT department, a service level agreement (SLA) has to be written. This SLA should cover procedures such as

- Backup and recovery,
- Archive and restore,
- Storage and long term archive of data, and
- Disaster recovery.

This SLA will cover the minimum service levels agreed together with performance metrics so that they can be monitored for effectiveness.

System Documentation

Documentation

The documentation supplied with the CDS application or system (both hardware and software), user notes and user standard operating procedures will not be discussed here as it is too specific and also depends on the management approach in an individual laboratory. However, the importance of this system specific documentation for validation should not be underestimated. Keeping this documentation current should be considered a vital part of ensuring the operational validation of any computerized system. The users should know where to find the current copies of documentation to enable them to do their job. The old versions of user SOP, system and user documentation should be archived.

Standard Operating Procedures

Standard Operating Procedures are required for the operation of both the CDS applications software and the system itself. SOP are the main medium for formalizing procedures by describing the exact procedures to be followed to achieve a defined outcome. Procedures have the advantage that the same task is undertaken consistently, it is done correctly, and nothing is omitted, and that new employees are trained faster (20). The aim is to ensure a quality operation.

The FDA Guidance for Industry on Computerized Systems in Clinical Investigations provides a minimum list of SOP expected for a computerized system (21). This list is reproduced below but has been edited for a CDS operating in a GLP or GMP laboratory.

- System setup/installation (including the description and specific use of software, hardware, and physical environment and the relationship)
- System operating manual
- Archive and restore (including the associated audit trails)
- System maintenance

- System security measures
- Change control
- Data backup, recovery, and contingency plans
- Alternative recording methods (in the case of system unavailability)
- Computer user training and roles and responsibilities of staff using the system

Note that this is a generalized list of SOP and more procedures may be required if the operating environment is more complex. Conversely, some of the procedures above could be condensed into a single SOP with more scope. The key issue is that all areas for the operation and maintenance of the system are controlled by procedure.

Validation Summary Report
The validation summary report brings together all of the documentation collected throughout the whole of the life cycle and presents a recommendation for management approval when the system is validated. The emphasis is on using a summary report as a rapid and efficient means of presenting results as the detail is contained in the other documentation in the validation package; see chapter 11 for more details.

MAINTAINING THE VALIDATION STATUS DURING OPERATIONAL LIFE
After operational release comes the most difficult part of computerized system validation: maintaining the validation status of the system throughout its whole operational life. Look at the challenges that will be faced when dealing with maintaining the validation of a CDS or indeed any system; some of the types of changes that will impact an operational CDS are as follows:

- Software bugs will be found, and associated fixes will be installed.
- Application software, operating system, plus any software tools or middleware used by the CDS will be upgraded.
- Network improvements: changes in hardware, cabling, routers and switches to cope with increased traffic and volume.
- Hardware changes: PCs and server upgraded or increases in memory, disk storage, etc.
- Interface to new applications, for example, spreadsheets or LIMS.
- Expansion or contraction of the system due to work or organization reasons.
- Environmental changes: moving or renovating laboratories.

All of these changes need to be controlled to maintain the validation status of the CDS. In addition there are other factors that impact the system as well from a validation perspective, such as

- Problem reporting and resolution,
- Software errors and maintenance,
- Backup and recovery of data,
- Archive and restore of data,
- Maintenance of hardware,
- Disaster recovery (business continuity planning), and
- Written procedures for all of the above.

In this section, the number of measures will be discussed that need to be in place to maintain the validation status of a chromatography data system.

Change Control and Configuration Management
Changes will occur throughout the lifetime of the system from a variety of sources such as

- Upgrades of the CDS software,
- Upgrades of network and operating system software,

- Changes to the hardware: additional memory, processor upgrade, disk increases, etc., and
- Extension of the system for new users.

This is the key item from the installation of the system to its retirement. Changes must be controlled. From a regulatory perspective there are specific references to the control of change in both the OECD consensus document (22) and EU GMP regulations (13).

Change control was implemented through a SOP that defined the procedure for change control. A change form was the means of requesting and assessing change.

- The change requested was described first by the submitter.
- The impact was assessed by the system managers and then approved or rejected by management.
- Changes that were approved were implemented, tested, and qualified before operational release.

The degree of revalidation work to be done was determined during the impact analysis. Changes that impacted the configuration (hardware, software and documentation) were recorded in a configuration log maintained within Excel.

Operational Logbooks

To document the basic operations of the computer system a number of logbooks are required. The term logbook is used flexibly in this context; the actual physical form that the information takes is not the issue, rather the information that is required to demonstrate that the procedure actually occurred. The physical form of the log can be a bound notebook, a pro forma sheet, a database or anything else that records the information needed, as long as security and integrity of the records (paper or electronic) are maintained.

Backup Log

The aim of a backup log is to provide a written record of data backup and location of duplicate copies of the system (operating system and application software programs) and the data held on the computer. The backup schedule for the disks can vary. In a larger system, the operating system and applications software will be separated from the data that are stored on separate disks. The data change on a fast timescale that reflects the progress of the samples through the laboratory and must be backed up more frequently. In contrast, the operating system and application programs change at a slower pace and are therefore more static; the backup schedule can therefore reflect this.

For smaller systems, such as personal computers, the data and programs may be located on the same disk and partitioned by the directory structure. If the backup software is capable of performing selective backups then the comments in the paragraph above apply. However, if there is little sophistication the whole disk may have to be backed up routinely. Again, for PC systems this may be an area to evaluate closely before buying. An alternative is a PC network, where the programs and data are held on a central server and can be backed up more efficiently and effectively than stand-alone systems.

Some of the key questions to ask when determining the backup requirements for the CDS are as follows:

- How long should the time between backups be? This can be answered by considering how much data the laboratory can afford to use. If it is up to a week (most unlikely), then the backups can be weekly but typically it is daily. If criticality determines that no data can be lost, then shadowing or duplicate disks are with RAID technology may be appropriate.
- Who is authorized to perform backups and who signs off the log? The laboratory manager in conjunction with the person responsible for the system should decide this. The authorization and any counter signature required should be defined in a SOP.

- When should duplicate copies be made for security of the data? This question is related to the security of data and programs. Duplicate copies should be part of the backup procedure at predetermined intervals. The duplicate copies should be stored in a separate location in case of a hazard to the computer and the original backups located nearby. Duplicate backups are also necessary to overcome problems reading the primary backup copies.

Problem Recording and Recovery

During the operation of a computer system, boot up, backup or other system functions, it will be inevitable that errors may occur. It is essential that these errors are recorded and the solution to resolve it also written down. Over time, this can provide a useful historical record to the operation of the computer system and the location of any problem areas in the basic operation.

Areas where this may be the case may be in peripherals where a print queue has stalled. This is relatively minor, however there may be cases where the application fails because of a previously undetected error. In the latter case, there is a need to for link the error resolution to the change control system.

Software Error Logging and Resolution

As it is impossible to completely test all of the pathways through CDS software or any software (1), it is inevitable that errors will occur during the operation of the system. These must be recorded and tracked until there is a resolution. The key elements of this process are to record the error, notify the support group (in-house or supplier), classify the problem and identify a way to resolve it.

Not all reported problems of a CDS will be resolved, they might be minor and have no fundamental effect on the operation of the system and may not even be fixed. Alternatively a work around may be required which should be documented, sometimes even retraining may be necessary. Other errors may be fatal or major, that mean the system cannot be used until fixed. In these cases, the revalidation policy will be triggered and the fix tested and validated before the CDS can be operational again.

Maintenance Records

All quality systems need to demonstrate that the equipment used is properly maintained and in some instances calibrated. Computers are no exception to this. Therefore records of the maintenance of the CDS need to be set up and updated in line with the work carried out on it. The main emphasis of the maintenance records is toward the physical components of a system: hardware, networking and peripherals; the software maintenance is covered under the error logging system described above.

If the hardware has a preventative maintenance contract, the service records after each call should be placed in a file to create a historical record. Also any additional problems that occur that requires maintenance will be recorded in the system log and there will need to be cross-references to the appropriate record there.

Many smaller computer systems have few if any preventative maintenance requirements but this does not absolve the laboratory from keeping records of the maintenance of the system. If a fault occurs that requires a service engineer to visit, then this must be recorded as well.

On sites where maintenance of personal computers is maintained centrally for reasons of cost or convenience, maintenance records may be held centrally. The remit of the central maintenance group may cover all areas of a site or organization including regulated or accredited as well as nonaccredited groups. It is important for the central maintenance group to maintain records sufficient to demonstrate to an inspector of the work they undertake. As defined in EU GMP annex 11 (13), the third party undertaking this work should have a service agreement and also have the curriculum vitae of its service personnel available and up to date.

Disaster Recovery

Good computing practices require that a documented AND tested disaster recovery plan must be available for all major computerized systems. It rarely is. Failure to have a disaster recovery plan places the data and information stored by major systems at risk, the ultimate losers being the workers in the laboratory and the organization.

Disaster recovery is usually forgotten, or not considered, as "it will never happen to me." The recovery plan should have several shades of disaster documented. From the loss of a disk drive: how will data be restored from tape or backup store and then updated with data not on backup, through to the complete loss of the computer room or building through fire or natural disaster.

Once the plans have been formulated, they should be tested and documented to see if they work. Failure to test the recovery plan will give a false sense of security and compound any disaster.

Revalidation Criteria

Any change to a CDS should trigger consideration if revalidation of the system is required. Note the use of the word "consider." There is usually a knee-jerk reaction that any change means that the whole system should be revalidated. One should take a more objective evaluation of the change and its impact before deciding if full revalidation is necessary.

Firstly, if revalidation is necessary, to what extent is it required to test (23): a software unit, module or the whole system? There may even be instances where no revalidation would be necessary after a change. However the decision must be documented together with the rationale for it.

Therefore a procedure is required to evaluate the impact of any change to a system and act accordingly. One way to evaluate a change is to review the impact that it would make to data accuracy, security and integrity (9). This will give an indication of the impact of the change on the system and the areas of the application affected. This allows the revalidation effort to target the change being made.

SYSTEM RETIREMENT

System retirement occurs at the end of the life cycle of any computerized system; however there are no directly stated regulatory requirements for formal system retirement. System retirement is typically considered when a CDS including file formats, operating systems and hardware platforms become obsolete. The key issue is that the process must be controlled and the records generated by the system maintained for the duration of the records retention period mandated in the respective predicate rule and any corporate policies. See the book and articles by McDowall for further information about this process (1,19,23).

REFERENCES

1. McDowall RD. Validation of Chromatography Data Systems: Meeting Business Needs and Regulatory Requirements. Smith RM, ed. RSC Chromatography Monographs Series. Cambridge: Royal Society of Chemistry, 2005.
2. Dyson N. Chromatographic Integration Methods. Smith RM, ed. RSC Chromatography Monographs Series. 2nd ed. Cambridge: Royal Society of Chemistry, 1998.
3. Burgess C, Jones DG, McDowall RD. All you wanted to know about A/D converters but were too afraid to ask. LC GC International 1997; 10:791–795.
4. Parental Drug Association. Validation of computer-related systems, technical report number 18. Journal of the PDA 1995; 49.
5. Concord Laboratories. FDA Warning Letter, July 2006.
6. Able Laboratories. 483 observations, July 2005.
7. McDowall RD. Qual Assur J 2006; 10:15–20.
8. Kornbo C, McDowall RD. Exploiting the benefits of electronic signatures with a chromatography data system. Scientific Computing and Instrumentation, January 2002.
9. International Society for Pharmaceutical Engineering. Good automated manufacturing practice guidelines version 5. Tampa, Florida, 2008.
10. McDowall RD. Validation of spectrometry software: critique of the GAMP good practice guide for validation of laboratory computerized systems. Spectroscopy 2006; 21(4):14–30.
11. Food and Drug Administration. Guidance for industry, General principles of software validation, 2002.
12. Institute of Electronic and Electrical Engineers. Guide for developing software requirements specifications, standard 1233–1998.

13. Commission of the European Communities. Good manufacturing practice for medicinal products in the European community, annex 11—computerised systems. Brussels, 2007.
14. McDowall RD. Effective and practical risk management options for computerised system validation. Qual Assur J 2005; 9:196–227.
15. 21 CFR 211. Current good manufacturing practice regulations for finished pharmaceuticals, revisions as of April 1, 2002.
16. Spolana. FDA Warning Letter, October 2000.
17. McDowall RD. Validation of spectrometry software: the proactive use of a traceability matrix in spectrometry software validation, part 1: principles. Spectroscopy 2008; 23(11):22–27.
18. McDowall RD. Validation of spectrometry software: the proactive use of a traceability matrix in spectrometry software validation, part 2: practice. Spectroscopy 2008; 23(12):78–86.
19. Donath J, McDowall RD. LC-GC Europe 2005; 453–464.
20. Hambloch H. In: Stokes T, Branning RC, Chapman KG, et al., eds. Good Computer Validation Practices: Common Sense Implementation. Buffalo Grove. Illinois: Interpharm Press, 1994:113–140.
21. Food and Drug Administration. Guidance for industry, Computerised systems in clinical investigations, 2007.
22. Organisation for Economic Co-operation and Development. Consensus document on principles of good laboratory practice applied to computerised systems. Paris, 1995.
23. Browne D, Thompson T, Mole D, et al. The prospective validation of a MS data system used for quantitative GLP studies. J Validation Technol 2002; 8:250–259.

21 | Case Study 3: Laboratory Information Management Systems

Christopher Evans
GlaxoSmithKline

Laboratory information management systems (LIMSs) are widely used in the laboratories of pharmaceuticals and related industries. The LIMS is typically based on client server technology supported by a relational database management system (RDBMS) as a storage repository. They can be used to manage and process large amounts of GxP[a] electronic analysis data, locally within a laboratory, or company wide between sites and will typically provide additional functionality used for increasing laboratory efficiency.

LIMS is now a mature technology and many packages are commercially available. Therefore, this chapter will concentrate on the selection, configuration, and implementation of a LIMS in a pharmaceutical manufacturing environment, rather than the development of in-house software for laboratory management.

LIMS provides an automated, efficient, and regulatory compliant means of managing electronic data produced in the laboratory, and its introduction into the laboratory is therefore encouraged by quality assurance (QA)/quality control (QC) laboratory managers. The implementation of a LIMS should also reduce the possibility of errors because of to personnel performing repetitive scheduled tasks. There are increasing demands from the regulatory authorities for data integrity, data security, and risk-based validation of LIMS in-line with the current regulations and guidelines (1–4), with LIMS as a focus of interest for pharmaceutical regulators.

The modern day commercially available LIMS packages may be configured to meet the needs of most laboratory analysis and data handling activities. This configuration is made up of software modules designed to provide standard features (e.g., sample log in, test management, work allocation, results entry, data processing and reporting). As this is the case there is now little need to develop bespoke/custom software to meet normal laboratory needs. Areas of bespoke/custom software development are often limited to programming the reporting package to extract and print information from the LIMS to meet specific business owner needs, and developing interfaces between the LIMS and other business systems.

The success of fully integrating the LIMS within the laboratory relies, in part, on robust, reliable interfaces between the core LIMS and a wide range of analytical instrument and system interfaces, for example, simple instruments such as those producing a single data point, complex instruments such as those producing spectra or result files, and complex computer systems such as chromatography data systems and manufacturing resource planning systems (Fig. 21.1). This will involve the interfacing of different computer platforms and analytical instruments, and gives inherent difficulties in providing a compliant risk-managed solution as required by the pharmaceutical industry regulations.

As LIMSs are basically configurable software packages, the quality of the design, coding, and documentation provided by the supplier is critical to obtaining an acceptably validatable LIMS. The LIMS suppliers have developed an appreciation for the unique requirements of the pharmaceutical industry resulting in an improvement in the system development life cycles they follow. As part of the decision of which LIMS supplier to use, a thoroughly documented evaluation of the supplier's practices should be performed.

---

[a]GxP—the combination of Good Manufacturing Practice (GMP), Good Laboratory Practice (GLP) and Good Clinical Practice (GCP).

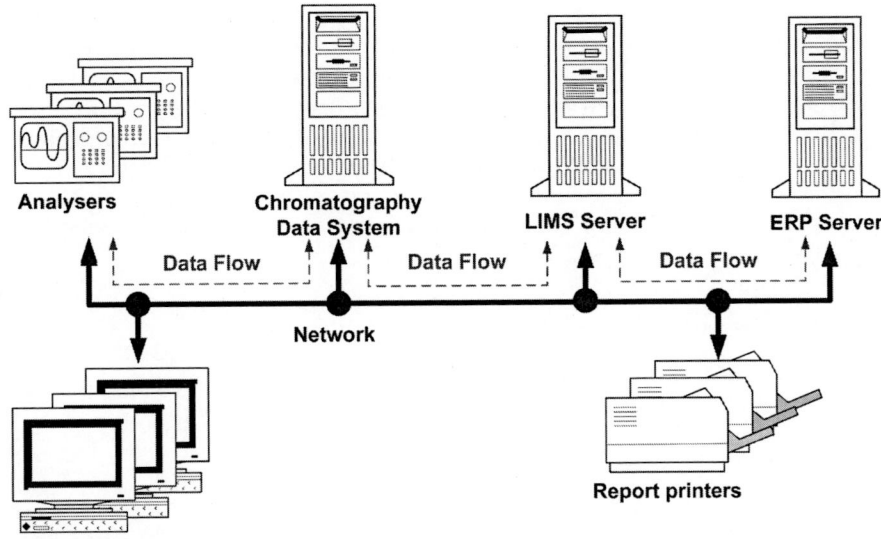

Figure 21.1 Basic LIMS structure. *Abbreviations*: LIMS, laboratory information management system; ERP, enterprise resource planning.

LIMSs are used to collect, store, and report data, which can be used to provide the final verification that a pharmaceutical product may be released to market or may provide data to be included in regulatory submissions. As this is the case an incorrectly implemented LIMS poses potentially serious risks to patient safety, and therefore the validation of LIMS is of critical importance to the compliance profile of the pharmaceutical company.

This chapter reviews the areas that are critical to the successful validation of a LIMS, which is based on risk to patient, and suggests an approach that represents good practice for validation of this type of application in a pharmaceutical manufacturing environment.

LIMS PROJECT PERSONNEL

Fundamental to the successful application of a risk-based approach to validation of a LIMS, and ongoing maintenance, is to have experts with full understanding of the laboratory business process that is to be automated by the LIMS, the proposed technology and clear understanding of their roles and responsibilities:

- *Business owner*: the person requiring the implementation of the validated LIMS. Business owners will have varying levels of expertise, experience, and resource availability to contribute to the implementation and validation; however, they have overall responsibility and accountability for ensuring that the LIMS is validated and maintained in a validated state.
- *Supplier*: the company that sells the LIMS application. As the developer of the LIMS, it is expected that the supplier has in-depth technical knowledge of the product and has experience in its implementation. The need to interface to other systems is typically a requirement for a LIMS. It is expected therefore that interfaces for widely used products (e.g., chromatography data systems, analytical equipment, MRP systems, etc.) will be readily available from the supplier.

    Where there is a need to develop additional functionality or interfaces to nonstandard systems, it is recommended that the business owner works closely with the supplier to ensure that any necessary recommendations are taken into account. The development of these functions or interfaces may be performed by the supplier (for inclusion as a standard feature for a future release) or by the integrator or business owner.
- *Integrator*: the company (or internal group) engaged by the business owner to configure and implement the LIMS, and also where required, to develop interfaces and

business owner specific code. Use of an integrator is optional, dependent on the services provided by the supplier and the level of expertise of the business owner.

- *Support team*: the group responsible for the ongoing maintenance of the LIMS following implementation. This group will provide a service to the business owner for managing daily operation, fault management, backup and restoration, and access management. It is important that the operation of this group is performed under the control of a formal agreement with the business owner (e.g., service level agreement, SLA), to ensure there is a clear understanding of the services to be supplied and the responsibilities of the support team.

LIMS FUNCTIONALITY AND TECHNOLOGY
Functionality

A typical LIMS would have the following functionality related to the management of the sample life cycle:

- Log-in of samples and receipt tracking
- Assignment of tests to samples
- Production of analyst worksheets and schedules
- Real-time data input
- Interfacing with analytical equipment and chromatography data systems
- Performance of configured calculations
- Monitoring of out of specification results using use programmed limits
- Review and reporting of analytical results
- Comparison of analytical results with specification
- Maintenance of sample status
- Basic statistical analysis of analytical information
- Audit trail of events linked to results

In addition, a typical LIMS would have the following functionality related management of the laboratory:

- Test method and specification management
- Management of analytical equipment calibration schedules
- Management of the stability program
- Export of data for statistical analysis
- Configurable reporting of laboratory performance metrics

A useful description of LIMS functionality can be found in ASTM Standard E1578 (5).

Most modern LIMSs are configurable applications built around sets of standard modules. In order for the LIMS application to be acceptable for use in the pharmaceutical industry, these modules should have been developed to industry quality standards, and have multiple implementations for which references are available.

To ensure that the software comprising the validated LIMS is controlled, it is good practice to avoid installing other software (e.g., word processors, drawing packages, e-mail) on the LIMS client PCs. There may, however, be the need for the use of some standard software packages, for example, spreadsheet packages, where the capability for data extraction from the LIMS to spreadsheets[b] is available. The client PC configuration including the LIMS client and the additional packages should be used during the validation testing to ensure that there are no conflicts between the applications. These packages may be updated from time to time (usually by the IT department); therefore this must be taken into account when assessing the needs for ongoing validation maintenance.

---

[b] Note! It is important to understand that the use of spreadsheets to extract data, where this data is to be used for GxP purposes, is only acceptable if the process for extraction and the use of the spreadsheet containing the data are validated and controlled/secure.

Technology

A complicating factor in the validation of a LIMS is the wide variety in the scale of LIMS implementations. A laboratory management system can vary from a PC-based application in a single laboratory to a client/server-based system running across multiple sites on a company wide area network with shared access to servers and multiple interfaces to other business systems.

The development of browser-based interfaces to replace the more traditional client software is a recent innovation and should provide the basis for a simplification of the validation activities relating to the configuration of the client PCs.

The increased use of wireless or web-hosted access to systems is also a factor that needs to be considered, in terms of security of data. The approach to validating these technologies will need to be integrated into the validation strategy.

Because of the critical nature of the data stored in the LIMS database, data integrity must be assured. Current technology allows for disk mirroring to ensure that any database additions and changes (and therefore changes to electronic records) are copied to additional, and possibly remote, locations. The function and use of the mirrored copy for backup, for disaster recovery and for system start-up after shutdown must be understood and tested as part of the validation.

DEFINING THE LABORATORY BUSINESS PROCESS/REQUIREMENTS

Definition and documentation of the laboratory business process that is to be implemented, and documenting this process in a business requirements document is the first step in the LIMS implementation process. These business requirements are not system specific and define, at high level, what will be automated and what will continue to be performed manually. The business requirements related to automation will be used to define the system requirements for the LIMS, and the financial justification for implementing a LIMS.

The definition of the laboratory business process and the business requirements related to automation should be the basis for a risk assessment. This assessment will be the vehicle for determining which functions of the LIMS have the potential for posing risk to patient safety, and the level of that risk. This will be the foundation for devising the risk management strategy that will input to the validation strategy for the LIMS.

The business case for automating a laboratory business process will normally be based on the ability of the LIMS to

- Increase the efficiency of the laboratory while reducing costs, and possibly resources
- Provide accurate and reliable data to support product release, development activities, or submissions to regulatory authorities
- Decrease the risk to patient safety and regulatory compliance risk through automation of a manual laboratory business processes.

The implementation of a LIMS should not be performed without fully assessing the impact of how it will be integrated into the laboratory business processes, and how it will impact existing laboratory staff ways of working.

It must be clearly understood, and covered in the business case, that the implementation and validation of the LIMS is only the initial component of the cost of the system through its life. Initial validation is just the beginning of an ongoing commitment to the maintenance of the LIMS in a validated state. This includes the commitment to the ongoing cost of maintenance of the system, infrastructure, and support services.

The Regulatory Influence

Today regulatory authorities accept and indeed encourage the principle of using computer systems for laboratory data management and for controlling release of product to market, and it has now become the norm for laboratories to use LIMS. There have been no recent changes in the overall expectations for the validation of this type of IT system other than a general expectation that the approach should be based on risk.

Table 21.1 Regulations Applicable to LIMS

U.S. Code of Federal Regulations Title 21, Part 211 Clause 68, and Title 21 Part 11	U.S. Code of Federal Regulations Title 21, Part 58, GLP for nonclinical laboratory studies	European Union Directive, 91/356/EEC Annex 11, April 2008 Draft
Personnel	Personnel	Personnel
Building/facility	Facility management	
Equipment design	Equipment design	System software
Equipment maintenance	Equipment maintenance	System
Standard operating procedures	Standard operating procedures	
Record retention	Record retention	Data
Calibration		
Change control		Change control and configuration management
Testing	Testing	User testing and system's fitness for purpose
Backup files		Backup, migration archiving and retrieval
Data storage		Data storage
Hard copy records		
Ongoing evaluation		
Security		Security
Electronic records		Electronic records
Electronic signatures		Signatures

Abbreviations: LIMS, laboratory information management system; GLP, Good Laboratory Practice.

Regulatory inspections of laboratories often audit the use of LIMS. Typical areas of regulatory interest include the following:

- Change control, including software, configuration, and documentation, and the cumulative effect of change
- Validation strategy:
  - Documentation of specification and design
  - Documentation and extent of qualification and testing
  - Traceability between specifications and tests

- Management of data, assurance of data integrity, and record retention
- Management of access to the LIMS (i.e., user profiles, providing and removing access)
- Failure to validate user created GxP reports
- Ineffective standard operating procedures (SOPs), including security practices and backup and restore
- Use of unvalidated spreadsheets, for data collection or extraction from LIMS

Application of the current regulations and how the authorities interpret them is not always consistent. It is therefore important to develop and document a well-thought-out strategy for applying a risk-based approach to validation to ensure that all aspects of the various regulations are addressed and the validation status of the resulting installation is easy to maintain. There is general guidance within the regulations covering computer systems but in the main it is not specifically aimed at LIMS. It must nevertheless be clearly understood that however general the regulations and guidelines, regulators will have their own specific interpretation of them. Table 21.1 indicates the coverage of the applicable parts of the regulations.

THE GOOD PRACTICE AS THE BASIS FOR LIMS VALIDATION

Good Practice is the basis for implementing computer systems in laboratory applications and is now well established and based on many years of experience. The LIMS is simply another use of the power and flexibility of client/server or browser accessed computerized systems, and therefore LIMS should be validated in a similar manner to other IT-based systems utilized within the pharmaceutical industry. Specifically a LIMS is deemed to be a form of automated system, and it is therefore recognized that the development and implementation of a LIMS should be conducted in-line with a defined life cycle with the focus of validation effort on the basis of the outcome of risk assessment, such as the quality methodologies based on those detailed in Good Automated Manufacturing Practice (GAMP) (1).

As prework for the validation project, time, and resources (and therefore money) must be expended to ensure that the potential risks posed by the LIMS to patient are understood, and are assessed from a regulatory and business perspective. This activity should be executed as early as possible, most definitely prior to purchasing the LIMS. Many projects have had problems because of lack of appropriate funding, and consequently validation difficulties, because of the project personnel not understanding sufficiently early in the project life cycle, what needs to be done, and the discovery at a late stage that the system does not meet the all the needs of the business.

It is the responsibility of the pharmaceutical manufacturer to ensure that they are using a "quality" supplier for the LIMS package and a suitably qualified system integrator for implementing the package. This is easy to state but difficult to do; there are many suppliers and integrators in the market who will advise that their solutions and practices are the best and have passed many regulatory inspections. The pharmaceutical manufacturer must therefore take appropriate steps such as performing supplier evaluations, which will normally require an audit of the LIMS package supplier (and the integrator if separate from the supplier), before any decision regarding which LIMS to implement is made.

Where the pharmaceutical manufacturer has a requirement not met by the supplier's application, there may be a requirement for development of customized software. If the LIMS supplier is to be used as the developer the development life cycle and ongoing support arrangements should be agreed. The LIMS supplier may wish to include the customization in a future release of the LIMS to increase the functionality of the package. Care should be taken to ensure an approach is taken to testing of the customized code commensurate with the risk it may pose to patient safety before implementation as it may not have been used previously in a production environment.

VALIDATION LIFE CYCLE APPROACH

The validation life cycle is typically split into a number of phases that are all executed following a life cycle approach documented in a validation plan (Fig. 21.2).

It must be clearly demonstrated that the quality assurance (QA) function of the pharmaceutical manufacturer is endorsing the validation strategy for the implemented LIMS as documented in the validation plan. In its simplest form this will require the QA representative to authorize the validation plan and the subsequent key documents produced to support validation (e.g., requirements specification, qualification documentation, reports, etc.). Table 21.2 shows the linkage between project activities and the qualification activities, which will be under the control of the validation plan.

Laboratory Business Process and Business Requirements Definition

The main purpose of this phase is to ensure that the business process is fully understood, and the business requirements for automating the business process using the LIMS are defined and documented. This will be the means for providing a justification to senior management that the LIMS is a cost effective and compliant solution. The business requirements will be the basis for applying for funding for purchase and implementation of the LIMS.

The business owner of the LIMS must understand the benefits and the drawbacks of implementing a LIMS prior to purchase. The business owner will also need to understand the longer-term responsibilities for maintaining the LIMS is a validated state.

GxP Risk Assessment

Part of the early decision-making process is to assess which parts of the LIMS are to be subject to full validation and which are to be covered by good IT practice. To define this, a risk assessment should be performed. This risk assessment will then need to be revised during the

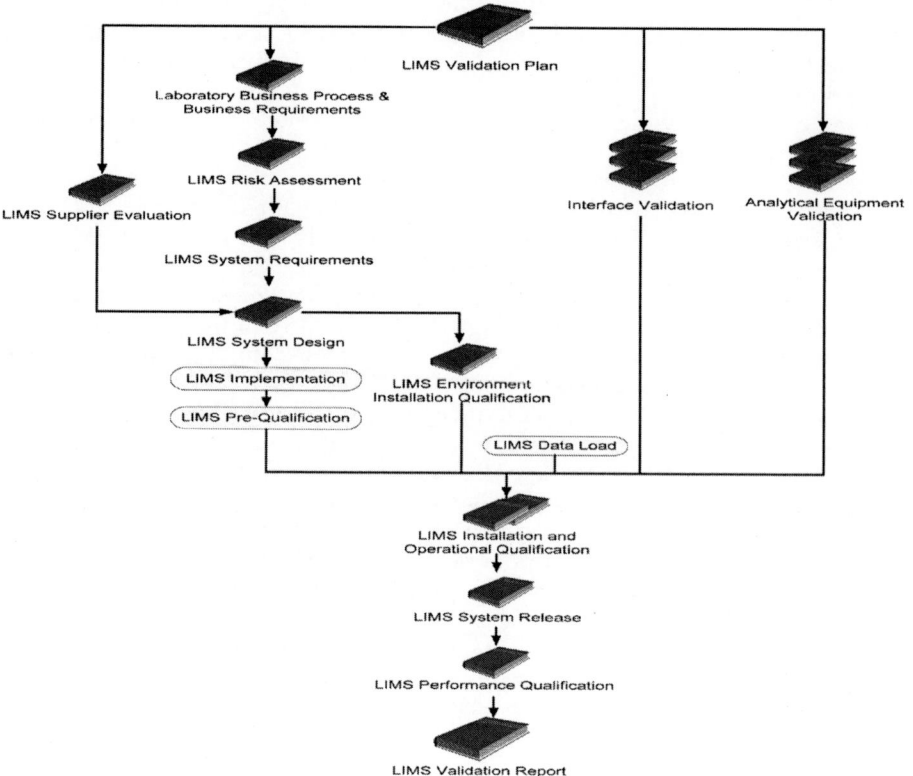

Figure 21.2 Validation life style. *Abbreviation*: LIMS, laboratory information management system.

project as new information becomes available. As changes are made to the risk assessment this may impact the validation activities to be or already performed.

The use of the LIMS within the company computing architecture may mean that it is utilized for both GxP activities that could pose high medium or low risk to patient safety and record integrity and non-GxP related activities. The risk assessment for the LIMS ensures that the validation effort is focused toward those system GxP elements that pose the most risk to patient safety (i.e., the high and medium risks). The justification for this risk-based approach must be based on good science and business process/product understanding and will focus on the GxP criticality of the data in the LIMS database, the use of the data, and the sources of the data. The impact of regulated record integrity on patient safety should also be considered

Table 21.2 Validation Phases and Project Activities

Validation phases	Project activities
Laboratory business process and business requirements definition	• Business process definition • Business requirements and business case development • System requirements development (initial draft) • Project planning (and supplier quality planning)
GxP risk assessment	• Perform GxP risk assessment
Validation planning, system requirement definition and supplier/integrator evaluation	• Validation plan production • Definition of risk-based validation strategy • System requirement specification development (version 1) • Supplier assessment/audit and system/integrator selection

Table 21.2 Validation Phases and Project Activities (*Continued*)

Validation phases	Project activities
Design and configure LIMS package	• Functional design specification production • Detailed design specification production for hardware and LIMS package (including interfaces) • Design review • Install and document the development and Validation/test environment • Configure LIMS package • Configuration review
Design and code bespoke/custom software	• Program specifications • Develop bespoke/custom software • Source code review • Unit testing • Design review
Prequalification activities	• Installation of hardware for the production system (computer system hardware; servers and client PCs) • Installation of software (operating system, Standard software modules and bespoke/custom software) • Implementation of infrastructure (e.g., networks and communication links) • Install and document the production environment • Implementation of interfaces • Develop test data sets
Installation qualification	• Verify correct installation of hardware and record as initial configuration • Verify installation of software and record as initial configuration • Verify correct installation of network infrastructure and record as initial configuration • Verify correct installation of interfaces • Verify import of test data • Verify availability of supporting Supplier Documentation • Verify equipment environment is suitable in terms of temperature humidity and electrical/magnetic interference

Operational qualification and system release	• Load and verify static data
	• Functional testing of the business process including
	  • Sample life cycle management
	  • Reporting
	  • Access control
	  • Audit trail
	  • Interfaces
	  • Security
	  • Backup and restore
	  • Develop and verify business continuity and disaster recovery plan
	• Implement support system (e.g., help desk, incident management)
	• System release releasing LIMS for use in the production environment

Performance qualification	• Ongoing verification of functionality • Performance monitoring by support organization
Validation reporting—authorization for use	• Validation reporting

Validation maintenance—operational compliance	• Ongoing support and integrity checks
	  • Disaster recovery and business continuity
	  • Backup and restoration
	  • Access control
	  • Change control
	  • Incident management
	  • Configuration management
	  • Document management
	• Performance monitoring by support organization

Abbreviation: LIMS, laboratory information management system.

when assessing risk. The risk assessment will have the benefit of reducing the amount of validation effort required where system functionality poses low or no risk to patient safety; functionality posing low risk to patient safety may simply be implemented in accordance with good IT practice.

Typical GxP critical data elements include:

- Batch/lot number
- Item number
- Shelf life
- Tester identification

- Material specification
- Sample number
- Sample date
- Sample time

- Sample status
- Test methodology
- Retest days
- Test results

This approach must be methodical to ensure that relevant data and the operations using that data in the LIMS is not overlooked.

It must be accepted, however, that risk assessments can be subjective and therefore rely heavily on the experience and knowledge of the assessors. The team performing the risk assessment should, therefore, consist of the business owner, or personnel with in-depth knowledge of the business process, a person with knowledge of the regulatory requirements for LIMS (typically someone from quality assurance), the project manager, (as he/she will be responsible for implementing the agreed risk management strategy, which is the output of the risk assessment), and technical personnel with knowledge of the proposed technology.

Approach to Risk Assessment

There are a many of approaches that may be used in assessing risk, but all have the same general principles (Fig. 21.3), which are

- To assess consequence to patient safety of a failure of the system,
- To assess likelihood of a failure occurring and possibility of its detection in a timely manner,
- And using this information to define level of potential risk posed by system functionality if a failure were to occur.

This overall list of risks and their rating makes up the risk profile for the LIMS.

The LIMS, in its role of providing data for batch sentencing and therefore release of product to market, obviously has a potential for high risk to patient safety, if the functionality associated with this part of the laboratory business process does not work correctly. There are, however, also functions in the LIMS such as reporting of laboratory performance metrics and work scheduling, which are simply management tools and have no impact on patient safety. As there is a finite amount of time and resource available for any validation project, it is very important that the major validation effort is focused on the areas of high or medium risks to patient safety rather than areas that pose low or no risk.

This approach fits in well with the current regulatory expectations of applying risk-based approach to the validation of systems.

The output from the risk assessment will be a report that defined the risk profile on which the risk mitigation strategy will be based. The strategy for mitigating these risks may be documented in the risk assessment or directly in the validation plan. This strategy should

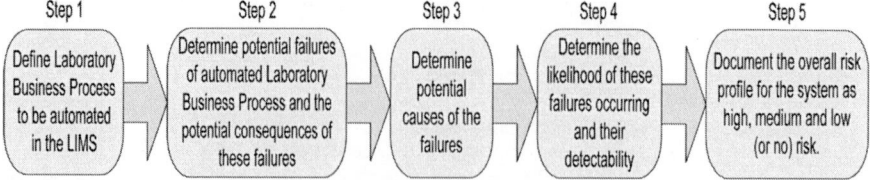

Figure 21.3   Creating a risk profile. *Abbreviation*: LIMS, laboratory information management system.

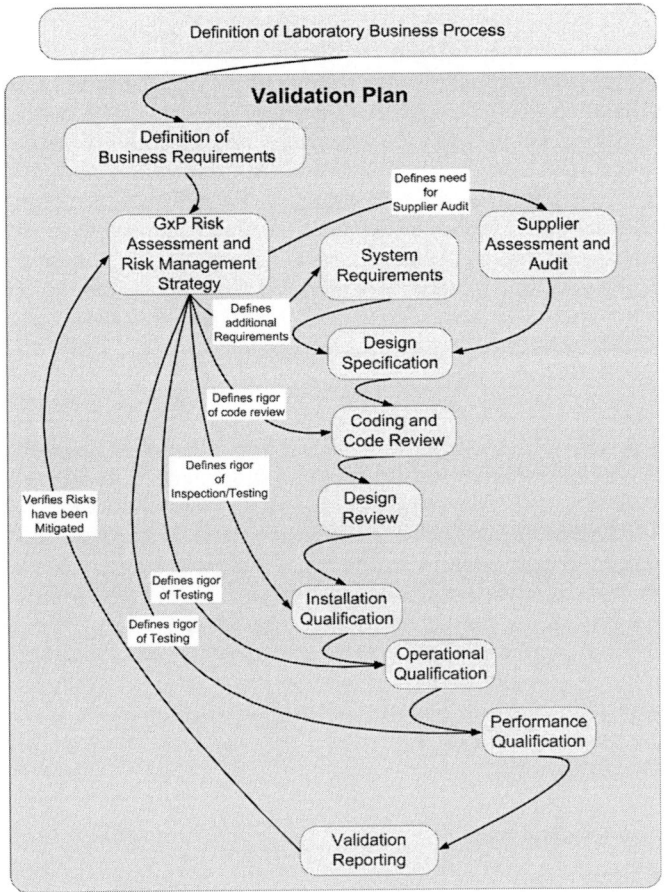

Figure 21.4 Impact of risk assessment on validation activities.

identify where additional validation effort is required and where a less rigorous approach may be applied and will impact a number of the validation activities (Fig. 21.4).

The outcome of the risk assessment may indicate that to mitigate a risk there is a need to provide additional functionality (e.g., forced double checks of data entry, increased redundancy, etc.). These additional requirements can be built into the business/system requirements at this early stage, hence gaining the benefit of gaining a clearer understanding of the overall cost of the proposed system and reducing the potential for additional cost and rework at the later stages of the project.

Validation Planning

The validation plan documents the scope of the risk-based validation for the LIMS. This plan must clearly define the boundaries of the validation activities, the required deliverables/ activities, and the responsibilities of each contributor, and will identify what is within the boundary of the LIMS in terms of hardware, software, and interfaces.

The validation plan will define the following:

Responsibilities of Personnel/Organizations

The success of any project of this type will rely on personnel having the right experience in terms of LIMS and GxP expectations. These personnel will have the responsibility for identification and mitigation of risks via the risk assessment, authorization of validation documentation and executing the validation activities within the scope of this plan. The responsibilities of the developer, supplier, and integrator must also be defined and documented in the plan.

Scope of Validation

The boundaries of the validation project must be defined to ensure that there is full coverage; consideration should also be given to clarifying responsibilities for interfaces between the LIMS and other systems. For example, will the analytical equipment or chromatography data system interfaces be validated as part of the project? Will supplier evaluations be required? It is very important at this stage to determine what is within the scope of the LIMS validation plan and what will be validated under other associated validation plans. The validation of the implementation of the LIMS within the laboratory should be managed as a complete entity to ensure that all parts of the LIMS are developed and validated to the appropriate standards. This may be achieved by the use of a validation master plan (VMP) for all the laboratory processes and information management. The validation plan for the LIMS and any associated plans for other interfaced systems would be referenced in, and be under, the control of the VMP.

Standards/procedures

To ensure consistency, standard and formal processes should be used in the project and the procedures/standards to be followed will be documented in the validation plan. It is strongly recommended that the company have defined procedures governing the validation and ongoing maintenance of the LIMS. This will be the basis for providing assurance to regulators that the company has a clear and understood approach to validation. Company standards and procedures will typically cover production of validation documentation, project governance (e.g., documentation management and issue/risk management), and validation maintenance (e.g., change control, configuration management, incident management, and business continuity planning). Where the supplier is to provide documentation supporting validation, this should be created either under the pharmaceutical manufacturers' procedures or, if the supplier's procedures are of an equivalent standard (as verified as part of supplier assessment), under the suppliers' procedures.

Risk-Based Validation Strategy

The strategy to be followed for validation of the LIMS must be clearly defined in the validation plan and must be commensurate with the risks posed by the system to patient safety. This approach should be driven by the risk profile of the LIMS and will focus more of the validation effort of the areas that are defined as the highest risk, and less on the area of lowest risk.

This plan should provide the project team and the LIMS supplier/integrator with sufficient detail to allow the appropriate documentary evidence to be produced to support the finally handed over compliant system. It is important that the validation plan (or at least the relevant information) be issued to the LIMS supplier/integrator to ensure that no surprises regarding the validation expectations at a later stage in the project. It is acceptable for the supplier and integrator to use their own development/implementation methodology, providing that the business owner has

- Understood the interface points and gaps between the methodologies;
- Satisfied him/herself that the development/implementation methodology followed by the supplier and integrator meet the company quality standards.

Where a LIMS implementation project is split over a number of different locations (e.g., different laboratories or different sites), but is designed to provide an integrated solution it may be appropriate to produce individual validation plans and associated documentation, for each of the LIMS location. Where an installation is implemented and supported on a company corporate basis it may be more appropriate for the VMP and the resulting validation documentation for the core LIMS to be managed from a central location. In such cases validation plans and associated validation documentation supporting the local implementation should be maintained on site.

The laboratory VMP mentioned in the preceding text should act as an umbrella document for the whole project, identifying the scope of each individual validation plan. Using this method of breaking down the documentation will allow a much more focused approach to be taken to the production of documentation, and also simplifies the problems normally found in obtaining validation document authorization across sites and geographical regions. A further advantage of using individual validation plans is that the individual LIMS instance may be put into operation, as soon as the testing and installation is complete without having to wait for the final validation report covering the whole project. Also when periodic reviews are performed on the LIMS again this may be performed on the local instance of LIMS without needing to review all other instances.

As the validation plan will be produced at such an early stage, it will inevitably only encompass the initial understanding of availability of personnel and project scope, so it will need to be further developed during the project. The project team and LIMS supplier/integrator must appreciate that this is a live document that will be developed during the project and will continue to be live until the LIMS validation report has been approved.

Supplier and Application Selection

The LIMS application should be chosen from an analysis of the application capabilities against the business/system requirements and conformance to the GxP regulations for the market to which the pharmaceutical products will be sent. This initial choice of application will lead to a preferred supplier for the project. For the supplier's system to be appropriate for use in a regulated laboratory, the supplier must have followed good software development practices during the development and in-house testing of the LIMS application, and have experience of supplying to the pharmaceutical industry. Failure to use good software development practices can lead to a requirement for the purchaser to perform significantly more testing to achieve a validated LIMS implementation. Such testing will require knowledge of the underlying structure of the LIMS database, and this information is typically not released by the LIMS application supplier.

Assessment of the supplier's software development practices can usually be achieved by performing a supplier audit. Experience shows that to address deficiencies, any actions from audit observations should be agreed with the supplier and included in the purchase and support contracts for the system.

System Requirements Definition

The purpose of the system requirements specification is to collate the detailed business owner requirements for how the LIMS application is to operate in both the laboratory and the company environment; this will be an expansion of the business requirements. There may of course be areas in the system requirements that are wishes rather than essentials and also some requirements that are GxP critical and/or identified as part of the risk assessment.

These System Requirements will form the basis for the configuration and testing of the standard application, and any additional development of functionality that is required.

Following approval, the first version of the system requirements may be used for obtaining quotations and proposals from the proposed LIMS supplier/integrators, and as a result this version will act as the foundation for the project. Time spent in ensuring that the system requirements are complete, structured, and clearly understandable will deliver large savings of time and effort (and therefore costs) at later stages of the project and will also reduce the chance of project failure.

It is unfortunately human nature to wish to spend as short a time as possible on this phase of the project as it does not seem to produce anything but documentation, when the business owner wants to see something more tangible. This is an oft repeated mistake that is likely to come back to haunt the project in the later stages and result in project overspend and dissatisfied users.

The feasibility of any specific LIMS product meeting the requirements of a laboratory (both technical and maintainability) must be fully assessed prior to purchase. This assessment process must take into account the supplier's proposed technical solution ongoing maintenance support (including future development) as well as any company strategy for laboratory management. The decision as to which LIMS supplier to use is of major importance and should be based on a detailed understanding of both the supplier/integrator and the proposed LIMS product. It is therefore essential at this early stage in the project to build a relationship with the

LIMS supplier/integrator to use the knowledge of their technical personnel in the development of the system requirements.

The approach of bringing the supplier on board at this early stage will assist in the structuring of the system requirements; the supplier will gain all he needs to know to provide a meaningful proposal for a technical solution and how his design documentation is to be structured. It would also be possible at this stage for the business owner and supplier to be able to agree who will be supplying the documentation that will support the validation, and which methodology is to be followed. It is not common for this sort of close alliance to be established in these types of project, but the potential for a successful and efficient project delivery makes putting some effort in this direction worthwhile.

The structuring of the system requirements to identify uniquely the numerous requirements will allow the project to demonstrate traceability of each of the GxP requirements through the design and testing process, which is a regulatory expectation.

The agreed system requirements should clearly identify the following:

- System size and capacity (e.g., numbers of items, client PCs, user interfaces, peripherals, data storage capacity, etc.)
- Integration of the LIMS with other systems and interfaces (e.g., MRP II, chromatography, use of LAN/WAN)
- The number of and types of pieces of analytical equipment to be interfaced to the LIMS. There will also be a requirement to define the means of communication with these pieces of equipment; this communication link may be typically via RS232 or network interface
- System performance and availability targets
- Documentation requirements (e.g., user manuals, technical manuals, software listings, database structure diagrams, etc.)
- Requirements for data presentation, records, and reports
- Requirements for data manipulation, for example, calculations
- Requirements for the use of spreadsheets for the collation or manipulation of data
- Details of external reporting and statistical analysis packages
- Details of supplier change control and patch documentation

The system requirements are the responsibility of the pharmaceutical manufacturer; however, it is highly unlikely that the company will have sufficient LIMS experts that can execute the project. It is strongly recommended that the future LIMS users provide their input to ensure that there is buy-in for the design and functionality of the solution. This applies in particular to the client PC user interfaces (e.g., display graphics, user entry screens) and reporting mechanisms.

Where expertise on design is required in the form of LIMS supplier/integrators, or consultants, these personnel should be included within the project team, but they must work closely with the business owner. Using the right people at this time will enable the requirements of the future LIMS users to be converted into a functional design document that will form the basis for a successful configuration of the application. When the system requirements have been agreed, which may take several iterations, it will act as the source document for the design and design review process. Any change to the system requirements will have an impact throughout the rest of the project, and the later any changes are introduced, for example, after the design/configuration/testing has begun, the more significant the effect will be on the project cost and time scales.

Supplier/Integrator Evaluation

Because of the expanding nature of the laboratory information management market there are a growing number of suppliers with offerings in this area. Main stream LIMS suppliers are backed up by supplementary application suppliers offering connectivity software, work flow solutions, archiving solutions, reporting packages, implementation expertise, etc. The LIMS market is also sufficiently large and diverse to support segmentation into specialist applications, for example., manufacturing versus research, highly configurable versus limited configuration packages, multi lab versus single lab solutions.

It is essential to conduct a supplier evaluation of the proposed LIMS supplier and integrator as part of the supplier selection process. It is important that this evaluation be conducted prior to placing an order for the LIMS, as it may be the case that there are some major concerns regarding the ability of the supplier/integrator to deliver a LIMS that meets the business owner's quality standards.

Where the supplier evaluation or risk assessment has indicated that there is a need for additional verification of the capability of a supplier or integrator this should be via a structured supplier audit that will focus on the areas of potential weakness.

Supplier/Integrator Audit

The purpose of a supplier/integrator audit(s) is to allow the pharmaceutical manufacturer to review documented evidence of the application of the supplier quality management system (QMS) throughout the development of the LIMS package. The supplier audit will also confirm that the supplier is capable of delivering the correct standard of software engineering and development documentation for the LIMS.

It has been shown many times that ISO9001: 2000 accreditation does not necessarily mean the supplier is capable of developing software and hardware products to meet business owner quality requirements. This can be because the LIMS supplier/integrator is accredited for the implementation and integration of LIMS, rather than the development of software. In these cases, the only way of gaining confidence in the approach taken in the software/ hardware development, core product testing, and change control processes is to visit the LIMS supplier/integrator's premises. Where a supplier is accredited to ISO9001: 2000 TickIT this will give the pharmaceutical manufacturer greater confidence that a defined life cycle will be followed as it is specifically aligned with the production and ongoing development of software.

When an audit is conducted the supplier audit report will be the evidence created by the pharmaceutical manufacturer, documenting any issues that were identified during the audit and providing recommendations for any corrective actions. It is expected that agreed corrective actions will be implemented as part of the approval of the LIMS supplier/integrator for the project, and will form part of the commercial contract between the two organizations.

The audit is a critical step in the validation of the LIMS supplier, as it allows the pharmaceutical manufacturer to evaluate key testing and validation deliverables from the supplier.

It must however be realized that despite performing an audit and identifying that there are shortfalls in the supplier's development methodology (e.g., the supplier does not perform documented code reviews, only high-level functional testing is performed on code changes, etc.), the LIMS application will already exist (except for any specific bespoke functionality that the business owner has requested). As this is the case, changing the supplier practices (e.g., to perform documented code reviews) will only impact future developments, which mean that a decision on whether to use the existing LIMS application and what to do to mitigate the supplier shortfalls in a key part of developing the strategy for managing risk. These decisions should be part of the risk management strategy and documented in the validation plan.

Design Specifications

Although the pharmaceutical manufacturer has defined the system requirements, they are normally generic and not technically specific to any particular LIMS application or supplier. The purpose of the design specifications is for the supplier to translate requirements into a specific technical solution, which will fulfil the requirements of the business owner. The structure of the design documentation, detailing the physical, functional, and performance criteria for the LIMS should be directly traceable back to the system requirements to allow the supplier to confirm that all requirements have been addressed. This may be achieved by the use of a cross-reference (traceability) matrix, which identifies each of the system requirements, references the location in the design documentation where the requirement has been addressed and also holds justification for any requirements excluded or not to be fully implemented. The design documentation will usually consist of a functional design specification (FDS) to provide a high-level overview of the proposed LIMS with more detailed specifications for the hardware, software, and configuration. The design documentation should typically specify the following:

- System overview description with diagrams (a concise nontechnical description of the functionality and operation of the LIMS covering all interconnected systems), which for complex systems may be a separate document providing a description of the system, which may be used to describe the system to a regulator.
- System architecture with details of hardware with diagrams
- Data architecture with details of the sources, uses, and ownership of each static data item
- System landscape showing the interactions of the various instances of the application
- Interfaces with diagrams (including networks, systems, and analytical equipment)
- List of software modules with details of versions, standard software products (e.g., spreadsheet packages), etc.
- System performance requirements (e.g., timing, memory storage, availability, and spare capacity)
- Number of users, user hierarchy, and response time required from the LIMS
- Software design documentation
- Communications and network protocols (e.g., RS232, TCP/IP, Ethernet)
- Methodology for data storage, backup, and retrieval
- Server environmental conditions including power supplies
- Client PC user interface software configuration details and terminal emulation requirements
- Peripheral devices (e.g., printers and backup devices)
- Functional descriptions of the LIMS operations
- Business continuity (contingency) plans, maintenance, and operating procedures
- Application software and reports
- Security and access control methods

Each of the elements within the FDS should have a unique identifier for traceability through the testing phase of the project. The functions identified should be in sufficient detail to allow meaningful tests to be produced. The results of these will be recorded in the qualification test documentation.

Where bespoke/custom software is produced software module design specifications are required.

Design Review

Design review is a formally documented and structured process of confirming that the LIMS design is both fit for purpose in terms the requirements of the users/business and also meets regulatory expectations. This is all part of the process of building quality into the design. Where functionality has been identified as high or medium risk in the risk assessment, additional effort should be focused in considering the software and hardware associated with this functionality.

The process begins with a review of the agreed business and system requirements against the proposed FDS, this is to identify any shortfall in the proposed design. The outcome of this review should be formally documented and will act as a key milestone in agreeing that the proposed design meets the expectation of the business owner. Following such agreement, the detailed design activities for the LIMS may commence.

The information used in this initial design review is the system requirements, the risk assessment, the LIMS supplier/integrator's proposed design specification, the company's standards, the pharmaceutical regulations, and any local regulations. The intention of the ongoing design review process is to monitor the development of the design throughout the project and as such does not end until the LIMS is in operational use. Each change to the design following the initial design review should be given full consideration of its potential impact on the risk assessment and the quality of the final LIMS installation. Elements that makeup the reviews of the detailed design will cover the execution of configuration reviews, source code reviews for user created applications (e.g., user customized reports and spreadsheets), which utilize data from the LIMS, GxP assessments, etc. It is important to note that the validation plan must clearly document the methodology to be followed and the documentation to be produced supporting the design review activities.

During the design review process a key requirement is to evaluate the system against the requirements for ongoing use and maintenance. For example, if the LIMS utilizes analytical equipment interfaces that must be isolated for maintenance, the LIMS must be designed to allow for maintenance activities without there being any effect on the operation of the rest of the system and any impact on the use of the LIMS.

The design review process begins as soon as the first version of the system requirements has been produced or after the LIMS supplier has been selected whichever is the earlier, and then effectively continues until the implementation of the LIMS in terms of hardware, software, database configuration and beyond. The design review process should then become a standard part of any ongoing modification to the LIMS through change control. The objective of the design review process is to ensure that a quality system is installed and maintained, with all identified risks mitigated to an appropriate level, thus giving a system in which the pharmaceutical manufacturer has a high level of confidence. Quality can only be built into the design, configuration, and implementation of the LIMS; it cannot be tested in afterward. Therefore, if the project does not take the opportunity to apply a structured and formal design review process it may result in an expensive and time-consuming exercise to redesign the LIMS if, as could easily happen, the installed LIMS does not meet the business owner's needs and /or regulatory expectations.

Rigorous testing has the potential for finding fundamental faults with the design of the LIMS, which may mean there is a requirement for redesign at a late stage in the project. The resulting redesign may negate design reviews that have already been performed as well as having a knock on effect on any testing that has already been performed.

The whole point of applying a life cycle approach is to provide a structured, controlled, and confidence-building approach to the development and validation of the LIMS and associated analytical equipment and related system interfaces.

As a result of this process in-depth testing can be focused where high or medium risk to patient safety has been identified via the risk assessment, thus reducing the overall amount of testing and documentation required to support validation.

There is a direct relationship between the quality of the documentation produced during this review process and the quality of the finally installed system. This is because clear, concise, and accurate documentation describing the required implementation will allow the project team to generate test scenarios most closely matching the system use and boundary conditions.

Application Development

To provide a controlled, structured environment for the development, testing, validation, and ongoing use of the LIMS, it is important to set up server-based environments at an early stage in the LIMS project. These environments are typically:

Development Environment

This is the area where the project team are able to configure existing standard LIMS software modules or, where required, create new software modules to meet the specific requirements of the agreed design. This area may also be used for piloting of proposed LIMS software module constructions to determine the most suitable application of the LIMS for the business owner. This environment is sometimes called a "sand pit" as it is expected that the developer may need to try out different means of implementing the agreed design, not all of which will be successful, this is sometimes called rapid application development. These trials are often not formally documented and are used as a means of evaluating the LIMS product. With modern LIMS applications the functionality required by the business owner is usually available in the standard modules and therefore most of the developer's activities will revolve around implementation of a configured solution that meets the business owner's needs.

This environment will have restricted access to enable the development personnel to create and store versions of custom software modules and instances of the LIMS. There will need to be extra control on access to the version controlled standard LIMS software modules so that they cannot be inadvertently modified, and this would normally be achieved by maintaining a secure library maintained under a configuration management system. Control of the library will normally be by personnel independent of the development team who will be responsible for managing and maintaining the module library on behalf of the LIMS owner. It is also normal practice for the developer to carry out informal testing of modules and proposed configurations of the LIMS in this environment to ensure that no obvious errors are present. When the module or instance of LIMS has been accepted by the owner as being ready for formal testing it will be transported to the validation/test environment.

Validation/Test Environment

This is where the final version the software is tested, through structural testing and functional/ stress or checklist testing, prior to releasing the LIMS instance for use in the production environment. This environment will be strictly controlled and will only be used for the validation/qualification activities. The environment, hardware, software, data, and configuration should be an accurate representation of the production environment. The testing must be formally documented.

At the time of validation testing, training of the LIMS users will be required and operational documentation available, for example, SOPs. All data sets used for the validation testing of the LIMS will be retained under configuration management as reference information supporting validation. Following the successful completion of the validation exercise a decision will be made by the project quality assurance representative for the LIMS software to be released to the production environment.

Production Environment

This is where the validated version of LIMS is made available for the users. This environment must only be accessible to trained users of the LIMS. Release of software from the development/test environments to the production environment must be strictly controlled. The production environment will be used only for "live" LIMS data and activities.

Environment Location

The environments may be set up either on separate servers or the same server; however, it is normal practice for the development and validation/testing activities to be performed on a server specifically dedicated to the original development and testing and ongoing support. Where separate servers are used, confirmation of equivalence between the validation and production environments (software and hardware) will be part of the validation exercise.

Configuration Management System

The flow of software will in all cases be from the development to the validation/test environment and then finally the production environment. Modifications and development of new software modules occurs only in the development environment. This discipline must be maintained to ensure that modules already in the validation/test environment are not modified. This could cause a mismatch between the library copies of the modules, held in the configuration management system, and the tested modules. The configuration management process is key in controlling the make-up of the different environments and the movement of modules between environments.

It should be possible to determine what versions of software modules, hardware components, and documentation used to make up the LIMS throughout the design, development, testing, and use of the system. As such configuration management should cover:

- Hardware components (e.g., servers, PCs, interfaces to laboratory equipment, etc.),
- Software (e.g., operating system, database, application software modules, data sets, bespoke/custom/standard software modules, etc.)
- Documentation (e.g., configuration records, test records, validation protocols, validation reports, etc.).

In the context of the initial project to implement the LIMS, configuration management is a means of ensuring that when the system is validated there is a specific record of the make-up of the instance of LIMS in use. If this information is not available, changes to modules as a result of test failures cannot easily be assessed to determine any impact on testing that has already been performed. For ongoing maintenance there is a similar need to understand the make-up of the LIMS components. This is a process of management/control that is expected to be in place as part of the ongoing support activities for any IT GxP computerized system.

Source Code Review

Although it is no longer common practice to write bespoke/custom software solutions when implementing a LIMS it may be necessary for software modules to be developed or modified to meet a specific business need. In such cases, the new or modified module should be developed in accordance with a software development life cycle.

As part of the software development life cycle custom/bespoke, or modified software, should be subject to source code review (SCR). This is another means of building quality into the LIMS. When conducting an SCR the code should be reviewed against the agreed design and the developers programming standards. Particular emphasis should be applied to the bespoke GxP software deemed as posing a high or medium risk to patient safety. For GxP software posing a low risk to patient safety or no risk it may be considered that an informal review of the code is all that is needed.

Following the completion of the SCRs a review report or a number of review reports will be produced to summarize the review findings. These reports may identify actions to be taken by the LIMS supplier/integrator, or the personnel responsible for performing the validation of the LIMS. These actions may range from identifying specific testing to revisions of the source code due to deviations from programming standards or to correcting the code errors.

Test Strategy

The testing of the LIMS is the final intensive activity that is performed prior to release of the LIMS into the production environment. A test strategy should be developed to consider what testing is required to mitigate the potential risk of LIMS failure impacting patient safety to build a high degree of confidence that the LIMS will continually operate as per design. This strategy is crucial in determining the amount and depth of testing that is appropriate to assure the implementation meets current regulatory expectations. The test strategy should also examine the testing requirements for the business continuity and disaster recovery plan. It is recommended that the pharmaceutical manufacturer execute the test strategy with assistance from the LIMS supplier/integrator.

The approach to testing should be based on risk and formally documented in a test plan that identifies the rigor of the testing and the documentation that is to be produced to act as evidence that the implemented LIMS meets the business/system requirements and regulatory expectations (Table 21.3). Testing will typically cover two main areas:

- The first is module functionality and stress testing to demonstrate that the required LIMS functionality that has identified as potentially high, medium, or low risk to patient has been successfully implemented to meet the design expectations.
- The second area is integration testing to verify that the integration/interfacing of the LIMS to other systems has been successful.

Table 21.3 Approach to Risk-Based Testing

Risk level	Test case content
Low risk	• Test are statements of the test to be performed—not how to do it • Results are tester evaluation rather than evidence collection
Medium or high risk	• Additional steps included to provide the desired sequence of tester actions • Only evidence required is that which is proving the objective of the test is being met (includes negative testing). Indicate when to collect and label evidence (e.g., screen shots) • For high risk perform additional testing to prove the operation of the system over an extended period and range of data sets

As part of the risk assessment all LIMS functions (e.g., manual data entry, automated data entry, and report generation) should have been assessed to evaluate the effect that they could have on the data that will support regulatory submissions and release of product to market.

Prequalification Activities

To ensure that all the interconnection of system hardware and loading of software has been successfully completed on the target system, prior to validation, the installation must follow written test plans and be formally documented.

Because of the inherent complexity of demonstrating a LIMS meets expectations and the logistics of arranging for all the analytical equipment interfaces to be present at this time it may well be the case that some simulation of data inputs or connections to other systems needs to be used. As this will be a deviation from the expected approach it will need to be justified in the test plan.

Supplier's installation work must be formally reviewed and approved by competent personnel, and the pharmaceutical manufacturer must also ensure that the completed integration testing documentation is of a quality suitable to support validation. Where supporting validation, the supplier's work should be preapproved by the company QA, and the results reviewed by the pharmaceutical manufacturer and a summary report produced.

The summary report will identify any issues raised, and the pharmaceutical manufacturer will need to review this information to determine if the LIMS is fit for purpose and therefore suitable for moving on to the installation qualification. Confirmation of acceptance of the LIMS will normally be given only if all but minor issues have been resolved.

Data Load

As part of implementing a LIMS there will be a need to populate the LIMS database with the relevant fixed data. There are two types of data dynamic data and static data:

- *Dynamic Data*: Information related to individual samples or lots, for example, test results, sample status. This will be required if the implementation is an upgrade to an existing LIMS. The process should enable the extraction, cleansing, verification, loading and maintenance of the "dynamic" data from the system being replaced. Dynamic data should be held securely under change control.
- *Static Data*: Information that is not related to individual samples or lots, for example, test methods, specifications, calculations, field formats. This process should enable the collection, cleansing, verification, loading, and maintenance of the "static" data, for example, specification limits, calculations, test regimes, etc. Static data sets should be maintained in a configuration control system.

To be sure that this data is loaded correctly it must be achieved via using a validated tool, be proven to be correct via an approved statistical technique, or a 100% verification of the data. Data transfer may not be as simple as copying data from one database to another, as there may well be different data fields or different field formats between the two systems. Data cleansing and archive may also be necessary.

Where the LIMS is installed to replace an existing LIMS or is implemented to replace manual or semiautomated laboratory information management activities, there will be a need to transfer already existing data from another system or from manual paper-based sources to the LIMS.

Test Data

Test data will be needed for use during the development testing and operational qualification. Test data should cover all of the data types and values that will be expected in the LIMS and ideally should be based on a copy of data from a live LIMS. Test data sets should be managed under configuration management and retained as part of the validation deliverables.

Installation Qualification

The installation qualification (IQ), as its name suggests, is a testing/inspection process that is designed to confirm the compliance of the LIMS installation with the agreed design and also includes the development/validation system if separate from the production system. This IQ should cover all aspects of the hardware, firmware, software, documentation, environment (development, validation/testing, and production), and infrastructure for the installed LIMS.

The IQ activity is the formal documenting of the installation hardware, software, configuration, static data, infrastructure, and interfaces to be used for the development and production LIMS in their final location. The IQ will be performed in two distinct parts, the initial installation of the development and testing system and the IQ of the production system. The IQ activities for the development and validation/testing system environments are required prior to performing the operational qualification activities in the validation/test environment. The IQ of the production environment is required prior to commencing performance qualification.

Attempts are often made to "rush in" to starting the IQ with the consequential problem of failure of tests/inspections. This not only has the effect of introducing retests and delays, but it also does not give a regulator, who may inspect the system, confidence that the LIMS has been designed and installed by a quality aware organization. As there is ample opportunity for a preinspection to be performed by the LIMS integrator, it is the case that the validation personnel are simply rechecking and recording something that has already been checked; therefore, no failures should be experienced. It is essential that the integrated project team are aware of the contents of the IQ test/inspection protocol to be used in the IQ, to ensure that there are no surprises for the testers when this activity takes place.

The IQ protocol will also cover the physical environment into which the LIMS and interface equipment is installed, which may need to be controlled in terms of temperature/humidity, electrical interference, physical access, etc. In addition the provision of services will need to be assessed, for example, electrical supplies.

The typical contents of an IQ protocol are shown in Table 21.4.

The IQ is also the vehicle for confirming that the design review process has been successfully completed. It is essential that all installation and infrastructure issues are resolved prior to the completion of IQ. It is therefore useful to provide a check within the IQ protocol, which looks back at the results of the design review process and integration testing and assesses the effects of any issues raised regarding the compliance of the installed system.

Installation Qualification Report

Following the execution of IQ there may be exceptions, where the LIMS has not been installed in full accordance with the approved design. There are often issues due to hardware, documentation, drawings, environment, interfaces, installation, or even missing components of the LIMS. These will result in failures in the IQ checks as the LIMS does not comply with the requirements specifically noted in the IQ protocol acceptance criteria. Any failure will be recorded in an installation qualification report and a decision as to what to do next

- To fix the noncompliance prior to commencing operational testing,
- To accept that the corrective action may be deferred,
- To accept that the issue will become a permanent feature of the LIMS installation.

Table 21.4 Example IQ Content

Subject	LIMS hardware	LIMS software
Software versions for operating systems, utility software, application software, firmware		√
Configuration of development, testing/validation and production environments		√
Licences for core software and layered software products		√
Hardware platform details with unique identification (e.g., serial numbers)	√	
Labeling of hardware platform equipment (including interface cabinets)	√	
Diagnostics self-test for analytical equipment interface	√	
Network compatibility of peripherals (printers, PCs, etc.)	√	√
Power-up/power-down tests	√	√
Installation of services, e.g., power supplies	√	
Installation of internal wiring and marking up of cabling for maintenance, confirmation that no disconnected wiring is present	√	
Computer room environment testing (temperature, humidity, radio frequency interference, electromagnetic interference	√	
Network connections to LAN/WAN	√	√
Security access testing	√	√
Draft user procedures	√	√
Provision of SLAs	√	√

Abbreviations: LIMS, laboratory information management system; SLAs, service level agreements; LAN/WAN, local and wide area networks.

To make this decision the project team will need to assess the situation and justify the course of action to be taken. If the failure does not affect the operational testing, (e.g., a specification is incorrect), this could be corrected in parallel with the operational testing. Otherwise, the exception must be corrected and retested prior to moving on to the next phase of qualification.

The installation qualification reports are milestones in the project that completes the IQ and records acceptance that the next phase of the project can commence.

Operational Qualification

When the pharmaceutical manufacturer is happy that the installation of the LIMS and the testing/validation environment has been satisfactorily completed, the project will move on to the stage where the functionality of the system can be demonstrated.

There is a need at the beginning of this phase to review any issues raised during the design review process, and also IQ. Any issues that will have an effect on the operational testing must be resolved prior to the commencement of OQ.

Operational qualification (OQ) is the vehicle for providing documentary evidence of the demonstrated functionality for the integrated LIMS. The OQ consists of a series of tests based on the LIMS FDS and the rigor of this testing will be commensurate with the risk posed to patent safety by the LIMS functionality being tested.

It is widely accepted that a validated system must not use data from an unvalidated system through an interface where this data is used for GxP purposes. The LIMS project must, therefore, take into account the need for the validation of all data sources (e.g., analytical equipment, chromatography data systems, etc.) and the interfaces used to obtain these data. The validation of interfaces follows its own validation life cycle with the integrity and security of data being of primary importance.

Operational Qualification Protocol

The OQ documents must cover all GxP relevant functions in sufficient detail to provide the pharmaceutical manufacturer with a high level of confidence that the LIMS operates in accordance with the agreed design. This testing should be performed to a minimum standard that is in accordance with Good IT Practice, this will be sufficient for the testing of functionality determined, via risk assessment, to be of low or no risk to patient safety.

Where the tests are for functionality that is either low or no risk to patient safety the expectation for test case detail and gathering of evidence is reduced. For example, a simple checklist style approach may be used with minimal evidence gathering needed. For this type of testing, it may be cost effective to consider using the supplier testing documentation, providing that it meets the expectations for Good IT Practice.

Where the risk assessment has indicated that there is functionality in the LIMS that could pose a high medium risk to patient safety, the rigor in the testing should be increased above the expectations of Good IT Practice.

Where the risk is assessed as being higher it is necessary to increase the rigor of testing both in terms of the test method to be followed and to ensure that specific evidence of correct operation is provided. In these cases, the supplier documentation may still be used however in these cases it would be an expectation that the supplier complies with the quality management system used by the pharmaceutical manufacturer, as this is the documentation that is most likely to be reviewed by regulators.

It is recommended that the structure of the OQs should match that of the FDS to provide a simple mechanism for demonstrating that all of the functions of the FDS have been tested. Where this is not a possible mechanism for demonstrating traceability of requirements through to design and on to testing is required, for example, via the use of a requirements traceability matrix. The OQ may have one or more functional test scripts, for testing each of the functions that are uniquely identified in the FDS. Table 21.5 identifies typical OQ protocol contents.

At the OQ stage there will be a need for the pharmaceutical manufacturer to ensure that the appropriate management systems in terms of procedures and business continuity (contingency) plans have been assessed and confirmed as suitable. In some cases, it may be that as part of the OQ process draft versions of the procedures and business continuity/disaster recovery plans are developed with the final version of these documents being issued following the completion of OQ.

Standard Operating Procedures

SOPs must be written to cover operational activities. The SOPs should be written by personnel knowledgeable in the low-level detail of the LIMS and should be detailed enough for the user to work without reference to other personnel or documentation, or memory. By using the OQ

Table 21.5 Example OQ Protocol Contents

Subject	Core LIMS	Custom code
Special configuration functions	√	√
Testing of bespoke/custom software		√
Signal diagnostics from linked analytical equipment interface	√	√
Special calculations and algorithms	√	√
Verify operating manuals including challenge testing	√	√
Verify sample data and system parameters, i.e., check against source records to ensure the accuracy of data within the RDB. This involves checking of data loaded manually or via automated upload from legacy systems.	√	√
Software backups and restoration of data	√	
Training Records of users	√	√
Routine maintenance/calibration routines	√	√
Data upload and migration checks	√	√
Data integrity checks, e.g., range checks, validation of inputs	√	√
Communication driver tests	√	
Database structure and population	√	
Disk shadowing demonstration (where fitted)	√	
Archive and retrieval of documents and records	√	
User results input and displays	√	
Draft user procedures	√	√
Analytical report generation	√	
Audit trail verification	√	
Demonstrate features supporting the use of electronic records and signatures	√	

Abbreviations: OQ, operational qualification; RBD, relational database.

as a means of formally testing the SOPs, any issues of detail should be identified. The first version of the SOPs must be authorized and issued prior to the start of performance qualification (PQ).

SOPs are also required for maintenance activities for the LIMS such as backup and restoration, change management, etc.

Operational Qualification Report

Following the execution of the OQ tests there may be issues noted that the LIMS does not function in accordance with the approved design. As for the IQ report the project team will need to review these failures and determine a plan of action or a justification for moving on to the next phase.

Following the successful completion of the LIMS OQ as documented in the OQ report the LIMS software is ready for PQ to commence.

At this stage, it is normal practice to issue the first version of the validation report of a form of notification that the system is to be released for use in production. This report/ notification summarizes all of the qualification activities to this point. Authorization of this report/notification indicates that the LIMS application is ready to be promoted into the production environment and made available to the trained operators for live use.

Performance Qualification

Although the term PQ is more of a process qualification activity there will be a requirement to monitor the ongoing operation of the LIMS in terms of system performance and user interaction, for which I am using the term PQ. There may also be some functionality that can only be proven in the live environment (e.g., the interaction with other systems via interfaces). The PQ typically consists of

- Monitoring the system performance through execution of the life cycle for each sample type, using the user SOPs, in the production environment,
- Confirming that the coverage of SOPs is complete,
- Monitoring of system incidents and failures,
- Monitoring fault calls from users of the LIMS, and
- Monitoring of change requests.

The PQ documents that the integrated LIMS system and interfaces perform effectively and reproducibly using live data and user interaction in the production environment. As with the previous phases the first part of the PQ will be to determine that there are no outstanding issues from the design reviews and IQ and OQ reports that need to be addressed prior to the start of PQ.

The length of time over which a complex LIMS will be subjected to this PQ testing is typically 6 to 12 weeks, but each system will need to be assessed on a case-by-case basis to determine if this is an appropriate period. Where a LIMS is found to have a high level of incidents and user problems it may be that the agreed PQ period will need to be extended.

It is a European expectation that if a LIMS takes over from a manual system then the new LIMS and the manual system should be operated in parallel for a period of time. This period is expected to be long enough to confirm that the LIMS is fit for purpose to take over the manual system. This activity may form part of the PQ activity.

Performance Qualification Report

At the end of the PQ a report will be produced that summarizes the continued operation of the LIMS for the initial period following going live. During the period of the PQ the LIMS is expected to have stabilized in terms of user support requests, system performance, and changes implemented as a result of the qualification process.

LIMS Validation Report

The final validation report for the installed and validated LIMS reviews the results of each of the preceding validation phases. It is effectively verifying that the activities and deliverables defined in the validation plan have been performed/delivered, and that all the risks as identified in the risk assessment have been mitigated to an acceptable level. This report will act as a summary of the overall validation status of the entire LIMS. There may have been validation reports associated with individual items of analytical equipment and perhaps the core LIMS itself. This final validation report should cover:

- Summary of the outcome of each of the validation activities,
- A summary of any outstanding issues associated with the LIMS,
- Time scales for future periodic reviews of the LIMS validation status,
- Details of the justifications from the pharmaceutical manufacturer for any deviations from the original validation plan,
- Details of any constraints or limitations of use,
- A statement that the LIMS is fit for purpose.

The LIMS validation report will be an input to in the first periodic review to confirm that any issues recorded have been successfully addressed.

ONGOING VALIDATION MANITENANCE/OPERATIONAL COMPLIANCE
Ongoing maintenance of the LIMS validation status requires a suitable support service to be in place. This support will consist of a LIMS manager/owner and appropriate SOPs, support teams (e.g., help desk) and SLAs where appropriate. The LIMS manager/owner will be responsible for controlling any changes to the system, interfaces, LAN/WAN architecture, LIMS functionality, and the data held within the database.

Responsibilities
The LIMS manager/owner is typically responsible ensuring that the daily administration of the entire LIMS (core LIMS database, LIMS servers, peripheral devices, for example, printers, user PCs, etc., networks) is performed. The support service must respond to user requests and problems in agreed timescales and is in effect providing a service to the laboratory. The duties of the LIMS manager/owner will include the following:

- Addressing user problems
- Adding and removing users
- Controlling user privileges
- Managing upgrades to the core LIMS, standard software packages, and operating systems
- Managing SLAs with the LIMS supplier

The LIMS manager/owner will be responsible to the company for ensuring that the LIMS validated status is monitored and maintained during ongoing operation. Where there is a need to make a change as a result of component failure, upgrade, LIMS development, or the implementation of new or modification to existing static data, a change control system must be followed. Maintaining access control (both physical and electronic) to the LIMS and interfaces is essential in ensuring the maintenance of the validation status over the lifetime of the system.

Change Control
Change control of the LIMS hardware, software, data, and associated documentation (SOPs and operating manuals) is necessary to prevent the system becoming unmaintainable. A good change control system will allow the LIMS manager/owner to determine what changes have been made to the LIMS, when they were made, and what effect they had. It is not acceptable for changes to be made to the LIMS functionality without the impact of the change being assessed against the current validated status, and also current GxP. If the change is necessary and impacts on the validated status then appropriate revalidation must be performed. This may result in rerunning one or more tests from the IQ, OQ, or PQ or may, in the worst case, result in a reevaluation of the fundamental design of the LIMS.

The LIMS change control system must record:

- Details of the change
- Justification/business case
- Authorization for the change
- Assessment of the effects of the change on GxP
- Details of the outcome of the design review
- Date when the change was requested and the date when it was implemented
- Testing performed to verify the operation and reconfirm the validated status and details of the test results

Where a change is required to the LIMS hardware due to the failure of a component there are two possible scenarios. This first is that the failed component is no longer available and a new design must be installed. In this case, an impact assessment will be required to assess the effect on the rest of the LIMS, followed by the normal testing approach. The second is that the component is a standard offering from the supplier and is therefore a "like for like" replacement. In this case, simple testing of the functionality of the replaced component is all that is required.

Where the change has an impact on the existing functionality of the LIMS or provides new functionality, it will be required to be assessed following the risk assessment process. The outcome will be an amended LIMS risk assessment that may result in an increase, or in some cases a decrease, in rigor of the required testing.

The change control system will be utilized in the maintenance phase of the LIMS life cycle; however the level of details of the review and the rigor of the testing should be the same as was used in the original validation process. The testing must therefore be carried our by competent qualified personnel, and the records of the testing retained as part of the LIMS documentation supporting the validation status.

Configuration Management
To ensure that the LIMS remains in a validated state the management of configuration must continue for the life of the system. The support team for the LIMS should be able to show the versions of each LIMS modules that are currently implemented within the application, and be able to define the module versions of the LIMS at any point in the past, so that it could be reconstructed (in case of regulatory concerns). Only the module versions that have been confirmed as current will ever be a part of the LIMS application are to be promoted to the production environment for validation and live use.

Upgrades to the LIMS
Following the installation of LIMS the core software will continue to be developed by the LIMS supplier to fix known bugs and implement new features. This means that the LIMS manager/owner will be routinely advised that there is a need to change to the latest version of software. In some cases, the LIMS supplier SLA may be linked to the installation of software upgrades. As a consequence of this the LIMS manager/owner will be responsible for the review of the impact of any upgrade taking into account the effect on the existing validation documentation as a result of the making this sort of change. Any change to the validation status is likely to involve updating documentation as well as the software and subsequent testing. The approach to updates should be as follows:

Software Updates (Patches)
Assess software updates for compatibility with the existing software. In terms of system or standard software products there is normally a "bug fix" list and details of new, modified, and removed features. Supplier's documentation should include an analysis of the impacts of the patch on their core system and this should be used to assess the extent of testing required to validate the patch implementation. Any testing that needs to be performed will need to be incorporated into the LIMS validation documentation.

Hardware Updates

Assess new hardware for compatibility with the existing hardware. Should any differences be identified the impact of these differences should be assessed against the existing LIMS design intent and appropriate inspections/testing performed as part of installation.

Documentation Updates

The document changes will normally be as a result of modifications to the LIMS hardware, software, or operating methods. The design documentation and SOPs supporting the maintenance of the LIMS should be updated to reflect any changes. It is not acceptable for any documentation that is to be utilized for maintenance purposes to be out of date.

It is also important that in all the updates described earlier, the quality function, within the pharmaceutical manufacturer's organization, provide sign-off that any change has been performed in a manner that maintains the validated status of the LIMS.

LIMS Backup and Restore

Backup and restore applies to the application code, the configuration, and the static and dynamic data. The objective is to be able to recover the system following a crash or other catastrophic event.

The LIMS must be backed upon a regular basis to maintain the security of the database. The regulatory inspectors will not accept that data within the database has been lost because of the failure of the server or other incidents (e.g., fire or flood). To prevent losses the pharmaceutical manufacturer is responsible for implementing a reliable, robust, and documented backup regime. The frequency of backups must be assessed as part of the GxP risk assessment process, and determination made of the risk posed by the potential loss of data the outcome of this assessment may be to implement redundancy features in the server to address the potential risk of data loss (e.g., in the form of disk mirroring). Automation of the backup regime is acceptable providing that it is validated, this is normally performed as part of the operational testing phase of the project.

A robust and secure process must be implemented for storage, renewal, and eventual destruction of backup media. It is strongly recommended that the backed up copies of the data be stored in a location remote from the server (e.g., another building or the site security gatehouse). This process should be documented as part of the backup and restoration SOP. It is important that it can be demonstrated to the regulators that appropriately trained competent personnel are managing this process and that there is evidence to confirm that the backups have been performed to schedule, and any issues/failure of backups have been investigated and resolved.

To demonstrate that the data backed up is capable of being restored in the event of a breakdown, the business owner organization must implement a periodic test of the restore procedure. This procedure will demonstrate that the data restoration is possible and that the procedures covering this activity are effective. However, this testing must not compromise the integrity of the production data and it is recommended that the testing take place on an off line version of the LIMS, for example, the test/validation instance.

Data Archiving

Archiving is required to manage growth of the data within the LIMS database. The policy and process for archiving must be documented and validated to ensure that any removal of data from the LIMS does not impact data integrity or increase the risk of data loss. It is important that it can be demonstrated to the regulators that appropriate personnel are managing the transfer of data to the archive. To meet the requirements of the FDA Electronic Records and Signatures regulation 21 CFR Part 11, the archiving process must meet the company retention periods for GxP critical data.

LIMS Access Control

The implementation of an effective access management process is required to comply with the regulatory expectations for access to, and security of, data and to ensure that only authorized personnel can approve analytical results and create certificates of acceptance (CofA) for use in batch sentencing decisions.

The pharmaceutical manufacturer will be responsible for managing the access control of the LIMS. This is normally accomplished through software protection, for example, passwords and log-on accounts but may also take the form of protection through physical restrictions, for example, locked-up or restricted areas. The management of this function should be in accordance with a formal SOP. The use of passwords and high-level accounts must be strictly controlled to prevent security breaches. Typical examples of control should be:

- Restricted number of high-level users
- Unique IDs to provide traceability of personnel making changes, decisions, or signing electronic records
- No shared user identifiers
- QA should authorize all users performing GxP activities
- Where electronic signatures are used they must be the equivalent of hand written signatures
- Training of users in security expectations

Business Continuity/Disaster Recovery Planning

Business continuity/disaster recovery plans define the controls that minimize the impact of temporary or long-term loss of all or part of the LIMS. The extent of planning will be determined by the criticality of the LIMS with respect to the GxP operations that it controls or monitors, and the data that it manages. Business continuity, in the form of standby systems and manual ways of working, should be considered during the development phase for any highly critical computerized systems. However, it is also vital that plans are established that assume the inevitable, that is, what can go wrong will go wrong. Plans will need to define the requirements for system archive, periodic backup, restoration procedures, and SLAs. Additionally, plans must address the method of system (which may mean temporary use of a system at a remote location) and data recovery, and define the manual operations that may need to be applied in the interim until the LIMS is reinstated. Where manual processes have been performed it is also important to manage the retrospective loading of any relevant data into the LIMS.

Training

Training of LIMS users is a key requirement that is likely to be the subject of a regulatory inspection. It is therefore essential that a training program be organized as part of the LIMS implementation and be maintained as part of the ongoing maintenance of the LIMS. Training is required not only for the LIMS users but also for the in-house support and development staff. The need for training not only applies to the pharmaceutical manufacturer's personnel but also, through audit, to the LIMS supplier/integrator's personnel and the validation personnel if they are independent of the LIMS supplier/integrator. As part of the supplier/integrator evaluation the pharmaceutical manufacturer must assure that competent trained personnel are to be utilized on the project. It is recommended that evidence of this training be provided for reference in the project validation documentation, perhaps in the form of staff resumes or copies of training records.

Periodic Reviews

The validation integrity of the LIMS must be periodically reviewed to ensure that ongoing support systems are effective. The review process should be designed to identify trends that may indicate noncompliance with support procedures or weakness in the original validation exercise. The review should further examine the original test data sets to determine their applicability to the current computer system configuration and duty. The review shall determine if there is a need for further validation or revalidation of the existing LIMS installation.

These reviews should typically include an assessment of

- System performance
- Maintenance records
- Backup records
- Risk assessment

- Change control records
- Access privileges
- Network integrity

- SOPs
- Fault/incident reports
- Supplier audit follow-up

CONCLUSION

This chapter has reviewed a risk-based approach that may be taken to the validation of a typical LIMS that will meet the expectations of the regulatory authorities. If the LIMS is to be utilized to support pharmaceutical manufacturing laboratories then the system must be subjected to validation.

A key aspect of any validation exercise is to apply the appropriate amount of rigor applied at each stage of the LIMS project to ensure that the implemented system is fit for purpose. This means that the personnel that are to use the LIMS have a system that meets the needs of the company in terms of functionality and capacity, and that the system meets the regulatory expectations for this type of it system. The modern thinking of focusing on risk to patient and using a risk assessment for identifying risk allows this thinking to be built into the validation approach. This gives the potential for an increase in compliance by focusing on the highest risks while providing an overall cost saving by reducing the amount of effort spent on lower risks, as this is a benefit to the company as well as to the patient.

In order for risk-based validation to be successful it is essential that the personnel involved in the process are knowledgeable and experienced in the validation of LIMS and the application of risk-based thinking. This may mean that there is a need to ensure that the LIMS supplier/integrator can provide suitable validation personnel or that external specialists may be needed.

Once a validation project is complete the system must be maintained under change control. The responsibility for this is the LIMS manager/owner who will also be responsible for ongoing validation maintenance. Ongoing validation maintenance not only covers the issues of upgrades, new developments, maintaining validation documentation, and managing all aspects of change control, but also ensuring that existing and new personnel are fully trained and competent in their roles in the LIMS.

REFERENCES

1. Good Automated Manufacturing Practice (GAMP®5) Version 5, 2008—A Risk Based Approach to Compliant GxP Computerised Systems. International Society of Pharmaceutical Engineers (ISPE).
2. Rules Governing Medicinal Products in the European Community, Volume IV Pharmaceutical Legislation—Medicinal Products for Human and Veterinary Use—Good Manufacturing Practices, including Annex 4 and Annex 11.
3. Title 21 Code of Federal Regulations (21 CFR Part 11) Electronic Records; Electronic Signatures. Final Rule Published in the Federal Register.
4. United States Code of Federal Regulations Title 21: Part 58 'Good Laboratory Practice for Non-Clinical Laboratory Studies'.
5. American Society for Testing and Measurement—Standard E1578. 'Standard Guide for Laboratory Information Management Systems [LIMS]'.

22 | Case Study 4: Clinical Systems

Chris Clark
NAPP Pharmaceuticals

Guy Wingate
GlaxoSmithKline

Current legislation and regulatory guidance for the management and conduct of clinical trials is undergoing significant changes and the European and U.S. regulatory bodies are increasingly focusing greater attention on the compliance of the pharmaceutical industry to these regulations (1–3). Recent observations noted during both regulatory inspections and company vendor audits have indicated that one area of critical noncompliance, and a potential barrier to successful license applications, is that of the development, implementation, management, and controls applied to the use of computerized systems in the Good Clinical Practice (GCP) environment. Companies invest large amounts of time, resource, and finance into the process of developing, investigating, documenting, and registering new products, a process that can take upwards of 10 years to result in a successful launch to the marketplace. There are many stages during this process whereby the new product can fail, for example, not demonstrating adequate/beneficial therapeutic value, the presentation of adverse side effects, or not capable of being formulated into a delivery system suitable for mass production.

The above-listed risks are well understood by those experienced in the new product development process, and they place great reliance upon the collection, manipulation, and presentation of data. It is that very reliance on data that places yet another question of risk in our pathway. The risk raised here is that presented by the failure of our computerized systems involved in these processes to infer adequate integrity and security to this hard won data. Regulatory submissions based on data that has been handled and/or managed by systems that cannot demonstrate adequate controls around integrity, and security will, in most likelihood, be rejected by the assessing authority as being unreliable. This avoidable scenario, which inevitably results in loss of revenue through the need to repeat expensive clinical trails and the consequent delays in getting a product to the marketplace, can be prevented by the application of sound computer systems validation practices.

COMPUTERIZED SYSTEMS AND THE CLINICAL TRIALS PROCESS

There are a wide variety of computerized systems involved in the clinical research process, from small standalone desktop systems to more complex enterprise-wide systems. Furthermore, they can often be categorized into four functional subgroups. Table 22.1 provides some examples of such widely divergent systems:

GENERAL VALIDATION REQUIREMENTS

Clinical studies may be supported by computer systems in a number of ways from data capture, data processing, production control, and document management. Some systems may be complex, others simple. Some systems may be custom made, others based on commercial off-the-shelf (COTS) products. Whatever the character of clinical computer system, the same basic GCP/GLP principles apply. All computer systems that play a part in the conduct or support of clinical studies intended for regulatory submission therefore need to be validated. It is vital that such systems manage clinical data reliably and securely.

Validation of clinical research computer systems should demonstrate that the computer system is suitable for its intended purpose (4). Validation is achieved through a life cycle approach to computer system development, operation, and maintenance. The various

Table 22.1 Examples of Clinical Systems

Functional subgroup			
Data capture	Data processing	Production control	Record management
Subject information systems	Data management	Clinical trails supplies production	Protocol management
Interactive voice response systems	Randomization systems	Analytical instrumentation	SOP management
Clinical trial data collection systems	Clinical trial review tools	Inventory system	Electronic publishing and regulatory submissions
Medical device measurement systems	Statistical analysis systems	Labeling system	Electronic document management systems
Optical character recognition Systems	Pharmacovigilance systems	Environmental monitoring	Training records
Bar code readers	Electronic data transfer systems	Product tracking/ monitoring	Archive repository

international GCP/GLP requirements also emphasize the importance of data integrity. This covers data input, manipulation, output, and archiving.

This chapter is based on the work published by ACDM/PSI (5). General GCP/GLP computer validation requirements are reviewed. This is followed by a summary of key topics for special attention for a selection of common systems found in the clinical environment.

The basic computer compliance requirements for development and installation of clinical systems can be summarized as follows:

- Validation of data processing software prior to use
- Auditing suppliers of software-based systems
- Assessment of the investigator site prior to the start of the trial, including investigator-supplied software-based systems considering such issues as validation activities (planned and completed), level of understanding of GCP requirements for use of computerized systems, statement of level of compliance with 21 CFR Part 11, the presence of remediation plans if required, and calibration and maintenance procedures

The capture, processing, and retention of data should be carefully defined and managed. The ICH GCP Guidelines (3), which recognize that clinical trials data can take many forms (paper, optical, electronic), indicate that the sponsor has specific responsibilities regarding the handling of electronic data and/or the use of remote electronic trail data systems (subsect. 5.5.3). Such responsibilities include the following points:

- The system is validated
- SOPs are in place covering usage
- Maintenance of an audit trail for data changes
- Adequate security systems in place to prevent unauthorized access
- Control of user access rights
- Data backup and recovery procedures
- Maintenance of blinding during data entry and processing

U.K. GLPs suggest systems that organize, tabulate, subject the data to statistical or other mathematical procedures, or which otherwise manipulate or analyze electronically stored data, should permit the retrieval of original data entries (4). All network communication links used for data transfer should be considered as potential sources of error and controlled appropriately.

OECD regulations have further identified the following operation and maintenance requirements (6):

- Procedures for operation and use of computerized systems (hardware and software), and the responsibilities of personnel involved

- Procedures for security measures used to detect and prevent unauthorized access and program changes
- Procedures and authorization for program changes and the recording of changes
- Procedures and authorization for changes to equipment (hardware and software) including testing before use if appropriate
- Procedures for periodic testing for correct functioning of the complex system or its component parts and the recording of these tests
- Procedures for the maintenance of computerized systems and any associated equipment
- Backup procedures for all stored data and contingency plans in the event of a breakdown
- Procedures for the archiving and retrieval of all documents, software, and computer data
- Procedures for monitoring and auditing the compliance of operational computer systems

Where system obsolescence forces a need to migrate electronic raw data from one system to another then a process must be validated to ensure integrity (4). If such migration is not practicable then the raw data must be transferred to another medium (e.g., paper, microfiche) and this verified as an exact copy thereof, prior to any destruction of the original electronic records (4,7).

RESPONSIBILITY OF GCP/GLP QUALITY UNIT

The quality assurance organization has no mandated role in the development of computer systems other than defining QA functional requirements (8). Once the system has been validated, accepted, and installed then QA will be responsible for monitoring data collection until its reliability is confirmed in accordance with SOP, compliance of user SOPs, training, and security policies (8). Any performance problems should be communicated to the responsible management personnel in a timely fashion. QA should monitor corrective actions and unscheduled downtime records.

The British Association for Research Quality Assurance (BARQA) has interpreted international GCP/GLP regulations and expects QA personnel to (9)

- Conduct GCP/GLP awareness training, validation training, and change control training
- Review and approve validation and change control procedures
- Review quality plans and key validation documents (i.e., validation plan, requirements, test plan, test results, acceptance, record retention (archiving), change control
- Advise projects on software development
- Review changes (individually or part of periodic review process)
- Conduct system audits (including system development, software, operation and use)

In addition QA commonly provide general consultancy and advice on the interpretation of regulatory requirements for computer compliance.

Data Capture Systems
Subject Information Systems
Electronic diary cards are portable, hand-held systems designed to be programmed according to specific protocol requirements and are used by patients to record directly information on their condition and/or medication consumption during a particular study. They should be specified and designed so that they are highly prescriptive since they are used in a relatively uncontrolled environment (e.g., subject's home). Specific considerations for the validation of electronic diary cards are:

- Suitability for use by the target patient population
- E functionality, for example, time of data capture, checks for logical consistency, data
- Confirmation and auditability, provision of investigator signature
- Supplier auditing

- Usability, robustness, and integrity of both software and hardware
- Tamper-proof software, that is, modification for other purposes should not be possible
- Power backup in the event of expiry or removal of batteries
- Security, controlled by password, including access restrictions and integrity of data
- Transfer of the diary data to the host database, including any data modification, annotation or processing occurring before, during, or after the transfer
- Documented training of site personnel and individual patients.

Systems are also required to record whether all dispensed medication for a clinical trial can be accounted for at the end of the study. For each subject in the trial the amount(s) of dispensed medication are compared with the amount(s) consumed and the amount(s) returned. The returned supplies are then destroyed and certified as such. The amount(s) dispensed may come from a pharmacy system, and the percentage consumption within a dispensing interval could be derived as a measure of subject compliance. Specific points to consider during validation are therefore:

- The incorporation of any derived data algorithms
- Electronic transfer from and to other systems

Interactive Voice Response Systems
An interactive voice response system (IVRS) is a communications platform based on the telephone network used to coordinate key clinical trail activities and provide real-time information for study managers. By utilizing the telephone network, the system provides for a direct connection between clinical trial patient and the central study-specific database. This permits the collection of data in response to preprogrammed prompts from the system, ensuring the recording of key trail events and the provision information critical to the successful conclusion of the trail. Most IVR systems are individually tailored to each specific study based on requirements defined by the sponsor.

Specific considerations for the validation of IVR systems are:

- Formally agreed and documented sponsor requirements
- Formally documented design specifications
- Validation planning for sponsor specific project
- Supplier auditing
- Traceability between design and testing
- Formal change control system for data and system
- Validation of the data transfer process to the sponsor database

Medical Device Measurements
Medical devices used to take clinical trial measurements must comply with medical device regulatory requirements. This covers design controls and software validation (10). Another case study in this book deals with medical device validation (see Chapter 36, Case Study 18).

Optical Character Recognition
OCR systems recognize images as α-numeric data, as if the data had been entered directly from a keyboard. They do this via recognition engines, operating by template matching, feature extraction, neural networking, or a combination of these approaches.

There is an explicit reliance on operator involvement in the verification of the captured data, whereby the software presents the operator with uninterpretable input image for manual resolution. Validation needs to take account of all dimensions of the system, testing with a sufficiently varied selection of input image. Specific considerations for the validation of OCR systems are:

- Supplier auditing
- Reliability, calibration, and maintenance of scanners

- Correct identification by the system of the type and number of scanned input forms
- Functionality, for example, substitution, learning capacity, verification
- Reliability of interpretation of the specified field images by the recognition engine
- Handling of indeterminate data
- Training and competency of the operator
- Transfer to the host CDMS or analysis package

Bar Coding Systems

A bar code is a pattern of dark bars separated by spaces. The bar code is read by passing a beam of light over it. Light is absorbed by the bars and reflected by the spaces. The differences in reflection are sensed by the scanning device (e.g., light pen, hand held scanner, flat bed scanner) and converted into electrical signals corresponding to the widths of the bars and spaces which can then be decoded into the numbers and letters represented by the bar code. There are a number of different bar coding standards.

Specific considerations for the validation of bar coding systems are:

- The system used for creating the bar code labels, for example, acquisition of number, conversion of number to bar code
- Print quality of the bar code, for example, specks of ink in the spaces, edge definition of the bars, and print contrast between the bars and spaces
- Presentation of the bar code to the scanner, that is, creased labels, protective covering
- Robustness and maintenance of the scanning device
- Verification of decoding
- Control of the re-use of pre-printed bar codes.

Data Processing Systems

Data Management Systems

Clinical data management systems can be used in a wide variety of applications. At the study level, it should be possible to set up the database efficiently to allow easy access to the data. Where the functionality is available, points to consider during the validation include:

- Entry screens function as expected (e.g., range checks, lookup tables, autoencoding using the appropriate dictionary, derived data calculations)
- Entry screen fields correctly relate to database fields, fields are correctly defined in terms of format (e.g., character/numeric, length)
- Entry of confirmed missing values is possible

Data entry should include identification of individuals using combination of user-ID and password at the start of the data entry session. Automatic log-off is appropriate for long absences of individuals during operator sessions.

The validation of data capture should include:

- Verification, if part of the transfer process, should result in discrepancies between two manual entries being correctly identified, and their subsequent resolution should result in one correct entry on the database.
- The transfer process should enable single entry of certain data, for example, electronic laboratory data.
- Data should be loaded into the correct location, that is, table and field.
- The transfer process should detect duplicate records.
- The user should be notified of nontransferred data.
- Any data identifiers should be correctly assigned.
- The date and time of initial loading of each data item to the database should be recorded by the system, that is, the audit trail should commence at initial loading.
- Any auto encoding, if part of the transfer process, should function as expected using the correct dictionary for each coded variable.

Data checking should include edit checking, plausibility checking, range, and consistency checking. Any data derived should be validated. At the system level points to consider include:

- Libraries of standard data checks should be accessible
- Standard data checks should be adaptable for specific needs
- Study-specific data checks should be possible
- Study-specific checks should be correctly incorporated, with standard checks, into the study-specific editing functionality
- The checks should be executed correctly, that is, the correct checks should be applied to a data item at the appropriate time
- Data items accepted following a failed data check should not fail again unless the data change
- Failed data checks should remain flagged until resolved
- At the study level the set of specified edit checks should be tested using tailored dummy data to ensure the absence of false-positive and false-negative failures.

The management and use of the system, and related reference data (e.g., laboratory reference ranges and coding dictionaries), should be controlled by standard operating procedures. Such procedures should include taking data extracts, possibly as predefined reports. Extracts should be validated to demonstrate they correctly identify, combine/merge, and report data requested. At the system level the functionality of a reporting system should ensure that

- Template programs are available for easy adaptation,
- Study-specific programs can be easily developed,
- Program development takes place in a separate environment from the use of validated programs,
- Documentation of output should include source program, date and time generated, user, page number, and total number of pages,
- All programs, and subroutines or macros called within programs, used to produce formatted output from the CDMS for clinical reports should be validated.

Validation should ensure that the facility for locking/securing the database prevents unauthorized write access. The unlocking of a database should be strictly controlled by an SOP.

Randomization Systems

Randomization systems, which are usually used by statisticians and/or pharmacy staff, may provide any of the following:

- A list of random numbers
- A code-break envelopes
- A packaging labels for drug supplies
- An electronic file of the patient treatment codes to be incorporated into the study database after it has been locked.

Validation of the randomization system should be rigorous as randomization codes and code breaks, and their security, are key to maintaining the integrity of any clinical trial. The codes and code breaks, generated prior to the start of the trial for the packaging of medications, will not be linked to the data until the end of the study when the clinical database has been locked, which may be several years after the codes were produced. The following points should be considered during validation:

- The source of the core random number generator and its validation status
- The ability to reproduce the randomization schedule
- Storage of randomization codes and code breaks and access control

- Backup and restoration procedures and their regular testing
- Linkage to the clinical database management system, administrative system and drug supplies accountability system

Clinical Trial Review Tools

Validation of computer aided review tools, which may be used by in-house or regulatory reviewers to explore the project database on a read-only basis, should address both the system and the project-specific aspects.

The underlying code of the generic shell that comprises the tool should be developed according to the software development life cycle. Depending on the degree of sophistication of the system, testing should cover the following areas of functionality:

- Selection of compound
- Selection of trial
- Display of raw data
- Subsetting of data
- "Point and click" cascading menus (i.e., increasing or decreasing the level of detail or subsetting)
- Search facilities
- Display of graphical results
- Linkage between annotation facilities (for sponsor and/or reviewer) and related data
- Transfer of data between different software systems (e.g., from the SAS package to a spreadsheet)
- Analysis and reporting.

Testing for correct project and study set-up should demonstrate that the data have been loaded into the system correctly. This will involve checking:

- Completeness, correctness, and consistency of the labels and formatting
- Correct functioning of the screens
- Consistency of the viewed data with the project database
- Consistency of reports and views of data output to the screen with clinical trial report tables, listings, and original.

The different ways of viewing data may be too numerous to test exhaustively. Validation requirements, therefore, need to be realistic to ensure an appropriate level of overall confidence.

Statistical Analysis Systems

The statistical software systems used for analysis of clinical trial data can range from custom programs for specific statistical techniques to COTS packages. Such packages (e.g., the SAS system, SPSS, S-Plus) provide the user with a library of statistical procedures (e.g., analysis of variance, regression, generalized linear modeling, nonparametric methods), which can be accessed either by using the native programming language or by selecting the required options from the package's user interface.

It is generally considered that there is no requirement for validation of statistical packages such as the SAS system as entities in their own right. Nevertheless, any custom program written using the package's native programming language should be validated.

The supplier-supplied installation tests should be performed and documented to ensure that the software is functioning correctly within the specific operating environment. In addition, a suite of supplier supplied programs, test data, and results can be a valuable aid to validation. Repetition of all these tests should be considered each time there is a change to the operating environment.

These include one-off and standard programs and macros developed using either a nonstatistical programming language (e.g., Fortran) or the native programming language of a COTS statistical software package (e.g., SAS programs, SAS macros, SAS/AF applications).

It should be shown that statistical procedures and functions (e.g., SAS PROCS), supplied as part of a COTS product, are used correctly within the context of the program. Software that automates the data analysis process across a number of clinical trials should be validated in the same way as other supplier or custom (bespoke) systems. However, the validation requirements for trial specific, one-off programs written using COTS package native languages are reduced. Specific issues to consider during validation are:

- Statistical competency of the developer
- Precision and rounding errors
- Handling of missing data values
- Handling of unequal (unbalanced) treatment groups
- Handling of ties in nonparametric analyses
- Facilities for checking the underlying assumptions of the statistical model
- Facilities for excluding outlying observations from analysis
- Printing of intermediate values during calculations
- Statistical competency and training of the users
- Operating environment and conditions

Pharmacovigilance Systems

Pharmacovigilance systems capture, store, process, maintain, classify, and report adverse event data. Any such systems generating reports for regulatory authorities (e.g., expedited reports, periodic safety updates) and the interfaces into them from a variety of sources should be validated. Specific considerations when validating these systems are:

- Reconciliation of adverse event data from the clinical trial database and other
- Sources, with the pharmacovigilance database through electronic interfaces
- Development of programs to generate reports for regulatory authorities, for example, expedited and periodic reports
- Assurance that all cases known to the system have been appropriately reported in the appropriate time frame
- Electronic transfer to regulatory authorities

Electronic Transfer of Data and/or Software

Clinical data and/or software may be transferred electronically, by diskette or direct line, on a routine basis from investigator sites, contract research organizations or central laboratories to the company (and vice versa), between different company locations, between computer systems within a location, and from the company to regulatory agencies.

Specific considerations for the validation of electronic transfers are:

- Internet, intranet, and/or other communication technologies (e.g., group ware, modem-to-modem, cellular technology)
- Externally owned lines
- Communication medium (e.g., diskette)
- Security (e.g., encryption, passwords, virus protection, "fire walls")

Specifications of the transfer file:

- File format (e.g., ASCI I, comma delimited)
- Size of file
- Number of records
- Linkage of comments to numeric data
- Recovery following interruption of transmission
- Corruption during transfer
- Consistency of electronic file with source
- Backup and disaster recovery in both the sender and the receiver locations

Production Control Systems

These computer systems should be validated to the same standards as expected for manufacturing control systems. Reference can be made to other case studies in this book as applicable:

- Clinical trials production (e.g., kilo laboratory production systems),
- Analytical instrumentation,
- Inventory systems,
- Labeling systems,
- Environmental monitoring,
- Product tracking/monitoring.

Record Management

Protocol Management

Protocol management may include any or all of the following features:

- Controlled protocol authoring
- Electronic storage of protocols or data, either scanned in or created electronically
- Controlled distribution of protocols to, and retrieval by, multiple users
- Review and/or approval of protocols, for example, within a work flow component
- Publishing of approved protocols
- Index generation
- Retrieval of indexed documents

The validation issues include:

- The life-span and characteristics of the storage medium used, including the frequency and type of testing required
- The security levels of the protocols, including process-specific security such as that used for electronic signatures
- Version control of protocols including audit trail
- Validity period of printed/published protocols, for example, SOPs
- The qualifications, training, and competency of users
- Indexing functionality

In most cases, there will be a requirement for a protocol to be appropriately approved and signed-off. Options include the scanning and storage of the signed document, scanning, and storage of the signature(s) associated with an electronic document and the use of electronic signatures. It is important to define the "master" version (i.e., as paper or electronic). Signatures should be verified and stored with associated protocols.

SOP Management

Management systems for SOPs should establish and validate:

- Workflow for approval of SOPs
- Electronic records and signatures
- Facility for user requests to change an SOP
- Storage of forms and templates in original software
- Storage of previous, current, and under-revision versions
- Controlled distribution.

Specific validation issues include:

- Access security, especially write access to approved SOPs
- Documentation of notification of new/revised SOPs to all appropriate staff

- Version control
- Integrity of the system, especially when replicated across servers
- Control of printed versions of SOPS

Issues to be considered during validation include:

- Testing of all possible routes to ensure that a document does not become suspended within the system
- Testing of parallel tasks to ensure that the result of those tasks is the same regardless of their sequence in real time
- Linkage and preservation of electronic annotations
- Corruption of the master document by annotations
- Printing of document and annotations.

Other validation requirements that may be applicable have been discussed under protocol management systems.

Regulatory Submission and Electronic Publishing Systems

Electronic regulatory submissions combine components from specific systems, for example, computer-aided review tools and electronic document management systems (EDMS). Electronic publishing systems assemble electronic documents and images into electronic dossiers. The validation requirements of the publishing system over and above the requirements for each component system should be assessed.

Electronic Document Management Systems

Document management refers to procedures or systems designed to exert an intelligent control over the creation, management, and distribution of documents. EDMS may include any or all of the following features:

- Controlled document authoring
- Electronic storage of documents or data, either scanned in or created electronically
- Controlled distribution of documents to, and retrieval by, multiple users
- Review and/or approval of documents, for example, within a work flow component
- Publishing of approved documents
- Archiving of documents for completed projects

Typically, EDMS may need to address a range of issues including version control, access control, organization and management, workflow, imaging, publishing, document reuse, indexing, and searching.

The document types that may be stored within the system may have a wide variety of file formats and sources and range from just key documents to the totality of documents generated for a project. For each type of document stored in the EDMS, it is important to define the "original" version (i.e., as paper or electronic).

The validation issues for EDMS include:

- The life span and characteristics of the storage medium used, including the frequency and type of testing required
- The security access of the documents and the system, including process-specific security such as that used for electronic signatures (see sect. 9.10)
- Version control of documents including audit trail
- Continuing readability of documents through technological changes, for example, the use of portable document format (PDF) file type
- Validity period of printed/published documents, for example, SOPs
- The qualifications, training, and competency of users
- Indexing functionality

Optical images may be produced by scanning in a paper document or a faxed image into the system. Apart from general configuration and installation requirements, specific validation considerations should include:

- Procedures for calibration and maintenance of scanners
- Definition of the master record, that is, paper version or electronic image
- Routing of images to appropriate locations
- Interfaces with other systems
- For fax-to-image, correction of transmission errors
- Readability of retrieved images
- Image quality prior to destruction of the original document
- Search and sort capability

Training Record Systems

Regulatory authorities go not generally inspect these systems, instead they inspect individual training records. Such systems should however be validated to ensure their reliability and performance. Specific validation issues to be addressed include (11):

- Testing record retrieval times
- Backup and recovery procedures
- Electronic record/signature controls

Archive Repository

The repository may range from a specific directory on a server, with a work group password protection, to a software package controlled database repository implementing full database security controls. There may also be a requirement to produce and store multiple renditions of a document within the repository. Specific validation issues include:

- Continuing readability following software upgrades
- Integrity of the document during conversion
- Production, storage, and retrieval of multiple renditions of a document
- Storage of signatures associated with documents

An important part of any electronic archive system is a policy or SOP that will affect the validation effort, including:

- Which documents should be kept in hard copy form and which may be kept only in their electronic form
- How long documents are maintained on the system
- How long hard copies of documents are kept
- The need for off-site electronic backup when hard copies of documents are destroyed
- The possible uses of the documents, including whether they may be required by a court of law
- Access to the archive
- Storage criteria for electronic media and any special considerations, for example, refreshing tapes/disks

E-MAIL AND INTERNET TECHNOLOGIES ISSUES

Clinical data and/or software may be transferred electronically, by diskette or direct line, on a routine basis from investigator sites, contract research organizations or central laboratories to the company (and vice versa), between different company locations, between computer systems within a location, and from the company to regulatory agencies.

Today the most common method of direct line transfer is likely to be an e-mail message attachment via an Internet link. This is fraught with several issues, the prime of which are:

- Lack of audit trail data
- Concerns over system administration

- Robustness of e-mail systems
- Security during data transfer

As a result of these issues, the specific considerations for the validation of electronic transfers and the use of e-mail and Internet technologies are:

- Internet, intranet, and/or other communication technologies (e.g., group ware, modem-to-modem, cellular technology)
- Externally owned lines
- Communication medium (e.g., diskette)
- Security (e.g., encryption, passwords, virus protection, "fire walls")
- Specifications of the transfer file:
- File format (e.g., ASCI I, comma delimited)
- Size of file
- Number of records
- Linkage of comments to numeric data
- Recovery following interruption of transmission
- Corruption during transfer
- Consistency of electronic file with source
- Backup and disaster recovery in both the sender and receiver locations
- Validation status of systems at either end of the transmission tunnel

ELECTRONIC RECORDS AND ELECTRONIC SIGNATURES

This case study will not discuss electronic record/signature controls in detail, as these are presented in detail elsewhere in this book. Nevertheless key points are summarized below.

US 21 CFR Part 11 Regulation

Since it became effective, there has been much discussion about the impact of FDA 21 CFR Part 11 (12), in particular in the manufacturing arena, with some progress being made toward achieving full compliance. However, it is generally accepted that such progress has been slow to commence within the clinical trails domain. Even today, several years after the rule became effective, it is more likely that any assessment of a computerized system for compliance with 21 CFR Part 11 will result in a number of issues being identified as falling short of what is required.

In general 21 CFR Part 11 does not introduce anything radically new to the debate about what requirements should be placed upon systems used for GCP processes. Most of the requirements detailed within the rule are essentially good IT systems and electronic records practice, with some additional emphasis placed upon controls for electronic signatures. What should be recognized is that compliance with the rule is not simply one based on a technological approach. Many compliance issues are related directly to putting in procedural systems to support any technology introduced. Hence, there is not an unreasonable expectation by the FDA that many of the procedural systems will have already been introduced within organizations, and, where gaps exist, plans for remediation will have been drawn up.

Specific 21 CFR Part 11 considerations in the validation of clinical research computerized systems include:

- Changes to data/software
- Audit trail—design and integrity
- Audit trail—paper versus electronic
- System controls for electronic signatures
- Event logging—date and time stamp synchronization
- Procedural controls for granting access and permissions
- Company security policy and action on detection of fraudulent activity
- Personnel training/understanding of electronic signature authority/responsibility

Other International Regulatory Requirements

Complete copies of records must be available for inspection, review, and copying by regulatory authorities.

Audit trails should be generated to log the creation, modification, and deletion of electronic records. From this information, it should be possible to reconstruct the electronic records as they existed at any date and time in the past. It should be possible to associate all changes to data with the person making these changes. Audit trails therefore need to be time-stamped with date and time changes were made. Computer systems should provide for the retention of full audit trails to show all changes to the data without obscuring the original data (4). Consequently, audit trails need to be protected such that no direct modification of the stored information can be made.

In addition to the requirements defined in 21 CFR Part11, the legal requirements surrounding the attachment of a signature to an electronic record is controlled under EEC legislation, in the form of Directive 1999/93/EC—a Community Framework for Electronic Signatures (13). The most common form of electronic signature is through the applying of unique combinations of user-ID and password. It is recommended that passwords be changed at established intervals (4). Biometric signatures can also be used. When considering the application of signatures to electronic records it is worth considering the actual nature and GCP criticality of the action. In many cases, a signature is being applied to indicate the completion of an event or the attainment of a milestone. In many less frequent cases is the signature actually being applied to indicate compliance or achievement of a GCP requirement. The distinction here is to be able to distinguish whether the signature is being used for identification or authentication purposes. This distinction can be clarified by this definition, provided by Julian Ashbourn (14):

> *Authentication* refers to the verification of a claimed identity. In other words, the user wishes to log on to a network or service and claims to be a certain person.
>
> *Identification* seeks to identify a user from within a population of possible users, according to a characteristic or multiple characteristics, which can be reliably associated with a particular user without an identity being explicitly claimed by the user.

The distinction should be made between identifying signatures that must be incapable of being repudiated in a court of law (e.g., GCP critical) and those that are not critical enough to warrant the nonrepudiation safeguards.

REGULATORY INSPECTION

Inspectors will normally want to identify those computer systems involved with the particular clinical study under investigation. They will be interested in data capture, processing, and retention. User interaction will also be a key topic of interest in regard to how the computer system assists making decision, collation of study data, and submissions. Examination of validation documentation and methods of testing may be requested for specific functions. Occasionally demonstrations of specific functionality might be requested. Security access controls and general administration SOPS on the other hand are often discussed. The main focus is likely to be data integrity, and computer system operation and maintenance.

As an example topics covered at a recent GCP inspection by the FDA of a pharmaceutical manufacturer in North America included:

- Study-specific data entry system—validation, supplier audit, change management, and test protocols
- Data entry system security—virus protection, access management, disaster recovery, archive and retention
- Electronic records/signatures—assessments and follow-up plans
- Laboratory information management systems—data transfer from clinical systems to networked data management applications

A further example involves an FDA investigation related to a specific submission. In this example, the investigators not only made observations related to the data handling systems employed directly by the pharmaceutical company, but went on to investigate the systems employed by a contract research organization employed to collect data related to the study. Their resultant observations made it clear that the FDA expected the sponsor organization undertake due diligence when it comes to employing such organizations and failure to exercise such diligence could result in the issuance of a "483" observation. The same investigation also reviewed the procedures employed for the electronic transmission of electronic data between the CRO and the sponsor, resulting in observations related to the failure to adequately safeguard the security and integrity of the data that was subsequently part of the submission.

No matter how well a pharmaceutical manufacturer believes it conducts validation, it will count for nothing unless during an inspection the regulator understands what has been done and can easily find his way around supporting documentation. To this extent a key feature in any validation exercise is inspection readiness.

Seven key elements that must be ready for inspection are listed below; others can be added as appropriate to how a pharmaceutical manufacturer wishes to manage regulatory inspections.

- Inventory of systems
- System/project overviews
- Validation plans/reports and reviews
- Presentation slides
- Internal briefing papers
- Document map
- Trained Personnel

Using terminology that the various regulatory authorities are familiar with will help enormously for example align to GAMP®5 (15). Try to avoid use of company specific jargon. IT staff especially tend to freely use company specific acronyms and terminology. Time should be taken to explain topics during an inspection and to prepare IT staff on what to expect during an inspection. They are typically not familiar with regulatory inspections, but with more and more clinical systems using databases and/or client-server technology, there is a much higher likelihood that IT staff will be required to support inspections.

REFERENCES

1. FDA. Computerised Systems Used in Clinical Investigations, Guidance for Industry. Rockville, MD: Food and Drug Administration, 2004.
2. EU. Good Clinical Practice for Trials of Medicinal Products, 1991.
3. ICH. Guideline for Good Clinical Practice, ICH Harmonised Tripartite Guideline, International Conference on Harmonisation of Technical Requirements for Registration of Pharmaceuticals for Human Use, 1996.
4. UK Department of Health. The Application of GLP to Computer Systems, The Principles of Good Laboratory Practice, United Kingdom Compliance Programme, London, 1995.
5. ACDM/PSI. 'Computer Systems Validation in Clinical Research: A Practical Guide', Version 1.1, December 1998.
6. OECD. Principles of Good Laboratory Practice to Computerised Systems, Paris: Organisation for Economic Co-operation and Development, 1995.
7. FDA. 21 CFR Part 11 Electronic Records; Electronic Signatures—Scope and Application, Guidance for Industry. Rockville, MD: Food and Drug Administration, 2003.
8. DIA. Computerized Systems Used in Non-Clinical Safety Assessment: Current Concepts in Validation and Compliance (known as Red Apple II), Drug Information Association, 2008.
9. BARQA. Regulatory Compliance and Computer Systems, Conference Proceedings, 1997.
10. U.S. Food and Drug Administration. General Principles of Software Validation; Final Guidance for Industry and FDA Staff. Rockville, MD: Food and Drug Administration, 2002.
11. Gallup D, Beauchemin K, Gillis M, et al. Selecting a Training Documentation/Record-Keeping System. PDA J Pharm Sci Technol 2003; 57:(1).
12. FDA. Electronic Signatures and Electronic Records, Code of Federal Regulation Title 21, Part 11. Rockville, MD: Food and Drug Administration, 1997.

13. Directive 1999/93/EC of the European Parliament and the Council of 13th December 1999 on a Community Framework for Electronic Signatures.
14. Ashbourn J. Biometric Definitions. Available at: www.ntlworld.com/avanti/authentication.htm, 2000.
15. ISPE. GAMP®5: Risk-Based Approach to Compliant GxP Computerised Systems. Tampa, Florida: International Society for Pharmaceutical Engineering. Available at: www.ispe.org, 2008.

23 | Case Study 5: Control and Monitoring Instrumentation

Peter Coady
Pfizer

Tony de Claire
APDC Consulting Ltd.

INTRODUCTION

Measurement and control instrumentation must be applied and maintained in support of computerized system validation. Unsatisfactory instrumentation can cause significant operational problems. This case study embraces the design and validation of both standard and intelligent instrument applications, and briefly discusses special instrument systems (shutdown systems and analyzer packages). Particular attention is focused on producing auditable documentation in compliance with GMP requirements (1,2).

The "approach" is based on good practices applied to primary/bulk process control systems and is equally applicable to secondary manufacturing automation and to support utilities. It ensures that the "user" is in control of the measurement instrumentation and other in-line devices applied within the GxP environment. In general, the levels of detail addressed would also be of benefit in documenting the technical aspects of IT systems and Infrastructure and associated cabling where appropriate standards and practices have not been followed.

Instrumentation is the vital link between the manufacturing process and the control system. Instruments are the eyes (i.e., transmitters, sensors) and limbs (i.e., actuators, positioners) of a process control system and enable it to perform the actions that were once performed by operators and/or laboratory technicians. The number of instruments that can be involved with the dynamic operation of a pharmaceutical manufacturing process can be large (2000–3000 items). In most cases, they are remotely installed from the control room environment and operate unsupervised except during routine maintenance work.

The correct operation of any computer control and data acquisition system is totally reliant on the information it receives from the instrumentation. If an instrument should malfunction, data integrity and the predefined control actions will be affected. Therefore, it is essential that an instrument is carefully chosen to be fit for purpose (i.e., correct type, size, materials, accuracy, repeatability, reliability, documentation, etc.) to enable confidence to be gained in its ability to perform its intended function (3–5).

SUPPLIER SELECTION

Instrument suppliers are generally selected on the basis of company or site "standards." Where there is no stated preference or knowledge of a particular instrument, selection can be made using a technical evaluation and a tender process. Companies that supply special instrument systems (e.g., shutdown systems, analyzers), companies that provide design/validation consultancy, and subcontractors (engineering design contractors, site installation contractors, panel manufacturers) should be prequalified to determine their suitability, from both an engineering and a commercial perspective, to receive the tender inquiry documents. Specialist instrument suppliers and consultants would generally be subject to a Supplier Assessment prior to any order being placed. Similarly, suppliers wishing to be included on the preferred vendor list for the site will also be subject to a Supplier Assessment.

A key element of supplier assessment is evidence and contractual commitment to a level of calibration/test procedures and record documentation that supports front-line measurement and control instrumentation.

Supplier Assessments should be conducted by suitably trained/qualified personnel against the applicable ISO 9000 (6) series standard, with special reference being made to software quality guidelines [e.g., ISO 90003 (7), TickIT (8)] for instrument systems involving software. The software assessment may cover the development of both application software and core (operating system level) software, depending on the type of system to be supplied. The decision to perform a Supplier Assessment, any follow-up assessments, and the type of assessment to be performed (e.g. on-site audit, postal audit, etc.) can be based on a documented risk assessment as detailed in Appendix M2 of the GAMP®5 Guide (9). Follow-up assessments should be considered for critical systems.

INSTRUMENT APPLICATION DESIGN

The design of instrument applications is an interactive process centered around an instrument schedule (sometimes referred to as a "loop schedule") and process data. The instrument application design process is shown diagrammatically in Appendix 23.A and its chronological position, in relation to an overall project development life cycle, is presented in Appendix 23.B.

The instrument schedule is generated from an approved set of process and instrument diagrams (P&IDs) and is used to identify all instruments associated with a manufacturing process. The instruments in the schedule are grouped in tag number (loop) order along with an identification of all associated documentation on which they appear.

Process data are provided by the end user and include critical process parameters and data that are generated from, or with reference to, the end user's pharmaceutical product manufacturing specifications. In consideration of a risk-based approach to validation (9) and good practice calibration regimes (10), the GxP criticality of each control and monitoring instrument should be identified on the instrument/loop schedule to form a verifiable reference to critical parameter definition in process and/or system design specifications. This level of record could also be used to identify "business-critical" measurements that may not be identified as direct GxP critical measurements.

A HAZOP study (sometimes known as a hazard study or hazard analysis) is normally carried out on the P&IDs during the design phase to determine where potentially hazardous conditions could occur during the operation of the process and the circumstances that lead to them. The results of the HAZOP study are used to generate safe working practices and the selection of suitable safety devices including rupture disks, safety relief valves, dedicated safety shutdown systems [solid state or programmable logic controller (PLC)-based systems], and hard-wired (relay-based) trip systems.

The types of documentation produced during the instrument application design process are listed below, and the content of each document is described in more detail in Appendix 23.C.

- Instrument schedule
- Instrument specification/data sheets
- Cable block diagrams
- Cable schedule
- Pneumatic tubing schedules
- Termination drawings
- Miscellaneous label schedules
- Field panel specification and drawings
- Field junction box drawings
- Electrical hookup (loop or wiring diagrams) drawings
- Pneumatic hookup drawings
- Process hookup drawings
- Instrument layout (location) and cable/tubing routing drawings
- Earthing schedules and drawings
- Miscellaneous drawings (control room/instrument room layouts)
- Instrument installation specification
- Package plant instrument specification
- Electrical and instrumentation interface panel
- Any special instrument specifications and wiring diagrams

Because of the interactions between the various types of instrument design documentation, and the sharing of input information, many of the documents can be produced in parallel. Each document should contain the required content (Appendix 23.C), presented precisely and completely, to enable that part of the plant to be constructed and/or maintained in a proper and auditable manner.

MANUFACTURING INSPECTIONS

Equipment that is either high cost or complex is often inspected during manufacture by the customer, an engineering design contractor, or a third-party inspection agency. During the inspection visit, the equipment is checked against its design specification, and the manufacturing/fabrication progress is assessed against the delivery program. Any problems found are recorded on an inspection report form, along with agreed on corrective actions; and copies are given to the manufacturer and the inspector's employer/customer, as applicable.

CALIBRATION

Before any instrument can be calibrated, it is essential to know the context of the measurement and the impact that it is capable of having.

There needs to be a system of calibration management, which defines what activities are required to be carried out, when, by whom, and why. There should be a prioritized, effective method for meeting the calibration needs of the pharmaceutical industry and satisfying the requirements of the regulators.

The importance of being in control cannot be overemphasized, and making the correct decisions depends largely on whether the correct information is acquired. This information is often gathered through measurement and it is, therefore, critical that it is both reliable and accurate.

For example, it is essential that during the steam sterilization of a fermenter, clean saturated steam at a temperature of $121°C \pm 2°C$ is maintained for the appropriate amount of time, as temperatures of $121°C$ or greater are required to kill off the bacteria. In manufacturing, however, temperatures in excess of $123°C$ may denature the medium within the fermenter. Instrumentation used for monitoring and controlling the temperature should, therefore, be both reliable and accurate.

The calibration of process control instruments may drift, because of the complexity of their construction and the environmental conditions in which they operate. This drift may be significant within a short time frame and accounts for the requirement for periodic recalibration.

Concentrating resources in the most important areas will maximize their effectiveness. The process of risk and criticality assessment will help in achieving this and in reducing costs.

This document provides education in the importance of calibration and the need to prioritize. The assurance of control and compliance within stated parameters can be achieved only through the implementation of effective calibration management. Simply put, if a parameter is not measured, control cannot be assured.

The importance of the measurement error depends on what is being measured, that is, assurance that a temperature measurement instrument is accurate to, for example, $\pm0.5°C$ during the critical step of a process is more important than an instrument measuring the temperature of an office where a $\pm3°C$ error is acceptable.

The key message is that the performance of the instrument is such that there is high confidence that the process has remained within the desired range, that is, indicated value plus possible error will remain within desired limits.

An assessment needs to be made on the criticality, accuracy, and stability of all instruments that can be calibrated to give a high degree of confidence that the process has remained within the desired range, and accuracy within tolerance.

CRITICALITY AND RISK ASSESSMENT

All instrumentation used for pharmaceutical manufacturing, whether directly or indirectly involved with the process, should be individually assessed against their specific use to establish their criticality to the process and therefore determine the precise calibration requirements.

The Criticality Impact and Risk Assessment is an essential part of the project phase of introduction of new equipment and must be completed prior to handover of the equipment to routine manufacturing operations. The outcome of the assessment will be the calibration requirements for the individual instruments such as the range, tolerance, and calibration frequency. This information will be used to populate the calibration register and calibration methodology.

The criticality assessment should consider the impact of the instrumentation on product safety, identity, strength, purity or quality plus plant/environmental safety or business efficiencies. The assessment should be performed by a multidisciplinary team including, but not limited to, representatives from production (process owner or expert), quality assurance, and engineering.

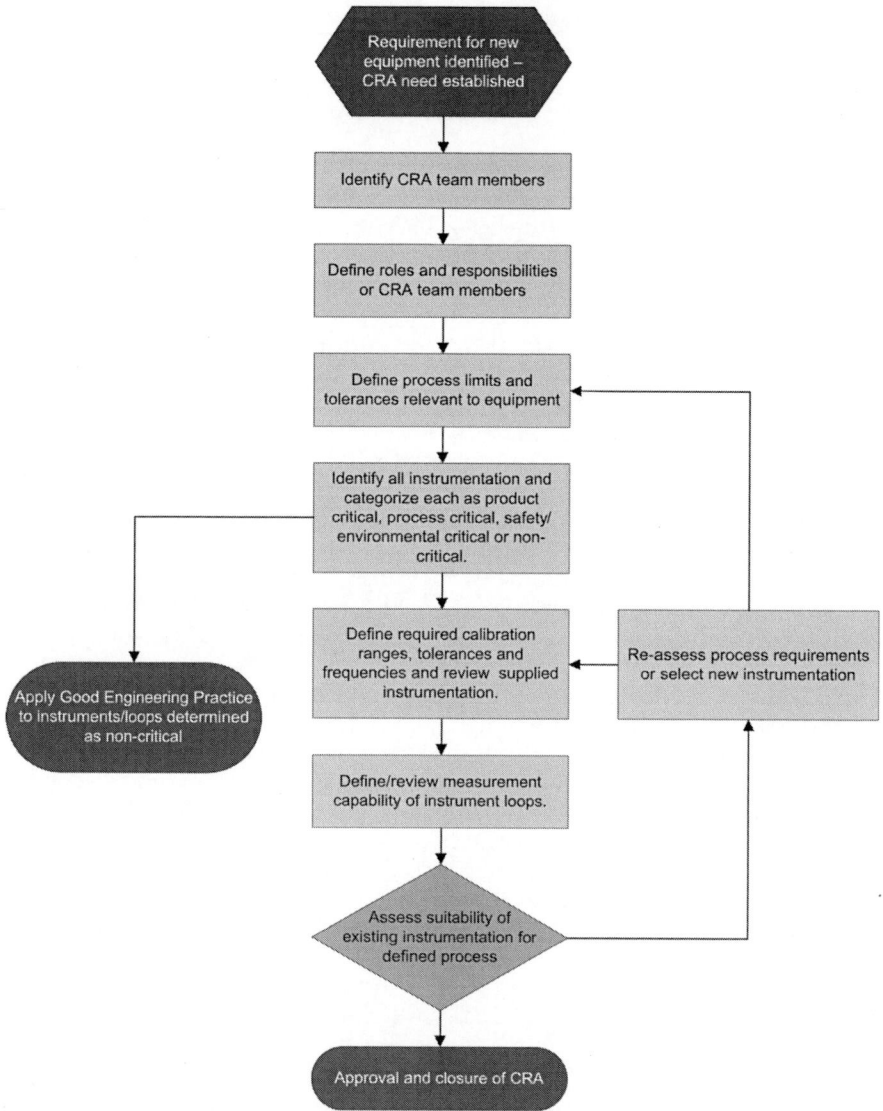

KEY REGULATORY REQUIREMENTS

The following statements are related to critical instruments, they outline the regulatory requirements for a successfully managed instrument calibration program.

1. A method of identifying the criticality of instruments used in production and control should be in place, documented and maintained.

2. Control, weighing, measuring, monitoring and test equipment that is critical for assuring the quality of intermediates or APIs should be calibrated according to written procedures and an established schedule.
3. Equipment calibrations should be performed using to a nationally, or internationally, recognized standards traceable to certified standards, if existing.
4. Records of these calibrations should be maintained.
5. Each instrument should have a permanent master history record.
6. The current calibration status of critical equipment should be known and verifiable.
7. Instruments that do not meet calibration criteria should not be used.
8. Deviation from approved standards of calibration on critical instruments should be investigated to determine if these could have had an impact on the quality of intermediate(s) or API(s) manufactured using this equipment since the last successful calibration.
9. All instrumentation should be assigned a unique number and all product, process, and safety critical instruments should be physically tagged.
10. Calibration frequency and process limits should be defined for each instrument.
11. All electronic systems used for calibration management for systems, which affect products for supply to the United States must comply with the Food and Drugs Administration (FDA) "21 CFR Part 11, Electronic Records, Electronic Signatures."
12. Calibration measuring standards should be more accurate than the required accuracy of the equipment that is being calibrated.
13. All instruments used should be fit for purpose.
14. Balances and measuring equipment of an appropriate range and precision should be available for production and control operations.
15. Measuring, weighing, recording and control equipment should be calibrated and checked at defined intervals by appropriate methods. Adequate records of such tests should be maintained.
16. There should be documented evidence that personnel involved in the calibration process have been trained and are competent.
17. A documented change management system should be established.
18. These statements should be read in conjunction with the following documents:
    - 21 CFR Part 211, "Current Good Manufacturing Practice for Finished Pharmaceuticals"
    - 21 CFR Part 11, "Electronic Records, Electronic Signatures"
    - "Rules and Guidance for Pharmaceutical Manufacturers and Distributors," Medicines and Healthcare Products Regulatory Agency (MHRA), 2007

PREDELIVERY TESTING AND CALIBRATION

A detailed account of the calibration life cycle processes can be found in the GAMP Good Practice Guide on Calibration Management (10).

Calibration and Test Equipment Requirements

The manufacturer should possess test equipment to enable all manufacturing tests and inspections to be performed. All test equipment used by the manufacturer must have a standard of accuracy better than the stated accuracy for the instrument(s) to be tested. The amount and nature of the scale of the improved accuracy should be based on the results of the Criticality and Risk Assessment results. All applicable test equipment must have a valid calibration certificate issued by a calibration laboratory that is certified to either a national or international standard [e.g., United Kingdom Accreditation Service (UKAS)] for calibrating the specific types of instruments concerned. This equipment must also have a calibration certificate showing as found and as left results and the same level of sign off. There should also be documentary evidence of traceability and the uncertainty budget applied.

MEASUREMENT UNCERTAINTY

Once the process proven acceptable ranges, process tolerance and Instrument ranges have been defined then consideration can be given to the uncertainty that will be intrinsic to the measurement loop as a consequence of Instrument Accuracy, environmental conditions, calibration standards and procedures.

The sciences of metrology, measurement uncertainty, and statistical analysis can be complex and should be considered and scaled on the basis of risk. For the majority of cases however the key contributors to uncertainty in the loop will be the accuracies of the loop components themselves combined with that of the Secondary Standard employed to calibrate them. As these are usually defined as part of a class B estimate of uncertainty then a simple Root of Sum of the squares of all the error values can be utilized to characterize the total uncertainty or likely maximum error in the loop. It is then quite common for the loop uncertainty to be multiplied by a coverage factor of two sigma (two standard deviations) to generate the calibration failure limits for the critical loops. In most situations this coverage factor is adequate to ensure that the "design space" for the process is not compromised by measurement system capability.

For noncritical loops an informed engineering decision is all that is required to establish calibration failure limits since by definition these are low risk and have no impact on critical aspects of the process. Calibration of these devices is therefore only designed to establish that the components are operating correctly and the data can be relied on to make non-GMP/regulatory decisions. Good Engineering Practice however dictates that the rationale used for deriving these limits should be documented and that the failure limits should be recorded on the instrument data sheet for future reference.

Factory Testing

The manufacturer should have written test procedures and should test all equipment supplied against these procedures prior to delivery. The manufacturer's testing should comprise the physical checking and operational and functional testing of instrumentation (e.g., valves, transmitters) and equipment (e.g., panels, analyzers) to be supplied. All tests must be fully documented, the results recorded, and the appropriate test sheets signed off by an authorized person. Calibration and test documentation should be made available for the Pharmaceutical end user's records.

Factory Calibration

Factory calibration should be carried out against the instrument specification/data sheets supplied. The calibration activities should address the following areas, as applicable:

- Process operating ranges
- Instrument operating range(instrument calibration capability)
- Required process accuracy
- Repeatability
- Hysteresis effects
- Switch set points
- Condition of switching (e.g., on a rising or falling measured variable)
- Switch action (e.g., open or close on fault condition)

In most cases, the manufacturer's own calibration procedures should be acceptable, as they form part of the manufacturer's own performance guarantees and/or quality certification. If special calibration is required, then companies with custom-built test facilities must be used (e.g., a magnetic flowmeter that requires a specific certification will require calibration in a flow rig that has been certified by an approved certifying authority).

DESIGN REVIEW

Design review [also called design qualification (DQ)] is the name given to the technical and quality assessment of the instrument application design engineering process, construction documentation package, vendor documentation, factory inspection report forms, and calibration data. The purpose of the design review is to verify through defined procedures and support documentation that the individual items of instrument application design have been designed and approved so that they meet the needs of the customer and the contractor's project and quality plan. The findings of the design review and the documentation inspected should be formally recorded, and a design review report should be produced.

Manufacturing Documentation Requirements

The manufacturer should provide copies of the following documents prior to the delivery of the equipment. It should be specified that documents are written in the local national language wherever possible.

Factory Calibration Certificates

The calibration certificate should include the following information:

- Instrument/device serial number
- Instrument specification/data sheet number
- Loop tag number
- Model number
- Certificate numbers of the test equipment used for the calibration
- Date of calibration and the validity period of the calibration certificate
- Calibration data
- Limits of uncertainty (for critical instrumentation)

Further information on calibration certificates can be found in the GAMP Good Practice Guide on Calibration Management (10).

Other Equipment Test Records

Copies of the test records for panels and associated instrumentation/devices should be provided. Test records can include the following:

- Copies of the signed and approved test record sheets
- The manufacturer's own quality system compliance documents
- Electromagnetic compatibility (EMC) directive (11), declaration of conformity certificates (self-certification) for equipment containing European Community (EC) CE-marked equipment (e.g., panels)

Hazardous Area Approval Certificates

Copies of hazardous area approval certificates for all applicable equipment should be provided. The approval certificates are issued by national or international approvals bodies, including the following:

- British Approvals Service for Electrical Equipment in Flammable Atmospheres (BASEEFA), which, along with other European approvals bodies, provides certification to the EC ATEX user directive (12)
- Factory Mutual Research Corporation (FM) and Underwriters Laboratories Inc. (UL), U.S.A.
- Canadian Standards Association (CSA)
- Standards Australia

Material Certificates

Material certificates are normally only required where the materials of construction were specified on the instrument specification/data sheets to comply with process or environmental requirements (e.g., valve bodies and trims).

Construction Documentation Package

The construction documentation package comprises the instrument installation specification, drawings, and documentation listed in Appendix 23.C, "Instrument Installation Specification." During this phase, and thereafter, it is important that the drawing register for the manufacturing process is maintained on-site and that only the latest revisions of drawings are used.

The instrumentation construction documentation package may also reference supplementary information contained in the following documentation:

- Process design documentation [e.g., P&IDs, engineering line diagrams (ELDs), utility line diagrams (ULDs)]
- Piping design documentation (e.g., piping isometric drawings)
- Process unit design documentation (e.g., vessel connections)
- Mechanical design documentation (e.g., package plant vendor installation drawings)
- Computerized control system documentation (e.g., termination and interconnecting cabling drawings)

EQUIPMENT DELIVERY, INSPECTION, PROTECTION, AND STORAGE

On arrival at site, instrumentation and equipment should be checked against the delivery note, checked for damage, and then either preinstallation tested and installed or put into a suitable store until required. Any discrepancy or damage should be recorded and reported to the supplier through the contractual channels established for the project. Rejected items should be stored and controlled separately from accepted items.

Instrumentation and equipment (e.g., panels, junction boxes) that cannot be installed on delivery must be housed in a properly constructed and conditioned store and protected from dust and moisture. Completion of control rooms should be programmed to permit the installation of panels immediately on receipt to minimize handling. If the control room heating system is not in operation, temporary heaters must be installed to ensure that the panels and instrumentation are kept within acceptable temperature and humidity limits.

Throughout the construction period, instruments that are not provided with housings must be adequately protected by covering with heavy duty plastic bags of an approved type or by applying more robust protection where necessary. The protection of instruments and the provision of covers is generally the responsibility of the installation contractor, and should be controlled.

SITE PREQUALIFICATION
General

All instruments should, wherever possible, be subject to a preinstallation test; this test should commence as soon as practicable after the receipt of the instrument on-site. The object of preinstallation testing is to ensure that each instrument has been supplied in accordance with its specification, is functionally correct, and is in working order. Where the preinstallation test is not specified or where circumstances prohibit carrying out the prescribed test, the installation contractor must propose a suitable test method for approval by the customer.

The tests should be performed as described below and with due consideration given to the manufacturer's recommended test methods. Adjustments must be carried out in accordance with the manufacturer's instructions. Any deviation from this must be approved by the customer and supplier prior to testing. All testing should be carried out in accordance with authorized test procedures. All tests must be fully documented, the results recorded on the appropriate preformatted test sheet(s), and the test sheets signed off by an authorized person.

Instrument testing should preferably be carried out in a calibration workshop. However, instruments that form part of an integrated system or control panel may be tested in the control room or instrument room after installation, using portable test gear and/or simulation equipment. All instruments that require calibration must be calibrated in both the upscale and downscale directions and, if necessary, adjusted until their accuracies are within the limits stated by the manufacturer. On completion of the tests, the instrument must be suitably cleaned and protected in accordance with the manufacturer's recommendations.

A detailed account of the calibration life cycle processes can be found in the GAMP Good Practice Guide on Calibration Management (10).

Preparation for Site Testing

The following checks should be carried out before preinstallation testing commences:

- All test equipment used by the installation contractor must be approved by the customer and/or engineering design contractor, and must have a standard of accuracy better than the manufacturer's stated accuracy for the instrument(s) to be tested. Test equipment must have a valid calibration certificate issued by a calibration laboratory that is certified to either a national or international standard (e.g., UKAS) for calibrating the specific types of instruments concerned.
- The instrument must be checked for damage (e.g., damage to doors, linkages). Any such damage must be rectified and approved before any tests are attempted.
- The data plate on the instrument must be checked for agreement with the information contained in the appropriate instrument specification/data sheet.
- A suitable means must be provided for simulating the required process conditions, and test gauges or meters must be made available with a sufficient degree of accuracy for the tests to be performed.
- The instrument to be tested should be mounted in the correct plane on a rigid and vibration-free stand or structure.
- The manufacturer's "Instruction Book" must be made available.
- All tests must simulate as closely as possible design process conditions.
- Tests must not be carried out on electronic instruments until an adequate warm-up period has elapsed. Wherever possible, instruments must be energized for at least 24 hours prior to testing.
- The instrument to be tested must be properly prepared prior to testing by the removal of any shipping stops and the installation of any miscellaneous components (e.g., charts, calibration fluid/material, oil).

Test Status Indication

On completion of each site test, the stage reached in the testing procedure should be clearly indicated. A typical method would be affixing to each instrument or installation a colored label conforming to the following code:

- Blue: preinstallation tested
- Yellow: pressure tested
- Green: cables tested
- Red: precommissioned
- White: test failed (a written message may be added giving the reason for failure)

This identification must be shown on all components in the loop, thereby making all personnel aware of the current status of any instrument and its installation. The label can also include the date of the test if this needs to be tracked.

Connecting an Energy Supply

The following procedure is common to all instruments that require an energy supply source and that generate a signal output.

Pneumatic Instruments

Connect the air supply and adjust the air supply regulator to the correct setting (e.g., 1.4 bar for a standard transmitter with an operating range of 0.2–1.0 bar, 20 pounds per square inch gauge (psig) for an operating range of 3–15 psig). Connect the output to a suitable test gauge via a capacity chamber (approximately 0.5 L capacity).

Electronic Instruments
Connect a suitable power supply. Connect the output to a suitable test meter, preferably a digital voltmeter, installed across a current dropping resistor.

Connecting a Signal Generator for Process Simulation

Connect a signal-generating source with an accurate indicator to the sensing device, together with the means of isolating and regulating its output. The type of signal generator required will depend on the type of signal to be simulated and should conform to recognized instrument calibration standards and the supplier's instructions. When hazardous fluids or gases are involved (e.g., oxygen or ammonium nitrate), suitable safety precautions must be observed and the test method must be agreed on with the customer.

Site Calibration

The calibration procedures adopted on-site must be agreed on with the customer and conform to recognized industry instrument calibration standards and the supplier's instructions. These procedures must be applied to all in-line instrumentation, loop instrumentation, local controllers, analyzers, and so on. Where the control and monitoring instrumentation is integrated with a computerized control system and where factory tests have been carried out, the installation calibration procedure should be agreed on with the customer.

Calibration checks are usually carried out on analyzers by injecting known samples into the sample conditioning systems. This must be determined for each type of analyzer, by reference to the manufacturer's handbook or by consultation with the instrument vendor, and it must be agreed on with the customer. Complex analyzer systems usually require specialist personnel from the analyzer manufacturer to assist in precalibration and commissioning and are generally outside the scope of the installation contractor's responsibility and experience.

The calibration results should be recorded (a sample instrument preinstallation calibration sheet has been provided for reference purposes in Appendix 23.D) and included with the site test records. Further information on calibration records can be found in the GAMP Good Practice Guide on Calibration Management (10).

Instrument Mounting and Accessibility

Each instrument to be installed must be inspected to check that its data plate agrees with the specification and that it has been preinstallation tested, if applicable, as described in the preceding sections. The instrument should then be installed in its intended location on brackets, a subpanel, a mounting post, or a pedestal, ensuring that it is leveled, plumbed, and firmly secured. The installation must follow good instrument installation practice and the supplier's instructions, and the instrument should be protected from damage until it is put into service.

Indicating instruments and instruments requiring adjustments should be accessible for observation and servicing from the floor level, walkways, permanent ladders, or platforms. Where possible, actuated valves should be accessible from the floor or permanent platform level.

Instrument Piping and Tubing

The installation and pressure testing of air supply piping, transmission/signal tubing, and process impulse piping must conform to, and be checked against, the instrument installation specification, the construction documentation package drawings, and recognized industry instrument standards.

Cable Installation and Testing

The installation and testing of signal transmission cabling and power cabling must conform to, and be checked against, the instrument installation specification, the construction documentation package drawings, and recognized industry standards. Immediately after cables have been laid and before connection, all electric and electronic instrument wiring must be checked for polarity, continuity, and insulation resistance between the conductors and between the conductors and earth. These tests should be carried out before final loop tests and should comply with industry standards and statutory regulations.

Coaxial cables used for data highways must be tested using sine-wave reflective testing techniques. Fiber optic cables must be tested for clarity, intensity and echoes (caused by fractures or kinks in the fiber optic) using an appropriate signal generator and receiver. Hazardous Area circuits involving intrinsically safe (IS) instrumentation connected with copper cables must be tested [e.g., loop impedance, inductance, inductance/resistance (L/R) ratio] in accordance with the manufacturer's instructions and approved by the customer. Wireless networks must be tested for signal strength and frequency and, if applicable, distribution pattern (for data privacy purposes).

Wiring that connects field instrumentation to a computerized control system should be isolated from the computerized control system (e.g., at the control room/marshaling cabinets) during cable testing to safeguard against damage due to incorrect connections. Isolation from the computerized control system may be provided by the use of isolating (e.g., knife edge) field terminals and/or by disconnecting the computerized control system input/output (I/O) wiring termination blocks.

After the tests have been completed, the cables should be identified with a colored label, which clearly indicates its test status (see above). Wiring should be reconnected on completion of cable testing and recorded as such.

Loop Testing
General Requirements
The object of loop testing is to ensure that all instrumentation components in a loop are in full operational order when connected together and are in a state ready for process commissioning and validation [operational qualification (OQ) and performance qualification (PQ)]. Loop testing also encompasses the integration of instrumentation with any associated computerized control systems.

The procedure to be adopted in carrying out these tests is detailed below but, in general, the completed loop should be tested as one system and, where necessary, adjustments should be made to ensure that the loop is fully operational as a system and is correctly calibrated. Associated alarms and process interlocks/trips must be checked during loop testing.

The loop test results should be recorded (a sample instrument loop check sheet has been provided for reference purposes in Appendix 23.E) and included with the site test records. Checks for mechanical/electrical completeness are recorded using the upper section of the sheet and the dynamic loop test results are recorded on the lower section. The test results sheets will provide the documentary evidence essential for installation qualification (IQ). Representatives from the installation contractor and/or the customer will witness the final loop tests and countersign the test sheets. Any tests not witnessed must be accompanied by written confirmation from the customer that witnessing has been waived.

Loop testing of remote control loops is a two-person exercise, with one person located in the field and the other in the control room or instrument room. Each person must be provided with an adequate means of remote communication (e.g., field telephones or two-way radios) as approved by the customer.

Loop testing of instrumentation and any associated computerized control and monitoring system should encompass the interfaces with electrical equipment (via the electrical and instrumentation interface panel) and its related operation. Loop testing must not be carried out on electronic equipment until an adequate warm-up period has elapsed. Where possible, equipment should be energized for at least 24 hours prior to testing.

On completion of loop testing, it is recommended that all control devices/functions are left set with the correct control action and with a 100% proportional band setting. Derivative and integral functions must be set at their minimum time values.

Loop Testing Procedure

The following test procedure should be carried out to test the correct operation of field instrumentation and equipment installed in a control loop, and to provide the necessary documentary evidence (test records) to satisfy the requirements of the IQ protocols:

- Inspect the loop and set air/electrical supplies where appropriate. Check in particular that control valve air supply pressures are set in accordance with the instrument specification/data sheet.
- For electronic loops, check polarities, measure the loop impedance, and make the necessary compensating adjustments. The compensating adjustments on smart instruments can be made using either a handheld terminal or directly at the instrument. Smart instruments, if supported, can also be adjusted using an instrument configuration page on a computerized control system.
- Transmitter output signals equivalent to 0%, 50%, and 100% of the instrument range should be generated, either manually (e.g., hot oil bath, dead weight tester) or by applying appropriate signals at the field terminals to check the response of all other instruments and control valves in the loop. Instrument zero settings and calibration checks/adjustments should be made as necessary.
- Switch the loop controller to manual operation and, by applying the appropriate output signals, ensure that the control valve(s) stroke correctly. Valve positioner gauges should also be checked during this stage.
- Apply an input signal to the loop controller equivalent to 50% of the instrument range and adjust the output of the manual pneumatic regulator to 50%. Adjust the loop controller set point to 50% and, by switching the auto/manual transfer switch, check for "bumpless" transfer. Using the manufacturer's instructions, adjust where necessary until a satisfactory "bumpless" transfer is achieved.
- Check all alarm and trip actions by varying the loop controller input signals and adjust as necessary.

After the tests have been completed, the loop controller must be switched to manual operation and then identified with a colored label that clearly indicates its test status (see above) and the loop setup/test sheet should be signed off by an authorized person.

Testing Computerized Control and Monitoring System Loops

Before this level of testing commences, the instrument loop must be prepared in accordance with the loop testing procedure outlined above. All test methods must be agreed on with the customer.

The operation of instrumentation must be checked from the field to the control room operator interface/graphics display unit or local controller I/O registers, as applicable depending on the type of system installed, and vice versa. The operation of control instrumentation (e.g., control valves, actuated on/off valves) should be checked by energizing each control system field output from either the control room display or local controller, as applicable, and observing and recording the results.

The operation of monitoring instrumentation (e.g., transmitters, switches) should be checked by either injecting a suitable signal at the field instrument terminals or by installing the instrument in a comparator (e.g., a hot oil bath, dead weight tester). The result received by the control system on the control room or local display or local controller I/O register, as applicable, should be recorded. Any problems should be reported to the company/companies responsible.

The results of all tests must be recorded on loop test sheets. A sample copy of a loop test sheet is provided in Appendix 23.E. As each test is completed, the tested item must be identified with a colored label that clearly indicates its test status (see above).

Testing Safety Interlocks

Safety interlocking and shutdown systems require detailed test procedures based around the design documentation (e.g., cause and effect charts, binary logic diagrams) that must be formulated and agreed on in advance.

Site Modifications and As-Built Drawings

Any modifications, exceptions, or additions to the construction documentation must be confirmed in writing by the customer before such work is commenced by the installation contractor. This work must be properly organized, with clear definitions of responsibilities for checking and approval, and must be undertaken following a strict change control procedure that identifies all documentation affected by the change. Existence of dedicated engineering change control procedures and their relationship to any site-level change control process should be clearly documented and both should be regularly audited.

The installation contractor should assist in the provision of as-built drawings that may be completed by others (e.g., the engineering design contractor) at the end of the contract. To assist in this exercise, the installation contractor must keep a set of printouts of all documentation that must be marked up as and when changes are agreed.

QUALIFICATION
Installation Qualification

On completion of installation, all documentation associated with the installation, calibration, and testing of the field instrumentation, along with any associated computerized control system documentation, should be collated by the project manager/engineer ready for IQ. Most of this documentation will be in the pre-"as-built" condition at this stage of the project and may, therefore, contain site (red line) markups.

The documentation should typically include the following:

- Construction documentation package (see Appendix 23.C, "Instrument Installation Specification")
- Instrumentation and loop test records and reports
- Instrument calibration certificates (10)
- Change control notices and supporting forms
- Manufacturer's operation and maintenance (O&M) manuals

IQ should be carried out using a written protocol (13) that is completed during the inspection of the installation and its documentation. The protocol should typically describe how the documents inspected will be marked to show their IQ acceptance status, as well as the acceptance criteria for the items inspected, and it should have space to record the reference numbers of the documents seen and where they can be located.

Operational Qualification and Performance Qualification

General

OQ verifies that the control and monitoring instrumentation, as integrated with the process equipment and any associated computerized control system, meets the operational and functional requirements defined in the instrument application design documentation and/or computerized control system user requirements specification (URS) (13). PQ verifies that the control and monitoring instrumentation, as integrated with the process equipment and any associated computerized control system, meets the operational and functional requirements defined in the instrument application design documentation and/or computerized control system URS, and produces pharmaceutical product consistently to specification (13).

OQ and PQ should be carried out using written protocols that are completed during the functional testing of the manufacturing process. The protocols should typically describe how the tests will be carried out, as well as the acceptance criteria for the items tested, and they should have space to record the reference numbers of the documents seen and where they can be located. The principal reference document for the OQ testing of the instruments will be the instrument site commissioning test procedures and results. Some points to consider during OQ and PQ are described below.

Recalibration

There may be a large time gap between the IQ, OQ, and PQ phases for instrumentation associated with particular manufacturing processes due to the site construction program (e.g., unavailability of utilities, panels). As a result, some control and monitoring instrumentation may need to be recalibrated prior to commencing OQ, and possibly again prior to PQ, depending on their calibration frequencies. It would be advisable to recalibrate critical instrumentation anyway to ensure its status is known prior to OQ and PQ.

Recalibration must be carried out to agreed standard procedures using calibration test equipment that is traceable back to national standards. All calibration tests must be fully documented, the results recorded, and the sheets signed off by an authorized person. Calibrated instruments must be provided with a full calibration certificate that details the test results and their limits of uncertainty. It is at this point that the data obtained on the calibrated status is entered into maintenance/calibration management system. A detailed account of the calibration life cycle processes can be found in the GAMP Good Practice Guide on Calibration Management (10).

Testing

The customer should be kept informed of all site tests and when they are to occur by the supplier/contractor so that arrangements can be made for the end user to attend for witnessing purposes. All testing must be fully documented using test record sheets and witnessed and signed off by the customer and the supplier. Copies of all test result sheets/test records and reports should be reviewed and approved by the supplier/contractor and suitably qualified customer representatives. Any document revisions necessary during OQ and PQ must be progressed through the site document management system and implemented under a formal change control procedure.

Handover

The completion of OQ marks the end of the site construction (installation and commissioning) phase for the project, and the manufacturing process is formally handed over to the customer by the signing of a handover certificate prepared by the engineering design contractor. The handover certificate should be accompanied by an as-built issue of the construction documentation package bound in a series of O&M manuals, along with all test results, calibration certificates, manufacturer's reference manuals, and so on, to enable the manufacturing process to be properly maintained for the rest of its operational life. Some of the information contained in the O&M manuals (e.g., calibration certificates) may be removed and placed in a centralized maintenance system for ease of control.

Reporting

Testing occurs at many levels in a project, as described in the preceding sections of this case study, starting with design phase verification activities (design reviews) and ending with the testing associated with IQ, OQ, and PQ on-site. All reviews and tests must be fully documented using appropriate methods (e.g., minutes of design review meetings, test record sheets), and must have the required initiating, checking, and approval signatures. These documents, along with change control documentation, calibration certificates (10), and so on, comprise the formal records generated to provide the necessary evidence to support validation.

PERIODIC REVIEW

Periodic and scheduled reviews of the manufacturing process, including the control and monitoring instrumentation and any associated computerized control system, must take place at designated intervals from the time it is handed over to a site until it is replaced and/or decommissioned (13), to verify that it continues to be capable of producing quality product to specification. The purpose of a periodic review, with regard to control and monitoring instrumentation, is to verify that it has been maintained in a validatable condition.

Typical activities supporting the periodic review phase for instrumentation include establishing a recalibration program (10,14) and conducting routine maintenance activities as part of a planned preventive maintenance scheme. All maintenance activities must be carried out under a formal change control procedure and any associated testing must be fully documented using test record sheets.

NOTES ON SPECIAL INSTRUMENTS AND TECHNOLOGIES

A number of special instrumentation systems may be required for a project. This section briefly describes some typical systems used in the industry.

Emergency Shutdown Systems

Emergency shutdown (ESD) safety systems are relatively new to the pharmaceutical industry but have been used for many years in the petrochemical and other industries. They provide an independent, reliable (high integrity) method for protecting plant and personnel from situations that could lead to a dangerous occurrence (e.g., runaway reactions, handling of dangerous substances). A typical application would be on a solvent recovery plant or tank farm for the isolation of solvent feeds to the process. ESD systems are often designed by the instrument discipline as part of trip system design and can be either solid state or PLC based.

Solid-state systems comprise a number of cards, each of which contain a number of independent logic channels (i.e., AND, OR, NOT function blocks) and special functions (e.g., timer blocks) that are hard-wired to provide the shutdown logic required. The cards are monitored and controlled by dual redundant CPU cards and are interfaced to the field by high integrity/availability relays to prevent problems with switch contact welding and/or malfunctioning due to lack of use. The technical systems (cards, CPUs, etc.) are subject to inspection and testing by an independent certification body [e.g., the German Technischer Uberwachungs Verein (TÜV) in Europe] and issued with an approval certificate.

PLC-based systems are similar to solid-state systems—they use the same field interface cards and have the same level of certification—but the logic functions are performed by software. They are also easier to interface to other systems (via serial communications) to notify the system of a failure. The PLC used is subject to rigorous source code (machine level) inspection and testing by an independent certification body (e.g., TÜV in Europe) to examine the safety integrity of the code, including all possible failure paths. The PLC system (application software and hardware) should be developed and tested using a formal, life cycle methodology (9,13).

The shutdown logic required is usually determined during a HAZOP and represented on either cause and effect charts or on binary logic diagrams. Cause and effect charts have a spreadsheet (matrix) type of presentation and show for each identified failure condition (e.g., high temperature alarm) the safety status required for all affected equipment. Binary logic diagrams perform a similar function but show the physical connections between each logic block in diagrammatic form. More detailed versions of the binary logic diagrams showing card numbers, addresses, and so on, should be provided by the system supplier as part of the documentation package.

Analyzer Packages

Analyzer systems are now becoming more widely used in the pharmaceutical industry at all levels of product development (primary, secondary, and R&D). Some typical systems include mass spectrometers, gas chromatographs, near infrared (NIR) systems.

Each analyzer should have a URS and should be supplied with a detailed design specification, installation and maintenance documentation, and operator instruction manuals. Some special considerations include the following:

- The sample loop: The sample (fast) loop for off-line analyzers must bring a representative sample to the analyzer in the shortest acceptable time so that, accounting for the analysis time itself and the output response time of any corrective systems, the system response time is achieved.
- Sample conditioning: The need for sample conditioning (heating, cooling, drying) must be addressed and consideration given to the effect this could have on the system response time.
- Analyzer results: The format of the data produced by the analyzer must be specified.
- Communications: The ability of the analyzer to send data or signals to other systems must be specified. Raw analysis data may need to be transmitted to another system for further analysis and/or presentation. Further, the analyzer may also monitor for the

occurrence of critical situations (e.g., the start of an exothermic reaction, high solvent content, high particulate count), in which case there may be a requirement to communicate (via communications link or hard-wiring) to a shutdown system or circuit.

• Validation issues: The analysis data usually form part of the manufacturing batch records. Consideration must be given to a Supplier Assessment, and this will include a formal software assessment if the analyzer is controlled by a computer system. The design and testing of the analyzer computer system (application software and hardware) will need to follow a formal, life cycle methodology (9,13).

Intelligent Instruments

Intelligent instruments are widely used in the industry and include smart transmitters, loop controllers, chart recorders, machine monitoring systems, and fume cupboard controllers. These instruments contain embedded software in the form of nonuser programmable firmware that is configured either by "filling in the blanks" or by entering high-level statements. These devices are considered to be "black boxes" in validation terms and can be classified under one of more of the Software Categories defined in the GAMP Supplier Guide (9) depending on the level of sophistication of the embedded software.

Although these devices are not subject to the normal rigor of application software validation on a project, the manufacturers of these devices are still expected to have formal documented methods and records in place for developing and testing the software, for controlling and reporting changes due to bug fixes and upgrades, and for the configuration management of both the software and firmware products. Information on these and other areas may be obtained for the project validation records either by a manufacturer assessment or by signed statements from the manufacturer in response to a questionnaire.

Intelligent devices can often be supplied which conform to the Fieldbus^{TM} standard (15) for data acquisition, device control, and configuration. Fieldbus is a multidrop network standard that allows both digital (serial) and analogue (4–20 mA) signals to share the same cable without interference. The main points to consider for validation are data integrity and security within the network (i.e., evidence to ensure data are not corrupted when being transmitted within the network between devices or to a computerized control system). This evidence should include formal documentation and records of the data transfer/communication standard used, and error and diagnostic checks.

In addition to the more commonplace monitoring instruments and measurement devices, technology continues to present the industry with a range of "super" instruments. Equipment such as robotics, spectrophotometers, bio-instruments, vision/imaging systems, color recognition, particle monitoring, and other sophisticated analyzers could be considered as in-line instruments with embedded software. With these devices, more complex calibration and qualification issues will have to be addressed in addition to the manufacturer's design and testing methods described above. However, it should be recognized that the fundamentals of controlled selection, calibration, testing, and documentation remain the same.

RETROSPECTIVE VALIDATION

This case study has concentrated on the design of new instrument applications subject to prospective validation. The validation of existing instrument applications will necessitate a retrospective validation approach. The main difference between the two approaches is in the use of historical design information to prove that the current installation is properly documented and maintained.

Retrospective validation invariably involves the verification of any existing documentation, and the generation of missing information/documentation. As the plant/process is existing, some of the documentation involved with the original site installation can be ignored, provided it does not contain information needed for operation or maintenance. Maintenance and effective records are the main considerations when deciding on whether a particular document is or is not required.

Considering the construction documentation package listed in Appendix 23.C, "Instrument Installation Specification," the minimum level of documentation that should be in place for retrospective validation is as follows:

- Instrument schedule
- Instrument specification/data sheets
- Cable schedules
- Pneumatic tubing schedules
- Field panel drawings (e.g., layout, wiring, termination, and piping drawings)
- Electrical hookup (loop or wiring diagrams) drawings
- Pneumatic hookup drawings
- Earthing drawings/details
- Electrical and instrumentation interface panel wiring diagrams
- Special instrument specifications and wiring diagrams
- Shutdown/safety system logic diagrams
- Junction box layouts

The following certification, operation, and maintenance documentations will be required:

- Instrument calibration certificates (10)
- Instrument hazardous area classification certificates
- Materials certificates or other method of material verification
- EMC directive (11), declaration of conformity certificates (self-certification) for equipment containing EC CE-marked equipment (e.g., panels)
- Manufacturer's O&M instructions

The following supplementary documentation provided by other disciplines will also be required:

- Process design documentation (e.g., P&IDs, ELDs, ULDs)
- Mechanical design documentation (e.g., package plant vendor drawings)
- Computerized control system documentation (e.g., termination and wiring drawings)

In conducting retrospective validation the identification and recording of the GxP critical parameters and data is essential, and the instrument/loop schedule can be a recognized "control document" for such purposes. As with prospective validation, this also provides a basis for the risk assessment for each critical measurement.

CONCLUSION

Instrument application design is one of the most extensive activities, in terms of the number of types of instrument and the associated paperwork, and requires a logical, carefully controlled approach—a change to an instrument tag number or the addition/deletion of an instrument can affect up to 10 different documents. However, with good document control, it is possible to produce a system of documentation that will allow instrument applications, both simple and complex, to be under control, and hence validated and easily maintained. This case study has described the instrument application design process commonly used in the industry and how this can be used to support validation activities.

ACKNOWLEDGMENTS

The authors would like to thank Mark Foss, Engineering Manager, Boehringer Ingelheim Ltd., U.K. for his value review comments regarding current calibration practice. Mark is currently leading the ISPE/GAMP Special Interest group revising the GAMP Good Practice Guide on Calibration Management.

REFERENCES

1. U.S. Code of Federal Regulations title 21, part 210 (last amended April 2000). Current good manufacturing practice in manufacturing, processing, packaging, or holding of drugs; part 211 (last amended April 2000). Current good manufacturing practice for finished pharmaceuticals.
2. Medicines and Healthcare products Regulatory Agency. Rules and guidance for pharmaceutical manufacturers and distributors (Orange Guide), 2007.
3. U.S. Code of Federal Regulations title 21, 211.68(a).
4. U.S. Code of Federal Regulations title 21, 820.72.
5. Medicines and Healthcare Products Regulatory Agency. Rules and guidance for pharmaceutical manufacturers and distributors (Orange Guide), part II, sections 3.34, 3.40, and 3.41, 2007.
6. Geneva: International Organisation for Standardisation. BS EN ISO 9001:2008. Quality management systems—requirements.
7. London: British Standards Institution. BS ISO/IEC 90003:2004. Software engineering—guidelines for the application of ISO 9001:2000 to computer software.
8. BSI DISC TickIT Office. TickIT Guide (issue 5.5). A guide to software quality management system construction and certification to ISO 9001:2000, issue 5.5. London, November 2007.
9. International Society for Pharmaceutical Engineering. GAMP®5: a risk-based approach to compliant GXP computerized systems, 2008.
10. International Society for Pharmaceutical Engineering. GAMP® Good Practice Guide: Risk-Based Approach to Calibration Management. 2nd ed. (draft 2009).
11. Electromagnetic compatibility directive 2004/108/EC for the European Community (EC). This was transposed into UK Law by the EMC regulations 2006 (SI 2006/3418), which came into force on 20 July 2007.
12. European Community ATEX ("atmosphere explosibles") directive. This is based on two EU directives: the ATEX workplace/user directive (1999/92/EC) and the ATEX equipment directive (1994/9/EC). The requirements of EU directive 1999.92/EC were transposed into UK law in the form of DSEAR (the Dangerous Substances and Explosive Atmospheres Regulations, 2002). The requirements of EU directive 1994/9/EC were put into effect by the Department of Trade and Industry's Equipment and Protective Systems Intended for Use in Potentially Explosive Atmospheres Regulations 1996 (SI 1996/192).
13. Coady PJ, de Claire AP. Best practice engineering for validation of process control systems. Pharm Eng 1995; 18–30.
14. U.S. Code of Federal Regulations title 21, 211.160(b)(4).
15. ANSI/ISA. ANSI/ISA-50. Fieldbus standard for use in industrial control systems. Multiple part standard.

OTHER RECOMMENDED READING

– "Vendor Documentation Requirements for New Instruments." This document has been sponsored by the ISPE C&Q COP and is to be published in the 2009 edition of the ASME-BPE standard.
– GAMP® Good Practice Guide: Validation of Process Control Systems (VPCS). October 2003 (currently being updated), ISPE.

Appendix 23.A Overview of the Instrument Application Design Process

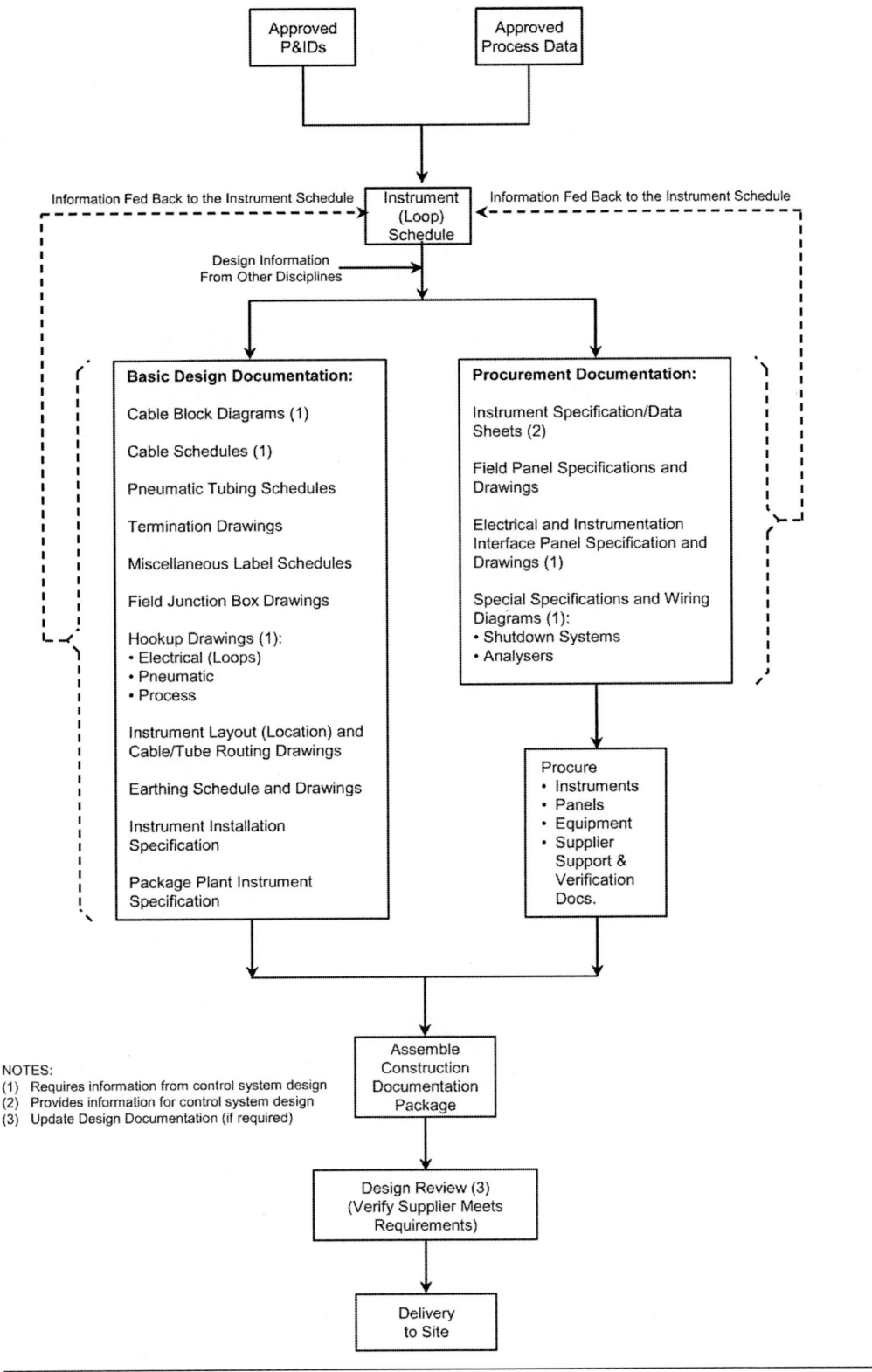

Approved P&IDs

Approved Process Data

Information Fed Back to the Instrument Schedule

Instrument (Loop) Schedule

Information Fed Back to the Instrument Schedule

Design Information From Other Disciplines

Basic Design Documentation:

Cable Block Diagrams (1)

Cable Schedules (1)

Pneumatic Tubing Schedules

Termination Drawings

Miscellaneous Label Schedules

Field Junction Box Drawings

Hookup Drawings (1):
• Electrical (Loops)
• Pneumatic
• Process

Instrument Layout (Location) and Cable/Tube Routing Drawings

Earthing Schedule and Drawings

Instrument Installation Specification

Package Plant Instrument Specification

Procurement Documentation:

Instrument Specification/Data Sheets (2)

Field Panel Specifications and Drawings

Electrical and Instrumentation Interface Panel Specification and Drawings (1)

Special Specifications and Wiring Diagrams (1):
• Shutdown Systems
• Analysers

Procure
• Instruments
• Panels
• Equipment
• Supplier Support & Verification Docs.

NOTES:
(1) Requires information from control system design
(2) Provides information for control system design
(3) Update Design Documentation (if required)

Assemble Construction Documentation Package

Design Review (3) (Verify Supplier Meets Requirements)

Delivery to Site

Appendix 23.B   The Instrument Application Development Life Cycle

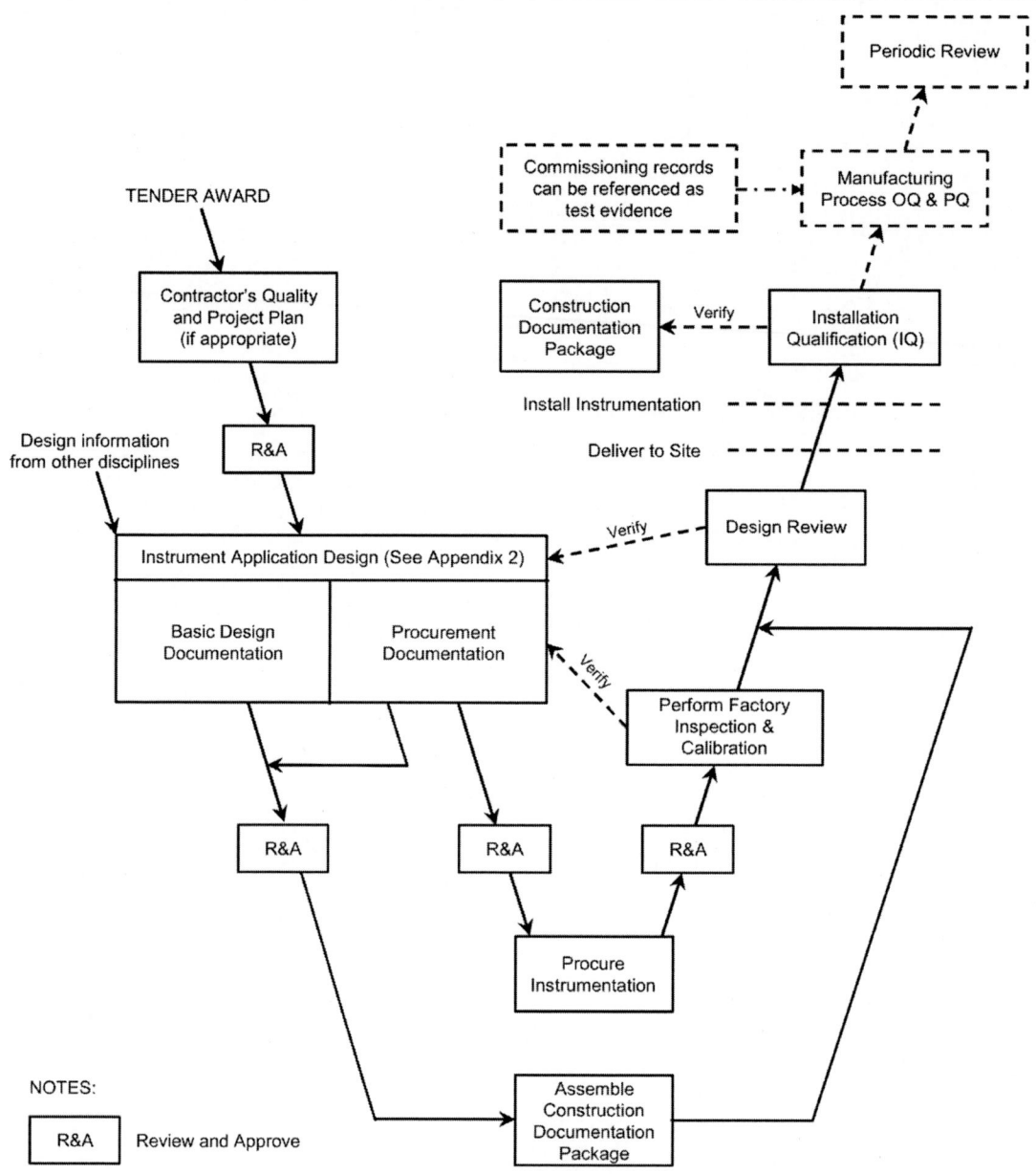

APPENDIX 23.C: INSTRUMENT-RELATED PLANT DOCUMENTATION

This appendix describes the typical contents of instrument application design documentation used by the industry. It is essential that the system of documentation used is suitable for its purpose, properly implemented, and auditable in order to support validation and future maintenance activities.

Document Control

Each document (drawing, schedule, specification, etc.) should contain a title block or front page (as appropriate) that contains the following information:

- Document title
- Document number
- Document revision number or code
- Project name and number
- Site name
- Plant area (if applicable)
- Type of document (e.g., instrument schedule, loop diagram, emergency shutdown system specification, temperature transmitter specification sheet)

Each document should contain the following change control and author information:

- Reason for issuing the document (e.g. "issued for customer comment," "updated in accordance with change note 123")
- Name of the document originator or modifier, as applicable, and the date of completion
- Name of the document checker and the date checked
- Name of the document approver and the date approved

Each document should contain cross-references to other design and/or reference documents, as applicable. Examples include the following:

- Process design documentation (P&IDs, process data specifications, etc.)
- Instrument design documentation (schedules, drawings, and specifications)
- Customer site or company standards

Instrument (Loop) Schedule

An instrument (or loop) schedule lists all instrumentation on the project, grouped by its unique tag (loop) number. For each instrument, the instrument schedule will typically provide the following information:

- Unique tag number
- Service/duty description
- Equipment description/type
- Location (e.g., pipe, process unit, or panel number)
- Manufacturer
- Requisition number
- Process design drawing number (P&ID, engineering line diagrams, utility line diagrams, etc.)
- Specification/data sheet number
- Electrical hookup (loop or wiring diagram) drawing number
- Pneumatic hookup drawing number
- Process hookup drawing number
- Control system I/O address or tag number
- Notes/comments

Instrument Specification/Data Sheets

Instrument specification/data sheets provide the technical specification and design data for each unique type of instrument on the instrument schedule. They are used for purchasing the equipment, for providing design information for other disciplines (e.g., the definition of instrument tag number, signal, and range that are essential for the design of any associated computerized control system), and as the basis for calibration data.

Similar items can be included on the same specification sheet in separate columns, and identical instruments can be listed by tag number under common specification details. There are three main classes of instruments: pipe mounted, process unit (e.g., vessel) mounted, and field/panel mounted.

Each specification sheet should contain the following instrument data:

- Unique tag number
- Instrument type
- Supply voltage/pneumatic supply details (as applicable)
- Electrical/pneumatic connection type (as applicable)
- Signal type (e.g., 4–20 mA, 0.2–1.0 bar, serial, Fieldbus, wireless)
- Type of mounting
- Range of instrument (calibration range) or switch set point values
- Materials of construction of wetted parts (i.e., parts in contact with process or utility media)
- Control characteristics
- Other requirements (e.g., smart or standard instrument, Fieldbus)

Each specification sheet should contain the following environment information:

- Process connection details (e.g., chemical seals, capillary lengths, flange rating)
- IP rating of the housing (e.g., weatherproof, dust tight)
- Type of hazardous area protection (e.g., IS, explosion proof)
- Requirement to meet EMC regulations (11)

Each specification sheet should contain the following process data:

- Process fluid/material
- Engineering units
- Working range (of all the process variables affecting the measurement)
- Maximum range (of all the process variables affecting the measurement)
- Fail-safe mode
- Duty requirements [applicable to modulating control or actuated on/off valves (e.g., trip service or normal operation)]

Each specification sheet should contain the required documentation needs, for instance, the following:

- Factory calibration certificates (full certificate or batch certificate of conformance)
- Testing/calibration equipment stipulations (e.g., traceable to national standards)
- Manufacturer's operation and maintenance manuals
- Approval certificates for equipment in hazardous areas
- EMC directive (11), declaration of conformity certificates (self-certification) for equipment containing European Community CE-marked equipment (e.g., panels)
- Layout drawings showing overall dimensions
- Electrical schematic wiring and/or pneumatic connection diagrams
- Valve sizing calculations
- Number of copies of each document required

Each specification sheet should contain the required calibration (10) and testing needs. Consideration should be given to how any associated computerized control system will be tested to ensure conformance to the user requirements specification. Typical information to be considered includes the following:

- Reviews of algorithms and calculations (smart instruments, loop controllers)
- Representative testing across the full operating range, including range boundaries
- Testing of alarms and hard-wired interlocks
- The need for specific test data, conditions, or equipment
- Records of test results
- References to specific calibration and testing procedures

Cable Block Diagrams

Cable block diagrams show in schematic form the panels (e.g., field panels, computer system cabinets, marshaling racks) and large items of equipment (e.g., packaged plant), the cables connecting them, and any intermediate junction boxes through which cables are interconnected. Each diagram should contain the following:

- Blocks uniquely representing instruments, combined systems (e.g., panels, PLCs), and large equipment items (e.g., analyzers)
- Interconnecting lines representing cables, referenced by unique cable numbers
- Clear representation of the location of each block in the plant

Cable Schedule

A cable schedule is a list of all field instrument cables in identification number order and is used for allocating cable numbers and providing installation information. Each schedule should contain the following for each cable:

- Unique cable reference number
- Cable source, location/routing, and gland type
- Cable destination, location/routing, and gland type
- Number of cores/pairs
- Cable ID/color code
- Estimated route length and route identification, if possible
- Cable specification/data sheet number
- Identification of special circuits (e.g., IS circuits)
- Any appropriate notes on segregation requirements for installation purposes (e.g., distance from electrical power cables)
- Other information if required (e.g., drumming details, termination details at each end)
- Notes/comments

Pneumatic Tubing Schedules

A pneumatic tubing schedule is a list of all field instrument pneumatic tubing (single and multitube) in identification number order and is used for allocating tubing numbers and providing installation information. Each schedule should contain the following for each tube/multitube:

- Unique tube reference number
- Tube source, location/routing, and bulkhead fitting type
- Tube destination, location/routing, and bulkhead fitting type
- Number of tubes
- Tube ID/color code (if applicable)
- Estimated route length and route identification, if possible
- Tube specification/data sheet number
- Notes/comments

Termination Drawings

Termination drawings are used to allocate cables to particular sets of terminals inside panels (e.g., field panels, system cabinets, marshaling racks), junction boxes, and plant equipment. Each drawing should contain the following for each core:

- Unique cable reference number
- Core reference number
- Terminal number at source and source location
- Terminal number at destination and destination location
- Core outer sheath color/number
- Core size
- Any appropriate notes on segregation requirements (e.g., IS circuits, analogue signals, security)

Miscellaneous Label Schedules

A miscellaneous label schedule is a schedule of all labels used on the project (e.g., instrument labels, junction box labels) and their specifications. Each schedule should contain the following information for each label:

- Equipment reference
- Dimensions of label
- Dimensions of character height
- Engraving details (including line separators)
- The numbers of each type of label required
- Label material type (e.g., Traffolyte, stainless steel)
- Label colors (e.g., Traffolyte sandwich colors)

Field Panel Specification and Drawings

Each specification should contain the following panel data:

- Panel reference
- Details of mounting or fixing
- Materials of construction and surface finish
- IP rating (e.g., weatherproof, dust tight)
- Type of hazardous area protection (e.g., increased safety, air purged)
- Requirement to meet EMC regulations (11)

Each specification should refer to all associated drawings, for example,

- Front of panel layouts, including label information,
- Interior layouts, including label information, and
- Internal wiring/pneumatic tubing drawings.

Each specification should refer to appropriate standards, for instance,

- General panel/junction box specifications and
- Cable/tubing standards.

Each specification should refer to test requirements and certificates, covering

- Electrical power and/or pneumatic checks and
- Functional checks.

Field Junction Box Drawings

Field junction box drawings can range in complexity from a simple cable termination box or cabinet, to a local control station containing electrical/pneumatic equipment. Complex junction boxes may require information similar to that described above for instrument panels (e.g., specifications, drawings, test certificates), depending on the level of complexity involved.

Junction boxes used for cable termination/routing require the following drawing details:

- Junction box reference
- Details of mounting or fixing
- Materials of construction and surface finish
- IP rating (e.g., weatherproof, dust tight)
- Type of hazardous area protection (e.g., increased safety, air purged)
- Interior layouts, including any label information
- Internal wiring details (cable and core identification, cable entry and glanding, etc.)

Electrical Hookup (Loop or Wiring Diagrams) Drawings

Instrument loop diagrams are the key design documents for instrumentation systems and are also the main troubleshooting tool for fault finding and maintenance activities on-site. Single loop diagrams, showing all items in the loop, tend to be the industry standard. Each loop diagram should contain the following functional information:

- All instrument items in the loop.
- All control items in the loop, including loop controllers and computerized control system I/O card address and termination details. The I/O details would need to be provided by the computerized control system designers.
- The cables connecting each applicable item in the loop along with their identification numbers and, for multicore cables, pair number/identification.
- Termination details (terminal block identification and terminal numbers).
- Electro-pneumatic equipment/circuit information (e.g., control valves and other electro-pneumatic valve circuits) is usually shown on the electrical loop diagram.

As applicable, each loop diagram should contain the following additional information:

- Hazardous area classification zone, temperature classification, and gas group.
- Instrument type and tag number.
- The diagram should show the power flow from its source, often in a control room, through the various locations of any intermediary junction boxes, barriers, or switches, and out to the load and back. The different locations through which each cable passes (e.g., control room, interface room, field panel, junction box) should be represented clearly so that the full extent of the different environments is conveyed precisely.

Pneumatic Hookup Drawings

Pneumatic hookups are the pneumatic versions of instrument loop diagrams and serve a similar purpose. They are produced for control loops that are purely pneumatic. For loops with a small pneumatic element (e.g., control valves and other electro-pneumatic valve circuits) the pneumatic circuit is usually shown on the electrical loop diagram. Each drawing should contain the following:

- All filter/regulation sets, piping, valves, and so on relating to the pneumatic hookup
- Details of all components and their connections/fixings
- Air supply pressure before and after controlling devices and at strategic points in the circuit
- Installation notes

Process Hookup Drawings

Process hookups are drawings showing the physical method for connecting and mounting instruments connected directly to process piping and equipment. The drawings are used for costing purposes by installation contractors and for the subsequent installation of instrumentation. Each drawing should contain the following:

- The instrument and its method of connection to the process piping/equipment
- Instrument mounting arrangement

- All piping and fittings required to install the instrument (material takeoff)
- Installation notes

Instrument Layout (Location) and Cable/Tubing Routing Drawings

Layout and routing drawings are floor plans of the building/process area showing the locations of process equipment and instrument panels. Each drawing should contain the following:

- The "spot" location of each instrument and its tag number
- The location of all instrument panels and junction boxes and their identification numbers
- Main instrument tray/ladder rack routes for cables and tubing and all tray/trunking branches (the identification numbers of the cables/tubes on each route and branch should be shown for installation purposes)
- The size and specification of each instrument tray/ladder rack and its height above finished floor level
- Elevations, as necessary, to help prevent potential clashes with other items of tray work/ladder racking and with other services (e.g., pipework and heating, ventilation, and air conditioning ducting)
- Separation distances from other cables and other services, on both plan and elevations, where compliance is necessary for safety reasons (e.g., intrinsic safety), for functional reasons (e.g., analogue signal integrity), or for electrical interference reasons
- Details of any transit frames or holes required through walls or floors to permit the installation of cables and tubing

Earthing Schedules and Drawings

Earthing schedules and drawings detail the preferred methods for earthing various types of equipment and its associated earth testing information. The documents should address the following (as applicable):

- Static earthing
- IS circuit(s) earthing
- Computer earthing

The earthing drawings should show how equipment should be connected for each of the above situations. Each item to be earthed should have an entry in the appropriate earthing schedule that details the following information:

- Plant item to be earthed and plant area location in which the item is to be found
- Earth bond size and location of earth bar to which the item is connected
- Earth test results (resistance in ohms) and date tested

Miscellaneous Drawings (Control Room/Instrument Room Layouts)

There is usually a need to provide additional drawings to clarify certain aspects of the design. A typical example is the provision of layout drawings for instrument/interface rooms and for control rooms to show the location of key items of equipment and their relationship to other equipment that may already be installed.

Instrument Installation Specification

An installation specification should be prepared for issue to prospective installation contractors for tender. The specification forms the technical section of the construction documentation package that should be appended to the specification. Each specification should contain, or provide detailed reference to, the following information:

- Regulations and codes of practice
- Company/work rules

- Competence levels required of the installer
- Safety standards required of the installer
- Good practice guidelines for the installation of instrumentation
- Good practice guidelines for the installation of cabling (including cable and core identification)
- Good practice guidelines for the installation of instrument piping/tubing
- Good practice guidelines for calibration management (10)

Each specification should incorporate a brief scope of work and refer to the construction documentation package that should be appended. The construction documentation package will typically comprise the following drawings and schedules:

- Instrument schedule
- Instrument specification/data sheets
- Cable schedules
- Pneumatic tubing schedules
- Termination drawings
- Miscellaneous label schedules
- Field panel specifications and drawings
- Field junction box drawings
- Electrical hookup (loop or wiring diagrams) drawings
- Pneumatic hookup drawings
- Process hookup drawings
- Instrument layout (location) and cable/tubing routing drawings
- Earthing schedules and drawings
- Miscellaneous drawings (control room/instrument room layouts)
- Electrical and instrumentation interface panel drawings
- Any special instrument wiring diagrams

Each specification should refer to the following test requirements and certificates:

- Loop test sheets
- Site calibration check sheets (10)
- Earth loop tests
- Insulation tests
- Completion certificates

Package Plant Instrument Specification

The package plant instrument specification details the instrumentation requirements and standards to be applied to any equipment packages. Typical equipment packages include water for injection systems, chiller packages, gas scrubber packages, tablet presses, packaging machines, freeze dryers, autoclaves, and so on. Each specification should contain, or provide detailed reference to, the following information:

- Regulations and codes of practice
- Company/work rules
- Safety standards required of the supplier
- Good engineering practice guidelines for the installation of instrument apparatus
- Good engineering practice guidelines for the installation of cabling (including cable and core identification)
- Good engineering practice guidelines for the installation of pneumatic piping/tubing
- Good practice guidelines for calibration management (10)

Each specification should contain details of the local power supplies available and refer to the necessary drawings and schedules, for example, the following:

- Supply voltage and air pressure
- Cable block diagrams describing the interface with the rest of the plant
- Cable schedules detailing interconnections with the rest of the plant

Each specification should refer to test requirements and certificates, for example, the following:

- Instrument hazardous area certificates
- Calibration certificates (10)
- Requirement to meet EMC regulations (11)

Electrical and Instrumentation Interface Panel

The electrical and instrument interface (interposing relay) panel is the method used to interface low voltage control systems (e.g., PLCs, distributed control systems, shutdown systems, loop controllers, etc.) to electrical switch circuits for operating and monitoring items of electrical equipment, including equipment packages. The documentation requirements for the panel comprise a panel specification, panel layout drawings, and wiring schematics (see above).

Special Instrument Systems

A number of other instrumentation and control equipment may be required for a particular project. These could include

- ESD systems,
- Analyzer packages, and
- Laboratory instrumentation (mass spectrometers, gas chromatographs, etc.).

Each item would generally require a detailed specification and installation documentation. The standards and documentation detailed in the appropriate sections above should be applied to this equipment. In addition, if the equipment involves the use of a computerized system, the system (application software and hardware) should be developed and tested using a formal, life cycle methodology (9,13).

Abbreviations: P&ID, process and instrument diagrams; I/O, input/output; IP, ingress protection; EMC, electromagnetic compatibility; PLC, programmable logic controller; IS, intrinsically safe.

Appendix 23.D   Sample Instrument Preinstallation Calibration Sheet

INSTRUMENT PREINSTALLATION CALIBRATION SHEET

CLIENT: PLANT:

CLIENT'S PROJECT No.:   CONTRACTOR'S PROJECT No.:

INST. TAG No. _____   SERVICE: _____

TYPE: _____   MODEL No.: _____   SERIAL No.: _____

MANUFACTURER: _____   ORDER No.: _____   SPEC. No.: _____

SIGNAL RANGE: _____   DIAL/CHART RANGE: _____

PHYSICAL CHECK: PROCESS CONN. CORRECT   ☐   PNEU./ELECT. CONN. CORRECT   ☐*

BODY MATERIAL CORRECT   ☐   RANGE/SPAN CORRECT   ☐

ELECT. SUPPLY SETTING CORRECT ☐   AIR SUPPLY SETTING CORRECT   ☐

GENERAL CONDITIONS SATISFACTORY ☐   ANCILLARY EQUIPMENT SUPPLIED ☐

_____

_____

_____

SHIPPING STOPS REMOVED   ☐

CALIBRATION CHECK:

INPUT		READING OR OUTPUT					
		RISING			FALLING		
%SPAN	ACTUAL	ACTUAL	%SPAN	ERROR%	ACTUAL	%SPAN	ERROR%
0							
25							
75							
100							

MAKER'S QUOTED ACCURACY ± _____%

CONTROLLER CHECK: CONTROL MODE: PROPORTIONAL ☐   INTEGRAL ☐   DERIVATIVE ☐

ON-OFF ☐   DIFF. GAP ☐

CONTROLLER ALIGNMENT CORRECT ☐   AUTO/MANUAL CORRECT ☐

SETTINGS: CONTROL ACTION: DIRECT ☐   REVERSE ☐   DIFF. GAP _____%

ALARM SETTING _____   TIME DELAY SETTING _____

LIMIT SWITCH SETTING: HIGH _____   LOW _____

OUTPUT LIMIT SETTING: HIGH _____   LOW _____

CORRECTIONS: AUTOMATIC TEMPERATURE CORRECTION RANGE _____

S.G./DENSITY CORRECTION SETTING _____

ZERO ELEVATION/SUPPRESSION SETTING _____

THERMO-COUPLE BURNOUT DRIVES: UPSCALE ☐   DOWNSCALE ☐

SHIPPING STOPS REFITTED ☐

* ANCILLARY EQUIPMENT LIST:

_____   _____

_____   _____

REMARKS: _____

_____

CHECKED BY: DATE: WITNESSED BY: DATE:

ACCEPTED BY: FOR: DATE:

INSTRUMENT TAG NO.

Appendix 23.E   Sample Instrument Loop Check Sheet

INSTRUMENT LOOP CHECK SHEET	
CLIENT:	PLANT:
CLIENT'S PROJECT No.:	CONTRACTOR'S PROJECT No.:
LOOP No.: _____	SERVICE: _____
LINE OR EQUIPMENT No.:	PIPE I.D.:

MECHANICAL/ELECTRICAL CHECKS

MEASURING ELEMENT:	INSTALLATION CORRECT	☐	LOCATION CORRECT	☐
	ISOLATING VALVES CORRECT	☐	MATERIALS CORRECT	☐
	TAPPING(S) POSITION CORRECT	☐	ORIFICE DIAMETER: _____	
IMPULSE CONNECTIONS:	CORRECT TO HOOK-UP	☐	MATERIALS CORRECT	☐
	PRESSURE TESTED	☐	TEST PRESSURE: _____	
	STEAM/ELECT. TRACED	☐	LAGGED	☐
FIELD INSTRUMENT(S):	INSTALLATION CORRECT	☐	AIR SUPPLY CORRECT	☐
	WEATHER PROTECTED	☐	POWER SUPPLY CORRECT	☐
PANEL INSTRUMENT(S):	INSTALLATION CORRECT	☐	AIR SUPPLY CORRECT	☐
	SCALE/CHART CORRECT	☐	POWER SUPPLY CORRECT	☐
CONTROL VALVE(S):	INSTALLATION & LOCATION CORRECT	☐	SIZE & TYPE CORRECT	☐
	STROKE TESTED	☐	POSITIONER CHECKED	☐
	LIMIT SWITCH(ES) SET	☐	I/P TRANSDUCER CHECKED	☐
AIR SUPPLIES:	CONNS. CORRECT TO DRAWINGS	☐	BLOWN CLEAR & LEAK TESTED	☐
TRANSMISSION PNEU:	LINES INSPECTED, BLOWN CLEAR & LEAK TESTED	☐		
ELECT:	INSULATION CHECKED - CORE TO CORE	☐	CORE TO EARTH	☐
	CONTINUITY CHECKED	☐	LOOP IMPEDANCE CHECKED	☐
	EARTH BONDING CHECKED	☐	ZENER BARRIERS CORRECT	☐
TEMPERATURE LOOPS:	T/C OR R/B CHECKED	☐	CABLE TO SPECIFICATION	☐
	CONTINUITY CHECKED	☐	LOOP IMPEDANCE CHECKED	☐
GENERAL:	SUPPORTS CORRECT	☐	TAGGING CORRECT	☐

CHECKED BY: DATE:	WITNESSED BY: DATE:

LOOP TEST:

	TRANSMITTER INPUT	TRANSMITTER OUTPUT	LOCAL INST. READING	PANEL INST. READING	
MEASUREMENT	0				
	50%				
	100%				

	CONTROLLER INPUT	TRANSDUCER OUTPUT	VALVE POS'NR OUTPUT	CONTROL VALVE POSITION	
CONTROL	0				
	50%				
	100%				

REMARKS: _____
_____

CHECKED BY: DATE:	WITNESSED BY: DATE:
ACCEPTED BY: FOR:	DATE:

INSTRUMENT TAG NO.

24 | Case Study 6: Process Control Systems

Roger Buchanan
Eli Lilly

Mark Cherry
AstraZeneca

INTRODUCTION

This chapter discusses industry and regulatory trends in assuring the "quality" of process control systems and how applying validation and qualification does not mean more or less effort, rather it is about taking a view of the "whole" process and "right sizing" your approach to ensure the system meets the business need, and is fit for its intended use.

Industry is looking for a highly streamlined approach to validation while still protecting patient safety, product quality and maintaining data integrity. The International Conference on Harmonisation (ICH), Food and Drug Administration (FDA), American Society for Testing and Materials (ASTM) E2500 standard, and GAMP®5 all give guidance encouraging industry to apply a risk-based approach and to remove none value adding activites, see Figure 24.1 (1–4).

Quality by design (QbD) is an initiative of the FDA it is intended to integrate the quality process through research, development, manufacturing and distribution. Correct implemetation of QbD improves speed to market, reduces product variation, improves operating efficiency and reduces costs at all stages of the process.

Quality risk management (QRM) is a systematic process for the identification, assessment and control of risks to the quality of pharmaceutical products across the product life cycle.

It should be understood that taking a more streamlined approach to validation and qualification does not mean the following:

- Taking chances with a patients well being
- Making do with insufficient time, money or people
- Providing an excuse not to do the right things

It should not automatically be assumed that there is a significant difference in cost/resources between integrating commisssioning and qualification of equipment with any associated process control system compared with validating them seperately. This does not mean that benefits cannot be realized, but achieving them needs a thorough understanding of the practices required, in the prevailing culture, that is, role of engineering/IT versus quality groups, to assure the system is validated for its intended use.

The level of validation and qualification should be based on a documented rational that identifies the risk and criticality the automation system shall have on the patient safety, product quality and data integrity. The term "verification" is being increasingly used as a umbrella term that encompasses the different approaches to ensure systems are fit for use.

BASICS OF PROCESS CONTROL SYSTEMS

Examples of process control systems range from relatively simple programmable logic controllers (PLCs), to more complex supervisory control and data acquistion (SCADA) and distributed control system (DCS) applications. PLCs can be embedded into equipment (so-called skid mounted) or stand-alone systems.

Programmable Logic Controllers

PLCs are microcomputers that have input/output (I/O) connections that enable them to communicate with external devices, usually in an industrial environment. Communication includes monitoring a state, detecting a change of state, activation, or deactivation. Devices

Figure 24.1   New process assurance standards.

include actuators, switches, thermocouples, and other instruments. PLCs can be programmed to process the incoming signal and, if required, respond with an appropriate output signal to enable control.

The use of PLCs provides an inherent flexibility for the automation of process equipment, facilities and utilities/services They can be used as an independent control system either embedded into or remotely linked to one or more items of process equipment.

Many manufacturing facilities may benefit from single control applications using a PLC module, that is, intallation of a preassembled unit (PAU) such as water deionizer for a sterile washer unit or autoclave (Fig. 24.2).

Expanded automation can be achieved by networking PLCs together or using them as slave stations to a central SCADA system or DCS.

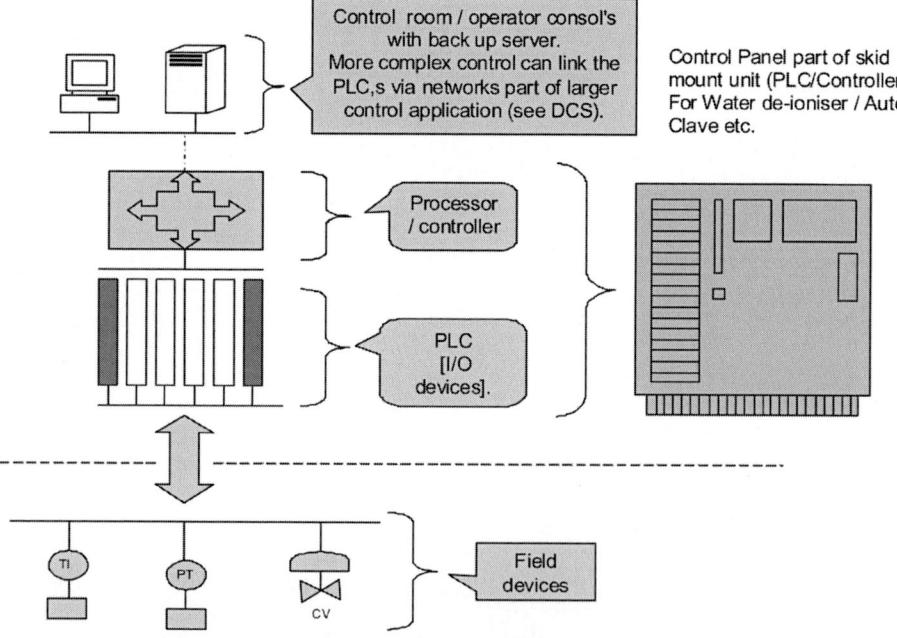

Figure 24.2   Basic programmable logic controller layout.

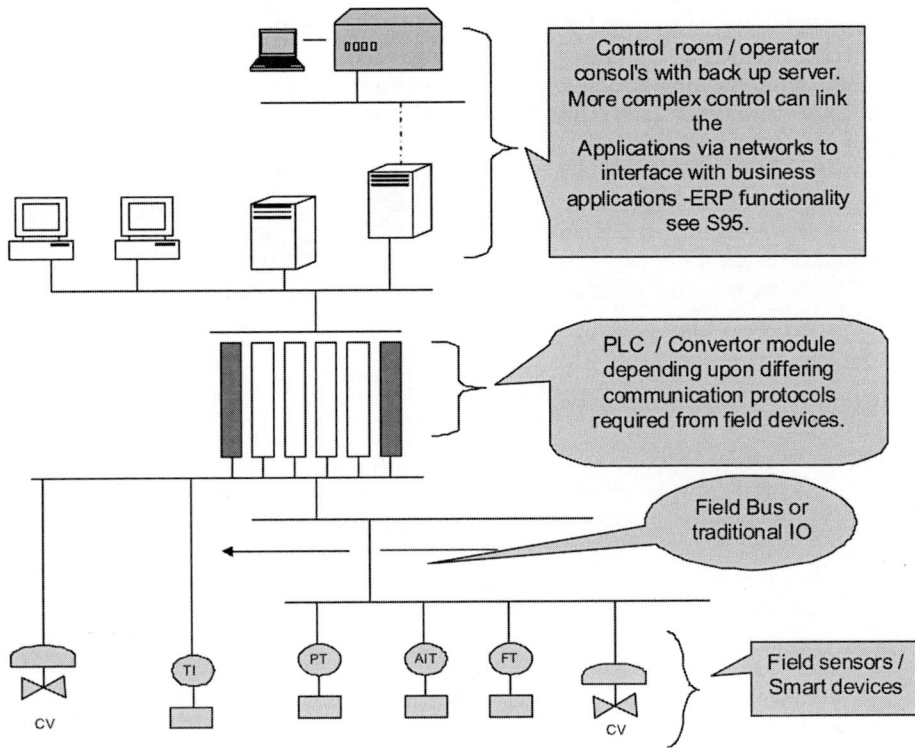

Figure 24.3 Basic supervisory control and data acquistion layout.

Supervisory Control and Data Acquisition System

Depending on the nature and use of the SCADA system, the system may contain many, if not, all of the following elements (Fig. 24.3):

- Hardware
- Operating system
- Network system
- Database management system

The control system and instrumentation may be embedded into items of plant equipment. Some organizations split SCADA systems into computer systems and instrumentation dividing the validation work between engineering support and plant operation. An advantage of validating the automation as a whole is that it avoids the management interface between two separate validation projects, the necessary agreement on exactly where instrumentation ends and the computer begins, and the determination of responsibilities and accountabilities for segments of the qualification.

Distributed Control Systems

The use of distributed control systems in the pharmaceutical industry is mainly (although certainly not exclusively) within bulk API manufacturing plants. The scale of DCS can vary tremendously: systems may cover multiple manufacturing facilities, or just a single process stage. A stand-alone system controlling a single process unit (e.g., a single PLC/ SCADA system for a packing machine) would not normally be considered a DCS, however multiple PLCs controlling a process stage connected to a SCADA system could be considered a DCS.

The main areas of functionality provided by a DCS are typically as follows:

- Operator interface: The provision of graphical and textual information on the plant status, also providing the operator the ability to control plant devices, either directly or by automatic sequences
- Sequence control of process operations and recipe/batch management and tracking
- Alarm and device interlocking (often in addition to separate hard-wired systems)
- Event and alarm recording, and historical trend recording of process variables
- Control of analogue process variables (e.g., temperature, flow, pressure, etc.)
- Interface to embedded control systems provided as part of packaged plant units such as filters, driers, centrifuges, etc.

Figure 24.4 illustrates how the various systems may interact or fall under a International Society of Automation (ISA) S95 control hierarchy.

Design Philosophy

Modern process control systems for batch process frequently make use of the ISA S88 standard for batch control (5). This provides a layered, structured approach to the system architecture, configuration and provides both a high degree of automation and flexibility within the process control system. The generic, modular nature of the architecture also provides opportunities for streamlining the validation process. Most examples of S88 implementations are based on new green field plants/installations; this case study considers the application and validation of an S88 based solution both to an existing facility (brown field application) and to a new (green field) facility.

The S88 approach to batch control provides a framework for the architecture of the system, having essentially four layers of control that operate on plant units (e.g., reactors, filters, driers) within a process cell (a collection of plant units within a facility or used for a process stage).

The control module layer is the lowest level and defines how field devices (e.g., valves, pumps, controllers, etc.) interact with the process control system. Phases are at the next layer and describe small (often generic) sequences (e.g., fill, transfer, initiate temperature control, etc.) that operate on a unit. At the next layer up the hierarchy, phases may be combined into unit operations to perform more complex functions (e.g., distillation, crystallization, etc.). The top layer is the procedural layer and this generally defines how unit operations are combined across plant units for the overall process. A feature of S88 is the ability to generate equipment modules; essentially common arrangements of control modules to provide a specific function (e.g., skid mounted temperatures control units for reactors or valve/pump arrangements for transfer routes).

S88 is complemented by the international standard for the integration of enterprise and control systems known as ISA S95 (6). The standard consists of models and terminology which can be used to determine which information has to be exchanged between systems for sales, finance and logistics and systems for production, maintenance and quality. This information is structured in unified modeling language (UML), which are the basis for the development of standard interfaces between enterprise resource planning (ERP) and manufacturing execution systems (MES).

The S95 standard can be used for several purposes, for example, as a guide for the definition of user requirements, for the selection of MES suppliers and as a basis for the development of MES systems and databases.

Unified Modeling Language
The UML is a graphical language for visualizing, specifying, constructing, and documenting the artifacts of a software-intensive system. UML offers a standard way to write a system's blueprints, including conceptual business processes and system functions as well as programming language statements, database schemas, etc.

Enterprise Resource Planning
ERP in its *simplest form is a means to integrate all departments and functions across a company via a single integrated software program (computer network/system), this will then allow departments to share and communicate all functions and needs.*

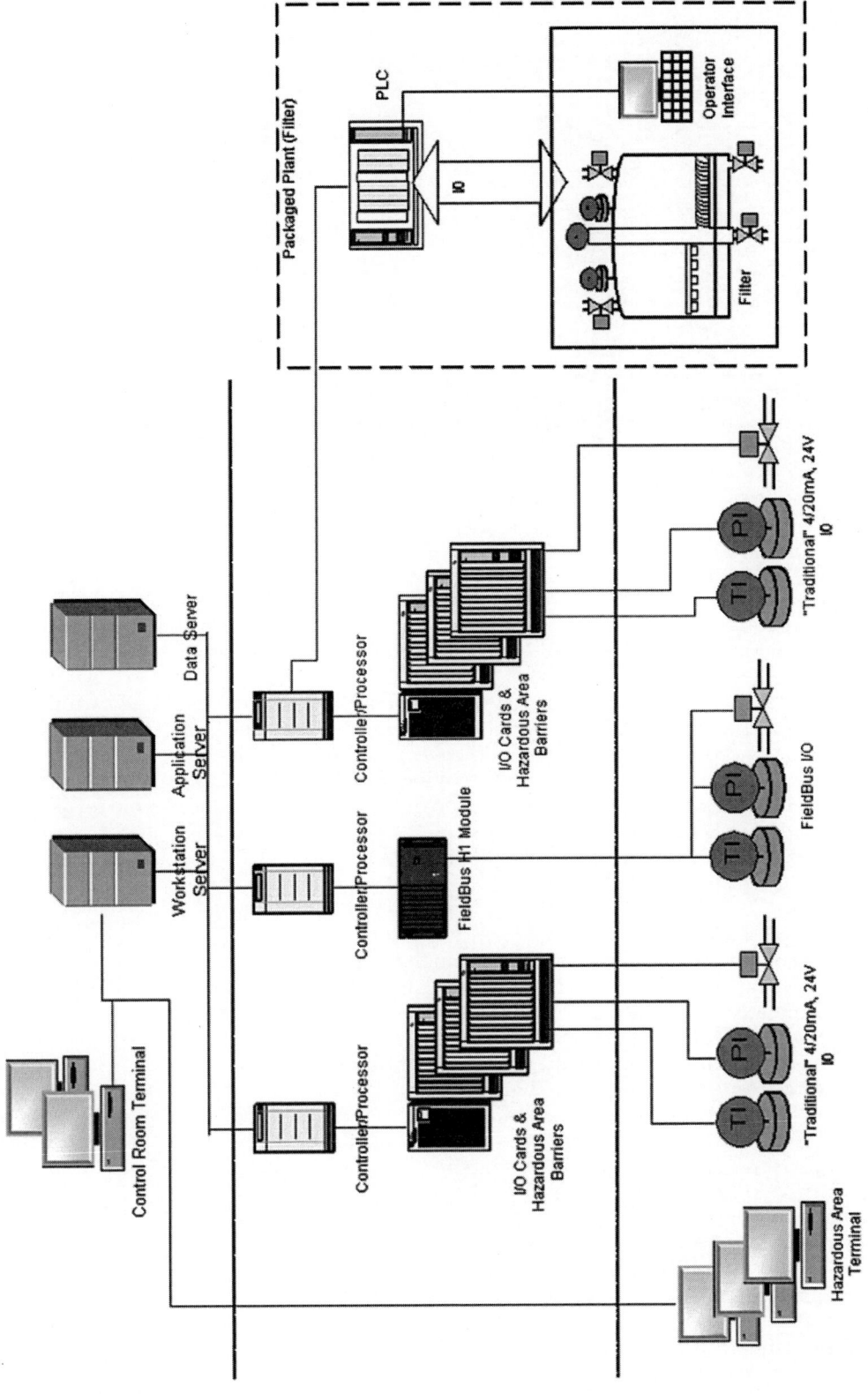

Figure 24.4 Basic distributed control system layout.

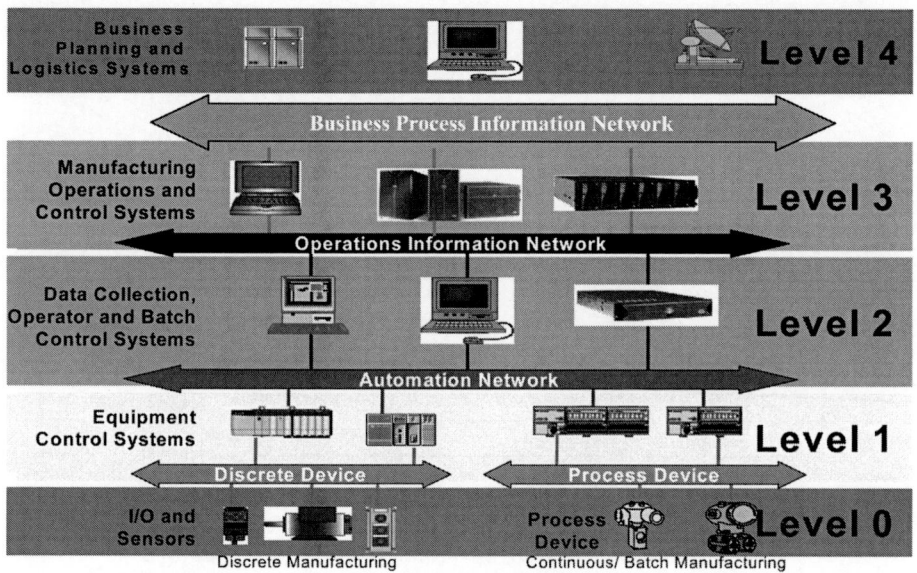

Figure 24.5 International Society of Automation S95 model.

Manufacturing Execution Systems

MES have the ability to deliver information that enables the optimization of production activities from start of order to finished goods. As long as the data is correct and accurate, MES can utilize this information and will guide, initiate, respond to and report on plant activities as they occur.

Figure 24.5 provides a diagrammatic representation of how the structures within the automation processes are linked. This model tries to illustrate how the S95 architecture brings the applications of processes control methodologies together, from the basic instrument field devices through to PLC to SCADA and DCS moving into the more complex network devices such as Data Historians and ERP/MES applications. The higher level of the S95 model tends to move into the IT application.

Note some consider the level 0 to be plant based actions/SOP processes, and place the I/O sensors/devices under level 1 also.

PLANNING FOR SUCCESS

If the business or supplier does not get a clear understanding of what is actually required then system requirements will change during the project life cycle, which can have major impact to the project from a cost and time perspective. Consequently it is very important to allow time within a project for communication and requirement gathering aided by a detailed "risk" assessment. The inability to understand what the user needs due to poor requirements and/or communications results in poor design. This in turn will lead to poor coding and configuration such that the system will not perform as intended. This is normally picked up during testing, which is the last phase toward "go live" and the most costliest point in the project to find major errors! Effective risk managemetn and validation planning are key to success.

Risk Management

It is important to understand the failure modes of the process control system and how the system interacts with the production/business processes. Key to success is getting the right people involved. Considerations should be given to the following:

- Risk facilitator (*Individual who understands the risk methodology applied and controls the time at each session*)
- *Subject matter experts* (SME)

○ Engineering (*software*)
○ Engineering (*process*)
○ End users

- QA
- Business [*custodian and owner (if possible)*]

Once there is a understanding of the basic risks to the business process the team can move into more detailed analysis of risks associated to the requirements. Typically this done thhrough conducting a HAZOP or FMECA study. During this process it may/can be benificial to have a classification process to distinguish risks, for example, the requirements could be identified on a scale of low, medium, and high.

High: These should be identified as critical, having a direct impact on business and public health, and consequently require greater levels of specification and detailed testing.
Medium: These should be identified as having potential impact on the business and public health, and require basic but complete specification and testing.
Low: These should be identified as very low impact to the buisness and public health, and require only high-level specification and testing.

A test strategy should emerge from the risk assessment. This is benificial not only to the business but also in that it can be used to demonstrate to any inspectors that the business understands what its critical process parameters (CPP) are and that the level of validation employed is comensurate with risk posed. Depending on how the data is stored within the system a further analysis may be needed for the management and control of eletronic records and electronic signatures (ERES).

Validation Master Plan

Within the pharmaceutical industry it is common practice to identify a manufacturing site's overall approach to validation in a site validation master plan (VMP). The VMP outlines how process control fits into this overall program and references the applicable sources of goverance (i.e., policies, procedures, standards). The U.S. FDA does not formally require a VMP, but inspectors may ask what the validation approach is and how specific systems are validated. Within the EU GMP directive annex 15 formally requires a VMP and as such the VMP should always be in place.

A site VMP will normally be supported by multiple individual validation plans covering different systems and facilities. A relatively straightforward planning hierachy involves the creation of a single VMP for an entire manufacturig site, supported by a number of validation plans, which exist for specific systems or projects. All of these documents are the responsibility of the pharmaceutical end user company, and need to be version controlled and approved by management and QA. It should be remembered that the VMP is a high level summary document so it is important that the overall approach is documented, taking care not to go into too much detail which may be included in the individual validation plans as illustared in Figure 24.6.

Validation Plan

The validation planning process should be started as early on in the system design life cycle as possible, ideally following the initial risk assessment on patient safety, quality and data integrity have been performed on an initial understanding of the user requirements (Fig. 24.7). The resulting validation plan should address the following:

- All the deliverables that are required for the successful implementation of the project such as Functional Specification (if COTS then reference to the supplier's own specification should be made) (Appendix 24.A), testing (if supplier is providing majority of test support again this needs to be referenced and supported by evidence of "management of the supplier").

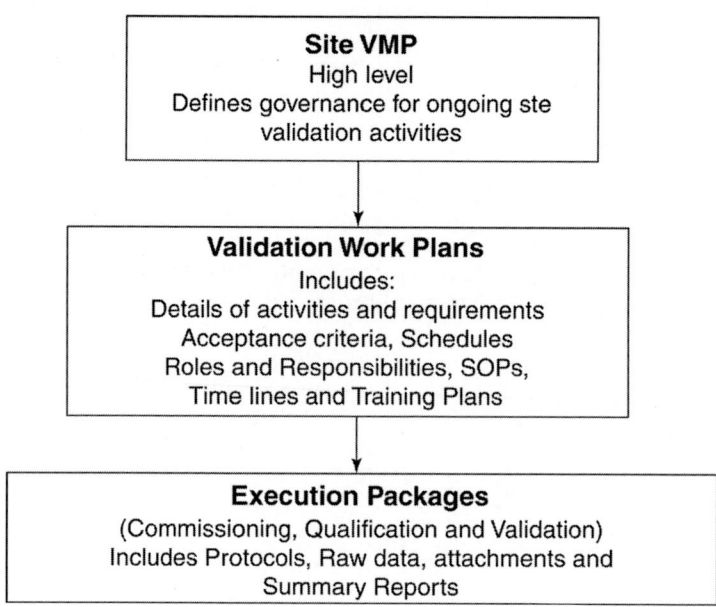

Figure 24.6   Validation hierarchy.

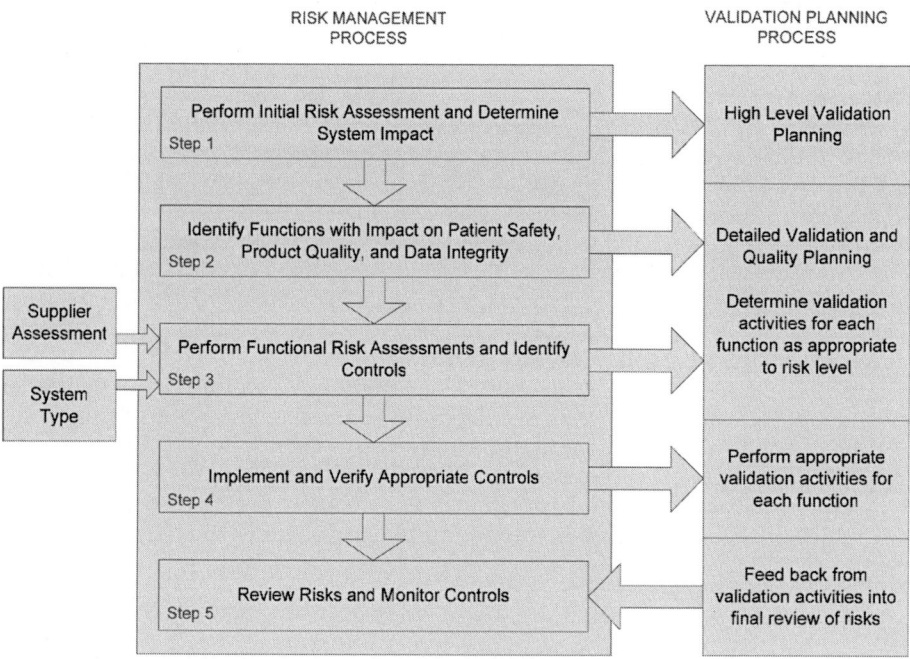

Figure 24.7   Validation planning and risk management.

- Regulatory expectations based on the nature of any impact on product quality, patient safety or data integrity. It is important to be able to demonstrate that the level of validation activities that are required for the system are based on how the process control system will interact/impact the manufacturing processes. If a classification tool has been used then reference this in the plan.
- The plan should also identify what activities are in-scope and out of scope (e.g., if the system is not required for ERES functionality then this should be stated, or if the sytem

is to be validated/qualified by the supplier (majority of project deliverables outsourced) then this needs to be identified and more emphasis placed on "controls" of the vendor).
- Considerations should also be given to maintenance activities for the control system once it has been handed over to the business such as. security administration, change control, disaster recovery planning (DRP), business continuity planning (BCP).
- Clear roles and responsibilities are identified.

The validation plan may need revising in the light of supplier assessment and information about how the supplier will implement the defined user requirements.

It is timely to reminder ourselves of recent FDA guidance (3):

Validation should be based on the software's complexity and safety risk—not on firm size or resource constraints. The selection of validation activities, tasks, and work items should be commensurate with the complexity of the software design and the risk associated with the use of the software for the specified intended use. For lower risk devices, only baseline validation activities may be conducted. As the risk increases, additional validation activities should be added to cover the additional risk. Validation documentation should be sufficient to demonstrate that all software validation plans and procedures have been completed successfully.

Figure 24.7 describes the relationship between validation planning and risk management.

Traditional Validation Approach

There are several well know approaches to system/software development including waterfall life cycle, spiral life cycle, and rapid application development (RAD). Figure 24.8 endeavors to illustrate the current approach commonly used in the development stages of process control systems. Figure 24.8 should not be taken to mean a sequential stepped development proccess where one step must be completed finished before the next can begin. Rather it is quiet acceptable (indeed often preferential) to allow activities to proceed in parallel for efficiency so long as an interdependnecy does not compromise the quality of the final process control system.

The traditional approach continues to be compliant with regulatory requirements but should be "right sized" on the basis of the risk assessment decribed earlier during validation planning. A well thought out system development approach should utilize methods such as prototyping and simuation testing to help mitigate anticpated/common issues (e.g., users do

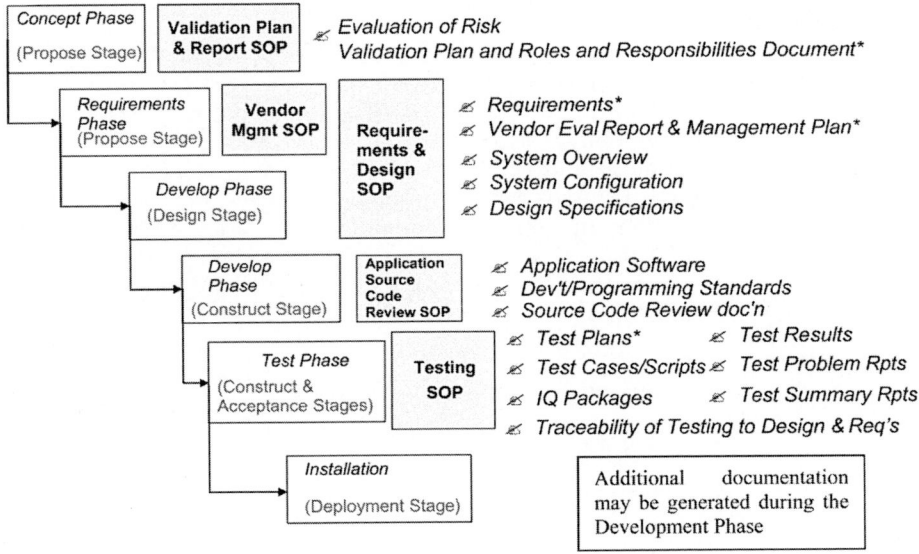

Figure 24.8 Traditional validation.

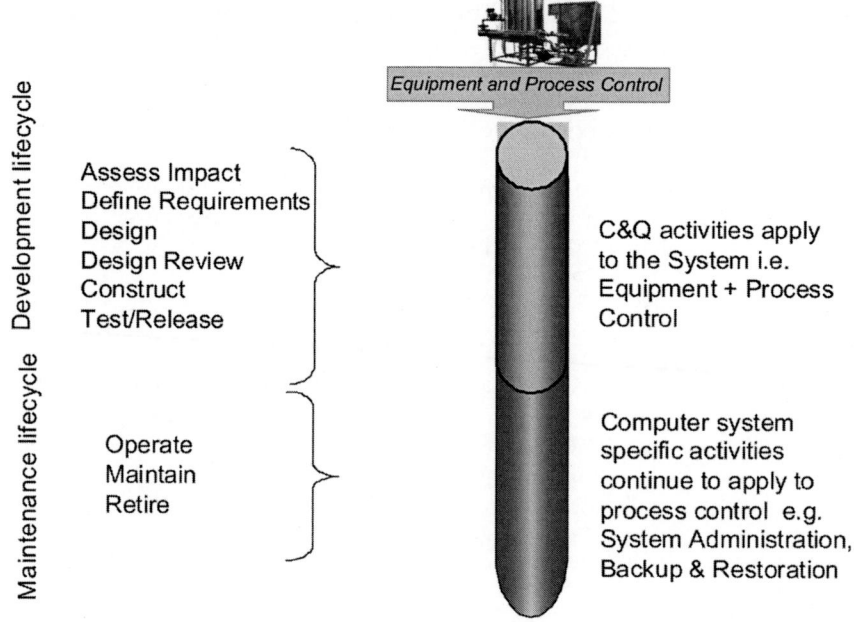

Figure 24.9 American Society for Testing and Materials E2500 approach.

not really know what they want until they see it, and production equipment downtime is very limited for system deployment).

A common frustration is that duplication can happen is with the vendors, we need to ensure if they have designed and tested the systems functionality users do not repeat those tests but rather just simply reference the evidence provided by the supplier. Another area of frustration is in project management, that is, if the project is running with one set of contractors (*i*) performing calbration of equipment, then installation qualification (IQ) of that equipment and contractor is (*ii*) performing commissioning, and as part of commissioning, they then repeat (*i*)—where is the value? Industry with guidance from the regulatory agencies needs to ensure it fully utilizing its resources and time and does not duplicate activities!

Commissioning and Qualification (Verification) E2500

New industry guidance such as ASTM E2500 is encouraging the industry to combine activities and not treat validation as an entirely independent exercise and hence remove much of the frustration about duplicating activities described earlier. ASTM considers the process control and equipment both as elements of unified system and thereby promotes an integrated approach to commissioning, qualification, and validation. Fundamental to this concept is that the focus should be on verifying process capability and not just system/equipment functionality. Figure 24.9 shows how commissioning and qualification (C&Q) can be combined with process control system qualification. It is worth noting that it is recognized that close integration does not always make sense such as when process control is not tightly coupled to equipment, in which case it is better to address the process control and eqipment seperately.

SYSTEM DEVELOPMENT

No matter what methodology is applied be it traditional validation or C&Q verification, we still need to demonstrate that the process control system is fit for purpose.

Requirements

Understanding our business processes are the first steps in development of requirements.

- What is to be automated and what is manual operations?
- How do they all interact with manufacturing process steps?

- What are the CPP?
- What are the critical quality attributes (CQA)?

Once we have established the business processes we can then look to define what is the best process control methodology. Typically requirements are grown from the very early conceptual stages of the project, by the business requesting an outline specification. Although not at a great detail this high level assists in the understanding of initial capital expenditure and resource needs. This information is then used to generate a more detailed definition of requirements often referred to as the user requirement specificaiton (URS). GAMP®5 describes the URS as "The extent and detail of requirements should be commensurate with risk, complexity, and novelty, and should be sufficient to support subsequent risk analysis, specification, configuration/design, and verification as required." A checklist of URS contents is provided in chapter 6 of this book.

It is good practice to make use of existing standard models when capturing requirements, particularly if the expectation is that the implementation will follow those models. For example, many process control systems exchange information with some form of enterprise data management or resource planning system. If that is the case, then the functionality required of the system should conform to the functional hierarchy given in ISA 95 part 1 and the boundary between the enterprise system and the control system defined. The required data flows across that boundary should then be defined.

It is important to consider the management and control of any electronic records and electronic signatures (see also chap. 13). PLC systems are not normally required to store electronic data that is required by the regulatory agencies. Care must be taken if your system is used to monitor process variables of critical areas, that is, sterile environments because this may well be consider a regulated electronic record, especially if the system is configured to store such data. Although they do not retain raw data (temperatures, pressures, flow rates, etc.), PLCs often store other important process data (recipes, control parameters, etc.) or configuration data (I/O scaling factors, etc.), the integrity and security of which may be important for the assurance of product quality, efficacy or patient safety. It is common for SCADA/DCS systems, meanwhile, to store data that is classified as regulated electronic records. Consequnently adequate controls must be put in place to satisfy regulatory requirements. Many systems adopt "historians" as the application to store critical data so we are now moving into the more advance levels of the recently discussed ISA 95 model.

A good URS will be clear and concise. The supplier or in house expert must be able to design the system and for the end user to be able umabigiously test these requirement to ensure the system has been designed corectly. Traceability of requirements through design to testing will be key to demonstrating complete requirements coverage.

Design

In the design process, the requirements are translated into a logical and physical representation of the software and hardware to be implemented. Design documentation must address how each requirement will be achieved. The design must cover the following:

- Functionality (e.g., alarm management, user and administrative security, safety, interlocks, recipes, interfaces, etc.)
- Hardware (e.g., number of controllers, controller loading, work stations, redundancy, network, interfaces between automation and equipment)
- Software (e.g., logic structure, type of programming language, parameters, purchased applications, etc.)

The level of detail for design documentation should be appropriate to what is needed to program or configure the system. This is part will be determined by the programming standards being used. The use of standards will improve system development, facilitate reusable source code, and aid system maintenance. Good design documentation can also assist the end user to rebuild the system in case of a disaster. Chapter 8 discussed programming standards in more detail.

Design reviews should take place to confirm the design is complete and meets the intent of thee URS. Such reviews can also be used to develop future testing strategy. The level of review should be based on the complexity of the system and the risk assement discussed earlier.

Testing

The testing of process control systems involves commisioning and qualification. The ISPE define commissioning as (7): "A well planned, documented, and managed engineering approach to start-up and turnover of facilities, systems and equipment to the end user that results in a safe and functional environment that meets established design requirements and stakeholder expectations." Qualification follows and tests focus on key functionality and critical components. Qualification should build on the contribution of commissioning so that the combined package meets the regulatory requirements and the overall approach process is run efficiently, minimizing duplicate testing and focusing on the.

Testing comprises IQ, operational qualification (OQ), and performance qualification (PQ), which collectively may be referred to as verification. IQ and OQ checklists are provided in Appendixes 24.B and 24.C. OQ is sometimes split into two parts.

- OQ1: Verification of basic functionality and calibration of instrument loops and verification of alarms/interlocks, followed by verification of process sequences and loop tuning—often using water to simulate the process
- OQ2: Process sequence verification and further loop tuning using solvent simulations and then commissioning batches

OQ2 can be considered part of PQ. The particular categorization nof tests does not matter so long as it is clearly explained. During this stage of testing, only very minor issues should be apparent with the system.

Testing cannot be reviewed without traceability to requirements.

- How do you know all requirement and design elements are tested?
- How do you know that acceptance criteria are correct?

Practical experience clearly indicates the need to avoid complex requirements traceability matrices (RTM). To simplify RTM, it is suggested that particular requirements are cross-referenced in their respective individual test cases with test protocols, and then the RTM is limited to tracing requirement to sections of test protocols. It might also be appropriate to ask suppliers to trace all their factory acceptance tests (FAT) to requirements but then limit OQ traceability to requirements dealing with higher risk functionality.

Utilization of Suppliers

Through good planning the user should look to maximize supplier involvement throughout the system life cycle to leverage knowledge, experience and documentation this will hopefully reduce the amount of activities they have to perform and avoid any potential duplication. Supplier documentation should be assessed for suitability, accuracy and completeness. There should be flexibility regarding acceptable format, structure, and documentation practices. Supplier documents do not have to adopt user templates for instance. Considerations specific to process control systems include the following:

- The degree to which the process controls system is "embedded" within the process equipment. For a packaged system, the knowledge of the process equipment (and, in some cases, of the process) may reside with the supplier rather than the end user. The supplier may therefore already have details of process capability onto which the end user needs to map the required design space. Such suppliers typically have increased responsibility within the risk management process and more involvement in process validation as well as system verification activities. The appropriate split of responsibility between supplier and end user needs to be considered at an early stage together with the appropriate level of supplier assessment.

- The complexity of the supply chain; process control system equipment is often produced by one or more suppliers (each of whom may, in turn, base their offering on standard packages or firmware produced by their own subsuppliers), and integrated and configured by another supplier. This needs to be taken into consideration when planning supplier assessment (does a supplier assess their own subsuppliers adequately or is additional assessment required from the end user) and when planning ongoing support for the system.
- The modular nature of many process control systems, both in terms of hardware and the control application, facilitates the progressive acceptance of the system and the opportunity to make use of supplier test documentation.

ALARM MANAGEMENT

A key aspecy of process control system functionality is alarm maanagement. Alarm management should prevent or minimize undesirable events and their resource implications, and review of alarm data supports continous improvement. Successful alarm management will mitigate what is often the reality of underperfoming alarm systems that are unhelpful or even ignored.

Sites/facilities should document the philosphy/strategy for the life cycle management of alarms produced by process control systems. ISA 18.02 "Management of Alarm Systems for the Process Industries" describes a life cycle activity to ensure effectiveness and robustness (Fig. 24.10):

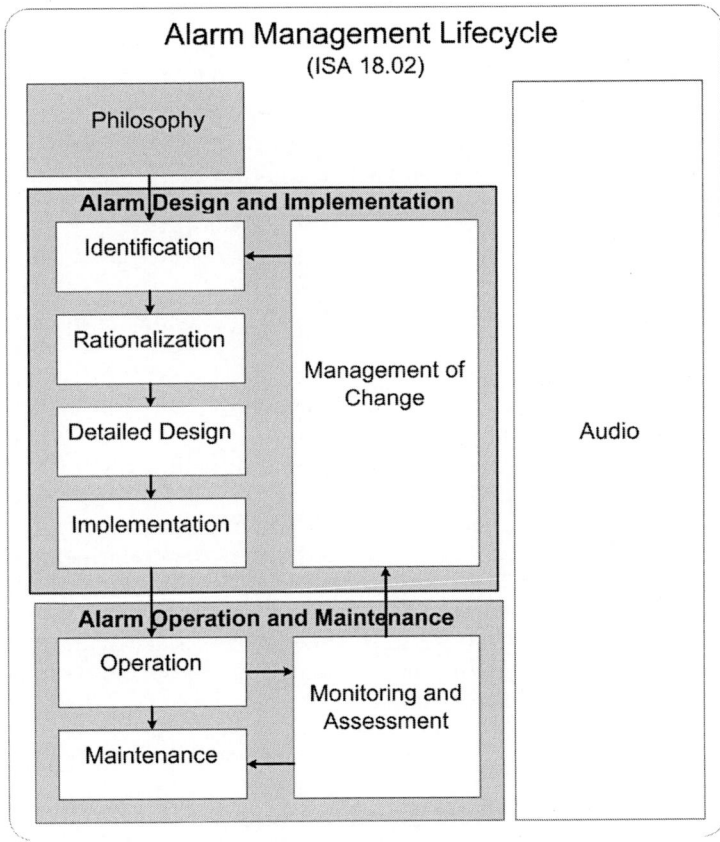

Figure 24.10 International Society of Automation Alarm Management (ISA 18.02).

Alarms should be defined to indicate important abnormal events that require a timely response. Alarms should not be confused with messages or other system generated notifications associated with normal operations. Care should be taken to ensure alarm terminology does not lead to unecessary confusion. For instance if the term "warning"is used it should be clear whether it signifies an alarm of lower priority or whether it is just a form of sytem notification. Further information on allocating alarms can be found in Engineering Equipment and Materials User Associations (EEMUA) publication number 191:99: Alarm Systems—A Guide to Design, Management and Procurement.

Alarm/event handling needs careful consideration during design of the process control system. It is important to define which process parameters are critical and require the right level/sequence of response to enable corrective actions to be put in place if the process being automated starts to deviate from its normal operating range. The proces control system needs to be designed to anticipate the event before it leads to a "critical" situation the product being manufactured has to be rejected because of an unacceptable excursion from specification. The way in which alarms are prioritized relies on risk assessment based on the related process parameters. In Europe a two-tier approach is common involving what is called commonly called critical and warning alarms. Note that use of the word critical does not mean out of control as is sometimes inferred by the term in the United States.

Critical Alarms

Critical alarms are generated whenever a process parameter has reached a level/limit at which there is the potential for an impact, for example, to the environment, safety, health and product SISPQ or equipment. The generation of a critical alarm requires system users to take appropriate action to bring the process under control, in accordance with a documented Alarm Response procedure. In addition, supervision must investigate the alarm event to determine if there has been an impact on, the environment, health, or safety (EHS). A critical alarm can have one, two or three categories as discussed shortly.

Warning Alarms

Warning alarms are generated whenever a process parameter has reached a level that requires system user intervention to prevent a critical alarm condition being reached. The generation of a warning alarm requires system users to take appropriate action to bring the process under control, in accordance with a documented alarm response procedure. No formal investigation is required.

Critical Alarm Categories

Critical alarms within a process control system can also be typically aligned to one or more of the following categories:

EHS: These alarms indicate trasgressions related to parameters concerned with the protecting the environment, as well as the health and safety of facilities/equipment plant and personnel. These alarms must be brought to the attention of the EHS organization.

GMP: These alarms indicate transgressions related to parameters concerned with product quality/patient safety, for example, CPP that affect CQA. These alarms must be brought to the attention of the quality organization (the clear identification of GMP critical alarms is key in supporting this action).

Process: These alarms indicate trasgressions related to parameters concerned with business operational needs (e.g., capacity. efficiency, avaiability). These alarms must be brought to the attention of the manufacturing team.

Alarms Actions

Alarms must always be actionable, in other words the alarm should result in the someone (normally the operator) taking action to counter the abnormal condition. Some example actions are shown in Figure 24.11.

Tag name	Description	Relevant operating conditions	Target SP	Alarm value	Operator actions to be taken
AI-xxx-A	Txxx Pool Tank pH	After pH adjustment and during Transfer Out	8.1 pH	LoLo = 8.0 pH HiHi = 8.2 pH	1. Take grab sample to confirm tank reading. 2. If grab sample confirms out-of-range pH, consult MSandT support. 3. MSandT support will consult data and advise if buffer can be further pH adjusted, or if buffer is to be discarded and a new batch made up.
TI-xxx-A	Txxx Pool Tank Temperature	During Transfer Out	5.0 °C	LoLo = 2.0 °C HiHi = 8.0 °C	1. Check the Cold Room temperature control and HVAC are operating correctly. 2. Check the T-xxx Agitator Equipment Module is operating correctly. 3. Check the temperature loop, TI-xxx-A. 4. Contact MS and T if necessary.
TI-1A	Final Product Freezer 1 – Hot Point	All the time	−20 °C	LoLo = −23 °C HiHi = −17 °C	1. Check alarm is not due to an open door, or recent addition of ambient material. 2. If freezer contains material, move to alternative freezer and update logs.

Figure 24.11 Example of GMP/CCP alarm actions.

Safety-Related Systems

Process control sytems will often support functionality that is critical to the safety of plant personal and process equipment. Where alternative (often hard-wired) safety/interlock systems are installed, consideration should be given to mirroring their actions within the DCS, or providing a status input into the DCS. In this way there is less chance that an activated hard-wired trip will be misinterpreted on the DCS.

Unless absolutely necessary, automatic recovery routines within DCS configuration should be avoided as these can be confusing to plant operations staff, often involve complex additional configuration and hence testing and validation. For sequence logic, it is often sufficient to have just a Hold state and Emergency Stop state to address abnormal operating situations. Key aspects to address include the following:

- What conditions would be notification only?
- For what alarm conditions would the impact depend on what stage the process was at?
- What alarms should initiate a hold condition on a sequence?
- During a hold sequence what state should I/O devices move to?
- How can an emergency stop be initiated?
- What state I/O devices should move to in the event of an emergency stop?
- How can adjacent process units be dealt with when a hold or emergency stop is initiated on a plant unit?

Further information on alarm philosophy can be found in International Standard IEC61508 (3), and ANSI/ISA-84.00.01-2004; Application of Safety Instrumented Systems for the Process Industries.

OPERATION AND MAINTENANCE LIFE CYCLE

Maintenance typically adds up to be a significant cost over the life of a process control system. If correct procedures and processes are put in place to maintain the system in a validated state then less errors and failures will occur resulting in less downtime and consequential loss of manufactured drug product to the business. Figure 24.12 summarizes typical maintenance activities to consider for a process control system.

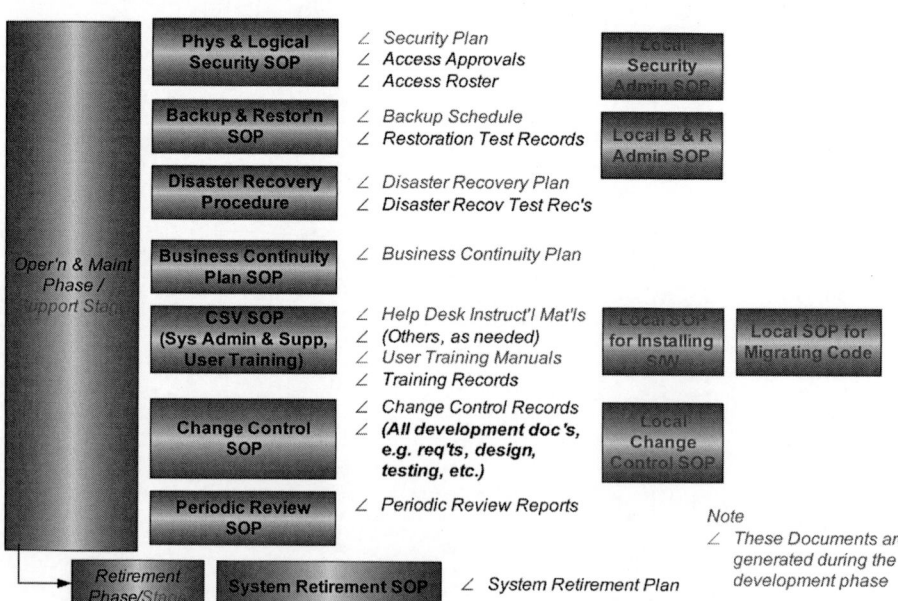

Figure 24.12   Typical maintenace activities.

Security

Management should address both physical and and logical security. The more critical the automation system the more comprehensive access controls you should look to implement. Physical controls include the use of card swipe access; log of entry and departure with date/ time and ID reconition. PLC cabinets should be secured for authorized access only. Cabinets should be locked to prevent ingress of dust particals. If a fire should occur then fire prevention system should activate and if a water system is used then the locking of the cabinet doors can assist in prevention of water ingress. Backup systems should be in place in case of electrical failures. Uniterruptable power supplies (UPS) and battery backups should be on a routine maintenance schedule to ensure correct operation/switch over on power failure. Temperature and humidity controls may be required dependent on the area the hardware devices are stored.

Logical (computer enabled) controls should include limiting access to critical control parameters to authorized access levels only. Dependent on the type of process control system there should be a ability to configure the system to different access levels—for example, operator, engineer, administrator. Consideration needs to be given to how the data is used recorded by the process control systems and if requirements for ER/ES should be implemented. Managemnt of electronic records and signatures is discussed more in chapter 13.

Backup and Restore

SOP should be in place that defines the backup restore process. Where the system is not connected to the LAN and is backed up onto temporary electronic media such as CD or USB then care must be taken to ensure these are stored in a secure and clean area to prevent unauthorized access and damage due to potential static buildup. Backup media should be stored in a separate location to the process control system it is supporting to ensure the ability to restore the data in case major physical damage to the area where the process control system exists. Agreements should be put in place that identifies who is responsible for the backup/ restore process and at what frequency for systems connect to LAN. A key prereqisiti to all of the above is an agreemnt on exactly what needs to backed up and this should be determined on the basis of criticality and risk assessment.

Disaster Recovery/Business Continuity Plans

SOP should be developed that describe what process and actions are required if a disaster should occur. A fire in the PLC room has the potential for major distruction and personnel should know who to contact what actions are required to try and protect the production process. It should be clear when a any manual paper based process neds to be invoked and consideration should be given to ensure data captured in this method is entered back into the automation system once the system is back up and running.

System Administration and Training

All personnel who use the process control system must be trained to ensure they are qualified to complete the role they are assigned to. More complex training may be required for engineering personnel who are providing system administration.

Change Control

Change control must be applied when changes are made to the process control system or associated equipment. If the change is a like-for-like change, that is, exact component to component change, then this should be covered by preventaive maintenance. A change that impacts the functionality of the control system or requires a new/differnet component should be implemeted under a controlled change process. It is important in the development of the change that the appropriate business, engineering and quality personnel are involved in the review, approval, and implementation process.

Periodic Review

Periodic reviews should be conducted to ensure the process control system is operating as intended, is fit for purpose, and satisfied current regulatory expectations. If during the review the system has had a high number of changes, deviations, or issues then this should raise a need to identify what is causing this trend and does it impact the system dependnability and compliance. Access log reviews should always be completed on the system, usually annually, but for the more critical systems every six months may be more apppropriate.

DECOMMISSIONING

When a system reaches the end of its operational life, an essential consideration before destroying all hardware, data and associated documentation are the record retention requirements.

Even for a system that is deemed not to contain electronic records, there will still be a requirement to retain validation and associated (e.g., specification, change control, etc.) documentation for the retention period following completion of manufacture of the final batch of product.

For systems deemed to contain electronic records, then provision must be made for ensuring that all such records archived remain secure and can be retrieved for the required record retention period. Appendix 24.D lists some options that could be considered.

ACKNOWLEDGMENTS

The author would like to acknowledge the help and assistance provided by Victor Christie (Eli Lilly), Des Keates (Eli Lilly), and Dr Guy Wingate (GlaxoSmithKline) in preparing this case study. In addition, the authors would like to recognize the ongoing discussions on this topic with colleagues in the GAMP Special Interest Group for process control systems who, under the leadership of Mark Cherry (AstraZeneca), are currently updating ISPE/GAMP Guide to Process Control Systems (8).

REFERENCES

1. International Society for Pharmaceutical Engineering. GAMP®5: Risk-Based Approach to Compliant GxP Computerised Systems. Tampa, Florida, 2008. Available at: www.ispe.org.
2. International Conference on Harmonisation. Quality Risk Management. Q9 Document, Technical Requirements for Registration of Pharmaceuticals for Human Use, 2005. Available at: www.ich.org.
3. Food and Drug Administration. General Principles for Software Validation, Final Guidance for Industry and FDA Staff. Centre for Devices and Radiological Health, 2002.
4. American Society for Testing and Materials. E2500-07 standard guide for specification, design and verification of pharmaceutical and biopharmaceutical manufacturing systems and equipment, 2007.
5. International Society of Automation. Batch Control Part 1: Models and Terminology. International standard S88.01, 1995.
6. International Society of Automation. International Standard for the Integration of Enterprise Control Systems S95, Parts 1–5, 2000, 2001, 2005.
7. International Society for Pharmaceutical Engineering. Baseline Guide, Commissioning and Qualification. Tampa, Florida, 2001. Available at: www.ispe.org.
8. International Society for Pharmaceutical Engineering. GAMP® Good Practice Guide: Validation of Process Control Systems. Tampa, Florida, 2003. Available at: www.ispe.org.
9. Cherry M. Case study: Distributed Control Systems. In: Wingate G, ed. Computer Systems Validation: Quality Assurance, Risk Management & Regulatory Compliance. Boca Raton, FL: CRC Press, 2004.

Appendix 24.A   Example Functional Specification Checklist

Content	Generic/ specific	Comments
Control modules model	Generic	Define the generic types of control modules, their functionality, alarm attributes, and faceplate displays
Graphics model	Generic	Sets display standards, colors layout, etc.
Alarm and security model	Generic	Sets standards for alarms and interlocks; describes user profiles and system security settings
Batch model	Generic	Describes how the batch executive interfaces with units and phases, how batch reporting is to be configured
Equipment model	Generic	Specifies the functionality of generic equipment modules such as temperature control units
Unit model	Generic	Describes each phase in structured English, describes how the unit hold and emergency stop function
Unit 'nnn' specification(s)	Specific	For each plant unit defines the unit type, phases, control modules, equipment modules, hold and emergency stop states, interlocks, and transfer routes
Procedural specification	Specific	For each process stage, describes the unit operations and phases together with the recipe values of parameters to be passed to phases
Hardware specification	Specific	Defines the hardware and network architecture including server specifications, uninterruptible power supplies, input/output subsystem and operating system, and application software versions

Source: From Ref. 9.

Appendix 24.B   Typical Installation Qualification Tests

Area	Test	Scope
Hardware	Confirms that all hardware has been supplied and installed in accordance with specification	Servers Workstations Network hardware Controllers I/O racks I/O cards Barrier systems Field instruments Power supplies UPS
Cabling	Confirms that all cabling is installed and labeled in accordance with drawings and/or specifications	Network cabling Data highways Power supplies Cabling to I/O racks Field cabling
Power-up/diagnostic checks	Confirms that all systems power up correctly	Servers Workstations Network hardware Controllers
	No error messages are present from system diagnostics.	I/O racks I/O cards Barrier systems Field instruments
	EMI/RFI checks (usually susceptibility to rather than emission)	Power supplies UPS
Software	Confirms that all softwares have been loaded and versions are correct	Operating systems Application software Configuration Bespoke code

Abbreviations: I/O, input/output; UPS, uninterruptible power supplies.
Source: From Ref. 9.

Appendix 24.C Typical OQ Tests

OQ phase	Test
OQ1	Confirms that all control modules operate "end to end" from the operator displays to the field—verifies whether correct device status is indicated on the graphics, and that the graphics are a correct representation of the plant configuration.
	Confirms that all analogue instrument loops are correctly calibrated (over their entire measurement range) from the field instrument to the DCS displays any other indicating devices. Where process trending functionality is included, this may also be verified during this test.
	Verifies correct operation of alarms, trips, and interlocks.
	Verifies correct operation of process sequences during water simulations—tune control loops; ensures that all process paths are tested including any "emergency stop" and "hold" conditions; verifies correct operation of any recipe management and batch data recording during these tests.
	Considers any stress tests that could not be performed during off-site testing—e.g., verifies how long a UPS will support the system following power failure.
OQ2	Verifies correct operation of process sequences during solvent—fine tune control loops.
	When there is confidence that the system is operating satisfactorily, and process validation is ready, product may be introduced into the plant and commissioning batches processed—such tests are conducted against the batch processing sheet, and are often shared tests for process and computer validation.

Abbreviation: OQ, operational qualification.
Source: From Ref. 9.

Appendix 24.D Archiving Options for Electronic Records

Option	Advantages	Disadvantages
Migrate all data to the new system.	Ready access to information; no requirement to maintain old hardware; if same manufacturer, then likely to be a standard upgrade/migration path	Migration method needs to be secure. Not an option if the old system is not being replaced.
Migrate all data to standard format, e.g., PDF.	Ready access to information; no requirement to retain old hardware	Migration will need to be validated—possibly bespoke software required to transfer records. Unlikely to support secure transfer of electronic signatures.
Retain sufficient elements of the old system to enable continued data retrieval capability.	Low-cost option (at least initially)	As time progresses, continued support of the legacy system will become more difficult and expensive.
Migrate old records to paper.	No requirement to retain an electronic system	Regulatory risk—unless this can be clearly demonstrated as the "last resort;" will not be acceptable for records with electronic signatures.

Source: From Ref. 9.

25 | Case Study 7: Manufacturing Execution Systems and Electronic Batch Records

Peter Bosshard, Michael Schneider, and Robert Fretz

Hoffmann-La Roche

The pharmaceutical industry is obliged to document carefully each step in the drug manufacturing process. Electronic batch recording systems (EBRSs) automatically create electronic batch records by collecting all the data that comes from the different manufacturing equipment, test results from in-process control (IPC), and other sources. To keep the batch record at reasonable length, only exceptions and deviations are reported, and the rest of the batch record is normally not reviewed. It can be considered that if a manufacturing process and associated materials are operating within specifications then there is no need to do a special review. The risk-based thinking behind this approach is to focus on the points that potentially could have an effect on the quality of the pharmaceutical product. This approach of review by exception (RBE) is acceptable if the manufacturing process, equipment, and systems are fully validated.

The following case study gives an insight into the business needs leading to the introduction of a manufacturing execution system (MES) including EBRS functionality and how such a system could be implemented and validated. The case study describes the system and the realization concept, the system specification, and the approach used for computer system validation. Risk management allows to focus all the efforts for development, testing, and use to the most critical aspects of a system. For the rest of this case study, the term MES will be used for both the MES functionality and the EBRS functionality.

BUSINESS CASE

The main objective of the MES application is to deliver the following benefits:

Time Savings

- Data is directly downloaded from the enterprise resource planning (ERP) system into the MES and therefore only one bill of material needs to be maintained
- Batch record review can be performed immediately, not delaying the following production steps (e.g., packaging).
- The review process of the electronic batch records is improved by concentration of the focus of the review to any exception (RBE)
- Discrepancies (if any) are easier to analyze if they are all listed automatically by the computer.
- Recording the entire process and control data allows the easy performance of investigations for failure analysis and production optimization.
- Batch records can easily be sent by electronic mail, provided that this function is validated and secure (no changes possible to the record). This is important in case product batches are exported to some countries which have these requirements.
- All signatures can be done electronically.

Improvement of Process Control

- The checking of a second operator in critical steps (e.g., dispensing) is replaced by a validated automated control.
- The completion of every production step, the corresponding results, and procedure parameters can be checked immediately.
- The sequence of the production steps can be configured in the MES, thus managing the manufacturing process (if necessary) is easier.

- The operator is guided through the manufacturing process step by step.
- The state of the production equipment is controlled. This means, for example, that a production vessel can be used only if it is clean, and a balance can be used only if it is calibrated.
- The successful process validation and preventive maintenance is monitored and controlled.
- All equipment and materials can be fully traced.

Improvement of Product Quality

- Boundary checking of process parameters is performed automatically. This means that whenever manufacturing data are outside the specified limits (alert and IPC), the application reports this discrepancy (on screen or by electronic mail). This increases the certainty that any irregularities are detected properly. Thus, it gives management the possibility for immediate reaction, so that timely and cost-saving corrective actions can be taken.
- The automatic data capture of the relevant process reduces input errors.
- Every batch record must be accessible at least for the shelf life of the drug [21 CFR 211.198 (1)]. The introduction of the new system gives the possibility to store the batch records electronically on various storing devices, such as optical disks. Compared to paper-based batch records, electronically stored records need much less room and increase the safety of the data.

Return on Investment

The return on investment (ROI) must be attractive for a project to be supported. Table 25.1 contains an exemplary business case that was published in 2004 (2).

Table 25.1 Return on Investment—Five-Year View—Capital Investment and Expense Reduction

	Year 0	Year 1	Year 2	Year 3	Year 4	Year 5	Total
1. Capital investment							
System costs							
Software	−500,000	0	0	0	0	0	−500,000
Services	−800,000	0	0	0	0	0	−800,000
Platform: Hardware/ software	−150,000	0	0	0	0	0	−150,000
Reduced working capital	0	0	0	0	0	0	0
Total	−1,450,000	0	0	0	0	0	−1,450,000
2. Expense reduction							
Assumed CAGR	10%						
From department spreadsheets							
Labor	0	1,128,586	1,241,445	1,365,589	1,502,148	1,652,363	6,890,130
Materials	0	419,500	461,450	507,595	558,355	614,190	2,561,089
Variable overhead	0	85,080	93,588	102,947	113,241	124,566	519,422
Other direct costs	0	90,125	99,138	109,051	119,956	131,952	550,222
Displaced costs of current System maintenance	0	10,000	11,000	12,100	13,310	14,641	61,051
Total	0	1,733,291	1,906,620	2,097,282	2,307,010	2,537,711	10,581,915

Abbreviation: CAGR; Compound Annual Growth Rate.

PROJECT IMPLEMENTATION STRATEGY

The MES application needed to manage an existing production process, from active ingredient to the final dosage form, that generated between 4000 and 5000 batch records per year, consisting, on average, of 10 pages each. These records were reviewed manually by the responsible pharmacists, the plant supervisor, and quality assurance personnel.

This project was divided into two phases to ensure management control of cost, time, and resources.

Phase 1

- Administration of the general data
- MES for bulk production (granulation, ointment, syrup, and sterile solution manufacturing)
- Data processing of the IPC and the environmental monitoring tests results

Phase 2

- MES for the sterile filling plants, the capsule filling plant, and the tablet compressing plant
- Administration and controlling of the maintenance data of the production equipment
- Administration of the personnel training records
- Automatic data capture from production and testing equipment
- Controlling of the filter test procedure

REGULATORY REQUIREMENTS

The functions of the planned computer system must satisfy the requirements of Good Manufacturing Practice (GMP) regulations. GMPs in the European Union (EU-GMP) and in the United States [Code of Federal Regulations (CFR)] require the recording of batch-specific information during the production stages [EC GMP 4.17 (3) and 21 CFR 211.188 (1)]. The manufacturing procedures and the batch records must be properly reviewed and electronically signed in conformance with 21 CFR 11 (4). Then the product is released for further processing [e.g., packaging in compliance with 21 CFR 211.192 (1)].

VALIDATION

The validation was performed according to the "Roche Computerized System Validation Policy and Guidelines." This comprised the definition of a validation plan, the performance of the planned activities, and the creation of a validation report. The scope of the validation activities was defined by GMP analysis (also known as a GMP assessment). The system requirements were analyzed regarding their GMP relevance.

The ISPE GAMP Group has recently issued a good practice guide (GPG) on the implementation and validation of MES (5). Although not available in time for use in this project, it does provide a useful resource for others embarking on a new project.

RISK MANAGEMENT

The above topics clearly illustrate that there is a significant opportunity to reduce operational risks resulting from the application of an MES. The MES supports compliant documentation and helps make manufacturing processes more reliable and predictable.

GAMP®5 describes in more detail how to adopt a risk-based approach to the validation of computerized systems (6). Risks should be progressively managed through the life cycle of the computer system development.

Risk Assessment

Risk assessment is an important step to identify the degree of the validation effort. Risk is calculated as the product of the probability of an incidence multiplied with the possible impact of the consequences. The risk assessment for this MES was performed in two steps. One general assessment was in order to determine the system risk and a second assessment was performed on the functional level during the development and building of the system. This functional risk assessment was then guiding the level of detail of the testing process.

For this MES all modules were identified as relevant for GMP. However, for the testing the functions were classified into three classes, one was direct product influence, for example, interfaces to the balances and the other was indirect influence on the product, for example, maintenance and training. The third category was the one without relevance to the product like performance reports, capacity utilization reports, etc.

Risk-Based Approach to Testing

The standard functions were taken as qualified by the supplier. This was verified during the supplier audit. The core system test therefore covered only the changed/customized functions. According to the findings of a supplier audit, all the standard functions not tested in the standard system test were included into the MES core system test in addition.

Risk Acceptance and Communication

The risk-based approach was outlined in the validation plan that was signed by all the stakeholders. The results were then later signed in the validation report, before the release of the system.

Risk Monitoring

The risk monitoring was implemented as part of the periodic review that was instituted during the go-live of the system.

Example of the Functional Risk Assessment for the Function RBE

Generally, risks need to be analyzed on a functional level taking into consideration also the corresponding business process. The function "Review by Exception" is a concept that was introduced with GAMP®5 (7).

This example of a risk assessment is based on a failure mode effect analysis (FMEA) where risk priority numbers (RPNs) are calculated as a product of severity, frequency, and detectability of potential hazards.

Table 25.2 illustrates the factors.

The schema for the FMEA is then to describe the process, identify the process steps, and the process sub steps. On this level the potential failures are listed and their RPN is calculated according to the above table.

So for the process step batch record review, the following substeps could be identified:

Process step: Prepare batch record

Sub process step: prepare dispensing part

Failure 1: page of dispensing part is missing—human error

Frequency: 6 (moderate, happens approximately once in 1 month)

Impact: 3 (major, effects, which do not cause a serious risk to health no side effects, but patient or inspector can observe the defect). GMP: consequences, which indicate system problems of process and handling; which might impact also other batches or products

Detectability: 2 (regularly detected: failure detected by procedure in place)
RPN: $6 \times 3 \times 2 = 36$

Failure 2: signatures or data entries missing—human error
      Frequency: 6; impact: 3; detectability 2
      → RPN 36

Table 25.2 Example Risk Factors

	Severity (consequences)	Meaning
Factor		
10	Critical class 1 (catastrophic)	Safety: Effects, which are potentially life-threatening or could cause a serious risk to health or a temporary health problem and/or might trigger a potential recall
		GMP: Close down of site or drug shortage and/or consequences do affect quality and regulatory compliance of a product
6	Critical class 2 + 3 (critical)	Safety: Effects, which could cause illness or mistreatment but are not covered by equivalent examples of "rating catastrophic"
		GMP: Consequences, which indicate systematic errors GMP systems or product registration
3	Major (marginal)	Safety: Effects, which do not cause a serious risk to health no side effects, but patient can observe the defect.
		GMP: Consequences, which indicate system problems of process/handling, which might impact also other batches/products
1	Minor (negligible)	Safety: Effects/complaints, which do not cause a risk to health.
		GMP: Consequences, which effect local daily operations
Frequency		
7	Frequent	Once in a week or more
6	Moderate	Once in 1 month
5	Occasional	Once in 1 year
3	Rare	Once in 10 years (e.g., once in life cycle of the system)
2	Unlikely	Once in 100 years (e.g., once in life cycle of a site)
1	Improbable	Once in 1000 years or less (e.g., once in life time of the factory or less)
Detectability (ability to find the failure)		
5	Normally not detected	Failure very likely to be overlooked, hence not detected (no technical control, no manual or visual control)
4	Likely not detected	Failure may be detected, e.g., audit as spot check
2	Regularly detected	Failure detected by procedure in place
1	Always detected	Failure immediately identified

Failure 3: wrong data entered – human error
→Frequency: 5 (once a year); Impact: 3; Detectability: 4 (likely not detected)
→ RPN 60

Failure 4: batch record sent to a wrong person for review—human error
Frequency: 5; impact: 3; detectability 1 (this is always detected, because there is no review, if the batch record is missing

→ RPN 15

Substep: prepare granulation part of the batch record

One risk for mistakes that occur during the review could look as follows:

Substep: review all values for compliance with the limits

Failure: limit violations are overlooked and not recognized—human error

Frequency: 7; impact: 6; detectability: 4
→ RPN 168

With the introduction of an EBRS many human errors can be eliminated. Specifically with the validated automated review of any exception (RBE) the RPN is 28 times smaller

For the RBE in an EBR

Substep: review all values for compliance with the limits
Failure: limit violations are overlooked and not recognized

> Frequency: 1 (improbable); impact: 6; detectability: 1 (process stops)
> → RPN 6

Of course all the other failures that can be contributed to "human error" are reduced too and as a result we have a very positive risk balance.

SYSTEM SPECIFICATION

Figure 25.1 gives a systematic overview of the various functional modules of an MES. In the annex to this chapter, there is a list of the most important functions of an MES. These may serve to group the systems requirements.

General Requirements

- Definition of authorizations for the specific functions of the system to ensure the appropriate workflow at the production line
- Reporting system for fast information regarding encountered discrepancies from the specified limits
- Identification of the materials used with unique identifiers (such as raw material, intermediates, filters, spare parts)
- Interface to the production planning system (MRPII)
- Identification of the production staff (attributes, resources, and training)
- Identification of the production equipment (status, cleaning, calibration)
- Identification of the types of production rooms used (cleaning, sterile, control, etc.)
- Identification of the desktop workspace (workstation, screen, bar code reader, etc.)

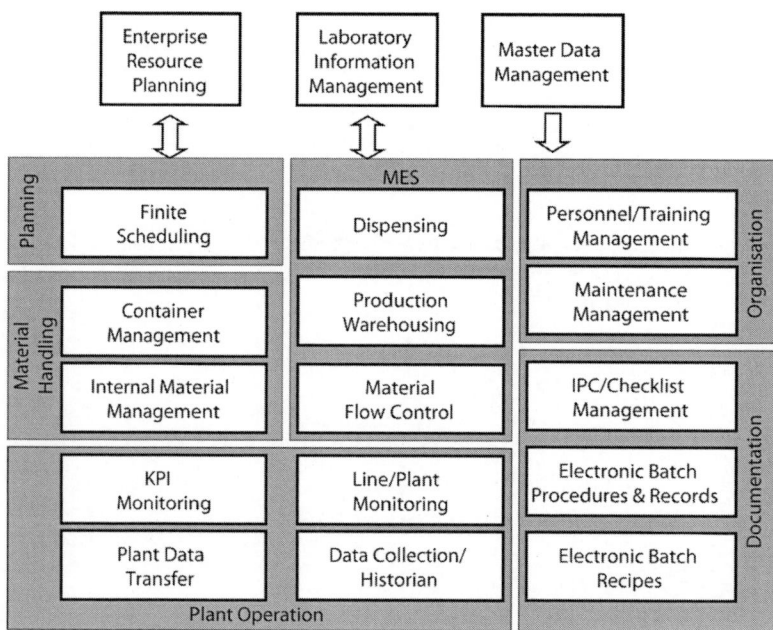

Figure 25.1   MES functionality.

Bulk Production

- Definition of master production protocols [21 CFR 211.186 (1)], through which the operator is instructed (through individual operating procedures) how the various production steps must be carried out.
- Creation of the master production record, by copying the valid master production protocol, before production is started.
- Filling in data by workers turns the master production record into the batch production record.

In-Process Control/Environmental Monitoring Test

- Environmental monitoring tests
- IPC
- Validations and calibrations

Electronic Records and Signatures in a Hybrid Environment
Even though the goal of the system is to provide fully electronic signatures and recording capability for all required entries, it is still necessary to have a possibility to work with hybrid solutions. Hybrid solutions are necessary to keep specimens such as samples of the packaging material used [see EU GMP 4.18 g (3)] or some printouts for older equipment that is not yet capable to be connected to the MES system. In case there is a specimen or handwritten signature that needs to be attached to the batch record the corresponding item is identified in the electronic batch record with a unique identifier that allows to locate the item later on if evidence is required.

SUPPLIER SELECTION
Today there are several competitors working in this field of MESs. Some of the best known are Consilium, Rockwell, SAP, Siemens, and Werum. A careful evaluation of the capability and prospectus of different systems was performed.

DESIGN AND DEVELOPMENT
The MES application was designed in accordance with company procedures and in accordance with industry standards.

Design Standards
Today many standards are existing that help to design and develop MES. Prominent standards include

- *ANSI ISA S95:* Clear separation between planning and production level with association of roles functions and responsibilities. Definition of the functionality of an MES in the form of UML and XML schemes.
- *ANSI ISA S88:* Batch control guidance in two parts. Part 1 defines reference models for batch process control.
- *ANSI ISA TR99:* Security technologies for manufacturing and control systems. Set of recommendation for the efficient utilization of electronic security technologies and plans for the production an process layer. This standard was developed by world famous cyber security and automation experts.

Data Architecture
Before starting with the project, the data available electronically were stored in several databases. Database update was time-consuming and difficult; data analysis even needed different program interfaces. Therefore, the database for the new system had to be a central uniform database for the whole production plant. The bill of material is fed from the ERP as well as the general planning. The LIMS is another important interface for the input of analytical data that is important for the calculation of correction factors for the content of API.

System Build

The system was built by the supplier. All corresponding standards and metholdologies were assessed during an audit of the supplier. Configuration and programming was performed according to the user requirements and the system delivery specification. Activities included:

- Definition of the programming standards
- Ensuring the independent functioning of each individual software
- Module
- Program description (source code)
- Source code review

In addition, computer hardware aspects needed appropriate management:

- Definition of the hardware components used
- Definition of the necessary installation procedures
- Ensuring the proper integration of software and hardware

QUALIFICATION

The finished program was tested by the developer, using unit and integration tests. In addition, installation qualification (IQ) of the hardware and the operational qualification (OQ) of the complete system (hardware and program) were performed on-site. In the test phase, the user acceptance tests were performed. The goal of these tests was to verify whether the completed system was performing according to the user requirements. The successful test completion was documented in the validation report, confirming that the system had been validated for use in daily business. Also included in this phase was the development of various operating procedures. The activities carried out are summarized below.

GMP Analysis

The user requirements must be analyzed regarding their risk potential and GMP relevance. The analysis determines if a function

- Has influence on the pharmaceutical technical quality
- Affects the medical safety of the drug
- Has influence on the data that become part of the registration documents
- Is critical for another important reason

Verification Test Strategy

A precondition for any testing is the availability of a document that correctly specifies the functions of the system. It is the basis for the verification test specification. In the case of the MES, this was the user requirement specification (URS). The major testing was done as black box testing by the supplier. White box testing was limited to the modules ranked as most critical in the GMP analysis. This included the formulas of active ingredient strength and the verification of algorithms in the source code (8).

Verification Test Procedure

A first step toward the validation of the system was the development of a testing procedure to be used for the different software modules and future revalidation. This procedure defines how test plans must be specified, how the tests are performed, and how they are documented. The goal of the testing is to establish documented evidence that the system is performing according to the specifications.

Defining the Verification Test Specification

Once the functions to be tested had been completely identified based on their GMP relevance, they were added to the verification test specification.

For each test case, the following information was added:

- Verbal description of the goals that a specific test must achieve
- Detailed description of the test procedure
- Definition of the required test data and the expected results
- Definition of the test protocol
- Listing of any related documents referred to in the tests

The following areas and functions were tested during the validation of the MES:

- *Daily usage of the system,* including authorizations and security, Windows™ menu control, error reporting, and communication with other devices (e.g., peripherals)
- *Characterization and handling of materials,* including the definition of raw materials, products, and auxiliary material (e.g., packaging)
- *Workflow at the production line,* including production steps, line type, line status, and line schedule
- *Production equipment and locations,* including definition and current state of buildings, production areas, and individual workstations
- *User specifications,* including the definition of user characteristics, groups, responsibilities and privileges, training, and scheduling
- *Control of auxiliary materials and equipment,* including characteristics and calibration of computers, containers, scales, and so on
- *Supplier management,* including the analysis of the vendor's quality management system (QMS) and its ability to deliver the requested system
- *Lot data control,* including the identification and maintenance (corrections, deletions, restrictions) of the lot data and content/ingredients calculations
- *Further testing,* including areas such as material storage, MRP II, cleaning protocols, product content, and archiving

Corrective Actions

Any discrepancy from the expected results that were encountered during the testing had to be analyzed for its relevance and documented. Problems that prevented GMP-conform usage of the system had to be corrected immediately.

Special Testing

Stress testing. A stress test regarding the data volume was performed by expanding the database to the possible volume of a half year's production records. The acceptance criterion was that it would still be possible to use the system with reasonable response times.

Client interrupt testing. It was tested to determine what happens in the case of a sudden breakdown of a client's personal computer (PC). The acceptance criterion was that the database would not be corrupted. Data not saved properly, should, thus, be rolled back automatically to the latest secure version of the data.

Automated Testing

Splitting the project into two phases made it necessary to integrate several modules sequentially. This process led to frequent revalidation activities. To reduce the testing time and to prevent typing errors during test execution, an automated test tool was used. However, this tool did not deliver all the benefits initially expected, because manual editing (programming) of the generated scripts for the testing tool could not be avoided completely and proved to be very time consuming. Such editing was necessary because

- Windows™ objects (e.g., buttons, menus) reacted differently than the testing tool expected;
- Scripts had to be commented to ease later editing, such as the insertion of additional test cases; and

- Date- and time-related functions, which are quite frequent in batch recording systems, led to problems during testing. For example, running a test script on a Friday, creating a production order for the following day, triggered the ("unexpected" question of whether the production order really should be started on a Saturday or the next Monday. Such questions would not arise Sunday through Thursday.

Furthermore, the execution of the test scripts was halted because of Microsoft Windows™ problems. Finally, maintenance of the test scripts became difficult once the system had been handed over to the business, since the know-how required to maintain and rerun the test scripts was no longer available.

Example

The following example (Table 25.3) describes how testing was done for a specific function of the system. The situation described in the example is such that within the MES, menus can be specified for the combination of

- Organization (plant, production line, etc.);
- Workplace type (weighing, drying, sterile conditions, etc.);
- User group (= access level), where each user group has access rights for its own or all lower access levels.

Table 25.3   Example Test Case

Function	Log-in of person 1 on the workplace defined by organization (plant) = B1 and workplace type 2 = AT2
Test procedure	1. Fill out log-in mask using the test data
	2. Fill out workplace-identification mask (log-in2) using the test data (see Fig. 25.2)
	3. Check if the menu displayed corresponds with the expected result (see Fig. 25.3)
Test data	4. Log-in mask: user name = TESTP1, password = XXXX00
	5. Log-in2 mask: workplace type = AT2, organization = B1
Expected result	No menu should be displayed, because there is no menu defined for user group 1 at the workplace B1/AT2 (see Fig. 25.3)
Test documentation	Printout of the workplace-identification mask and printout of the main menu mask (sign-off with initials plus date)

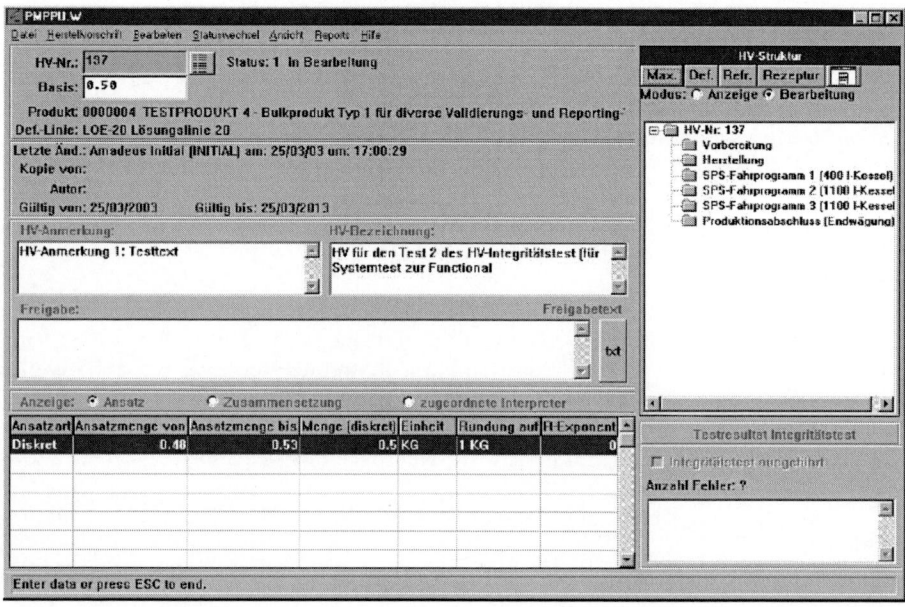

Figure 25.2   MES data screen.

(A)

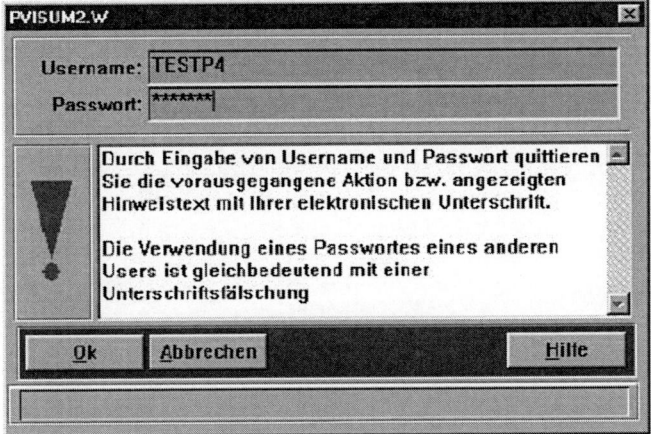

(B)

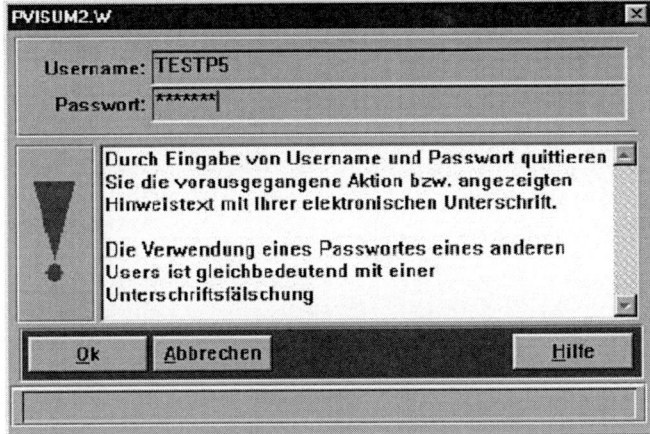

Figure 25.3 MES test screen.

A validation database was set up. This validation database contains always a defined amount of data and leads to precisely predictable results. For example, in this validation database, person 1 was set up with user name = TESTP1, belonging to user group 1 with the access level 1. The PC on which the validation testing was run was set up as workstation WS0001, belonging to the organization B1 with the workplace type AT2. There is no menu prepared for organization B1 with workplace type AT2 and user group level 1.

MAINTAINING OPERATIONAL COMPLIANCE

Once the system was successfully implemented and in operation the validated status of the system was maintained, using the necessary technical and organizational procedures.

- Change and configuration management involves procedures that control and report the implementation of changes that may affect the validation status of a system. This includes the tracking of problem handling, resulting from fault reports or change requests, to their solution. Change management ensures that the configuration of the system is identifiable and reproducible (4).
- Periodic system reviews are performed to verify that the system is operating as specified (performance, disk space), that it is properly administered (e.g. authorizations), and that the documentation is accurate.

Currently, the system is still under implementation and revalidation is done frequently—each time a system module is handed over to operations. Experience with ongoing validation of the system has yet to be gained.

EXPERIENCE OF INSPECTIONS

The MES is frequently a topic during inspections because of the modern technology. Most of the inspectors review the validation report. Sometimes the traceability matrix that is used to link the tests to the user requirements was also reviewed as well as some exemplary test cases. There were never any major observations.

According to literature the following observations are reported relating to production systems:

- Validation of the computer software used to operate and save data generated during manufacturing processes does not demonstrate or challenge security features. The ability to save changes made to processing data by operators was not challenged or documented. Audit trail capabilities were not activated, and the abilities to override a batch and delete an entire file were accessible at the supervisory level. Also, a master list passwords for all operators and users of the software was made globally available in the validation report (9).

Comment: Security and audit trail are very important characteristics of a modern MES. Most likely the system that led to this observation was not configured correctly and the fact that the master password was explicit in the validation report shows that this validation was not performed against modern standards such as GAMP.

- The computer control system software used to control coating operations in coating rows do not ensure that correct product recipes, which contain pre-established process parameters, are used for the corresponding products. For example, during manufacturing of a particular batch a recipe was downloaded by the supervisor for control of the process. Manufacturing orders for tablets require the use of recipe name although process parameters are different, including atomizing air pressure, spray rate, inlet air temperature, etc. (10).

Comment: It is very important that the operators are trained to use only the correct recipes for the correct product.

- Input and output verification from the computer, related systems of formulas, and records or data are not checked for accuracy. Specifically, the firm's computer system has not been validated to ensure the accuracy of formula input and output calculations and quantity changes to formula ingredients (11).

Comment: Any formulae used in a computerized system has to be verified during validation. Manual input of critical data has to be checked accordingly and whenever possible should be replaced by automated input over a validated interface.

- There are no procedures/system established to assure accountability for original batch records when removed from the files. Notably, the original batch record for lot could not be located. (11)

Comment: The data of the system should be protected by backup. There should be a clear procedure for disposal of old records, that is, batch records that are no longer needed after shelf life plus one year according to 211.180 a) (1).

- The validation of the software, which controls allocation of materials for production from the warehouse, did not include verification that the system would correctly calculate the amount of material required when the assay concentration of the material was not 100% (12).

Comment: Any formulae used in a computerized system has to be verified during validation. This includes also verification of formulae that are used to calculate the correct amount of material according to the lab result. Otherwise the data should be checked by a second operator.

- The firm's computerized control of the step has not been adequately qualified and/or validated in that
  a. The control program algorithms were not described and tested to establish their correct functioning.
  b. No data have been collected and documented during the validation runs to establish that the automated controls actually monitored and controlled the process as intended by the process owner.
  c. Monitor and control data during the batch process are not captured and included as part of batch records. There is no information available as to the actual time line information for the critical control steps (12)

Comment: During the specification phase (a) all algorithms used should be described, (b) afterward during the testing these requirements should be tested, and (c) all data relevant for the batch should finally be in the batch record (c). For the review it should be sufficient to only check the exceptions (see review by exception)

CONCLUSIONS
Realization of the Expected Benefits
High Degree of Automation Required
To be efficient, electronic batch recording must eliminate more than half of all manual collection, analysis, and review work. Otherwise, there will not be any significant time reduction.

Creation of a Central, Uniform Database
To ensure the efficiency and effectiveness of the system, all data required for batch review and release should be stored in one central database, eliminating the necessity of interfaces between different databases.

Efficient Validation
Availability of a Company Policy on Computer System Validation
Validation was performed according to the Roche company policy on computer system validation. A validation plan was established, the activities were performed and documented according to this plan, and the results were entered in the validation report. As a result, the product (drug) manufacturing process is better documented and analyzed, thus contributing to the safety of a product.

Analysis of GMP Relevance and Risk Assessment
The validation approach should be based on the analysis of the GMP relevance for each function of the system. The results should then be used to define the test strategy and the corresponding test cases.

Proper Vendor Selection
It is important to select a vendor who is capable of delivering the system in compliance with the requirements of the computer system validation policy. This ability should be verified by proper vendor selection and auditing.

Ensuring Ongoing Validation

Once the system is in a validated state and handed over to operation, it is important to ensure that the system remains in such a state. The available operating procedures must take into account that the unit operating the system does not have profound system knowledge, which the project team had.

REFERENCES

1. U.S. Code of Federal Regulations Title 21, Part 211 (Revised as of April 1, 2008). Good Manufacturing Practices for Finished Pharmaceuticals. Available at: http://www.accessdata.fda.gov/scripts/cdrh/cfdocs.
2. Anderson L, Burbage D. Estimating ROI for Automated Clinical Materials Management Systems, Pharmaceutical Engineering Supplement September/October, 2004; 8–11.
3. The Rules Governing Medicinal Products in the European Union. Good Manufacturing Practices. Volume 4, March 2008 edition, European Commission, Directorate General III—Industry, Pharmaceuticals and Cosmetics. Available at: http://ec.europa.eu/enterprise/pharmaceuticals/eudralex/homev4.htm.
4. U.S. Code of Federal Regulations Title 21, Part 11 (Revised as of April 1, 2002). Electronic Records and Signatures. Available at: http://www.accessdata.fda.gov/scripts/cdrh/cfdocs/cfCFR.
5. ISPE. GAMP® Good Practice Guide: Manufacturing Execution Systems. Tampa, Florida: International Society for Pharmaceutical Engineering. Available at: www.ispe.org, 2009.
6. ISPE. GAMP®5: Risk-Based Approach to Compliant GxP Computerised Systems. Tampa, Florida: International Society for Pharmaceutical Engineering. Available at: www.ispe.org, 2008.
7. Appendix S2, Electronic Production Records, Good Automated Manufacturing Practice GAMP, Version 5, ISPE, 2008.
8. FDA. Software Development Activities, Technical Report, Reference Materials and Training Aids for Investigators. Rockville, MD: Food and Drug Administration, 1987.
9. GMP-Trends, 15. Nov 2004.
10. GMP-Trends, 15. Sep 2004.
11. GMP-Trends, 1. Jan 2005.
12. GMP-Trends, 1 Sep 2006.

ANNEX: MOST IMPORTANT FUNCTIONS OF AN MES
Master Data Management

- Master data management
    work centers
    items
    storage
    site and shift calendars
    users administration

- Version-controlled master data
    routings
    bills of material
    production instructions
    SOPs
    procedure operations
    material safety datasheets

- Electronic signatures

Labor Management

- Personnel/facility master data
- Staff qualification, training administration
- Plant-/GMP- and SOP-related instructions
- Site and shift calendars

Electronic Batch Records

- Interactive on-line process control
- Electronic signature with password and user ID
- Gather data from processing equipment via PLC interface
- Automated generation of the batch record

Process Visualistaion and Data Acquisition

- Order processing according to the pull principle
- Real-time process monitoring and process control
- Collection of resources consumption data to calculate yields
- Utilization calculations
- IPC management
- Line set points control

Archive

- Search and reporting functions for all production data
- Long-term batch archive
    preferably online
    otherwise in a conserved format (PDF, TXT, TIFF)

- External archive management
- Retention period minimum 15 years (European Liability Law)

Detail Production Planning

- Checking of dates and resources
- Interface to ERP
- Order sequencing
- Optimization of set-up times
- Resource requirements and availability (qualification status)
- Personnel placement plan
- Simulations
- Monitoring of order progress
- Electronic planning board

Warehouse Management

- Interfaces to ERP
- Goods receipt/issue
    on material, batch and container level
    RFID interface
    EAN code management

- Point-of-identification control
- Warehouse control functions
    Cold rooms
    Humidity and temperature control

- Transportation management
    Automated guided vehicles
    High-rack warehouse

- Inventory functions
    Chaotic storage
    FEFO principle

Dispensing and Weighing

- Identification of containers and materials by RFID-scanners or barcode readers
- Order explosion
- Weighing schedule
- Recipe-based weighing
- Interfaces to balances
- Interfaces to ERP

IPC

- Management of IPC, that is, check plans, check points, checklists
- Statistical process control
- Continuous monitoring and documentation of environment conditions for batch documentation
- Automatic, acquisition of analog and digital measured values
- Visualization, monitoring, and trends, alarms, and alarm history
- Reports, evaluations

Performance Analysis

- Real-time and trend-oriented production monitoring system
- Management information (MIS) with on-line information about current order and process status
- Characteristic performance and productivity values
- Operational reporting
- Mean time to failure
- Equipment utilization and downtime

Maintenance Management

- Tracks and directs to maintain the equipment and tools
- Logbook
- Integration with ERP Plant Maintenance
- Preventive and breakdown maintenance

26 | Case Study 8: Building Management Systems

John Andrews
Andrews Consulting Enterprises Ltd.

Mark Foss
Boehringer-Ingelheim

The control systems associated with building environmental management, typically known as building management systems (BMSs), have always presented a difficult challenge to those responsible for the validation. Although the equipment involved may well be straightforward to validate on their own, the control systems themselves have presented a more difficult challenge. This has been because cGMP and noncritical facilities are generally housed in the same building. The control systems therefore have generally been mixed, thus making it very difficult and expensive to validate. Segregating the control system between cGMP and noncritical is also very difficult because the air-handling equipment and other such equipment may be common to both facilities. This has led to the common generalization that "BMSs can not be validated," the common get-out clause; however the regulators are not convinced.

BMS SCOPE AND DEFINITION

BMS can be a collective noun for a range of computerized systems including programmable logic controllers (PLCs), supervisory control and data acquisition (SCADA) systems, distributed control systems (DCSs), outstations/controllers and instrumentation (Fig. 26.1).

Instrumentation and Devices

Instrumentation and devices communicate measurements and status information to the control and monitoring logic usually in the form of digital and analogue inputs. Such information is interpreted by the control logic to deduce control actions that are used to refine the process control. The calibration and tuning of such instrumentation is critical to the accuracy of the process control.

Control

Control is typically provided by assembling standard control functions, for example, P&ID, Start/Stop, etc. into the required control scheme. Control and calibration parameters are inputs to the control scheme that establish process characteristics, process timings, and responsiveness of the control scheme.

Monitoring

Feedback from instrumentation influences the control scheme, which will respond to maintain process parameters within configured limits. Integrated and/or independent monitoring functions scale and check inputs against preconfigured statuses and limits, setting alarm conditions when deviations are detected.

Control and Calibration Parameter Management

Control parameter management enables users to change control parameters to achieve the desired characteristics of the process, for example, temperature and humidity set points, tolerances, time spans, alarm limits, etc. Such parameters are usually entered via graphical user interfaces or via control panels.

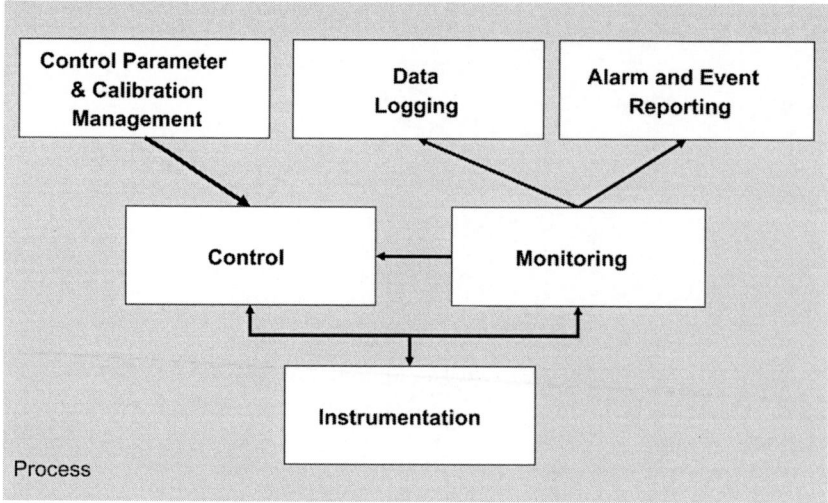

Figure 26.1   BMS functional model.

BMS FUNCTIONALITY AND REGULATORY REQUIREMENTS

The objective of the BMS is to centralize the monitoring, operation and management of a process area or unit. One of the criteria for installing such a system is that critical process and building environmental parameters, such as room pressures and temperatures, can be maintained and recorded. Another benefit of installing a BMS is that energy is used in a more efficient manner and costs are, thereby, reduced. In the process of meeting these objectives, the BMS has evolved from a simple relay and timer-based system into a fully integrated microprocessor controlled system with many features such as environmental optimization, PID (proportional, integral and derivative control) with full data recording, and archiving facilities. Figure 26.2 presents a typical BMS layout.

Environmental control in drug manufacturing facilities has drawn increased attention from the FDA and other regulatory authorities in the 1990s. The U.S. Code of Federal Regulation for Good Manufacturing Practice says in section 46 that (1)

1. Adequate ventilation shall be provided;
2. Equipment for adequate control over air pressure, microorganisms, dust, humidity, and temperature shall be provided when appropriate for the manufacture, processing, packing, or holding of a drug product; and
3. Air filtration systems, including prefilters and particulate matter air filters, shall be used on air supplies to production areas when appropriate.

European GMP directives and associated regulatory guidance have very similar expectations (2).

REGULATORY CONCERNS

Having established that the process should be the focus of risk assessment, it is possible to review regulatory citations over a number of years to identify regulatory concerns associated with typical processes controlled and monitored by BMS.

U.S. FDA Warning Letters

Table 26.1 summarizes example citations from U.S. FDA Warning Letters. While not all of the Warning Letters listed indicate the use of a BMS, the processes and equipment referred to are typical of those associated with BMS.

When reviewing regulatory citations, care should be taken, as the context within which the citation is raised is not always apparent. Several of the citations indicate the importance of

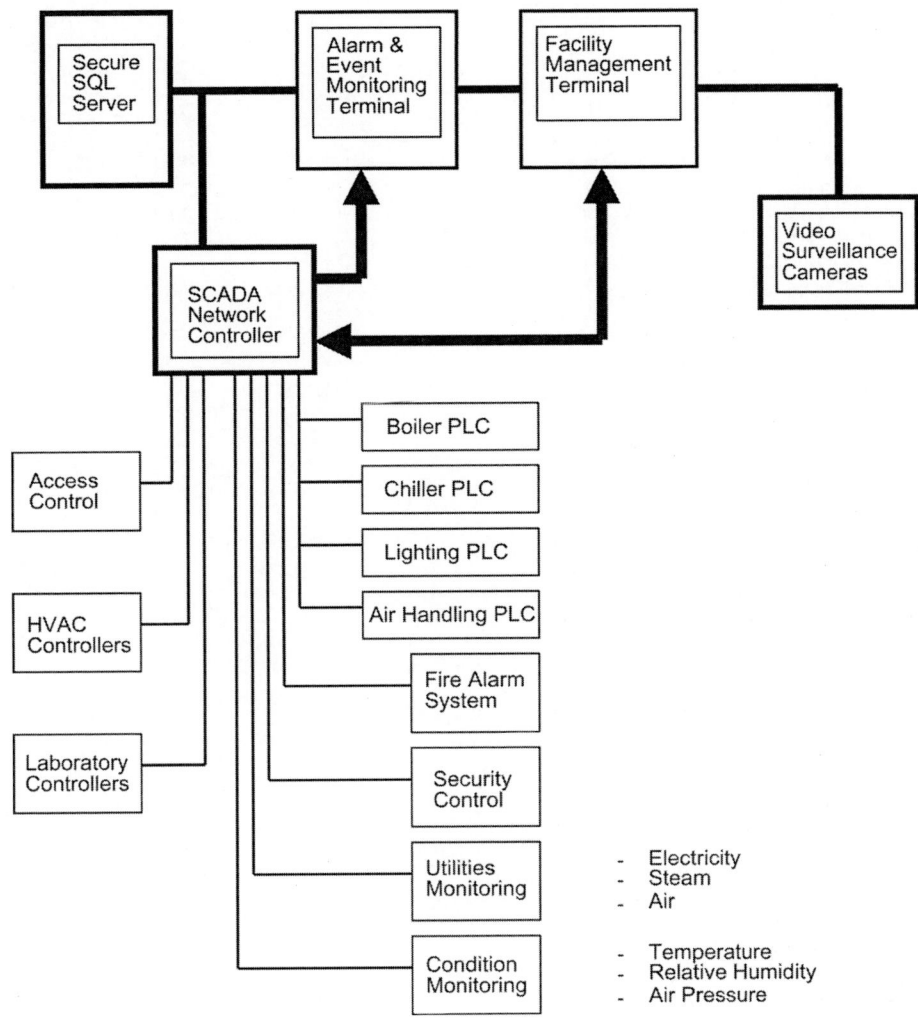

Figure 26.2 Typical BMS layout.

monitoring key (environmental) parameters against predetermined limits such as temperature, humidity, and differential pressures. Although predominantly focused on monitoring, some of the citations indicate a lack of validation\qualification of controls, even if, the criticality of the cited systems to product quality is not clear from the citation.

Access to specific European regulatory observations is restricted; however, anecdotal evidence from industry forums, such as ISPE GAMP, suggests that a number of companies have been challenged by European regulators about control of their BMS.

BMS COMPLIANCE STRATEGY

There has been some debate on the appropriate compliance strategy for building management systems. Regulatory guidance has suggested that BMSs used to control the environment for aseptic manufacturing have a critical impact on drug quality and should be validated (3). ISPE commissioning and Qualification Baseline® Guide further suggests that BMS applications with indirect impact on drug quality do not require validation and that documented Good Engineering Practice (GEP) is sufficient (4). This requires that there is no critical product quality dependency on the BMS and that qualified/validated independent monitoring systems are performing those functions critical to making decisions about the quality of product. Table 26.2 summarizes a suggested way forward.

Table 26.1 U.S. FDA Regulatory Citation Examples

Citation	Clause (where stated)	Warning Letter date
"Qualification and control of the ambient temperate and accelerated temperature stability rooms is inadequate …"	211.166	July 01, 1999
"The alarm system that communicates, records, and controls alarms such as air balance and temperatures for production, warehouse, and testing areas lacked validation documentation"	Not stated	January 2001
"No evidence that your firm investigated temperature failures that occurred for the incubators and refrigerators"	211.22(a)	January 16, 2001
"There is no written procedure in place for, nor is there any testing performed, for the environmental monitoring…"	211,160(b)	
"Failure to validate equipment, for example … Failure to document the rationale behind established alarm times to monitor the specified differential air pressures within the manufacturing areas"	211.168	March 02, 2001
"You have failed to validate the HVAC system used to control temperature and relative humidity in your manufacturing and warehouse areas"		August 14, 2001
"No formal specifications for temperature or humidity have been established for these areas."		
"You were noted to have portable chart recorders for monitoring of temperature and humidity in suites 1 and 2 and one recorder was noted in the warehouse."		
"A wide range of temperatures and humidity was noted in our review of the data from the monitored areas …"		
"IQ and OQ which support …. production …. The list of deviations include … replacement of HVAC systems and its control system."		February 15, 2002
"IQ and OQ which supports coating … list of deviations and resolution plan include failing acceptance criteria for temperature and pressurization flow direction … updating control system … converting to a … control system"		
"IQ and OQ which supports coating … list of deviations and resolution plan include failing acceptance criteria for temperature, HVAC alarm and interlock testing …"		
"No documentation of the validation of the air handling system or the water system used in production"	211.42	October 10, 2002

Table 26.2 BMS Compliance Strategy

BMS implementation	Quality dependency	Compliance strategy
Aseptic manufacturing	Critical product quality dependency on control and monitoring of BMS regardless of any independent monitoring	*New BMS*: Expectation is to validate entirely new BMS implementations[a] *Existing BMS*: Qualification of existing BMS should be reviewed and revised as necessary. Implement and validate independent monitoring. Note that independent monitoring is not needed if BMS is validated[a]
Nonaseptic manufacturing	Product quality dependency on monitoring	Implement and validate independent monitoring, review as necessary for existing BMS. Adopt good engineering practice for BMS.

[a]Validatable BMS are now available as commercial off-the-shelf products unlike a few years ago.
Abbreviation: BMS, building management system.

Aseptic Manufacturing BMS

Product quality is critically affected by BMS control and monitoring for aseptic manufacturing such as parenterals. Reliance on alarming out-of-specification environment conditions is insufficient to support high-integrity product. Independent monitoring does not relieve the basic reliance on BMS operability and should therefore be validated.

Nonaseptic Manufacturing BMS

Product quality is dependent on monitoring manufacturing environmental conditions; there is no critical product quality dependency on BMS environmental control. This scenario allows the implementation of validated independent monitoring (4). Independent monitoring systems can be complex or simple depending on monitoring requirements. Highly toxic, terminal steriles, and inhalation manufacturing often have sophisticated monitoring requirements involving multiple environmental parameters. Independent monitoring in these cases is best served by implementing an SCADA system. Oral dosage, liquid, and topical manufacturing that have much simpler monitoring requirements are probably best served by stand-alone chart recorders.

APPROACH TO VALIDATION

Fear of FDA intervention certainly is a compelling reason for a company to validate its environmental controls. Accomplishing business goals may be a better reason. According to the Landis Division of Siemens Building Technologies, Inc. (5), "It just makes good business sense to make sure the facility operates as designed to ensure quality products are consistently produced." The operations manager for Siemens explains it this way "Aside from the risk to the life and health of employees, the cost of product failure due to not meeting quality standards can be very high. Years ago, humidity, pressure and temperature were not considered part of quality control. Today, it is realized that controlling the environment boosts the production yield. It's not just the process that must be validated."

However, is validation still required for everything? If more than one building is to be constructed, all processes that must be validated by GLP (Good Laboratory Practice) or cGMP could be segregated to the same building and noncritical facilities housed in the other. If critical and noncritical areas are mixed within the building, the critical processes could be segregated to one area. Do offices, research and development labs, storage areas, and corridors really need to be validated? It may not be considered necessary. Finally, are all the hardware components critical (some may well have direct impact on quality, whereas others may have indirect or no impact in the way that interacts with the process/product)?

Hardware and software change control must be addressed early on, because it will affect the entire process. If thermistors are specified and then sealed behind drywall during construction, calibration will be a very expensive and time-consuming process (they must be replaced when they are out of specification). Resistance temperature detectors (RTDs), which can be calibrated in place and have field-replaceable parts, may be a more cost-effective solution in the long run, even though the initial cost is higher. If the software change control procedure requires revalidation with every minor modification, updates will be very difficult and costly. One should remember that the maintenance staff must live with the change control procedures for the life of the facility. Flexibility should be built in and subcontractors must be trained on the correct maintenance procedures.

Risk Analysis

The compliance requirements for BMS systems should be commensurate with how they are used. A risk analysis can be performed to determine whether the parameters controlled and monitored by a BMS application have direct or indirect impact on drug product quality (processing, storage, and distribution). Should the analysis reveal that the BMS is controlling and monitoring any parameters with a direct impact on product quality, there are two courses of action:

- Validate the BMS and
- Relieve the BMS system of its critical function.

The risk assessment process can be divided into two steps. The first step would be to evaluate the impact of a "system" on the product quality. The second step of the process would be to evaluate the criticality of the components in the direct and indirect impact systems, as they relate to product quality.

The determination of system impact as direct or indirect, and the result of criticality assessment, should be documented. Review and approval from quality assurance personnel is expected.

The application of this process helps to ensure that if validation route is chosen, the appropriate resources are applied to the parts of the system that have the potential to affect product quality. Secondly, it provides the rationale to focus qualification effort on quality related functionality, while still ensuring compliance for the product(s).

Impact Assessment

Review the system within the project, define the boundaries, and perform the system assessment outlined in Figure 26.3. The systems should be identified as

- Direct impact,
- Indirect impact, or
- No impact.

The no impact systems fall out into a classification where it makes good business sense to apply GEPs. The direct impact and indirect impact systems enter a preparation phase where the component lists of the systems are prepared for the second step of the process if the validation option is chosen.

Component Criticality Assessment

The second step of the risk assessment involves the criticality assessment of the system components. The components will be either critical or noncritical.

It may be easier to begin by creating a list of all the instruments, equipment, components, etc., in the direct and indirect impact systems to perform the criticality assessment. A series of questions and discussions may take place to evaluate each component and its associated

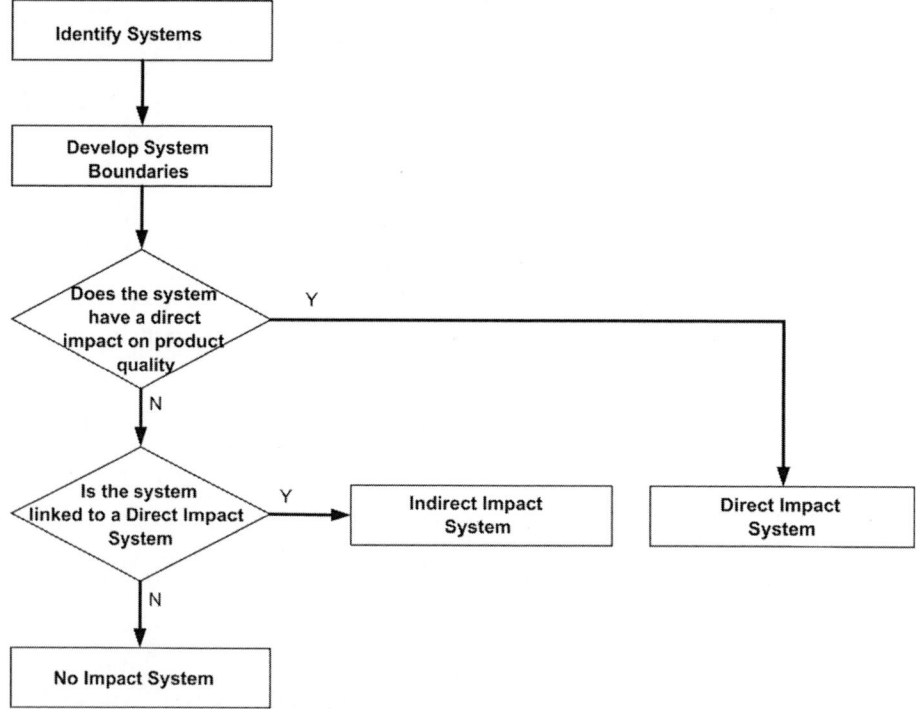

Figure 26.3 Impact decision tree

Table 26.3   Example Criticality Questions

1.   Is the component used to demonstrate compliance with the registered process?
2.   Does normal operation or control of the component have a direct effect on product quality?
3.   Will failure or alarm of the component or the associated functionality have a direct effect on product quality or efficacy?
4.   Is information from this component recorded as part of the batch record, lot release data, or other GMP documentation?
5.   Does the component (e.g., sensor) come into contact with product or product components?
6.   Does the component control critical process elements in such a way to affect product qualify without independent verification of the control system performance?
7.   Is the component and associated functionality is used to create or preserve a critical status of a system?

control requirements. A checklist may be used to track whether a component and its control features are critical during this process. Considering that there may be a significant number of components and functionality, the method of documenting the process should be determined in advance.

Table 26.3 includes example questions to help determine component criticality. Specific products may require additional considerations.

Component Criticality Versus System Impact
The results of the criticality assessment may then be checked against the matrix shown below. This matrix represents the relationships between systems and system components. Components are permitted to exist in three of the four boxes, and cannot exist in the lower left box. The relationships and their interpretations need to be understood before progressing with the assessment process.

Figure 26.4 illustrates a summary of the impact assessment process described so far additional points to note are:

- Indirect impact or no impact systems should not contain any critical components or associated functionality.

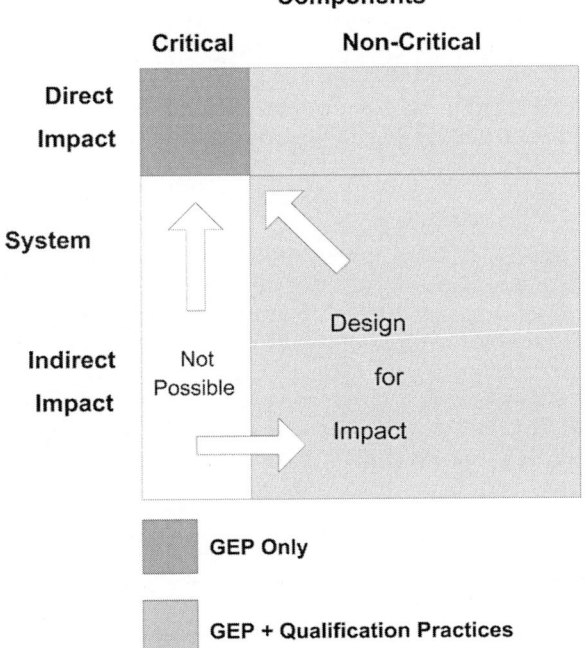

Figure 26.4   Criticality determination.

- Direct impact systems may well have both critical and noncritical components and associated functionality. The noncritical components and associated functionality can be reviewed and tested with a lower level of scrutiny.
- "Design for impact" reduces the scope of the components and functionality that are subject to a focused validation effort, allowing appropriate focus on the components and functionality presenting the greatest risk to produce quality.

"Design for impact" is the term used to describe the practice of making conscious design decisions with respect the impact a systems and it is associated functionality has on quality. By careful design the number of direct impact components and functionality can be reduced, thus reducing any unnecessary qualification and validation effort.

It is essential that this process is documented and approved (or at least reviewed) by quality assurance (QA) personnel. The process of assessment would normally be conduced by a team of qualified staff including representatives from engineering, production and quality. The documented conclusions will be used to position the validation approach that will be described in the validation plan.

Control System Considerations

Now that it has been established that BMS applications and their associated hardware should be designed in anticipation of potential impact they may have on the quality of the product the control system validation can be considered. Factors to consider include:

1. Are the hardware and software platforms supporting the BMS application suitable for validation?
2. Does the supplier implementing the BMS application have the capability to meet computer validation requirements? Where necessary consider an alternative supplier.
3. Is there a need/capability to interface with legacy systems? Interfaced legacy systems should be validated where GxP data is passed to the BMS or to any independent monitoring system, including the network/data link.
4. Can the BMS application be subdivided such that a discrete part of the BMS can applied to product critical areas?
5. Could a separate validated BMS be provided to product critical areas?
6. Is local support available at the level required for validation?
7. The degree of compliance in regard to the use of electronic records and signatures using technical and procedural controls (ref. 21 CFR Part 11 etc.).

The review should be formally documented and the system then made subject to ongoing change control. There must be no uncontrolled creep in the original quality assurance role of the BMS.

Independent Monitoring

GMP critical control input/output points typically in order of 5% to 10% of total input/output points. This has lead many pharmaceutical manufacturers to consider the use of validated independent monitoring systems for the GMP critical control points and hence alleviate validation of the control system to a GEP activity on the basis of qualification (4). Independent monitoring systems range in complexity:

- Chart recorders are the simplest devices (0–30 input points), being industry standard (GAMP® software category 3). Validation requirements are based on recording model and version numbers, complemented by the necessary calibration and commissioning of alarm signals.
- Data loggers are more complex than chart recorders (managing typically 30–300 input points), and while industry standard systems are available there is usually considerable configuration and/or bespoke programming. Validation is based on the combination of GAMP® software categories 1, 4, and 5, covering operating systems,

configurable software packages, and bespoke programming. A complete life cycle approach is therefore required, including archiving the raw data.

- SCADA systems are more complex again than PC-based data loggers (typically managing in excess of 300 input points). They may directly monitor or supervise a number of monitoring PLCs. The software has a similar character to that associated with data loggers and the validation approach should be the same. The scale of the validation work will be greater, however, than data loggers because of the increased complexity of the system.

Independent monitoring systems used to implement the key quality assurance controls must be validated (whether they are complex supervisory control and data acquisition (SCADA) systems, or simple chart recorders). For an independent system to be accepted as a validated alternative in the monitoring of critical parameters, the system must be able to manage key quality assurance functions. Such functions include the following, but are not necessarily limited to:

- Controls for maintaining set points
- Functions for alarms and alarm logs
- Functions for trending over short and long term
- Preventative maintenance including calibration
- Access controls for security purposes
- Data interpretation and management

In those applications where all key quality assurance functions are managed through independent monitoring system(s) then provision and maintenance of the controlling system can be managed through GEP. It is important to remember, however, that to avoid validating the control system the independent monitoring system(s) need to keep records of all critical drug manufacturing parameters, for example, air changes, temperatures, air exchanges per hour, pressure differentials. This data may be used to support batch records, regulatory submissions, or QA investigations for out of specification incidents. Regulatory requirements for electronic records should also not be forgotten.

Data Interpretation and Management
Any difference in monitored values between the validated independent system and GEP/qualified control system would act as a trigger for investigation, in advance of any routine calibration or performance check of the validated system. Decisions about product quality must be driven by data generated from the validated independent system where an independent system philosophy is followed. The data used to support these decisions must be archived.

GOOD ENGINEERING PRACTICE
The application of GEP differs considerably from organization to organization. GEP has been established for many years and requires the robust specification, design, commissioning, and testing of systems to ensure that they are fit for their intended use.

ISPE Baseline Guides refer to the need for documented design, design review, commissioning plans, and test records. In this sense GEP is not significantly different to qualification other than in the rigor applied to the development of documents and extent of quality assurance input during the commissioning and test process. This said, the Baseline Guide to Commissioning and Qualification recognizes the importance of the quality assurance role. Dependence on GEP should therefore be viewed as the application of professional practices to ensure that systems satisfy their predefined specification and not as an opportunity for inadequate design and testing of systems.

VALIDATION LIFE CYCLE
Typically, a BMS is a mixture of software categories; it is important therefore that the validation plan identifies what are the categories of software, which make up the system, as well as incorporating the results of the system and component risk assessment onto the overall

validation strategy. Another important element that feeds into the validation plan is the result of the supplier audit. The validation life cycle presented here in consistent with GAMP guidance and related case study material (6–8).

Validation Plan

As stated earlier the validation plan is a crucial document. From experience, the best method to create the plan is to set up a small team, consisting of the user, system expert, and quality assurance representative. The plan will include the results of the risk and software category assessments as well as any additional requirements determined by the supplier audit. The plan will state what documents are required, when they will be produced (i.e., what order) and by whom. The validation plan will state what must be done to confirm that a system will be considered validated.

Supplier Audit

As the BMS supplier generally supplies to the construction industry, they have little experience with cGMP and the resulting requirements for validation. Therefore, it is essential that a supplier audit is performed; this has a number of advantages:

- Defines the appropriate software life cycle method to be followed.
- Enables gaps in existing management system and documentation to be addressed early in the project life cycle.
- Builds relationship between client and supplier
- Clarifies uncertainties
- Educates supplier in customer specific validation requirements
- Identifies what follow-up activities may be necessary

A postal audit could be used to assist with audit planning and in some case, when the supplier is known to the client, may replace the need for a full audit. But this should only apply if the client has an on-going relationship with the supplier. An example postal audit can be found in the ISPE Validation of Process Control System (VPCS) Guide (8).

The key to the process is to understand the system that is being proposed. It is a good practice for the auditor to spend time reviewing the user requirement specification (URS) and the system descriptions and understanding what software categories exist for the proposed system. This should be followed up, with the postal audit checklist. This will also provide valuable information to enable the auditor to plan the audit. Available information should be used to customize the audit checklist to address the specific issues that are relevant to both the supplier and proposed project. Consider, for example, a system that includes hardware and software, where some of the software is custom, other parts are configurable and others are part of a standard package. The auditor will need to establish how each part of the system will be developed and how the build phase will be controlled. There may even be more than one supplier. The auditor would need to split up the main elements and examine how each part of the system will be built.

A flow chart indicating the activities of the auditor or audit team is given in Figure 26.5. The following sample supplier questionnaire can be used as the basis of the supplier audit, postal audit, and audit checklist.

- Summary of product/service under audit
- Is the supplier registered to ISO 9001:2000, or TickIT:2000. If so which parts and when.
- Product and service development
- Use of subcontract suppliers, etc.
- Contract reviews
- Specifications
- Software life cycle method
- Verification of purchased material
- Testing, with deviation management
- Change control for documents and software

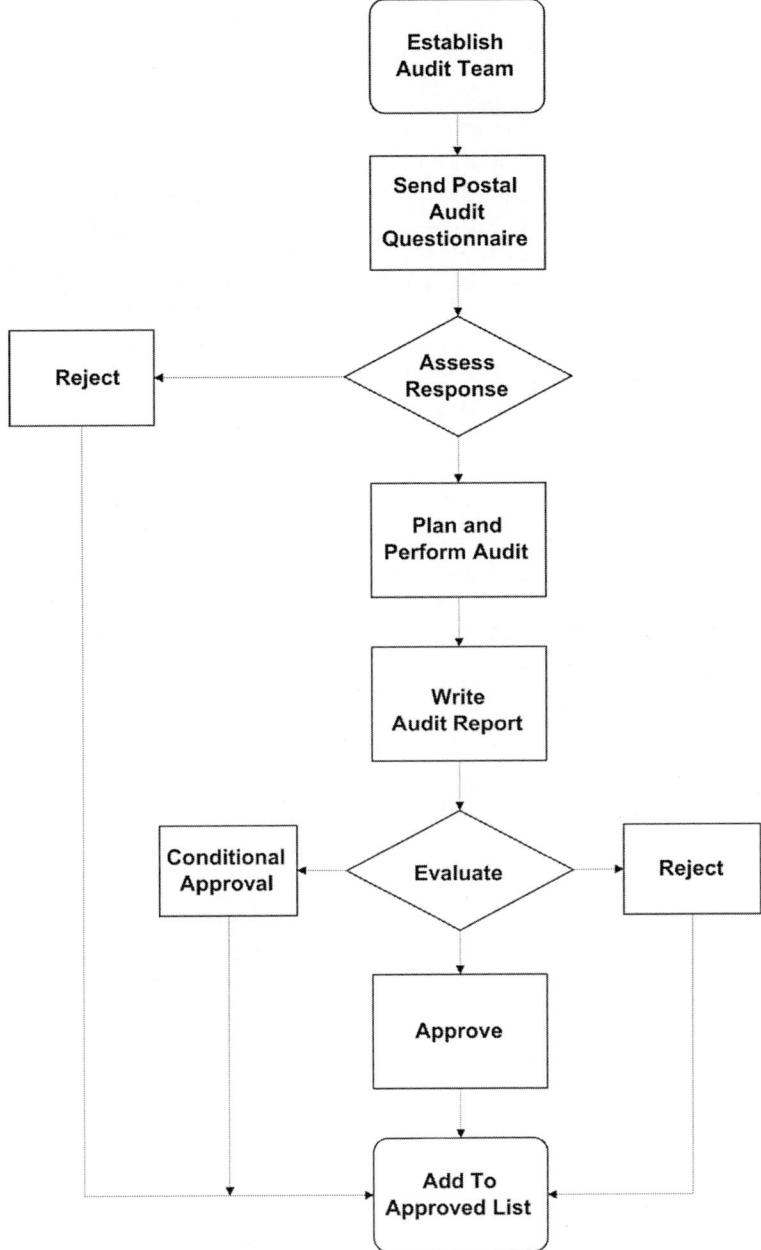

Figure 26.5 Audit process.

- Training
- Support and maintenance
- Support procedures and activities
- Fault reporting

The standard software packages and the custom elements of the system will require a similar review against the standard audit checklist. It is important to use the checklist as an aid to planning the audit, not to drive the audit. Remember *if you fail to prepare, prepare to fail*.

When choosing a BMS supplier, look for experience in the validation process as a prerequisite. A close working relationship can save time and money beyond the initial cost of installation. A primary criterion for choosing a BMS supplier should be the ability to provide support for the life of the facility. Their attitude shouldn't be one of walking away after commissioning.

System Specifications

System specification documents are required but often are not in the format that fits with the traditional GAMP "V-Model." In these circumstances, it is important that the user defines the required functionality in the URS. Making sure to define the objectives and separation requirements for all quality related areas, components, and associated functionality. However don't be too specific with stating what the control limits that must be achieved during the commissioning stage, better to specify some example limits and state that during the bedding-down period of the systems use, during the subsequent qualification phases, these limits will be confirmed and documented. Continue to use the legacy system in parallel with the new system during the bedding-down period, if possible.

Remember the user should get involved as early as possible and look at what the desired end result will be, not just the "correctness" of the specification. The URS is not always exactly what he wants, what he wants is not always what he gets, and what he asks for is not always what he needs.

It isn't typical to receive a separate functional and design specification; these are often included as part of the building's environmental control package. If this proves to be the case, then it is not necessary to rewrite these specifications. The recommended approach is to create a matrix that references those parts of the environmental control package that are relevant to the computer system validation requirements for the critical functions. Also ensure that this matrix is mapped back to the URS to ensure that the URS is met. This matrix will form the basis of the requirement traceablility matrix (RTM), which is intended to assure that all requirements have been addressed, the functionality is appropriate, consistent, meets predefined standards, and that the system is appropriately tested. The functional/design specification or the relevant sections of the environmental control package should typically address:

- Control and monitoring required for each specific area of the building; this should be in accordance with predefined separation policy for critical areas versus noncritical areas. It should also address the tolerances for control and monitoring accuracy.
- Field instrumentation requirements, which are impact versus indirect and no impact components covering:

  - Measurement range
  - Measurement accuracy
  - Control loop dynamics
  - Exposure to corrosion
  - Vibration
  - Hazardous area requirements, if any
  - Index of Protection (IP rating) for cleaning
  - Accessibility
  - Calibration requirements
  - Maintainability and spares

- Outstations locations and network requirements. These should be capable of overseeing the whole network with response times suitable for the process controlled. The presentation of data that takes account of the number of users on the system and the various levels of technical information required.
- Data and records production and how they are handled meeting regulatory requirements (e.g., 21 CFR Part 11).
- Presentation of data may include the following:

  - Dynamic mimic displays
  - Graphical trend plots

- Printed reports
- Data processing and how this handled
- Alarm set points and handling

Design and Development

The BMS functionality is constructed mainly from GAMP®4 software category 4, standard system modules configured to the user's specific requirements; therefore little software development is necessary. However, care must be taken to ensure that the configuration work and any code testing is independently reviewed and documented. Another aspect to consider is how the "standard modules" are to be configured, a sample recompiling of source code elements may be required. Also if there is a need for custom coding this should be treated as GAMP®4 software category 5, and full development validation will be needed.

Some aspects of validation are unique to HVAC control systems. Although the controls are one of the last things to be fitted, they must not be planned last. The user must make many decisions before the controls are installed, and there should be qualification meetings early in the process. Quality can't be tested into a process. It has to be designed into each system.

System Build

Build is normally part of the construction phase. It should be remembered that generally speaking, standard components from the suppliers preferred range will be used so no special build is required, except for purpose built marshaling and display cabinets that will require design and build drawings. Special care must be exercised when using intelligent instrumentation and the associated bus type communication networks. The desired functionality should be documented and design qualification (DQ) on these components should be undertaken.

The HVAC controls for critical (validated) areas should be grouped in specified field panels. One may want to label these panels, "Critical Process Controls: Please Follow Change Control Procedures," or something similar. This will prevent the necessity of having to validate noncritical controls. Electrical supplies and other utilities must also be evaluated. One may need a UPS (uninterrupted power supply) for critical field panels and PC (personal computer) workstations to continuously monitor critical equipment, such as refrigerators, incubators and particle counters with the BMS.

Factory Acceptance Testing

It is recommended that as much factory testing as possible should be carried out before delivery of the control system to site. To a large extent this will be limited to simulation of the input and output elements (i.e., it is not connected the building services and instrumentation yet). Full testing can occur once installation is complete. The extent of factory acceptance testing (FAT) is governed by how rigorous the simulation can be designed. Remember it is worth investing time at this stage to fully challenge the system to reduce the time and effort if faults are found once the system is installed on site.

On-Site Testing

After the HVAC mechanical equipment and controls are installed, the process should begin with a point-to-point checkout of every component (i.e., verifying that every input and output device is connected to the proper terminals). If formalized then this method would reduce cost and time by utilization of the commissioning documentation to support validation. For example, commissioning checklists can be referenced in the installation qualification (IQ). The alternative is to do them separately and duplicate a lot of paperwork. If calibration is required, the procedures and documentation must be referenced in the validation protocols. Once IQ is satisfactorily completed, start-up of the HVAC system can begin in accordance with the company's SOPs. The mechanical equipment must be up and running before operational qualification (OQ) can begin. This is where verification is done to ensure that the various mechanisms operate as intended (e.g., when the room thermostat calls for heat, does the hot water/steam valve open?).

Performance qualification (PQ) must be carried out by the user. This is where verification is done to ensure that all systems work together under as-used conditions to meet the URS. Do room temperature, humidity, and pressure stay in spec with production under way and people entering and leaving the facility? All systems must be operational to complete PQ. Cooperation between the various contractors (mechanical, controls, etc.) is vital to completing PQ in a timely and cost-effective manner. As discussed earlier, the user and the designer must sit down at the beginning of the project and determine critical (validated) and noncritical areas. Don't waste resources and money validating noncritical areas.

Maintenance

Change control procedures should address such issues as scheduling and documentation of maintenance and recertification of calibrated sensors. How will one ensure that a calibrated sensor is available if one fails, or that the control program changes stick to standard formats? This is the nature of BMS change control. The following quote from the Proposed Changes file of the cGMP Web site emphasizes the FDA's viewpoint: "To preserve the validated status of a process, measures must be taken that will allow any significant process changes to be recognized and addressed promptly. Such change control measures can apply to equipment, SOPs, manufacturing instructions, environmental conditions, or any other aspect of the process system that has an effect on its state of control and therefore on the state of validation."

An auditor must be able to evaluate the current status of a facility on the basis of the owner's documentation and compare it to the specifications, but the processes also have to work smoothly and allow improvement.

Reporting

The validation plan should require an assessment of the project success and the validation report should present, or refer to, the evidence to support this. It is normally the case with a BMS project that the documentation is too large to attach to the report. It is therefore sensible to present the documentation in a list form with the location of each document referenced. This report must also clearly state that the system is validated and is approved by the user and QA personnel.

IMPLEMENTING NEW SYSTEMS OR ENHANCING EXISTING SYSTEMS

URSs are important irrespective of the GMP criticality of any system. For a complex centralized BMS, there may be several "users" including engineers, system owners, data owners, QA, etc. It is important that the needs of all stakeholders are captured in the URS The URS must clearly define the relationships between the BMS (and independent monitoring systems where implemented) and the *process(es)* being controlled and monitored.

It is important that each requirement be categorized to define whether the requirement is safety critical, GMP critical (direct impact), business critical or otherwise. The categorization of the requirements will help determine the most appropriate approach to implement each requirement, that is, GEP or validation.

Figure 26.6 illustrates that the decision to apply GEP or qualification/validation is determined by combining the likelihood of product impact with the ability to detect and where possible correct failure. Guidance on approach to GEP, qualification and validation can be found in the following guidelines:

- ISPE Baseline Guide Volume 5, Commissioning and Qualification
- ISPE GAMP®5
- ISPE GAMP® Good Practice Guide, Validation of Process Control Systems

SPECIFYING NEW BMS FUNCTIONALITY

The design of the BMS will determine the ease with which the BMS can be implemented, quality assured, and managed in its operational life. Requirements specifications, design and operational controls may consider:

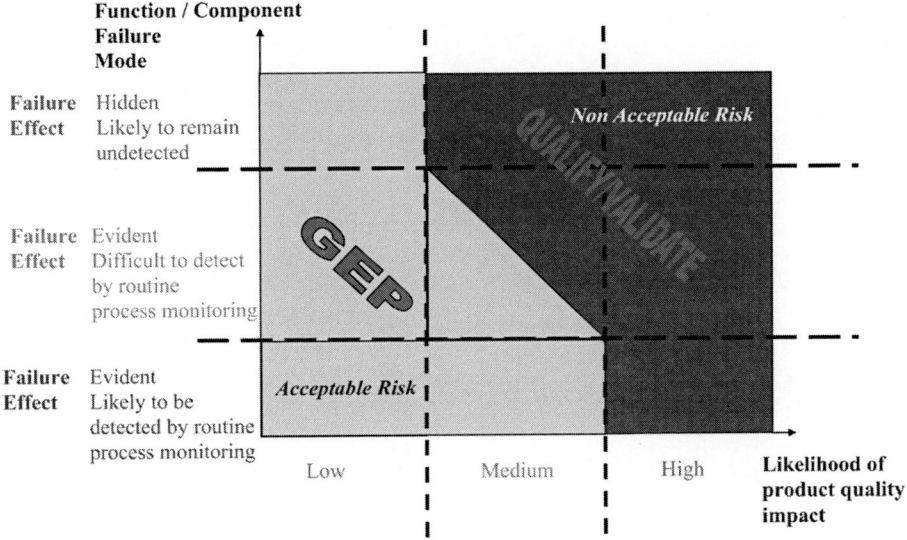

Figure 26.6   GEP versus qualification/validation.

LEGACY SYSTEMS

As with new systems, risk assessment will determine the need to implement additional quality assurance measures for existing systems. Where the risk assessment determines components of the system to be high risk, it still may be difficult to achieve current industry standards for a number of reasons including:

- Systems may not have been developed to current quality system expectations (supplier and manufacturer quality standards)
- Old technologies may be in place that do not allow for implementation of current control requirements
- Original design may be difficult to modify
- The scale of technology upgrade required may not deliver acceptable cost/benefit returns

In such cases, mitigation controls including procedural and where cost effective, technical controls should be considered. Such controls may include:

- Redesign (where practical)
- Implementation of appropriate automated and/or manual monitoring regimes at a frequency commensurate with risk (including where appropriate validation and calibration)
- Implementation of available logical security features (e.g., password controls, review and reorganization of security access rights)
- Introduction of physical security controls where possible (e.g., locked cabinets, tamper evident labels, etc.)
- Implementation of procedural security measures where technical controls are inadequate (e.g., periodic password change, periodic review of control parameter and alarm settings, control of access to programming devices, periodic review of I/O override settings)
- Recommissioning of areas of the system in accordance with ISPE Baseline Commissioning and Qualification Guide
- Review and adjustment of calibration schedules

The degree and nature of mitigation controls will obviously depend on the current status of the BMS.

ELECTRONIC RECORDS AND ELECTRONIC SIGNATURES REVIEW

Where the BMS (and/or associated monitoring system) is determined to be GMP critical, an assessment of electronic records and electronic signatures impact should be made. BMS (or associated monitoring system) holds data for a variety of reasons including business management, engineering maintenance, and GMP decision making. The context within which such records are used determines whether they are regulated electronic records or otherwise.

European (Chapter 4, Annex 11, PIC/S), U.S. (21 CFR Part 11), and Japanese (MHWL Guideline) regulations and guidance should be considered as appropriate when determining management controls. Table 26.4 defines typical data held by a BMS (and associated

Table 26.4 BMS Requirements/Design/Operational Considerations

Requirements aspect	Examples	Objective
System partitioning	Segregate GMP and non-GMP functionality and I/O	Avoids conflict between GMP and non-GMP functions and I/O. Enables boundaries to be put around GMP critical aspects of BMS
	Segregate GMP and non-GMP databases	Avoids conflict between GMP and non-GMP aspects of the system. Not always possible or desirable for operational reasons, e.g., viewing multiple databases
	Independent local network for BMS	Segregates critical aspects of the system and avoids conflict with non-GMP operations. Security is also easier to manage
Alarm handling	Alarm management strategy defines alarm prioritization that clearly differentiates product quality alarms from maintenance alarms, tolerance alarms, and system alarms	Critical process alarms are clearly differentiated from other alarms. Process alarms easily identified. Defined within alarm management strategy
	Separate alarm printouts/logs for product quality alarms	Those alarms requiring quality assurance review are differentiated
Security	Outstation security controls Restricted access to maintenance functions. Workstation multilevel security access Ability to synchronize security settings across the BMS infrastructure	Prevents inadvertent modification of critical control parameters. Minimizes risk of false readings, e.g., through forced I/O settings Ensures that roles can be differentiated and appropriate controls applied, e.g., engineering administrator, quality assurance, users, data stakeholders Changes to one workstation reflected across all workstations to ensure consistency of access from different points
Planned enhancement capability	Built in expansion to allow for easy addition of control and monitoring points	Easier validation/GEP of upgrades. Reduces pressure to combine GMP and non-GMP functionality
	Backward compatibility to enable controlled upgrade	Easier to maintain validated/GEP status following upgrade.
Testing	Simulation tools	Enable setting of I/O and status conditions to facilitate controlled testing (note these features can also be detrimental to validation if not provided under appropriate access control)

monitoring systems) and rationales for electronic records compliance determination or otherwise. The requirements for electronic signatures used within BMS are no different for any other computerized system.

The risks associated with potential BMS electronic records should be determined in parallel with the BMS functional risk assessment (Table 26.5).

Table 26.5 BMS Electronic Records Considerations

Data type	Use	Electronic records determination (direct impact, indirect impact, no impact)	Comments
Historical data logging			
Critical process measurements	Support regulatory decision, e.g., batch release	Direct impact, if used for batch release and investigation	None
	Support regulatory investigation, e.g., product adulteration	Indirect impact, if used for maintenance purposes	
Energy usage profile	Determine alternative energy strategy or report energy usage	No impact	None
Equipment failure and performance status	Condition-based monitoring	No impact	If alarm and event histories are used to influence changes in maintenance strategy they may be reviewed by inspectors and therefore will be electronic records
Alarm and event logging			
Critical parameter deviations	To determine process deviations	Direct impact, if alarm logs are retained in electronic history files to support future investigation	If alarms simply annunciate and are then printed, no electronic record exists.
		Indirect impact, if alarm logs used for maintenance.	If alarms are saved to removable storage media, then such media should be managed to prevent unauthorized change
Equipment failure and performance alarms	To determine need for maintenance	No	If alarm and event histories are used to influence changes in maintenance strategy they may be reviewed by inspectors and therefore will be electronic records
Control and calibration parameter management			
Calibration settings	To ensure accuracy of instrument and equipment feedback	Indirect impact	GMP decisions are not made on calibration parameters. Change control or operational procedures should be used to manage calibration changes. Calibration parameters should be secure from unauthorized or inadvertent change

(Continued)

Table 26.5 BMS Electronic Records Considerations (*Continued*)

Data type	Use	Electronic records determination (direct impact, indirect impact, no impact)	Comments
Critical process set points, control actions, and alarms	To establish required control scheme	Indirect impact	Parameters should be subject to change control/configuration management. Parameter should be secure from inadvertent or unauthorized change Regulated decisions are made on process performance rather than input parameters.
Noncritical process set points, control actions, and alarms	To establishing required control scheme	No impact	Non-GMP or indirect impact processes
Control			
Control logic	To ensure consistent and accurate performance of process to stated specification	Indirect impact, control logic is software that should be validated or subject to GEP to demonstrate that the system is fit for purpose	Although links have been made between software and electronic records, GMP decisions are not generally made by reviewing software as such, configuration management should be adopted
Instrumentation			
Any readings or records held by instruments are typically transient.	Readings and status of process sent to control logic	No, assuming measurements are transmitted to control and monitoring system and not retained and used within instrument, i.e., they are transient in nature	Instruments may be configured to establish operating ranges and control parameters. Such configuration should be subject to configuration management

BENEFITS DELIVERY

A major pharmaceutical company installed a building manufacturing system (BMS) in one of its sterile powder vial filling manufacturing suites. This would replace an old building service control system with independent monitoring via manual readings from fixed gauges. The BMS would control and monitor manufacturing suites, preparation areas including changing rooms, service areas, offices, corridor, and refreshment room. The implementation followed the strategy described in this study.

The first step was to assess the system for impact, that is, which parts are direct impact, indirect impact, and no impact. The purpose of this strategy was to decide how to apply separation of the control system and air handling equipment for the different areas. It was clear from the assessment that the impact areas were the manufacturing suites and changing rooms and the indirect areas was the preparation area. The no-impact areas comprised of the service areas, offices, and corridor and refreshment room.

The design therefore called for two control systems and associated hardware, which would require validation to be applied to one system with the other following the principles of GEP. The next step is to decide on the extent of the validation for the quality critical system with direct and indirect impact.

A further assessment of the quality critical system components and associated functionality was then required. The resulting list of direct impact components (and associated functionality) covered temp control, humidity and pressure differential between manufacturing areas to changing rooms, and preparation areas to offices, restrooms, and corridors. This then allowed for validation challenge testing of critical functions associated with the impact components and associated functionality. This was followed by validation confirmation testing of all other BMS control functions for indirect impact components (and associated functionality) on the BMS controlling the preparation areas including changing rooms.

The principle of GEP for the BMS system controlling offices, restrooms, and corridors was adopted. The project milestones were then planned and actioned in accordance with the combined (cGMP/GEP) quality plan. The supplier was audited and commissioned with the understanding that they must participate in the risk assessments. It was agreed that savings in project costs would be shared; however the company's QA audit group would assess the whole project and fines could be applied if breaches in quality were detected.

The project saved 40% of the original estimated validation effort, and the whole project was completed under budget. Risk analysis delivered real benefit while maintaining compliance.

CONCLUSIONS

This GPG has illustrated some of the complexities in managing BMS and the issues faced when determining the most appropriate quality assurance strategy for the BMS. The key points raised by this GPG are summarized as:

- BMS architectures vary from low complexity PLC systems to complex centralized, networked systems
- BMS implementations have often evolved over a long period of time with each extension to the system being implemented to technologies and standards available at the time
- Replacement of older aspects of the system to use current technologies and meet current industry standards may not provide an appropriate cost/benefit balance
- Risk assessment is essential to determining criticality of BMS and any associated monitoring systems (manual or automated)
- Risk assessments must focus on the probability that a BMS-controlled process will impact product quality
- Validated and/or calibrated independent monitoring systems can reduce the reliance on BMS for GMP decision making and enable a balanced cost/benefit approach to BMS quality assurance
- Consideration should be given to validation/qualification of potentially high criticality aspects of the BMS controls (e.g., sterile environmental controls)
- GEP should be applied as a minimum quality assurance standard for indirect and direct impact BMS systems. GEP represents a professional engineering approach to assuring a system is fit for intended use (10)

ACKNOWLEDGMENTS

The author would like to acknowledge comments and contributions of Dr Guy Wingate, GlaxoSmithKline.

REFERENCES

1. U.S. Code of Federal Regulations Title 21, Part 210, Current Good Manufacturing in Manufacturing, Processing, Packaging, or Holding of Drugs; Part 211, Current Good Manufacturing Practice for Finished Goods.
2. European Union Guide to Directive 2001/83/EC, Computerised Systems, Annex 11 of Rules and Guidance for Pharmaceutical Manufacturers and Distributors.

3. Pharmaceutical Inspection Co-operation Scheme. Good Practices for Computerised Systems in Regulated GxP Environments, Pharmaceutical Inspection Convention, PI 011-1, Geneva, 2003.
4. ISPE. Qualification and Commissioning Baseline® Guide, First ed.. International Society for Pharmaceutical Engineering, 2001. Available at: www.ispe.org.
5. Landis, Validating Building Control Systems, Siemens Building Technologies Inc.
6. Wingate GAS. Computer Systems Validation: Quality Assurance, Risk Management and Regulatory Compliance. Boca Raton, FL: CRC Press, 2004.
7. GAMP Forum. GAMP Guide for Validation of Automated Systems (known as GAMP®5). International Society for Pharmaceutical Engineering, 2008. Available at: www.ispe.org.
8. GAMP Forum. Validation of Process Control Systems, Good Practice Guide. International Society of Pharmaceutical Engineers, 2003.
9. ISPE. GAMP Forum GAMP® Good Practice Guide, Good Engineering Practice. International Society for Pharmaceutical Engineering, 2008.
10. International Society of Pharmaceutical Engineers, GAMP® Good Practice Guideline Use of Building Management Systems in Regulated Environments—BMS Special Interest Group, 2007.

27 | Case Study 9: Engineering Management Systems

Chris Reid
Integrity Solutions Limited

ENGINEERING MANAGEMENT SYSTEMS

Effective and efficient utilization of assets by life science research, development and manufacturing organizations is fundamental to the early delivery of new products to market and to satisfying customer demand once those products have been approved by the relevant regulatory authorities. A carefully designed engineering strategy is essential to optimizing and maintaining system reliability, capability and performance consistency; that is, assets should

- Be available when needed and should not fail during use,
- Function consistently to predefined performance criteria, and
- Meet performance criteria without undue stress, risk of failure, or reduced asset life.

The continuous improvement of asset reliability, consistency and capability, either mutually or simultaneously, is the basic objective of the engineering strategy to reduce operation and maintenance costs and ensure patient safety.

The engineering management systems (EMS) comprises a number of computerized systems that support engineering processes and engineering information management. At the core of the EMS are maintenance management, electronic document/records management and building/environmental management systems, condition monitoring, and computer-aided design (CAD).

KEY CONCEPTS
Process Understanding

The criticality of the EMS is determined by the criticality of the managed assets. For example, if only office facilities and environments are managed, the criticality will be less than if clinical and manufacturing systems are managed by the EMS. Performance monitoring and maintenance activities should be more greatly controlled as the risk to patient safety and clinical study integrity increases.

The EMS will typically maintain systems at two levels; firstly to minimize the risk and/or detect potential breakdowns and secondly to maintain and detect deviations from process performance. A system component failure does not necessarily lead to an immediate process performance failure.

Performance criteria should be defined for maintained systems and should be specified completely and accurately with full process understanding. Performance criteria stated as "Maintain room temperature at 18°C" are loose and ambiguous and may be more accurately specified as follows:

Temperature range:	18°C to 22°C
Control accuracy:	Set point ± 1°C
Acceptable excursions:	<5°C for less than 10 minutes

Failure mode and effect analysis (FMEA) or other similar techniques are used to determine required system performance criteria and potential failure modes on which the engineering strategy is built.

Asset Hierarchy and System Concept

Typically, the assets of a regulated company are managed as a hierarchy comprising (Fig. 27.1)

- Sites,
- Buildings,
- Rooms/areas/zones, and
- Systems (utilities, building services, process systems).

This hierarchical structure provides the foundation for maintenance planning, reporting and engineering information access.

This chapter focuses on "systems" as they are the most diverse and complex asset in the asset hierarchy and are the primary focus of continuous improvement strategies. Vessels, pumps and valves, which although are critical components, do not in isolation deliver the functionality and process performance. It is the integration of such components into a system that enables the designed system performance to be delivered and maintained.

A HVAC system delivers air to class 10,000, temperature to 20°C ± 2°C, and humidity to 50% RH ± 5% RH. The failure of a component although important, only becomes critical if performance is lost. Figure 27.2, differentiates between performance loss and system failure.

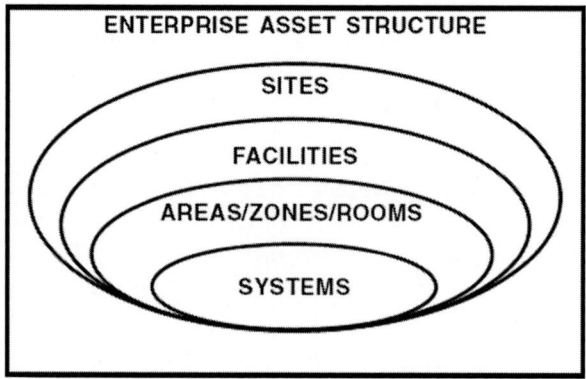

Figure 27.1   Asset hierarchy.

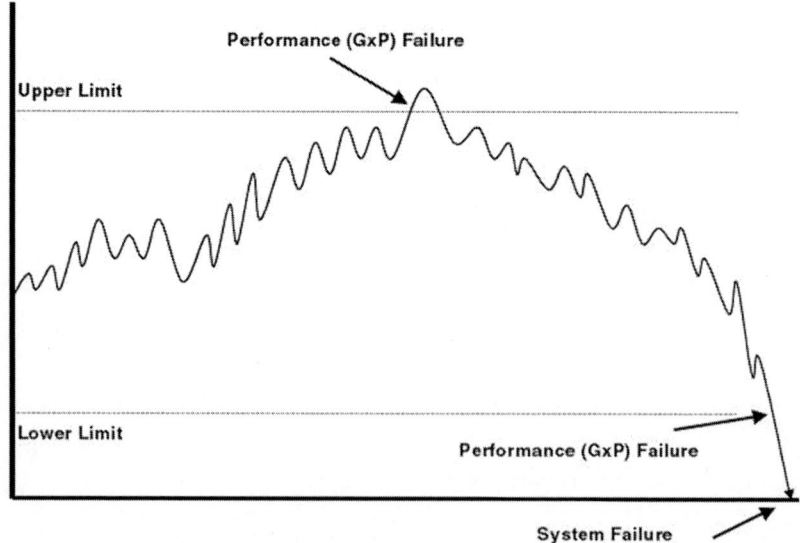

Figure 27.2   Performance monitoring.

GxP compliance is lost once the performance deviates from the predefined operating range, which is long before the system totally fails.

It is essential that system construction, functional, performance and maintenance information is maintained against the system so that it is accessible, maintained and useful in improving the engineering strategy.

ENGINEERING INFORMATION STRATEGY

The foundation for continuous improvement is "information," which should be established at the start of the asset life cycle with the definition of business need in measurable terms, that is, performance criteria, without which there is no basis for design, testing, operation, maintenance, compliance and consequently continuous improvement. Figure 27.3, defines a simple asset life cycle depicting the creation of critical asset management information at each phase.

Asset management information provides the foundation for other regulated documentation used in clinical and manufacturing operations as shown by Figure 27.4. The integrity of clinical study results or manufacturing operations can be jeopardized if the systems used to control drug stability testing or the manufacturing process are not operated and maintained in accordance with their predetermined performance criteria.

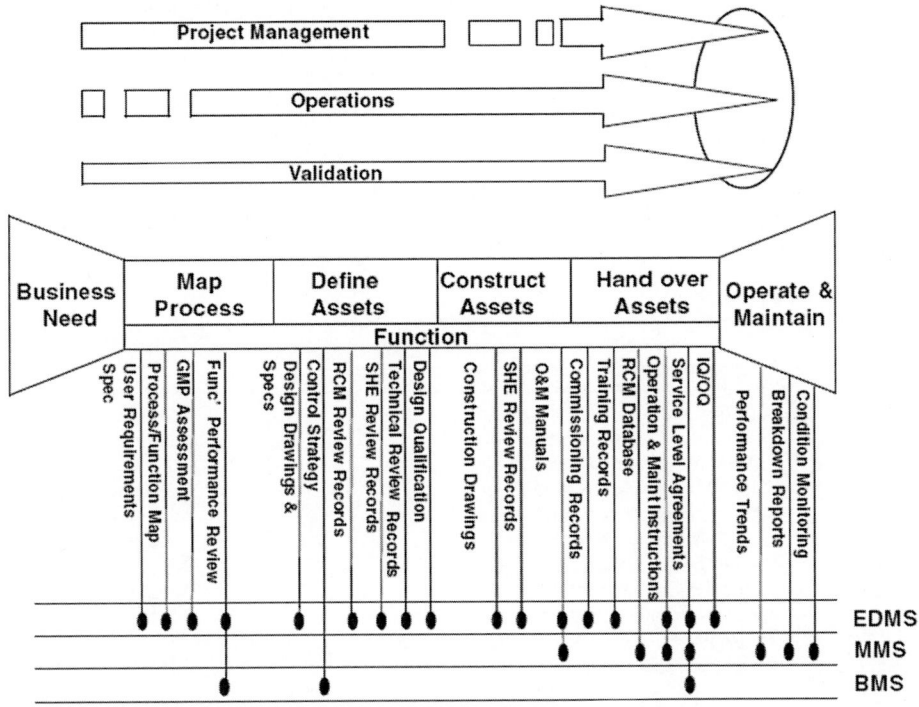

Figure 27.3 Asset life cycle.

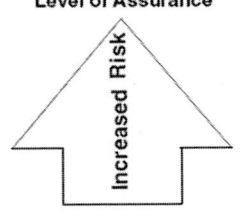

Level of Assurance

Increased Risk

Functional Criticality

A	Operational Safety
B	GxP Critical
C	High Business Impact
D	Minimal Business Impact
E	None

Figure 27.4 Functional criticality.

The regulated company should establish internal information standards that define minimum engineering information content, structure and format to ensure that the information can be

- Imported into the relevant EMS system,
- Maintained, and
- Readily and effectively accessed.

Delivering information to the required information standard, in a consistent format and in a timely manner is a considerable challenge, especially when multiple suppliers are involved and where those suppliers have traditionally delivered information to their own and sometimes unsatisfactory standard.

Electronic Records and Electronic Signatures

All electronic records need to be evaluated to determine their criticality and controls established to assure integrity, authenticity and where appropriate, confidentiality of information contained within. The controls required for each record should be determined by the criticality of the information maintained by the system (1–4).

Electronic information should be controlled to minimize the risk of unauthorized or inadvertent alteration. System security controls should ensure that only authorized personnel can access and maintain electronic information. Electronic audit trails are useful in providing a history of changes to maintained information. Table 27.1 provides examples of electronic records that may be maintained within the EMS and potential impact on patient safety in the event that information integrity is lost.

The need for authorization, review and approval of each electronic record in response to regulatory or internal quality requirements needs to be determined. Where records are signed electronically, controls should be established to ensure the integrity of applied electronic signatures.

Table 27.1 Example Electronic Records

System	Typical records	Potential impact on patient safety
Engineering database	Data sheets	Medium
	Calibration data and records	Medium–high
	Materials of construction data	High
	Asset configuration data	Medium–high
	Asset performance data	Medium–high
	Engineering drawing data	Medium
Building management system/environmental monitoring system	Asset performance data	Medium–high
	Process deviation alarms	Medium–high
	Environmental control deviation alarms	Medium–high
	Trends of critical process and environmental parameters	Medium–high
Maintenance management system	Maintenance schedules	Medium
	Maintenance instructions	Medium
	Maintenance activity reports	Medium
	Supplier records	Medium
	Noncompliance and breakdown reports	Medium
	Stock inventory	Medium
Electronic document Management system	Maintenance procedures and work instructions	Medium
	Engineering drawings	Medium
	Design and construction specifications	Medium
	Maintenance strategy records (e.g., reliability-centered maintenance records)	Medium
	Calibration procedures	Medium–high
	Training records	Medium
	Operation and maintenance manuals	Medium

Note: Where patient impact is indicated as "medium–high," the actual risk is determined by the nature of the processes being supported by the engineering asset.

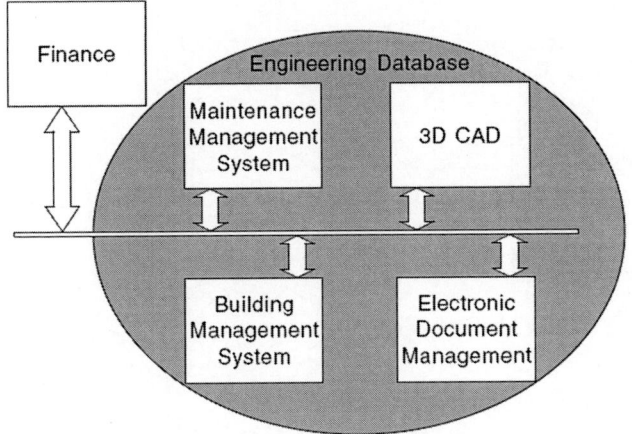

Figure 27.5 Engineering management systems architecture.

ENGINEERING MANAGEMENT SYSTEMS ARCHITECTURE

The EMS is not a single information system, rather a collection of integrated systems providing engineering control and monitoring functions. Figure 27.5 provides a high level representation of a typical EMS architecture. The Engineering Database is the hub of the architecture, providing a repository for information related to each system. The nuclear power and oil and gas industries have taken the lead in the development and utilization of "intelligent" databases. Such databases provide automated links to the maintenance management, electronic document management system (EDMS), and CAD system to provide single point access to data. The general principle that documents and information should be structured to minimize the number of occurrences of data is paramount to reducing the burden of information maintenance and the risk of regulatory noncompliance.

Access to information is as equally important as the controlled maintenance of information. For example, when a critical alarm is reported by the building management system (BMS), the engineer needs to promptly respond to instructions, service level agreements (SLAs), SOPs, and engineering drawings.

Functional Architecture of the Engineering Management Systems

There is obvious functional overlap between the systems described in Figure 27.5, for example, the MMS will provide some degree of document control, which obviously overlaps with the EDMS. We should therefore describe the generic functionality associated with the EMS architecture. Figure 27.6 provides an overview of a typical maintenance process.

Maintenance Planning

Maintenance tasks are either implemented proactively to "prevent" or minimize the chance of failure or reactively to "correct" a situation following failure. Preventive maintenance plans are derived from the FMEA process that defines the tasks and task frequencies required to maintain system "reliability," "consistency," and "capability." Having applied the FMEA process to determine the maintenance strategy, it is essential that the system

- Schedules planned maintenance at defined intervals,
- Identifies relevant maintenance task schedule (consistent with FMEA requirements),
- References the SLA defining system performance criteria and system criticality,
- References documents, drawings, databases supporting the maintenance tasks, and
- References instructions and SOPs to ensure controlled and consistent execution of maintenance tasks.

Corrective Maintenance, conducted following a failure, should be carried out in a similar controlled manner however; in this instance the engineer should manually construct the

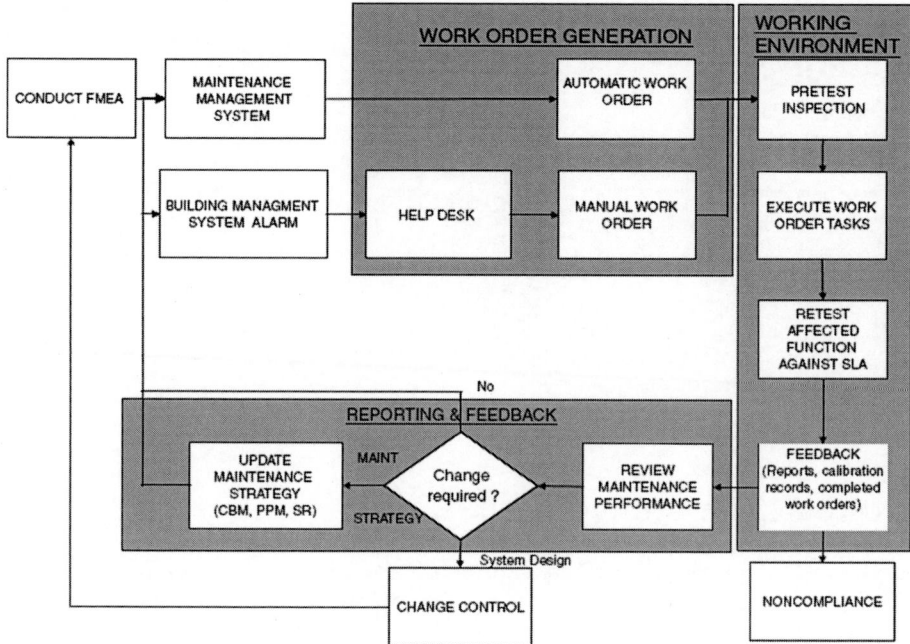

Figure 27.6 Operation and maintenance process.

maintenance task schedule following investigation of the problem. The risk to GxP compliance therefore increases because of the manual intervention. The structured organization of the EDMS and MMS in particular, to ensure that SOPs, documentation, drawings and information are easily accessible through the relationship to a specific system, is essential.

Operation and Maintenance Implementation
Work Instructions and SOPs define the operations required to start-up, operate, monitor, shutdown and maintain systems. These Work instructions and SOPs should be controlled to prevent inadvertent and unauthorized modification and to ensure access to only the latest revision of the document. Work instructions and SOPs should therefore be held in accessible, but secure areas that are periodically backed up and archived. Information system access should be controlled by a hierarchical security system that constrains system operations in accordance with the role, responsibilities and competency of the user. Access to and modification of the information supporting such work instructions and SOPs, for example, engineering drawings and specifications should equally be controlled.

System Control and Performance Monitoring
The BMS and Condition Monitoring systems provide control and performance monitoring functionality to ensure that performance criteria are met and performance deviations detected. Monitoring of process variables using available technologies such as BMS, ultrasonic and vibration analysis may often be deployed to predict pending failures, enabling corrective action to be taken before performance is lost.

Maintenance Reporting
Work Instructions and SOPs controlling maintenance operations should ensure that the maintenance engineer records all performance measures, observations and maintenance tasks in a consistent manner, for example, calibration records containing calibration parameters, calibration procedure, reference to calibration equipment, name of engineer, date of calibration, next due date, etc. Where automated condition and performance monitoring is employed, the integrity of the recorded data is a potential GxP issue.

Records generated by the asset management process should be used to bring about continuous improvement to increase the effectiveness and efficiency of the engineering strategy. All changes arising from the review should be controlled and documented.

EMS FUNCTIONAL RISK ASSESSMENT

The EMS provides important functionality that minimizes the risk of patient impact, however most functionality is considered to be medium risk as the failure does not necessarily lead to a process failure, process failures may be detected by other functions of the EMS and typically procedural controls should also be in place to control execution of maintenance activities. Table 27.2 provides example impact assessment of key functions of the EMS.

Table 27.2 EMS Functional Risk Assessment

Asset management function	Risk scenarios	Potential impact on patient safety
Planned Maintenance (PM)	• Planned work orders not generated in accordance with schedule.	Medium
	• Planned work orders include incorrect work instruction.	Medium
	• Obsolete work instructions linked to PM work order.	Medium
	• PM life cycle not followed (open, assigned, awaiting spares, in progress, complete, closed).	Low–medium
	• Record of Maintenance Activity not Created.	Medium
Corrective maintenance	• Record of maintenance activity not created.	Medium
Spares procurement and management	• Maintenance spares not procured against correct specification.	Medium–high
	• Maintenance spares procured from unapproved supplier.	Medium
	• Spares inventory inaccurate.	Low–medium
	• Incorrect spares assigned to a system/work order.	Medium
	• Spares held in different location to the location indicated by system.	Medium
Resource management	• Incorrect resources assigned to work order tasks.	Low–medium
Cost management and capital planning	• Incorrect budget forecasts and cash flow forecasts.	Low
	• Actual costs exceed planned costs.	Low
Fault analysis	• Faults not recorded against correct system or system component.	Medium
	• Fault trending does not correctly identify persistent failures.	Medium
Asset performance monitoring	• Alerts not generated in accordance with defined alert limits.	Medium–high
	• Performance trends do not highlight potential performance failures.	Medium

(Continued)

Table 27.2 EMS Functional Risk Assessment (*Continued*)

Asset management function	Risk scenarios	Potential impact on patient safety
BMS/EMS alarms	• Alarms not generated in accordance with defined limits. • Alarm and event logs not generated.	Medium–high
BMS/EMS trends	• Measurement logging not configured at appropriate intervals.	Medium–high
	• Measurements not logged at configured intervals.	Medium–high
Maintenance/service records	• Unauthorized personnel modify record of maintenance activity.	Medium
	• No audit trail of maintenance activity record changes.	Medium
	• Maintenance activity record not electronically signed.	Low–medium

Abbreviations: BMS, building management system; EMS, engineering management systems; PM, Planned maintenance.

REGULATORY IMPACT
GxP Assessment

As previously discussed, the FMEA process can be used to determine the consequence of process system failure. FMEA can in turn be applied to the functionality of the computerized systems comprising the EMS.

There are a number of considerations when determining the criticality of EMS functionality.

1. The criticality of the process systems being supported, for example, office environment control systems and/or direct patient impact systems.
2. The consequence of failure of the EMS function itself, for example, the delayed issue of a preventive maintenance work order does not necessarily mean that the process system will immediately deviate from defined performance parameters.
3. The failure may be detected by other functions of the EMS, for example, BMS process alarm.

The GxP assessment is conducted in accordance with FMEA principles. Figure 27.7 provides a high level representation of the GxP assessment process. The flowchart is supported by standardized questions that challenge the impact of the EMS function on GxP compliance. Typical challenges will include the following:

"Will the total or partial failure of the information system lead to"

*   Loss of or interruption to process system performance,
*   Failure to conduct critical maintenance activities in accordance with a predetermined schedule,
*   Use of superseded or wrong maintenance procedures or system documentation,
*   Incorrect maintenance/failure/performance reporting,
*   Incorrect chronological reporting of maintenance tasks, and
*   Loss or corruption of operational/maintenance data?

The above is not an exhaustive list but provides an insight into the extent to which EMS functions can impact GxP.

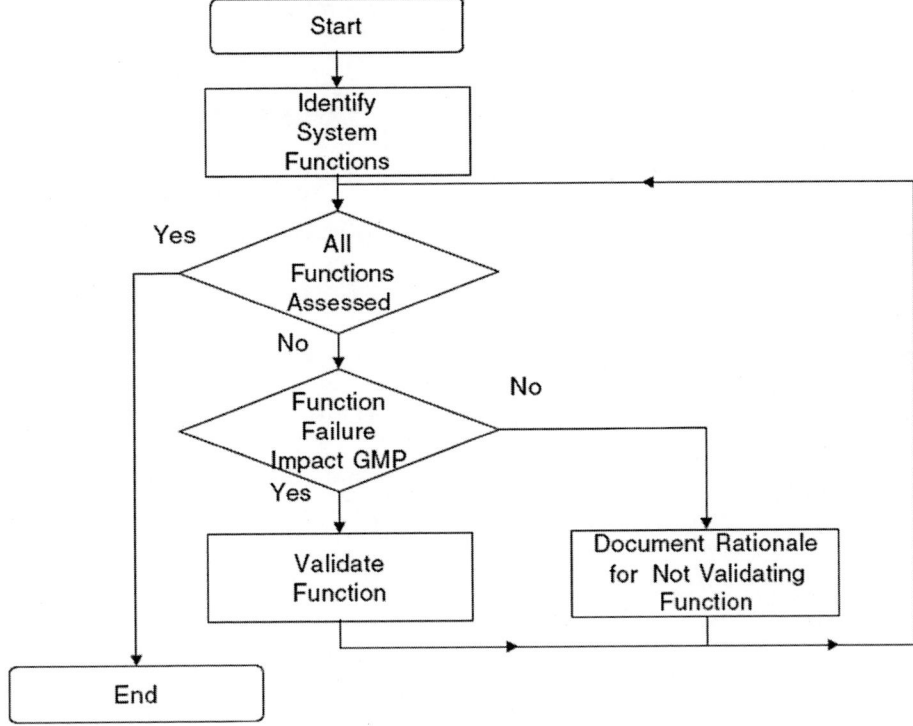

Figure 27.7   GxP assessment.

EMS VALIDATION APPROACH

Figure 27.8 depicts a typical validation life cycle from validation planning to validation reporting and ongoing support. Regulated users should be most heavily engaged in the early and late phases of the life cycle, namely planning, specification and testing. Suppliers typically have primary responsibility for design, build, configuration, development and acceptance testing.

It is however, the fundamental responsibility of the regulated company to assure themselves that the computerised systems they use have been developed in a quality manner, consistent with the expectations of the industry regulators (5–10). A detailed discussion on validation life cycles can be found in "Validating Automated Manufacturing and Laboratory Applications" (11).

Validation Plan

The regulated company should develop a validation plan (VP) to define the validation strategy for the implementation of the EMS. The principle of risk-based validation is well established across industry. The factors influencing the level of effort required to verify systems comprising the EMS include the following:

- Process/requirements/functional criticality (in particular patient impact)
- Degree of configuration and/or customization required to implement the system
- System size and functional complexity
- Supplier capability and availability of quality documentation

The identified risks determine the following:

- Resource requirements (internal and external)
- Engagement of business quality assurance in the project
- Need and approach to supplier assessments

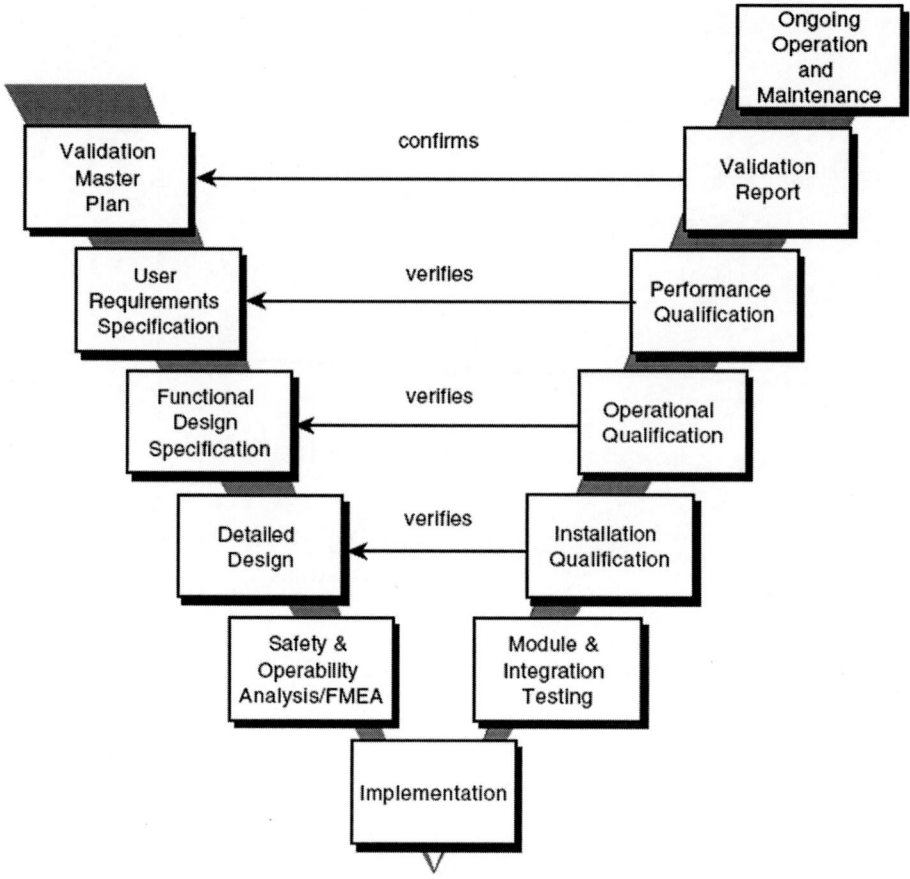

Figure 27.8   Validation life cycle.

- System development/implementation life cycle (for different aspects of the system, for example, custom, configured, out of the box)
- Number of in project risk assessments (process risk assessment, functional risk assessments, project risk assessments)
- Documentation set (expanded, omitted, combined based on risks)
- Documentation standards
- Documentation review and approval requirements
- Use of supplier documentation
- Data archive and migration requirements

GAMP®5 (12) recognizes that not all systems are implemented in accordance with a standard "V" model. It may be necessary to adopt and iterative approach to system implementation where requirements, design and configuration are implemented in situ and based on prototyping. This approach is commonplace but should be controlled to ensure that requirements, design and configuration are complete, traceable and reviewed by relevant parties. A requirements traceability matrix is a good tool for ensuring requirements, design, configuration and testing are aligned.

The project team should be trained in the expectations of the VP to ensure that all roles, validation controls, documentation and assurance requirements are fully understood.

The VP should clearly identify supplier responsibilities. Where supplier documentation is to be utilized, the process verifying such documentation should be defined.

Requirement Specification

A number of different stakeholders will contribute to requirements definition. Such stakeholders may include

- System owner,
- End users,
- IT,
- Technical (engineering),
- Quality assurance,
- Validation, and
- Safety, health, and environment.

Depending on the criticality, complexity and current understanding of the computerized system, the method by which the requirements are captured and documented may differ from project to project. Possible methods employed in the definition of requirements include

- Discussions and meetings,
- Prototyping,
- Workshops, and
- Supplier demonstrations.

As identified by GAMP®5 (12) a range of skills and body of knowledge is required to establish effective requirements including process knowledge, business systems knowledge, and technical knowledge.

Depending on the complexity and phasing of the project, a number of requirements documents may be developed. Such documents should reflect

- Project phasing,
- Systems comprising EMS, and
- Different supplier organizations.

Workflow diagrams, process flowcharts, state transition diagrams or other suitable approach should be considered when documenting engineering processes to be supported. Process definitions should

- Define key process steps,
- Identify regulated steps,
- Identify manual interventions,
- Define process interfaces, and
- Define data inputs and outputs.

Functional requirements underpinning the business processes should be documented.

Functional requirements should be

- Stated in a clear, concise and testable manner,
- Grouped according to topic (e.g., communication interface, security and permissive concept), and
- Uniquely referenced to support traceability.

Each requirement should be categorized according to criticality. The following categorizations are useful:

- Critical to patient safety
- Essential for legal/regulatory compliance
- Business operation impact
- Essential to health, safety or environment
- Low business impact

Requirements documents should be reviewed and approved by key stakeholders (e.g., engineering management, business quality assurance, IT). Following approval, changes to requirements should be subject to change control and/or other relevant document management procedures.

Providing functional specifications (FS), configurations specifications, or equivalent documents describing the processes, functionality and operation of the system are in place, Requirement specifications (RS) need not be maintained post go live. RS, however, should be maintained up to go live and should fully reflect the released system.

Documented traceability between requirements, functional design and testing should be established. Table 27.3 presents a typical index of contents for a RS.

Table 27.3 Typical Contents of the Requirement Specification

Functional and performance requirements

- Functional and performance requirements related to engineering processes, e.g.,
  - Asset hierarchies and numbering
  - Planned maintenance
  - Corrective maintenance
  - Work instruction management
  - Control of asset data
  - Maintenance reporting
  - Breakdown monitoring, alarms, and reporting
  - Spares procurement and assignment
  - Asset and maintenance cost management
  - Condition monitoring
  - Process performance measurement and trending
  - Nonconformance management

Operational environment
- Number of sites implementing the system
- Number of anticipated users (per site or department)
- Number of concurrent users (per site or department)
- Operator workstations and locations
- Work patterns

Information and electronic record handling requirements
- Information received from other systems
- Information required by other systems
- Information supporting reports
- Information displayed on screens

Data entry and capture
- Required data entry and capture (e.g., from instruments and devices)
- Data formats and structures
- Data entry validation requirements
- Recovery following data entry failure
- Mandatory fields
- Confirmation on saving, e.g., electronic signatures

Data migration
- Volume of data
- Type of data
- Data conversion
- Data archiving

Human interfaces
- Special interface requirements (e.g., devices)
- Number of operator screens
- Environmental considerations (e.g., safety)

Table 27.3   Typical Contents of the Requirement Specification (*Continued*)

Communication interfaces
- Data to be transferred
- Frequency of transfers and transfer triggers
- Volume of data to be transferred
- Transfer method
- Communications synchronization (e.g., polling or continuous transfer)

Reporting requirements
- Information to be contained on reports
- Requirements for custom report development
- Additional information required on reports (e.g., company logo, watermarks)

Security
- User account access controls
- Access levels required
- Audit trail requirements
- ERES requirements (e.g., detection of altered records)

IT infrastructure
- Operating system platform
- Hardware infrastructure platform
- Data storage platform
- Infrastructure security requirements
- Intranet standards

Safety and reliability
- Implication of a system failure on process and/or products
- Implication of a system failure on human safety
- Acceptable and unacceptable software and hardware failures
- Security standards
- Data entry violation procedures
- Fail safe operation
- Redundancy of hardware and/or software
- Standby and manual control operation
- Start-up and shutdown procedures
- System reaction to process events
- Demand on independent systems

System flexibility
- Software expansion/enhancements
- Hardware expansion/enhancements
- Performance tuning
- Storage capacity enhancements
- Interfaces to and impact on other systems/environment in the future

Support and maintenance requirements
- Documentation, manuals, and training
- Fault rectification (e.g., patches)
- Upgrades
- Help desk support
- On site support services

Constraints and assumptions
- Internal and external standards
- Project phases, milestones, and timescales
- Risk management requirements during system changeover (e.g., parallel running of legacy systems)
- Integration requirements (e.g., legacy systems)
- System reliability and consistency (e.g., number of permitted system outages of a defined duration)
- Procedural constraints (e.g., statutory and regulatory obligations, legal issues, working methods)
- Maintenance (e.g., ease of maintenance, expansion capability, likely enhancements, expected lifetime, long-term support requirements)

Supplier Assessment

A number of suppliers and/or development consultants may be engaged in the project. The need for and approach to supplier assessment is determined by the criticality of the implementation. Suppliers may be assessed by a number of methods including

- Postal audit,
- On-site audit, and
- References from other companies that have implemented similar solutions.

Postal audits are useful in gathering initial information that can be used to focus subsequent on site audits or to narrow down the number of potential suppliers. The outcome of the supplier assessment is used to determine the following:

- Fit of proposed solution to business needs
- Existence and adherence to quality management systems
- Pharmaceutical experience
- Technical competence
- Support infrastructure
- Long-term viability of supplier
- Availability of documentation, including testing that can be leveraged by the regulated company

Where a site audit is required, the audit can focus on the main areas of concern raised by the postal audit, thus reducing the duration of the audit and enabling a more in depth review of the main areas of risk.

Auditors should assure themselves that there is sufficient documentary evidence available to demonstrate that quality controls, appropriate to regulated industries, are in place and are being routinely applied.

Supplier Quality and Project Plans

Where a supplier quality plan exists, it will often be used as a guide to the audit. Specific activities and documentation stated in the quality plan should be reviewed against their controlling procedures.

The project plan should be reviewed to ensure that all tasks have been defined and that time estimates and resources appear adequate. Project plans should be maintained and reviewed regularly to identify potential slippages that may impact delivery milestones or quality tasks.

Functional Specification

The FS is the supplier's response to the regulated company's RS (Table 27.4). The FS will indicate how the "out-of-the-box" application will be implemented to reflect the approved RS. The FS should reflect business decisions that influence how the system is configured and/or developed to meet the RS.

The FS should be fully traceable to the RS, clearly demonstrating that all RS clauses have been met. Where the FS deviates from the RS, a rationale for the potential impact of the deviation should be provided.

Infrastructure Design Specifications

The infrastructure design specifications will define the infrastructure software and hardware platform supporting the EMS architecture. The regulated company may impose corporate standards to ensure compatibility with other installations on the site. In many instances the supplier will simply state the hardware requirements and allow the regulated company to procure the hardware, especially when the EMS architecture comprises a number of systems from a variety of suppliers. Table 27.5 provides the typical contents of the IDS.

The installation qualification (IQ) should verify the major critical components stated in the infrastructure design specification.

Table 27.4 Typical Contents of the Functional Specification

- System overview
  - System description
  - Schematic of system architecture including interfaces to other systems

- Process definitions
  - Diagrammatic representation of business processes implemented within information system, e.g., maintenance management, change control, document management, etc.
  - Work flows

- Functional definition (flowcharts, narratives)
  - Function inputs
  - Function objectives
  - Functional performance
  - Function outputs

- Record management
  - Identification of key records
  - Backup and restoration
  - Audit trails

- Data migration strategy
  - Migration of information from legacy systems
  - Import of information from manual systems

- User interfaces
- Communication interfaces
  - Data transfers between systems

- System reliability and performance
  - Failure mode recovery
  - Minimum process performance requirements
  - Minimum data storage requirements
  - Redundancy
  - Hardware upgrade paths

Table 27.5 Infrastructure Design Specification

- System architecture diagrams
- Layout and wiring diagrams and drawings
- Main component specifications
- System interface specifications
- Minimum processor performance
- Minimum memory requirements
- Minimum storage requirements
- Peripherals
- Interfaces (communications cards, network connections, cabling, speed)
- Settings (switch settings, firmware configuration)
- Environment (temperature, humidity, RFI, UV, electromagnetism)
- Electrical supplies (UPS, earth requirements, filters, etc.)
- Define relevant standards (safety, electricity, etc.)

Software Design Specification

Software design specification (SDS) relating to the standard application design should be verified where necessary by the supplier assessment process. Where custom software needs to be developed to supplement the standard software of the EMS, SDS should be created to define the structural and logical design of the software. The SDS should provide sufficient detail to enable unambiguous implementation of the software.

Software Coding Standards and Software Code Reviews

Software code reviews (SCR) confirm that implemented software has been developed to appropriate standards and identifies potential coding errors that might not be identified by testing alone.

SCR are the responsibility of the system developer as they require technical competence and understanding of system design. Code review records associated with the core product software should be examined during supplier assessments. Custom code developed to provide additional functionality during system implementation should be reviewed if it is complex, critical or developed by less experienced developers. The scope and content of Software Coding Standards is discussed more in chapter 8.

Configuration Management

The primary objectives of configuration management are to

- Support problem investigation,
- Enable disaster recovery,
- Support change management, and
- Support release management.

An electronic and/or paper record of configuration parameter settings against each release of the system should be maintained.

It is essential that business decisions leading to the setting of system parameters are recorded so that subsequent configuration changes do not adversely impact previous business decisions.

Configuration records may comprise any combination of the following:

Electronic backup	Used to recover configuration to a known point following system failure
System-generated configuration report	Generated after IQ/operational qualification (OQ) to provide a baseline of validated configuration parameters
Configuration specifications/log books	Provide a record of configuration changes post go live and business decisions leading the parameter settings

All configuration settings should be traceable to the release of the system in which they are implemented.

All custom software should be subject to change control and version control. Software code management tools should be used to ensure that code is secure and that changes are traceable. It should not be possible for developers to simultaneously modify the same code.

System environments should be controlled to ensure that development, test and production operations are segregated and appropriately controlled. Test environments should be representative of production environments (although not always identical) to ensure integrity of the testing and increase confidence that the implemented systems will operate correctly post go live.

Supplier assessments should ensure that supplier development processes provide adequate control over software including version control, change histories, access security, safeguards over simultaneous modification by multiple developers, release control and build controls.

Supplier Testing

Supplier development and system release testing should be verified by the supplier assessment. In addition to such testing, the supplier may conduct acceptance testing to demonstrate the system meets the approved design and requirements.

The regulated company may verify such testing rather than repeat it during IQ, OQ, or performance qualification (PQ). Qualification may include review of supplier test records or may involve witnessing supplier testing. The greater the criticality of the functionality under test, the greater the degree of qualification.

When leveraging supplier testing, it should conform to good testing practice as described in this chapter. Further, it is beneficial to include supplier testing when establishing requirements traceability to ensure that the required scope and depth of testing has been achieved.

Installation Qualification

The IQ is the responsibility of the regulated company. The IQ should define methodical inspections that verify the hardware and software installation against the design. Each inspection is verified against unambiguous acceptance criteria. The results of the inspection should be recorded on a test result sheet, referenced to or contained within the IQ protocol. An inspection is deemed to have "passed" if all the acceptance criteria set forward have been satisfied. Table 27.6 provides the typical contents of an IQ.

Table 27.6 Typical Contents of an Installation Qualification

- Installation plans/procedures
  - Satisfactory execution of installation procedures

- Software installation
  - Correct execution of installation media for application software, reporting software, database tools, etc.
  - Application of patches/service packs
  - Correct assignment of setup parameters

- Database setup
  - Correct installation of database tools
  - Correct configuration of database architecture

- Inspection of critical hardware components
  - Servers in correct locations
  - Minimum processor performance
  - Minimum memory capacity
  - Minimum storage capacity
  - Printers
  - System interface setup
  - Printer driver installation

- Networks
  - Connection
  - Network addressing
  - Server synchronization (e.g., date and time)
  - Network conflicts

- Electrical installation
  - Cable connections
  - Electrical testing
  - Uninterrupted power supplies

- Input/output
  - Outstation connection and configuration
  - Field device connection and calibration
  - Diagnostic checks

- Documentation
  - User manuals
  - Technical documentation
  - Availability of user SOPs
  - Availability of disaster recovery plans

- Training
  - Availability of training plans

Operational Qualification

The OQ should be integrated with the system acceptance testing (SAT) normally conducted by the supplier. The OQ verifies the functionality of the system within its normal operating environment. The OQ protocol clearly defines the methodology by which the tests should be conducted and the acceptance criteria used to determine the outcome of the test. Figure 27.9 provides an example of an OQ test script. The OQ should reasonably challenge the operating boundaries of each function (although never to destruction). For example,

- Schedule a maintenance task on a date before the last maintenance task,
- Ensure that alarms operate on defined boundaries,

OQ Reference:	001. Recipe Save and Retrieve
Prerequisites	Recipe to be created shall not exist.
TestEquipment/ Simulators/ Harness	None
Function/Purpose:	To ensure that Recipes are correctly saved to file FDS Ref: 3.1.2
Method:	1. From Recipe Edit Screen select 'Create New Recipe' 2. Enter a value in each field 3. Print screen and verify each field against entered values 4. Select 'Save Recipe' 5. Exit Recipe Editor 6. Re-enter Recipe Editor 7. From Recipe Edit Screen select 'View Recipe' 8. Enter Number of newly created recipe 9. Confirm that values are correct
Acceptance Criteria:	1. Recipe Edit Screen is Displayed 2. New Values are accepted, values out of range are not accepted 3. Printed fields match enter fields 4. Message 'Recipe Save Ok' is displayed 5. Main Menu is displayed 6. Recipe Editor is displayed 7. Prompts for recipe number 8. Recipe is recalled and displayed 9. Values match those on printout from step 3
Results/ Observations	
Acceptance Criteria Achieved (write clearly YES or NO)	
Tested by:............................. Date:	Witnessed by:.......................... Date:

Figure 27.9 Test script.

Table 27.7 Typical Content of the Operational Qualification

- Functional operation and performance
- Failure processing, reporting and recovery
- Multi-site challenge
- Operating boundaries (e.g., data entry)
- Network interfaces
- Serial communication interfaces
- Start-up and shutdown
- Security and access
- Data storage and retrieval
- Error reporting
- Backup and restoration SOPs
- User SOPs
- Contingency plans
- User and system administrator training (delivery)

- Attempt to modify a locked record,
- Attempt to use quarantined spares, and
- Attempt to procure spares from an unapproved supplier.

Table 27.7 provides the typical content of the OQ. As with the IQ, all protocols should be pre and post approved and tests independently reviewed. Tolerances on the acceptance criteria should be in line with design and should not be so wide as to guarantee success.

Data Migration

Data Migration controls maintain the integrity of data transferred from existing systems to replacement systems. The system owner should define the data to be migrated on the basis of the age, value, and criticality of the data. Consideration should be given to how nonmigrated data should be retrieved once the legacy system has been retired. Company, regulatory and legislative requirements for data retention should be met.

Some systems provide tools that automatically analyze data to ensure that the data is still compatible with database rules and to ensure that referential integrity is maintained.

Data structures in the replacement system may not be fully compatible with data structures in the existing system or manual records. The data migration manager should define the data transformation and manipulation requirements. Consideration should be given to the following:

- Mapping of database tables in existing and replacement systems
- Population of additional table fields in the replacement systems (especially mandated fields)
- Differences in field formats and sizes
- Manual data manipulation requirements
- Data validation rules for existing and replacement system
- Conversion of formats and numbers

Consideration should be given to how nonmigrated data should be retrieved following retirement of the legacy system. Options include but are not limited to

- Archiving to a neutral file format (e.g., PDF, XML),
- Archiving to paper, and
- Retention of existing software and hardware (not recommended!).

Performance Qualification

PQ has two primary sets of objectives.

- Verify consistent process operation and performance

- ○ Implemented processes correctly operate for "real" scenarios (e.g., processing a planned maintenance task from start to finish)
- ○ Interfaces between implemented processes function correctly
- ○ Processes can be operated using assigned roles

- Verify adequacy of system controls and system stability including:
  - ○ Fit of implemented requirements to actual engineering needs
  - ○ System failures and errors
  - ○ Number and scope of system changes
  - ○ Effectiveness of SOPs
  - ○ Effectiveness of user and support training
  - ○ Effectiveness of supplier service agreements

As much PQ testing as possible should be executed prior to go live. It may be necessary to execute some PQ testing (or monitoring of actual system use) during early system operation. System controls and stability can only be monitored post go live and will typically extend three months beyond go live.

The outcome of the PQ may indicate that there are areas of weakness in the design or implementation of the application, in particular if high levels of design changes or functional failures are observed within a localized area. In such instances, it is necessary to reconsider to adequacy of the OQ and further testing may be required to determine the root cause of the failures.

Verification

The term verification is now widely used to describe the systematic approach to verify that systems are fit for intended use. Verification encompasses all methods by which an organization ensures that stated requirements have been satisfactorily implemented such as review, inspection, commissioning, supplier testing and qualification. The key implication of the introduction of the term is that validation strategies can be more intelligent, effective and efficient by through leveraging and verifying traditional life cycle activities (e.g., commissioning and supplier testing) rather than blindly repeating testing and other activities under the qualification banner (12). Qualification becomes a process of reviewing or auditing commissioning and supplier testing.

Validation Report

The validation report (VR) responds to the VP, providing a summary of the actual approach taken and the documentation produced. Any deviations from the approach prescribed by the VP should be justified and the impact of the deviation assessed. Where the deviation is not acceptable, a corrective action plan should be formulated to address the issue. It may be possible to implement manual procedures to overcome the issue in the short term to enable the system to move into the operational environment while the issue is being addressed.

The VR should clearly demonstrate that a suitable operating environment has been established including procedures to control documents, changes, backup, archive and restoration, security and that appropriate service contracts have been established (12,13). Where residual risks are low or the risk can be addressed before any potential impact, the system can be recommended for operational use.

Operational Environment

The validated status of the computerized systems comprising the EMS should be maintained during the operational life of the system. It is therefore essential that procedures are developed to control operation and maintenance of the system post go live. Such controls should include the following:

- System operation
- Change control
- Incident and deviation management

- Backup and restoration
- System upgrade
- Disaster recovery and contingency plans
- Service level management
- System repair
- Corrective and preventive action
- Periodic review

IT infrastructure supporting the application must also be maintained in a state of compliance (14).

SUMMARY

This chapter has hopefully demonstrated that effective implementation of an engineering strategy is dependent on an integrated suite of computerized systems including maintenance management, document and records management, building and environmental monitoring, condition-based monitoring, and CAD.

The criticality of such systems is determined from the criticality of the scientific and manufacturing systems supported by the EMS coupled with the impact of failure of the EMS functionality on those systems.

The information systems referred to however, are only tools and a clear understanding of performance expectations and failure modes of the assets being maintained is essential. Further, establishing a culture within the Asset Management organization that ensures accurate and complete performance data is collected and continuously evaluated to identify engineering strategy improvements is of paramount importance.

REFERENCES

1. U.S. Food and Drug Administration. 21 CFR Part 11. Electronic Records; Electronic Signatures.
2. U.S. Food and Drug Administration. Guidance for Industry, Part 11. Electronic Records; Electronic Signatures—Scope and Application, August 2003.
3. International Society for Pharmaceutical Engineering. GAMP® Good Practice Guide: a Risk Management Approach to Compliant Electronic Records and Electronic Signatures, 2005.
4. European Union. Guide to Good Manufacturing Practice Used for Medicinal Products EU GMP, Chapter 4. Documentation.
5. European Union. Directive 2003/94/EC Good Manufacturing Practice in Respect of Medicinal Products for Human Use and Investigational Medicinal Products for Human Use, 2003.
6. European Union. Directive 2001/20/EC Good Clinical Practice in the Conduct of Clinical Trials on Medicinal Products for Human Use, 2001.
7. U.S. Food and Drug Administration. 21 CFR part 211. Current Good Manufacturing Practice for Finished Pharmaceuticals.
8. U.S. Food and Drug Administration. Guidance for Industry, Computerized Systems Used in Clinical Trials, April 1999.
9. European Union. Guide to Good Manufacturing Practice Used for Medicinal Products, Annex 11. Computerized Systems.
10. European Union. Guide to Good Manufacturing Practice Used for Medicinal Products, Annex 15. Validation.
11. Wingate GAS. Validating Automated Manufacturing and Laboratory Applications. Buffalo Grove, IL: Interpharm Press Inc, 1997.
12. International Society for Pharmaceutical Engineering. GAMP® Risk Based Approach to Compliant GxP Computerized Systems, 2008.
13. Pharmaceutical Inspection Cooperation Scheme. Good Practices for Computerized Systems in Regulated "GxP Environments," PI 011-02, September 2007.
14. International Society for Pharmaceutical Engineering. GAMP® Good Practice Guide: IT Infrastructure Control and Compliance, 2005.

28 | Case Study 10: Desktop Applications Including Spreadsheets

Arthur (Randy) Perez
Novartis Pharmaceuticals

INTRODUCTION

The GxP environment has evolved in the past 15 to 20 years from a condition where computerized systems were largely limited to lab instruments and large centralized applications to one where not only is there a computer on every desk, but users are savvy enough to use them in innovative ways to make their lives easier. The amount of data that life science companies process and store has multiplied tremendously. This is partly due to the fact that we can now collect, process, and store far more data far faster, and partly due to growing data requirements from increased regulation by all sorts of governmental authorities. It is small wonder then that today's more sophisticated users employ the desktop tools provided to them to help make sense of this vast datasphere.

The catch, of course, is that if they are using these desktop tools to support regulated processes, there may need to be some controls in place that they are ignoring, or more likely simply have not recognized. Laws and regulations that can affect desktop applications include requirements related to GxPs, data privacy, protection of intellectual property, legal discovery, etc. This chapter will focus on satisfying GxP requirements.

Depending on the way that desktop applications are used, there are often validation requirements associated with them. The catch is that those within the company responsible for monitoring and oversight of the validation program may be unaware of programs that reside on one or a few desktop PCs. Simply identifying all of the desktop applications at a company that may need some level of validation and/or controls may actually be more challenging than putting such measures in place. Regulators, however, seem to find such applications with regularity, and have cited firms for using these tools without requisite controls:

FDA Warning Letter, 2001:

- "Failure to have an adequate validation procedure for computerized spreadsheets used for in-process and finished product analytical calculations."
- "Failure to use fully validated computer spreadsheets to calculate Analytical results for in-process and finished product testing [21 CFR 211.165(e)]. For example, the computer spreadsheets used to calculate analytical results have not been validated." (1)

FDA Warning Letter, 2003:

- "...spreadsheet software used in manufacturing has not been validated for the purpose of generating a worksheet for formulation of reagents. No documentation was found to establish or verify corrections to the program."(2)

Where does the solution lie? Probably somewhere short of the "fully validated computer spreadsheets" cited above. The approach to computer validation using risk management principles has been presented in depth in such guidelines as *GAMP®5* (3). By their very nature desktop applications are likely to be substantially less complex than large server-based applications, and therefore easier to validate. Conversely, however, managing them to a state of adequate control can be more challenging.

DEFINITION OF DESKTOP APPLICATIONS

As the name implies, a large majority of desktop applications reside on a user's PC. The most common desktop applications are:

- Spreadsheets, built with tools like Microsoft Excel®
- Databases such as Microsoft Access®
- Word processing like Microsoft Word® with macros

However, the principles in this chapter can also be applied to applications written by any end-user that may reside on either a desktop PC or possibly on a server. It is the nature of the application that is the focus of the classification. Examples that merit the same approach to compliance but do not fit easily into the classifications above include:

- 4GL statistical analysis programs written using tool such as SAS®or
- 4GL "ad hoc" database queries, for example, using a tool such as SQL®
- Simple programs in 3GL language, for example, BASIC, VB®, C++

Common characteristics define issues around desktop applications.

- The number of users can vary widely
  - A one-off on a single PC
  - Multiple copies on several PCs
  - Accessed/downloaded by many users from a server

- Usually developed by an "amateur"
- Usually poorly documented (if at all)
- If tested, rarely challenged
- Versions often uncontrolled
- Security often lax
- Risk seldom considered in deployment

A key issue is to recognize that for all practical purposes, most of these must be considered as *applications*. Some are very simple, but they are still applications, and to meet regulatory requirements it is imperative to recognize them as such and to take *appropriately scaled* steps to ensure the same level of confidence in them as is enjoyed by their larger server-based brethren.

SPECIAL CONSIDERATIONS FOR SPREADSHEETS

Spreadsheets are analogous to Swiss Army® knives in the sense that they provide a remarkably versatile tool set that can be applied to solving a wide variety of different tasks. A spreadsheet might be any of the following:

- A simple document used as evidence to support GxP compliance
- A substitute for a calculator
- A template for a simple, repetitive data manipulation task
- A complex data processing tool
- A database

The very flexibility that makes spreadsheets so useful, however, introduces a significant level of regulatory compliance risk if proper controls and protections are not employed, and in some cases such controls may not actually be possible. Like a screw driven with a hammer, a solution that appears strong may actually fail when stressed. Just because you have a hammer (or a spreadsheet) does not mean that it is the right tool to use.

Spreadsheets as Documents

The use of a spreadsheet as a medium for generating and possibly storing GxP records is generally as straightforward as using a word processor. Spreadsheets may simply be a better document choice than a word processor in some cases. For example, spreadsheets provide an excellent platform for sortable and printable compiling lists.

Although validation is not needed, other controls are required exactly as for any other document. Copies can be printed and signed for retention as paper records. If the document includes calculations, these should be shown (most spreadsheet tools have a keystroke sequence that can display the calculations in the cells rather than the results); a copy of this can be printed as well. If the records are to be stored electronically, all appropriate controls that would be expected for any other document would apply as well; this could include some regulatory requirements such as those specified in FDA's 21 CFR Part 11 (4) or the EU's directive 1999/93/EC (5). If electronic signature is contemplated, similar considerations apply.

Spreadsheets as Calculators

It is fairly common to use a spreadsheet to do calculations that would have been done using a hand calculator several years ago, for example, for processing lab results prior to entry into a lab notebook. When used for this purpose, the analogy should be followed to the fullest. The calculations should be documented exactly as if a hand calculator were being used. No additional controls are necessary. There is no need to show that the calculations are accurate, any more than there would be with a calculator. The key is showing that the calculations that were done were the right calculations.

Spreadsheets as Templates

A very common use of spreadsheets is for automation of repetitive tasks, using a blank copy each time the task is executed. Templates can be of varying complexity, ranging from simple forms to sophisticated data manipulations. The resulting completed template is often a document that is retained as evidence of successful completion of the task (see sect. "Spreadsheets as Documents").

Validation requirements will depend on complexity, and should follow guidelines suggested in section Risk-Based Validation and Control Strategies.

Spreadsheets for ad hoc Data Processing

When faced with the need to manipulate a large amount of related information to solve a unique problem, it is very common to place data in a spreadsheet to take advantage of the tool's ability to facilitate such tasks as sorting, performing various statistical operations, etc.

Since validation is geared toward demonstrating reproducibility of a process, it does not make sense to validate spreadsheets that are "single use," that is, those that are going to be repeatedly used as part of a standard process. However, a small amount of verification can support compliance. The simplest approach is to retain the spreadsheet itself in a secure manner as long as the data analysis needs to be retained. If inspected, the original electronic evidence is thus available for examination.

If the records are retained on paper a little more effort is recommended. Using sorting functions have no compliance impact, as none of the data are actually changed. If complex arithmetic manipulations are used, the key compliance aspect is to show that the proper calculations were done. Printing a copy of the spreadsheet with the cell calculations visible is adequate to prove how the calculations were done.

Spreadsheets as Databases

Again due to their ease of use and the familiarity of many users with spreadsheets, only moderately sophisticated users have the ability to create spreadsheet databases. This is probably the highest risk type of spreadsheet application from a regulatory standpoint, stemming from

- Data integrity,
- Protection of the database structure and "code,"
- No support for audit trails.

The first two factors above are exacerbated by the fact that the data and the application are inseparable. Every change to data or "code" requires saving a new copy of the file. To make matters worse, users will typically save the updated file over the previous version.

When contemplating a spreadsheet database the following questions should be considered:

- Concerning the nature of the data:
  - Are records in scope for Part 11 (required by a predicate rule) or another e-record regulation?
  - Does the data require an audit trail?
  - Is the data just too sensitive to keep in a spreadsheet?
    - Are there intolerable risks if a laptop with this information is lost?
    - Is there valuable intellectual property at risk?
    - Are there additional regulatory risks, for example, HIPAA or data privacy laws?
- Concerning adequate protection of the of the application:
  - How will access be restricted to authorized users?
  - What are required controls to restrict improper data entry?
  - What will be done to prevent unauthorized data or code modification?
- Concerning management of the application:
  - How will change control processes be enforced?
  - How will it be version controlled?
  - How will backup and archiving be handled?
  - Who will manage security, and how?

In most cases where regulated data is involved and a database is required, it is advisable to develop it in a proper database tool. Even a desktop database like Microsoft Access offers significantly better control over many of the issues noted above than does a spreadsheet.

RISK-BASED VALIDATION AND CONTROL STRATEGIES

The approach to validation should, as with any application, be based on applying effort commensurate with risk. Figure 28.1 illustrates that as risk increases, the relative amount of effort expended on validation and the level of control required for the application increases. Thus, a spreadsheet that is a major tool for analyzing a clinical study merits significantly more attention than one that is used to track attendance at training sessions.

One factor that makes assessing risk for desktop applications easier is that most address only a single function within the business process. For example, consider a laboratory process that involves collecting HLPC data and then running trending analysis on a spreadsheet constructed from a template. There are two systems involved. The HPLC system controls several critical to quality (CTQ) parameters:

- Injection volume
- Eluent flow rate
- Eluent mixture
- Calculation algorithm

Figure 28.1 Controls and validation effort commensurate with risk.

The spreadsheet, on the other hand, addresses only one CTQ requirement: the calculated results based on output from the HPLC. Assessment of risk is thus simpler, as only one risk scenario needs to be considered: what is the consequence of an incorrect result from the calculation?

Approach to Evaluation of Risk

Risk to Patients

For all practical purposes the approach to the evaluation of the potential risk to patient population if a desktop application fails can be handled exactly as would be done for any other type of application. Therefore, a spreadsheet that performs a calculation used in a decision to release product for human consumption would have the same outcome as an incorrect analysis from a lab instrument: potential release of adulterated product, and hence high potential impact on patient safety. A spreadsheet used to calculate the average yield of a production campaign would have lower impact, as that information is primarily of interest for productivity, although it still has some GMP impact because low yield can be indicative of quality problems.

Complexity of Desktop Applications

A distinguishing factor for desktop applications is that the complexity is a much more readily assessable factor than for a system like that controlling the above HPLC. For the latter, complexity is staggering. There will probably be trillions of potential logic paths through the application. For most desktop applications this number will be orders of magnitude lower. Indeed, for many applications it will be just one, which makes assessing desktop applications for complexity fairly easy.

Complexity has long been used as a factor in developing validation strategy. For example, the GAMP categories as defined in *GAMP®5* are actually based on a combination of complexity and ubiquity. Accordingly, *GAMP®5* proposes an effective modification of the category concept for desktop applications (Fig. 28.2) (6). An extremely important concept that this picture helps to illustrate is the concept of risk as a continuum. There are no risk steps; risk cannot be compartmentalized to discrete values of high, medium, or low. There will be "high mediums" and "low highs" that will be closer to one another than to a "pure" high or medium risk.

For the purpose of defining complexity of desktop applications, note that the commercial platforms (e.g., Microsoft Excel) are considered category 1 (infrastructure software), just like an

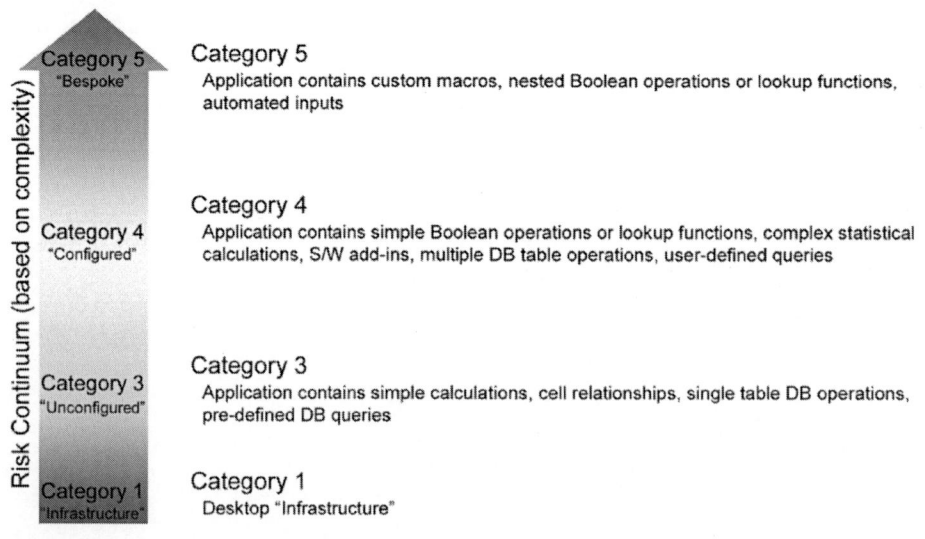

Figure 28.2   *GAMP®5* categories modified for desktop applications.

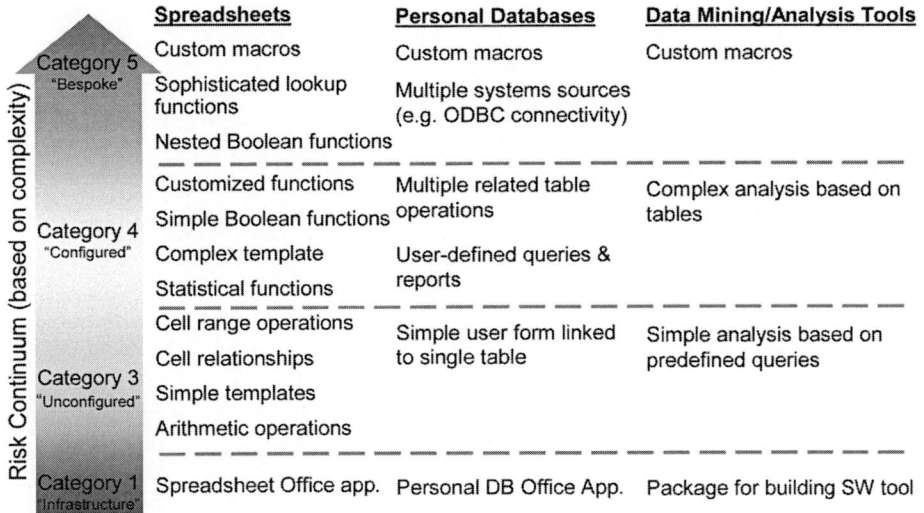

Figure 28.3 Application of modified GAMP categories to common desktop applications (picture© ISPE 2008).

operating system. This makes sense as the software provides the operating platform for the spreadsheet analogously to the way the Windows® operating system provides the platform for an HPLC control and data system.

The simplest desktop applications other than spreadsheet documents (see sect. "Spreadsheets as Documents") are considered comparable to category 3, unconfigured software. While configuration is not really relevant, the analogy holds because these desktop applications represent the simplest use of the tools, for example, Excel doing simple arithmetic or a single uncomplicated Access database table. Note that some of the spreadsheets that fall into this category may also be treated as documents for retention purposes, but that they are not "pure" documents because they do some math, and there will be some associated validation activities.

The introduction of Boolean operations (IF, AND, OR, etc.) or multiple database tables provides the next level of complication, as now the application may have multiple potential logic paths. A similar argument can be made for many of the statistical operations that are not part of the basic infrastructure package, but which can be downloaded and installed as add-ins. These elevate the classification to category 4 (parallel with configured software).

Category 5 includes the most complex desktop applications, characterized by features such as macros, nested Boolean operations or lookup tables, and direct input from other applications. The complexity associated with macros deserves some further explanation. Macros are in fact subsidiary computer programs, usually in a fully developed 4GL language such as VB, that reside within the desktop application. Macros often have logical branching themselves, and can be very significant in size and complexity.

Figure 28.3 shows some examples of the application of the GAMP categories to some common type of desktop applications.

Scalable Validation Approach

With the two issues of risk to patients and complexity understood, a simple overall risk picture can be developed by plotting these two factors as shown in Figure 28.4. As the classification of a system moves to the right on the x-axis or up on the y-axis, the overall risk of the application is higher and the commensurate requirement for validation and operational controls increases.

Validation Planning

Not all desktop applications will require a validation plan. Typically, this will not add much value for low risk applications, and can be forgone in such cases. Even for moderate- to high-risk applications the plan can be quite abbreviated, describing the approach to validation in a

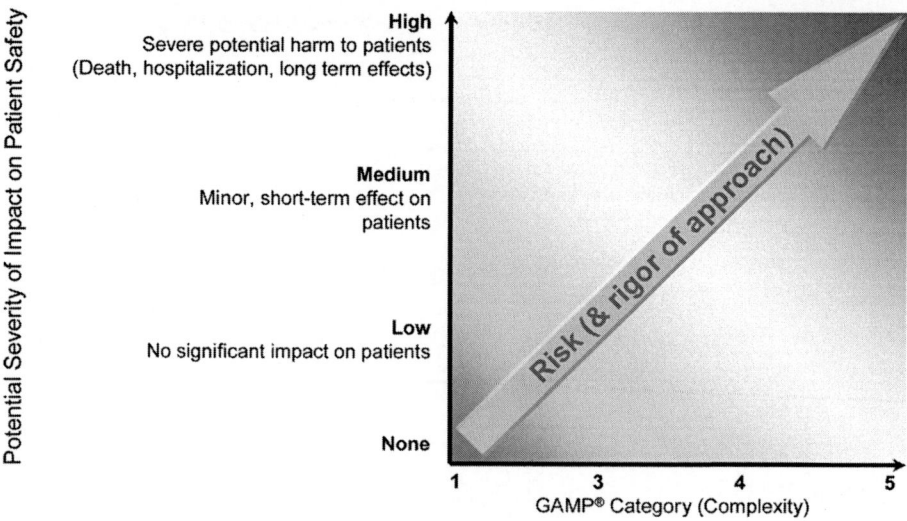

Figure 28.4   Overall risk of a desktop application as a factor of potential impact to patients and complexity.

URS ID#	Requirement
1	Record batch yield, API purity, content and identity of 2 largest impurities
2	Calculate campaign results for average and standard deviation of the above
3	Show yield results for night shift, day shift, and overall

Figure 28.5   Sample URS for a simple spreadsheet.

single page or less. Some firms may choose to develop a brief form to capture the relevant information, which should center on intended use, the assessed risk, and the approach to verification.

Specifications

Since even the most complex desktop applications are simple when compared to their larger brethren, some shortcuts to documentation are justifiable and reasonable. Like any other application, specification documentation is appropriate to describe what the system is supposed to do. However, it can be very brief. For instance, a user requirement for a spreadsheet used to track and trend a production campaign could be as simple as the example shown in Figure 28.5. If the calculations to be done are complex, they should be defined in the user requirement specification (URS). Special calculations that are needed should also be specified. If it is desired to calculate yields achieved by the night shift, this should be noted.

Functional specifications represent the transition between what the user wants and what the programmer will implement in the final application. Since the user is the programmer, there is really no point to a functional specification. Design specifications define *how* the system will be implemented. It is not necessary to test the calculations to verify that the arithmetic is done correctly by the tool, but it is necessary to demonstrate that the calculations built into the application are the *correct* calculations; this is what passes for a design specification for simpler spreadsheets. A very simple approach to demonstrating that the calculations are correct is to retain a copy of the application that shows the calculations, as shown in Figure 28.6. (This view can be obtained in Excel via the keystroke combination < ctrl ' >.) Note that, in this example, a simple Boolean operation is used (SUMIF), but since this does not really constitute a program branch the list of calculations is adequate.

	A	B	C	D	E	F	G	H	I
1	Batch ID	Date	Shift (d,n)	Yield (%)	Purity API (%)	Impurity 1 ID	Impurity 1 (%)	Impurity 2 ID	Impurity 2 (%)
2									
3									
4									
5									
6									
7	Avg Yield			=AVERAGE(D2:D6)					
8	Avg Yield Night Shift			=SUMIF(C3:C6,"=n",D3:D6)/COUNTIF(C3:C6,"=n")					
9	Avg Yield Day Shift			=SUMIF(C3:C6,"=d",D3:D6)/COUNTIF(C3:C6,"=d")					
10	Yield Std Dev			=STDEV(D2:D6)					
11	API Avg Purity			=AVERAGE(E2:E6)					
12	API Purity Std Dev			=STDEV(E2:E6)					
13	Impurity 1 Avg %			=AVERAGE(G2:G6)					
14	Impurity 1 Purity Std Dev			=STDEV(G2:G6)					
15	Impurity 2 Avg %			=AVERAGE(I2:I6)					
16	Impurity 2 Purity Std Dev			=STDEV(I2:I6)					

Figure 28.6 Sample design specification for a spreadsheet.

It is a fairly common practice of spreadsheet developers to hide rows or columns of cells that contain information that is critical to the way a spreadsheet works, but which the users do not need to see, such as lookup tables. Figure 28.7 shows a simple spreadsheet automation of the *GAMP*®*5* risk assessment tool (7). The calculation in cell I9 depends on the information in columns L-S, which are hidden in the spreadsheet available to users. Design documentation without this critical reference information would be of no value.

Effect of Macros on Strategy
Macros can be classified as two types:

- Simple
  - Constituting keystroke automation, geared toward improving usability and reproducibility
  - Usually recorded; generating simple macros does not require programming skill
  - Example: a macro that clears several data entry fields in preparation for the next set of calculations.
- Complex
  - Subprograms within the spreadsheet that include multiple potential logic paths
  - Usually programmed by someone with knowledge of the macro language (e.g., VB)

For simple macros, a copy of the macro should be retained (electronically or on paper). For applications that have embedded macros or other features that provide multiple potential logic paths it may be appropriate to show a flow chart.

	L	M	N	O	P	Q	R	S
10	Do not change or delete these cells!							
11			Probability					
12	Impact		L	M	H			
13	L	1	3	3	2		H	3
14	M	2	3	2	1		L	1
15	H	3	2	1	1		M	2
16			Detectability					
17	Risk Level		L	M	H			
18		1	H	H	M		H	
19		2	H	M	L		M	
20		3	M	L	L		L	

Figure 28.7 Hidden information critical to the function of the spreadsheet is a key part of the design specification.

Testing Strategy

In all cases, the rigor of documentation and test strategy should be linked to the assessed risk of the application. Applications that fall in the lower left quadrant of the risk continuum shown in Figure 28.4 will have much looser requirements than those in the upper right quadrant.

For simple cell operations (arithmetic, comparison of values, Boolean operations that do not generate program branching, etc.) there is no need to demonstrate that the spreadsheet software does the calculations correctly. However, there should be documented verification that they are the correct calculations needed to achieve the business goal. Two common ways of achieving this are:

- Manual verification that the cell generates the proper result with known data
- Verification by a third party that the cell formulae are correct

Whatever method of verification is selected, the results need to be documented. This will usually involve printing results to paper or PDF and retaining them as part of the validation package.

For more intricate operations like complex macros or the nested lookup tables in Figure 28.7, a predefined test script should be developed, including definition of the test data set. This can be incorporated in the validation plan.

Validation Reporting

With the possible exception of extremely complex and high-risk desktop applications, it is generally unnecessary to generate a separate report summarizing the validation activities. Since the document set is reasonably small, it should be sufficient to retain a well-organized package of the plan (if applicable), specifications, verification records, and documentation showing the processes used to control the application (analogous to system management SOPs for major applications).

Control of Desktop Applications

Security

Considerations of scalability can be applied to security of desktop applications as well, but the principal factor must be the nature of the information rather than the complexity of the application. In some cases, it may just be unwise to place sensitive information in a desktop application. The recent wave of data privacy breaches illustrates the vulnerability of such information. A laptop PC or CD or flash drive with sensitive information can be lost or stolen. It is not uncommon for a large pharmaceutical firm to lose an average of one laptop per day. Information on PCs can also be vulnerable to loss of information via malware in peer-to-peer networks.

Security issues specific to desktop applications include:

- Data integrity
  - Many desktop applications do not support user-specific passwords
  - Audit trailing of many desktop applications is not possible; spreadsheets are especially problematic for audit trailing
  - Data is often stored in the same file as the application "code"
  - Data that should be locked may be editable during subsequent sessions
  - Users can copy spreadsheets and save new and potentially altered versions
  - Some spreadsheets may fall under e-record rules like 21 CFR Part 11
  - All data used to support GxP decisions will have data retention requirements
- Access to application
  - Access to "code" (e.g., calculations, logic)
  - Version management can become a challenge as code is "tweaked"

Some level of security control is possible using the features of the desktop application; users developing the applications need to understand whether the available controls are adequate based on risk. For example, it is possible to lock all of the cells except for those needed for data entry, which at least partially mitigates the access to "code" issue. Applications can be protected by a password, although the password will have to be shared by users. Access management can also be augmented through operating system level controls and administrative procedures. If a record must be generated in a format that cannot subsequently be edited, printing to PDF is an option. PDFs can also support some secure digital signature technologies, for example, public key infrastructure (PKI). Printing and hand-signing a record might also be an option, although that would require managing paper records, which is becoming a less palatable option for many firms. Finally, one option that can mitigate many of the above concerns is the management of the records in a secure electronic document management system (EDMS).

Change Control

End-user applications require management of change just like any other. The nature of the control mechanism should be based on risk and complexity. Databases and spreadsheet templates require version control. Possible technical solutions include the use of an EDMS or software code management tools. The latter, used for version control of code segments by programmers, offer a less expensive but still effective solution that supports check-in/check-out processes, often with an audit trail.

CASE STUDIES—SIMPLE EXAMPLES OF DESKTOP APPLICATIONS

The following examples are hypothetical desktop applications based on the assumption of a risk level as illustrated in Figure 28.8. Readers may feel that the assessed risk level is inappropriate for the application as described. However, a company's risk tolerance plays a large factor in assigning risk. Regulatory trends, business exposure, and factors unique to a particular product can all play a part in shifting a point within the risk continuum.

1. Spreadsheet "infrastructure" (e.g., Microsoft Access or Excel)
   Low impact, low complexity
   - Version control
   - Evidence of correct installation
2. Simple spreadsheet template for arithmetic calculations of tablet weight distribution for product release

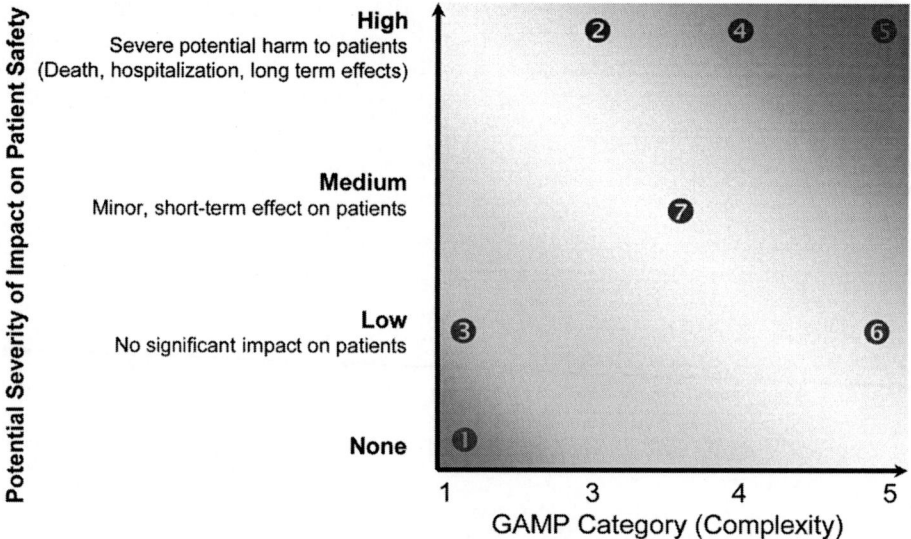

Figure 28.8   Assessed risk for example desktop applications.

High impact, low complexity
– URS, traceability, verification that the calculations as entered are the correct calculations
– Change control
– Security against unauthorized change
– Security limiting access to authorized users

3. Spreadsheet for tracking attendance at training sessions
    Low impact, no complexity
    – No functionality requiring specification or verification (basically it is a document)
    – Standard controls for a GxP document. Part 11 controls apply if GLP; 21 CFR 58.29(b) is the predicate rule requiring maintenance of GLP training records

4. Spreadsheet template for statistical analysis of toxicology studies, no macros

    High impact, medium complexity
    – GAMP category 4 approach
    – Validation plan, combined URS/FS/DS, traceability, documented testing
    – Change control, version control, back-up processes
    – Security against unauthorized change
    – Security limiting access to authorized users

5. Spreadsheet template for statistical analysis of clinical studies, including VB macros
    High impact, high complexity
    – GAMP category 5 approach
    – Validation plan, combined URS/FS/DS, traceability, documented testing, summary report
    – Change control, version control, back-up processes
    – Security against unauthorized change
    – Security limiting access to authorized users

6. Spreadsheet for statistical analysis of manufacturing data for statistical process control within validated ranges
    Low impact, high complexity
    – Abbreviated GAMP category 4 approach
    – Validation plan, combined URS/FS/DS, abbreviated documented testing
    – Change control, version control, back-up processes
    – Security against unauthorized change
    – Security limiting access to authorized users

7. Desktop database tracking disposition of printed labels
   Medium impact, low to medium complexity
   – Abbreviated GAMP category 4 approach
   – Validation plan, combined URS/FS/DS, documented testing
   – Change control, version control, back-up processes
   – Security against unauthorized change
   – Security limiting access to authorized users

CONCLUSION

The status of desktop applications at many firms is a bad news/good news situation. The bad news is that at most health science companies their desktop applications have generally been "under the radar" and have therefore received less attention to their validation needs than is appropriate. The net result is that many firms will have some catching up to do. Adding some urgency to developing a corporate plan for handling such applications is that in the absence of any policy regulators have tended to hold unvalidated desktop applications to the same standards as more complex server-based system systems, so the risk of a severe regulatory observation is higher.

The good news is that most end-user applications are quite simple and bringing them to a compliant state will not be onerous. Scalable and simplified approaches will usually be possible. The toughest part of reaching a state of compliance will probably be finding all relevant desktop applications.

Understanding the issues and evaluating compliance early can help avoid using the wrong tool. Spreadsheets, while handy, are not advisable as a platform for GxP database applications. Recognizing the risks and understanding their impact can be of benefit in developing an efficient approach to validating even the most complex desktop applications.

ACKNOWLEDGMENTS

I would like to thank my wife Jeanne for proofreading this chapter and providing suggestions for improving readability, and Guy Wingate for final review and comments.

REFERENCES

1. Available at: http://www.fda.gov/foi/warning_letters/archive/g1485d.htm.
2. Available at: http://www.fda.gov/foi/warning_letters/archive/g4452d.htm.
3. International Society of Pharmaceutical Engineers. GAMP®5: A Risk-Based Approach to Compliant GxP Computerized Systems, 2008.
4. Title 21 Food and Drugs. Food and Drug Administration. Department of Health And Human Services. Available at: http://www.access.gpo.gov/nara/cfr/waisidx_98/21cfr11_98.html.
5. EurLex. Directive 1999/93/EC of the European Parliament and of the Council of 13 December 1999 on a Community Framework for Electronic Signatures. Available at: http://eur-lex.europa.eu/LexUriServ/LexUriServ.do?uri=CELEX:31999L0093:EN:HTML
6. International Society of Pharmaceutical Engineers. GAMP®5: A Risk-Based Approach to Compliant GxP Computerized Systems, Appendix S3 (Desktop Applications), 2008.
7. International Society of Pharmaceutical Engineers. GAMP®5: A Risk-Based Approach to Compliant GxP Computerized Systems, 2008.

29 | Case Study 11: Databases

Arthur (Randy) Perez
Novartis Pharmaceuticals

INTRODUCTION

The FDA *Glossary of Computerized System and Software Development Terminology*, published by the agency in 1995, quotes an ANSI (American National Standards Institute) definition: a database is "a collection of interrelated data, often with controlled redundancy, organized according to a schema to serve one or more applications. The data are stored so that they can be used by different programs without concern for the data structure or organization. A common approach is used to add new data and to modify and retrieve existing data." Perhaps a more prosaic definition, and one that is probably much more readily understood by a large majority of computer system users, quality assurance organizations, and regulatory authorities, is simply that a database is a compilation of related data that is needed to support some activity.

The pervasion of electronic tools like the Microsoft Office Suite's Access®, Word®, and Excel® have placed the ability to build databases at the fingertips of many people who would never dream of trying to build an application using a more sophisticated tool like Oracle®. Users of such applications can make the mistake of not considering regulatory compliance. A perusal of the FDA Warning Letters and 483s from the last few years, however, shows that the agency, like other regulatory authorities, does inspect such applications.

The following list summarizes regulatory expectations that can impact databases. Many of these are specifically addressed the FDA's Final Rule on Electronic Records and Electronic Signatures (21 CFR 11) and FDA guidance that allows risk to be a factor in the selection of appropriate controls (1).

- Verification of data load process.
- Limiting computer access to authorized individuals.
- Protecting data from unauthorized modification and destruction.
- Use of authority checks to determine if the identified individual has been authorized to use the system or device, or to access or perform a particular operation.
- Changing passwords periodically.
- Use of time-outs of terminals to prevent their unauthorized use while unattended.
- Use of security measures to protect against natural system failures.
- Use of time-stamped audit trails. The audit trail provides the capability to reconstruct the data that has been modified in order to prevent the previously entered data from being obscured.
- Use of record revision and change control to maintain configuration management.
- Use of operational checks to enforce permitted operational parameters such as functional sequencing.
- Use of device (location) checks to determine whether the physical source of the data or e-signature is valid.
- Facilities for electronic signatures where required by application.

Examples of recent FDA citations for noncompliance of database applications include the following (2):

Regarding the deviation database currently maintained by the Quality Unit as an Excel spreadsheet file for monitoring the status of deviations and investigations:

- The firm has failed to put in place procedures defining or controlling the use of this database.

    The firm has failed to validate this database. [FDA 483, 2002]

Software such as Excel, Access, and Word used to create and maintain databases (rejects, complaints, and concessions) and electronic documents is not validated. [FDA Warning Letter, June 2000]

No security system to prevent unauthorized changes to computer database used to print labels. [FDA 483, 2001]

Due to the common password and lack of varying security levels, any analyst or manager has access to, and can modify, any HPLC analytical method or record. Furthermore, review of audit trails is not required. [FDA Warning Letter, 2006]

DATABASE ARCHITECTURE
Records and Fields

The common characteristic of even these simplest of databases and of the most complex of their cousins is the concept of records. A record comprises the smallest collection of related data elements that is typically retrieved by a search. Again quoting the FDA glossary, a record is "a group of related data elements treated as a unit. [A data element (field) is a component of a record; a record is a component of a file (database).]"

A good visualization is to think of a record as a line in a table, although it is not that simplistic in even moderately complex relational databases. In the Excel database cited in the preceding text, a record might consist of a name of an investigation, a product batch number, a date the table entry was created, to whom the investigation was assigned, and a status such as "in progress," "under review," or "completed." None of these data elements mean much in isolation, but when considered together they constitute an important collection of information.

As noted in the definition in the preceding text, the individual data elements that make up records are typically referred to as fields. In the example above, the fields are the name, the date, and the status of the investigation that is the subject of each record.

Database Management Systems

When discussing validating databases, it is important to distinguish between a database and a database management system (DBMS). The DBMS is the layered software that provides the tools to build and use a database. For example, Oracle and Microsoft Access® are two examples of a DBMS often used in pharmaceutical companies.

Types of Databases

The simplest kind of databases is a flat file. (FDA glossary: A flat file is "a data file that does not physically interconnect with or point to other files. Any relationship between two flat files is logical; e.g., matching account numbers.") In a flat file, searching for records is essentially a brute force task. In effect, the computer looks at all of the stored information in order to determine what records fit the criteria of the query. This is not very efficient, and is highly impractical for large amounts of data. Performing a sort on a spreadsheet, or using the Edit/ Find function is a way of doing this sort of search.

A more sophisticated database design is relational. This design makes use of defined relationships between data to vastly increase the efficiency of data retrieval. Their popularity is largely attributed to their relatively simple data model:

- Data is presented as a set of relations.
- Each relation is a data table.
- Columns within a table are data attributes.
- Rows represent entities possessing attributes.
- Tables have a set of attributes that when taken together uniquely identify each entity (a key) (3).

For example, consider the Excel table noted in the 483 referenced earlier. While it is not part of a relational database, it serves as a model for a table that could be part of it. Figure 29.1 shows how the table is organized. In this table, each investigation is an entity. There are six attributes that define the entity, but only two of these are needed to define the key by which unique investigations are identified. There may be duplicate values for attributes, but it is not possible for

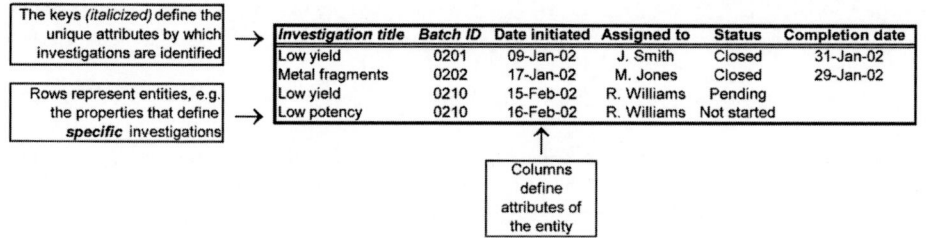

Figure 29.1 A database defining status of investigations into manufacturing discrepancies.

independent investigations to have duplicate values for all of the keys. As can be seen in this example, there are cases where the title is duplicated and cases where the batch ID is duplicated, but in no case can the title *and* ID both be duplicated.

During a search, keys can be used to combine data from this table and others, as long as all of the keys are identified and values are supplied. For example, another table in a relational database might contain information concerning the product that was being made. This might have batch ID as the sole key, so a search on investigation title and batch ID could provide all data shown in Figure 29.1 plus the identity of the product by finding the information in the two tables based on values provided for the keys.

Oracle and Access are two examples of relational database management systems. SAP®, which has thousands of tables, is an example of a large application that makes use of relational database technology. There are two other types of databases, hierarchical and network, but these are less common and generally limited to the mainframe world, and will not be discussed herein. This chapter focuses primarily on relational databases.

The kind of information managed, whether it is sales data, electronic documents, clinical trial data, or recipes for a manufacturing execution system, is fairly independent of the database type (although no one would build a flat file database for any of these). The choice of relational versus hierarchical versus network is primarily dependent on business needs.

APPLICATION DEVELOPMENT ISSUES
Validation Approach
When we speak of validating a database such as Oracle, we talk about validating the database *application*, not the DBMS. In the updated edition of GAMP®5 database managers are listed as Category 1 (infrastructure software), and validation requirements are limited to demonstration of correct installation and that they are managed under a state of control. This means that change control, configuration management, and related system management processes are paramount (4).

Many database applications can be built directly by user companies, while many others will be incorporated into commercially available applications by software suppliers. SAP, for example, uses the Oracle DBMS as its database engine.

Validation of the database application will typically be based on either a configuration approach (Category 4) or custom/bespoke software (category 5). GAMP®5 recommends the following activities for this category of software (4):

- Evaluate supplier.
- Conduct full validation life cycle.
- Record version (and configuration of environment).
- Verify operation against user requirements.
- Manage to a state of control.

User Requirements
As with any other system that must be validated, the starting point for a database is with a strong user requirements specification (URS). It is very important for the users to truly understand the data they want in the database, and the relationships between various data

elements, including the keys they want to be able to use to relate the data. Users do not need to understand the underlying design, or even the theory behind relational databases. That can be left to the designers, but users do need to be able to explain how their data are related.

It is crucial in this step to thoroughly understand and document the nature of the data, whether the database is being designed and built by an individual or by an internal IT organization, or whether a third-party package is being purchased to meet the need. Although the latter case is unlikely to involve creating the relationships between various data elements, there would probably be some level of configuration involved, such as naming and defining fields.

Database design is very dependent on being able to define hierarchies and relationships of data. For example, if you were designing a database to document GMP training, a logical way to define records would include assigning certain attributes to employees, such as name, employee number, department, and date of hire. None of these data mean particularly much if they are separated from the employee's identity (the name, or conceivably the employee number), which would be one of the keys analogous to those discussed in section "Types of Databases." However, when you put them all together you can draw certain conclusions about the training required.

The mode of data input must be specified. Will the system have to accept input from another system such as a laboratory instrument data system, or will it have to take direct input from sensors, for example, thermocouples? Will it be interfaced to another database, such as an enterprise resource planning (ERP) system that must determine a batch status residing in a laboratory information management system (LIMS) before allowing a raw material to be allocated to a manufacturing step? Will there be direct entry of data via keyboard or barcode scanner?

Required automation features of the application must be specified in the URS. Does the database need logic or arithmetic functions to populate automated fields? For example, when controlling product release a database may need to keep a status field as "quarantine" until five other fields have acceptable values, in which case the status switches to "released"; *or* until any one field has an unacceptable value, which switches status to "rejected".

In cases where not all of the data required for a record will be available when the record is created, the users must know whether this should affect the search and reporting capabilities of the system. For example, in view of the FDA rule 21 CFR Part 11, hypothesize a scenario wherein only part of the record is entered initially. When data entry is finally completed should this show up the initial commitment of the record, or as a change to the whole record in the audit trail, or should the audit trail be granular enough to show that this is the initial entry in those particular fields? Definition of this user requirement needs to be based on a combination of factors that will include the business process, risk, and any specific regulatory requirements.

The URS needs to record any special requirements for searching. For example, do the users want to be able to search on partial strings? Should searching on the name John retrieve John Smith, Edward John, and Mary Johnson, or only one of the possible subsets of this? Do they want to be able to use wildcard characters? Do they need to be able to search ranges between numerical values? Do they want to be able to refine earlier searches by applying new criteria to a prior search result?

If there are common reports that the system will be asked to generate, these should be specified in the URS. If reports need to be generated without user intervention, that needs to be noted as well.

For GxP databases data integrity issues will always be viewed as critical by regulatory authorities. It is important for those who will build the database on the basis of this URS to fully understand the implications of regulations such as Part 11. For example, deciding whether an audit trail is required and developing the design approach for it will be dependent on a number of factors on which the users can shed light. This might include considerations such as how often data will be changed after the initial record is committed, how they envision audit trails should be available on-line and printed, or whether there is a requirement for electronic signatures. It is important that quality assurance/regulatory compliance be involved in the assessment of such issues, both to ensure that the database will be designed and built in a compliant fashion and to ensure that this is done in a manner consistent with corporate compliance standards.

Functional Specifications

In the functional specification (FS), the database architect converts the information gleaned from the user requirements into a definition of what the application will actually do (but not how it will do it). A good URS is prioritized, because users tend to ask for the world, while management is only willing to fund a rocky islet. The FS must be based on, and directly traceable to, the URS. Prioritization is removed at this stage because the FS represents what will be designed. This definition of the functionality defines the majority of the basis for operational qualification (OQ).

Design Specifications

Entity Relationship Diagrams

Relational database design is certainly unlike other forms of programming, but there are tools available to help with the task. Since the relationship between data elements is a key consideration, the entity relationship diagram (ERD) is an important one of these tools. There are many conventions for documenting ERDs; the one depicted below is one of the simpler examples to understand.

The ERD is a basis for developing the database tables that define the relationships between data elements and as such is an important factor in determining how to challenge a database during validation testing. Each of the entities is defined by one or more associated tables. Figure 29.2 shows an ERD for a training database. In this scheme, each department has a one-to-many relationship with training requirements and a one-to-many relationship with employees, that is, there are many employees in a department. The employees, in turn, have a one-to-many relationship with courses taken (training history). Similarly, the training department's course inventory has a one-to-many relationship with both the departmental requirements (the training department offers many courses for each client department) and each employee's training history (each employee takes many courses). The database tables will be set up to make optimum use of these relationships. For example, the names of the courses an employee needs to take can be determined by taking the intersection of the department and the inventory, which gives the courses required by the department, less the courses already taken by the employee. As can be seen from the example, this database is extremely simple in that none of the tables will require more than three columns, yet the database will be able to track and report a reasonably complex (and important) regulatory activity.

When compiling validation documentation, it is also worth noting that visual aids are quite valuable tools during regulatory inspections. Confucius may have been thinking of this when he noted that one picture is worth 1000 words.

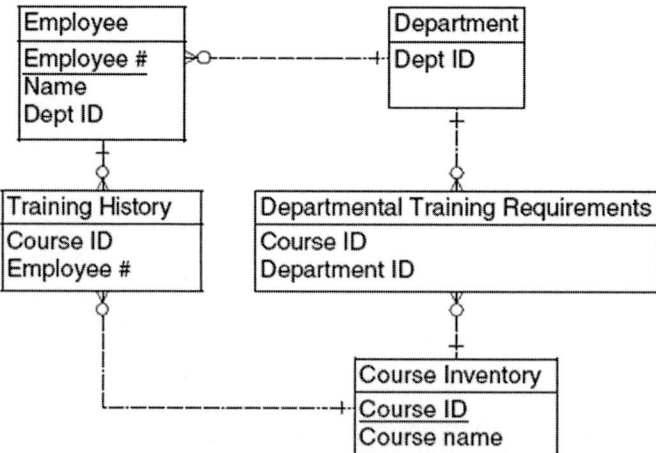

Figure 29.2   Entity relationship diagram for a simple training database (Crow's foot technique).

Field Definition

It is important that database fields be defined properly to fit the type of data expected to be entered in that field. A common point of failure for poorly designed databases is an inability to handle unexpected data. One of the aims of database validation is to demonstrate that this is not a problem. It is obviously best for the design to preclude such problems, thus avoiding heavy reliance on less reproducible factors like training and administrative processes.

An example of where field definition can improve the database integrity is simply date format, which is especially critical for data shared between organizations in the United States and other parts of the world. If a database query is entered for records between 1/4/09 and 7/4/09, should the database interpret this as January 4 to July 4 or April 1 to April 7? If the field is defined to display dates in a dd-mmm-yy format, 04-Jul-09 will be unambiguously interpreted.

Similarly, it is much more difficult to validate databases that store important data in free text fields. Searches on free text often have to have more precise parameters than is reasonable to expect, and programmers cannot be expected to anticipate every free-text query. For example, a free text search for "blue and green" probably would not find "green and blue," or even "blue and green" unless the database developers were prescient. However, if there are two fields for color, it is fairly easy to build logic that will search for both colors regardless of order. Defining fields properly can improve database performance too. Free text fields are also generally slower to search.

Prudent requirements planning and design will help to preclude user errors. If a field should accept only positive integers below 9, specifying and constructing the field in that manner may prevent a large portion of inadvertent data entry errors. Care should be taken when specifying boundary values, as this has often been a regulatory hot button. Validation efforts must include challenging this type of design feature, because systems often fail at boundary values. An example of where this is critical might be a LIMS database, where batch release is dependent on an assay of 98% to 100%. It is critical that the acceptance criteria be set knowing whether 98 and 100 are values that should be passing or failing. It is possible that only values of 99 would allow batch release if the programmer used ">" and "<" instead of "$\geq$" and "$\leq$". Boundary testing is one important fashion in which validation practitioners can demonstrate that a system does not fail (as opposed to demonstrating that it works, which is much easier.)

Finally, it is important that the design recognizes those fields without which the record is meaningless, and these fields should be mandatory. This clearly includes the keys that are used to relate the tables, but that may not be the limit of critical data. In the training database above, if you do not accurately store the courses taken our model would obviously be pretty useless, but that relates to data entry and not design, because it is possible the employee record will be set up before he has taken any training. However, the records would have no meaning without the identity of the employee (what is being tracked without this?) or without the department (there would be no record of what is required), and these are things that can be made mandatory when setting up the record. A record should not be allowed to be saved missing such critical information. Validation testing must verify that there are no meaningless records in the database.

Data Input Interfaces

The design must account for the origin of the data going into the database. There are few issues for systems whose only input source is keyboard, but as systems grow more electronically integrated this becomes less common. An important consideration when designing a database that is to be validated is ensuring that incoming data is of adequate integrity. The GI-GO philosophy (garbage in—garbage out) applies in the world of validation as well. However, it must be recognized that as integration of systems becomes more commonplace it cannot be assumed that all input to validated systems will originate from validated systems. For example, consider a certificate of analysis for sodium chloride that is transmitted directly from the supplier to an ERP system. If a company were to demand that a supplier of salt validate its systems, they would probably have to explain the concept of validation, after which the supplier would enjoy a hearty laugh. The highway department probably uses more salt in one winter than a pharmaceutical company will use in several decades. This does not mean that the supplier's certificate of analysis is not trustworthy. Rather it means that there needs to be some research to determine data integrity.

Interface design should have some checks to ensure that data transmitted electronically meets expectations for completeness. For example, if one were using some electronic device to record attendance at a training session, for example, an ID-card reader, it is imperative that the incoming record of attendance be associated with a course ID. Validation must verify that such transfers work properly.

Audit Trails

In view of 21 CFR Part 11 audit trails have become a standard part of many database designs. When specifying any database that will include GxP data, it is important to recognize that specifications for Part 11 compliance must go beyond a statement that the application "must be Part 11 compliant." It is also important to recognize that unless a predicate rule specifically requires an audit trail the 2003 FDA guidance on Part 11 also provides leeway for a user company to define whether one is required. This decision is based on risk and should include consideration of:

- Risk to public health:
  - Is the data so sensitive that patients could be placed at risk in an inappropriate change were not recognizable?
- Risk related to the business process:
  - Is it necessary to recognize the change history to ensure integrity of the process?
  - Are operators of the system making frequent changes?
  - Are there multiple system operators who might make changes to a record?
- Risk related to supporting processes:
  - Would an audit trail be a critical part of the investigation of process deviations or some other ancillary process?

Not all designers or suppliers of commercially available solutions will have intimate familiarity with the regulation, and even those who think they do may have a different concept of compliance than the user company. Ergo users must be specific in their expectations as to what they deem to be a compliant solution. If an audit trail is required, key considerations to ensure compliance should include requirements to:

- Ensure that all data entries, modifications, or deletions are identified with the user, date, and time of the action. This must be based on a good understanding of the underlying predicate rules. For example, there may be a requirement for a motivation field in addition to the information noted earlier [GLP regulations require this reason for change (5)].
- Ensure that modifications or deletions do not obscure any of the previous values for the changed fields.
- Ensure that the audit trail is irrevocably tied to the record and will be retained for as long as the record.
- Ensure that reports can be generated for regulatory review, both electronically (to screen and file) and on paper.

Validation testing must demonstrate that the audit trail as implemented successfully meets the documented Part 11-based audit trail criteria.

Electronic Signatures

If a database is to employ electronic signatures, Part 11 is again a guide for design. Key considerations that should be recorded as requirements and challenged in validation testing are:

- In all displayed manifestations of the signature (both on-line and printed) the signature must display the name of the signer (not just a user ID), the unambiguous date and time of the signing, and the reason for signature (e.g., approval, executed the task, etc.)

- The signature must be irrevocably linked to the signed record. It cannot be excised and applied elsewhere, and it must be invalidated if the record is changed subsequent to the signing.

Validation planning must ensure that these points are documented and challenged during testing.

Security

Databases comprise a class of application that often requires multiple levels of security. For example, it is possible that a business process may necessitate keeping the ability to modify existing records distinct from the ability to enter original records. Large database applications like ERP systems have many roles defined, and virtually no one should be able to enter, manipulate, or delete data across the whole system. Role-based security schemes are required, where appropriate, by Part 11. In any case, general users should never have the same level of access and edit/delete privileges that a database administrator would have. All access levels need to be challenged in validation testing. Role-based security may be built upon the ability to access certain tables and views in the database, and this may be a reasonably complex mechanism, so understanding the database design is quite valuable in developing a validation strategy for security.

Testing and Qualification

Test Planning

Validation test planning for a database application has the same two principal foundation blocks as for any other type of application: the traceability matrix and a risk assessment process. The traceability matrix is a tool that both ensures the test plan challenges everything that needs to be challenged, and properly maintained, and enables a firm to demonstrate to regulators that each current specification has a corresponding successfully executed test.

A judiciously applied risk assessment process is an important tool that can provide essential guidance at a number of key project junctures. It may be appropriate to use a variety of risk assessment techniques in one project. For a good example of one of these techniques see *GAMP*®5 (4).

The first risk assessment, and generally a very easy one to execute, is an assessment in conjunction with user requirements analysis that determines whether the database has any GxP bearing and thus requires validation.

If packaged solutions are being considered, another assessment should be conducted prior to selection of the supplier. Much of this assessment will be based on results from a supplier evaluation. In addition to a critical look at supplier quality systems it pays to understand the database design process. Sometimes, in an effort to cut costs or meet tight timelines a supplier may move from one DBMS to a newer one (e.g., DB2® to the most recent version of Oracle) but not update the design. Such a practice can even go back more than one generation of the application. This may manifest in problems of incomplete compatibility and lead to such troubling problems as orphan data after deletion, etc. Especially in view of Part 11, this can lead to questionable data integrity. Such risks, and the potential associated increase in the complexity and the amount of work required for the validation, should be carefully considered. The rigor and extent of validation testing is one lever that can be applied to the problem of a poor supplier quality system. (The same issues would of course apply if similarly dubious practices were employed in an internal database design project.)

A third level of risk assessment enables validation planners to justify the extent of testing. First a determination of functionality that is actually critical to patient safety (as manifested in product quality and data integrity) is required. Only these critical to quality (CTQ) functions require further assessment. Functions that are not CTQ should be tested according to expectation of good practice.

The next level of assessment should look at each of the CTQ functions of the database and assess them for likelihood of failure (based on a combination of design quality and complexity, user ability, and frequency of use); failure consequence (based on patient or worker safety and regulatory and/or business impact); and the likelihood of detection before serious consequences arise. Mitigation strategies, which often result in increased or decreased depth of testing, can then be developed. Other strategies that might be considered include (but are by no means limited to):

- Enhancing automated error checking processes for critical database entries, such as by requiring a data entry confirmation or building acceptance criteria logic into critical fields
- Requiring a confirmation of critical entries by a second operator
- Developing procedural checks, etc.

Any of the above strategies will have an impact on test planning, as they all are intended to reduce unacceptable risk to a tolerable level. Regulators have consistently demonstrated that they believe that there needs to be demonstrable evidence that everything that should have been tested has been, and that the depth of testing needs to be justified. The traceability matrix provides the former, and a sound risk assessment practice satisfies the latter.

These tools should not be considered limited in application to the initial validation effort. By keeping the traceability matrix up to date it becomes (and will remain) an important tool for assessing the impact of changes to the database, and both it and the risk assessment process are still important test planning tools as part of change control. Of course, sound change control practices are absolutely imperative for keeping an application validated.

User Acceptance Testing
User acceptance testing (often equated with operational and/or performance qualification) of the application is also required against the functionality of the application's specification.

- Reports
- Calculations
- Data entry processes
- Search processes
- Logical access to the application
- Role-based security challenges
- Back-up and recovery processes
- Archive and restore processes
- Interfaces to other systems

There should also be testing within the business process, challenging the application in terms of everyday usage patterns.

Traceability

As with any other type of application, one of the keys to developing adequate testing will be thoroughly traceable specification documentation. Database design must be built from the FSs, and the FSs must be predicated on the user requirements. Beyond addition to general adherence to this fundamental concept, significant attention must be paid to the details as to how each element of design relates to both the previous and the next level of specification. Figure 29.3 shows an example of how the design of the training database module discussed previously might be derived from a simple requirement to report on outstanding training needs. This single requirement leads to a set of four FSs, which in turn are elaborated into many more design elements. As noted in earlier understanding the design is important for developing test challenges, and a good traceability matrix is an aid in this. The matrix should provide transparent traceability from URS through FS and DS to actual test cases.

The above scenario is fairly readily applicable to bespoke systems, but for commercially obtained systems traceability will often go from URS to configuration (as applicable) to test.

In both of the above cases, the critical validation consideration is traceability of the CTQ functionality. It is good practice to have comprehensive traceability, but it is obviously most critical where there is the most risk.

Specification ID	Specification
URS-1	System must be able to document training required by each employee
FS-1	System will track all departmental training requirements
FS-2	System will track employee department
FS-3	System will track courses taken
FS-4	System will be able to report difference between courses taken and courses required
DS-1	Employee table defines name and employee number
DS-2	Department table defines Department ID
DS-3	History table includes employee number and Course ID
DS-4	Requirements table lists Department ID and Course ID
DS-5	Training inventory lists Course ID and Course name
DS-6	Department to employee one-to-many relationship
DS-7	... and so on

Figure 29.3 Traceability example: all of the functional specifications and design specifications in this table are derived from the user requirement labeled URS-1.

Documenting traceability requires significant rigor in order for a traceability matrix to be useful. Further, the traceability matrix must be maintained throughout the life of the system to support change control. Using manual methods such as the table in is barely manageable for a small application like this training database; for larger systems it becomes nearly impossible. Fortunately, there are a number of automated solutions to this problem available.

It should also be noted that the ERD is an important part of the design specifications. It would be conceivable that the design specification referencing URS-1 and FS-1 through FS-4 would be the ERD and some field definition information such as defining the employee number as a positive six-digit integer. The important consideration is that a programmer be able to unambiguously interpret whatever DS is provided to build the right database.

The key to developing and maintaining traceability is making certain that there is a process to ensure that changes to any of the referenced documentation is evaluated for potential propagation to other levels of specification. Developing and adhering to this process is an important validation activity and a regulatory expectation. The activity of planning validation testing should be built around such a matrix.

OPERATION AND MAINTENANCE
Backup/Recovery
Given Part 11's orientation toward data integrity, the FDA has stressed in Warning Letters that they consider backup to be an integral part of protecting data integrity. For databases, the principal concern is protecting them against corruption, which can result from a variety of causes. When validating databases two points need to be kept in mind:

1. The back-up process needs to be defined, and most importantly it needs to fit the business scenario. For example, if a database is used very infrequently, let us propose hypothetically only in the month of January to close out year-end activities, it is inappropriate to use a process of monthly backups. This is because data centers typically retain only about four back-up copies, overwriting the oldest copy when it is time to do the fifth backup. In such a scenario, if this database became corrupted in March the last good back-up copy would be overwritten in July, but the corruption would not be discovered until the following January, at which point the data would

be irretrievably lost. A much better strategy is to do annual backups in February, saving them for several years. It saves the IT department work, it saves the expense of storing worthless copes, and most importantly it provides a much more reliable data protection scheme.

Similarly, if a database were extremely heavily used daily, monthly backup would be inadvisable because failure in the third week would force the restoration of a copy missing a large amount of data. There are back-up strategies that can adapt to such scenarios while minimizing the burden on the IT group, such as daily incremental backup (backing up only what has changed from the previous day) with weekly or monthly full backups.

Even scheduling needs to fit business processes. For example, if the business process calls for large batch jobs to be run overnight, and IT is counting on nights to run backups, this issue needs to be addressed.

2. Backup and restore must be tested, and this testing too needs to be within the scope of the business process. Quite frequently validation teams simply note that their application and data are on machines that are supported by the data center, and that these are already within the scope of existing back-up processes. By not testing the restore process they are exposing themselves to potentially unpleasant surprises when a recovery of backups becomes necessary after the system is placed in production and the process does not work as expected.

When changes are made to a database that may affect the back process, the change control process needs to ensure that back-up processes are adjusted accordingly. Business customers will not be happy if they request a recovery operation and they are told that an empty directory has been backed up for a year because files were moved without telling the storage management team.

It is important to remember that by their very nature databases are constantly evolving. This gives a slightly different flavor to decisions regarding back-up strategies that would be the case with a system like a chromatography data manager, where the data is for the most part static.

Archive and Restoration

While they are often lumped together in discussion of system management issues, it is important to understand the difference between backup and archival to understand how each relates to validation and maintaining the validated state. While some companies do it, it is the wrong concept to retain back-up copies for the length of the archive period. It is also terribly inefficient, since the back-up tapes will have the application and operating system in addition to multiple redundant copies of any data that needs to be archived. Figure 29.4 shows a comparison of the properties of a backup versus archival.

As discussed in chapter 15, Part 11 has made archival problematic, especially in the world of preclinical and clinical data management where retention requirements may amount to decades. This means that it is virtually impossible to avoid archiving data, if only because it is impractical to keep obsolete hardware and applications running ad infinitum.

In this light it is imperative that the archival strategy and validation effort for a database consider metadata. If metadata is incompletely copied, records restored from archive will not be properly retrievable and/or reported. Ergo it is imperative that the database functionality be challenged again with restored records after it has been found acceptable with "normal" data.

Change Control

The principal issues around keeping a database validated are in essence the same as for any other type of applications. Change control procedures need to ensure that all changes are assessed for impact on the database (and interfacing systems). Decisions regarding the extent of testing should be based on a risk assessment.

Properties of Back-up	Properties of Archival
Periodic copying of the data, applications, possibly even the operating system	1. Periodic copying of data 2. Retention of old versions of application software
Intent is to protect system against unforeseen problems by retaining an image that can be recovered after problem resolution.	1. Intent is to remove low-value data from the system (to provide long-term protection of the data and possibly enhancing database performance and usability) 2. Intent is to retain obsolete software versions in case data needs to be reconstructed and this cannot be done on a newer release
Short term storage of full copies	Long-term storage of selected data / programs
Backed up data stays in the live system	Archived data deleted from live system
Media often recycled; few worries over media life	Media life a critical concern

Figure 29.4 Differences between archival and backup.

Infrastructure and Layered Software

To be considered validated, a database (or any other application) must be running on a qualified infrastructure. This includes servers, and as applicable, network(s) and workstations. For an infrastructure to be considered qualified, the support organization must have current, approved documentation describing its configuration, and evidence that demonstrates that it has been properly built and appropriately challenged. There must be a reasonable, documented, and approved mechanism for handling change and problem resolution. Compliance requirements for IT infrastructure are discussed further in Case Study 15 (chapter 33).

Security

In addition to the role-based security described earlier it is necessary that operating system level security be enforced. It does little good to have sophisticated application level security if a user can access a controlled directory through the operating system and employ standard tools to modify or delete data. This means that this level of security should be included in planning for the security-oriented validation testing.

Especially in electronic signature databases, passwords must be controlled, enforcing periodic renewal. This process must not be too frequent, however, else users may try to simplify their lives by using inappropriately simple passwords, or worse, writing them down. A good practice is to control user IDs should not be recycled after a user leaves the company. This helps to ensure that all ID/password combinations are unique. If password aging is managed by the application, this needs to be verified during testing. If it is handled administratively, one of the activities the validation team needs to plan for is the verifying that the procedures have been developed and properly implemented.

DECOMMISSIONING

Decommissioning databases usually entails some decisions regarding the fate of the data within the database. Regulatory, legal, or business concerns often require retaining the data past the time when the cost-benefit ratio justifies keeping a database active. This may mean expending considerable effort migrating the data to another database or to a format that can be handled by a generalized archiving tool (see sect. "Migrating Databases").

Once the decision has been made regarding what to do with the data, a formal decommissioning notification should be prepared and signed by IT, the system owner, and by quality assurance. Two potential scenarios merit further discussion:

1. If the data is indeed no longer needed, the decommissioning letter should note this. Once the documentation is complete IT should delete all instances of the application, all copies of the data, (including archives), and all supporting documentation (including validation documents). Users should destroy relevant documentation. Firms must remember the costs and implications of legal discovery processes if they are tempted to retain data that should really be destroyed. The decommissioning letter should be retained in accordance with appropriate legal and regulatory expectations.
2. If the data is not being destroyed because there are business regulatory, or legal reasons to retain it, where data will reside and any special tools or procedures required to access it should be noted in the decommissioning letter. The letter should be retained with the validation documentation until such time as the data can be destroyed in accordance with the guidelines in the preceding paragraph.

MIGRATING DATABASES

Migrating databases becomes a major hurdle for companies using electronic record systems when that data needs to remain electronically processable. Given the rapidity with which technology becomes obsolete, it is naive to assume that database systems that are state-of-the-art today will exhibit any greater longevity than their older siblings. Under this assumption, there are two logical routes to retaining electronic data. The concept of retaining obsolescent hardware and software can be rejected out of hand because of the expense of retaining it. The only realistic alternative, often unattractive in its own right because of the complexities involved, is migrating old data to new database applications. *GAMP®5* includes an appendix discussing data migration (4).

It is important to preserve or translate as much of the original form and format of the data as possible, and that includes metadata. A database can contain a tremendous amount of metadata such as audit trail information, electronic signatures, relationships between database tables, definition of field characteristics, etc. It is necessary to consider this metadata as part of the data set; failure to do so could inhibit searchability and reporting after migration, make modification of old records problematic, or possibly even result in loss of the integrity of the records. Validation tasks associated with data migration must be geared to demonstrate that neither of these circumstances prevails.

Validation testing for migrated records should include testing where similar examples of migrated and freshly entered data are challenged in a similar fashion and the test results compared. It can be the case that metadata is lost or otherwise affected during the migration, in which case otherwise identical records may behave differently in such tests.

Finally, firms intending to migrate data must remember to include already-archived data in their migration plans. It is possible that complications may arise with this archived data. For example, data archived from earlier software releases of the database might be readable through the current version, but it could be that minor differences in metadata could render this older data unreadable after migration if these differences are not specifically addressed in the migration process.

CONCLUSION

The principles of validating databases are essentially the same as they are for any other computer system. Key issues are having good user requirements, developing traceability while generating the functional and design specifications, test planning based on risk assessment, documenting everything thoroughly, and maintaining that documentation to reflect the current state of the system after it goes live.

However, understanding how the database is designed and keeping it in mind throughout the project will have a significant impact on the effectiveness of testing and on the ease with which the system can be supported after implementation. Extending this understand to how metadata affects the records will make maintenance and eventual retirement a smoother operation.

ACKNOWLEDGMENTS

I would like to thank the following people for their contributions to the preparation of this chapter: my wife Jeanne for proof reading and suggestions for improving readability, Philip "Flip" Rutledge for help with the finer points of entity relationship diagramming, and Guy Wingate for final review and comments.

REFERENCES

1. Guidance for Industry: Part 11, Electronic Records; Electronic Signatures —Scope and Application, FDA, 2003.
2. Warning Letters issued by FDA since 1997 can be found on their website at http://www.fda.gov/foi/warning.htm. While there are many examples related to back-up and most do not mention databases specifically, the following examples all serve to illustrate Agency concern with the concept of back-up as a critical part of guaranteeing data integrity: Apheresis Technologies, November 1999—In relation to a spreadsheet used as a database: "There is no documentation covering Excel application software, or any procedures instituted covering the protection of electronic records or *an established back-up system*." Glennwood LLC, May 1999—Regarding a chromatography data system (which includes a database of chromatography results): The software allows for overwriting of original data. There are no written procedures for the use of passwords, levels of access, *or data back-up*. Hydro-Med Sciences, February 1999—"There are *no procedures for backing-up data* files and no levels of security access established."
3. Barman D. "Dilip's Brief Introduction to Relational Databases." Available at: http://www.cs.unc.edu/Courses/wwwp-s98/members/barman/databaseLesson/.
4. ISPE. GAMP®5: A Risk-Based Approach to Compliant GxP Computerized Systems, 2008.
5. FDA regulation 21 CFR Part 58.130(e) states: "...Any change in automated data entries shall be made so as not to obscure the original entry, shall indicate the reason for change, shall be dated, and the responsible individual shall be identified."

30 | Case Study 12: Electronic Document Management Systems

Robert Stephenson
Pfizer

INTRODUCTION

Document management covers the preparation, review, approval, issue, change, withdrawal and storage of documents. Good practice (GxP)—for example, as described in GAMP®5 Appendix M9 (1)—must be applied consistently throughout the full document life cycle for a regulated organization to be able to demonstrate the level of control now demanded by regulatory bodies. Electronic document management systems (EDMS) are a class of applications specifically developed to support these requirements. This paper describes an approach to successful implementation of such systems.

What is a document? The IT Infrastructure Library® (2) defines a document as "information in a readable form." This provides a helpful distinction between a "record"— which may be just a value contained on a database table—and a readable "document," but is not sufficiently detailed to further inform this discussion. More helpfully, the International Society for Pharmaceutical Engineering (ISPE) (web-based) glossary (3) defines "documentation" as "any written or pictorial information describing, defining, specifying, and/or reporting of certifying activities, requirements, procedures or results."

This definition immediately feels closer to the intended meaning of the term used so widely throughout the current GxP regulations and associated guidances, which determine how activities within the healthcare sector are performed. It is these types of documentation that are primarily intended to be in the scope of use of any EDMS.

Documentation is inevitably a critical asset to any organization as it can contain critically important information that, if lost or corrupted, may not be able to be recovered. The security of information may also be a consideration to protect the intellectual property of the organization from unauthorized access. A "holistic" quality management system (QMS) approach within regulated organizations, such as described in International Conference on the Harmonisation of Technical Requirements for Registration of Pharmaceuticals for Human Use (ICH) 10 (4) also requires effective and demonstrable controls over documentation such as those offered by EDMS.

Traditionally documentation has meant paper. For GxP documentation, legacy systems have been developed to manage paper and ensure that the right pieces get to the right place at the right time. These have been generally based on the multicopy approach with controlled copies being distributed to known locations and being withdrawn as required. This approach is likely to be expensive, time consuming and error prone; paper copies can be easily lost or damaged and strict controls are required to ensure that controlled copies are kept consistent with each centrally held master copy.

The problems associated with the management of paper can be overcome by implementing a comprehensive and effective EDMS.

ELECTRONIC DOCUMENT MANAGEMENT SYSTEMS

EDMS are, therefore, concerned with the management of the life cycle of specific information, which can be presented in human readable form. EDMS are designed to control and retain documents throughout their life cycle—from creation to archiving or destruction, and at all stages in between.

A word processing package used only to prepare a document for use in "hard copy" paper format would not in itself be considered to be an EDMS. However, if the same word processing package was integral to a system in which a document is created, reviewed, approved, viewed,

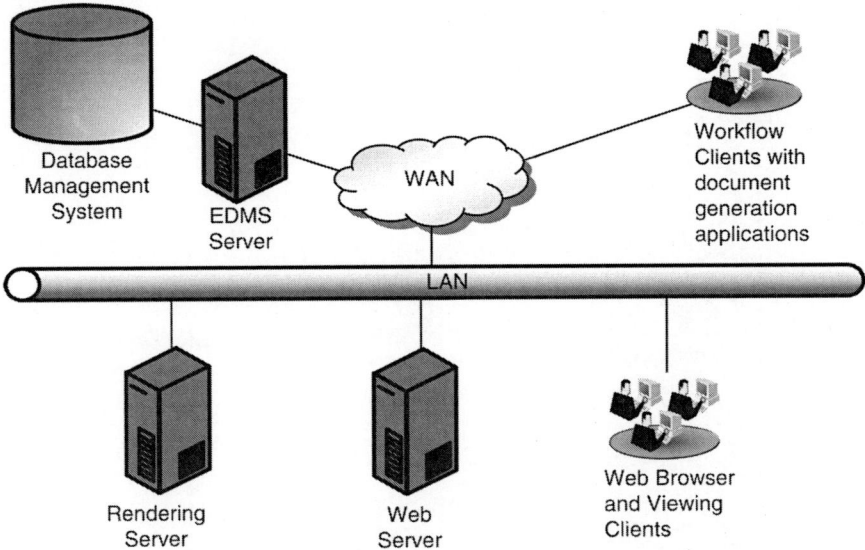

Figure 30.1 Example of a system architecture.

superseded, and archived, then it would be a significant component of that EDMS. However, as indicated by the ISPE definition above, the term document cannot be restricted to the output of a word processing package alone since documents can now contain content in a variety of formats including diagrams, spreadsheets, pictures and video. Similarly documentation can exist on a wide variety of media, including hard copy, electronic, magnetic and optical.

A search of the Internet will quickly identify many proprietary EDMS solutions. Some vendors have particularly geared their products toward the pharmaceutical industry and have built in much of the functionality required to meet GxP regulations. For this reason EDMS applications are generally designated as configured products/GAMP software category 4 (1), which means that configuration of the system based on specific customer requirements will be required. This process is discussed in more detail in later sections. For some applications customization of code may also be required which will require a more rigorous approach to specification and verification activities appropriate to GAMP software category 5.

The needs of the organization and the agreed scope of use of the system will determine the type of EDMS to be implemented. Systems can be local or, more commonly, distributed throughout the organization using local area networks (LANs) and/or wide area networks (WANs) to maximize the sharing of information and benefits. This distribution is increasingly organization wide. Achieving network compliance is covered in Case Study 15 (chapter 33) of this book.

Increasingly web-based interactive EDMS packages are available, although older client-server based systems are still in use. Figure 30.1 provides an example of EDMS system architecture for a multiuser distributed system.

EDMS can be configured in many different ways to support the way documents are managed. Figure 30.2 shows an example life cycle for a document such as a standard operating procedure.

THE REGULATORY ENVIRONMENT

All Regulations governing the discovery, development and manufacture of healthcare products demand that appropriate procedures and records are in place. Regulatory inspections use this documentation as the primary source of evidence of compliance or otherwise. However the regulatory environment is currently in a state of transition, the broad context in which many regulations are being reconsidered and revised is a desire [initially expressed through the FDA's "Pharmaceutical Current Good Manufacturing Practices (cGMP) for the 21st Century: a Risk-Based Approach" initiative (5)] to enable pharmaceutical manufacturers

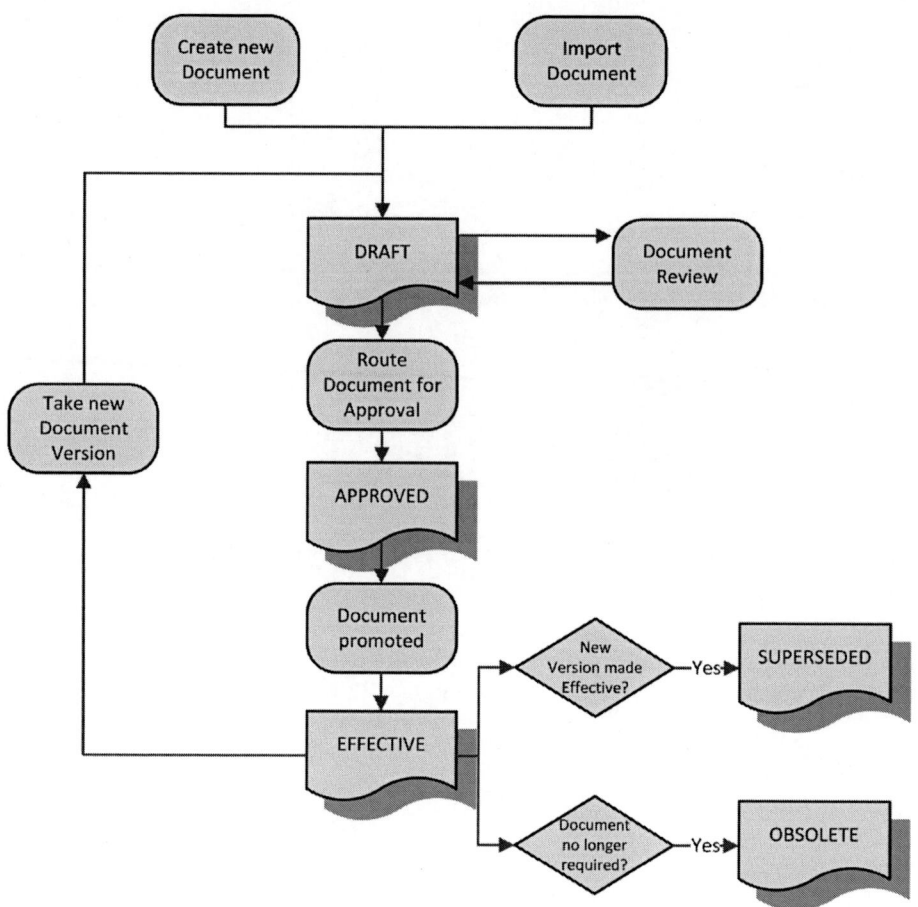

Figure 30.2 Example of document life cycle.

to be more innovative and to be encouraged to adopt modern quality management techniques and systems into their operations, including the application of "state-of-the-art" pharmaceutical science and risk-based approaches. This initiative means that new classes of documentation such as quality system documentation and risk assessments may become subject to regulatory review and, as a result, the scope of use of EDMS may need to be extended to cover them.

Although only applicable to systems subject to inspection by the FDA, the most influential regulation specifically relating to electronic records (and therefore to all regulated documentation maintained in electronic form) is FDA 21 CFR Part 11 (Electronic Records; Electronic Signatures) (6). The rule, issued in 1997, drove a "step change" in the functionality of electronic systems with respect to record integrity controls and the need to achieve "irrefutable" electronic signatures when applied to electronic records (i.e., legally equivalent to handwritten signatures applied to paper documents). Unfortunately, 21 CFR Part 11 also had the unintended consequence of triggering many costly "assessment and remediation" programs throughout the industry which were, in retrospect, of limited benefit.

At the same time that the FDA announced their major review of the cGMP regulations in 2003 a "Part 11, Electronic Records; Electronic Signatures—Scope and Application" document (7) was issued which contained "non binding recommendations" intended as a corrective measure to bring a more reasonable and considered approach to the application of the 21 CFR Part 11 regulation. This initiative was taken in advance of a promised comprehensive revision of the rule. While the first objective has been achieved, at the time of writing, the revised 21 CFR Part 11 rule has yet to be issued.

Other regulations of relevance to EDMS include the Eudralex Rules, specifically "chapter 4: Documentation" and "annex 11: Computerised Systems." The European regulators have drafted revisions to both chapter 4 and annex 11 (8) "in response to the increased use of computerized systems and the increased complexity of these systems," which have generated a great deal of comment during public consultation from April to October 2008.

When the revised regulations will be published and what they will contain remains uncertain but it is hoped that regulators will adopt a consistent approach which clarifies what controls are expected from computerized systems (including EDMS) but does not define specific technical solutions which may be short-lived and put unnecessary constraints on the industry and its suppliers.

Many of the requirements explicitly required by part 11 are implicitly stated in European and other non-U.S. regulations and directives. Current thinking by international regulatory agencies is presented in the influential PIC/S guidance document, Good Practices for Computerised Systems in Regulated "GxP" Environments (9).

EDMS IMPLEMENTATION

Implementing an enterprise wide, intersite or site-wide EDMS is a major undertaking requiring significant investments of time, resources and money. Existing document management practices must be carefully analyzed to determine what the critical business requirements are. An EDMS offers the capability to streamline document management processes and procedures across diverse business areas. A major project initiative with wide representation following a well-structured life cycle approach will most likely be essential to ensure that the benefits to the organization are maximized. Opportunities for the EDMS to be interfaced with other systems to deliver an integrated set of business solutions should also be considered.

Life Cycle Approach

A comprehensive risk-based life cycle approach to the development, implementation and support of the system should be adopted to ensure that a fully acceptable EDMS is delivered to the organization with effective and efficient use of internal and external resources. A life cycle model such as that described in GAMP®5 (1) is strongly recommended. This divides the life cycle into phases:

$$\text{CONCEPT} \rightarrow \text{PROJECT} \rightarrow \text{OPERATION} \rightarrow \text{RETIREMENT}$$

Concept

Following the principles contained in GAMP®5 it is important that the output of the initial strategic overview of the project is documented in a suitable concept document. The concept document should be at a sufficiently detailed level to provide the decision makers in the organization with an overall understanding of what will be delivered and what level of business commitment (resource and funding) may be required to deliver it. For example, the "concept" of an EDMS could vary from providing a simple system to control standard operating procedures at a single site to a major initiative such as developing an enterprise wide document management solution intended to manage all electronic records, whatever their format (text, picture, video). The concept document must be suitably endorsed before detailed project work commences.

An initial risk assessment should be performed, either at this stage or early in the project phase, to establish and document the overall GxP determination for the system. Given the critical role of documentation in demonstrating regulatory compliance it is anticipated that the EDMS as a whole will be deemed "GxP regulated" and that more detailed risk assessments will be required during the project phase to determine functional risks and identify required controls.

Project

Following the agreement of the concept the implementation of the EDMS moves into the project phase.

Project Team
A project team should be appointed. The project team should consist of a core group of representatives from the key project areas. This should include the following:

- A project sponsor who ensures that the team is resourced appropriately
- A project manager to coordinate team activities and provide progress reports to the sponsor and senior team
- User group representatives who will be subject matter experts (SME) in the documentation practices and needs of the organization
- Quality unit representatives who will assure quality and regulatory compliance throughout the EDMS project
- IT systems representatives to provide technical expertise (EDMS and Infrastructure)

This team may be supplemented by SME from external suppliers once the project is underway. The project team should meet regularly to review quality practices and set up regular communication sessions with the wider user base to keep them aware of all facets of the project and to ensure that any critical reviews and decisions can be made in a timely manner.

Specification and Verification
One of the major factors to be considered when implementing a GxP compliant EDMS is how to address the specification and verification of the system. An approach based on the life cycle approach described in GAMP®5 (1) is appropriate. Figure 30.3 shows how specification and verification documentation relates to typical project activities. For those EDMS functions determined to be GxP-relevant validation activities will be required as described below.

Specification activities
User requirements specification A comprehensive analysis of requirements is a very important first step in the project as it forms the basis for the definition of the system, supplier selection and the approach to verification and validation. It should be the mechanism by which the users have the opportunity to express their needs. Prioritizing needs into, for example, mandatory (high priority), beneficial (medium priority), and "nice to have" (low priority) provides the project team with an indication of how to weight the requirements during supplier selection. The output of this step will be a detailed requirements document or user requirements specification (URS).

Examples of EDMS-relevant items covered in the URS are as follows:

- *Operational requirements*
  - Functional/regulatory
    - Document creation
    - Document management throughout its life cycle
    - Ability to update, withdraw or archive documents
    - Ability to route documents for review and signing
    - Ability to take new versions/retain old document versions
    - Ability to print or prevent the printing of controlled documents
    - Access and permission controls based on job functions
    - Audit trail requirements
    - Use of electronic signatures
    - Signature manifestation
    - Signature/record linking

  - Technical
    - Access control and security requirements
    - Speed of access to the system and document retrieval times
    - Presence or absence of hypertext linking

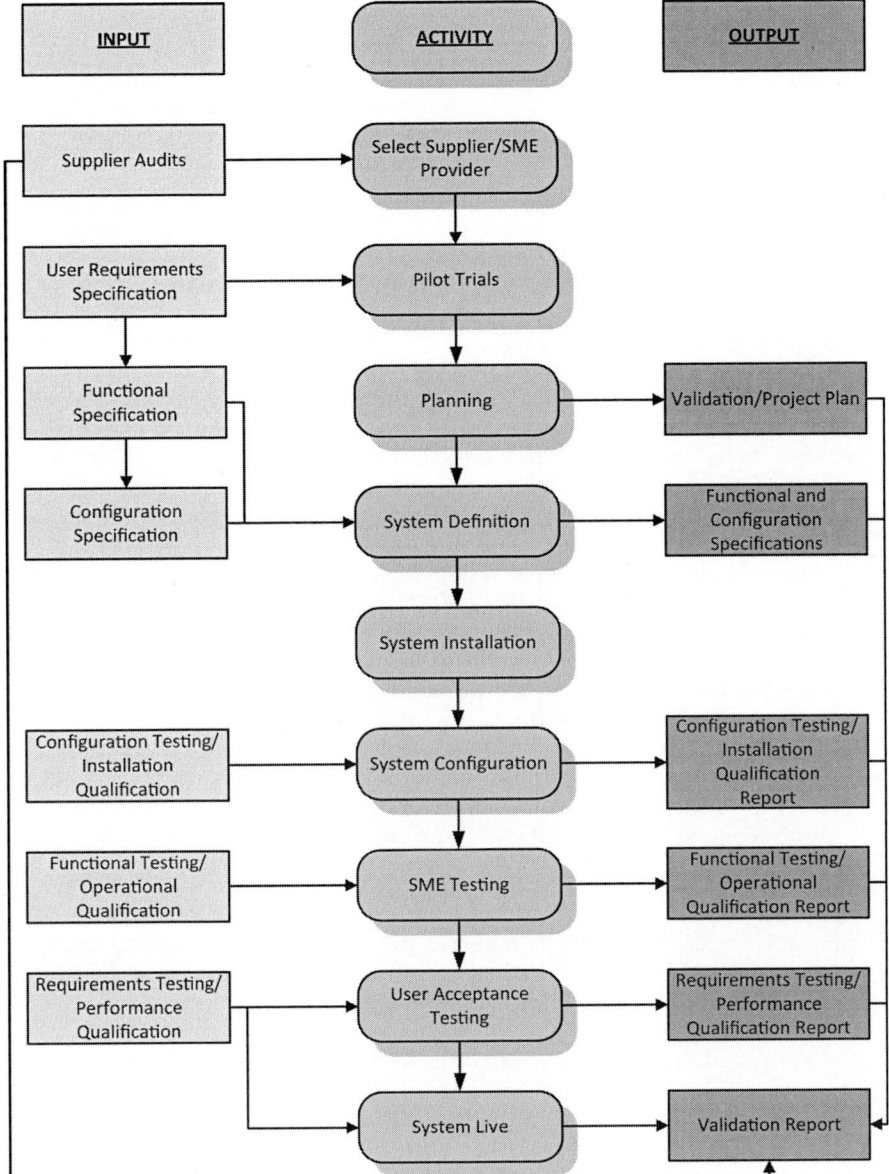

Figure 30.3 Project phases with validation deliverables.

- o Backup and recovery
- o Archive requirements
- o Disaster recovery

- Interfaces
  - o Human machine interfaces (HMI) required and their design
  - o Interfaces to other computer systems

- Environment
  - o Physical security
  - o Impact on existing IT infrastructure

- *Constraints*
  - ◦ System availability requirements
  - ◦ Need to transfer legacy data
  - ◦ Future use/growth of system
  - ◦ Potential number of users of the system

The URS should then form the basis of the evaluation criteria for suppliers of EDMS.

Selection of a supplier/subject matter expert provider   In large multisite organizations the system to be implemented may be governed by a corporate standard. This gives advantages both at the implementation stages and throughout the system's life in terms of the potential availability of individuals with knowledge of the product within the organization—SME. This may simplify the selection of project team members and help to reduce the overall project cost.

Where a new supplier is being selected a number of factors need to be considered and a detailed supplier selection process may be undertaken. Evaluation may consist of gathering information from various sources such as the suppliers themselves and companies who already have the EDMS installed. Trial demonstrations and evaluation visits should then be arranged. Where possible a fixed duration trial (pilot) should be set up on site for the project team and user representatives to run through some scenarios of how the system may be used (see below).

A supplier audit is essential to determine if the system has been developed (and will continue to be managed) within the framework of an adequate QMS and to good software engineering standards. A positive supplier audit gives the user confidence that the GxP critical functionality of the system can be validated and that, once implemented, the supplier's activities will not adversely affect maintenance of a demonstrable state of control throughout the operational phase.

The supplier selection process may result in the selection of a system that requires customization for it to fully meet the URS. If this is the case then a systems integration partner with the ability to provide external SME may be required to support the implementation. Where appropriate the QMS of the systems integrator should also be audited to ensure that their methodology will not compromise the validation approach.

Pilot trial   It is recommended that a short trial of the most probable supplier's system is piloted with the involvement of a representative cross-section of the user community. A pilot system is, by definition, temporary and will not necessarily be built to the same standards as the actual EDMS, however it should be configured to meet the URS requirements as closely as possible. The pilot trial is an important preparatory step as it should

- Provide "hands-on" evidence to the users that the system will meet their requirements as expressed in the URS,
- Allow the team to form a better understanding of what the EDMS in question will provide,
- Assess possible configurations,
- Identify any potential functional shortcomings or technical problems, and
- Expose the users to the system interfaces, thus allowing any problems in this area to be highlighted at an early stage.

Pilot Trials can also be used to compare possible suppliers. The decision to stage trials of more than one system should not be taken lightly since conducting a useful pilot requires considerable resources and time on the part of the team, plus the setting up of a pilot system by each supplier may involve a significant cost.

The completion of the pilot trial phase will result in a decision to continue or abandon the EDMS project. If the decision is to continue then the URS may be further refined in light of the experience gained, both adding and deleting functionality and expanding detail as necessary. It must be remembered that changes to the URS will affect the supplier's quotation, which must be resubmitted on the basis of any revisions.

Planning   Completion of the pilot trial phase is an appropriate time to prepare a detailed validation plan for the project. There is a case for preparing a version of the validation plan at the start of the pilot but the effort may be abortive if the project is subsequently abandoned.

Completion of the validation plan will usually accompany the preparation of a detailed project plan. Quality unit representatives on the project team should ensure that key validation activities are included and that adequate resources are assigned.

To ensure that business needs are met and ensure full compliance with current regulations any EDMS system will require a comprehensive URS and an equally detailed functional specification (FS). It is therefore crucial to build traceability controls into the project documentation from the very start. Tools such as documentation matrices and requirements traceability matrices should be used to keep track of the necessary interrelationships throughout the system life cycle.

Functional and configuration specification. The next stage of the project phase is to expand the basic URS requirements into how the organization wants that functionality to be implemented. Users must be involved in this stage, both to give them ownership of the product, and, more importantly, to obtain the benefit of their experience and knowledge of documentation management. This process must also involve EDMS system SME to facilitate user group discussions. The involvement of SME helps the team to achieve a realistic and practical set of functional requirements by identifying key deliverables from the system while advising against functionality that is unworkable or which will require significant amounts of customization. Discussion groups can also highlight inconsistencies between requirements of different user groups, especially when the system is designed to be used by all departments rather than by, for example, a dedicated documentation management function. One example of an area of potential conflict is print protection where one department may want free access to printing, whereas another group requires that printing of certain documents be strictly limited using system controls. Hence, a compromise must be reached which provides the required functionality in a manner that complies with GxP requirements and which is acceptable to the quality unit.

From the above discussions a detailed FS can be prepared. The FS must be cross-referenced with the URS to ensure that it encapsulates all of the users' requirements.

At this stage a Functional Risk Assessment should be performed and documented to identify which functions have the potential to adversely impact on patient safety, product quality and data integrity. These functions will need to be qualified during the functional testing stage. For other functions a verification process according to good engineering practice (GEP) is appropriate.

The creation of the FS is therefore critical in terms of validation as the operational qualification (OQ) test scripts will be written against this document, ensuring that all GxP-related functionality is tested against a preapproved qualification protocol.

The FS should describe the individual functions of the EDMS and define the required data. User interfaces should also be described. Example elements of an EDMS FS include the following:

- Performance criteria and system availability
- Document database structure
- Document types supported
- Document work flows
- Viewing capability
- User group configuration
- Access control, for example, passwords
- Audit trail definition
- Search features
- On-line help
- Interfaces to other systems
- Training requirements
- Maintenance functionality
- Error handling

Once the FS has been agreed the EDMS SME can use this to prepare a detailed configuration specification (CS) from which the EDMS can be configured. This will include both the hardware and software configuration.

EDMS configuration. Following system build and installation according to the supplier's protocols and according to the regulated organizations compliance standards the EDMS should be configured from the CS. The project team should review the operation of system once it has been configured to ensure that it meets the business requirements. Process redesign and reconfiguration (performed under project change control) at this stage will significantly reduce user dissatisfaction, delays and cost compared with similar activities after the system has been implemented.

To enable the system to be developed and user tested simultaneously it is useful to create separate instances of the EDMS, one for the SME to configure (development system), and one for users to try out the functionality during its development (test system). Once the development of the system reaches the implementation stage then an instance of the EDMS in which to perform validation testing is required. This may be the test EDMS put under strict change control to prevent unauthorized changes, or more usefully, a separate validation instance.

Upon completion of functional testing the "live EDMS" instance can be created and implemented after conducting requirements testing [performance qualification (PQ) and user acceptance testing] to ensure that it behaves identically to the validation instance EDMS.

The provision of several EDMS instances may require significant infrastructure resources, as such, it may not be possible in all cases. However, it is essential to have an EDMS instance in addition to the live system to be able to investigate, correct and verify any faults found—without endangering the integrity or availability of the live system. This process requires strict software\configuration version control to ensure that the various systems are using the appropriate version.

Where bespoke components are developed comprehensive unit and structural testing should be carried out by the SME. If this is carried out properly, the number of faults found during subsequent testing should be significantly reduced. This testing should be documented according to GxP guidelines such as those published by GAMP/ISPE (10). The test scripts and evidence of testing must be handed over to the customer in a formal handover meeting on completion of the installation phase. The quality unit must be present at this meeting and accept the documentation as satisfactory.

Verification activities
Configuration testing: installation qualification   For configured systems the configuration testing may be a phased process. For configuration items affecting GxP critical functionality an installation qualification (IQ) protocol should be developed to confirm that the EDMS is configured as intended. Evidence may be collected during the installation and configuration process or retrospectively once successfully completed.

Checks on the following may be included:

- The installed hardware
- The installed operating system software
- Environmental conditions
- Support procedures
- Maintenance agreements

The second phase is the installation of the configured software. This may include checks on the following:

- Installed core software
- Installed configuration
- Installed bespoke code (if appropriate)
- Software licenses
- Support procedures
- Maintenance agreements

A summary configuration testing/IQ report should be issued to record successful completion and to document any unplanned deviations and their resolution. This will confirm that the system is in a suitable state of control for functional testing to commence.

Functional testing: operational qualification

All functionality as defined in the FS should be tested to demonstrate that the system is fit for purpose, GxP critical components should be tested against preapproved OQ test scripts.

This will involve taking a document through all GxP-related work flows from start to finish and, if the work flow is a continuous cycle, then at least two cycles should be tested.

Example elements of OQ testing are as follows:

- Administration
- Security of access to the system
- Security surrounding system functionality
- Creation of draft document
- Document review and commenting
- Document approval and release, including use of electronic signatures
- Document rejection
- Document made effective, that is, in use
- Document superseded
- Document withdrawal
- Document revision—content
- Document revision—template or format
- Importation of legacy documents
- Viewing of documents
- Controlled printing of documents
- Production of all "GxP" reports
- System robustness
- Components such as the system for creating a rendition, for example, MS Word to portable document format (PDF) format, and the software for this process

It is equally important to identify and test for functionality that should not happen as well as checking that the system works as expected, for example, in a work flow that involves parallel review followed by approval the test should check that the document is not forwarded to the Approver until all the parallel reviews have been completed. Unless the users have been provided with a test EDMS this will be the first time the full system is available to the project team and so there may be a tendency to propose enhancements to the system as it is tested. The project team must be clear whether such enhancements, or in the systems integrators language "functionality creep," will be implemented with the associated cost and delays or whether only major concerns, such as noncompliant GxP functionality, will be corrected. If faults are found during OQ it is better to complete the full protocol, if possible, so that all required changes are identified and resolved prior to running the OQ again.

When a document is prepared in an application such as MS Word and then rendered into another format, for example, PDF, there is a possibility that a particular character in a particular font or symbol set will not be able to be rendered by the rendition software. In this case the software will substitute its best guess. It is, therefore, necessary to verify each character of each font or symbol set used to ensure that it is accurately rendered by the system in use by creating a document in the font/symbol set in question, rendering it and comparing the characters on the two documents. This verification is required for all native application software fonts or symbol sets, which will be rendered by the rendition server and should be part of the OQ.

A summary OQ Report should be prepared following OQ testing highlighting outstanding issues, their criticality to the project and assigning responsibilities with a time scale for completion. The PQ phase cannot begin if there are unresolved critical issues from the OQ tests.

Requirements testing: performance qualification

PQ may consist of testing the system in a "live" environment with a restricted user base but using the system as envisaged when rolled out to all users. Following approval of a PQ report acceptable to the business owner, system owner, and quality unit the EDMS can be made available for use.

Alternatively, PQ may be used to assess the system after it goes live, checking system attributes that cannot easily be tested as part of the OQ. Testing here may include the following:

- System availability including ability to log on and access documents
- System access times and document retrieval times with the full user base, network traffic and expected number of concurrent users
- Performance of the infrastructure
- Ability of the users to use the system
- Number of incidents and change requests
- Password management

A summary PQ Report should be prepared, again identifying issues from the testing, their criticality to the project and assigning responsibilities with a time scale for completion. This type of qualification may be termed ongoing assessment or performance monitoring.

At the completion of testing and when the system is ready to go live the tested EDMS must be promoted from the validation to the production system (if differing EDMS are used). Handover of the EDMS to the users for operational use usually requires some specific activities such as setting up desktop icons for connecting to the system, setting up user passwords and so on. This should all be recorded.

User procedures   A validated system must have written procedures that have been formally reviewed, approved and issued. These procedures should be reviewed (by someone from the intended user base who has an appropriate level of expertise in document management and who has been trained on the EDMS) prior to approval by the quality unit. In addition controlling procedures for the system administration function must also be established.

Database population   Early in the project it must be decided whether existing documents will be imported onto the EDMS. If the decision is to bring documents into the system there are a number of ways of doing it. For example, either the electronic files can be imported into the EDMS or hard copies of the document can be scanned and the resulting file imported. Generally, it will be necessary to employ a mixture of methods particularly where old documents on obsolete word processing packages are involved or where not all of the electronic files are available. For GxP critical documents a verification program should be established to ensure that the version of the document in the EDMS is a true representation of the regulated document. For electronic files it is possible that the way they have been managed has not been to the same standard as that for the management of the paper system. Care must be taken to ensure that the correct document, that is, the current approved and issued version, has been imported into the EDMS and that the file has not been corrupted or changed.

For scanned images the verification of the document in the EDMS should check the following:

- It is the current approved and issued document.
- All the pages of the document are present and in the right order and orientation.
- There are no erroneous pages.
- The image is legible.

For imported electronic files the verification of the document in the EDMS should check the following:

- It is the current approved and issued document.
- It has not been modified, for example, a user has started to produce the next version using the file for the current approved version.
- It is in a verified text (see section on validation of fonts and symbol sets).
- Any symbols or special characters have been correctly rendered (see section on validation of fonts and symbol sets).

Both of the above types should involve checking the document in the EDMS against the current approved document. This should incorporate a check on the accuracy of the attributes entered on importing the document.

The importance of this exercise cannot be over emphasized. If the system contains incorrect information, there could be GxP compliance issues and, from a practical perspective, users quickly become disillusioned with systems if they cannot rely on the information they contain. Hence, it is important that the information is also maintained during the implementation phase to ensure that any documentation updated in the hardcopy system is also updated in the EDMS.

In addition to verification of each individual imported document a check should be made to ensure that all required documents have been imported. Failure to do this could result in critical documents being missing from the EDMS. As part of this final check the documents should also be checked to see that the EDMS contains the current version of all the documents in case documents have been updated since import. On completion of this verification step the management of the documents in question should then be transferred to the EDMS.

Training Acceptance and the continued use of a system are reliant on the perception of the user base of its usefulness. Training is key to helping the user to have a positive impression and ensure that they know how to use the functionality that they require in their job function.

The timing of user training will depend on whether all users will use the system immediately it goes live or whether there will be a phased roll out of users. The former obviously demands that all user training is completed prior to implementation of the live system, whereas, the latter means that each individual user must be trained before being allowed access to the live system. Training should use the procedures that will be available for the system. This not only checks the procedures to determine if they are correct and that they are easy to follow, it also familiarizes the users with them. If the user base is large it may be useful to train a group of people who can provide on-the-job support to their colleagues.

Training is also required for the administrators of the system so that the EDMS can be maintained. The EDMS SME (internal or externally sourced) usually provide this.

Ongoing training is also needed to retain the validated state of the system, for example, for new users and refresher training for current and lapsed users. All training must be documented. Increasing use is being made of Computer-based training for EDMS users as this can be provided "on demand" and without the use of scarce SME resource.

Validation report At the completion of the implementation a validation report is required to summarize all of the validation activities. It should summarize the outcome of each of the steps identified in the validation plan and review the progress of any outstanding actions from the IQ, OQ, and PQ reports. There may also be issues from the supplier audit or the review of compliance of the systems integrator against their own QMS that may need to be assessed. The report should then assign a validation status to the EDMS.

At this point a date for when a periodic review of the system should be conducted—determined using a risk-based approach according to the policy of the regulated organization—should be established and documented. If the project is being implemented in phases the report may defer assigning a review date until the completion of subsequent phases.

Operation

Getting to a validated state requires significant expenditure of time and money. As well as being required for regulatory compliance it makes good business sense to maintain the system under control throughout its operational life. A formal set of procedures and systems are required. These should include:

Change management process is a system that manages change, whether they be changes to hardware, version changes of the core software or local configuration changes, is critical to maintaining control of the system. The identification of categories where changes can impact the system can help to decide the degree of verification/validation effort required.

- *Access security*: Control of access to the hardware, software and to the system via the user interface is very important. Access to system administration functionality should be controlled, particularly where a user performs significant events such as the creation or modification of user accounts.
- *Incident management*: An easy to use system for users to report unexpected events with the system is an important monitoring tool. This should be supported by an effective CAPA process to ensure that corrective and preventative actions are taken.
- *Business continuity planning*: In the event of system unavailability, for example, this is particularly important if the EDMS manages the instructions on how to make product. Paper copies of the instructions may need to be held with some mechanism to prove that they are official copies and are true representations of those that are held on the system.
- *Disaster recovery planning*: Required in the event of a major failure to the server or other crucial elements of the system. A risk assessment should be performed to determine the criticality of the system to the business. The higher the degree of risk the more comprehensive the plans should be to quickly restore the system. This adds to the cost. Disk mirroring, platform mirroring, backup strategy, identifying a business partner who will provide a similar platform in an agreed time frame, should all be considered. As well as considering actions in the event of the unavailability of the main platform, recovery actions due to failure of other crucial elements of the computer infrastructure, such as networks, should be included.
- *Backup and media storage strategy*: How it is done, records to demonstrate that the procedure is being followed, how the backup can be restored and shelf life of the storage media all need to be considered. It is also essential to prove that the restore procedure works before it is required.
- *Support and maintenance agreements*: With the support providers (IT infrastructure, application vendor, internal administration SME) should be established and periodically reviewed to ensure that the service delivery continues to meet the needs of the user organization.
- *Periodic review*: This will include reviews of the change and incident management methodology and an assessment of the cumulative effect of any changes. Training, procedures and records and any outstanding actions from the validation report or previous reviews will also be reviewed.

RETIREMENT

Retirement of an EDMS system may be triggered by the decision to migrate to a new system or by external events affecting the organization (sale, closure, etc.). Since GxP and other documents critical to the organization are involved careful consideration must be given to which records must be retained, for what period and how this will be achieved technically. The effort required to retire the system should not be underestimated and should be carefully controlled using a system retirement plan to define the approach to be taken.

A detailed discussion of this topic can be found in GAMP®5—Appendix M10: System Retirement (1).

SUMMARY

EDMS provide an invaluable tool to manage regulated and business important documentation throughout their life cycle and to share information within the organization in a way that ensures effective review and approval, and that current versions of critical documents are always at hand.

In the healthcare sector, verification and validation processes are a prerequisite to their use for GxP purposes. The validation methodology used is similar to that used for other information management systems. As with all systems, the more attention that is devoted to the design, validation and ensuring that the users will be happy to use the system the greater will be the benefit to the business.

ACKNOWLEDGMENTS
The author would like to acknowledge the first edition of this case study by Richard Mitchell and Roger Dean of Pfizer Ltd., Sandwich, Kent.

REFERENCES
1. International Society for Pharmaceutical Engineering. GAMP®5: A Risk-Based Approach to Compliant GxP Computerized Systems, 2008. Available at: www.ispe.org.
2. IT Infrastructure Library® (ITIL®) V3 Glossary. Available at: http://www.itil-officialsite.com.
3. International Society for Pharmaceutical Engineering. Web-Based Glossary. Available at: www.ispe.org.
4. International Conference on the Harmonisation of Technical Requirements for Registration of Pharmaceuticals for Human Use. Pharmaceutical Quality System—Q10. Available at: www.ich.org.
5. Food and Drug Administration. "Pharmaceutical Current Good Manufacturing Practices (cGMPs) for the 21st Century: A Risk-Based Approach" Initiative, 2004.
6. Food and Drug Administration. U.S. Code of Federal Regulations, Title 21, Part 11. Electronic Records; Electronic Signatures, 1997.
7. Food and Drug Administration. Electronic Records; Electronic Signatures—Scope and Application, Final Guidance, Rockville, MD, 2003.
8. Draft Eudralex Rules. Chapter 4, Documentation and Annex 11, Computerised Systems.
9. Pharmaceutical Inspection Cooperation Scheme. Guidance on Good Practices for Computerised Systems in Regulated "GxP" Environments (PI 011-3), September 2007. Available at: http://www.picscheme.org.
10. International Society for Pharmaceutical Engineering. GAMP® Good Practice Guide: Testing of GxP Systems, 2005. Available at: www.ispe.org.

31 | Case Study 13: Enterprise Resource Planning Systems

Guy Wingate
GlaxoSmithKline

MISSION IMPOSSIBLE?

> I want a system to replace all the standalone pockets of automation handling the warehouse, purchasing and materials management systems. The system is to interface to my laboratory management systems and provide Internet links with my customers and suppliers. ... I expect to be able to reduce my inventory, reduce lead times, reduce IT costs, improve regulatory compliance, and install a system which is flexible to change.

The answer, besides do not we all want this, is the implementation of an ERP system? Well, ERP does offer a solution but it is no panacea. Indeed there are examples where the cost of a poor implementation of an ERP system has lead to a pharmaceutical company going out of business.

An additional requirement in the pharmaceutical industry is the need to fulfill the requirements of good practice (GxP) regulations that impact the use of computer systems. This requirement is often expressed as an afterthought to the quote at the beginning of this paper: "Oh, and by the way, I want the project to fully comply with GxPs but at a minimum cost."

The GxPs (covering good clinical practice, good distribution practice, good laboratory practice, and good manufacturing practice) necessitate the ability to demonstrate that a drug product can be consistently made to its specified quality criteria. Failure to satisfy these regulations can result in a regulatory authority refusing to accept pharmaceutical products made using the computer system concerned. Lost sales revenue for a single top selling drug could exceed €2 million per day. An ERP system usually coordinates operations across an entire site or sites. Deficient application, operational error, or system malfunction could potentially affect the manufacture of all the products using the ERP system. A single medium sized secondary manufacturing site may have an associated annual drug sales revenue in excess of €1 billion. Factoring up this rule of thumb over a number of drug manufacturing sites quickly demonstrates how "super critical" ERP systems are.

It is vital that when a company commits to the implementation of an ERP system that it does so knowing how critical project management with compliance is.

PROJECT APPROACH

The implementation of GxP is often referred as validation and the well-known regulatory authorities such as the U.S. Food and Drugs Administration (FDA) and the U.K. Medicines Healthcare Products Regulatory Authority (MHRA) have given guidance on what they expect to see during an inspection. A life cycle approach should be adopted in the implementation project, and care taken to ensure that after cutover the system is maintained for ongoing compliance. The GAMP®5 Guide (1) provides general industry guidance but this must be adapted to the needs of an ERP project whose activities will typically include the following:

- Project initiation
- Supplier selection
- Install development system
- Define business processes
- GxP assessment
- Conduct conference room pilot
- Review legacy data and data upload

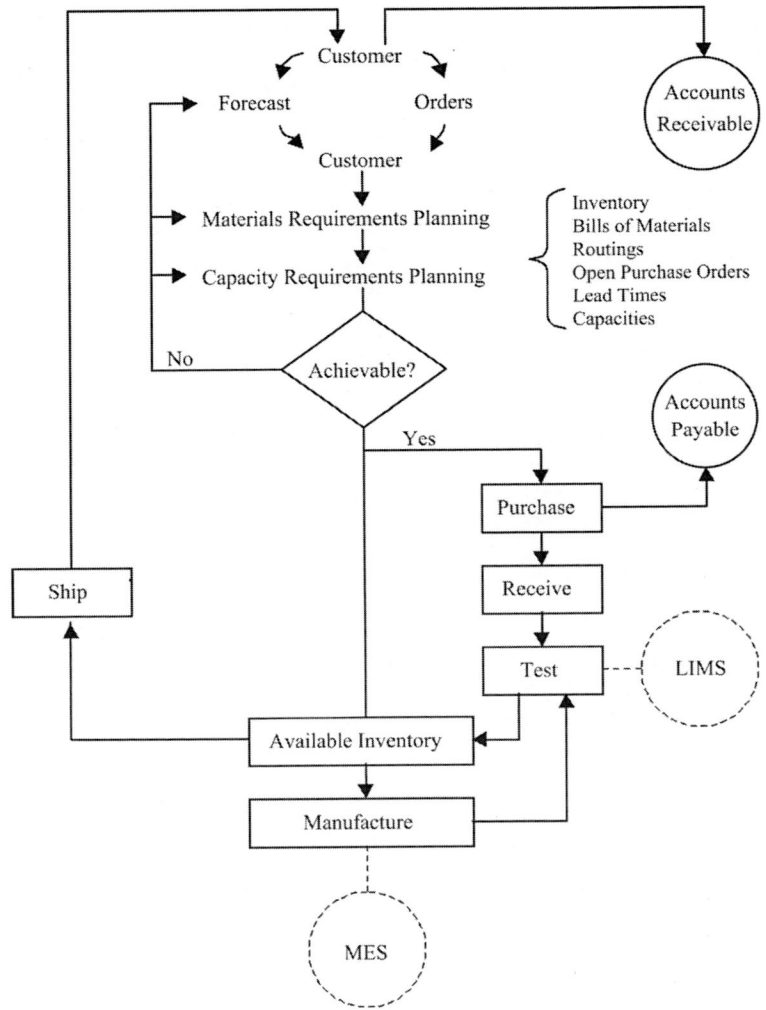

Figure 31.1   Overview of ERP functionality. *Source*: From Ref. 1.

- Readiness review and go live
- Performance improvement

Project initiation will scope the business processes and systems integration needs for the system within what is usually referred to as a user requirements specification (URS) or "to be" document. The typical operability of an ERP system is shown in Figure 31.1. Project initiation will also include project planning, budgeting and project risk analysis. For the pharmaceutical industry it is also the point at which quality assurance practices begin. A validation master plan (VMP) will need to be prepared to identify the project process, procedures to be adopted, personnel requirements, roles and responsibilities, documentation to be delivered, and milestones showing the rollout of the project to completion. In essence the principles of ISO 9000, but recognizing that validation for GxP goes beyond such quality management systems. The VMP may itself reference a number of validation plans covering specific aspects of the ERP system. A project quality plan may also be produced.

It is likely that selection of the supplier has occurred by default rather than choice, for instance SAP R/3 is the clear market leader for ERP systems. Nevertheless, a supplier audit should take place and include the original vendor(s) of the software product suite being used, and any system integrating companies taking responsibility for delivery of whole or part of the system. This audit needs to assess the confidence that a pharmaceutical manufacturer can place in the quality of the software and hardware products used in the ERP implementation. The GxP regulatory authorities hold the pharmaceutical manufacturers directly accountable for such quality and where there are deficiencies or insufficient evidence of quality they expect the pharmaceutical manufacturer to remedy the situation. This may involve working directly with the supplier to improve their quality management system, or working through a third-party consultancy.

Defining the business processes to be implemented by the ERP system, either within the system or in conjunction with other interfaced systems, is a key task. SAP refers to this activity as blueprinting. This will involve reviewing the current ways of working and perhaps embarking on a program of change in these working practices—business process reengineering (BPR). An overview diagram showing how top level business processes fit together should be prepared along with diagrams illustrating the operability of the main functional elements (business processes) making up the ERP system.

Once the business processes have been agreed a GxP assessment can be conducted. This should address those operational aspects of the system that impact the quality of finished pharmaceutical products and will include supplier details, batch records, laboratory quality control records, batch release, and recall. Example GxP impacting functionality in an ERP system is given in Appendix 31.A. Experience suggests that perhaps between 25% and 50% of ERP functionality[a] is GxP critical (2,3). The GxP operational aspects will form a focal point during any GxP regulatory inspection. It is very important to document why these aspects and not others are deemed GxP impacting, and where these operational aspects are defined and tested. Remember that inspectors will keenly challenge the distinction on non-GxP impacting so form a robust assessment and err on the side of caution as to whether something is rather than is not GxP impacting. This determination will bring focus to the validation exercise.

Almost immediately a system will need to be installed to provide development and testing environment. A separate system is usually installed later to provide a go-live production environment for cutover.

Setup of the various system environments must be managed. Documentation must be developed to describe the hardware platform and installed software, including any network infrastructure. Hardware architecture design documentation should be prepared. A diagram should be included to illustrate the geographic distribution of any client-server hardware. Client-server software also needs to be defined. Clients are often referred to as either "thick" or "thin" depending on whether they require substantial or minimal application related software. Client-server software can be considered to consist of the following (Fig. 31.2):

Operating system: Operating system independent of the client or server application. GAMP level 1 software requiring version to be recorded (e.g., UNIX OS).

System software: Standard software specific to intended use of client (e.g., desktop utilities) and server (e.g., network and database utilities). GAMP level 3 software requiring the version to be recorded and operability confirmed (e.g., Oracle Database and Microsoft SMS).

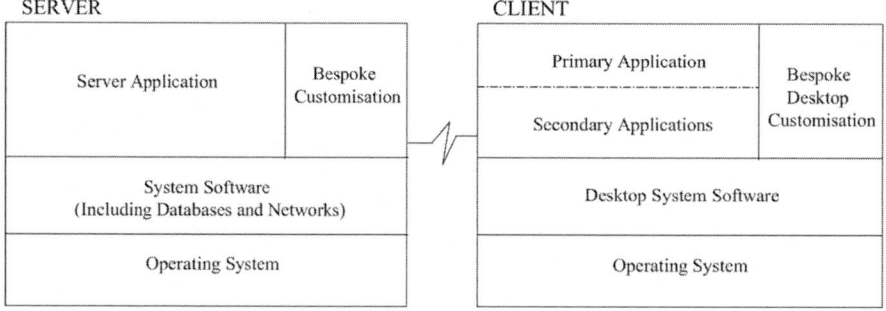

Figure 31.2 Client-server software schematic.

---

[a]SAP R/3 modules: CO (costing), FI (finance), MM (materials management), PP-PI (production planning—process industries), QM (quality management), and SD (sales and distribution)

Server application: Application software products such as the ERP software product. GAMP level 4 software requiring a supplier audit, validation of the configuration, and confirming the operability of the standard element of the software (e.g., SAP R/3). There may also be some standard software, GAMP level 3 requiring the version to be recorded and operability confirmed (e.g., third-party utilities provided with the server application).

Client applications: A client may be used for more than one application (e.g., ERP, LIMS, and EMS). Each application will have an associated file set providing what is often referred to as it is graphical user interface (GUI). File sets are usually built into standard client setups. Individual files may include some element of configuration. GAMP level 3 (e.g., Windows NT) and GAMP level 4 software require the version to be recorded, operability confirmed, and any configuration validated. Supplier audit requirements are usually satisfied as part of the server application validation.

Bespoke customization: Bespoke programming (e.g., macros defining reports and forms, and interfaces especially written for the client or server). GAMP level 5 software requiring a supplier audit and validation of the bespoke code (e.g., SAP R/3 ABAP form and report programs, and Microsoft SMS client scripts).

Robust server architectures are required to provide a dependable service to what may number hundreds or even several thousand clients. Basic configuration management of the server(s) and network are expected. It is also important to define client builds (sometimes referred to as the desktop) and maintain them under configuration management. Client builds should have their applications integration tested to check there are no conflicts. It is quite common for clients to run multiple applications and it cannot be assumed that conflicts will not occur, even between standard application products. Automatic desktop configuration tools should be validated in their own right.

An installation qualification (IQ) is needed to define and execute tests to verify successful installation of the hardware platform and resident software. As the project ramps up the system is likely to require expansion to cope with a larger user base in which case the IQ must be revised. The IQ should normally include the following:

- Inventory and configuration checks for the hardware platform (clients, server, and network).
- Inventory check of software used.
- Check whether all vendor-supplied manuals are present and correct.
- Check whether necessary SOPs are available.
- Make environmental checks in computer room housing hardware platform on power supplies, backup power supplies, temperature, and humidity.
- Check physical security mechanisms.
- Check system boot-up diagnostics.

Following on from the URS a system definition consisting of functional design specification (FDS) needs to be collated. The URS does not necessarily specify the chosen ERP system, and if this is the case, the FDS will need to introduce and overview the selected ERP system. The FDS will define the URS business processes at a transaction level. Referenced documentation published by the supplier defining the standard ERP software product and its functionality should be retained and maintained with the current version of the ERP software used. It is important to identify those functions of the standard ERP system that are used and specifically document which functions are not being used.

Process flow diagrams should be considered as the basis for SOPs developed for the transactions implementing the business processes as they are generally easy to understand and can be designed to highlight user interaction and interfaces to other systems linked to the functionality provided by the ERP system. SOPs, forms and reports must be drafted and under version control ready for piloting in what is sometimes referred to as a Conference Room Pilot. Appendixes 31.B to 31.F present some typical business processes for procurement, production planning, production, sales and distribution, and finance with associated example SOPs.

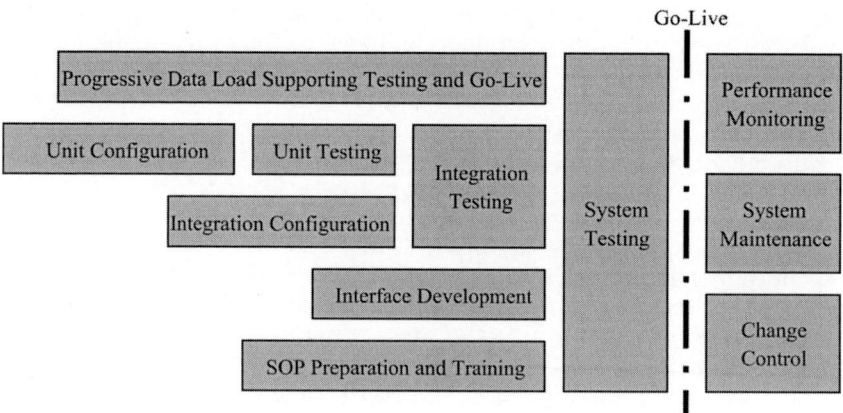

Figure 31.3 Schematic ERP testing plan.

Once defined the business process transactions can be configured within the development environment of the ERP system. There are normally instances when it is easier to amend the business process to fit the standard functionality of the ERP product software than to make a customized bespoke modification. Any bespoke modifications, like the interfaces, must be fully documented in design specifications, test specifications and test records. One important aspect to avoid during configuration is to set up the system to accept default user entries. There have been several recalls within the pharmaceutical industry because a user failed to recognize that a default entry on their ERP system was incorrect. It is always a good idea to have positive user confirmation of key data entry or decision points. If defaults are still required then make them fail safe, that is, default entry on product sample status should be "reject" and require positive selection of alternatives such as "retest," "rework," and "pass." Figure 31.3 indicates how configuration can be split into unit and integration activities. The pace of ERP projects usually brings pressure to begin testing as soon as possible and this can be facilitated by testing unit configurations and then as a follow-on activity their integration. Perhaps as much as 80% of the configuration activity can be attributed to unit configuration.

The completed FDS should be verified that it is consistent within itself and with SOPs implementing the business process transactions, and that it fulfills the requirements of the URS. The activity is often referred to as a design review or design qualification (DQ). The use of a requirements traceability matrix (RTM) should be considered to demonstrate how URS elements are addressed in the functional specification and design documentation. This RTM can later be extended to trace test specifications and results.

When the business processes have been implemented they are transferred to the testing environment. A key prerequisite to testing is data load. Ensuring the integrity of data is a must—garbage in, garbage out! It is important to review legacy data and new data entry requirements in readiness for testing and cutover to the production environment of the ERP system. Not all data from the replaced legacy systems needs to be transferred to a new ERP system. Decommissioning and archive of legacy data must be carefully considered. Some pharmaceutical companies have tried to distinguish between critical and noncritical data (see Table 31.1) and set different data accuracy requirements for each. In reality, there is little difference when it comes to user and customer satisfaction. All data should be checked for accuracy and if its integrity does not pass then cutover must not occur. Transport mechanisms for data load must be validated to provide assurance of data integrity. The distribution of data, once loaded into the system, and its control must also be defined.

Testing in a conference room pilot, referred to at this stage as operational qualification (OQ), can largely be limited to "black box" functional testing where a standard system is used without modification as long as the supplier audit determines a high confidence in the embedded quality of the system. If a system is customized or a supplier audit notes significant

Table 31.1 GxP Data Elements in ERP Systems

Batch information	Assets	Bill of materials
Batch number	Purchase order number	Items
Batch status	Contract of supply	Quantity per
Dates of manufacture		unit of measure
Expiry dates	**User**	Conversion factors
Quantity/potency	Name (and password)	Work centers conversion
Factors	Security access	Yield factors
Approval restrictions		approval
	Customer orders	
Item	Shop order number	**Shop order**
Item number	Customer order number	Quantities
Item classification	Customer addresses	Receipt date
Location		Transactions
Type	**Supplier**	
Quality/potency	Quality approval	
Shelf life and retest interval		

issues with the supplier's quality development of the system then "white box" structural testing should also be conducted. Either testing should include challenge tests to verify the system can detect within reason operator error and bad data. As with the IQ there should be a preapproved protocol before testing begins, and test records must be collated. At this stage in the OQ the system is still under-refinement and any changes must be logged and necessary retesting carried out. The size of ERP systems means that conference room pilots are often organize to exercise certain areas of functionality, based around the pharmaceutical company's organization or the ERP systems standard functionality, so it is important not to forget to rigorously test the integration and interfaces interconnecting these areas. OQ tests will include, but are not be limited to, the following:

- SOPs implements transactions (Appendixes 31.B to 31.F)
- Verifying the processing of batch and laboratory records
- Challenging user interactions
- Testing accurate data manipulation and presentation (e.g., rounding errors and number of display digits)
- Verifying backup and recovery procedures
- Checking product recall processes

The OQ should also include system performance tests to confirm the system can cope with high numbers of active users and large volumes of data.

Education and training programs should be established for the project team and the end-users. Education is based on presenting principles while training is based on practical *hands-on* tutorials. Course modules must be documented, their content approved, and delivered by authorized trainers. Staff training records, including those for contractors, should be maintained to track attendance on courses. The use of competency questionnaires to verify learning should be considered. Mere attendance does not necessarily imply that an individual has understood and taken on board course material.

There may be a significant training requirement associated with the new ERP system. The conference room pilots provide an opportunity to train users in new SOPs and hence reduce the need for separate training events. Organizational structures and ways of working often alter with the implementation of an ERP system, and with large scale training requirements many pharmaceutical companies employ change management consultants as well as ERP specialists. Pharmaceutical manufacturers should anticipate training requirements as demonstrating the competency of staff is a key aspect of the GxP regulations.

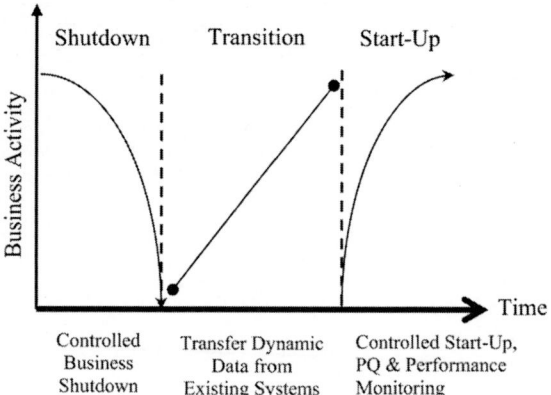

Figure 31.4   Business cutover period.

Successful OQ means that cutover of the system into live operation can be considered. Other issues affecting cutover are whether all procedures and software have been frozen and issued, whether all tests have been completed, whether all tests have passed, project documentation is complete, and that relevant business managers are in themselves confident on a successful cutover. It must be stressed that in the pharmaceutical industry cutover must not be allowed if the projects GxP-related validation documentation is not complete. A formal "go/no-go" decision should be taken to document the cutover decision with signed approvals. It is advisable to incorporate this within an Interim Validation Report authorizing cutover of the system from a regulatory compliance standpoint.

The cutover process can be considered as comprising three main stages (Fig. 31.4). First, the business operations must usually be shut down in readiness for the decommissioning of existing systems and the switch to active use of their replacement system(s). This can be a complex management exercise if many existing systems are being decommissioned. The next cutover phase involves dynamic data upload, which cannot by the nature of the data occur earlier. Static data will have been loaded earlier usually in the OQ phase of the project. Finally, a controlled start-up of operations can begin. Back-out plans and procedures should be put in place in case a major problem occurs during the cutover period.

For many implementations cutover is the point of no return! It is vital that the ERP system is ready for cutover before the cutover is authorized. Validation case studies on ERP, MRPII, EDMS, LIMS, and warehouse business systems have stressed the importance of cutover management (4–6). There is often considerable pressure to cutover on time and a reluctance by individuals to be the first to say that their aspect of the system implementation is not complete and ready for cutover. As far as possible an open and honest culture should be established. It is better to delay a cutover and take corrective actions than cutover on time and live through operational difficulties directly attributable to not being ready for cutover. Hindsight is a wonderful thing, but not when you are unable to release drug products to customers. Try and use terms such as breakpoint to describe the time at which a decision to go forward or not is taken. Breakpoint implies work will stop if criteria are not met. Referring to milestones does not have the same impact, and avoid using terms such as "drop-dead" date as its very emotive—who drops dead the organization for going live when the system was not ready, or the messenger who brought this to the attention of senior management? In any event it is wise to develop contingency plans (sometimes also called business continuity plans) and challenge their feasibility before cutover just in case . . .

After cutover the performance of the ERP system should be monitored and evaluated. This is sometimes referred to as performance qualification (PQ). The PQ protocol should be prepared identifying key performance metrics such as the following:

- Successful batch release in live environment
- Number of new change requests
- Number of outstanding change requests

- Number of help desk calls
- Changes to business processes
- Data accuracy
- User enquiries and retraining requirements
- System outage (partial or total)
- Security profile changes

Following a period of say three months from cutover it should be possible to demonstrate that the ERP system is enjoying a period of stable operation. A PQ report should collate data, possibly graphical, that can demonstrate these trends.

To conclude the implementation project validation report is prepared in response to the validation plan issued at the beginning of the project. It summarizes what went according to plan, and explains what did not go to plan. Amendments to the plan must be justified. Some issues may still be outstanding in which case forward audit trails to corrective actions must be made. The Validation Report must demonstrate that the ERP implementation is fit for purpose and can be used to support drug manufacturing. Because of the large number of documents often associated with these projects a library index or route map may also be useful to include either in the Validation Report, or more normally to reference in the report as a separate document.

ROLLOUT STRATEGIES

Within large corporate rollouts of ERP systems there is likely to be a core system configuration providing a company standard way of working. Individual sites will implement the preconfigured system with minimum variations to the standard core system. In this way the site implementations can share the standard core system documentation. Each site will normally have its own validation plan directing (and hence Validation Report responding to) site-specific validation activities and placing these activities in the context of the standard core system documentation so that when a regulator comes to inspect the system they will understand it from it from a site perspective. To date the vast majority of regulatory inspections of ERP systems come in from a site inspection.

Corporate rollouts are often phased: typically finance, inventory and warehouse in the first cutover, then customer services (sales and operations planning S&OP in the second cutover, and distribution), followed by production as a third cutover. However, some rollouts consist of a single cutover and as such are sometime referred to as "big-bang" events. There is a high risk associated with big-bang cutovers because of their complexity and the total dependence in the new system. Instituting interim procedures to bridge phased rollouts, however, also has risks and a balance must be struck to manage the issues posed between big-bang and phased rollout approaches. It is worth noting that it is often not practical to continue running original systems being replaced by the new ERP system for any length of time after a big-bang cutover. This position, however, as the GxP regulations stand, might be considered noncompliant.

MAINTAINING OPERATIONAL COMPLIANCE

Controls must be put in place to ensure the ERP system is maintained in a validated state. These controls will include change control (hardware, software, data and documentation), configuration management, desktop management, maintained list of users, security and access management, service level agreements for maintenance and repair, ongoing education and training, and periodic reviews. Business processes and SOPs must also be maintained. Information on these topics will be presented by other speakers at this conference, see also further reading.

The strategy toward version upgrades and bug fixes of the ERP product software should have been specified in the validation plan. Each software release should have been successfully regression tested and be market tested before it is used. The term market tested has not been defined but it has been suggested that this means that the version of a software product being used, or bug fix, has been released into the market for at least six months and that there are a large number of users of that particular software. The aim is to reduce the risk associated with installing new software that does not have a track record of successful operation. It has been well documented elsewhere how complex software such as that making up ERP systems can degrade as it ages because additional functionality and bug fixes can actually introduce more problems than they solve—the "software death cycle." Even so, it could be argued that not implementing bug fixes is negligent. Any upgrade should be conducted under change control. Release notes from the supplier will require review in conjunction with how the system is used to determine whether there is any impact. It must be recognized that revalidation of the ERP application may be needed.

Inevitably the ERP system will become unavailable to users from time to time. Planned maintenance activities and development activities can be managed to control their GxP impact. If the system crashes for any reason (e.g., server goes down), however, then the system will need to be recovered to a known controlled state. This usually involves rolling back the system to its last archived state. There is then a need to catch up data entry and processing to reflect what happened while the ERP system was unavailable. Data processing centers providing these services are subject to the same GxP requirements in this respect as site manufacturing the actual drug products. Backup and restoration procedures must be defined, tested and approved. The size of the catch up task will depend on how long ago the last archived backup was taken, the duration of the outage, and what data processing was achieved prior to the outage and subsequently during this period. Regular back-ups will reduce the necessary catch up effort but will require more stand-by hardware and storage media.

BUSINESS BENEFITS

The business benefits ERP systems can bring are well documented. Better warehouse and inventory control saved enough money during the first year of operation to pay for the ERP implementation in one Irish pharmaceutical company. Savings in this regard by other companies are not always so dramatic, but generally they are significant. Other pharmaceutical and healthcare companies may also see benefit of automation in reducing user error and speeding up data processing. This coupled with a need to validate their existing unvalidated ERP systems has lead many companies to replace the old with new systems. Whether it is possible to successful retrospective validation of ERP systems has been questioned during FDA inspections. The cost of retrospective validation (perhaps 20 times cost of original unvalidated ERP implementation) and the uncertainty of satisfying regulatory inspection have lead to a general replacement rather than fix solution being adopted in industry. Not that validation is all cost. As a consequence of validation better maintenance documentation should make modifications easier, faster, and hence cheaper. Indeed, one U.K. pharmaceutical and healthcare company managed to reduce its operation and maintenance costs by more than 75%.

ELECTRONIC RECORDS AND SIGNATURES

There has been much debate in the pharmaceutical industry over a U.S. regulation known as 21 CFR part 11 affecting electronic records and electronic signatures that became effective in 1997. Paper records have been the traditional medium for regulatory records demonstrating validation compliance. Electronic records are allowed but must be reliable and secure and facilitate bound signatures. Controls for electronic signatures are required so that they are legally equivalent to handwritten signatures. Do not assume that all electronic records and electronic signatures are compliant by default. The complexity of this issue has led many ERP systems to fall back on paper records as masters, using electronic versions as working copies. This approach, however, was until very recently acceptable to FDA, who enforces compliance with 21 CFR part 11 (7).

Some ERP vendors offer a special edition of their product to address pharmaceutical industry needs (e.g., SAP offers "PharmaPack"). Pharmaceutical manufacturers should work with the vendor of their ERP system to ensure such special editions do indeed meet their needs. Cilag Pharmaceuticals recently shared their experiences with SAP in this regard (8). Some of the issues they raised were as follows:

- Multiple entries being posted in an audit trail for a single change to an electronic record
- Some non-GxP records creating entries in GxP electronic record audit trails

- Large amount of information being processed and stored for audit trails was impacting overall system performance

Any pharmaceutical manufacturer implementing an ERP system must pay careful attention electronic records and electronic signatures, and justify their position in a discussion document so that they can respond if challenged by a GxP regulatory authority. Hybrid solutions based on adding procedural controls and possibly supplementary software may be necessary to establish compliance with the regulation.

REGULATORY INSPECTION

Regulatory inspection of the operational ERP system may occur several years after implementation and can be just as critical if the system is found to be noncompliant with the regulator's expectations for validation. Pharmaceutical manufacturers who have been subject to detailed scrutiny of the computer validation for their ERP system understand the need to properly validate. The financial consequences to a multi-national pharmaceutical manufacturer of noncompliance can be immense.

Pharmaceutical manufacturers implementing ERP systems should carefully consider briefing their appropriate regulatory authorities in advance of any potential inspection. Few regulatory inspectors would claim to be ERP experts and while they will understand the principles of computer validation they may not be familiar with an individual pharmaceutical manufacturer's validation philosophy for enterprise applications. Both parties will benefit from understanding each other's perspective. Any concerns or misunderstandings can then be proactively managed. Figure 31.5 presents a variant of the well known validation V-model for enterprise applications used in this paper but how does this fit individual implementations?

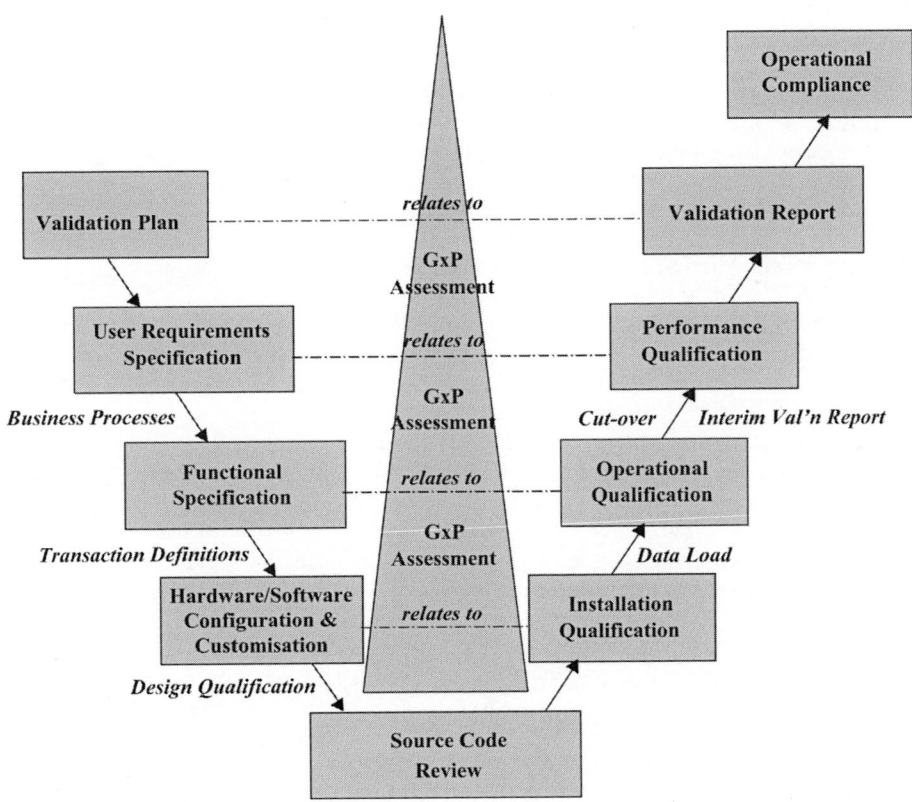

Figure 31.5 Validation life cycle for enterprise applications.

When a regulatory inspection does arrive it is likely to look at the ERP system as it used from a site perspective rather than from a corporate perspective. The availability of site-specific document sets including site validation plans and site validation reports will be key. Documentation for the core system should be readily accessible. Managers must consider who owns the ERP for the purposes of fronting an inspection, and access to staff who are knowledgeable on the core system and site application. Some staff will naturally move on to new jobs within and outside the company. It is very important to ensure a critical mass of knowledge about the ERP project is maintained within the company.

The RTM linking the specification, implementation and testing of functional aspects of the ERP system together with the GxP Assessments provide a very useful tool to assure that all aspects of the system, and especially the GxP impacting elements, have been successfully validated. It also provides a route map through the documentation set for the ERP system. There can be many thousands of documents and being able to quickly retrieve appropriate documentation during an inspection is very important. It is no good having done the validation if you cannot retrieve it for an inspector! The route map will also provide those preparing to receive an inspection a means of reminding themselves of project and document organization so that a knowledgeable and professional front can be presented during the inspection. A high level overview of a validation document set is presented in Appendix 31.G. As an aside, the benefit of having documentation presented in a common format and in neat labeled binders should not be under-estimated. Impressions count for a lot during an inspection—remember the pharmaceutical manufacturer is basically trying to demonstrate that the organization is in control.

INSPECTION CASE HISTORY

Here are some actual observations made by the FDA during an inspection of a SAP R/3 application at Solvay Pharmaceuticals (9). The observations were made by Thomas Arista and Robert Tollefson and recorded in an FDA 483. The SAP R/3 applications were inspected over a two day period shortly after go live.

- No final validation report (PQ still in process after cutover)
- No application version number mentioned in validation documentation
- No formal approval of vendor-supplied documentation
- Not all required SOPs in place with users as formally issued copies
- No list of current approved users with levels of access

Other comments raised during the inspection by the FDA were as follows:

- Distinguish between "validated state" and "authorized for use." The validation methodology did not have a rationale for justifying why the MRP II system was acceptable to support production and release after cutover but before the PQ was complete.
- Do not refer to internal audits in validation documents. The FDA were possibly concerned here that because the internal audits were presented as open to inspection. This could lead to any direct comment being made to the pharmaceutical manufacturer's senior management not being as explicit as it otherwise might be.
- IT departments will be part of future inspections. To date many IT functions with pharmaceutical manufacturing companies have not been subject to inspection.
- Project documentation should adhere to good documentation practice. There are often many thousands of project documents for ERP systems. They should all be subject to document (life cycle) management including following approval processes, indexing, and archive. Document management must cover the implementation project, operation and maintenance of the system.
- Key project and system documents should be available at sites. It is generally not practical to maintain a complete set of system documents at each site supported by an ERP system (remember there may be many thousands of documents). A complete set

of documents should be managed at a central location, with key stage and site-specific validation documents being formally copied to sites.

- Major milestones should be formally authorized by senior management. The successful implementation of an ERP system would normally be considered business critical. As such senior management would be expected to meet to agree progression between major work stages, possibly connected to the release of project budget. These authorizations to proceed should be formally recorded and retained.
- Focus effort on GxP relevant processes within system validation. The size and complexity of ERP systems means that full validation of everything is not practical. The adoption of GxP should therefore be considered for the whole system implementation with full validation of directly GxP impacting processes and functionality.

It is important to understand that these comments were passed on a limited audit of the application. The observations cannot therefore be considered comprehensive.

FURTHER INSPECTION EXPERIENCE

Another major pharmaceutical company has recently experienced a FDA review of its ERP system during a site GMP inspection, which included a subsequent dialogue involving three and a half pages of detailed follow-up questions. The keys aspects of the ERP system that FDA was interested in were as follows:

- Use of ERP in relation to steps in manufacturing processes. FDA interested in interfaces and interdependencies between ERP and other systems.
- SOPs for system use were reviewed. The FDA was also very interested in what constituted master data and how this was controlled and where copies existed these were contemporaneous with master data.
- Labeling processes examined in detail.
- Explanation of exactly what central project team did versus activities at multiple sites implementing the ERP system.
- Review of test strategy with interdependence of site and central responsibilities.
- Detailed review of site PQ and data migration activities.
- Finally, FDA was interested in the validation acceptance criteria and how validation deviations and incidents were handled.

Pharmaceutical and healthcare companies should take note that their ERP systems are highly to be inspected. Initial questions will be on how the systems are used followed as appropriate questions concerning validation. The level of questioning will be dependent on the experience of the inspector and the criticality of the aspect of operation that the inspector has any concerns over.

CONCLUDING REMARKS

This paper has outlined the basic approach to implementing and validating an ERP system. Managers must tailor their project approach to match their particular business organization, availability of in-house and external resource, scope and size of implementation. Successful implementation, validation and operation of an ERP system will also depend to a great extent on ensuring that the project does not (10) do the following:

- Unwittingly compromise the standard nature of a configurable standard software product by too much customization.
- Lose control of quality during what are often fast track projects. There are often large numbers of project staff from a variety of backgrounds, not all of which are necessarily conversant with either IT systems or validation.
- Unacceptably increase project risk by business process reengineering instead of limiting the implementation to current established ways of working.

- Ignore known shortcomings with the core supplier product (how they are tackled should be documented).
- Disregard the potential financial and validation impact of upgrading/integrating their current IT infrastructure (clients, servers, networks) to support the business system implementation.

In addition managers should closely monitor stress levels and morale amongst the project team. Illness amongst key team members can sorely hit a project's progress, an issue which becomes ever more critical toward cutover. The retention of staff has already been discussed in relation to fronting inspections, but key staffs are also necessary to maintain and further enhance the configuration and use of the implemented ERP system after cutover. It is quite common for a company implementing an ERP system to loose one third to one half of their original project team within a year of cutover. Reasons for the departure of staff are many: permanent staff taking highly paid contractor ERP positions elsewhere, individuals suffering from stress or uncertainty on how they will fit back into their own organization, individuals being poached by other companies embarking on an ERP implementation. It is very difficult to retrieve the situation when an individual has come to the point of leaving a company. It is better to actively manage to minimize the potential problem from the outset.

REFERENCES

1. International Society for Pharmaceutical Engineering. GAMP®5: risk-based approach to compliant GxP computerised systems. Tampa, Florida, 2008. Available at: www.ispe.org.
2. Gottschalk F. Validation of SAP R/3 and other ERP systems: methodology and tools. Pharm Technol Eur 2000; 26–30.
3. Hambloch H. An approach to risk assessment for IT systems. ISPE Conference on GAMP Concepts and Case Studies. Zurich, September 18–21, 2000.
4. Wingate GAS. Computer Systems Validation: Quality Assurance, Risk Management and Regulatory Compliance for Pharmaceutical and Healthcare Companies. Boca Raton, FL: Interpharm Press, 2003.
5. Snelham M, Wingate GAS. Validating Laboratory Information Management System. J Validation Technol 2000; 6(4).
6. Thompson D. Wiring up the Warehouse. Pharm Vis 2001; 62–66.
7. Food and Drug Administration. 21 CFR part 11. Electronic records; electronic signatures—scope and application, guidance for industry. Rockville, MD, 2003.
8. Smith C. Validation of SAP release. ISPE Conference on Principles and Applications of GAMP®4. Zurich, September 25–26, 2002.
9. Rakhorst W. Validation of SAP R/3: experiences with a big bang in an international environment. Institute of Validation Technology Conference on Computer & Software Validation. London, February 21–22, 2000.
10. Wingate GAS. Finding a way. Perform Chem Int 1998; 13(10):28–30.

Appendix 31.A Example ERP GxP Impacting Functionality

GxP impacting functionality	
Sales and operational planning	Master production schedule
	Capacity planning
	Routings
	Shop orders
	Bill of materials
	Materials resource planning
	Customer complaints
	Recalls
Supplier management	Purchase of materials
	Planned orders
	Order amendments
	Repetitive supplier scheduling
	Schedule batch release
Manufacturing and packaging	Issue of materials
	Manual allocation of materials
	Shop order release
	Shop order close
	Labeling
Warehouse/inventory management	Goods receipt
	Quality control inspection
	Movement of raw materials
	Location creation
	Movement of work in progress
	Movement of finished goods
	Returns to supplier
	Scrapping of materials
Quality control/quality assurance	Release of materials to production
	Quarantine of materials
	Scrapping of materials
	Testing of materials
	Retesting of materials
	Release of material (QP release)

Source: From Ref. 4.

Appendix 31.B Example Procurement Business Processes with SOPs

Business process	Standard operating procedure
Purchasing	Request quotation
	Create/change/approve purchase order
	Purchase item receipt
	Purchase order archiving
	Setup/change material/item details
	Setup/change supplier details
	Setup/change buyer details
Warehousing	Goods receipts from suppliers
	Goods returns to suppliers
	Customer returns
	Warehouse palletization
	Stock placement and removal
	Hazardous material handling
	Material requests/reservation/staging for production
	Goods received from production
	Return of unused or partially used materials
	Relocation movements within warehouse
	Stock accuracy checks (e.g., perpetual inventory)

(Continued)

Appendix 31.B   Example Procurement Business Processes with SOPs (*Continued*)

Business process	Standard operating procedure
Invoicing	Purchase item invoicing
	Invoice matching, approval, and payment
	Damaged goods processing
	Spend approval
	Quota arrangements
Quality management	Assign/revoke supplier approval
	Managing preferred supplier lists
	Quarantine materials/goods

Appendix 31.C   Example Production Planning Business Processes with SOPs

Business process	Standard operating procedure
Sales and operations planning	Planned order conversion
	Schedule creation
Demand management	Planning
	Display PIRs
	Validate customer demand data
	Planning hierarchy maintenance
	Review/maintain flexible planning data
	Identify demand/forecast changes
	Historical forecast creation/monthly table review
	S&OP demand review
MPS	Maintain batch run parameters
	MPS and exception message operation
	MPS review
	Run MPS manually
	Manage write-offs
	Customer and dependent demand review
MRP	Maintain MRP batch run parameters
	MRP and exception message generation
	MRP review
	Run MRP manually
	Manage write-offs
	Customer and dependent demand review
Capacity planning	RCCP setup
	RCCP execution
	Capacity evaluation review
	Capacity evaluation issue resolution
	Demand version creation
	Scenario creation and initiation
	PIR management
	Plan and capacity
	Purchasing and financial simulation
	Feedback to operative system
Forecasting	Forecast demand control from BPS
	Replenishment
	Purchase order and receipt control
	Sales order and dispatch control
	Inventory/material requirements control
	Forecast demand control
Maintain master data	Manage forecast master data
	Manage MRP master data
	Manage MPS master data

Abbreviations: MPS, master production schedule; MRP, material requirements plan; PIR, purchase item receipt; RCCP, rough cut capacity planning; S&OP, sales and operations planning.

Appendix 31.D Example Manufacturing Business Processes with SOPs

Business process	Standard operating procedure
Process order management	Process order creation
	Process order preliminary costing
	Process order approval
	Process order archiving
Manufacturing	Release individual/collective process orders
	Material ordering and staging
	Print shop floor documentation
	Task list and work center processing
	Issue material to process order
	Missing parts processing
	Product labeling
	Batch record processing
Packaging	Release packaging orders
	Material ordering and staging
	Print shop floor documentation
	Task list and work center processing
	Issue product and packaging materials to process order
	Missing parts processing
	Packaging labels and documentation
	Batch record processing
Labeling	Create/change/approve labels
	Print/reprint labels
	Reconcile labels
Waste management	Backflushing
	Rework
	Disposal
	Stock reconciliation
Quality assurance	Inspection checks
	QC sampling and sample labels
	Certificates of analysis
	Variance/defects reporting
	Change management
	Retest materials
	Changing shelf life of products
	Reassigning products
New product introduction	Annual product review
Maintain master data	Maintain BOM master data
	Maintain master recipes

Abbreviation: BOM, bill of materials.

Appendix 31.E Example Sales and Distribution Business Processes with SOPs

Business process	Standard operating procedure
Presales handling	Customer records
	Pricing data
	Customer purchase restrictions
	Product equivalence
	Batch allocation
	Foreign trade processing
Direct sale to consumer	Creating an inquiry
	Providing a quotation
	Order placement
	Shipping processing and tracking
	Transportation processing (including freight forwarding)
	Billing
	Intersite/company transfers
	Export sales and special documentation
	Contract/tender management
	Toll sales
	Free of charge sales
	Rush orders
	Bonus goods allocation
Third-party order processing	Creating an inquiry
	Providing a quotation
	Order placement
	Transportation processing
	Billing
Returns processing	Batch tracing
	Cancelled customer orders
	Recall processing
	Returns order processing
	Bill corrections (credit and debit notes)

Appendix 31.F Example Finance Business Processes with SOPs

Business process	Standard operating procedure
Asset management	Asset master record maintenance
	Asset acquisitions and capital expenditure
	Asset decommissioning
	Depreciation simulation (including tax)
	Asset calculation
	Asset revaluation and write-ups
	Insurance revaluation calculation
	Investment support
	Period end closing and reporting
Revenue and cost controlling	Actual vs. planned operating costs
	Cost center allocation
	Profit center balance sheets
	Creation and maintenance of internal orders
	Budget values and availability controls
	Commitment accounting
Product costing	Stock valuation
	Planning and comparison
	WIP valuation
	Variance calculation
General ledger	Account maintenance
	Posting of general ledger journals
	Period end processing (day, month, year)
	Financial reporting
Accounts payable	Supplier/vendor data maintenance
	Invoice verification
	Matching purchase order to invoice
	Down payments on purchase orders
	Settlement (payment) of supplier/vendor account
	Invoice and credit note processing
Accounts receivable	Customer data maintenance
	Invoice and credit note processing
	Customer down payments
	Bills of exchange
	Letters of credit
	Debt collection
	Customer settlement (payment) of account

Abbreviation: WIP, work in progress.

Appendix 31.G Example ERP Validation Document Set

Validation master plan	URS	Functional design implementation specification	Qualification	Validation report

• Project procedures (GAMP)
• System overview
• Supplier audit (application vendor, service providers)
	• URS with business process models
	• GMP assessment (risk analysis) part 1: business process evaluation
		• Functional specification
		  - Procurement
		  - Production planning
		  - Manufacturing
		  - Sales and distribution
		  - Finance
		• User procedures
		• GMP assessment (Risk analysis)
		  - Part 2: GMP tests for SAP R/3
		  - Part 3: GMP tests for associated interfaces
		• Configuration definition and management
		  - Software components
		  - Hardware components
		• Bespoke reports and forms and other custom code
		  - Software design
		  - Programming standards
		  - Source code review
		• Vendor manuals
		  - Standard software
		  - Standard computer room hardware
		  - Standard IT infrastructure
		  - Standard database
		• Training and education materials
		• Requirements traceability matrix
			• Design review (phased?)
			• Installation qualification (initial and upgrades)
			• Data load reports
			• Conference room pilot
			• Operational qualification (phased?)
			• Performance qualification (phased?)
				• Validation report
				• Glossary of terms
				• Document index and navigation aid
				• Updated training records
				• Support procedures (system management, change control, disaster recovery, security practice, performance monitoring, periodic review)

Abbreviation: URS, user requirements specification.

Case Study 14: Marketing and Supply Applications

Louise Killa
LogicaCMG

INTRODUCTION

The data held within marketing and supply applications are key inputs into the efficiency of the supply chain and provide the vital link that connects Information Technology to the physical world of raw materials, intermediate stock, finished inventory, business processes and people. What is done with that data—how it is collected, processed, communicated, stored or otherwise manipulated, determines its true value to any organization as Information Technology takes its place as an enabler for efficient enterprise coordination.

Contemporaneous control of these activities should also enable business benefits to be achieved such as increased productivity arising from more efficient use of key equipment and personnel, greater accuracy and elimination of common errors and the possibility of lowering stockholding levels without risk to customer service. However, organizations must ensure that such benefits are not achieved at the expense of regulatory expectations.

The computerized systems used by pharmaceutical organizations are expected to operate in accordance with their intended design to reliably and consistently support the regulated business processes. Whether it is agreeing artwork with the marketing organization, communicating medical information on licensed products, placing a purchase order for the supply of products, reading the bar code on the item, tracking its movement through the storage and distribution network, capturing relevant data at various transit points throughout the supply chain, or recording customer complaints, to retain the various licenses that are required to operate in the pharmaceutical markets around the world, the various regulatory authorities require to see formal evidence that these systems have been validated to confirm their suitability for use.

In recent years, there has been an increasing regulatory focus on the marketing and supply aspects of the pharmaceutical supply chain. While these activities can be carried out manually, it is more commonplace to find that a computer system is used to support them either wholly or partially (a hybrid system). The reliance on these systems as the sole mechanism of recording information means that they should be developed to an appropriate level of compliance and validated for their intended use to ensure they are able to provide a consistent output to support the regulated process.

This study outlines the considerations that need to be made when determining the validation activities that need to be undertaken on these types of systems which wholly or partially provide functionality to support the regulated processes outlined in Figure 32.1.

MARKETING APPLICATIONS

Marketing applications used within the pharmaceutical and healthcare industry include those computerized systems used to support international artwork and the provision of medical information supporting pharmaceutical and healthcare products released to market.

- *Artwork* needs to be agreed between the pharmaceutical, medical device or healthcare company and the marketing organization. Traditionally this has been achieved using fax and/or e-mail attachments. More recently the intranet and Internet have been used to facilitate such transactions in conjunction with electronic document management systems (EDMS).

  A variety of applications may be used to support the artwork process within an organization to ensure that it is able to rapidly respond to changes that are required to

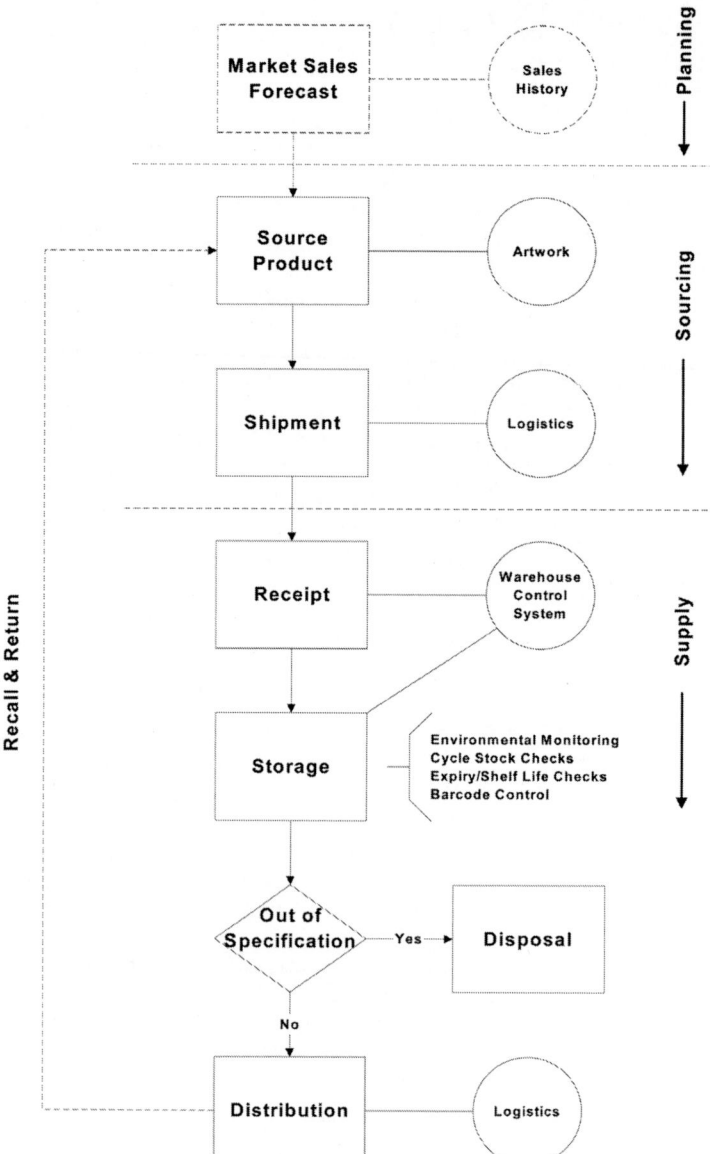

Figure 32.1 Context of supply application functionality.

meet appropriate market regulations. These range from different standard artwork generation packages, supporting templates, portable document format (PDF) technology and corresponding applications capable of reading and editing these files and the various e-mail and infrastructure applications used to facilitate the secure transmission of the artwork to the approved external studios.

Additionally, where an organization may find itself managing a significant amount of artwork changes and approvals, a workflow tracking system may be used to determine the status of a particular piece of artwork at any point in time.

When using workflow media, care needs to be taken to allow for the fact that colors that are displayed on a computer screen may actually look different if printed out on different printers, this can at times give false confidence regarding the acceptability of the material which might differ again in the final printed output.

All artwork produced should take into account the technical requirements of the relevant printing process and the materials used. Failure to do so could result in a process that could present problems if the validation of activities on supporting systems is required.

Because of its potential to have a considerable impact on patient safety, artwork processes generally involve several stages of critical checks within the pharmaceutical organization to ensure that the final copy is as error free as possible. The completion of these checks is recorded in supporting workflow applications at key checkpoint stages.

Depending on the level of technology involved in the supporting system, approval may use a hybrid form involving the appendage of a physical signature on printed output to authorize its progression to its next stage, or may use encryption technologies such as electronic signatures.

• *Medical information* supporting products released to market can exist in many different formats. The medical information produced for the finished product, whether in label format or as a package insert should correspond with the actual ingredients that the product is formulated to contain (1) and the method and dosage that corresponds with its license.

Regulatory authorities have also stated (2) that labeling is not limited to the immediate product container or package insert, but also includes all promotional material that is distributed and/or published in connection with a particular product or medical device.

The Internet has found increasing popularity as a means of publishing medical information about pharmaceutical and healthcare products including press releases, sales sheets, brochures and advertisements. Such statements should be backed up with process, controls and checks to ensure the information displayed is accurate and relevant to the product.

Statements made by, or on behalf of marketing organizations during promotional audio conferences will also attract regulatory attention (3) if they minimize crucial risk information and promote a drug or device for an unapproved new use.

Incorrect, misleading or incomplete artwork and medical information could lead to the inappropriate use of a drug or medical device in which case a recall of that product is usually required. These systems should therefore be validated as they can impact public health.

Other marketing applications that may fall within the scope of regulatory scrutiny are those used to record complaints received regarding drugs or medical devices, any other EDMS applications and systems that retain and develop documentation used in dossier submissions.

Not every marketing process supported by a computer system would fall within the scope of regulatory scrutiny. Computer systems such as forecasting applications, cost control systems and systems that are used to purely support only financial or accounting activities may not need to be validated at all. However, each of these systems is still expected by Industry regulators to undergo some form of formal documented regulatory assessment to determine whether or not this is the case.

SUPPLY APPLICATIONS

In addition to corresponding changes within manufacturing operations, in recent years the warehousing and distribution aspects of the industry have tended to become more automated. This has minimized the need for human intervention thereby eliminating many user errors and removed the reliance on paperwork for the operation of the plant and the management of the many specialist products and suppliers, leading to improvements in productivity.

The distribution requirement planning (DRP) process that is supported by supply applications focuses on the flow of goods through the downstream channels of the supply chain. Planning information regarding the demand for goods cascades from customers to licensed local distribution depots, and from these to a main warehouse and finally back to the supplier's factory (or external source of supply), taking into account ordering and delivery between the levels. These systems can be seen as a front-end extension to a manufacturing resource planning (MRP)II system.

The supply applications discussed in this study are those applications within the overall supply chain that are used to support the sourcing, receipt, storage, disposal and distribution of licensed products and may also be used to support product recall activities, should this activity become necessary. The principles discussed in this chapter would equally apply to applications that are used to support the receipt, storage, disposal and distribution of raw materials, intermediate products and material for clinical trials.

The application solution selected by an organization to support its supply process may consist of one integrated application that has different modules capable of handling all elements of the supply process. Alternatively, it may comprise a series of separate complementary applications that have been linked together by a series of system interfaces and business processes to provide an overall solution to the business.

The connection of these systems within and between organizations is key to the successful operation of the supply chain because it allows important information to be shared by suppliers, partners, distributors and even customers, who may view data to see the progress of their orders.

Typical systems that support the business in supplying its inventory to licensed wholesalers or distributors may include the following:

- A *procurement/purchasing system* that manages the sourcing of items from approved suppliers. This is a function that has traditionally generated considerable quantities of paperwork to communicate information from one function to another to facilitate action, to indicate requirements to suppliers and to obtain the necessary goods required by the supply chain on time and to specification.

  In recent years purchasing has been recognized as playing a more strategic, rather than a purely transactional role within an organization. The advent of more integrated purchasing software applications has facilitated integration with other applications to improve the performance of the overall supply chain.

  Goods and services should only be sourced from suppliers assessed as being capable of providing materials products and services that meet the standards of the organization. Regulatory authorities expect the organization to have a documented system for the assessment and approval of the suppliers appropriate to the type of item or service being sourced. If the procurement process followed is supported by a computer system that records the approval of the potential supplier, then the usage of this computer system is expected to be validated.

  Applications used to support the procurement process for items having a regulatory impact should clearly indicate the approval of a supplier and should prevent the selection of unapproved suppliers for products and services that have been deemed to have a regulatory impact. The validation of the system will be expected to demonstrate that this relationship is correctly established within the system.

  Any computer system responsible for supporting the shipment of the items into the supplying warehouse, or the return of goods (for whatever reason) is generally under the direct responsibility of the supplier of the goods and is likely to exhibit the characteristics of the distribution system outlined later on in this section. However, where the goods being shipped fall within the scope of regulatory scrutiny, the data provided by any shipment applications is expected to be created and maintained in line with regulatory requirements order to meet these requirements.

- An *inventory management system* that contains all relevant supplier details to enable the receipt of goods, checking against purchase orders placed, allocating suitable storage locations, any requirement to repackage or relabel to meet local market requirements, inventory status management, segregation of products following recall or return activities, stock rotation and associated checks, disposal and stock level monitoring and replenishment from the source of supply. This system would also record any status changes necessary to reflect the activities that are carried out as this stage of the supply chain including approved release by the qualified person (QP) or quality control (QC) function.

The complexity of such systems can vary greatly from large-scale turnkey solutions, to more simple PC-based packages. Systems available also include front-end applications capable of adding real-time communications, bar code scanning and advanced productivity management to existing legacy systems. Advanced systems often incorporate an alerting capability which is able to monitor and detect when specific programmed events need to happen (for example replenishments needing to be completed two hours before a picking cycle commences), or when expected events fail to materialize (for example, the late arrival of an expected urgent delivery from a particular supplier).

Depending on the requirements of the operation, this system may also include breakdown or aggregation of inventory received into smaller or larger stock-keeping units (SKUs) for onward distribution. It may also include interfaces to purchase order systems, labeling applications, automated materials handling systems such as sortation systems, stock location systems, automated storage and retrieval systems, radio data terminals (RDTs), radio frequency identification units (RFIDs), automated guided vehicles (AGVs) or conveyor systems and could involve integration with devices such as bar code scanners, balances, asset tracking solutions and microchip identification systems.

- A *warehouse system* encompassing all aspects of the management and maintenance of the storage facility in line with the relevant regulatory and local health and safety expectations. This will encompass maintenance of standards of cleanliness, sterile areas, temperature, humidity, pest control systems, physical security (including the restriction of access to controlled drugs). This system may interface with physical alarm systems for the building and other automated access control systems (e.g., swipe card systems).

- A *sales order processing system* that contains all relevant customer and inventory details to enable a sales order of the correct characteristics to be placed against the actual stock available to fully or partially satisfy the order. The characteristics of the contract between the company and each customer may differ greatly, for example some customers may only be licensed to receive a very limited subset of the total inventory held within the system, while others may be licensed to receive everything but a few specified inventory items available from the supply chain. In both cases, the system would be expected to contain functionality which allows each type of circumstance to be correctly set up to ensure that sales orders for onward wholesale suppliers are only raised for inventory that the supplier is licensed to hold.

    Additionally, the contract to supply different licensed customers may specify different supply terms for the same inventory item. For example, a pharmacy wholesaler may only accept deliveries of orders containing inventory with at least a 12-month expiry date, while a grocery wholesaler may be willing to accept inventory with a minimum six-month expiry date. Therefore the sales order processing system would be expected to differentiate between the stock rotation dates of the inventory made available to satisfy the orders placed by each type of customer.

    The Sales Order Processing system is likely to be one of the prime sources of information used should any product recall activity become necessary. Depending on the type of sales order processing system, it may have interfaces to other systems that electronically feed in sales orders to the organization and to automated or manual order picking systems.

- A *distribution system* to ensure that the items specifically picked to satisfy a sales order are successfully and safely delivered to the required destination with no deterioration in product quality or risk to public safety. The system should ensure that orders can be tracked throughout their distribution cycle to enable clear control and traceability to be demonstrated when required. This system may be required to interface with logistics management systems run within the organization or by third party logistics suppliers. This system may include functionality that interfaces with a separate proof of delivery

(POD) system and financial systems containing general ledgers managing the payments to distributors for services received and invoicing to licensed wholesalers for the goods they have ordered.

The distribution system may also be the initial point at which returns to the warehouse reenter the supply chain system, for example if goods are supplied to wholesalers on a sale or return basis. Depending on the functionality and interfaces of the system, upon completion of successful deliveries, it may provide confirmation of this activity back to the sales order processing system.

Any system used by representatives from manufacturing or distributing organizations to record the distribution of drug samples to licensed healthcare professionals upon request would also fall within this category.

REGULATORY REQUIREMENTS

The validation of marketing and supply applications encompasses exactly the same fundamental activities as any other validation exercise carried out on a system supporting activities within the supply chain that have a regulatory impact.

Addressing these activities should ensure that the software is developed, adequately tested and maintained to remove the likelihood of system failure, and to ensure that the system possesses the necessary resilience to ensure a rapid and complete return to operation in the unlikely event that a failure does occur.

Requirement to Validate Marketing and Supply Applications

Depending on its use, supply chain applications supporting the business processes of marketing, product sourcing, shipment, receipt, storage, disposal and distribution could fall under the scrutiny of various regulatory authorities covering the countries in which the organization operates and those to which it supplies products.

Within the European Union (EU) there are several regulations (4–10) that govern marketing and supply operations. Within the United States, there are different regulations (11–18) that govern marketing and supply operations, although the principles are broadly the same in covering the manufacturing, holding and distribution of products and medical devices and their associated records. Regulatory citations such as the ones below are not uncommon.

You have failed to validate XXXX computerised systems used for drug product distribution information. (19)

Failure to validate the computerised system used by your firm to track drug products from receipt through distribution, in accordance with 21 CFR 211.100(a). This computerised system is also used by your firm to generate the "Unique Barcode Labels" that are applied to cases containing drug products, and/or to the individual drug product containers, as part of your relabelling operations. (20)

Failure to exercise appropriate controls over and to routinely calibrate, impact or check automatic, mechanical, or electronic equipment used in the manufacturing, processing, and packaging of a drug product according to the written program designed to ensure proper performance (21 CFR Part 211.68) in that the Installation Qualification (IQ), Operational Qualification (OQ), or Performance Qualification (PQ) for the XXXXX was not performed. (21)

Prescription drug products stored at your firm, with temperature range controls from 68°F to 77°F, were not held in accordance with the label requirements to ensure the identity, strength, quality and purity of the drug products as set forth in 21 CFR Part 211.142(b). There were no reading logs or data of temperatures or relative humidity conditions of the warehouse since March 13, 2001 as required. The temperature indicator at your firm during the inspection indicated a storage temperature of 81°F. The air conditioning system is not run continuously and is turned off overnight, weekends and holidays. (22)

Failure to have a written individual record of major equipment cleaning, maintenance and usage [21 CFR 211.182]. There were no equipment logs for the bar code scanner, programmed label reviewer and the roll splicing equipment. (23)

Your firm failed to conduct quality audits at the intervals listed in your Quality Audit Procedure xxxxxxxx, to verify that the quality system is effective. (24)

Developers of applications that are intended to be used by organizations operating under the scrutiny of more than one regulatory body should ensure that the validation activities within their project lifecycle encompasses the requirements of all these organizations. This will ensure that each of the intended end user sites can successfully complete their on-site validation activity. Appendix 32.1 identifies some of the possible GxP business processes supported by marketing and supply applications.

Requirement to Follow Standard Operating Procedures and Quality Standards

Failure to adequately document standard working practices is a deficiency commonly cited by regulatory authorities (20,25–27). Recent FDA inspections have indicated that in some circumstances, no attempt has been made to cover this activity: "failure to establish procedures for the warehousing and distribution of stock" (28); "failure to establish written operating procedures for drug production and process control steps. For example, the receipt and handling of drug components and containers" (19); "failure to establish written procedures for the receipt, identification, storage, handling, sampling, examination, and/or testing of labelling and packaging materials; preparation and printing of labels, examination and review of labels; disposition of rejected labelling; issuance of labelling, reconciliation of quantities of labelling issued, used and returned; destruction of unused labels bearing lot numbers; and the 100% visual inspection of labels hand applied to drug products" (21); and "failure to establish written procedures for the monitoring of temperatures, humidity and the air handling system in the production area" (21).

The absence of written operating procedures is particularly relevant to the validation and operation of bespoke computer systems used to support warehousing and distribution processes. Regardless how PC literate an individual may be, it is unlikely that any staff members would be immediately able to use a bespoke warehousing and distribution application without some form of training. Standard operating procedures (SOPs) form an important part of this training. Failure to do so can also result in Warning Letter citations (19,25,27).

Electronic Record and Electronic Signature Requirements

Regulatory authorities only require pharmaceutical and healthcare organizations to satisfy any relevant electronic record and electronic signature (ERES) considerations such as those in the respective EU (7) and U.S. (11) regulations if the usage of the application has been deemed to support a GxP-critical process.

Where this is the case, the organization is expected to be able to provide evidence that it has assessed these applications against all relevant ERES requirements and initiated satisfactory remediation plans to address any issues identified by this assessment. It is essential that controls are established to ensure the authenticity and integrity of the electronic records and where appropriate, any associated electronic signatures used as the legally binding equivalent of a handwritten one.

Organizations should ensure that they are aware of exactly which activities they capture electronically within their marketing and supply applications would fall within the scope of any relevant ERES regulations. Records transmitted by electronic means such as fax, or word-processed documents that are subsequently printed, authorized and maintained as paper records may not always fall within this category. Mechanisms should be put in place for prospectively assessing the ERES capability of any new applications that they intending to commission for their use.

Even if an application does not have a GxP regulatory impact, it may still need to meet other local ERES requirements if this is a mandatory criteria for the organization for any other reason, for example to meet the expectations of local financial, legal or health and safety regulatory authorities. POD functionality using a component capable of capturing a recipient's signature with a stylus such as a handheld personal digital assistant (PDA) might be one such example of this.

VALIDATION LIFE CYCLE
Validation Determination

Authorities regulating healthcare companies expect regulated companies to have adequate documentation to support the GxP/non-GxP determination processes recorded for each of their systems. Also, for every system used to support a regulated process, a corresponding documented ERES assessment is expected to be present.

The business process that the system supports should provide the primary determination regarding whether or not any validation activity is required to be undertaken on the system. Each system that has been identified must be assessed to see whether it performs quality or business critical functions, whether there are sufficient controls in place to ensure its performance and whether it is required to be validated or not.

This assessment can only realistically be performed by representatives from the business community within the organization who have an understanding of the business processes undertaken. Other personnel who have specific expertise in the regulatory expectations surrounding that particular activity may assist staff. The assessment should consider the effect of the use of the system with respect to the following:

- Product purity
- Product identity
- Product efficacy
- Patient safety
- Regulatory submission process

In addition, organizations need to consider if the business process could be used as the basis for any regulatory discussion, even if it is not the primary purpose of the process.

Only when it is confirmed that the business process requires validation do the regulatory requirements for any system (or manual process) that supports it need to be considered in any further detail.

Validation Plan

Successful validation of computer systems cannot be built in as an afterthought and validation planning should commence as soon as possible after the requirement to validate the application has been determined. Retrospective validation is far more expensive and resource intensive than prospective validation with no guarantee that regulatory expectations can be satisfied at the end of it.

An overall validation master plan (VMP) for a marketing and supply application should be established. It is essential that the plan extends to encompass all of the components of application including hardware network, infrastructure, interfaces and other systems that are necessary for its successful and continued operation.

The user site should ensure that the scope of validation reflects the actual use of the system by the end users, and not just the intended use documented. For example, an off the shelf warehouse application may contain functionality developed to meet the requirements of many different companies, so could actually contain far more functionality than is actually required by the end users from one particular company. Users may select this functionality if they believe that it provides them with additional features that would enhance their business processes in preference to the intended documented usage that the organization intends to validate. Such functionality should be disabled, or users should be expressly prohibited from using this alternative functionality through SOPs and training.

Care should also be taken to ensure that GxP data is not being extracted from the application, manipulated in an uncontrolled manner by another application and then input back into the original application. Situations such as this destroy the data traceability necessary to meet regulatory requirements. Events such as this often arise because users have developed a local workaround to a system problem rather than reporting the issue and getting it resolved and fixed properly by the nominated support team.

Subordinate validation plans may be developed for individual user sites, or to some other logical business grouping. For example, a particular system type within the same site, such as all DRP systems, or the computer systems operated by a particular business unit or department within the supply chain covering multiple sites. Where services have been outsourced to an external third parties, the relevant aspects of the work of the third parties are required to be verified by a supplier audit of each party that forms part of the overall validation plan.

Validation plans should reference any governing VMP (or quality plan as appropriate) in addition to defining site-specific validation activities. Both site validation plans and central VMP should clearly differentiate central team accountabilities and deliverables from site validation accountabilities and deliverables.

Site validation plans should define where external supporting documents will be used in support of site validation activities. Central and site teams should agree relationships between their activities and documents and should work together to ensure that all required activities are covered. If the validation plan covers an identical usage of a system on more that one site, consideration should be given to developing templates to ensure that activities are performed in a consistent manner across the organization and to maximize use of central documents without duplicated site effort. Where responsibility for validation activities is shared in this way, each site should maintain an entry in their system register for those applications/products they use that are centrally developed/supported, with the central support groups maintaining a corresponding system register of sites using the GxP systems they support.

Configuration Management

A configuration management plan should be established to outline the process to be followed to ensure that configuration and version control is established for the software, development tools and supporting documentation that are used. This will enable accurate configuration baselines to be taken at specified points in the software development lifecycle.

It is important that electronic master copies of documentation are placed under configuration control to prevent accidental changes being made to current or future revisions of a document.

All staff should undertake GxP training so they are aware of the significance of the expectations placed by regulators on the accuracy and fitness for purpose of the software they are developing and the reasons behind why the development lifecycle needs to be undertaken without any unapproved or undocumented deviations.

User Requirements Specification

The user requirements specification (URS) for any system is typically written by the business community (users) and describes what it is intended that the system should do. This is a key factor for consideration in any system validation determination as it should indicate whether the business process the system is intended to support has a regulatory impact. The requirements outlined in the final URS document will subsequently tested by user qualification.

Site needs should be established and documented in central URS and site-specific URS as appropriate. Wherever possible a common set of user requirements should be developed between sites using the same system. Specific local regulatory requirements should be clearly stated in addition to any other statutory requirements necessary to satisfy other bodies such as the financial reporting requirements of the Inland Revenue within the United Kingdom.

Outsourcing systems development is becoming increasingly common and where this is the favored option, a version of the URS should be included in the invitation to tender (ITT) sent to potential Software Suppliers, and must clearly distinguish between mandatory requirements and requirements that may only be preferable if the design could accommodate it. The outsourcing organization must ensure that the companies that have been invited to tender clearly understand the implications of developing the software that is required to meet regulatory expectations and that the company has the necessary processes in place to support this.

Supplier Audits

Organizations that have decided to outsource key elements of their regulated computer systems should ensure that only approved suppliers capable of consistently supplying software products and services that meet regulatory expectations are used. Contracts placed with the supplier should include provision for ensuring immediate and ongoing regulatory compliance. This is no different to the expectations for suppliers satisfying other types of supply criteria within the pharmaceutical and healthcare supply chain (29,30) and regulatory authorities may request documented evidence to support the supplier selection process.

Regulatory authorities will also expect to see evidence that the software supplier has been audited against appropriate documented assessment criteria by suitably qualified individuals before the products are used in the production environment. This assessment is expected to be performed again at regular intervals after the initial assessment to determine their continued suitability to supply and support the software.

Central development and support teams should ensure the quality of suppliers supporting central activities have been assessed. Centrally organized supplier audits should be conducted in conjunction with regulatory groups that are familiar with the organization's operating model. Suppliers supporting local modifications should also be subject to supplier assessment.

Functional Specification

The functional specification (FS) for any system is typically written at a detailed level by the supplier (developer) and describes what it is intended that the system will do. The requirements outlined in the final FS document will subsequently tested during user qualification.

The signed FS forms the agreement between the technical staff and users that the requirements stated in the URS have been correctly understood and provides a level of confidence that the intended system will meet user requirements. If the development of the application is being outsourced, a preliminary version of the FS may be included in the supplier's response to the ITT, although the final version is expected to be prepared by the successful software supplier in conjunction with the user.

When considering the requirements specified for the functionality of a system, the planned local usage of any report writing and user configurable utilities present in the software should be assessed to determine whether or not this usage is required to be subject to local validation activities. The basic functionality of utilities of this type does not generally contain any GxP data types, although they should have some form of user controls placed on their use. However, the business processes that determine the way in which these utilities are used on site could make them subject to regulatory requirements, and if so, they would need to be validated.

Identification and Segregation of GxP and Non-GxP Data

Where the use of an application has been assessed as having a GxP impact, this does not necessarily mean that every module or area of functionality within the software has a regulatory impact.

It is acceptable for an organization to determine that they only need to validate a subset of the overall software used (that part with functionality that has GxP impact) and not the whole application if that software is distinct. Appendix 32.2 indicates the data within marketing and supply applications that typically have a regulatory impact. However, regulatory authorities will expect to see evidence of some form of impact or risk assessment, which has determined that this segregation will not present a risk to the successful operation of the system in a regulatory environment.

System functionality that is common to both regulated and unregulated application modules, such as system security and menu access should always be considered as having a GxP impact.

Not all of the data identified as having a GxP impact may be considered to be GxP critical by an organization and the organization may undertake a risk based approach and subsequently decide to validate only the data considered to be GxP critical in line with their own usage. If preliminary assessment identifies a significant amount of data as having a GxP impact, it may be of little benefit to proceed with a full GxP data segregation exercise, as it may involve less resources and overall effort to consider the whole application as having a GxP impact.

Regardless of whether the specific module has been determined as having a regulatory impact or not, all modules within an integrated application will be expected to follow consistent change control and configuration management processes.

Design Specification

The design specification should be a complete definition of the equipment or system in sufficient detail to enable it to be built and is written by the developer (or supplier). Unless the system developed is very basic, it is usual for the design to be broken down into several different documents. Where this is the case, regulatory authorities expect an overall summary of the system design to be generated showing traceability between these documents. A documented design review should also take place to confirm that the design meets the stated requirements.

Throughout the lifecycle of a project, it is common to find that this documentation evolves at the project progresses to incorporate subsequent approved changes or to correct design defects that are found during any phase of the testing process. One of the commonest regulatory failures cited regarding system design is that the software development documentation set is not kept up to date, or where it has been updated, does not bear any approval signatures.

The requirements outlined in the final design specification document will subsequently tested during user qualification to verify that the correct equipment or system is supplied to the required standards and that it has been installed correctly.

Software Coding

The development of validatable applications requires evidence that the code is an accurate translation of the intended design of the system. Inspections involving computer systems frequently involve an analysis of the coding work undertaken.

Every project should indicate the mandatory documented program coding standards, directory structure standards, file naming conventions and configuration management processes that are expected to be undertaken for all coding output. All of these documents should be readily available, either in electronic or hard copy, for all staff to consult as necessary.

All software development tools and the configuration management tool used should be assessed for suitability and fitness for purpose and this assessment should be documented.

All code should be subject to configuration and version control and where command files or compilers are necessary to facilitate the software build process, these should also be controlled in the same way. The lack of any obvious or consistent version control is often cited as a regulatory deficiency.

Once program code has been compiled, it is expected to undergo an independent documented coding review to confirm that it meets its design requirements and has been developed in accordance with the relevant coding standards. If these conditions are not met, the coding review should document the deficiencies and the code should then be reworked until it has been verified that these conditions have been met.

Development Testing

The testing of an automated system is required to be performed at several levels that should demonstrate that the controlling specification has been implemented correctly. Successful completion of a particular testing phase will therefore allow the project to progress to the next phase of the project lifecycle. Testing should also encompass the testing of the hardware and infrastructure as well as the software itself.

Testing should be fully documented, including approved, detailed test specifications, test cases and the results of testing in the form of signed test sheets and raw data to provide a complete record of the testing undertaken.

As part of the validation activity, testing should be independently checked to confirm that it demonstrates traceability back to the corresponding requirement and that it verifies that the requirement has been met. For example, unit testing should verify that the design specification has been satisfied, system testing should verify that the FS has been satisfied, and user acceptance testing should verify that the URS has been satisfied.

The validation plan should also clearly indicate the split of responsibilities between user and supplier. It should be noted that the supplier is normally involved in all levels of testing. The validation plan should state that an independent expert (often QA) will confirm that the testing is being carried out in accordance with the documented test strategy may witness testing. This is important where the software is developed externally, or where the development is taking place centrally on behalf of several sites.

User Qualification

User qualification activities involve the completion of the installation qualification (IQ), operational qualification (OQ), and performance qualification (PQ) activities.

The hardware and software IQ processes may be reasonably straightforward if the physical system is intended to be located on the same site and only used by that site. However, the introduction of centrally managed systems, shared service operations and data warehouses as a means of lowering costs and providing a more streamlined infrastructure within an organization may make the overall validation activity for the site more complex.

It is also common to find several different supply chain applications residing on the same host machine. This may impose additional complexity for the overall site validation activity should some of these have a regulatory impact and others not. Where an operating system is shared, this could result in non-GxP critical applications operating with additional controls not to compromise the activities of the GxP critical application.

Site validation activities should include configuration management, data load, and specification/design/testing of any locally developed specific site modifications. Validation of site-supported infrastructure should be incorporated within site activities.

During the OQ phase, tests should be conducted to verify that any wireless data solution employed in addition to hard-wired networks meets the roaming and resilience criteria documented in its corresponding requirements specification. Central testing should be used in support of qualification wherever possible. However, should the OQ have been run centrally, it may be possible for the site OQ activity to be waived, provided that the software has not been locally modified and the central test environment used has the same characteristics as the local site.

PQ is a site-specific activity but may be coordinated across multiple sites if appropriate. Issues raised during PQ should be reviewed by both the central support teams and the local site. In some instances, it may be beneficial for templates for validation documents to be developed by central teams to provide a consistent approach across sites.

PQ is heavily dependent on the business producing test cases on the basis of their current or intended SOP that correspond to the URS and provide the required operational and maintenance controls. It is important that these test cases are up to date and reflect all of the relevant agreed changes that have been implemented since the start of the project. The use of computer systems should be consistent with relevant company policies and relevant regulatory expectations for the business processes that they are intended to support.

Validation Report

To complete the project lifecycle, the project team should produce a validation report that aligns with the site validation plan. In addition to confirming site validation activities, validation reports should confirm the adequacy of all relevant central activities. A central validation summary report (quality report) should be developed reviewing the adequacy and release of each application/product version.

Independent experts may review the results of specific validation activities and the summary report may incorporate their findings and any corrective actions necessary following their review of the original activities.

OPERATION AND MAINTENANCE
User Documentation

Effective SOPs for the use of computer systems should be established by the business process owners to provide the correct operational and maintenance controls for the validated system. Following SOPs, plus any supporting detailed localized work instructions, should ensure that a state of control is established and maintained at all times while the system is being used to support a regulated process.

These SOPs should clearly define the inputs and expected outputs of the process covered by the SOP to provide objective measurement criteria to determine whether the process is being successfully operated or not. This state of control should then allow the required electronic records to be created and maintained by the system, in addition to the creation of any documentation specified as part of its requirements, as outlined in the traceability requirements diagram shown in Figure 32.2. Any relevant supporting process flows, system overview or network diagram documentation should also be reviewed and revised as necessary to reflect the new business processes if the requirements and SOPs for the system are updated in the future.

Particular care should be taken to ensure that suitable SOPs are in place to monitor finished inventory in line with regulatory expectations to ensure that it retains the purity, identity, efficacy and safety characteristics required by its licensed specification. If it does not remain within acceptable boundaries of this specification, then it will be considered to be "out of specification" and must be immediately allocated a restricted status such as "quarantined," followed by disposal in a safe manner in line with local health and safety and regulatory expectations.

The business should ensure that adequate SOPs are available and routinely tested to ensure business continuity can be maintained in the event that the system becomes unexpectedly unavailable. These SOPs should also outline alternative procedures that should be followed if the process is expected to continue in the event of failure of key system components such as bar code scanners that may not necessarily render the system inoperable.

In addition to system and technical configuration documentation created as part of the development of changes, it is essential that any user manuals issued to support the use of computer systems in conjunction with the local SOPs are updated whenever required. These manuals should display indication of review and subsequent approval and should also clearly indicate which revision of the software that they apply to. This is especially important if these user manuals are developed by external software houses.

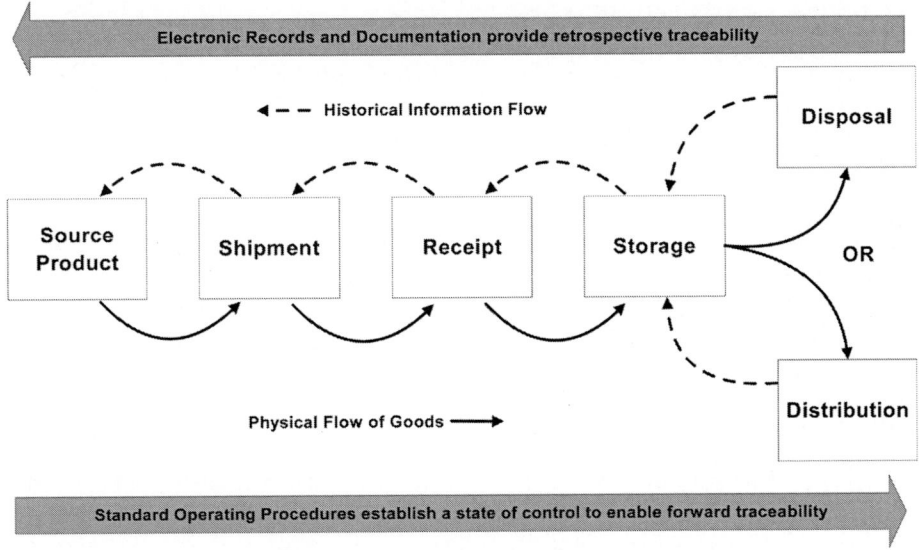

Figure 32.2 Traceability requirements for supply applications.

Training

The training of staff who operate systems is of key importance to the delivery of its expected benefits. Regulatory authorities expect companies to be able to demonstrate that they have evaluated the training needs for their staff (30). They also expect that all staff have received an appropriate level of training in the specific processes and systems they are expected to follow and operate respectively before they commence their work. Staff will also be expected to have the language capabilities to be able to fulfill their roles (31). In addition to any specialized training necessary to fulfill their role, general GxP training is required.

This training should be recorded in the training records kept for the individual. Training should be delivered by suitably qualified individuals who can demonstrate their current regulatory awareness (32). If external consultancies are used, evidence should be obtained of their qualification to deliver this training (5,6,17). Training is also required to be repeated at regular intervals so staff knowledge remains up to date.

Warehouses are one area within the supply chain where contractor staff are sometimes used on a regular basis. Contractor staff are often used at short notice to cover for sickness or other unplanned absences, or to provide additional resources at peak busy periods to supplement the existing operators. Another area that may involve the use of contractor staff is the logistics and transportation process that delivers the goods to the licensed wholesaler. In this instance, often the entire operation is subcontracted to a third party.

Both areas involve the handling and processing of finished inventory, with the possibility of contractor warehouse operatives also being involved in the packaging of the goods or assembly of packs. Therefore organizations must ensure that all contractor staff have received satisfactory process and GxP training appropriate for the tasks they are to undertake before they start work.

Organizations are also responsible for ensuring that any contract manufacturers, suppliers or third party distributors used are made aware of the requirement to ensure that their staff receive adequate GxP training. This requirement extends to IT staff who develop the systems used to support the organization's operation, even though they may not physically handle the goods themselves. Regulatory authorities make no distinction between the training requirements for permanent, temporary, part time or third party staff when it comes to training (32).

Companies should seek to attract and retain knowledgeable staff who will be able to contribute to the further development of supporting systems at appropriate points and also should the system's usage change at some point in the future. Where staff leave the organization, or move on to different roles, it is essential that knowledge transfer to replacement staff is properly effected and recorded to maintain a stable operation.

However, managers should be aware that if problems are subsequently encountered regarding the use of the system, and training has been eliminated as the root cause of the problem, then further training in itself is unlikely to solve the problem. Managers should remember that staff may need encouragement to apply the knowledge gained during their training properly.

Change Control

Like any validated application, subsequent changes to marketing and supply applications within a production environment should be controlled and documented to demonstrate that the validated status of the application has not been affected by the change being made. Change control and configuration management processes need to be documented (39) and should be considered as parallel activities, particularly during the assessment of the impact of the change.

Care should especially be taken regarding changes made to intranet sites, where situations could easily arise where the content of the site is changed without a full impact assessment being carried out on the effect of the change. The control maintained over the intranet site itself as well as its contents are crucial. The data displayed must be up to date, correct, accurate and maintained in accordance with documented procedures.

There have been numerous instances in recent years of pharmaceutical and healthcare companies operating via the Internet to publish medical information about drug products or devices without realizing the GxP relationship between the Internet information and the actual labeling of those products. Drug press releases have been defined as labels by regulatory authorities (13) and making false and misleading representations can have serious consequences for any pharmaceutical organization (33). Depending on the extent of the misbranding, this could cause a product to be considered as a drug, and legally prevent its marketing without a corresponding approved new drug application (NDA) (2).

Regulatory authorities such as the U.S. FDA have established routine monitoring and surveillance programs, operated by its Division of Drug Marketing, Advertising and Communications (DDMAC) and any violations detected will result in Warning Letters being issued without regulatory authorities actually visiting any of the organization's operations (3).

Care also needs to be taken regarding the consistency between the labeling of the product itself and any promotional statements being made by third parties such as licensed distributors on behalf of the manufacturer. Should these claims cause the product to be misbranded, it is the manufacturer who may receive the actual citation from the Inspector, in addition to the distributor (34).

Where changes are being made by a central development and support group, this group should ensure that changes to the system are communicated to all affected local sites and receive appropriate agreement prior to being actioned. This ensures that any local implications of the change can be considered prior to its development. The central group are responsible for ensuring that a change is completed in such a way that the validation status of the core application is not affected during the change control process.

Emergency Changes
Unless an emergency situation has developed that would result in a safety hazard, loss of product quality arising because of a system shutdown, sites should not install software in a production environment until all of the relevant validation activities have been completed. If such an emergency situation develops, a documented justification is required to indicate why they considered it necessary to install the software without completing the necessary validation activities. Regulatory authorities expect completion of any outstanding activities as soon as possible after the installation has taken place. A scheduled system change where the implementation plan does not allow enough time for it to be properly executed is generally not considered to be a sound justification for an emergency change. Indeed regulatory authorities will expect to see corrective action being put in place, wherever reasonably practicable, to minimize the likelihood of the risk that the emergency situation will reoccur in the future.

Security
The integrity of the data held within any system is essential to the successful operation of a regulated business process. This can only be guaranteed if effective security measures have been established to prevent any unwarranted access to the data, whether intentional or not. The generation of electronic copies of master production records without any apparent controls to ensure that their authenticity or data integrity will not be overlooked by regulatory authorities (25).

The security within an application may be a combination of application system security and the standard security features provided by a proprietary operating system. Care should be taken to ensure that the ERES audit trail required by an application is created to track any work undertaken in this area, especially the activities that may be performed by a System or database administrator without using the "front-end" graphical user interface (GUI) of the system.

A list should be readily available of system users and their access levels to various functions within the application. This list does not have to be available to all users via a formal menu option within the system, but should be easy for a competent System Administrator to generate.

Care should be taken where a system logon is given to a temporary member of staff to use for the duration of their employment (e.g., an contractor forklift driver engaged at short notice to cover for unexpected sickness in the warehouse). Regulatory authorities expect to be able to clearly trace who actioned a particular transaction, and the practice of using a generic log-on such as "contractor temp" used by different individuals on different days should not be encouraged.

Routine IT activities such as the resetting of user passwords should contain steps that require some form of authentication check on the person requesting the password reset.

A key security area that often gets overlooked is the removal of system access or specific privileges from a user's profile for an application when the privileges are no longer necessary. This is often done if an employee leaves an organization, but this often forgotten if someone moves to another job within an organization.

Regular reviews of system privileges should be encouraged to ensure that staff are not granted more privileges than their current role may actually require. The withdrawal of privileges should also include the removal of the ability to apply electronic signatures where a change in role no longer requires this capability.

However, managers should ensure that any deletion or alteration of user profiles does not affect the transactions held within the system. For example, while the system ID is active, a master file for a particular piece of artwork may show the name and system ID of the person who approved it, and thus meet regulatory expectations. However, if this information is sourced from a master file of current systems users, depending on how the design of this link has been established, deleting the user may end up removing the user's detail from the artwork record, thus taking a previously compliant record out of regulatory compliance.

Backups

To prevent the loss of critical GxP data, backups of systems should be taken at regular intervals, as documented in local SOPs. This prevents the physical loss of important system data or its accidental deletion by users. A backup also provides a basis from which the application can be restored in the event of an unforeseen event arising that results in the need to invoke the documented disaster recovery plan. Data that should be subject to backup activities includes system software, such as the operating system and any preconfigured software modules, application software, and configuration parameters. Within a regulatory context, examples of the type of operational data that would be subject to a backup process would be the critical data items identified from the list of data types shown in Appendix 32.2 of this chapter.

The frequency at which the backup is taken is largely dependent on how frequently the data changes. For example, backups of sales order processing systems that may process thousands of new orders per day are likely to be taken more frequently than backups of system configuration parameters, which are less likely to change on a daily basis. During the process, data is copied onto media external to the system and then stored in a secure offsite location in line with documented SOPs. The backup process should be tested at regular intervals and this testing should be documented.

Disaster Recovery

The support group for the system must ensure that adequate procedural documentation is in place to ensure that system continuity can be restored in the event of a system failure. Business continuity plans should be developed in line with relevant service-level agreements (SLA) agreed between the business and the supporting IT organization.

Regulatory authorities expect that the process that will be followed to recover a regulated application will be documented in SOPs and that training will have taken place at regular intervals to ensure that all applicable staff are aware of their responsibilities in relation to the business continuity plan. Evidence should be available to confirm that these processes are regularly tested to ensure the supporting SOPs are adequate to complete the activity or modified if these tests indicate that events during the test did not quite turn out exactly as planned.

This expectation is exactly the same for any third party systems used, for example logistics systems. Organizations can expect regulatory authorities to inquire about whether this topic was covered during the supplier audit phase.

Upgrades

All upgrades, regardless of category must be carried out in accordance with documented change control and configuration management plans. The upgrade of a validated system should be considered as a site, rather that system specific activity, as it is likely to involve several different aspects of the IT operation on site. System upgrades can be split into three broad areas.

- Upgrades of the application itself
- Upgrades of the supporting operating system
- Upgrades of the supporting infrastructure or hardware

An upgrade of any part of a previously validated system does not necessarily mean that full revalidation is required. The validation plan that addresses the upgrade should incorporate an impact assessment to determine the exact nature of the change, how much of the validated system will be affected by it and whether it would be within regulatory expectations to undertake a partial validation only. It is recommended that this assessment should take a documented risk based approach.

If the configuration item being upgraded is an established commercially available operating system (Category 1), then less validation activity is normally necessary. However, upgrades to configurable software (Category 4) or bespoke packages (Category 5) are likely to involve more validation work (35).

Decommissioning

The decommissioning of any supply chain application is an important regulatory process that often receives less attention than the commissioning of new applications. Decommissioning should be planned well in advance of the deployment of the new solution, and the plan should include a full assessment of the impact of the withdrawal of the system, including the effect of this on any external third party applications that interface with the application to be retired.

Early planning will allow for full consideration to be given to the best method of preserving any GxP data to allow it to be retrieved at any time in the future within its specified regulatory retention period. This may involve migrating legacy data into any application replacing the retiring application.

Data Archiving and Record Retention

From time to time it may be necessary to remove data or other electronic records from an on-line computer system to another durable secure location for long term off line storage. This removal also includes the removal of the metadata associated with the record, for example, the properties of a Microsoft Word document or a series of data definition and table relationships.

This may happen as part of a planned migration process when active records are moved from a legacy system to a new system. It may also occur in systems that have been operational for some time when inactive records are removed from a database to improve the performance of the existing system. A record transitions from an active to an inactive phase when its documented ageing or status requirements are satisfied and the record is no longer subject to change.

How legacy data should be retained is often a complex issue. The approach taken depends on many different factors, but the most important of these is usually whether or not the method of retention can be maintained throughout the required retention period. It is possible that that technology chosen may become unsupportable during this period and if this is the case, then an alternative method may have to be chosen either before the original archiving takes place, or while the data is in the archive itself. Depending on the circumstance, this may also include archiving to nonelectronic media.

The periodic review of the data contained within the electronic archive often tends to be overlooked. Organizations must ensure that they monitor the records for signs of deterioration over the record retention period, including the media used for storage, if this is being done electronically.

The business owner of a particular system is ultimately accountable for the electronic retention of business data. Support groups are often delegated responsibility for the electronic record retention of technical data such as operating systems, source code, configuration records and any other software necessary to operate the application itself including supporting documentation for these components.

Organizations are also responsible for ensuring that any contract manufacturers, suppliers or third party distributors used are made aware of the length of time they will be required to retain records of their activities either electronically, or on paper, in case these may be required in the future.

There are many other reasons why pharmaceutical and healthcare organizations are required to hold records, not just regulatory purposes. The record keeping requirements of local environmental, safety or financial regulations should also be considered when determining the length of time a record is required to be held. Whatever requirement stipulates the longest retention time is the one that should be adhered to and recorded to determine the disposal criteria.

Where records are retained centrally for use by more than one site, then all of the applicable regulatory, environmental, safety or financial retention periods for each site need to be considered when determining the record retention period required.

Records are not required to be held indefinitely. When the chosen record retention period has been met, they may be destroyed. The approach to record destruction will vary according to the design of their archive location, but should consider the maintenance of the data integrity within the archive system. The frequency of record destruction also needs to take into account factors such as operational cost, performance and ease of disposal.

FUNCTIONALITY ISSUES
Packaging and Labeling

All labels generated by supporting application should be printed and applied so as to remain legible and affixed during the customary conditions of processing, storage, handling, distribution and, where appropriate, use (10,12). Labels should not be released for storage or use until they have been inspected for completeness and accuracy by the designated individual (31).

Any labels generated by a supporting application for finished pharmaceuticals must ensure that the way in which the information is displayed on the label gives prominence to the active ingredients (36). Where a label for a drug or medical device has been generated by the supporting application, it is still expected to be accounted for in the same way as labels produced by other means (26).

Computer applications used to support repacking operations and subsequent relabeling should contain functionality that operates checks to verify that the integrity of the data relating to the batch is maintained. For example that repacked products from the same parent lot still retain the original lot number and that individual items do not display multiple expiration dates (37).

Where the final item being shipped consists of a collection of units the final pack may often comprise different items that each have different expiry dates and batch numbers. Records that are held within supporting computer applications should accurately reflect the lot number and expiry dates for all of the elements contained within each pack to facilitate traceability. Once the pack has been produced and entered into the system as an item in its own right, the expiry date shown in the system for the pack and on its external packaging should be the earliest component expiry date. This should be verified as a specific test case during OQ and PQ testing.

Inventory Cycle Checks

Regulatory authorities may request that a pharmaceutical organization can account for all inventory of a particular item even if a product recall has not been necessary (36).

Inventory management applications should have functionality and supporting SOPs to enable both planned and unplanned cycle stock checks and other stocktaking activities to be easily undertaken—the latter is especially important as it may form part of any product recall activities. This checking may be required to cover stock at several locations, not all of which may be under the direct control of the organization responsible for managing the recall.

Coordination is very important. The system should be configured to rotate stock either a first in–first out (FIFO) or first expired–first out (FEFO) basis.

Storage Conditions

Pharmaceutical distributors should ensure that the storage of finished goods meets accepted temperature and humidity conditions for both product quality and regulatory compliance. This requirement applies to all areas within a warehouse where temperature, humidity and airflow are required to be monitored such as specific hot or cold spots within the racking, loading bays or repackaging areas, areas of entry and exit and other restricted access areas (e.g., sterile or hazardous areas). Goods in transit should also conform with storage conditions.

Where a computerized system is used to support this activity, the validation of the computer system should include ensuring that any automated monitoring systems are functioning within the specified range (21). System alarms and other alerts should be adequately tested prior to use to verify that they would be triggered correctly. This testing should be periodically repeated to ensure the validated status of the system is maintained. It may be appropriate to undertake a warehouse mapping exercise to identify climatic variations. Validation activities should also verify that perimeter access systems are adequately tested to ensure that staff who should not be permitted access to restricted areas are actually prevented from gaining this access.

Stock Selection

Developers of regulated systems must ensure that the system incorporates functionality that prohibits the selection of stock with a restricted status to satisfy a sales order. An example of this would be a situation where the testing and release procedures for a particular stock item require appropriate laboratory determination of satisfactory conformance to its final specifications prior to release (38). This requirement should be stored within the computer system as an attribute of the particular item. The system should be configured in such a way that the stock cannot be released into the sales order processing system until the conformance of the product to this specification has been recorded in the system by a suitably authorized individual. Another example would be where the business operates a process within its supply chain that allows stock with an expired lot date to be received back into a warehouse via a credit note for final disposal. The supporting system must ensure that this stock is allocated a system status such as quarantine that will prevent it being reselected to satisfy another sales order. The system should also ensure that this stock can only be released for disposal under the authority of a QP. In both circumstances, regulatory authorities would expect to see evidence that this functionality has been adequately tested during validation.

Segregation of Stock

Organizations need to ensure that the functionality within supporting computer applications can be configured to align with the physical characteristics of any warehouse used. The application needs to be able to make clear distinctions between the available storage locations that can be assigned to items with a particular allocated status or storage characteristic. For example, the put-away algorithm logic needs to ensure that only appropriately restricted locations would be displayed as being available for the storage of any item with a 'Quarantined' status. Any automated selection of storage locations needs to take into consideration the type of items stored in any adjoining locations, following relevant approved industry guidance. For example it may not be acceptable to store antibiotics next to other items in case of contamination. OQ and PQ activities within the validation lifecycle would be expected to cover scenarios such as this in detail, in addition to ensuring that only authorized system users were permitted to authorize the removal of items from certain locations using an automated process.

Sample Management

Inventory records need to account for samples withdrawn for stability, expiry or shelf life checks. Depending on the allowable tolerances for controlled drugs, such sampling, if not recorded properly, can lead to unacceptable levels of unaccounted stock in the event of a product recall being initiated. Sample request or receipt forms may be transmitted photographically or electronically, for example by facsimile transmission (FAX) or electronic data transfer provided that the method of transfer meets the security requirements outlined in 21 CFR Part 203 (14). Computer systems used to manage or record the distribution details for samples held by Manufacturers' or Distributors' representatives also require validation.

Out of Specification

There are many aspects of the inventory storage process that can have a direct bearing on the characteristics of a product, for example the characteristics of a product may change if it is stored at an inappropriate temperature or passes its shelf life/expiry date. Therefore steps must be put in place to ensure that the application used to control this storage highlights areas of concern before they happen and does not directly or indirectly contribute toward the generation of an OOS result (36). Such steps may include ensuring that the warehouse control system only selects locations with appropriate storage conditions, or ensuring that stock is rotated following an agreed strategy. Functionality should exist within the application to provide information well in advance on those products whose shelf life is approaching the next check cycle or whose expiry dates is approaching. This should be established in accordance with the requirements specified by the business for the particular item. This should be tested during validation. OOS records generated by the computer application must be retained as part of the batch production, packing or control records (35) and are expected to be investigated (39).

Inventory Disposal

Where it is necessary for inventory to be disposed of, for whatever reason, this should be done in accordance with any relevant regulatory and local health and safety legislation. This will require certain information pertaining to the disposal of the lot/batches in question to be recorded in the appropriate system in accordance with all legal requirements relating to the type of drug being destroyed. The type of data that is required to be recorded could vary greatly in accordance with the legislation that governs its management. For example, the rules applying to the disposal of controlled drugs are generally far more stringent than those applying to the disposal of OTC drugs regardless of the regulatory body involved. The activity required to validate this functionality should include the record keeping requirements of all types of drug disposal that it is anticipated that the application will be required to support.

Artwork Transfer

Artwork transfer processes should be validated including storage of electronic files on servers. External artwork studios and printers are expected to be subject to a Supplier Audit and periodic review in the same way as any other supplier to the pharmaceutical and healthcare industry. All software, including fonts, used for artwork preparation should be licensed to the organization. An experienced market representative should proof read the hard copy printout of files (such as PDF) to ensure correct content. Once this has been confirmed, the artwork administrator should check that the printed copy, the electronic version, and the original artwork file are controlled and in alignment before engaging the print supplier. A final check is required (17) by QC to confirm that the packaging components delivered to site correspond with their intended specifications before they can be released for use. Validation needs to consider the regulatory requirements relating to any e-mail or Internet applications used in the artwork transfer process. If the use of these technologies is taking place within a closed environment, the amount of validation activity required could be considerably less than if these technologies were employed within an open environment (7,11). If an open environment is created by using e-mail to transfer artwork files to design studios, regulatory authorities will expect to see all aspects of this activity covered in the validation plan. Even if the final agreed hard copy is hand signed to confirm approval of the artwork, the system infrastructure used in the artwork process still needs to be validated because it controls the integrity of the artwork that has been approved and demonstrates the validity of the approvals process itself. The final approver is only one member of a group of experts that may have viewed this artwork electronically during its generation and approval process.

Product Recall

Validation needs to be completed for those systems that would support any product recall activities. Regulatory authorities expect that traceability is established throughout the supply chain to facilitate the recall of an entire lot/batch of a product if this becomes necessary to minimize, amongst other things, any adverse impact on patient safety. Sites may have a locally developed IT solution, a centrally supported IT solution or may have even subcontracted this activity to a third party as part of an outsourced Sales Order Processing activity.

Outsourcing of Sales Order Processing operations may add additional complexity to the product recall process. Depending on the overall systems solution being employed, the third party may not have direct access to the manufacturing systems meaning that in some cases a hybrid system may be involved. For example, a signed physical shipment note containing GxP data might travel from the factory with the goods, with its details being input into a third party system on arrival in the distribution warehouse. In this instance, it is likely that the third party system would be used to effect any recall and not the internal one.

If a particular lot is the subject of a recall and the information relating to the lot is not easy to retrieve, then it is possible that a regulatory issue could develop. An example of a more complex situation might be where the vendor lot number has been manipulated in some way and there no easy means of tracing back to the original vendor lot number. If a formal report has been set up within the application and validated to ensure it that it gives the ability to generate a listing of sales orders by the original vendor lot number, as opposed any different system allocated lot number, the information required for any recall activity could be easily obtained.

Regulatory authorities do not mandate any one specific process to be followed for product recall, but they expect the process selected by an organization to be adequate for the task (37) and for the recall to be carried out within a reasonable timeframe. Accuracy is also of key importance, organizations need to be able to account for all of the items that are subject to a recall, not just those nearest to hand. Both the relevant U.S. (17) and EU regulations (4,5) clearly state they expect the chosen process to be fully documented (27) and evidence to be available to indicate that the relevant staff have received training in it prior to it becoming effective.

Contemporaneous Data

Less automated systems that need manual input to confirm that an activity has been completed require data entry to be contemporaneous. If this were not the case an inspector might walk, for instance, round an available stock area of the warehouse, notice an unattended pallet on a floor in the picking area which is available for picking, and ask someone to show him where this pallet should be. If an inquiry is made into the system that shows that the pallet should be in a separate quarantined area, as outlined in a current SOP, then staff should expect to discuss this further with the inspector (25,40).

Applications Storing GxP Data in Multiple Languages

Over recent years there has been a distinct trend toward the introduction of global solutions that support the supply chain processes for different countries using the same production instance. This sometimes requires the same information to be stored in more than one language and the operator would be expected to select the language required from within the range of options specified within the application. It is a stated requirement of some regulatory authorities (12) that the translation must convey the meaning properly to avoid confusion and dilution. This is particularly important in the case of warning statements. Unless the marketing company has made a prior arrangement with the supplying site for translations to be obtained locally, any foreign language text that is required to maintained within the supporting application should be supplied by the marketing company.

Database fields that are used to record label information are often free text fields. Care should be taken when validating fields that contain entries in different languages because mistakes may not be apparent to systems testers or operators if they are not familiar with the language concerned. It is often the case that IT development and support staff may only be given the file containing the approved translation from a third party. Care should be taken to ensure that any specific character or accent characteristics required by the new language will be recognized within the new application and will therefore appear exactly as they are intended to on the screen or printed output. Responsibility for the final checking of the text remains with the market and this requirements should be incorporated into the validation plans for the pharmaceutical and healthcare manufacturer's commercial organization.

Presence of Additional Functionality in Externally Developed Software

Where an industry standard modular application is selected as a software solution, it may include functionality within the application that supports different business processes operated by other companies who use the same application. Unless this additional functionality can be disabled within the application, which is not always technically possible, there is a risk that this functionality may be accessed by end users even though the site has not validated this usage. This situation should be controlled by ensuring all end users are trained in the specific functionality within the application that they are expected to use to support their business processes and this training should be supported by SOPs that clearly state that no alternative processes are to be undertaken.

ORGANIZATIONAL ISSUES

Relationships Between Local Sites and Central IT Development and Support Groups

Regulatory inspections are not necessarily limited to sites where the application is operated. Some production instances may be centrally managed and supported by a particular IT Group. Other applications may only be centrally supported by a particular IT Group, while the control of the production instance may be managed locally by the site using the application, or physically managed on behalf of a particular region or user group by another site using the production instance. The scope of an inspection can be extended to cover central IT development and support groups. Central IT groups must therefore ensure that they are ready at all times to face any inspection.

Service-Level Agreements for Inspection Support

The local site is responsible for ensuring that the dependencies for inspection support are understood and accurately documented in all appropriate supply agreements or SLA. Depending on the complexity of the documents, consideration should be given as to whether it might be more appropriate to develop a dedicated SLA for inspection support activities. Where organizational changes are taking place within a company, care should be taken to ensure that continuity of support between the central IT support team and its dependent local sites is planned from the outset and is monitored to ensure that any unacceptable risks are mitigated.

Outsourcing Activities to Third Parties

Outsourcing activities may bring about cost efficiencies and allow pharmaceutical and healthcare organizations to specialize in the core activity in which it believes it has a competitive advantage. However, where outsource activities fall under the scope of regulatory scrutiny, this does raise the issue of the computer systems no longer being under the direct control of the pharmaceutical and healthcare organization.

It is essential in any outsourcing operation that there is a written contract covering exactly what is to be supplied and the standards expected for the supplied item(s). The may be complemented by SLAs for hardware, software and network support. The contract should also state that an adequate quality management system (QMS) is expected to be in place to establish the necessary controls to ensure that the software systems meet all of the stated requirements given by the outsourcing organization. Third parties should ensure that they clearly understand exactly what is permitted within the relevant regulations. A satisfactory Supplier Audit is not a substitute for the relevant validation deliverables such as design specifications.

For software development, documents such as URS and FS are the controlling specifications for the items to be delivered and as such form part of the contractual document set for the software. These should be signed by all parties before any subsequent development work is undertaken, and changes in requirements should follow a standard contract variation process and should not be agreed informally between staff from both parties. Organizations should also consider whether it would be beneficial to include the type of documentation that would be required in an inspection situation as part of any ESCROW agreement that is established. In the event that the third party unexpectedly ceases to trade, the pharmaceutical or healthcare organization would then have access to the relevant documentation to back up its systems development.

TECHNOLOGY ISSUES
Use of Bar Codes
Applications interfacing with electronic reading components such as bar code or microchip readers must be validated fully before they are used in the same way as any other application. The bar code for an item may be incorporated into the artwork of the packaging for the item in a standardized format (e.g., ISO/EAN). Validation should include the use of bar code reading components.

In most instances it is not practical for the actual lot number or lot expiry dates to be captured in the artwork itself. If the warehouse management system chosen is bar code driven, consideration may be given to producing a further bar code to record this information and affixing this to the items from the relevant batch. However, if this is to take place, care should be taken to select a location for the new label that does not obstruct or deface key information on the packaging and also to ensure that the risk of the new bar code becoming detached from the packaging is eliminated. Care should be taken when selecting the font used to print out the bar code. Bar code fonts have been known to create EAN/UPC symbols with serious design defects. The design of the font, an operator input, or a combination of both may sometimes cause problems. In addition, most fonts do not automatically calculate and add the check digits and other security features to bar codes expected by regulators. A separate application is usually needed to calculate the check digits first so that they are available to be added to the bar code created. Even if a bar code system is used within the supply chain to generate the lot number and lot expiry for each SKU, and thereby increase efficiency, both types of information still need to be present in a format visible to the human eye for the end user. This is usually achieved by punching or embossing this information on one or more of the primary, secondary, or tertiary packaging solutions used.

In addition to the placement of bar codes on product packaging and labels, some pharmaceutical manufacturers are now considering printing bar codes on individual drugs. The intention is to ensure that medicines used in hospitals are compatible with computerized systems used to support day to day operations to ensure that a patient gets the right medicine in the right strength at the right time. Data types that could be encoded onto the product to support this are the medicine type, its dosage, lot number and expiration date. In addition to reducing human errors, the codes would simplify recalls, investigations of adverse events and the purging of expired medicines from inventory.

Internet Applications
The increased use of the Internet in recent years as a means of communicating information on drugs and medical devices has provided considerable benefits to the healthcare industry. The Internet offers many possibilities in terms of graphic representation of data and the ability to publish information to a wider audience at a faster speed than by the use of traditional marketing channels. The overriding compliance consideration, however, is the accurate transmission of information to reader, whether the reader is a member of the public or a pharmaceutical and healthcare professional.

Website developers should consider the possibility that features visible with some Internet browsers may not always be visible if another browser is used. This can be difficult to fully scope during testing because it is unrealistic to expect testing to encompass all of the web browsers that are known to be available within the market place. It is suggested that the most practicable approach that could be taken would be to do the following:

- Ensure that the website and its development environment remain under configuration control at all times and changes to its content should be made in accordance with a documented change control plan appropriate to the technology in use.

- Establish and document the standards to be used for all website development following the same approach used for other coding standards within the software development lifecycle.
- Ensure that these standards outline the use of standard HTML/XML with no use of browser specific extensions.
- Ensure that standards specify minimal use of browser scripts (if at all) and plug-ins and should prohibit the use of any platform specific plug-ins.
- Incorporate any relevant considerations such as requirement for accessibility for users with visual handicaps and text only browsers into the layout and design of the web page.
- Consider clearly marking the information with the date/time on which it is valid, to distinguish it from the date/time it could be printed out by readers.
- Verify during the design review phase that the code produced contains valid logic. Appropriate tools such as HTML checkers may be used to assist in this process.
- Ensure that the test cases used for OQ and PQ should test a reasonable range of browsers currently available to determine that the information is displayed in a consistent format across this range.

This approach should ensure that the design of the website avoids creating warnings, hazards or other pertinent facts within medical information statements. If a different browser is being used that does not support the original technology used then important medical information may not be brought to the reader's attention.

As Internet information may be held in the 'cached' memory on servers other than those of the originating organization (for example, some public search engines), consideration should be given to clearly marking the information with the date/time on which it is was published. This is needed to distinguish it from the date/time it could be printed out by readers. This should be incorporated into the OQ and PQ testing undertaken before the website is released or updated.

Although not a regulatory issue, organizations intending to use the Internet as a method of communication should also consider taking control of the domain names they use and publicizing them as the official communication channels for the organization. Internet users may find it difficult to distinguish between information posted by an organization on its website(s) and other opinions stated by other organizations on different websites, especially if the website address on which the information has been placed is similar to the official website address. Available domain names can be legitimately registered that are extremely close to an official website address for the manufacturing or marketing organization. For example, two completely different organizations could register the same website, one using the ".co.uk" suffix and the other using the ".com" suffix.

Use of Spreadsheets

The use of spreadsheets within the pharmaceutical and healthcare industry is one area that is increasingly coming under greater regulatory scrutiny. Spreadsheets need to undergo an assessment exercise similar to other computer systems during their requirements phase to determine if their usage needs to be validated before they can be used. Authors and users should have different access profiles to prevent accidental over-typing of data. Spreadsheets that have a regulatory impact should be stored on servers with managed access control. This may be achieved using the "password to modify" option or by establishing NT access control functionality. Consideration should also be given to the effect of upgrading the spreadsheet package on the validation of the spreadsheet. An upgrade of functionality could affect any existing prerecorded macros present within the spreadsheet file, for example any calculations may not give exactly the same result that they did previously. Validation should take place to confirm that the upgrade has not introduced any undesirable features.

Radio Frequency Identification

RFID is a technology that looks to have the potential to make a major impact on the Pharmaceutical Supply Chain over the coming years in the fields of product and resource identification and tracking. RFID is creating innovative new business opportunities by making everyday objects and products intelligent and interactive. RFID is a unique technology that enables data to be transmitted from a micro silicon chip at very fast speeds and without the need for line of sight as is currently required by bar codes. RFID has been at work for several years in systems where fresh food products are tracked through the supply chain. RFID is robust and will survive in harsh environments where a bar code would normally be destroyed. As the technology has significantly decreased in size and cost since its introduction, it is now becoming cost-effective to place an RFID tag, consisting of a chip and antenna, on virtually any object, and allow that object to be identified uniquely and tracked accurately

Awareness of Technology Limitations

When selecting any application that is intended to be used to support a regulated process, organizations should ensure that they are fully aware of any limitations within the technology methods they are considering and whether these limitations may make it difficult to successfully validate the use of the chosen solution. Limitations may include, but are not limited to the following:

- Spreadsheet packages that have a limit on the number of characters that can be contained within one cell
- Databases that have limits on the total number of records the database can contain or on the maximum size of particular field types
- Graphics or drawing packages that may have a limit to the number of pages that can be created in a particular file
- E-mail applications that may truncate any attachments

In some instances it may not be apparent to end users that the particular limitation is reached, for example, the sender or recipient may have no indication that the attachment to the e-mail sent has been truncated. In other instances, existing information at the beginning of a file may be overtyped by information entered once the data limit for the file has been reached.

If there are any known limitations for any chosen application and the decision to proceed with the use of the application to support a validated process is made by the organization, then documented warnings should be given to staff so they are aware of the situations that they should avoid in the course of their work.

VALIDATION ISSUES
Awareness of Relevant Regulatory Expectations

In addition to ensuring that staff receive adequate and frequent training that meets the relevant regulatory expectations in full, it is important that they also receive training in the Inspection process itself so they are aware of how this process is likely to be carried out.

This should ensure that potentially awkward situations do not arise where they are reluctant to provide a regulatory authority with information that might reasonably expect to be discussed during an inspection (39). A refusal to provide relevant information to FDA for instance may contribute toward a Warning Letter being issued to an organization that might otherwise have been avoided (41). Such refusals are regarded as serious violations because they are deemed by regulatory authorities to hinder an inspector's ability to thoroughly and completely evaluate an organization's ability to make safe and effective drug products (23) and medical devices.

Change in Regulatory Status

Even if a system has previously been assessed as having no regulatory impact, it does not follow that this assessment will remain correct for the lifetime of the system. A purchasing system, for example, that has previously only been used to purchase non-GxP items, but subsequently used to purchase GxP items will now require validation even though the functionality remains exactly the same as documented in the previous assessment. All central groups developing and supporting a system are responsible for notifying their local user site if there are any planned changes in the use of the computer systems. The local site will then need to ensure that an impact assessment is undertaken to determine the effect of the change and whether its implementation means that a full or partial revalidation exercise is necessary.

Human Error

Where a system is not fully automated and requires transactions to be undertaken and subsequently confirmed by users, opportunities for human error arise. Human error in artwork is typically a major problem. A seemingly small error, often with decimal points, similar product names or minor inconsistencies in dosage information can have far reaching consequences for the organization concerned because it could jeopardize patient safety (36). Another accidental error could result in inventory being put in the wrong physical warehouse location. While it could be easy to locate the misplaced item(s) if the space is adjacent to the one that should have been used, the incorporation of functionality such as the requirement to input check digits or scan a unique bar code for each location is encouraged. This can then be used to automatically update the system to confirm that the put-away activity has been successful. A further common source of errors occurs when high bay racking locations are out of reach for the operator and as a result they are required to input the check digits or scan a bar code for the location that has been placed in a more accessible location as a substitute. Fully automated systems rather than hybrid ones are therefore recommended.

Reliance on Suppliers

Contracts placed with the supplier should include provision for ensuring regulatory compliance. This is no different to the expectations for suppliers satisfying other types of supply criteria within the pharmaceutical and healthcare supply chain (30,32) and regulatory authorities may request documented evidence to support the supplier selection process.

Regulatory authorities will expect to see evidence that the software supplier has been audited against appropriate documented assessment criteria by suitably qualified individuals before the products are used in the production environment. This assessment should be repeated at regular intervals after the initial assessment to determine their continued suitability to supply and support the software. Regulatory authorities will not have been party to the confidentiality agreements between the pharmaceutical or healthcare company and the supplier being audited. Consequently without due cause they will not expect to see the detailed findings of the audit itself. Instead regulatory authorities expect to see evidence that audits has taken place, including evidence of the management of any follow up activities.

Contracts to supply for software are expected to list the documentation to be provided as part of the contracted deliverables. With an externally sourced software product, these generally include documentation such as installation instructions, user manuals, system administrator manuals, configuration manuals, data definition guides and training materials, all of which are key documents for the validation process. Other documentation is likely to be proprietary and confidential to the software supplier for intellectual property reasons. Where this is the case, it is a regulatory expectation that the organization has undertaken a Supplier Audit to assure that the software product complies with regulatory expectations and is fit for its intended purpose.

Central development and support teams should ensure the quality of suppliers supporting central activities have been assessed. Centrally organized supplier audits should be conducted in conjunction with regulatory groups that are familiar with the organization's operating model. Suppliers supporting local modifications should also be subject to supplier assessment.

Effect of Subsequent Local Modifications to Globally Released Applications

Applications that have been globally released to different sites and then legitimately modified locally to meet specific reporting requirements for the local site only may require more validation than just the change itself. The receiving site may no longer be able to wholly rely on any development documentation provided by the central support group. This is because it may no longer correspond to its actual business use where the local modification has amended the functionality present in the software. Local sites should ensure that any documentation generated for the new modification adequately addresses this gap.

Process Improvement

For medical devices, inspectors expect to see a process improvement program in place encompassing all activities within the supply chain, including the systems that support it. Activities that would provide evidence that such a program is in place include the implementation of documented processes such as corrective and preventative action (CAPA) (41), self-inspection and internal audits (42). Organizations should ensure that where a SOP has been put in place to initiate a process improvement activity, the SOP is adhered to. For example, if a quality audit SOP states that all areas shall be subject to audit at least once every two years then inspectors will expect to see evidence that this has been the case. Failure to do so is likely to result in a citation similar to (24).

Within any software development lifecycle, the effectiveness of the change management process and the corresponding configuration management process are often indicators of whether or not any improvements need to be made. One opportunity for continuous improvement that is often overlooked in the software development lifecycle is the post project review process which concentrates on any negative business and technical issues that have arisen as a result of the deployment of the new software system. Project teams should also be encouraged invite members for other teams to take part in certain review processes to encourage two-way knowledge sharing on what they have both learnt from the execution of their respective projects. This could shorten delivery times for later projects and help teams avoid earlier pitfalls by learning about some of the activities and methods applied by other projects. The effectiveness of the process improvement program depends to a certain extent on the culture in which the information is being exchanged. However, if this process is not managed constructively, it could be easy for this to develop from an 'improvement culture' into a 'blaming culture'.

Use of Electronic Records and Electronic Signatures

The introduction of paperless systems has meant that specific consideration has to be given to the way that ERES are created and subsequently managed. The method and justification for preserving data on electronic media should be documented in sufficient detail and communicated to all relevant staff. This is important because there is generally no paper documentation to back these transactions up should insufficient electronic data be recorded or satisfactory data subsequently become corrupted. The electronic records created to support all transactions that fall within the scope of the various pharmaceutical regulatory authorities should generate satisfactory audit trails within the application that record the creation, amendment and deletion of these transactions. This is particularly important in the case of record deletions because the electronic record itself will no longer be present to support the transaction.

The approval process for artwork is one area where the growth in electronic communications technology could deliver significant reductions in lead-time to market for key information. Electronic transfers between the organization and external parties such as design consultancies and graphic studios means that the design of these processes needs to take special care to ensure that data cannot be amended in any way while it is being transferred between the parties involved.

IT staff should ensure that the boundaries between open and closed network components are clearly defined and there are methods specified for protecting data, such as access controls, firewalls and cryptographic techniques for data protection. The validation of open systems especially needs to be carefully planned to ensure the interest of all parties is protected.

Prospective Vs. Retrospective Validation

Wherever possible, validation should always be carried out prospectively before the application is used for the first time in a GxP context. It is acknowledged, however, that this is easier to achieve with a new business process/facility. If the warehouse facility or system, for instance, has originally been used for non-GxP purposes and is then used for GxP purposes, retrospective validation may be unavoidable and will need careful planning to be successful. Regulatory authorities often raise concerns regarding retrospective validation decisions, unless a sound justification for the decision can be demonstrated. Even if computer system validation has been attempted retrospectively there is no guarantee that any retrospective validation is going to be meaningful and therefore deliver a satisfactory outcome. There must be sufficient system documentation in place to demonstrate that the system has been developed against a formal QMS that takes into account regulatory expectations and good software development practices.

Global Impact of Regulatory Deficiencies

The ability to correct reported defects across an organization is an area that is coming under increasing regulatory focus with organizations being expected to provide a written commitment to regulatory authorities that they will prospectively correct similar potential or actual defects across other sites in their network. If the same known defect is found to be present in the same computer system on different operating sites, or on different computer systems supporting other operations on the same site, regulatory authorities could legitimately interpret this as a corporate pattern of bad practice (43). If the situation only applied to one circumstance on one site, a less severe censure such as an observation might be raised against the organization. Should the situation be found to apply to multiple circumstances, the organization might find itself in a situation where the inspector considers that there is no other appropriate option but to escalate the issue (e.g., FDA Warning Letter).

CONCLUSION

The validation activities required for any marketing or supply application, whether central or local, must have active and visible support from senior management to succeed. This sends a visible signal to staff that satisfactory validation of the software is a management concern and priority and that it should bring tangible benefits to the organization in the longer term.

Organizations should not wait for an inspection to detect or correct any issues arising regarding the validation of their marketing and supply applications. Regular monitoring of validation activities as they are taking place should be used as a basis to determine the effectiveness of the validation master plan. Any system related deviations or incidents should be investigated and used as part of the continuous improvement process that is a customary part of the SOPs that should be in place to support the software development lifecycle within all pharmaceutical and healthcare organizations.

ACKNOWLEDGMENTS

The author would like to thank Guy Wingate (Director, Global Computer Validation, GlaxoSmithKline), Don Valentine (Principal Consultant, Pharmaceutical Practice, Industry, Distribution & Transport Business Unit, LogicaCMG), Mark Lawson (Quality Support Scientist, Boots Healthcare International), and Paul Hetherington (Director, Emcee Squared Limited) for their contributory discussions and review comments to this work.

Louise sadly died during the preparation of the new edition of this book and is greatly missed by her friends, colleagues and family. The case study remains current and relevant which is a testament to her work.

REFERENCES

1. Food and Drug Administration. Warning letter issued to Vita-Erb Ltd. by the Kansas City district office, June 7, 2002.
2. Food and Drug Administration. Warning letter issued to Earth & Plant, Inc. by the Seattle district office, August 13, 2002.
3. Food and Drug Administration. Warning letter issued to Pharmacia Corporation by Division of Drug Marketing, Advertising and Communication. Rockville, MD, February 1, 2001.
4. EU directive 2001/83/EC on the community code relating to medicinal products for human use. Official Journal of the European Communities L311/67, 2001.
5. Pharmaceutical Inspection Co-operation Scheme. Guide to good manufacturing practice for medicinal products, document PH 1/97 (rev), December 2000.
6. Good distribution practice for medicinal products for human use 94/C 63/03. Official Journal of the European Communities L113, April 30, 1992.
7. EU directive 1999/93/EC on the community framework for electronic signatures. Official Journal of the European Communities L13/12, December 13, 1999.

8. United Kingdom Medical Devices Regulations. Regulation 2(1): S1 1994 No. 3017 (This U.K. regulation implements EU directive 93/42/EEC), 1994.

9. European directive 93/42/EEC concerning medical devices, class 1lb parts III and V.

10. Medicines Control Agency. Best practice guidance on the labeling and packaging of medicines. Her Majesty's Stationery Office, United Kingdom, December 12, 2002.

11. U.S. Food and Drug Administration, Department of Health and Human Services. 21 CFR part 11. Electronic records: electronic signatures.

12. U.S. Food and Drug Administration, Department of Health and Human Services. 21 CFR part 201. Labelling.

13. U.S. Food and Drug Administration, Department of Health and Human Services. 21 CFR part 202. Prescription drug labelling.

14. U.S. Food and Drug Administration, Department of Health and Human Services. 21 CFR part 203. Prescription drug marketing.

15. U.S. Food and Drug Administration, Department of Health and Human Services. 21 CFR part 205. Guidelines for state licensing of wholesale prescription drug distributors.

16. U.S. Food and Drug Administration, Department of Health and Human Services. 21 CFR part 210. Current good manufacturing practice in manufacturing, processing, packing for holding of drugs; general.

17. U.S. Food and Drug Administration, Department of Health and Human Services. 21 CFR part 211. Current good manufacturing practice for finished pharmaceuticals.

18. U.S. Food and Drug Administration, Department of Health and Human Services. 21 CFR part 820. Quality system regulation, medical devices.

19. Food and Drug Administration. Warning letter issued to Farouk Systems, Inc. by the Dallas district office, August 1, 2001.

20. Food and Drug Administration. Warning letter issued to Borschow Hospital & Medical Supplies, Inc. by the San Juan district office, February 21, 2002.

21. Food and Drug Administration. Warning letter issued to Pharmaceutical Distribution Systems by the Baltimore district office, January 3, 2003.

22. Food and Drug Administration. Warning letter issued to the Alero Corporation by the San Juan district office, January 16, 2002.

23. Food and Drug Administration. Warning letter issued to Eon Labs, Inc. by the New York district office, February 6, 2003.

24. Food and Drug Administration. Warning letter issued to Krieger Medical, Inc. by the New England district office, January 10, 2003.

25. Food and Drug Administration. Warning letter issued to Cardinal Enterprises, Inc. by the New England district office, December 7, 2001.

26. Food and Drug Administration. Warning letter issued to the Opti-Med Controlled Release Labs, Inc. by the Detroit district office, January 9, 2002.

27. Food and Drug Administration. Warning letter issued to Imperial Drug & Spice Corp. by the New Jersey district office, January 16, 2002.

28. Food and Drug Administration. Warning letter issued to Icon Laboratories by the Florida district office, February 4, 2003.

29. Food and Drug Administration. Warning letter issued to Purdue Pharma, Inc. by the New Jersey district office, November 9, 2001.

30. Food and Drug Administration. Warning letter issued to Hearing Aid Express by the Dallas district office, February 22, 2002.

31. Food and Drug Administration. Warning letter issued to Trusted Care by the New England district office, December 14, 2001.

32. Food and Drug Administration. Warning letter issued to PharmaScience Laboratories by the New Orleans district office, September 21, 1999.

33. Food and Drug Administration. Warning letter issued to West Agro, Inc. by the Kansas City district office, March 28, 2002.

34. Food and Drug Administration. Warning letter issued to Biogen, Inc. by the Center for Biologics Evaluation and Research. Rockville, MD, March 29, 2001.

35. Food and Drug Administration. Warning letter issued to CASA Lab, Inc. (d/b/a Walking Bird International, Inc.) by the New Orleans district office, January 2, 2002.

36. International Society of Pharmaceutical Engineers. GAMP—Supplier guide for validation of automated systems in pharmaceutical manufacture, version 4, December 2001.

37. Food and Drug Administration. Warning letter issued to Tom's of Maine, Inc. by the New England district office, November 14, 2002.

38. Food and Drug Administration. Warning letter issued to TYA Pharmaceuticals by the Florida district office, August 6, 2002.

39. Food and Drug Administration. Warning letter issued to the PureTek Corporation by the Los Angeles district office, February 10, 2003.

40. Food and Drug Administration. Warning letter issued to Hobart Laboratories by the Chicago district office, December 6, 2002.

41. Food and Drug Administration. Warning letter issued to Sountec, Inc. by the Dallas district office, November 14, 2002.

42. Food and Drug Administration. Warning letter issued to Minnesota Extrusion, Inc. by the Minneapolis district office, March 29, 2002.

43. The United States of America versus Schering-Plough Corporation and Schering-Plough Products, LLC (corporations) and Richard J. Kogan and Steven C. Chellevold (individuals) filed in the U.S. District Court, New Jersey, May 20, 2002.

Appendix 32.1   Possible GxP Business Processes Supported By Marketing and Supply Applications

This appendix lists possible marketing and supply business processes that may be supported by an application falling under relevant U.S. or European Union regulatory scrutiny. This list is indicative only and depending on the functionality offered by the chosen application, further regulatory conditions could also apply.

Business process	GxP relevance
Artwork	Organizations are responsible for ensuring that the text and graphic details of all printed packaging component artwork for all products and devices are accurate, complete and compliant with all known local requirements. Critical items of information should be located together on the artwork and appear in the same field of view where practicable. Where practicable, artwork for packs should include space for the placement of a dispensing label. Checks should be carried out to ensure that hard copy, electronic copy and original artwork file are all in alignment before engaging the print supplier.
Packaging	Packaging materials should be representatively sampled upon receipt and again before use. Records should be maintained for each shipment received of each different packaging material indicating receipt, examination or testing and whether the packaging has been accepted or rejected.
Labeling	All relevant information must be presented in a legible manner that is easily understood by all those involved in the supply and use of the drug or device. Labels should be printed and applied so as to remain legible and affixed during the customary conditions of processing, storage, handling, distribution, and where appropriate, use. No person other than the manufacturer, packaging organization or distributor should be identified on the label of the drug, drug product or medical device.
Medical Information	Marketing information should correspond with the terms of the license for the product. The information should not be false, lacking in fair balance or otherwise misleading. Only positive statements should appear on labeling to avoid ambiguity of the message, e.g., "for intravenous use only." Negative statements such as "not for intravenous use" should not be used. Information on a particular product or device should be consistent across all marketing channels. Marketing information for prescription drugs should detail all facts pertinent to the use of the drug including a true statement of information in brief summary relating to the side effects, contraindications and effectiveness of the drug.
Installation instructions (medical devices only)	Each manufacturer of a device requiring installation should distribute the instructions and the procedures for this activity with the device or otherwise make them available to the person(s) installing the device.
Supplier order placement	Orders should only be placed with suppliers who have been assessed and subsequently authorized to supply a product meeting documented acceptance criteria.
Received goods	Goods should be checked upon arrival to ensure the consignment corresponds to the order and that the goods have been checked for damage.
Materials handling	Prevent contamination or mix-ups during the course of receipt, identification, storage, packaging, labeling and quarantine operations.
Storage	Monitor and record environmental conditions such as temperature, humidity and air quality in accordance with predefined standards and procedures.

Appendix 32.1 Possible GxP Business Processes Supported By Marketing and Supply Applications (*Continued*)

Business process	GxP relevance
Stability testing (out of specification)	Assess the stability characteristics of drug products should be established and followed to determine appropriate storage conditions and expiration dates. This should include the periodic retesting of finished products during their storage. Similar written procedures should also exist to outline how the expiry conditions and dates for medical devices should be monitored.
Release of product	Where the release of batches for sale or supply is carried out using a computer system, the system should only allow a QP to release the batches and it should clearly identify and record the details of the person who released a particular batches.
Status of inventory	Goods should be assigned an accurate status that indicates their position within the supply chain at any particular point in time. Examples could include "received," "on hold," "quarantined," "available," "planned," "picked," "despatched," "in transit," "delivered," "rejected," "referred," and "awaiting life extension." Quarantined status may apply to goods shipped without certificate of analysis, returned products, damaged products, incomplete products, counterfeit products, expired products, misbranded and adulterated products or goods from an unauthorized supplier.
Inventory reconciliation	Losses, errors and inventory reduction following destructive testing should be reported and recorded. Stockholding inventories should be adjusted to reflect these activities.
Sales order processing	A sales order should only be raised in favor of persons who are authorized to hold and distribute the finished inventory or medical device.
Stock rotation	A process should be in place to ensure that the oldest stock should be distributed first—first in first out (FIFO) or first expired first out (FEFO).
Despatch	Finished inventory should be checked by a trained individual for identity, damage and to ensure they have been held under the correct storage conditions prior to distribution.
Distribution	The market supply planning undertaken should ensure continuity of supply in the event of an unexpected emergency situation occurring. For finished pharmaceuticals, distribution records should contain the name, strength of the product, description of the dosage form, name and address of consignee, date and quantity shipped and lot or control number of the drug product. For medical devices, distribution records should contain the name and address of the initial consignee, the identification and quantity shipped, the date of shipment and any control numbers used. Where a device's fitness for use or quality deteriorates over time, procedures should exist that ensure that expired devices or devices that have deteriorated beyond acceptable fitness for use are not distributed.
Record retrieval	Records that can be immediately retrieved by computer or other electronic means should be readily available for inspection during their retention period. Records kept at a central location apart from the inspection site and not electronically retrievable shall be made available for inspection within 2 working days of a request by an authorized official.
Complaints	Procedures describing the handling of all written and oral complaints regarding a drug product should be established and followed. A written record of each complaint should be maintained in a file designated for drug product or medical device complaints.
Product recalls	A documented process should be established to specify how product recall activity can be readily undertaken should this become necessary.
Product returns	Returned products are expected to be labeled as such and segregated from other stock to prevent reuse. Returned products should be destroyed unless examination, testing or other investigations prove that the returned drug product still meets appropriate standards of safety, identity, strength, quality or purity.
Product salvaging	Drug products and medical devices that have been subjected to improper storage conditions, or where the history of their storage cannot be verified, should not be salvaged and returned to the marketplace.
Product disposal	The destruction of defective or date expired products should be carried out in accordance with written procedures.

Appendix 32.2 Examples of Regulatory Data Types Within Marketing and Supply Business Processes

The table below provides examples of typical data types that may have a regulatory impact that can be found within various regulated marketing and supply applications. It should be noted that many of these GxP data types are common to more than one area of the marketing and supply process.

Process area/data type	Source product	Receipt	Storage	Disposal	Distribution	Marketing information
User control						
Security access	X	X	X	X	X	X
Name and password	X	X	X	X	X	X
Medical information						
Product/established name						X
Generic/proprietary name						X
Dosage form						X
Storage conditions						X
Sterility						X
Pharmacological/therapeutic class						X
Warning/hazard statements						X
Artwork details						
Storage conditions	X	X	X	X	X	X
Product/established name	X					X
Generic/proprietary name	X					X
Active ingredient	X					X
Dosage	X					X
Quantity/pack contents	X					X
Tamper evidence statement[a]	X					X
Product license details	X					X
Registration number	X					X
Bar code	X					X
Contact information	X					X
Packaging						
Packaging type	X	X	X			
Item/part reference number	X	X	X			
Item description	X	X	X			
Date of receipt	X	X				
Quantity received	X	X	X			
Supplier name	X	X				
Supplier address	X	X				
Examination/testing data		X	X			
Acceptance or rejection decision		X	X			
Labeling						
Product/established name					X	
Generic/proprietary name					X	
Dosage/quantity/pack contents					X	
Potency					X	
Lot/batch number					X	
Control number					X	
Date of expiry					X	
Handling conditions	X	X	X	X	X	
Storage conditions	X	X	X	X	X	
Installation instructions (medical devices only)					X	
Name of the manufacturer					X	
Manufacturer place of business					X	
Name of the packing company					X	
Packing company place of business					X	
Name of the distributor					X	
Place of business of the distributor					X	

Appendix 32.2 Examples of Regulatory Data Types Within Marketing and Supply Business Processes (*Continued*)

Process area/data type	Source product	Receipt	Storage	Disposal	Distribution	Marketing information
Inventory of samples						
Product/established name	X	X	X	X	X	
Generic/proprietary name	X	X	X	X	X	
Potency	X	X	X	X	X	
Number of samples received	X	X	X			
Name of sample recipient					X	
Address of sample recipient					X	
Date of sample distribution					X	
Number of sample units shipped					X	
Date of sample disposal				X		
Number of sample units disposed				X		
Supplier details						
Name of supplier	X	X	X	X		
Supplier address	X	X				
Address goods shipped from	X	X				
Address goods shipped to	X	X				
Purchase order number	X	X				
Supplier batch number	X	X	X	X	X	
Supplier control number	X	X	X	X	X	
Quality approval	X	X				
Lot/batch information						
Lot/batch number		X	X	X	X	
Lot/batch status		X	X	X	X	
Control number		X	X	X	X	
Date of expiry		X	X	X	X	
Date of receipt		X	X			
Quantity		X	X	X	X	
Potency		X	X			
Conversion factors		X	X			
Batch notes[a]		X	X	X	X	
Item						
Item number	X	X	X	X	X	
Item description	X	X	X	X	X	
Item notes[a]	X	X	X	X	X	
Location			X	X	X	
Type	X	X	X			
Quality	X	X	X			
Shelf life	X	X	X	X		
Retest days		X	X			
Bar code		X	X	X	X	
Purchase order						
Purchase order number	X	X				
Supplier	X	X				
Purchase order date	X	X				
Purchase order quantity	X	X				
Date of receipt		X				
Quantity received		X				
Unit of measure	X	X				
Supplier batch number		X	X	X	X	
Bill of materials						
Item number		X				
Item description		X				

(Continued)

Appendix 32.2 Examples of Regulatory Data Types Within Marketing and Supply Business Processes (*Continued*)

Process area/data type	Source product	Receipt	Storage	Disposal	Distribution	Marketing information
			Functional area			
Quantity		X				
Units of measure		X				
Conversion factors		X				
Work centers		X				
Potency		X				
Yield factors		X				
Critical process parameters		X				
Approval		X				
Process order						
Process order number		X				
Quantities		X				
Receipt date		X				
Transaction		X				
Inventory receipt						
Supplier name		X	X			
Purchase order number		X				
Purchase order quantity		X				
Quantity received		X	X			
Quantity outstanding		X				
Units of measure		X	X			
Conversion factors		X	X			
Carrier name		X				
Customer orders						
Customer order number					X	
Customer name					X	
Customer address					X	
Quantity ordered					X	
Quantity supplied					X	
Address goods shipped from					X	
Address goods shipped to					X	
Item number					X	
Item description					X	
Item notes[a]					X	
Lot/batch number					X	
Control number					X	
Distributor details						
Distributor name					X	
Distributor code number					X	
Distributor address					X	
Address goods shipped from					X	
Address goods shipped to					X	
Date collected					X	
Quantity collected					X	
Item number					X	
Item description					X	
Item notes[a]					X	
Lot/batch number					X	
Control number					X	
Recipient details						
Customer order number					X	
Customer name					X	
Customer address					X	

Appendix 32.2   Examples of Regulatory Data Types Within Marketing and Supply Business Processes (*Continued*)

			Functional area			
Process area/data type	Source product	Receipt	Storage	Disposal	Distribution	Marketing information
Shipping address					X	
Shipping notes[a]					X	
Date of despatch					X	
Date of receipt					X	
Returned goods shipment						
Customer order number		X			X	
Return goods note number		X			X	
Customer name					X	
Customer address					X	
Address collected from		X			X	
Address returned to		X			X	
Shipping notes[a]		X	X		X	
Date of return		X	X		X	
Date of receipt		X	X		X	
Quantity returned		X	X		X	
Lot/batch number		X	X		X	
Control number		X	X		X	
Date quarantined		X	X		X	
Reason for return		X	X		X	
Stock adjustment						
Item number			X	X		
Item description			X	X		
Item notes[a]			X	X		
Lot/batch number			X	X		
Supplier batch number			X	X		
Quantity disposed			X	X		
Date quarantined			X	X		
Date of disposal			X	X		
Reason for disposal			X	X		
Inventory transfer						
Item number			X			
Item description			X			
Item notes[a]			X			
Lot/batch number			X			
Date of transfer			X			
Quantity transferred			X			
Unit of measure			X			
Transfer from location			X			
Transfer from warehouse			X			
Transfer to location			X			
Transfer to warehouse			X			

[a]Batch, item, or shipping notes or other data type headings such as warning or hazard statements are often free text fields within an application and are purposely created to capture any comments, instructions or special conditions that need to be associated with the particular batch or item at all times. Care should be taken when validating the use of free text fields because these fields do not normally have any mandatory system verification placed on them to confirm that that data entered is of the correct data type (e.g., text or numeric values) before it is committed to the database. The only verification that can usually be tested is that data can be added, amended and deleted from these fields and that the modifications made are on one screen are correctly reflected in any subsequent screens. This factor is particularly important where large blocks of text are being entered, for example text entries for medical information systems supporting products released to market. If a site's particular usage of the application is determined to be GxP critical, then the system security programs controlling the integrity of any GMP or GDP data should also be subject to local validation activities, as shown in the table above.

33 | Case Study 15: IT Infrastructure and Associated Services

Chris Reid
Integrity Solutions Limited

Barbara Nollau
Abbott Vascular

INTRODUCTION

IT infrastructure comprises communication networks, hardware (e.g., servers, peripherals), operating systems (including service packs and patches), IT services, clients, infrastructure applications and facilities (e.g., data centers) that provide a platform for successful operation of business applications. Figure 33.1, provides an infrastructure architecture overview adapted from the ISPE GAMP infrastructure guideline (12).

IT infrastructure typically evolves over a period of time in response to new demand for application support, services and storage. As such, management controls should ensure that current design and configuration information is maintained in line with the current installation. Figure 33.2 illustrates how a number of projects may impact common aspects of the infrastructure, leading to historical design and configuration information becoming obsolete quickly.

A suite of records including network diagrams, configuration data, cabling diagrams and system inventories should be maintained to define the current configuration status of the infrastructure (Fig. 33.3). These records should be controlled so that access is limited and unauthorized changes cannot be made.

Changes to IT infrastructure are implemented in accordance with traditional lifecycles, change control procedures and/or operational procedures depending on the nature of the change. Management controls should ensure that the introduction, modification and retirement of software, hardware and services is evaluated, managed and verified against functional and performance criteria.

Validated business applications cannot be viewed in isolation of the IT infrastructure. Business applications are increasingly interfaced using both local and wide area data networks. IT infrastructure should be suitably qualified to support validated applications. Regulated companies should ensure that IT organizations are knowledgeable with current good practices pertaining to infrastructure management [e.g., GAMP (12), ITIL (13)] to ensure that investment in business application validation is not jeopardized. It may be argued that a system cannot be considered fully validated if qualification of the infrastructure is absent.

IT infrastructure compliance had a high profile amongst regulatory agencies in the early 2000s. Since then, greater understanding of IT infrastructure risks have been developed and controls implemented to ensure potential impacts to business application functionality, performance and data integrity are implemented.

Many regulated companies have implemented IT quality organizations that work closely with business quality organizations to bridge gaps in technology understanding and to ensure that quality practices are integrated into IT processes.

ROLES AND RESPONSIBILITIES

IT senior management should establish policies and standards that define the minimum controls to be applied in the management of IT infrastructure. Defined controls should be scalable on the basis of risk to business application performance and data integrity, and should be aligned with the company's overall risk tolerance level.

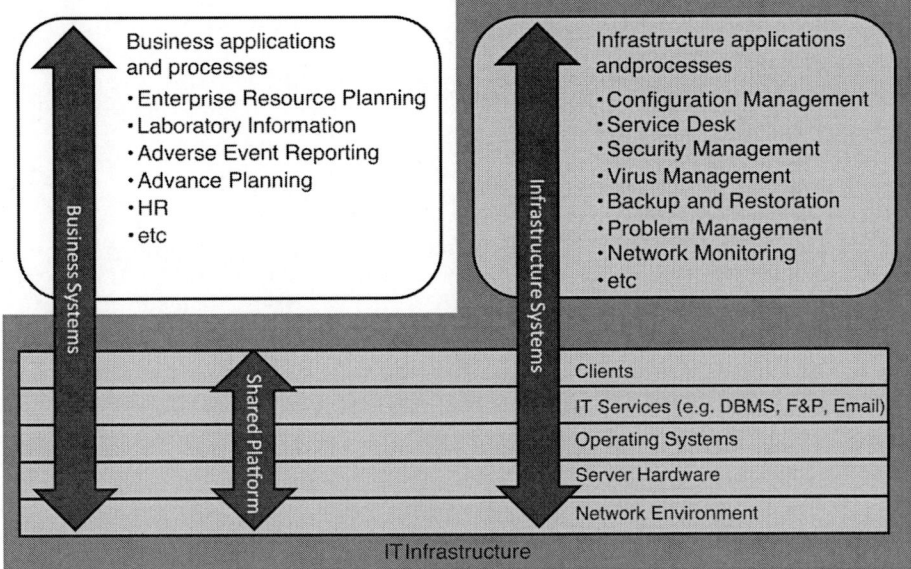

Figure 33.1 Infrastructure architecture—adaptation of GAMP®5.

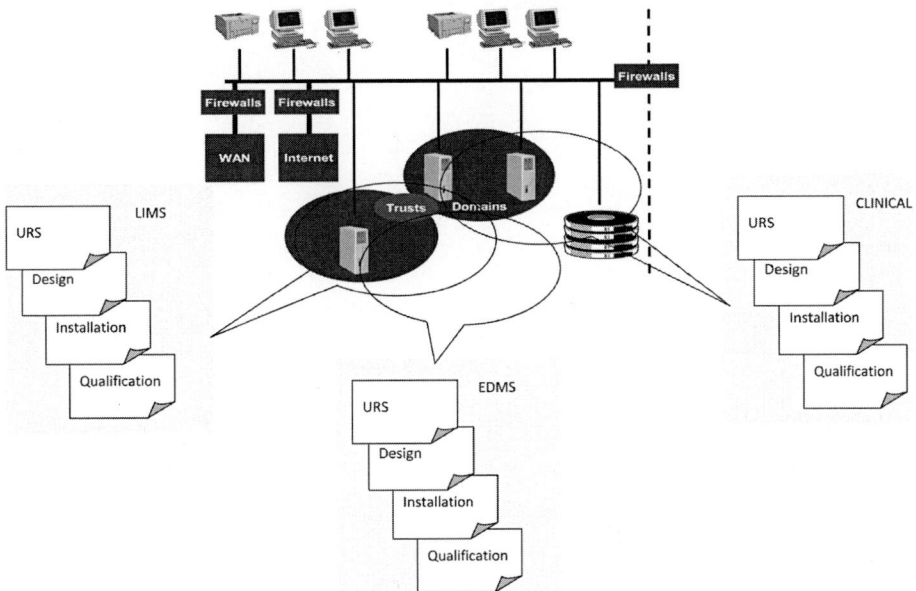

Figure 33.2 Impact of overlapping changes.

Implementation of such policies and standards should not be underestimated. There may be significant investment required in the implementation of support procedures, work instructions, infrastructure applications and training.

Clarity of accountabilities across IT and business organizations is essential. IT may comprise a corporate IT function and local business aligned IT functions with dotted line relationships to corporate IT management (Fig. 33.4). In such organizations, business aligned IT functions may assign higher priority to business direction than corporate IT direction. As such, corporate IT policies, standards and directives may be implemented with differing levels of enthusiasm and risk consideration across businesses. Accountabilities should take account

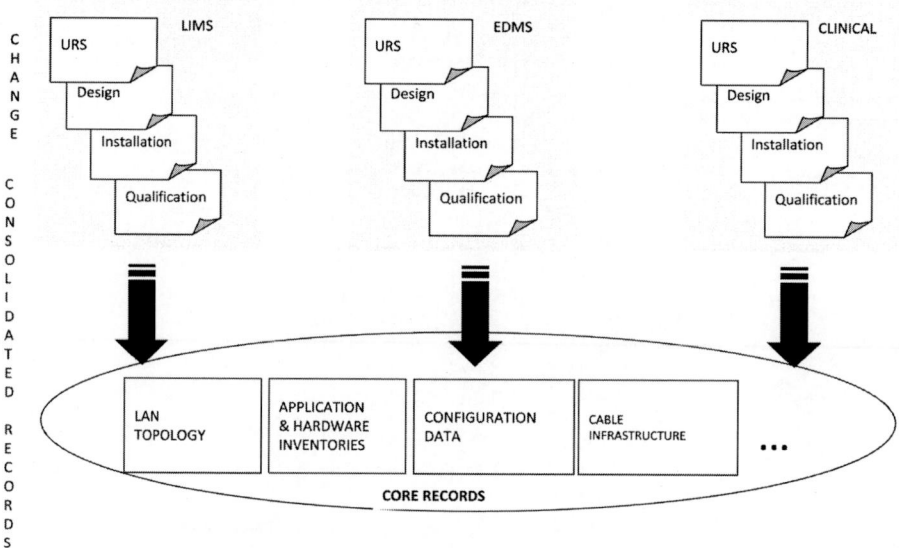

Figure 33.3   Consolidated design and configuration information.

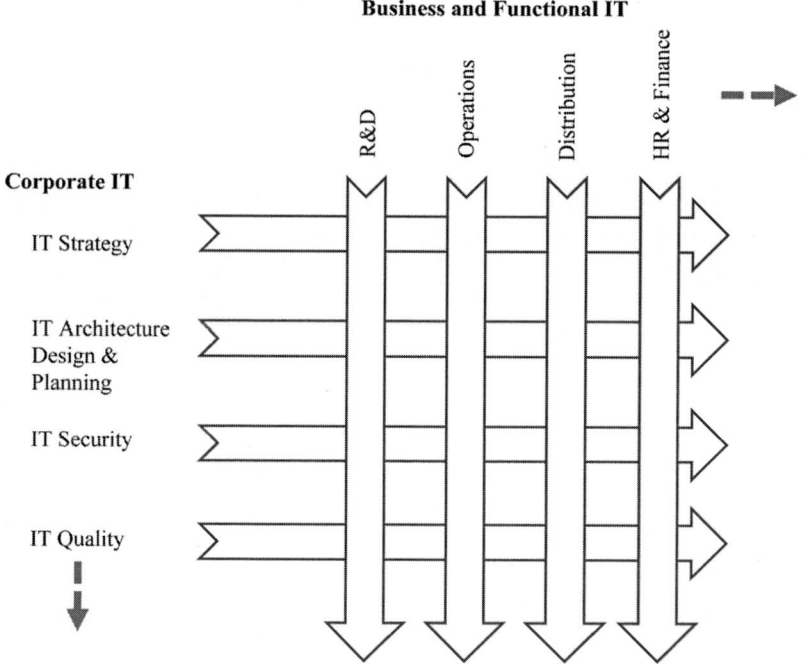

Figure 33.4   IT organization.

of such organizational structures and should always carry the necessary authority to execute assigned accountabilities. For example, corporate quality and security functions cannot be solely accountable for compliance if businesses have power of veto.

IT leadership should evaluate and implement IT infrastructure architectural and service requirements in response to essential business needs. Defined quality controls should be applied to the implementation, support, and retirement of IT infrastructure based on business and regulatory impact.

IT leadership should ensure the following:

- Standard operating procedures (SOPs) and/or infrastructure systems are implemented in response to IT infrastructure policies and standards.
- IT personnel are trained in the requirements of policies, standards and procedures.
- Personal objectives include quality accountabilities.
- Nonconformances against policies, standards and procedures are evaluated and addressed in accordance with risk.

Business organizations should ensure that IT infrastructure supporting critical and regulated business applications and data is appropriately qualified. Business organizations may conduct audits of IT to ensure that appropriate controls are in place and are being applied. Business quality assurance functions should work closely with IT quality functions to understand IT quality controls.

Business organizations should

- Define service level, functional, and regulatory requirements to IT so that appropriate resources, technical solutions and/or services may be established,
- Provide primary input to determination of business and regulatory risk,
- Establish and own business continuity plans and ensure that IT disaster recovery plans meet business need,
- Define data migration, archiving and retention requirements, and
- Communicate infrastructure issues and incidents to IT.

IT quality organizations support IT personnel in the implementation of risk-based quality controls. IT quality provides first line assurance that IT incidents, nonconformances, changes and risks are appropriately identified, evaluated and addressed.

IT quality should

- Monitor and communicate developments in regulatory expectations and industry practice to IT management so that the impact on IT operations can be determined and addressed,
- Support the determination of regulatory and quality impact associated with IT infrastructure changes and incidents,
- Support the implementation of relevant controls to address quality and regulatory requirements,
- Conduct audits and periodic reviews of IT infrastructure controls to ensure compliance with governing policies, standards and procedures,
- Ensure that outsourced IT service providers are evaluated to ensure compliance with relevant aspects of IT policies, standards and procedures,
- Support IT led projects, changes and initiatives to ensure that regulatory and quality requirements are understood and addressed, and
- Communicate the compliance status of IT infrastructure to business quality functions to ensure they are fully aware of compliance risks and associated preventive and corrective actions.

IT personnel deliver IT services and provide support in accordance with defined procedures and work instructions. It is essential that IT personnel understand and accept their accountabilities for quality rather than perceiving quality as a "quality organization" responsibility.

IT personnel should create and adhere to SOPs and infrastructure systems and should be encouraged to report IT incidents and propose quality improvements.

IT QUALITY MANAGEMENT SYSTEMS

IT leadership with the support of IT quality should implement a quality management system in accordance with industry standards such as GAMP®5 (12), ITIL (13) and ISO 9000 (15).

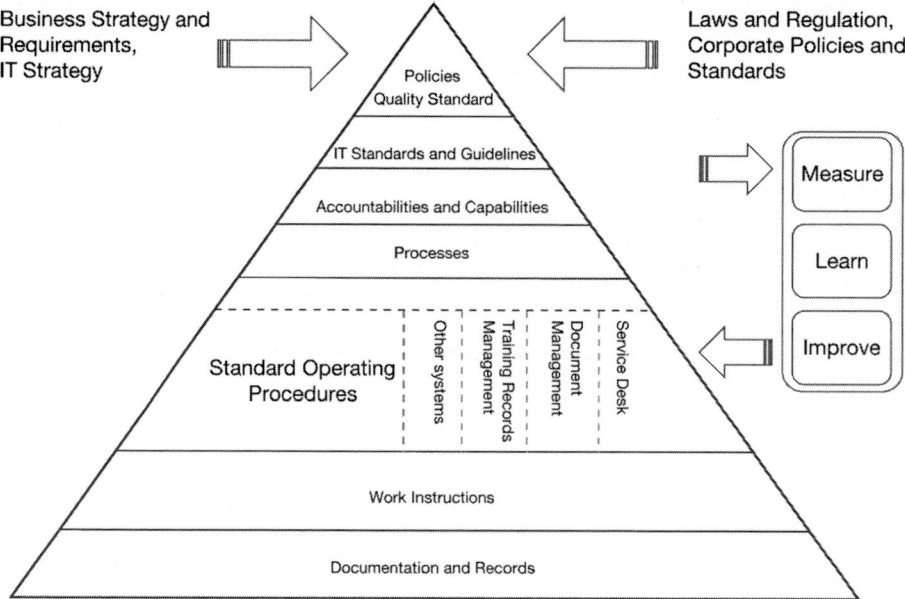

Figure 33.5 IT QMS structure.

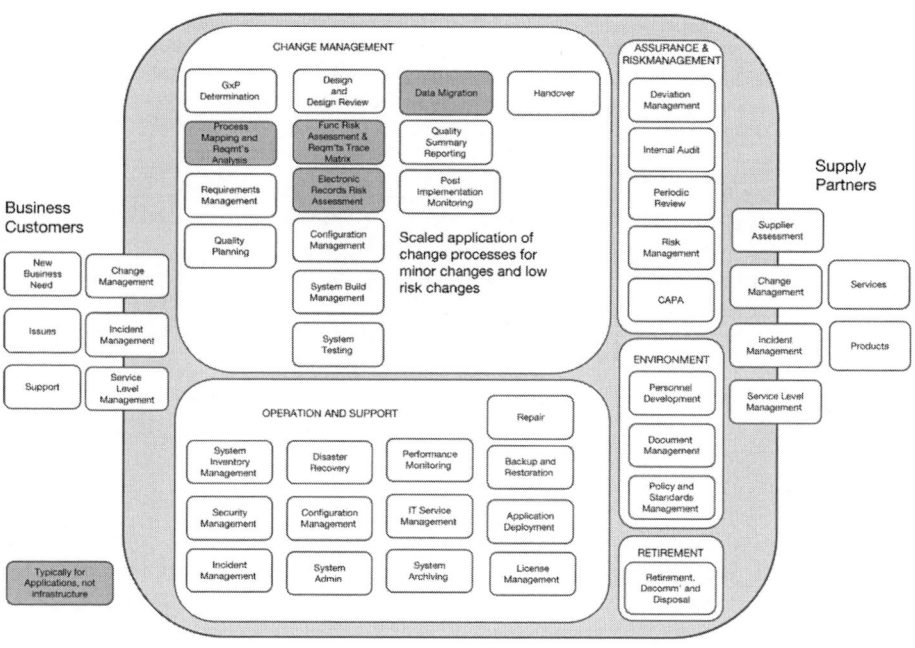

Figure 33.6 IT QMS processes.

Figure 33.5 defines a typical quality management system structure driven by organizational and company policies and standards. IT quality processes are typically defined within SOPs and/or IT infrastructure systems such as service desk. Detailed IT service activities such as operation of backup tools, patch panel management and infrastructure security profile configuration will be defined in Work Instructions.

Figure 33.6 provides a process view of the IT quality management system, identifying some of the key processes and procedures that will need to be established. Change control, incident management, and service management processes typically control the interface between IT and business (customer)/supply organizations.

IT INFRASTRUCTURE IMPLEMENTATION
Risk Assessment and Management
Risk assessments should identify the potential impact of proposed changes on business application functionality, performance and data integrity. Risk assessments should consider

- Criticality of impacted business applications and data,
- Scope and complexity of the change,
- Use of new technology, and
- Contingency provisions, system stability, and capability support exposure.

The level of control required to implement the change should be commensurate to the identified risks. From a practical perspective, it can be difficult to relate technical infrastructure changes to potential impact on business applications and data. As such, a phased roll out of changes to lower risk business environments (e.g., human resources) or extensive testing before roll out of the change may be more appropriate than theoretical risk assessments.

An upfront risk assessment may be conducted against primary infrastructure components to define likely impacts associated with component failures and to determine typical controls when implementing changes to such components.

Care should be taken to ensure that the evaluation of impact takes account of operating context (e.g., failure of a data storage device containing noncritical information will be different to the failure of a data storage device containing critical clinical data). Such risk assessments are useful when defining routine operational changes that need to be efficient and effective without considering the same risks each time the operational change is made.

Requirements Specification
Requirements Specifications are not generally required when purchasing one off standard infrastructure components, for example, router and server. However, requirements specifications are beneficial when establishing a new significant infrastructure design such as high availability facilities, replacement desktop projects, data center, clustered services (e.g., Oracle server farm).

Where IT infrastructure implementation is integral to a business application implementation project, the infrastructure requirements may be integrated into the computerized systems Requirements Specification, particularly if the infrastructure specifically supports that system. Otherwise, a qualification of infrastructure to include these elements can be referenced within the validation packages of all supported systems.

Common topics for consideration when documenting infrastructure requirements are defined in Table 33.1.

Applicable internal IT standards (e.g., preferred equipment suppliers, operating systems, security) should be referenced.

Constraints such as project milestones, and technology platforms should be detailed in the requirements specification.

Table 33.1 Infrastructure Requirement Specification Considerations

- Network organization (e.g., topology, location of data centers, communications centers, access locations)
- Performance (including response times, concurrent users, data volumes, availability)
- Storage (including capacity, scalability, redundancy, and data retention)
- Security (physical and logical)
- Environmental (including temperature, humidity, static, intrinsic safety, rack space, power supply, cabling, fire protection)
- Redundancy (requirement for dual or standby components)
- Backup (e.g., backup frequency, backup media, hot and cold backups, media management)
- Data recovery
- Fail-safe operation (such as uninterruptible power supplies, high availability)
- Licensing
- Desktops and servers (including minimum performance standards, operating system compatibility, configuration requirements, security requirements)

Table 33.2 Infrastructure Supplier Assessment Planning

- Are the components or services to be provided unusually complex?
- Do the components utilize new and relatively unproven technologies?
- Is there confidence in the supplier based on previous experience?
- What is the scope of the work to be carried out by the supplier?
- Is the supplier expected to perform qualification testing?
- Are there concerns as to whether the potential supplier has the necessary technical and/or quality capabilities to complete the work?
- Is the work being carried out under the control of the supplier quality systems without supervision?
- What is the criticality of the business applications, infrastructure components, data or services supported by the supplier?

Procurement and Supplier Assessment

Typically, IT infrastructure components are standard components. Such components should be procured from reputable, well-established companies with the ability to provide appropriate ongoing support.

It will not normally be necessary to conduct supplier assessments of standard component suppliers. However, where a supplier provides a critical service, for example, outsourced platform management, supplier assessment should be considered.

Quality assurance should be consulted prior to engaging the services of a new IT service provider. The decision to assess a supplier should be determined by the criticality of the components or services provided. Table 33.2 defines considerations when determining the need for and approach to infrastructure supplier assessments.

Depending on the nature of risks, the supplier assessment may take the form of one or more of the following: postal audit, on site audit, review of previous experience or references from previous clients of the supplier.

Quality Planning

A quality plan should be developed for the implementation of new major components of the IT infrastructure. For routine or minor modifications change control should suffice.

The quality approach is dependent on the criticality, complexity and scope of impact of the infrastructure components. Specifically, the quality plan should define the following:

- Objective and scope
- Quality approach appropriate to the scope and criticality of the infrastructure components
- Roles and responsibilities
- SOPs controlling specification, design, qualification and support activities
- Training requirements

IT Infrastructure Design

For major infrastructure, documented design specifications should be maintained. IT infrastructure design documentation should largely be dependent on the scope and complexity of the design, and extent to which supplier documentation and information can be leveraged (Table 33.3).

Infrastructure design documentation should be reviewed and approved by IT technical authorities.

Quality assurance should periodically audit IT infrastructure documentation to ensure appropriate and up to date documentation is in place.

Configuration Management

The primary objectives of configuration management are to

- Ensure consistency of configuration practices,
- Support problem investigation,
- Enable disaster recovery, and
- Support change control.

Table 33.3 Infrastructure Design Documentation Considerations

- System architecture diagrams
- Equipment inventories, schedules, and datasheets.
- Layout and wiring diagrams and drawings
- Main component specifications
- System interface specifications
- Minimum processor performance
- Minimum memory requirements
- Minimum storage requirements
- Peripherals
- Interfaces (communications cards, network connections, cabling, speed)
- Settings (switch settings, firmware configuration)
- Environment (temperature, humidity, RFI, UV, electromagnetic)
- Electrical Supplies (UPS, earth requirements, filters, etc.)
- Relevant standards (safety, electrical, etc.)
- Data center and telecommunications room layouts

Abbreviations: RFI, Radio-Frequency Interference; UV, Ultra-Violet.

An electronic and/or paper record of configuration parameter settings against for each configuration item should be established. Standard build items may be defined in preapproved configuration standards that are applied and verified each time a new build is implemented.

Configuration records may comprise any combination of the following:

Audit tools	Used to interrogate the current configuration of infrastructure components
Electronic backup	Used to recover configuration to a known point following system failure
Configuration specifications/record sheets/log books	Provide a record of configuration settings for installed components

Certain infrastructure equipment and components will contain numerous default configuration parameters of which only a limited number will be change. Configuration records need only document the parameters that have changed.

Installation Qualification/Verification

The correct installation and successful operation of infrastructure components in their intended operating environment should be verified through installation and operational testing.

Hardware and software components should be subject to installation verification to confirm correct installation, connection, configuration and conformance to minimum capacity and performance requirements.

Installation verification documents should define or reference clear instructions on how to install and configure hardware and software. Clear criteria should be set to demonstrate successful installation. If the installation is being performed by the supplier, installation verification must still be performed and documented, and any documentation received from the supplier/installer retained with the documentation package.

Low criticality components may be verified informally. Standard reusable installation instructions and verification steps may be defined for off-the-shelf standard components.

Disk cloning and automated deployment of configuration affecting multiple devices may be used in support of installation and configuration operations. The use of these approaches should be documented and verified to confirm that automated processes operate as expected. Table 33.4 defines typical design, installation and operational verification requirements.

Table 33.4 Design, Installation and Operational Verification Considerations

Computer rooms and data centers
- Work in progress equipment identified and located to prevent inadvertent misuse
- Adequate areas for tape storage and rotation in a controlled manner
- Appropriate charts/drawings (data center layout, racking, cabling and wiring, electrical)
- Environmental specifications and controls and evidence of adherence to them
- Monitoring programs and maintenance programs
- Physical and logical security specifications and controls and evidence of adherence to them
- Automated alerts for infrastructure failures
- Redundancy and backup systems (e.g., RAID, high availability systems)
- Entry logs

Networks
- Hardware installation checks, including visual inspection of components against the following:
  - Design specification
  - IT standards
  - Statutory requirements
  - Manufacturers' recommendations
- Network addressing
- Check that all equipment and materials are undamaged, and correctly installed (refer to installation records)
- Check for hazardous area requirements
- Capacity testing
- Software versions checked
- Electrical supply and interference testing
- Installation diagnostic testing
- Power on-off testing (blackout testing)
- Configuration/system testing on installation or replacement of equipment
- Simulated communication between network points
- Cable specification and definition
- Cable routing and redundancy, e.g., dual backbone
- Security

Client (desktops and mobile PCs)
- Document standard client configuration
- Document and verify application scripting and automated deployment processes
- Restrict installation of unauthorized software and downloads from internet
- Virus protection management
- Evaluate impact of client component upgrades on resident computerized systems
- Security policies
- Internet connection policies
- Management of coexisting applications

Mobile PCs specifically
- Use of remote access security when connecting from a remote location [e.g., virtual private network (including PC Firewall)]
- System clock change restrictions
- Screen saver password and lockout change restrictions
- Private use of laptops
- Virus protection management when not connected to network
- Security management when not connected to network

Servers
- Start-up and shutdown
- Virus protection
- Operating system version and patch status (service packs)
- Configuration settings
- Hardware switch and address settings
- Backup and restoration
- Security configuration
- Folder security settings
- Network addressing

Operational Qualification/Verification

Operational verification confirms that an infrastructure component or integrated design operates according to specification. Major infrastructure components supporting critical data and IT services should be subject to documented verification to ensure the following:

- Correct start-up and shutdown of IT servers and network components
- Correct start-up and shutdown of services (e.g., DHCP, DNS)
- Operation of login and other scripts
- Synchronization with other devices (e.g., automatic clock update)
- Accessibility to other required network areas
- Configuration of security policies
- Operation of infrastructure software (e.g., antivirus software, backup)
- File share and print services (user facing)

Operational verification should also challenge built in redundancy, high availability, hot standby, power failure, and alerts design.

Operational verification should be executed and documented in accordance with a preapproved protocol that defines verification procedures and acceptance criteria.

Quality Summary Reporting

A quality summary report should be generated for major projects, that is, when a quality plan has been established. The quality summary report should

- Confirm that all activities called for by the quality plan have been satisfactorily completed,
- Summarize the outcome of verification activities,
- Confirm that all documentation called for by the quality plan has been created, reviewed and approved, and identify all infrastructure documentation created by the project,
- Record deviations from the quality plan and provide an evaluation of the resultant corrective actions and risk,
- Identify any outstanding actions, along with persons responsible and timescales for completion, and
- Identify any sustaining activities/controls and responsibilities for carrying them out.

The quality summary report provides the authorization or otherwise for operational use of the implemented infrastructure. Residual risks should be evaluated to ensure that business operation and critical data integrity shall not be adversely affected.

OPERATIONAL CONTROLS

Performance Monitoring

Performance monitoring demonstrates that the IT infrastructure is able to operate under normal and peak load conditions. Performance Monitoring should determine

- Network failures,
- IT infrastructure component failures,
- Storage capacity utilization, and
- Network utilization.

Automated alerts to mobile phones, e-mail or pagers are commonly used to warn of pending or actual IT incidents. Identified performance issues and failure conditions should be managed in accordance with incident management procedures.

Configuration and Inventory Management

Initial configuration should be defined during the design of new infrastructure components and should be maintained throughout the operational life of the infrastructure.

An inventory of all major components of the IT infrastructure should be maintained. The inventory should include the following:

- PCs
- Servers
- Storage devices
- Switches
- Routers
- Performance monitoring devices
- UPS
- Environmental management equipment (specific to the infrastructure)
- Infrastructure software (e.g., database management software, virus software, backup software, network monitoring software, security management software)

Configuration information to enable initial build, setup and/or recovery of infrastructure components should be maintained. Such information may be documented in configuration standards (e.g., standard client build), configuration specifications (e.g., server/client/router setup configuration) and/or asset management databases as discussed earlier in this chapter.

IT infrastructure components are typically dynamic and subject to frequent change. As such, configuration information may be more readily maintained in an electronic information management system (e.g., a configuration management database) rather than paper based documents. Such tools should be subject to appropriate verification to ensure that the integrity and security of configuration management information is being maintained.

Where possible, configuration of IT infrastructure components should be subject to backup to support recovery following component failure.

Change Control

Change control procedures should be used to manage changes to the IT infrastructure configuration, with the exception of routine operational changes such as security management that are managed in accordance with specific operational SOPs.

All changes should be requested, planned, designed, implemented, tested and approved by personnel from relevant technical, business and quality functions.

Occasionally it may be necessary to carry out emergency changes and procedures should allow for this. In the case of an emergency change, verbal authorization should be obtained from defined authorities. The change should be documented and formally approved during normal business hours of the next working day. Emergency changes are only implemented when there is an immediate and severe risk to business operation or critical data integrity, and when the risk of not making the change immediately outweighs the risk associated with the lack of a prerequisite, formal review (e.g., virus threat, etc.).

Back-out plans and instructions should be considered for complex or high-risk changes. Because of the nature of emergency changes, formal back-out planning may not be possible, however consideration should be given to infrastructure recovery in the event of a change failure.

Like for like changes may be managed in accordance with routine repair/replacement procedures/work instructions that do not require the same level of authority to make the change. In the context of like for like changes, infrastructure equipment and components evolve rapidly. The definition of like for like equipment and components must be defined but should be practical when considering minor supplier upgrades to components that do not adversely affect function or performance.

Incident Management

Incident management procedures should be established to ensure that IT infrastructure failures, faults, and anomalies are documented and evaluated, and that appropriate corrective and preventative actions are taken. All unplanned events or interruptions to IT services should be notified to the IT function.

Information describing the Incident should be recorded (e.g., description of the incident, date and time, scope and impact assessment). Escalation procedures should also be defined. Immediate steps should be taken to limit the effect of any incident (e.g., shut e-mail down following a major virus threat).

IT management should ensure that incident records are reviewed periodically to identify trends and potential causes of persistent incidents. Appropriate actions (e.g., systemic corrective actions, alerting other business areas or sites, etc.) should be initiated as a result of such reviews.

Where an incident has a major impact on the user community, a dedicated incident manager should be nominated. The incident manager should convene an appropriate team of managers and technical specialists to ensure that timely investigations and actions are taken.

The incident manager should invoke disaster recovery and continuity plans where appropriate to ensure that critical business functions can continue to operate while the incident is being resolved. The business and quality assurance should be involved or at least informed when any incident poses a risk to business systems they own or govern.

System Administration

System Administration tasks should be defined in appropriate SOPs or work instructions. System administration tasks include but are not limited to the following:

- Check health and disk status of servers
- Check previous nights backup(s)
- Change backup tape(s)
- Check mail flow
- Check internet connection
- Check server event logs
- Check available hard disk space on servers
- Periodic reboot of servers
- Active directory (AD) housekeeping

Support and Maintenance

Service level agreements (SLAs) should be implemented with external service providers to ensure that required support services are defined, service management controls (e.g., SOPs) are in place and documentation and record keeping requirements are established.

SLAs should consider the following:

- Scope and definition of services to be provided
- Hours of provision
- Controlling procedures
- Skills, knowledge and training
- Governance controls, roles and responsibilities, interfaces between organizations
- Ownership of equipment
- Ownership of documentation and records
- Initial and ongoing surveillance audit requirements
- Support during regulatory inspections
- Service performance reporting
- Escalation procedures
- Contract exit

Service performance targets should be periodically reviewed and appropriate action taken to address deviations from agreed targets. SLAs are also useful when implemented between IT and the business to define the services provided to the business, for example, business application support, disaster recovery, backup and restoration, incident management, service desk, and system/service availability.

Data Management

A data management plan should be established for each server. The data management plan should define the following:

- Backup and restoration frequency based on criticality of data
- Data archiving requirements
- Data purging requirements
- Data integrity checking requirements
- Data copying controls
- Data protection of confidential information

Backup and Restoration

Documented Backup and Restoration procedures should be established. Backups should be scheduled in accordance with criticality of the business applications, data and configuration maintained on the IT infrastructure.

Backup procedures should be periodically audited and backups tested in accordance with a documented schedule.

The Installation and Operation of backup system hardware and software should be verified to confirm that the system is correctly installed and is operational.

Backup failures should be reviewed to determine risk and corrective action (e.g., rerun backup). Business system owners should be notified of major or extended failures.

Backup media should be stored in a manner and/or location that minimizes the risk of simultaneous loss of online and backup data. Backup media should be stored in facilities that minimize the risk of fire, water damage and theft.

Where third-party service providers are utilized, the capability and appropriateness of the service provider's resources, processes and facilities should be evaluated.

Disaster Recovery and Business Continuity

IT infrastructure disaster recovery plans support business continuity and should consider the following risks to infrastructure outage. Planned controls should be based on an analysis of such risks.

- Human error
- Fire, water damage, weather
- Power/electrical failure
- Failure to comply with policies and procedures
- Software or hardware failure
- Accident, absenteeism
- Sabotage, natural disaster
- Enforced site evacuation

IT should establish controls based on the analysis of risk to reduce the impact of significant IT incidents including the following:

- An ongoing review and challenge of the appropriateness of the backup and recovery procedures and strategy
- Effective security controls
- Failover and backup systems (e.g., uninterrupted power supplies, dual network backbone, and standby systems)
- Installation of intrusion detection and intrusion prevention tools and utilities
- The installation of failure detection systems such as network diagnostic monitoring and fire detection

IT infrastructure disaster recovery plans should consider the following actions:

- Immediate steps to be taken to minimize further impact following disaster detection
- Prioritization of recovery based on business impact and technical dependencies

- Actions to be taken to implement alternative working arrangements
- Roles and responsibilities for recovering the situation

IT disaster recovery plans should cater for alternative working environments and practices to support business continuity following a significant disaster scenario including the following:

- Connection of users to an unaffected part of the IT infrastructure
- Supporting manual processes until services reestablished
- Relocating essential business operation and technical infrastructure support staff to alternative premises
- Establishing alternative IT infrastructure away from the affected areas

Disaster recovery plans and business continuity plans should be tested in accordance with a defined schedule. Test frequencies should be determined by the identified threats and vulnerability of the IT infrastructure to such threats.

Disaster recovery plans, business continuity plans, supporting documentation, and procedures should be stored so that they are accessible during a disaster scenario and should be updated as required on the basis of the outcome of challenges.

SECURITY MANAGEMENT

IT infrastructure security management warrants specific consideration by this chapter. Security threats are a primary concern for business application performance and critical data integrity. Documented security policies should be established to

- Maintain the integrity of data stored and transmitted by the IT infrastructure,
- Maintain the availability of information assets and ensure business continuity,
- Maintain confidentiality of information, and
- Protect data from threats such as viruses and other malicious code, theft, and inappropriate remote access.

Regulated companies should implement and maintain appropriate physical security controls to mitigate the risk of theft or damage to information assets, unauthorized access to or erasure of proprietary information. The implementation of physical security may not be the direct responsibility of the IT function (e.g., some aspects of physical security may the responsibility of an estate management organization or landlord).

Access Controls

To prevent unauthorized access to information systems, formal procedures should be in place to control the allocation of access rights to information systems and services.

Multiple level access accounts should be defined (e.g., user account, system administrator account). The number of levels should be determined by the business need.

The privileges associated with system administrator accounts should be relative to their normal daily activities. Additional access rights required to complete nonroutine work should be granted and revoked on a needs basis.

When user accounts are locked out, only system administrators should be able to reset user passwords following confirmation of the user's identity. The only exception to this should be the use of self-service password reset functionality performed by the user himself.

Where possible the following basic user account controls should apply the following:

- User account access codes should be unique.
- A password that is written to a file or a database should be encrypted.
- Passwords should not be viewable by anyone, including IT personnel.
- Passwords should be obscured when entered.

- Password complexity rules should minimize risk of password determination by unauthorized personnel.
- Password ageing.
- User ID/password combinations should be restricted from reuse (i.e., cyclical passwords).
- Automatic lockout of a user should occur following a defined number of successive failed login attempts.
- A log of all security violations should be maintained by the system.
- It should be possible to lock a workstation while unattended and all users should be advised to do so while their workstation is unattended.

User access rights should be reviewed at regular intervals to ensure they are appropriate to the user's defined accountabilities. User access rights should be revoked when the user leaves the company, and should be evaluated for change when the user changes roles.

Policies and technical controls should be established to manage access to the internet and external networks. Firewalls should be implemented to detect attempted intrusions and virus attacks.

Users should not download software from the internet or other external networks without prior review and approval by IT management. Software will include freeware, tools, applications, screensavers, clipart, games and other similar items.

Remote connections should be via secure virtual private network (VPN). Technical solutions should ensure the following:

- Remote connections are terminated following termination of the user session.
- The identity and where appropriate location of the user is verified prior to connection to the network.
- Communication failures result in termination of the user session (assuming they cannot be immediately recovered).
- Normal user authentication procedures (e.g., user ID and password) should still apply.
- Remote connectivity by suppliers or similar entities should be managed and authenticated via callback functionality or similar safeguards.

Physical and Organizational Controls

Access to buildings and offices containing information assets should be defined and controlled. Barriers such as walls, card controlled entry points and manned reception desks should be used to protect areas that contain information and information processing facilities.

Intruder detection systems should be installed to detect unauthorized entry to facilities. Entry to a data center or computer room should be controlled. Persons who do not have the required security to enter the data center or computer room should be supervised by the relevant system owner, IT manager, or system/IT administrator. A log of access to the area should be kept.

Where possible telecommunications lines into information processing facilities should be underground or subject to adequate alternative protection. Network cabling should be protected from unauthorized interception or damage.

Cable and equipment markings should be used to reduce the possibility of errors, such as incorrect patching of network cables. Access to patch panels and cable rooms should be controlled and logged.

Care should be taken not to leave equipment holding business critical information in insecure locations. Mobile computers and other devices such as PDA's are designed to be mobile and used away from the regulated company's premises. However, policies should ensure these are kept secured by the user and any other equipment holding critical information should not be taken off site without prior permission.

Supplier security requirements should be defined in third-party SLAs and verified to ensure they are in place, operated and maintained.

Precautions are required to prevent and detect the introduction of malicious code, for example, viruses and unauthorized mobile code. Virus signature databases should be up to date.

Mobile computers should be scanned for viruses when reconnected to the network if the normal virus scanning period has been exceeded.

Removable devices (e.g., removable disk drives, CDs, DVDs), e-mail attachments and other methods of importing files into the IT infrastructure should be automatically virus checked. In the event of a virus infection, appropriate action should be taken to minimize the spread of the infection. This may include temporary shutdown of information assets.

System documentation (e.g., system specifications) containing information that would increase the possibility of a security violation should be controlled so that it is not readily accessible. Such information includes security setup information and database modification management procedures.

The clocks of all relevant information processing systems within an organization or security domain should be regularly synchronized with an agreed accurate time source. Clock synchronization should be automated where technically possible.

RETIREMENT AND ARCHIVING

Retirement plans should be established to define the controls for the Retirement of major aspects of the IT infrastructure. Retirement plans should ensure the following:

- Data is migrated, converted, or archived to alternative media or systems in accordance with business and regulatory retention requirements and that the data transfer is appropriately verified.
- Data that is to be destroyed is approved before destruction.
- Obsolete storage media is appropriately discarded to ensure that risks to information confidentiality are minimized, and that security and access authorization is maintained through the retirement process.
- The impact on the remaining IT infrastructure is evaluated and addressed, and that disruption to business applications and IT services is minimized.

LEGACY IT INFRASTRUCTURE

Legacy IT infrastructure comprises IT infrastructure components that are in operation and have not been implemented in accordance with the controls defined within this chapter.

For such IT infrastructure, some of the implementation lifecycle activities are not practical or possible (e.g., requirements specification, supplier assessment). Minimal information should, however, be established.

- Documented inventory of main infrastructure hardware and software components
- Network diagrams
- Collation of existing documentation supporting the configuration and verification of legacy IT infrastructure, including network performance error data, infrastructure capacity data (e.g., data storage), error diagnostic data, configuration information, administration, and operation data (e.g., manuals)

Diagnostic and other performance monitoring tools should be used to evaluate current system stability and performance. Where performance is unacceptable, design and/or operational changes may be required.

BASELINE INFRASTRUCTURE QUALITY ASSESSMENT

Prior to embarking on an IT quality improvement program, it may be beneficial to conduct a baseline quality assessment to determine the current practices, systems and capability in place. Where possible and practical, leveraging existing practice may be helpful in gaining commitment and buy in. The ability to leverage existing practice is however dependent on the effectiveness and adequacy of those practices.

The baseline assessment is conducted against IT standards and industry practice. The baseline assessment identifies current best practices and shortfalls. The IT organization should remain familiar with best practice to minimize effort required to raise standards and to quickly bring the organization to a common platform.

Shortfalls are reviewed and prioritized for remediation on the basis of the criticality and scope of the impact. Cross-site and regional teams should be formed to provide a consistent solution across the organization. This approach further enables cultural differences and regional approaches to be addressed by any delivered solution.

Appendix 33.A provides a questionnaire for assessing IT quality and compliance practices. In addition to asking specific questions as described in the questionnaire it is also useful to conduct interviews with key people from the organization including

- Technical architects (designers),
- Service managers,
- Service delivery,
- Outsource contract managers,
- Business representatives (customers),
- Security managers, and
- Quality assurance.

Interviews will generally help to identify particular quality issues that are causing frustration within the organization.

ELECTRONIC RECORDS AND ELECTRONIC SIGNATURES

IT infrastructure supports databases and files that contain regulated information. Such files and records are typically associated with business processes and business applications and many controls required to manage such records are established during the implementation of business processes and applications.

IT infrastructure controls defined within this chapter provide additional safeguards that ensure the confidentiality, authenticity and integrity of regulated records hosted by the IT infrastructure platform.

Electronic records specifically associated with IT infrastructure management include the following:

- Security configuration
- Electronic configuration management records
- Infrastructure specifications
- Electronic change control records
- Electronic LAN and WAN diagrams
- Standard operating procedures and work instructions
- Electronic diagnostic and fault reports (including virus alerts and security violations)
- Installation and deployment records
- Qualification records
- Specifications
- IT staff training records

Although GxP significant, such records do not have a direct impact on patient safety and therefore may require less stringent controls than, for example, batch records and pharmacovigilance records.

Electronic audit trails and electronic signatures are useful in maintaining IT infrastructure record integrity and where available in service desk, asset management, and document management systems should be implemented, although hybrid paper controls may also provide adequate control.

Appendix 33.B provides a high level assessment of IT infrastructure controls beneficial in the management of electronic records.

IT infrastructure controls supporting electronic record and electronic signature functionality of business applications (e.g., security) should be exercised when validating the business application.

Some IT organizations continue to manage IT infrastructure information using traditional paper systems. As such, design and configuration information is often not maintained as the effort required to update paper documents is significantly higher than the effort to implement a configuration change. The benefits that asset management, service desk and other similar tools bring in terms of control of infrastructure design and configuration information should be considered favorably against the risk of such systems not providing audit trail and electronic signature functionality.

INTERESTING ISSUES
Middleware
Middleware is typically software that connects two otherwise separate applications. For example, there are a number of middleware products that link a database system to a web server. This allows users to request data from the database using forms displayed on a web browser and it enables the web server to return dynamic web pages based on a user's request and profile.

The term middleware is used to describe separate products that serve as the glue between two applications. It is therefore, distinct from import and export features that may be built into one of the applications.

Middleware is typically GAMP (12) category 1 and is validated in situ with the business applications with which it is integrated.

Wireless Networks
Implementation of wireless networks is on the increase. In simple terms the fundamental concepts of this chapter will apply. However, with a traditional network, it is clear how cables are generally secured from tampering within secure physical buildings.

The risk associated with wireless networks and devices is more related to the risk of business sabotage and information confidentiality than unauthorized modification of regulated records.

Regulated companies should ensure that they can demonstrate authenticity, integrity and where appropriate confidentiality of data transmitted via wireless networks through the implementation of security features, public key infrastructure (PKI), and other techniques as appropriate. Further, controls should ensure that access cannot be readily gained to the IT infrastructure as a result of weak security controls.

Handheld digital wireless devices are used for remote communications access to intranet and internet applications. Examples include mobile phones, pagers, two-way radios, and smart-phones. Communications are facilitated through the wireless application protocol (WAP). Applications are written using wireless markup language (WML). Data entry through such devices needs to confirm the identity of the user and their authorization before accepting data input. Similar issues exist with remote data capture devices used in the production and laboratory environments.

Extranet
VPN offer cheaper web solutions. Remote access is achieved using a client with a browser connected to a corporate web address or universal resource locator (URL). A secure session, or tunnel, is established between the VPN server and the end user workstation. Internet service providers (ISPs) and telecommunications carriers are endeavoring to provide a managed extranet/VPN service to corporate subscribers. A major benefit of web-enabled applications is the ability to recover from disaster scenarios. Business continuity plans can actively make use of such applications.

Web-Based Applications
Web-based applications used in GxP critical processes are no different to other GxP applications and should be validated and subject to appropriate operation compliance procedures. However, web-based applications are in principle readily accessible by the general public and therefore should be subject to secure user access controls. Further, having gained access an application it should not be possible to gain access to other applications located on the infrastructure.

Enterprise User Directory

Some organizations are moving toward a single digital signature for all applications running on integrated infrastructure. Such systems overcome the need to have multiple user accounts with different user IDs and passwords and are generally easier to manage. However, there are a number of issues associated with such an approach that need to be managed including the following:

- Suspension of access to systems once the person leaves the company.
- Management of access to systems when the person changes their role.
- If security codes are breached, access is provided to multiple systems.
- Access rights should still be managed for each application as access levels for one system may not be appropriate for all systems.
- Need to be able to track which systems a user has access to for security investigation purposes.

E-Mail

The use of e-mail is often taken for granted; however, it is important to consider how e-mail is used to support GxP operations. The qualification of such e-mail systems poses fundamental problems around the lack of audit trails, administration, robustness, and security, particularly with data passing from open to closed systems.

Where e-mail is routinely used to communicate authorization and approvals, regulatory authorities will expect to see evidence that the authorization and approval mechanism is secure, robust and verified. In such cases PKI techniques and use of digital signatures will bring additional security necessary to ensure robustness and security of transfers.

It is important that policies and procedures are in place to ensure that information transmitted by e-mail is suitable for its intended purpose and that it is only used for that purpose. Records sent via e-mail for information purposes should not be used to demonstrate regulatory compliance or to fulfill regulatory requirements.

Care should be taken to avoid automatic purging of e-mails when transmitting electronic records for regulatory purposes.

E-mail communication is often used to request new user accounts or password resets. Care must be taken to ensure that security processes are able to confirm the identity of the individual and system owner authorization for access. Typical control considerations may include the following:

- IT knows the person making the request.
- System owner/line manager issues request, but account details sent to user.
- User ID and password issued in separate communications.
- Signed, scanned access request forms are used in conjunction with e-mail.
- Memorable words are used as additional security.
- Passwords reset after first log-in.

OUTSOURCING

Outsourcing agreements range from sourcing an external resource to manage internally owned infrastructure assets, to utilization of outsourced company-owned infrastructure. There are many objectives of outsourcing including but not limited to the following:

- Focus on core, value-adding business activities (making drugs!)
- Increased service delivery performance
- Cost reduction
- Increase access to expertise and technology
- Optimum utilization of assets and resource
- Simplified supply chain

It is important that quality and compliance considerations are integrated into outsource agreements. Regulatory concerns relating to IT infrastructure include but are not limited to the following:

- Failure to manage infrastructure changes
- Failure to document infrastructure configuration
- Failure to maintain asset and software inventories
- Failure to manage security
- Failure to manage network management tools
- Failure to train IT personnel in GxP requirements
- Inappropriate documentation controls
- Lack of LAN and WAN documentation
- Lack of defined roles and responsibilities
- Inadequate management and qualification controls
- Inadequate backup and archive controls

Outsource contracts and SLAs must clearly define quality and compliance expectations including the following:

- How will the client company demonstrate that they own the data hosted by the infrastructure service provider?
- What assurance processes are required to ensure that the service provider meets quality and regulatory requirements and is outsource company able/willing to comply?
- How will client and outsource company's quality processes, procedures and systems be interfaced?
- What are the client company and service provider responsibilities for incident management, change control and security management?
- Who will own documentation and records; are there any issues with shared documentation management responsibilities; whose document standards will apply?
- How will local site issues be managed and prioritized within a global contract framework?
- How will outsourced service quality be measured and reported?
- How will the outsource organization work with different client processes procedures and systems across regions and sites?
- How will the client company regain control of processes, documentation and information in the event that a new service provider is selected or services are taken back in house?
- How will outsource company maintain required skills and knowledge levels especially where there is a high staff turnover?

Service partners are typically selected through extensive "due diligence" processes that include supplier quality assessments. Due diligence processes solicit information from various sources to obtain a profile of the service partner's experience and capability. Supplier assessments should be conducted to ensure that appropriate quality organizations and systems are in place and that quality systems are adhered to and maintained. Supplier assessments should be conducted against the organizations that will actually deliver the services, not necessarily a corporate function.

From a quality perspective, one of the key elements to include in the contract is the issue of data ownership. Regulatory authorities will be keen to see who is accountable for regulated "processes" and "data." Table 33.5 defines typical quality considerations for an outsource contract:

Service provision must be monitored. Typically periodic reports are issued defining service performance against agreed service levels. Where global or major outsourcing agreements are in place, care should be taken to ensure that the reporting structure does not obscure significant quality and performance incidents at a site level.

Table 33.5 Outsource Contract Considerations

- Service definition, quality requirements, and performance requirements
- Quality organization structure and relationships
- Compliance expectations for internal policies and processes
- Definition of quality framework—quality management objectives and key process requirements
- Client and supplier QMS interfaces
- Training requirements
- Documentation and information responsibilities/ownership
- Performance monitoring expectations
- Auditing requirements
- Review and approval requirements
- Quality improvement requirements
- Security policies and standards
- Regulatory inspection support requirements
- Incident reporting and investigation

Periodic audits should be conducted to ensure that agreed policies, standards, SOPs and Work Instructions are being adhered to. A central auditing group should conduct audits where possible (or use the same audit tools and methods) to ensure that the audits are consistent from site to site and region to region.

Implications of contract exit must be considered early and must ensure that business application performance and data integrity is maintained. Depending on the nature of the outsource agreement, infrastructure design and configuration records may also need to be recovered by the regulated company.

THE CULTURAL CHANGE

Quality management within an IT environment can be as much about cultural change as it is about quality systems. A formal quality program can be viewed as an unnecessary and expensive constraint to innovation and agile working.

Before embarking on a cultural change program it is essential to first determine what a quality culture will look like when it is achieved and the current starting point. The change program should be carefully designed and planned to ensure that the objectives, end products and success criteria for the program are defined and bought into by senior management.

Some key indications of cultural change may be as follows:

- Appointment of an IT quality manager
- Quality demand reflected in IT budgets
- Internal development of pragmatic, compliant and value adding processes
- Increased consultation between IT and users
- Increased recognition of quality issues in decision making processes
- Ongoing improvement of processes and systems, driven by management and technical staff rather than IT or business quality

REFERENCES

1. Gindin S. Guide to E-Mail and the Internet in the Workplace. Bureau of National Affairs, 1999.
2. Wingate GAS. Validating Corporate Computer Systems: Good IT Practice for Pharmaceutical Manufacturers. Interpharm Press, 2000.
3. Crosson JE, Campbell MW, Noonan T. Network management in an FDA-regulated environment. PDA J Pharm Sci Technol, 1999; 53(6):280–287.
4. D'Eramo P. Computer qualification in an Internet age. GAMP concepts and case studies. International Society for Pharmaceutical Engineering Conference. Zurich, September 18–21, 2000.
5. Food and Drug Administration. "Off-the-shelf software use in medical devices," guidance for industry. FDA reviewers and compliance, September 9, 1999.
6. Organisation for Economic Co-operation and Development. GLP computer qualification, section 3(b) (ii), 1995.

7. U.S. Code of Federal Regulations. 21 CFR part 11. Electronic records and electronic signatures.
8. EU Guide to Good Manufacturing Practice for—medicinal products for human and veterinary use, including annex 11, Computerized systems.
9. Pharmaceutical Inspection Cooperation Scheme. Guidance on Good Practices for Computerized Systems Used in Regulated GxP Environments (PI-011-3), September 2007.
10. Food and Drug Administration. Guidance for Industry Part 11. Electronic Records and Electronic Signatures—Scope and Application, August 2003.
11. International Society for Pharmaceutical Engineering. GAMP®5: A Risk Based Approach to Compliant GxP computerized systems, 2008.
12. International Society for Pharmaceutical Engineering. GAMP Good Practice Guide, IT Infrastructure Control and Compliance.
13. IT Infrastructure Library (ITIL), ISO. Best Practice for Information and Communications Technology (ICT) Infrastructure Management.
14. International Organization for Standardization. ISO/IEC 27002:2005 Information Technology—Security Techniques—Code of Practice for Information Security Management.
15. International Organization for Standardization. ISO 9000:2005 Quality Management Systems—Fundamentals and Vocabulary.

Appendix 33.A Baseline Quality Assessment

Reference	Challenge
A	*IT management and organization*
A1	*Roles and responsibilities*
A1.1	Are IT management roles and responsibilities defined (e.g., remit, job description)?
A1.2	Are IT quality roles and responsibilities defined (e.g., remit, job description)?
A1.3	Is the organization documented (e.g., organization charts)?
A2	*Capability and competency*
A2.1	Are training plans in place for IT personnel?
A2.2	Have IT personnel received training in regulatory expectations (where appropriate)?
A2.3	Are training records in place to demonstrate that training has been delivered?
A2.4	Do training records document <ul><li>description of training,</li><li>date of training,</li><li>instructor, and</li><li>evidence of attendance?</li></ul>
A2.5	Do training records demonstrate that the attendee understood the training?
A3	*Internal organization interfaces*
A3.1	Are interfaces between IT and other infrastructure organizations defined?
A3.2	Are service agreements in place between internal infrastructure organizations?
A4	*External support organizations*
A4.1	Are contracts and/or service agreements in place for all external service/support organizations?
A4.2	Have external service/support organizations been assessed (e.g., audited) against contract requirements?
A4.3	Is service performance monitored against defined service levels?
A4.4	Have service providers been trained in your company's procedures where relevant?
A4.5	Have service providers been trained in your company's security policy?
A4.6	Are their controls in place to ensure that only authorized personnel from the service organization have access to your network and files?
B	*Quality systems*
B1	*General*
B1.1	Are IT projects managed in accordance with lifecycle project management systems that meet the requirements of industry standards and/or internal policies?
B1.2	Is there an overview document (e.g., quality manual) describing the quality management system?
B1.3	Is the quality management system periodically reviewed for its effectiveness?
B1.4	Are quality metrics in place to enable measurement of quality system performance?
B1.5	Are documentation and records management processes, systems and/or procedures in place?
B1.6	Are automated support systems compliant with regulatory and company requirements, e.g., SOP systems, configuration management, change control, etc.?

(Continued)

Appendix 33.A Baseline Quality Assessment (*Continued*)

Reference	Challenge
B1.7	Are infrastructure documentation standards in place including

B1.7   Are infrastructure documentation standards in place including

- planning,
- requirements specification,
- design specification,
- development,
- installation verification,
- operational verification, and
- report requirements?

B1.8   Does the QMS address operational processes, e.g.,

- change control,
- security management,
- backup and restoration,
- disaster recovery,
- archive and retention,
- help desk,
- client management,
- configuration management,
- system performance monitoring,
- maintenance,
- problem investigation, and
- retirement?

C   *Computer rooms and data centers*
C1   *Environmental conditions*
C1.1   Are computer rooms and data centers environmentally controlled? Environmental conditions include the following:

- Temperature
- Humidity
- Vibration
- Radio frequency interference
- Electromagnetic interference
- Electrostatic interference

D   *Infrastructure specification*
D1   *Hardware*
D1.1   Are inventories of hardware components in place?
D1.2   Are specifications, diagrams, or other documentation in place to describe the site local area network including

- Complete network layout of the site showing the backbone cable path and location of main network objects, e.g., hubs, servers, etc.,
- For each area or building, the location of each network component and cable path, and
- Network access points?

D1.3   Are documented configuration specifications in place for each network component (e.g., mainframes, servers, storage devices, transceivers, repeaters, bridges and routers, etc.)?

- Manufacturers details
- Location
- Addressing
- System performance (processor speed, memory, disk space, BIOS, etc.)
- Cards within component (including address)
- Configuration settings

D2   *Network organization*
D2.1   Are network trusts, domains, etc., documented (including access controls)?
D3   *Software and configuration*
D3.1   Is there an inventory of all network control and monitoring software/tools?

- Operating systems
- Communication protocols
- Performance monitoring software
- Virus protection
- Backup and restoration
- Software deployment tools (e.g., SMS)
- etc.

Appendix 33.A Baseline Quality Assessment (*Continued*)

Reference	Challenge
D3.2	Is there an inventory of all applications and data storage areas within the network?
D4	*Cable infrastructure*
D4.1	Are (internal or external) standards used to define cable requirements?
D4.2	Are cabling diagrams or specifications in place?
D4.3	Are cables tagged or labeled to aid identification?
D5	*Control of external connections*
D5.1	Are connections to WANs defined?
D5.2	Are controls in place to ensure that only authorized users can access the system remotely (e.g., secure ID or callback)?
D5.3	When a remote access link is terminated, is the user automatically logged off the network?
D6	*Electrical supplies*
D6.1	Are backup power supplies (e.g., UPS) in place to guard against power loss to critical components?
D6.2	Do electrical supplies conform to earthing, loading, filtering and safety standards?
D7	*Redundancy and fault tolerance*
D7.1	Have redundancy requirements been assessed, e.g., disk mirroring, RAID?
D7.2	Have requirements for automatic standby systems been defined?
E	*Infrastructure verification*
E1	Are critical hardware components, e.g., servers, storage devices, subject to installation verification?
E2	Are infrastructure tools, e.g., virus protection, backup, performance monitoring subject to installation verification and operational testing?
E3	Are computerized infrastructure tools and processes (e.g., change control, configuration management, access authorization, etc.) used?
F	*Network performance and fault management*
F1	*Speeds and capacities*
F1.1	Are procedures or automated controls in place to monitor network performance and capacities including • Speed, • Bandwidth, • Storage capacities, • Disk performance (e.g., fragmentation, thrashing), and • Address clashes?
F1.2	Are procedures in place for reporting, investigating and documenting network faults?
F1.3	Are event logs created and maintained in support of service performance monitoring?
G	*Data management, disaster recovery, and contingency plans*
G1	*Backup and restoration*
G1.1	Are procedures in place to assess backup requirements against business and regulatory needs?
G1.2	Have backup restoration procedures been formally tested?
G1.3	Are installed versions of operating systems, communication protocols, applications, etc., archived to facilitate recovery?
G1.4	Do backup procedures address the following? • Frequency of backups. • Physical labeling of media. • Review and retention of backup logs. • Periodic testing PF backups to verify that the backup procedure is functioning. • On-site and off-site storage of media. Full backups should be periodically stored off site. • Rotation of backup media.
G1.5	Do off-site backup storage considerations include • Location of facility and • Formal processes and controls over physical access to media both on a schedule and "on-request" basis?
G1.6	Do restoration procedures adequately address the retrieval of single and multiple files?
G2	*Archive*
G2.1	Are Retirement processes in place?
G2.2	Are processes in place for management of data deletion?
G2.3	Do processes, systems and/or procedures implement the requirements of company archive, records management, retention, and disposal policies?

(*Continued*)

Appendix 33.A Baseline Quality Assessment (*Continued*)

Reference	Challenge
G2.4	Do archive procedures include the following?

- Identification of archive media
- Management of archived media
- Documentation of records to be archived
- Retention periods
- Secure and safe storage of archive media
- Frequency of archiving
- Periodic evaluation of archive media
- Migration following system upgrades

G2.5   Do archive restoration procedures address

- Authorization to request records from archive and
- Procedure for performing restoration?

G3   *Business continuity plans*
G3.1   Are contingency plans in place to manage critical processes and maintain data integrity in the event of a failure?
G4   *External data management organizations*
G4.1   Are external organizations managing backup and archive facilities subject to appropriate controls including

         - contact/service definition,
         - audit, and
         - performance monitoring?

H   *Network access and security*
H1   *Security general*
H1.1   Are processes, systems, and/or procedures in place to address the requirements of IS security policies?
H2   *Physical security*
H2.1   Are servers and other critical hardware located in secure areas where access is controlled by key or other security device (e.g., card key)?
H3   *Logical security*
H3.1   Are responsibilities for security management defined?
H3.2   Are firewalls in place and documented to control access to the network?
H3.3   Are procedures in place to ensure that users are restricted to those parts of the network required to fulfill their defined role?
H3.4   Is virus detection software in place?
H3.5   Are controls in place to ensure that unauthorized software and files cannot be loaded into the network?
H3.6   Are procedures in place to detect and investigate potential security violations?
H3.7   Are user IDs based on two components and unique?
H3.8   Do user accounts automatically time out after a period of inactivity?
H3.9   Are user accounts disabled after multiple failed access attempts?
H3.10   Are users removed from the system when they leave the company or change jobs?
H3.11   Are there documented rules for password management including the following?

- Passwords should not be written down.
- Passwords should not be shared.
- Users should change their password upon logging into an account for the first time or following modification of the password by anyone other than the user.
- Enforcement of password complexity.
- Passwords should expire on a period basis.
- Policies in place to discourage reuse of passwords.
- Minimum password length should be 5 characters.

H3.12   Do procedures exist to manage cards and tokens including

- Issue of temporary and permanent cards and tokens, consistent with the security, account management and password procedures,
- The testing of their correct operation upon issue and periodically thereafter, and
- Cancellation in the event of loss?

H3.13   Are user access rights documented?

Appendix 33.A   Baseline Quality Assessment (*Continued*)

Reference	Challenge
I	*Configuration management*
I1	*Physical controls*
I1.1	Are development, test and production environments managed to ensure that software, hardware, and configuration integrity is maintained?
I1.2	Are GxP and non-GxP areas segregated or are GxP level controls applied to both?
I2	*Procedural controls*
I2.1	Are installation plans used to control the installation and verification of new hardware and software on the system?
I2.2	Are specifications, configuration statements and other documentation updated following changes to hardware and software?
I2.3	Do configuration statements document the following information for hardware and software installed on the network?
	• Item name or identifier
	• Serial number
	• Model or hardware type
	• Manufacturer
	• Item location
	• Storage devices
	• Operating system software, including version
	• Layered products, including version
	• Relevant application software, including version and the system owner
I2.4	Are controls in place to control access to system documentation?
I2.5	Are retention periods defined for system documentation in line with the site/function record retention schedule?
J	*Client management*
J1	Is the standard client defined?
J2	Are local extensions/configurations to standard client defined?
J3	Are processes in place to management to deployment of client applications?
J4	Are processes in place to audit client configuration?
J5	Is client configuration documented?
J6	Are processes in place to manage the build of new clients?
J7	Are processes in place to manage upgrades to the client?
J8	Are processes in place to maintain up to date virus protection?
K	*Service control management*
K1	Have service start-up and close down processes been defined?
K2	Have processes for implementing and communicating service restrictions been defined?
K3	Are facilities in place for fault reporting and tracking (e.g., help desk)?
K4	Are support services defined (e.g., first, second, third line support)?
K5	Are escalation procedures in place for management of service shortfalls?
K6	Are continuity plans in place to address critical service outage?
L	*Change control*
L1	Are change control procedures in place to manage changes to Network hardware, firmware and software, including impact assessment of any application affected by change?
L2	Do change control procedures require testing to be conducted when hardware or software is added, removed or modified within the infrastructure?
L3	Do change control processes include risk/impact assessment?
L4	Are IS/QA/user responsibilities defined for change control?
L5	Are patches, configuration changes, etc., subject to change control?
L6	Are changes tested/qualified?
L7	Do change control procedures address emergency changes?

Appendix 33.B   Electronic Records and Electronic Signature Infrastructure Challenges

Training and personnel
Is there adequately documented training, including on the job training, for the following groups? *11.10(i)*
• System administrators
• System developers
• IS/IT support staff

(*Continued*)

Appendix 33.B   Electronic Records and Electronic Signature Infrastructure Challenges (*Continued*)

Security

Is access to the system platform/network/architecture limited to authorized individuals, with their details recorded and maintained up to date? *11.10(d)*

Is there a group or role whose remit includes responsibility for platform/network/architecture security-both logical and physical? *11.300*

Documentation controls

Is there distribution and access control over infrastructure operations and maintenance documentation? *11.10(k)(1)*

Is the distribution of sensitive documentation, such as information on system security features, controlled? *11.10(k)(1)*

Change control

Do procedures/documentation exist for the design, installation, qualification and maintenance of the platform/ network/architecture/technical infrastructure components? Are these fully versioned and change-controlled? *11.10(k)(2)*

Is system documentation (e.g., design, installation, qualification, and maintenance documentation) available for the platform/network/architecture? *11.10 (k) (2)*

Is there a current inventory of all hardware and software components? *11.10(a)*

Is there a change control procedure to ensure that all hardware and software changes are properly documented? *11.10(k)2*

Is a change history maintained for this platform/network/architecture? *11.10(a)*

Policies

Is there a written policy that makes it clear that individuals are fully accountable and responsible for actions initiated under their electronic signatures in the same way as for their hand-written signatures and has this policy been communicated? *11.10(j)*

Is there a procedure requiring a formal investigation into suspected instances of electronic signature falsification? *11.10(j) 11.300(c)*

User ID/ID device and password controls

Is the identity of an individual verified before assigning a user ID, card, or token? *11.100(b)*

Is there a procedure to periodically check, recall or revise passwords, user IDs, cards, or tokens and to test that the latter function properly and have not been altered? *11.300(b), 11.300(e)*

Are there procedures that address the loss or compromise of user identification devices (cards/tokens, etc.) or passwords including electronic de-authorization, immediate and urgent reporting and rigorous control of temporary or permanent replacements? *11.300(c), 11.300(d)*

Is there a procedure for recalling, as appropriate, a user ID, card, or token in the event an individual leaves the position, the company, or is transferred? *11.300(b), 11.300(c)*

Is there a procedure assuring that repeated or serious attempts at unauthorized password usage are reported to organizational management? *11.300(d)*

Is it assured that the electronic signature is unique to an individual and cannot be used by anyone else—including system administrators? *11.100(a)*

Are user IDs assigned in such a way that they are never reused? *11/100(a) 11.300(a)*

Do passwords periodically expire? *11.300(b)*

Remote access

Is there an additional level of authentication for remote access? *11.200(a)i*

Are there additional controls around third-party remote access? *11.10(d)*

Clock settings

Is the time/date stamp applied by the system to any records reliable and can any alterations made to it be readily identified? *11.10(e)*

Open systems

Do the measures for open systems architectures, additional to all the above, assure authenticity, integrity and required confidentiality of records and signatures? *11.30*

Archiving

Do adequate procedures exist for the archiving and retrieving of media and data, which includes (where needed) archiving and storage of obsolete software and hardware needed to retrieve the electronic records through the required period of retention? *11.10(c)*

Are these periodically tested through the required period of retention? *11.10(c)*

Backups 11.10 (c)

Are backups performed on a regularly scheduled basis? *11.10(c)*

Is there documentation of regular backups?

Is the backup process periodically tested?

Is backup/archived media rotated to prevent degradation?

Disaster recovery

Do disaster recovery/business continuity plans exist for this platform/network/architecture? *11.10(a) 11.10(c)*

34 | Case Study 16: Internet/Intranet Applications

Winnie Cappucci
Bayer Pharmaceuticals

Ludwig Huber
Labcompliance

Arthur (Randy) Perez
Novartis Pharmaceuticals

INTRODUCTION

Internet applications have become ubiquitous for all types of businesses including healthcare. Two main distinctions are the use of websites for all types of online transactions and e-mails for exchanging messages with and without attachments. An example where the Internet can play an important role is shown in Figure 34.1. A pharmaceutical company outsources part of its clinical studies or laboratory analyses to a contract laboratory. The physical sample is sent to the contract laboratory, analyzed and the data sent back to the sponsor by e-mail with reports attached, or by data entry into a proprietary database.

Other examples for using the Intranet or Internet in the healthcare business are as follows:

- Release of batch approvals
- Electronic artwork transfer
- Remote approval of certificates of analysis at contract laboratories
- Updates, exchange and approval of training records and SOPs
- Administration of electronic patient records
- Billing information exchange between healthcare provider and insurance
- Telemedicine (remote surgery, diagnostics, imaging)
- Drug prescription online
- Electronic patient card
- Centralized and local patient data administration
- Tracking of drug product distribution
- Management of pharmaceutical sales forces
- Library research

When transporting clinical studies or any other data as mentioned above data traffic needs to adhere to:

Confidentiality: The contents of the data should only be accessible by authorized persons. The evolution of data privacy legislation opens the possibility for significant financial penalties should personally identifiable information be exposed.

Integrity: The data should be the exactly same at source and destination computer. Controls need to be in place and demonstrable, else the data may be open to question in the eyes of regulators.

Authenticity: The authenticity of the sender of the data must be guaranteed.

Nonrepudiation: Sender and recipient of the data cannot deny sending/receiving the data.

FDA's regulation on 21 CFR part 11 on electronic records and signatures requires records to be trustworthy (1), a word that combines all requirements as mentioned above.

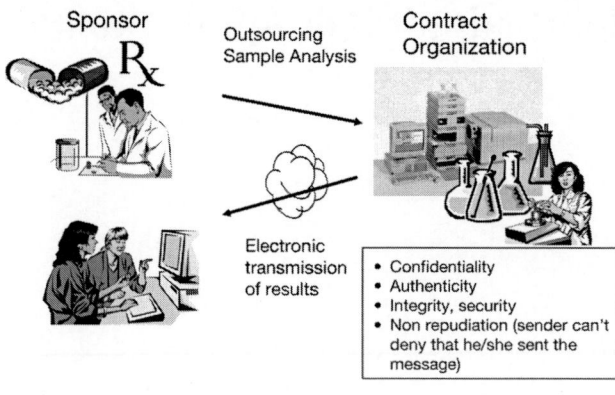

Figure 34.1 Example case study.

The Internet by its nature is an insecure and unreliable environment and therefore without special precautions is not compliant with the above-mentioned requirements. For example, the TCP/IP Internet communication protocol was designed to transmit and receive data correctly. It verifies correct delivery or if it detects a transmission error will retransmit until full delivery is successful. However, security beyond the confirmation of the data delivery is not a part of its design. Almost daily we hear in the news about hackers, viruses and worms, scam artists and online predators. For example, in 2007 the Open Web Application Security Project (OWASP) published an updated report about the top ten web application security vulnerabilities (2). The top ten vulnerabilities identified were as follows:

1. *Cross-site scripting (XSS)*: It occurs whenever an application takes user-supplied data and sends it to a web browser without first validating or encoding that content. XSS allows attackers to execute script in the victim's browser, which can hijack user sessions, deface web sites, possibly introduce worms, etc.
2. *Injection flaws*: Injection flaws, particularly, SQL injection, are common in web applications. Injection occurs when user-supplied data is sent to an interpreter as part of a command or query. The attacker's hostile data tricks the interpreter into executing unintended commands or changing data.
3. *Malicious file execution*: Code vulnerable to remote file inclusion (RFI) allows attackers to include hostile code and data, resulting in devastating attacks, such as total server compromise. Malicious file execution attacks affect PHP, XML, and any framework that accepts filenames or files from users.
4. *Insecure direct object reference*: A direct object reference occurs when a developer exposes a reference to an internal implementation object, such as a file, directory, database record, or key, as a URL or form parameter. Attackers can manipulate those references to access other objects without authorization.
5. *Cross-site request forgery (CSRF)*: A CSRF attack forces a logged-on victim's browser to send a preauthenticated request to a vulnerable web application, which then forces the victim's browser to perform a hostile action to the benefit of the attacker. CSRF can be as powerful as the web application that it attacks.
6. *Information leakage and improper error handling*: Applications can unintentionally leak information about their configuration, internal workings, or violate privacy through a variety of application problems. Attackers use this weakness to steal sensitive data or conduct more serious attacks.
7. *Broken authentication and session management*: Account credentials and session tokens are often not properly protected. Attackers compromise passwords, keys, or authentication tokens to assume other users' identities.
8. *Insecure cryptographic storage*: Web applications rarely use cryptographic functions properly to protect data and credentials. Attackers use weakly protected data to conduct identity theft and other crimes, such as credit card fraud.
9. *Insecure communications*: Applications frequently fail to encrypt network traffic when it is necessary to protect sensitive communications.

10. *Failure to restrict URL access*: Frequently, an application only protects sensitive functionality by preventing the display of links or URLs to unauthorized users. Attackers can use this weakness to access and perform unauthorized operations by accessing those URLs directly.

Should we neglect the advantages of the Internet because of the many problems that have been reported? The answer is resoundingly no.

The FDA recognizes the increasing use of the Internet and gives recommendations on how it can be used in an FDA regulated environment. FDA issued a draft guidance document in 2001 concerning validation of electronic record-generating systems. Although this guidance has since been withdrawn because many of the controls it recommends would not be universally applicable, a very pertinent statement within it remains valid (3):

We recognize the expanding role of the Internet in electronic recordkeeping in the context of Part 11. Vital records, such as clinical data reports or batch release approvals, can be transmitted from source to destination computing systems by way of the Internet.

There are a lot of security tools and technology available and offered as part of browser software or as add-ins or separate applications. There are also tools available that help comply with other requirements such as authenticity, data integrity, accuracy of data transfer and nonrepudiation. This was also recognized by the FDA in the withdrawn 2001 guidance (3):

"The Internet can nonetheless be a trustworthy and reliable communications pipeline for electronic records when there are measures in place to ensure the accurate, complete and timely transfer of data and records from source to destination computing systems."

Availability of tools does not necessarily mean that everybody takes advantage of them. This is where this case study will help. Its aim is to give guidelines on the steps to take to make the use the Internet trustworthy as required by regulations, such as FDA's 21 CFR part 11.

A grounding in some of the basic terms and technologies is required. These include open vs. closed systems as defined by 21 CFR part 11, FTP/IP protocols, cryptography, digital signatures, digital certificates, Public Key Infrastructures (PKI) and secure multipurpose Internet mail extensions (S/MIME). Although the details are beyond the scope of this chapter, there is a lot of reference material available. Three very helpful websites are RSA Security, http://www.rsa.com; Internet Engineering Task Force, http://www.ietf.org; as well as the Open Web Application Security Project (OWASP), http://www.owasp.org referenced above.

An excellent guideline on the validation of computerized systems can be found in *GAMP®5* (4). The qualification of network infrastructure and validation of networked systems is documented in a GAMP Good Practice Guide (5).

OPEN VS. CLOSED SYSTEMS

The Internet is a classical example of an open system in the definition of 21 CFR part 11 (1):

"Open system means an environment in which system access is not controlled by persons who are responsible for the content of electronic records that are on the system." As illustrated in Figure 34.2 Internet service providers (ISPs) have access to data, and worse, data can reside on intermediate servers in the transmission chain for an uncontrolled time, exposing it to malicious use. The bottom line is that persons who are responsible for the content cannot control access to any data transferred throughout its entire exposure on the Internet.

Section 11.30 of part 11 specifies requirements for open systems:

Persons who use open systems to create, modify, maintain, or transmit electronic records shall employ procedures and controls designed to ensure the authenticity, integrity, and, as appropriate, the confidentiality of electronic records from the point of their creation to the point of their receipt. Such procedures and controls shall include those identified in Sec. 11.10, as appropriate, and additional measures such as document encryption and use of appropriate digital signature standards to ensure, as necessary under the circumstances, record authenticity, integrity, and confidentiality.

DATA TRANSFER THROUGH THE INTERNET

Data are transmitted through the Internet by using TCP/IP communication protocols. A specific function of TCP/IP protocols is so-called packet switching. When a data file is sent through the Internet it is not sent in one piece. Instead the file is broken into packets that can be routed separately through the Internet, a process that is called packet switching. This is illustrated in Figure 34.3.

At the sending computer the files are broken into packets. They are sent through a LAN or modem and gateway through routers to the receiving computer.

The receiving computer reassembles the packets into a single file, which is identical to the original file. The advantage of this concept is that each packet can find the fastest way through the Internet. When one way becomes overloaded packages broken down from one file can be directed through different lines.

Computers on the Internet are identified through IP addresses. IP addresses of the sending and receiving computers together with some other information is included in a header created for each packet. This is the reason why all packets from one data file find their way to the same computer and can be reassembled again. The routers read the message headers and forward the packets to either another computer or to the gateway at the destination site.

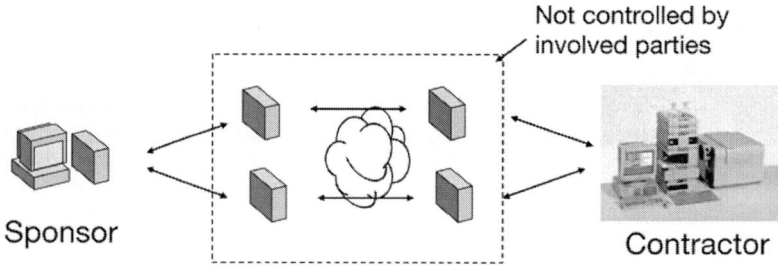

Figure 34.2   The Internet as an open system.

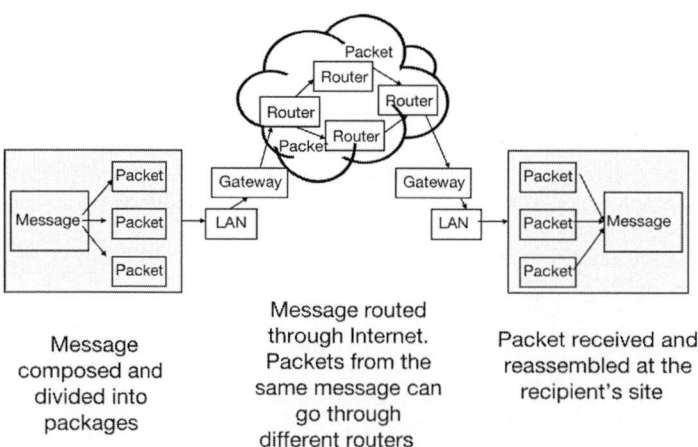

Message composed and divided into packages

Message routed through Internet. Packets from the same message can go through different routers

Packet received and reassembled at the recipient's site

Figure 34.3   Data transfer through the Internet.

CRYPTOGRAPHY AND DIGITAL SIGNATURES

As already mentioned the Internet is a classical example of an open system. TCP/IP communication protocols were not designed for security and file information can be accessed by both the personnel authorized by the ISP and unauthorized individuals who have gained illegal access. To prevent the information from being read by people who are not authorized to do so the information must be encrypted. Only a computer that "knows" how to unscramble the information (decrypting) can generate a readable record.

Encryption technology is used for online shopping to make credit card information invisible. If we want to encrypt the word "test," all we have to do is to convert each letter to the next letter in the alphabet. After doing this the word reads "uftu" and unless the reader knows the encryption mechanism (the "key") he does not understand the meaning. Of course in practice the encryption algorithms are far more complex.

Cryptography can be divided into two groups: symmetric and asymmetric encryption. In symmetric encryption the sender and receiver use the same key. This method is very fast but the sender needs to send not only the message but also the key, otherwise the receiving party cannot read the message.

Asymmetric encryption is also called public key encryption. Two keys are required: a private key and a public key. Usually, the sender encrypts the data with the public key and the receiver decrypts the data with a private key but it can also be the other way around. Public keys are frequently located on the Internet. Private keys are located in a secure area of the owners' computer.

HASH FUNCTIONS TO ENSURE DATA INTEGRITY

Cryptographic hash functions are used to verify data integrity. A cryptographic hash function is a procedure that utilizes a selected data string of any length and produces a fixed-length binary value (hash) as the output (fingerprint). The hash value should concisely represent the longer message or document from which it was computed. Therefore any change to the original data will change the hash value. A cryptographic hash has the following characteristics:

- It should be easy to compute.
- It should be hard to modify the encoded data without changing its hash value.
- It should be hard to find the encoded data that has a given hash value (one-way hash).
- It should not be feasible to find any two encoded data strings with the same hash value or to deduce any useful information about the source data from the hash value.

The sender's computer calculates the hash value and attaches the value to the message. The receiving computer uses the same algorithm, recalculates the hash value and compares the result with the value as attached to the message. Obtaining exactly the same hash means the file is the same. The probability that two different records generate the same message digest can be as low as 1 in 10^{87}, which is effectively zero (the highest estimate for the total number of subatomic particles in the entire universe is 10^{87}).

Neither cryptography nor hash values can ensure authenticity of the sender over the Internet. To do this we need digital signatures that combine one-way hash calculations and cryptography using a person's private key. A digital signature is an encrypted hash value that is appended to a message. The receiver uses the sender's public key to decrypt the message and calculates the encrypted hash value using the same hash function that the sender was using. When correctly implemented digital signatures are legally equivalent to handwritten signatures on paper and the signer cannot deny legal responsibility for the content of the document.

DIGITAL CERTIFICATES OF PUBLIC KEYS

With encryption, hash calculations, and digital signatures, as described above, there is still one open question: how can you be sure that the person whose name is in the message is really the sender? To ensure that a public key is assigned to an individual or organization we need electronic certificates that verify the identity of a user. They are also the basis for secure electronic transactions. The information that should go into a digital certificate is standardized in a protocol called X.509. A public key infrastructure consists of the following:

- A certificate authority (CA) that issues and verifies digital certificates. A digital certificate contains the public key or information about the public key.
- A registration authority (RA) that acts as the verifier for the CA before a digital certificate is issued to the requestor.
- One or more directories where the digital certificates (with their public keys) are maintained.
- A certificate management system.

When implemented correctly both sender and receiver can trust each other. An example of an organization that issues digital certificates is Verisign (www.verisign.com).

E-MAILS THROUGH S/MIME

Sending a text message via e-mail on the Internet is similar to mailing a picture postcard. The mailing process is inexpensive and quick, but your message is public and open for everyone to see. Anyone who happens to see the card can pick it up and read your message. E-mail is similar in that it can be easily read by a variety of off-the-shelf hacker applications. This leaves sensitive information sent by standard e-mail vulnerable.

For secure e-mail transactions the format of S/MIME was developed. It uses most of the technologies as described in previous sections of this chapter. E-mails can send encrypted e-mail messages including attachments, and allows users to use digital signatures. It uses the PKCS 1 RSA encryption, PKCS 7 cryptographic message syntax and PKCS 10 certification request syntax specifications, which are in effect the de facto industry standards. The life of an e-mail message is illustrated in Figure 34.4.

The sender writes the text and can attach documents. The sending computer generates a hash value of the message. Using either the private or public key of the sender the message and the attachments are encrypted and sent over the Internet to the receiving site. The recipient takes his/her private key plus the public key of the sender and decrypts the message. Using the sender's digital signature and certificate the recipient can verify who wrote the message before he/she reads the message and attachments.

VIRTUAL PRIVATE NETWORKS

Sending confidential information to third parties as shown in the example in Figure 34.1 is not the only need for companies; they also want to be able to send information within the organization, sometimes from remote locations. Within the office this is easily done via the corporate network. Access to a corporate network can be very well controlled as long as the

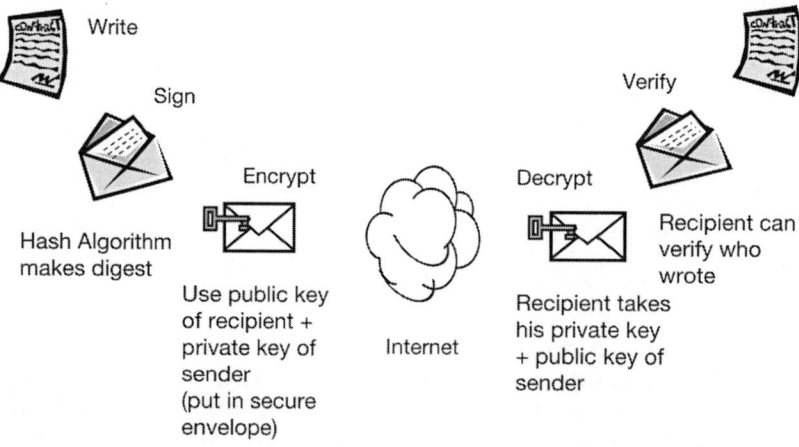

Figure 34.4   Life of a secure multipurpose Internet mail extensions e-mail message.

- Traffic is encrypted
- Remote site authenticated
- Multiple protocols supported

- Branch office
- Home office
- Business travelers
- Business partners

Token card +PIN

Figure 34.5 Virtual private network.

company uses transfer lines inside its firewall. However, there is also a need to communicate confidential data across the globe, for example, to remote sites, business travelers, home offices and trusted business partners.

In the past companies leased private telephone lines to build private networks. This is very expensive but thankfully it is no longer the only solution to secure data transmission from remote locations. With modern technology as described in previous chapters companies can now achieve very much the same level of security and confidentiality using public infrastructure and a concept is called virtual private network (VPN), which is illustrated in Figure 34.5. Data are encrypted when passing between the protected corporate network and a remote corporate user. Such users may be employees working in branch or home offices, business travelers, or even business partners who have been provided with the private key. VPNs can provide secure connection between computers over the Internet and are cheaper than private networks.

Access security is the major concern of VPNs. Static passwords are not secure enough. Access to VPNs typically requires dynamic password control through tokens. Tokens are often built in to company laptop computers, or may be separate devices. These devices provide additional information as part of the login process that verify the user is genuinely authorized to be on the network.

VALIDATION

It is not realistic to assume that all systems controlling communication through the Internet will be formally validated. However, the value of corporate assets accessible via a VPN will be so great that there is no question that a very solid state of control will be required as a business practice. Companies must certainly also be able to demonstrate this as part of a health authority audit.

Again quoting the withdrawn 2001 FDA's part 11 validation guidance states (3):

We recognize that the Internet, as computer system, cannot be validated because its configuration is dynamic. For example, when a record is transmitted from source to destination computers, various portions (or packets) of the record may travel along different paths, a route that neither sender nor recipient can define or know ahead of time. In addition, entirely different paths might be used for subsequent transfers.

When an application accepting transmissions via the Internet is validated, testing should include many of the same measures that would be taken when validating transmission of data between interfaced systems on the same network. This would include verification that the target application can detect and flag incomplete or corrupted transmissions, and if the intent is that transmissions will be encrypted, that it accepts only properly encrypted input, and that the decryption algorithm works properly to yield a readable record. If the transmitting application is also validated, that validation exercise should verify that the right information is getting into the file and that it is properly encrypted.

Acknowledgement of record transmission and reception may be desirable if the risk related to patient safety is high, for example, in monitoring of certain clinical studies. The best approach would be fully automated, although it could be adequate for a person to review transmission logs according to a defined and documented process. Any requirements for written acknowledgment should be risk based.

Validation of an Internet application comprises of six parts.

- Configuration management and documentation of systems at sending and receiving site(s).
- Validation of applications on source and/or destination computers. Depending on the nature of the applications, it may not be appropriate to expect one of the systems to be validated. For example, a purchasing system sending a purchase order to an ERP system is unlikely to be validated. Similarly, a supplier system sending a certificate of analysis to an ERP system may not be validated.
- If a browser is involved in the data transfer, test correct browser functionality and user interface.
- Verification of correct file transfer.
- Security testing.

Configuration Management

The objective of configuration management is to have detailed information on the system throughout its operational lifetime. Improperly configured web and application servers are prone to security problems and are a common source of vulnerabilities (2). SOPs should be available for both initial configurations as well as for planned and unplanned changes. Worksheets are useful to document initial configurations. This includes computer hardware, operating software with product name and revision number, application software with product name and revision number, network devices with product name and software/firmware revision number, documentation such as user manuals and configuration settings. Any changes should be documented following documented change control procedures.

Validation of Computer Applications at Sending and Receiving Site

For the validation of computer applications at the receiving site we recommend applying any lifecycle concept, for example, according to GAMP®5 (4). Validation activities should be well planned and documented in a validation project plan with validation activities, owners and time schedule. The validation approach and results should be documented in a validation report. Specific attention should be given to the design, configuration, testing, and control of the interface portion of the application, especially if, as may be the case for many Internet-facing applications (Fig. 34.6), data may be coming in a variety of formats. Also critical is authentication for access to the application. This would include both the acceptance of data transmission and the logging in of remote users.

Not all computer systems sending information over the Internet to validated systems will be validated. Most companies today probably take feeds from suppliers, whether directly their own applications or via e-mail attachments with information that will be uploaded to validated systems. On the basis of the principle of garbage in, garbage out (GIGO), it is the responsibility of the regulated company to ensure that external data that goes into their validated applications meets required standards for content and format. Without that assurance data integrity of the validated application may be questioned. The receiving company would need to do some validation work to ensure the feed can be taken cleanly, and will have to verify that the sender's processes for configuration management and change control are sufficiently robust to preclude unanticipated changes that break the interface.

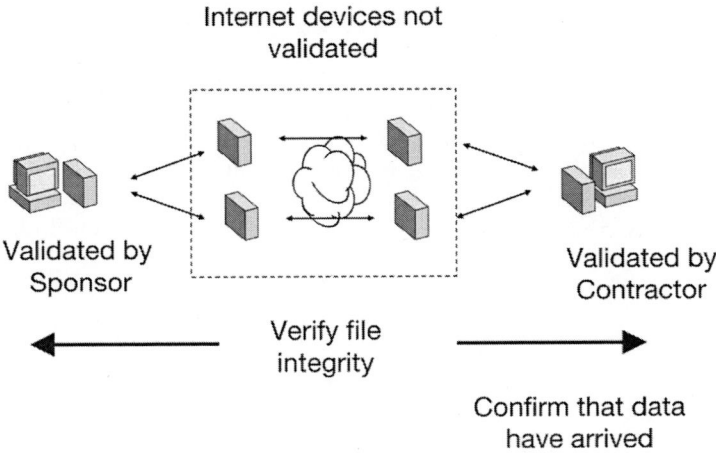

Figure 34.6   Validation activities of Internet applications.

When the input originates from validated systems, the onus on owners of the receiving system is likely to be a bit lighter, as the control processes at the sender should be reasonably robust. However, the owner of the system receiving the input still needs to ensure that they are notified of any changes at the originator that may affect the data feed. It is a very good idea to make review of the validation package part of the supplier assessment process.

Testing Browser Functionality and User Interface

User interface of web applications are typically accesses via browsers such a Microsoft's Internet Explorer, Firefox, or a similar application. While browsers are getting better and web standards seem to be converging, there are still cases where websites do not render correctly or different browsers behave differently (6,7).

Many applications may also require additional software add-ins at the browser end, such as Java applets, ActiveX controls, Flash players, etc. This can be problematic, as the user's company is likely to have one defined standard browser with a set configuration. Suppliers who are not bound by corporate standards and the strictures of validation may adopt new browser technologies or new versions of old add-ins more rapidly than the user company's standards change. This can force a company to stay with old software releases.

The compatibility of the standard browser configuration with the applications should be challenged. Some areas that may prove problematic include the following:

- Correct functioning of scroll bars
- Correct functionality of buttons
- Incorrect or dead hyperlinks
- Correct arrangement and functioning of frames
- Visibility of all information on the screen at different screen resolutions

Companies should be alert for situations where Internet-facing applications might have browser requirements that conflict with each other, with corporate standards, or with corporate policies regarding security.

Verification of Correct File Transfer

As with any interface between systems, verification of data exchange is important for web-based systems. When the file transfer occurs through the Internet, encryption is a complicating factor. In addition, if the file transfer is unpredictable (i.e., not a regularly scheduled transmission), there is an added challenge of verifying that all of the information the sender transmits actually gets to the receiver. A hash calculation approach can be used for this purpose, as could an answer-back function where the receiving system tells the sender that transmission was successful. Such a controls need to be design into the systems.

Validation activities related to file transfer should include the following:

- Verification that all sent files are received. This would include challenge testing of the controls to detect transmission failures.
- Verification that sent files are properly encrypted.
- Verification that received files are properly decrypted and are readable.
- Verification that received files are complete.
- Verification that unavailability of bandwidth does not negatively impact the transmission.
- Verification that file size (within predictable business norms) does not negatively impact transmission.

Security Testing

Security testing is critical for web applications. Aside from controlling unauthorized access within the company, the threat of hacker and malware attacks over the Internet is vastly greater than for applications whose communications are entirely within the firewall. Data privacy and protection of intellectual property are also paramount considerations for both a business and compliance perspective.

Tests scenarios should be set up test to challenge the following:

- Access to the VPN (not necessary for every application being validated, but it has to be done at least once)
- Password expirations
- Authorized access to certain areas
- Data integrity (see also previous section)
- Data filtering at firewalls
- Encryption (where appropriate)
- Granting and revoking of PKI (or similar) certificates to third party companies

Many companies run intrusion detection software as a routine control protecting the firewall. As part of the risk management strategy for a validated web application consideration should be given as to whether the application needs to be in scope of such a control.

Another good practice is penetration testing for perimeter hardening. This is testing in which independent testers/assessors attempt to circumvent or violate the security features of a system. They use all system design and implementation documentation, this may include system source code, configuration, manuals, and diagrams. They will work under the same restrictions and rules applied to users. The end result is an analysis of the system for any potential vulnerabilities that may result from poor or improper system configuration, known and/or unknown hardware or software flaws, and operational weaknesses in current process or technical controls.

DEVELOPMENT AND COMMUNICATION OF PROCEDURES

All of the controls discussed to this point are heavily dependent on the staff of the regulated company developing a culture that recognizes the risks of Internet connectivity. This requires development of procedures for good Internet practices and training on how to use the procedures day by day as well as enforcement of the procedures.

When online users should think with every click and keystroke about what they are doing and consider its potential impact on privacy and security.

Specifically, users need to recognize the risks related to the following:

- Downloading files from nontrusted websites (even YouTube videos could be infected with a virus)
- Opening e-mail attachments for which the origin is unclear
- Running programs not vetted by the IT department (or at least scanned for malware)
- Zipped files as potential malware vectors

In addition, IT departments need to develop a culture and standards that ensures the right level of security protection. It must provide the company with a reasonable level of protection while being responsive enough to allow users to achieve business goals (perhaps loading PC software, downloading browser add-ins, etc.) and to keep them from being tempted to circumvent the process. Aspects of this include but are not limited to the following:

- It must include a risk-based process for evaluating patches and rapidly applying those that expose the enterprise to intolerable risk (8).
- It must cover a wide variety of areas such as browser controls, registry protection, automated virus scanning of files to be downloaded and portable media like floppies, CDs, flash drives, etc.
- It must evolve, addressing both business needs and threats.
- It should identify "trusted sites" from which materials may be downloaded with a high degree of confidence in their safety. The list of trusted sites must be maintained on a timely basis for both additions and deletions.
- It must include safe handling of cookies.
- It should severely limits administrative rights to both servers and workstations. Restricting the ability to load software closes a major route for malware to infest a company's computers.
- It should recognize that on the basis of risk some computers may require tighter control than others (a computer interfaced to a clinical research organization and the clinical trial monitoring system probably should be more carefully controlled than a computer used by a sales representative).
- It should ensure the constant management of essential controls like update of virus definitions, intrusion detection monitoring, etc.

Procedures should be developed for using the technologies described in previous sections such as using digital signatures and certificates, sending and receiving e-mails with S/MIME and validation of computers at the sending and receiving site.

In addition, corporate policies should be available defining acceptable use of the Internet, whether it is used for regulated applications or not. These are procedures that should help to protect the computer and data against accidental or incidental attack from outside. The company needs to recognize that users will use both e-mail and the Internet for personal purposes as well as business, and policies need to reflect that immutable reality.

Policies and procedures are obviously not of much value if they are not followed. The internal audit program should include procedures on how to use the Internet.

Security procedures are also important but do not fit into the scope of this chapter. We recommend looking at literature, for example, NIST has published good guidance documents on IT security, for example, the NIST Security Guide for Interconnecting Information Technology System (9) and the Guide for Assessing the Security Controls in Federal Information Systems, 800-53A (10).

Microsoft also has good recommendations for security, for example, on the Windows operating system (11).

CONCLUSION

The bottom line for web-based systems is essentially the same as it is for non-web-based systems. This is well stated in the current FDA guidance on electronic records and signatures (12), which states a need for "documented evidence and justification that the system is fit for its intended use (including having an acceptable level of record security and integrity, if applicable)." Although the specific reference in this case is in regard to legacy systems, it reflects the general regulatory expectation for systems of any type.

ACKNOWLEDGMENTS

This case study was originally authored by Ludwig Huber and has been subsequently updated and revised by Randy Perez and Winnie Cappucci.

REFERENCES

1. Code of Federal Regulations Title 21. Food and Drugs, Part 11. Electronic Records; Electronic Signatures; Final Rule. Federal Register 62(54), 13429–13466.
2. Open Web Application Security Project (OWASP). Top 10 2007. Available at: http://www.owasp.org/index.php/Top_10_2007.
3. Food and Drug Administration. Guidance for Industry: 21 CFR Part 11. Electronic Records; Electronic Signatures Validation (draft). U.S.A., August (now withdrawn) 2001.
4. International Society for Pharmaceutical Engineering. GAMP®5: Risk-Based Approach to Compliant GxP Computerised Systems. Tampa, Florida, 2008. Available at: www.ispe.org.
5. International Society for Pharmaceutical Engineering. GAMP® Good Practice Guide: IT Infrastructure Control and Compliance. Tampa, Florida, 2005. Available at: www.ispe.org.
6. Mitchell R. Web standards on the edge. Computerworld, February 24, 2009.
7. Mitchell R. When good browsers go bad—and they all do. Computerworld, February 24, 2009.
8. International Society for Pharmaceutical Engineering. GAMP®5: A Risk-Based Approach to Compliant GxP Computerized Systems, Appendix S4, 2008.
9. National Institute of Standards and Technology. Security guide for interconnecting information technology systems, 2002. Available at: http://csrc.nist.gov/publications/nistpubs/800-47/sp800-47.pdf.
10. National Institute of Standards and Technology. Guide for assessing the security controls in federal information systems, 800-53A. Available at: http://csrc.nist.gov/publications/nistpubs/index.html.
11. Microsoft Security Central. Home page. Available at: http://www.microsoft.com/security/default.mspx.
12. Food and Drug Administration. Guidance for Industry Part 11. Electronic Records; Electronic Signatures—Scope and Application, 2003. Available at: http://www.fda.gov/cder/guidance/5667fnl.htm.

35 | Case Study 17: Medical Devices and Their Automated Manufacture

Guy Wingate
GlaxoSmithKline

Tom Ryan
Boston Scientific

A medical device is an instrument, apparatus, appliance, material, or other article, whether used alone or in combination, together with any software necessary for its proper application, which

   a. is intended by the manufacturer to be used for human beings for the purpose of
      i. diagnosis, monitoring, treatment, alleviation of disease,
      ii. diagnosis, monitoring, treatment, alleviation of or compensation for an injury or handicap,
      iii. investigation, replacement, or modification of the anatomy or of a physiological process, or
      iv. control of conception, and

   b. does not achieve its principal intended action in or on the human body by pharmacological, immunological, or metabolic means, even if it is assisted in its function by such means (1).

The definition's reference to software extends the scope of medical devices to include those based on programmable technology. Such devices can be extremely complex and consist of a large number of programmable elements. Figure 35.1 provides a schematic overview of a medical device's healthcare service: the delivery, control, and/or monitoring of medical treatment.

Medical devices require both product and process validation. Product validation is necessary to assure that they are designed and assembled consistently to assure a high quality of service. Process validation is necessary to assure that they are produced under a compliant regime of Good Manufacturing Practice (GMP). Particular validation requirements are laid down by European Directive 2001/83/EC (2) and the U.S. Code of Federal Regulations Title 21, Part 820 (3). Some medical device manufacturers in Europe additionally seek CE marking, which is a quality standard given to organizations who successfully pass a quality inspection by a regulatory agency.

Regulatory expectations also require that "when computer or automated data processing systems are used as part of production or the quality system, the [device] manufacturer shall validate computer software for its intended use according to an established protocol" (3). In addition, computer systems that implement part of a device manufacturer's production processes or quality system may be subject to electronic record and electronic signature requirements. Medical devices destined for the United States for instance will be subject to 21 CFR Part 11 (4). Example computer applications that typically require validation include medical device design tools, laboratory testing and analysis, product inspection and acceptance, production and process control, environmental controls, packaging, labeling, document control, and compliant management (5).

This case study discusses the particular issues affecting the automated manufacture and validation of a medical device involving the management and coordination of a number of suppliers:

- One medical device designer
- One technology development designer

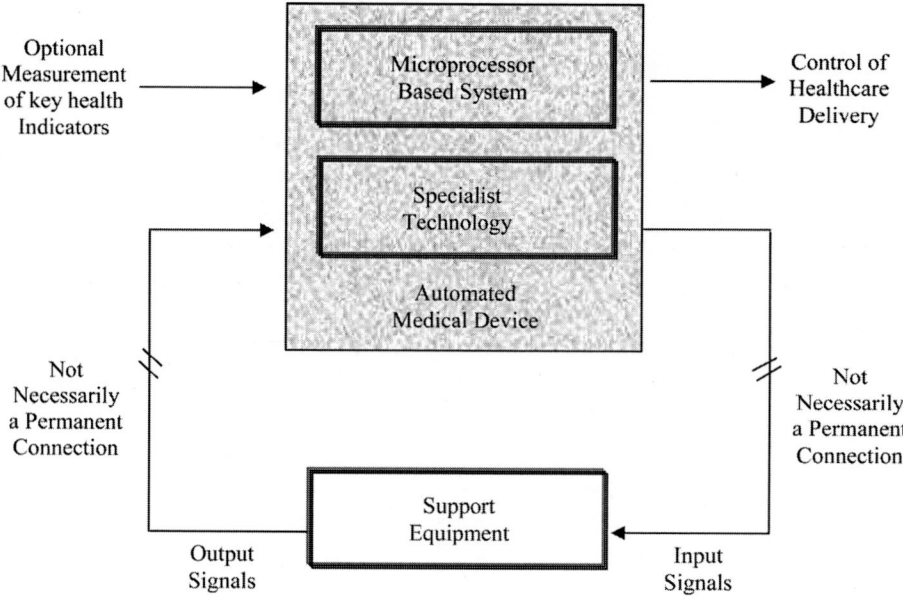

Figure 35.1 Automated medical device.

- Two supporting equipment designers
- One medical device manufacturer
- Four manufacturing process lines

The approach to validation is unchanged from that described in this book with one exception. Risk assessments for medical devices should focus on severity of instances of erroneous behavior. Risk assessments should not take account of probability of erroneous behavior as any occurrence may be critical. Electronic record and signatures are not discussed since this topic has already been discussed in detail in chapter 13.

VALIDATION PLANNING

The coordination of suppliers is vital, and, to this end, a validation master plan is often produced, referencing a number of validation plans for each element of the automated medical device. Suppliers should be encouraged to produce their own quality plans in response to the validation plan for their portion of the system, so that any inconsistencies and ambiguities can be identified and corrective actions instituted before change become too inconvenient and expensive.

The selection of suppliers is often limited because only one or two suppliers will generally have the capability to provide a particular item of technology used in the medical device or in its manufacture. It may not be the case, therefore, that the supplier to be used has any experience in validation. Indeed, many suppliers supporting medical devices are small organizations and have a limited opportunity to develop an in-house validation capability. If this is the case, then training should be given to ensure that the supplier has no misunderstandings about the expectations made by GMP regulators. Care must be taken not to always take capability "sales-speak" at face value. The use of external validation consultants and/or the ISPE GAMP® guides for computer validation can prove useful in training mechanisms (6). Most suppliers are keen to pick up new skills and will welcome the chance to enhance the competency of their staff.

Because of such uncertainties, each supplier should be audited at the start of the project to establish whether a Quality Management System (QMS) exists that will support the validation of the equipment or process. Where none exists, an agreement must be reached with each supplier as to what quality measures will be used. This should be outlined by each

Table 35.1 CE Marking and Validation Documentation Relationships

EU Directive 93/42/EEC, Annex III, Section 3	Validation documentation
The documentation must allow an understanding of the design, manufacture, and performance of the product and must contain the following items in particular:	
A general description of the type, including any variants planned	User requirements specification
Design drawings; method of manufacture envisaged; in particular sterilization; and diagrams of components, subassemblies, circuits, etc.	Drawings, hardware design, and software design specifications
The description and explanation necessary to understand the above-mentioned drawings and diagrams and the operation of the product	Functional specifications, operator's manual
A list of the standards referred to in Article 5, applied in full or part, and descriptions of the solutions adopted to meet the essential requirements if the standards referred to in Article 5 have not been applied in full	Validation plan, supplier quality Plans
The results of the design calculations, risk analysis, investigations, technical test, etc., carried out	Results of clinical trails; threats and controls; FMEA, software structure analysis; IQ, OQ and PQ testes; and fundamental science document
A statement indicating whether the device incorporates, as an integral part, a substance as referred to in Section 7.4 of Annex I and data on the test conducted in this connection	Relevant to medical device as a whole; not directly applicable to its automation
This clinical data referred to in Annex X	Clinical trial report
The draft label and, where appropriate, instructions for use	Relevant to medical device as a whole; not directly applicable to its automation

Note: Annex III, Section 3 of the European Union's medical Device Directive (2) holds the key area that maps CE onto computer validation requirements.

company in their quality plan. In our case, we decided that all companies concerned should comply with the ISPE GAMP Guide, as this outlines validation documentation that is suitable for the U.S. Food and Drug Administration's (FDA's) requirements and may also be used to support the European Union's CE mark accreditation (see Table 35.1). The device was to be released in both the American and European markets.

A particularly important area to be addressed during planning is the consistent use of terminology. The ISPE GAMP Guide (6) includes a lexicon of validation terminology that can prove to be a useful reference in this respect some practitioners prefer to include specific terms associated with a validation project and their definitions in the validation plans.

Intellectual property rights (IPR) should also be agreed to as part of the validation planning exercise. Particular elements of the automated medical device or its manufacture may be confidential to individual suppliers. Contracts must clearly define the terms and conditions affecting the supplier's rights.

REQUIREMENTS

Requirements should be developed for the medical device and automated equipment used to support its manufacture. The requirements should clearly state the intended use of software. Areas of special importance include allocation of system functions to hardware/software, operating conditions, user characteristics, and potential hazards (5).

The FDA recommends a software requirements specification document is generated. Regulators will deem software unvalidated without predetermined and documented requirements. The scope of requirements should cover (5):

- All software system inputs
- All software system outputs
- All functions that the software system will perform

- All performance requirements (e.g., data throughput, reliability, response times)
- How users will interact with the system
- The definition of all external and user interfaces
- The definition of internal system interfaces
- What constitutes an error and how errors should be handled
- The intended operating environment (e.g., hardware platform, operating system)
- Any ranges, limits, defaults, and specified values that the software will accept/reject
- Any potential hazards and design constraints (i.e., safety-related requirements)

Each requirement identified should be evaluated for accuracy, completeness, consistency, testability, correctness, and clarity (5). The U.S. quality system regulation requires a mechanism for addressing incomplete, ambiguous, or conflicting requirements [Clause 30c (3)].

In this case study, let us consider an insulin delivery system to aid diabetes (7). Diabetes is a relatively common condition where the human body is unable to produce sufficient quantities of a hormone called insulin. Insulin metabolizes glucose in the blood. The conventional treatment of diabetes involves regular injections of genetically engineered insulin. The problem with this treatment is that the level of insulin in the blood does not depend on the blood glucose level but is a function of the time when the insulin injection was taken. This can lead to very low levels of blood glucose (if there is too much insulin) or very high levels of blood sugar (if there is too little insulin). Low blood sugar is, in the short term, a more serious condition as it can result in temporary brain malfunctioning and, ultimately, unconsciousness and death. In the long term continual high levels of blood sugar can lead to eye damage, kidney damage, and heart problems.

An insulin delivery system might work by using a microsensor embedded in the patient to measure some blood parameter that is proportional to the sugar level (7). This controller computes the sugar level, judges how much insulin is required, and sends signals to a miniaturized pump to deliver the insulin via a permanently attached needle. Insulin delivery systems are likely to be software controlled. Figure 35.2 is a data-flow model that illustrates how an input blood sugar level is transformed to a sequence of pump control commands.

The requirements for the insulin delivery system would include specific patient safety needs such as

- A single dose of insulin shall not be delivered that is greater than the designated maximum dose,
- The daily cumulative dose of insulin shall not be greater than a designated maximum dose,
- An audible alarm shall sound when any device anomaly is detected, and
- Diagnostic messages should indicate nature of warning and remedial action required.

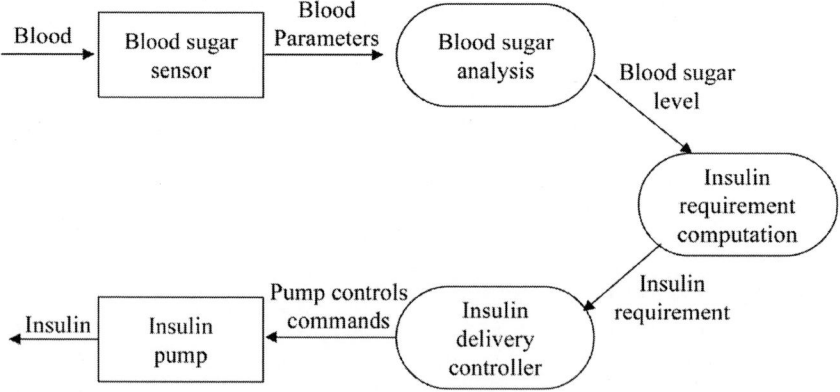

Figure 35.2 Example data flow for insulin delivery system.

A risk analysis is then required to confirm that these safety requirements can be met, that the medical device will not malfunction, and that a safe state is maintained in relation to patient health. For the insulin delivery system, a safe state is a shutdown state where no insulin is delivered. Over a short period this will not pose a threat to the diabetic's health (7).

RISK ANALYSIS (SAFETY CASE)

Risks should be identified that can result is system malfunction or failure. The consequences of failure should be analyzed, along with requirements to mitigate such malfunctions and failures. It has been suggested that risk, which might otherwise be evaluated through likelihood and consequence, should only be factored on consequence because of the social unacceptability of any known harmful impact a medical device might pose. In practice some allowance must be made for likelihood, but with a careful eye also on the probability of detection so that corrective action can be taken (8).

The process of risk analysis generally involves considering different classes of hazard such as physical hazards, electrical hazards, biological hazards, radiation hazards (where appropriate), hazards due to service failure. Each of these classes is then analyzed in detail to determine the acceptability of associated risks.

An insulin delivery system, for example, might have the following hazards and associated classes (7):

1. Insulin overdose (service failure);
2. Insulin underdose (service failure);
3. Power failure due to exhausted battery (electrical);
4. Machine interferes electrically with other medical equipment such as a heart pacemaker (electrical);
5. Poor sensor and actuator contact caused by incorrect fitting (physical);
6. Parts of machine break off in patients body (physical);
7. Infection caused by introduction of machine (biological);
8. Allergic reaction to the materials or insulin used in the machine (biological).

The hazards posed to a medical device associated with its manufacture should also be included. Complex arrangements may require multiple phases of hazard analysis.

The level of risk posed can then be determined and its acceptability considered (see Table 35.2). Risks should be attributed to be acceptable or unacceptable. Unacceptable risks require management. Acceptable risks require no further action.

It is rarely possible to completely mitigate a risk other than by somehow taking action to avoid the associated hazard in the first place. Instead risks need to be reduced so that they become "as low as reasonably practical" (ALARP). Remedial project actions should be specifically documented—this is sometimes referred to as the "safety case." Remedial actions may employ hazard avoidance strategies, introduce hazard tolerant design features, or apply specific project management controls, or a combination. Further information on risk management for medical devices can be found in ISO 14971 (9).

Table 35.2 Example Risk of Analysis of Identified Hazards

Identified hazard	Hazard probability	Hazard severity	Estimated risk	Acceptability
1. Insulin overdose	Medium	High	High	Unacceptable
2. Insulin underdose	Medium	Low	Low	Acceptable
3. Power failure	High	Low	Low	Acceptable
4. Machine incorrectly fitted	High	High	High	Unacceptable
5. Machine breaks in patient	Low	High	Medium	Unacceptable
6. Machine causes infection	Medium	Medium	Medium	Unacceptable
7. Electrical interference	Low	High	Medium	Unacceptable
8. Allergic reaction	Low	Low	Low	Acceptable

In the example, the first two hazards are software related within the medical device and will require attention as part of the design process. The remaining hazards meanwhile are not software related but can be countered by self-checking software that monitors the system state and alerts unsafe conditions. Warnings that alert detection of a hazard should be designed to allow an accident to be avoided by prompting some defined remedial action, for instance, power failure and incorrect fitting of the device. Monitoring software itself of course is safety-critical and will require validation.

DESIGN

Early on in the design phase it is important to identify and understand the impact of faults on the medical device so that controls can be incorporated as necessary. Fault tree analysis is often used to identify medical device fault scenarios (see Figure 35.3). Recommended design controls including any user procedures should be clearly logged. Arithmetic errors might for instance be mitigated though exception handling.

Design documentation will consist of hardware and software design specifications for equipment and process definitions for process lines. Where a feasibility study has been carried out on a certain technology, a fundamental science document will be generated to summarize the technology on which the medical device is to be based. Fundamental science reports will examine the use of specialist hardware and programming and the ability of these technologies to provide the necessary functionality. The use of particular technologies may be inhibited because they are deemed as unvalidatable or because the validation is too expensive. A justification of the validation approach to be used for different technologies must be documented and made available for inspection by regulatory authorities. It is highly unlikely that regulatory inspectors will be experts in the technologies being used; therefore, a step-by-step argument supported by validation evidence should be developed so that the inspectors can walk through the validation exercise to check its integrity.

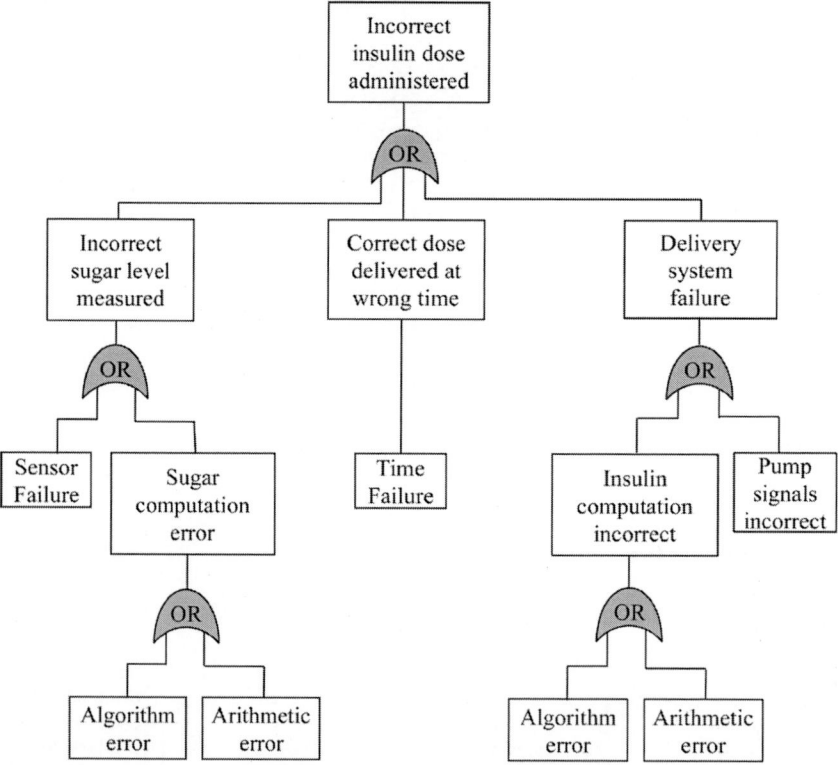

Figure 35.3 Example fault tree analysis.

DESIGN REVIEW

The FDA recommends that a traceability analysis be conducted from requirements to design, including the risk management documentation, to verify that the design is fit for purpose (5). A failure modes and effect analysis (FMEA) should be carried out to confirm that the hardware of the medical device and the supporting equipment cannot fail in an unsafe way. HACCP (hazard analysis and critical control point) has also proved useful for some medical device–manufacturing processes (10) but may have more limited use directly on software. The use of FMEA and HACCP should be documented, and any recommendations on redesign should be carried out before testing. A check needs to be made that ALARP risk mitigation results in an acceptable residual level of risk (refer back to Safety Case a few pages earlier).

SOFTWARE PROGRAMMING

Software written for the medical device and any automated manufacturing equipment should be prepared in accordance with established industry good programming practices. In particular, programs should be well structured and commented, and include headers giving details of version and change control. A source code review (sometimes known as a software structural assessment) should be conducted to verify the adoption of good programming practices and verify any critical algorithms, such fast Fourier transforms (FFTs) on digital signal processing (DSP) microchips, error handling by the software, and fail-safe or graceful degradation scenarios.

To address concerns that source codes for off the shelf (OTS) software may not be available, the FDA will allow "black box testing" as a validation method whenever source codes and design specifications cannot be obtained from suppliers (5). Alternatively a supplier audit may be used to document that the supplier has employed acceptable software programming practices.

ASSEMBLY

It is not uncommon for a company to design a medical device prototype and then have it built by a company specializing in electronic manufacture. Where this happens, as in our case, a number of suitable companies should be considered for the project and those short-listed audited for capability. Selection would depend on the quality system in place, previous experience in medical device manufacture, control of subcontractors, level of testing supplied, and the amount of in-house technical support available. It is advisable to ensure that a legal agreement, quality plan, and user requirements specification (URS) are employed to secure production standards for the device.

There might be a need with some prototype medical devices to reorganize the circuit board components to obtain a layout that will allow easy automatic assembly. Any changes must be noted in the hardware design specification, and the new board must be checked against the prototype for functional equivalence. The FMEA report should be consulted to see if there are any critical components on the circuit board. If so, the manufacturer will have to make special arrangements to ensure that component traceability exists from supplier to circuit board to user.

A typical six-step manufacturing process for a medical device would be:

1. To build circuit board. This will usually be done by automatic machines.
2. To check circuit board. The electronic manufacturer should have in-house expertise to program an automatic "in-circuit" tester for the board that ensures all components are in the correct place and that measures the correct value.
3. To load a test program into the device and place in a heat cycling oven. The heating profile and testing time should be agreed on between manufacturer and customer. It is important that the oven is temperature calibrated with a supporting certificate.
4. To load the current version of the validated software into the device.
5. To test the device on a suitable automatic test system.
6. To pack and ship to customers.

It is advisable to use a color-coded labeling system to keep track of the test stages. In our case, a yellow spot was added after successful heat cycling with test software loaded, and a green spot half covering the yellow spot once the validated software had been loaded and the board tested successfully; no device can be shipped unless both spots are present. The manufacturer will be expected to supply a certificate of conformity for the devices produced and packed with each batch.

Assembly of the medical device may be partially completed by suppliers before final assembly by the device's registered manufacturer. In such situations, the registered manufacturer is entirely responsible for the work and is expected to assign their own quality assurance staff to monitor and perhaps witness supplier assembly.

QUALIFICATION

The medical device and all supporting equipment and processes covered by validation plans will each require a test specification. These will be generated by reading through the URS, functional specification, threats and control analysis, and (where appropriate) the FMEA for each unit and determining which functions require testing. Critical areas should be examined very carefully, and a demanding set of tests drawn up to check each key area. If an automatic testing system is designed to test the medical device during manufacture, as in our case, the tests will need to check both the pass and fail routes of every test by using test medical devices with known hardware faults introduced.

Another point to remember is that where a software test harness is written to help test a medical device, this must be subject to validation (i.e., it must have functional, design, and test documentation to support it).

The tests will be broken into hardware acceptance tests (or installation qualification), system acceptance tests (or operational qualification), and equipment tests (or performance qualification). According to the size of the unit, all tests may be in one test document, or there may be three separate qualification documents. A test document will define the test philosophy and how the tests should be run. Each test will have a title, a reason for the test, an outline of any test equipment required, a description of the test, data to be recorded, and the test acceptance criteria. Calibrated test equipment must be supported with calibration certificates.

Each supplier will be responsible for generating its own test specifications, so it is important to ensure that all suppliers use the same standard for the test document. Reference to a common standard such as ISPE GAMP should be considered.

Another point to remember is that where a software test harness is written to help test a medical device, this must be subject to validation (i.e., it must have functional, design, and test documentation to support it).

Environmental Tests

Apart from qualification testing, a medical device should undergo a number of environmental tests, as outlined in standards such as BS 2011 (British standard for environmental testing), CISPR 11 (limits and methods of measurement of electromagnetic disturbance characteristics of industrial, scientific), ISM (medical radio frequency equipment), and BS4826 (British standard for packaging electronic equipment for transport).

Typical tests that might be carried out on a medical device under environmental testing are as follows:

- Temperature operating range test
- Humidity operating range test
- Pressure operating range test
- Impact test (including drop and bump)
- Vibration test
- Damp heat test
- Transport packaging test
- Radio frequency radiation test
- Radio frequency immunity test
- Reliability test

These tests would be carried out on the final manufactured product by a company specializing in this field.

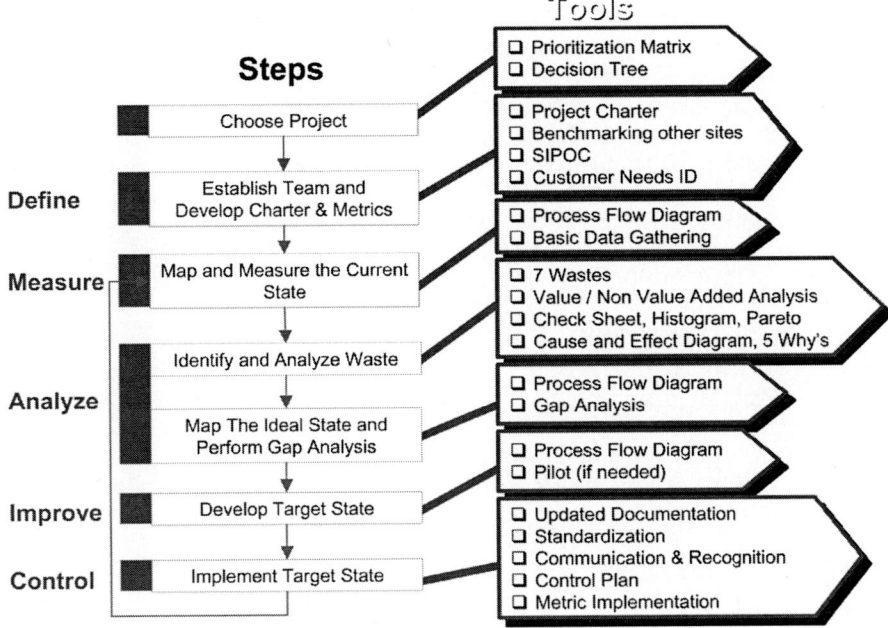

Figure 35.4   Validation documents.

REPORTING

Potentially, the number of validation documents associated with a medical device can be great because each subsystem of the device is subject to its own validation life cycle (see Figure 35.4). It is very important to collate these documents. They must be managed, otherwise it is likely with the large number of participants that some will be lost, which will severely compromise the validation. In our case, an agreement was reached, at the beginning of the project, as to what validation documents would be generated by which supplier. This information was recorded in a live document–tracking report containing the title of document, reference, who was responsible, required date, current status, and issue date. It was issued every four weeks and proved invaluable with over 100 documents under control. The document-tracking report ended with a document library index.

Once the document-tracking report shows that all validation documents called for by a corresponding validation plan are present, all of the documents are reviewed and a validation report is generated, concluding on whether the automated medical device is suitable for its intended purpose. Any outstanding issues must be discussed, and either corrective actions raised or justification for taking no action must be made. The validation report should be periodically reviewed to check that the results of any changes, corrective actions, or regulatory requirements have not altered the validated status of the medical device. If additional validation is required, then this should be planned and completed as soon as reasonably practical. Any aspects of the additional validation that directly impact its safety should be addressed immediately.

When the project comes to the end, and all validation reports have been completed, each report must be reviewed to ensure nothing is outstanding and that all parts of the project are satisfactory for use. This information is covered by the validation summary report, which should be the last validation document of the validation suite to be written.

CHANGE CONTROL

Change control must be established for the validation project and ongoing support of the medical device. Each supplier associated with the project is likely to have its own particular change control practice. These must either be linked and coordinated or a single change control procedure to be used by everybody must be enforced, to ensure effective logging and management of changes. In our case, a single coordinated change control procedure was established with all suppliers. Any supplier can make a change request, but it can only be authorized by all concerned parties. Once authorized, it may be carried out under the quality system of the supplier, with the test results being sent to all parties for approval. The medical device itself must also be subject to change controls. It should be given a serial and version number. All embedded software must also be under version control.

The FDA also will require evidence revalidation whenever a change is made to the software to determine the extent and impact of the change on the entire software system (5). Basic validation will be sufficient for low-risk devices.

MAINTENANCE AND DECOMMISSIONING

The robust operation of a medical device must be maintained throughout its operational service. Upgrades must be validated prospectively using the same basic life cycle as described earlier for the original medical device. Distribution records need to be maintained so that any medical device upgrade or product recall can be effectively conducted. Such records are also required when notifying withdrawal of support for a medical device and any decommissioning that might involve.

OPTIMIZING SOFTWARE VALIDATION

Validation can be thought of as a process whereby resources are applied in a manner to produce a validated process or product, with relatively high predictability in any step of the process. The software validation life cycle process for a medical device is a multistep operation as presented earlier in this case study. If well designed, the process will perform with predictable expenditure of resources and with predictable quality. However, it needs to be characterized in enough detail to identify the parameters that significantly affect measured output: cost and timelines and ultimately compliance. The FDA estimates poor quality design to be responsible for greater than 40% of product recalls in medical devices (11). By inference, this emphasizes the importance of designing a validation strategy appropriately.

Lean Sigma has been successfully applied to optimize the process of software validation. The Lean Sigma approach combines the methodology of Lean Manufacturing and Six Sigma, two distinct improvement processes that are complementary in nature. Lean is focused on eliminating non–value added steps and activities in a process. Six Sigma is focused on reducing variation from the remaining value-added steps. Thus, Lean is about making sure the right activities are engaged and Six Sigma is about making sure that the right activities are completed right the very first time to maximize predictability of outcome.

The well-known operational excellence approach known as DMAIC (define, measure, analyze, and control) was applied as set out in Figure 35.5. The objective was to (12):

- Identify key business issues and understand current performance levels,
- Achieve breakthrough improvement,
- Transform how day-to-day business is conducted.

Figure 35.5 illustrates a number of tools that may be applied in the DMAIC phases, it is not always necessary to employ them all. The tools include flowcharting, time measurement, resource tracing and value stream identification. Although the diagram focuses on some technical tools, it is important to note that Lean Sigma can be looked at as a marriage of social and technical systems (13); it is as much about people excellence as it is about technical excellence.

In the case study the define phase produced a validation process flow chart for the current approach (current state) to the software validation for a representative medical device to act as a baseline for improvement. The flow chart identified the functional levels, for example, engineering disciplines and the process loops, often repeat loops, taken to complete validation from the very first to the very last step. The Lean Sigma philosophy of focusing on problem definition was seen to align well with the validation requirement of characterizing the process both in terms of functional disciplines required and the intricacies of achieving the

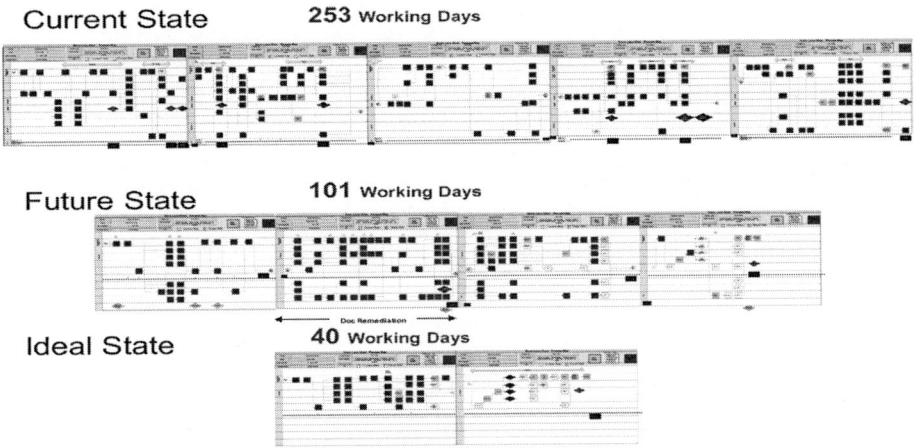

Figure 35.5   Applying DMAIC. *Abbreviation*: DMAIC, define, measure, analyze, and control.

milestone or phases in the current state. It developed a shared process understanding across the functions of the team. It confirmed the opportunity based on the amount of loops or repeat activities that seemed apparent in the actual workflow.

The measure phase provided detailed information on the performance of the life cycle process. The measure phase established the metrics that provided objective criteria for inference on the capability and stability of the life cycle process, for example, it could take up to 253 days to complete an upgrade to a process incorporating software.

The analysis phase involved analyzing the detailed information produced by the measure phase. This led to a strategy of a target and ideal state for the life cycle process. This led to a proposal that significant cycle time reduction could be obtained by just better deployment of the functional resources and by eliminating waste through reducing duplication of effort and task execution time with better project management organization; while at the same time optimizing compliance through reducing opportunity for error. On the basis of this strategy, for target and ideal state the man hours and amount of time were proposed to be reduced by 60% and 84% based on current state. This represented a huge potential saving, in terms of validation time and cost, and speed to bring the validation life cycle to completion.

The improve phase involved an experimental phase at the plant by running exploratory pilot studies deploying the disciplines set down by the target state. Its successful outcome meant that the strategy was proven worthwhile and should be implemented for future projects. The phase involved the adoption of standard work with the underlying principle that greater team collaboration at all stages leads to financial savings through the accomplishment of reduced life cycle timelines and without the usual repeat document approval review loop and wasted resource effort. Additionally, the improve phase endorsed the belief for the team that more life cycle improvement was possible and that the DMAIC model was a novel way to systematically uncover opportunities that would accomplish the ideal state and beyond without compromise to compliance. The improve phase proved that reduced phase duration was actually achievable in the order of 40% based on real projects. This was a significant improvement but somewhat less than the predictions made in the analyze phase. The Lean Sigma (DMAIC) approach clearly shows how to link and sequence individual investigative tools, this is one of the primary reasons for its success. The investigative process is probably more important than the individual tools deployed in any one phase of the project.

The control phase was identified with the broader initiative of holding the gains for future validation projects. In so far as the Lean Sigma DMAIC strategy controlled the whole project in the improve phase, it too was considered successful. The economic gains in terms of reduced validation efforts consequent on collaboration and project management structure were important to be extended to future projects, be the big or small (Fig. 35.6).

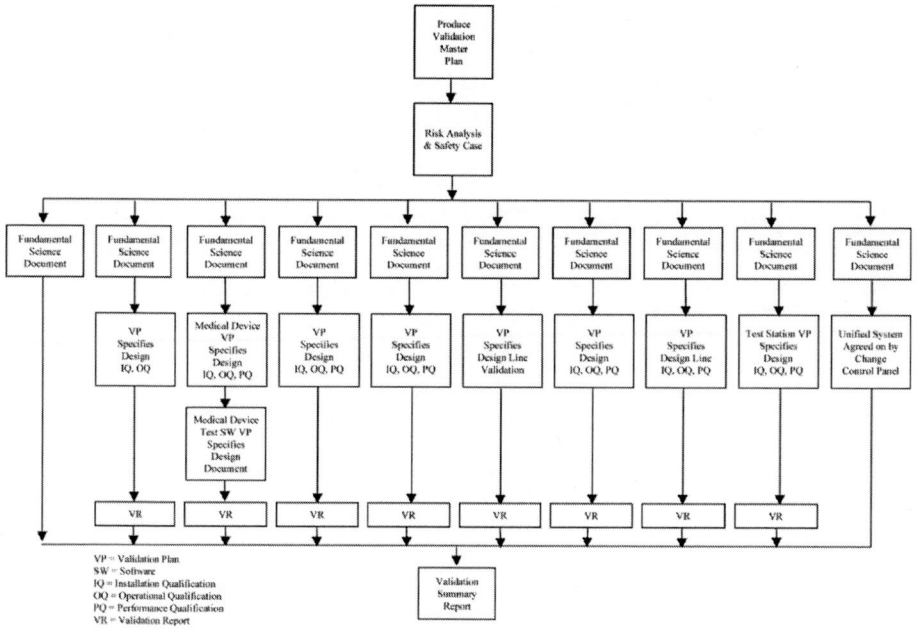

Figure 35.6 Organising validation activities

Few formal case studies have been published; however, recent experience from a large multinational pharmaceutical company suggests that typical costs of validation on information technology (IT) project can be reduced by in excess of 65% (14). The above findings based on this case study would indicate that opportunities associated with control systems in some cases may parallel IT projects.

INSPECTION FINDINGS

A selection of observations taken from a number of different FDA Warning Letters that reference software is provided below. It is not a comprehensive listing but rather it has been collated in support of the validation approach proposed by this case study.

Failure to validate computer software used to ensure the software will for its intended use. 21 CFR 820.70(i).

Failure to validate computer software for its intended use according to established protocol when computers or automated data processing systems are used as part of production or the quality system as required by 21 CFR 820.70(i).

Failure to validate computer software for its intended use according to an established protocol when computers or automated data processing systems are used as part of production or the quality system as required by 21 CFR 820.70(i). For example: your firm's [redacted] is computer controlled. It uses software programs to record data from measurements of the radius of curvature and corneal refraction of the eye. However, your firm has not validated the software and computer system used to record this data for its intended uses. Your firm has no documentation to assure that they perform as intended. Also, there is no validation and documentation of subsequent changes to the software.

Failure to validate processes that cannot be fully verified by subsequent inspection and test, as required by 21 CFR 820.75(a). For example, the complaint-handling software program, ultrasonic sealing procedure, leak-testing procedure, and injection-molding procedure have not been validated.

Your firm failed to validate several computer databases that are used for quality functions including your access database, your [redacted] software, and your MS Excel spreadsheet program as required by 21 CFR 820.70(i).

Your organization failed to document the selection and design specification of the catheter testing equipment, including the computer system, software, data acquisition hardware, and meters.

Failure to maintain procedures to ensure all purchased or otherwise received products and services conform to specified requirements [21 CFR 820.50]. For example, your firm failed to ensure that the supplier of the main computer board has documented all of the required test results to indicate the supplier's quality acceptance of the computer boards manufactured and delivered to your firm.

Your firm failed to establish and maintain procedures to control the design of the device to ensure that specified design requirements are met, as required by 21 CFR 820.30. For example, the software designed by your firm was developed without design controls.

Failure to validate computer software for its intended use according to an established protocol prior to approval and issuance, and document the results of these validation activities, as required by 21 CFR 820.70(i). For example,

A. The associated computer hardware and software used to identify incoming devices.
B. Software used to control the production and assignment of work orders and the control of master SOPS.
C. The software and hardware used to print labeling.

Failure to maintain procedures to ensure that the device design is correctly translated into production specifications. For example, the [redacted] software source code version 1.6 did not go through a formally documented design transfer process. The source code's electronic file transfer to the master chip before production release was not documented and the approved source code version 1.6 (hardcopy or electronic file) was not retained under document controls.

Software validation report not reviewed, approved, and signed.

No documented corrective and preventative action for software bugs found during retrospective validation. Validation testing revealed several responses that were unexpected and may potentially adversely affect the performance of the telemetry device. Yet these responses were not evaluated and addressed. These unexpected responses include the software acceptance of a new patient under an existing patients identifier without displaying an error message and four other unexpected responses documented in the validation document.

Failure to validate processes with a high degree of assurance where the results cannot be fully verified by subsequent inspection and testing, and have those processes approved and documented according to established procedures, as required by 21 CFR 820.75(a). Specifically, revalidation of the microprocessor software used in the XXXX has not been completely performed following an engineering change in.... Additionally, computer and/or automated data processing system software used in production and quality systems, including the use of electronic signatures has not been validated.

Failure to address and correct problems with software bugs/errors and defects identified during your retrospective software validation and retrospective risk assessment. You indicate these defects will be reviewed in....; however, you provide no justification to support your continued marketing of these products until such time as these defects and deficiencies are corrected or otherwise resolved. Please explain your reasoning in this matter and provide whatever documentation supports your position that these devices are safe to market.

CONCLUSION

Managing a large number of suppliers adds greatly to the administrative complexity of validation. This case study has identified some of the important issues. Other issues will arise on particular projects. The key to success is establishing a partnership between suppliers and the device's registered manufacturer.

It is acknowledged that development activities may be dispersed, occurring at different locations being conducted by different organizations. Regulatory authorities such as FDA hold the device manufacturer ultimately responsibility for ensuring validation is conducted and sufficient regardless of the distribution of tasks, contractual relations, source of software components, or the development environment (5).

Above all a team effort is required. Without it, validation is extremely difficult, if not impossible, to achieve.

ACKNOWLEDGMENTS

The authors would like to acknowledge the contribution of Bob Paige, formerly of ICI Eutech Engineering, in preparing the original version of this case study that has subsequently been updated and extended. The authors would also like to acknowledge the support of Boston Scientific in regard to the research work presented on optimizing software validation.

REFERENCES

1. United Kingdom Medical Devices Regulations. Her Majesty's Government Stationary Office, ISBN 0 11 042317 8, 2002.
2. European Directive 2001/83/EC concerning medical devices—class 1lb Parts III and V.
3. U.S. Code of Federal Regulations Title 21, Part 820, Good Manufacturing Practice for Medical Devices.
4. U.S. Code of Federal Regulations Title 21, Part 11, Electronic Records; Electronic Signatures.
5. FDA. General Principles of Software Validation; Final Guidance for Industry and FDA Staff. Rockville, MD: U.S. Food and Drug Administration, 2002.
6. ISPE. GAMP®5: Risk-Based Approach to Compliant GxP Computerised Systems. Tampa, Florida: International Society for Pharmaceutical Engineering. Available at: www.ispe.org, 2008.
7. Sommerville I. Software Engineering. 7th ed. New York: Addison-Wesley, 2004.
8. FDA. Guidance for the Content of Premarket Submissions for Software Contained in Medical Devices. Guidance for FDA Reviewers and Industry. Centre for Devices and Radiological Health, 1998.
9. ISO 14971. Medical Devices—Application of Risk Management to Medical Devices, 2007.
10. Jahnke M, Kuhn KD. Use of hazard analysis and critical control points (HACCP) risk assessments on a medical device for parenteral application. PDA J Pharm Sci Technol 2003; 57(1):32–42.
11. Lasky FD, Boser RB. Designing in quality through design control: a manufacturer's perspective. Clin Chem 1997; 43(5)866–872.
12. Nave D. How to Compare Six-Sigma, Lean and the Theory of Constraints: Quality Progress. Available at: www.asq.org, 2002
13. DeFeo JA. Six-Sigma: Road Map For Survival. HRfocus Newsletter, 1999 July.
14. Ryan T. Thesis Applying Lean-Sigma methodology to optimise the software validation lifecycle strategy in combination medical device manufacturing, M.Sc. thesis, Dublin Institute of Technology, 2009.

36 | Case Study 18: Blood Establishment Computer Systems

Joan Evans
ABB

As the first decade of the 21st century draws to a close, blood establishments are potentially challenged on a number of fronts; publicly funded facilities are under increasing cost pressures, and for the private sector competition is intense. While the regulators can be judged to be taking a more pragmatic approach to compliance (of which more later), public concerns about data privacy and data integrity are, if anything, increasing, as are "client" expectations as to responsiveness and level and quality of service provision by donor centers and other such facilities.

In seeking to address these and other issues, blood establishments have followed the trend observed across the pharmaceutical and healthcare industries, and use of computerized systems continues to increase rapidly. These systems now assist, manage, and in some cases, control, the analysis, creation, and management of critical records, for whole blood, blood components, and blood derivatives.

The range of facilities classed as blood establishments stretches from those covering the entire spectrum of collection, testing, processing, storage, and distribution of blood and blood components through hospitals, and nursing/care homes where blood is received for transfusion (see Fig. 36.1). The number of associated computerized systems is, therefore, enormous.

PC-/PLC-based systems may be used for laboratory analysis or manufacturing control equipment, for example,

- HPLCs, GCs, electrophoresis equipment, automated blood typing machines
- Manufacturing control equipment (e.g., PLCs on a hematocrit centrifuge)
- Environmental control or building management systems

Identification systems for blood and blood component/containers may be essentially manual operations or may extend to complex systems including cameras, barcode recognition software, and fully automated processes for barcode label generation, printing, and application.

While a small nursing home may rely on paper records or a simple spreadsheet or database application, larger facilities are more likely to use complex database applications for donor information management, blood inventory management, laboratory information management, or even a single system for what is grandly termed "center management."

Increased levels of automation and use of computerized systems are both welcome and appropriate; however, the constant challenge for the owners and operators of blood establishments is to make efficient and effective use of the technology at their disposal. Computerized systems are no universal panacea and need to be appropriately specified, designed and built, proven to meet their predefined requirements, and operated and maintained by trained and qualified personnel. It is therefore disappointing, but not necessarily unexpected, that deficiencies continue to be identified by the regulators for blood establishment computerized systems.

This chapter focuses on database systems for assisting in decision making and the management of data associated with blood products and their donors, but the principles and good practice outlined are applicable to all computer systems handling critical data with potential public health impact.

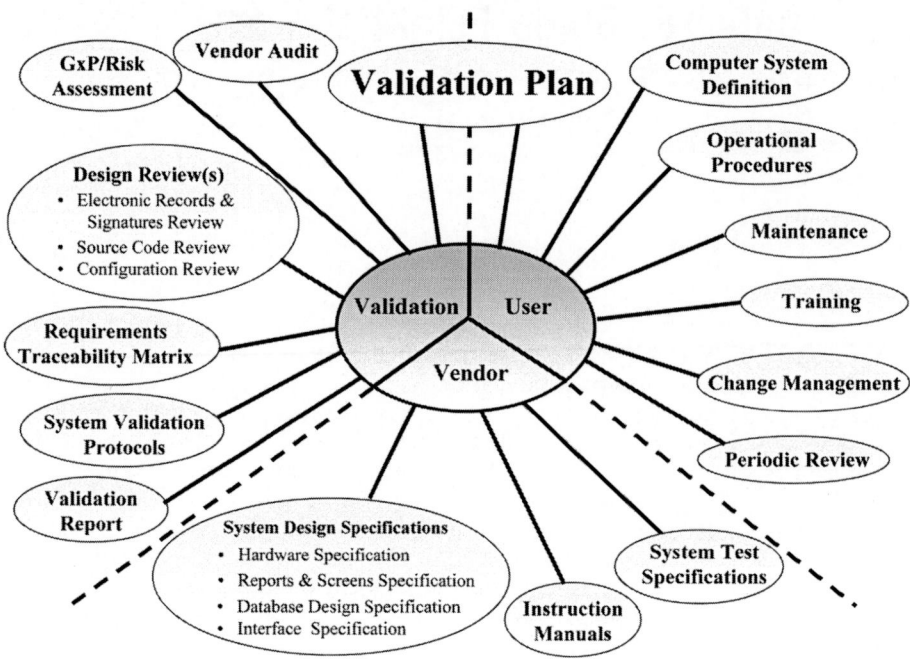

Figure 36.1   Computer system environments.

THE REGULATORY ENVIRONMENT

There has been much activity in this area in recent years. On the one hand, the regulatory burden may be considered to be reduced with increasing focus by the regulators on science-based risk management and acceptance of scaleable life cycle activities.

Since the FDA's final report on "Pharmaceutical cGMPs for the 21st Century—a risk-based approach" in September 2004, publications have included:

- ICH publications on pharmaceutical development (1), quality risk management (2), and pharmaceutical quality system (3)
- ASTM E2500 for pharmaceutical and biopharmaceutical manufacturing systems and equipment (4)
- GAMP®5—a risk-based approach to compliant GxP computerized systems (5)
- EU GMP Volume 4, Draft Annex 11 (6)
- FDA blood establishment computer system validation in the user's facility-draft guidance (7).

On the other hand, however, the issue of a number of EU directives and subsequent national implementation of corresponding regulations has resulted in the extension of requirements for

- Quality management system (QMS),
- Traceability,
- Record keeping,

to a number of facilities that hitherto had not been the focus of regulatory scrutiny.

Despite recent technology developments, examples of "missing data" and unauthorized transactions (thankfully, outside the healthcare sector) continue to be reported and leave the public, at large, skeptical as to the competence of either private companies or government agencies to manage or protect volumes of data.

For blood establishments concerns focus on donor/patient confidentiality, traceability, and, in particular, about potential for blood contamination. This has all resulted in increasing pressure on blood establishments to assure the consistency and reliability of their operations, and the security and data integrity of their critical records.

Regulatory Requirements

U.S. regulatory requirements governing blood establishments stem from three sources:

- Since blood and blood components are classed as drugs in the Federal Food, Drug, and Cosmetic (FD&C) Act, current good manufacturing practices (cGMP) as defined in 21 CFR Parts 210 (8) and 211 (9) apply.

Regardless of whether the software is developed in-house or is vendor supplied, blood establishments with computerized systems are required to perform user validation to ensure that the software meets its predefined requirements for its intended use.

- All software used to manufacture blood and blood components, to maintain data, to make decisions about donor (suitability) eligibility, or to release products for transfusion or further manufacture is classified as a device under Section 201(h) of the FD&C Act. Regulations 21 CFR Parts 800 (10) and 820 (11) therefore apply.

The device provisions such as registration as a device manufacturer, product listing, medical device reporting, compliance with the quality system regulation, and premarket notification 510(k) or application apply to the device software manufacturer (vendor or blood establishment itself). In the latter case the blood establishment must still register as a device manufacturer, list the products they manufacture, and meet the quality system regulation requirements of 21 CFR 820.

- Thirdly, as per the U.S. Public Health Service Act (12), blood, blood components, and blood derivatives are defined as biological products; therefore, 21 CFR Part 600 (13), Part 606 (14), and Part 610 (15) apply.

Other U.S. regulatory requirements are included in CFR 640 "additional standards for human blood and blood products" (16).

In addition to the predicate rules outlined in the preceding text, regulation 21 CFR Part 11 Electronic Records, electronic signatures (17) will also apply if computerized system electronic records/signatures are used for GxP purposes; that is, if electronic data held by a computerized system is used as the basis for a GxP decision then that data is considered as an "electronic record," and, as such, must comply with the stated requirements of 21 CFR Part 11. However, in the absence of a rewritten regulation, the key feature of 21 CFR Part 11 continues to be the emphasis it places on the necessity to meet predicate rule requirements for the reproducibility of critical records throughout their retention period and the preservation of integrity of their data. These are obviously issues that are fundamental to blood establishment operation and, in particular, to their database systems.

A number of blood establishments are also moving to use of nonbiometric methods, for example, application of user ID plus password for signature of blood establishment records online, in which case the requirements for *electronic signatures* must also be met.

For the European Union, general requirements are to be found in the Guide to Good Manufacturing Practice for medicinal products (6).

Specific requirements for blood establishments are contained in:

- EU Directive 2002/98/EC (18), which sets standards of quality and safety for the collection, testing, processing, storage, and distribution of human blood and blood components
- EU Directive 2004/33/EC (19), regarding technical requirements for blood and blood components

- EU Directive 2005/61/EC (20), regarding traceability requirements and notification of serious adverse reactions and events
- EU Directive 2005/62/EC (21), regarding community standards and specifications relating to a quality system for blood establishments.

The above EU regulations have been transposed into national legislation, for example, in the United Kingdom (22) via

- Blood Safety and Quality Regulations 2005 No 50
- Blood Safety and Quality (Amendment) (No 2) Regulations 2005 No 2898

effective from November 8, 2005 and

- Blood Safety and Quality (Amendment) Regulations 2006 No 2013

effective from August 31, 2006

Regulatory Inspections

For those facilities subject to U.S. regulation, even a cursory inspection of the FDA Web site contents confirms both the FDA's concerns about risks to public health, potentially resulting from blood establishment deficiencies, and the continued inability or unwillingness of certain owner/operators to apply good practice.

While a key feature of recent Warning Letters for blood establishments has been failure to maintain and/or follow written standard operating procedures (SOPs), a number of deficiencies related to computerized systems have been identified. Recurring themes include:

- *"Failure* to maintain records concurrently with the performance of each significant step in the collection, processing, compatibility testing, storage, and distribution of each unit of blood and blood components so that all steps can be clearly traced {21 CFR 606.160(a)(l), 606.160(b)(2)(i), 606.160(b)(2)(ii), and 606.160(b)(5)(iv)}."
- *"The Blood Bank lacks* quality control records to document the capacity of a shipping container to maintain proper temperatures while in transit ... Donor blood specimens ... had been transported from a mobile collection site to a blood bank laboratory in an unqualified cooler."

In addition to those highlighted in Warning Letters, a number of product recalls/medical device notifications can be traced directly to software deficiencies, for example,

- *"In the Blood Order Processing module* when, at least, two tests are added to a unit in the compatibility testing window, under specialized conditions, the system will save result(s) with incorrect records. Customers are asked to request the software correction via the client support web page."

The party with regulatory responsibility for compliance of computer systems to the above may be the blood establishment itself or the system supplier (as the manufacturer/distributor of the medical device, in this case the computer software).

While one might be tempted to apply the maxim "the more things change, the more they remain the same" regarding inspection results from U.S.-regulated facilities, the picture in the European Union has been quite different. In the United Kingdom, for example, the implementation of the above-mentioned recent EU directives in the form of the blood safety and quality regulations (22) has been a challenge for a number of establishments such as hospital blood banks seeking to demonstrate GMP compliance to a new inspection authority (in this case the MHRA). Typical comments from blood establishment owner/operators have been:

We can read the regulations, but don't understand them and don't know what to do in terms of meeting the requirements

The concepts of GxP and an appropriate QMS have been (in some cases painfully) absorbed, but the biggest challenge facing these establishments has undoubtedly been the identification and application of an effective strategy for their computerized systems. Many such facilities do not have resident computer system validation (CSV) resources and are potentially at risk of carrying on as before (the "ignore the regulators and they'll go away" approach) or of attempting to document and validate everything, regardless of risk (the "volume of evidence equals validation quality" approach).

Regulatory Reporting Requirements

All blood establishments are required to submit information describing their proposed computer system along with the establishment license application (ELA). Moreover, any "important" proposed change to a computer system or database is also reportable as a supplement to the original ELA.

THE APPROPRIATE VALIDATION FRAMEWORK

As noted earlier, a database application for a blood establishment may be a simple in-house developed tool (e.g., an MS Access application with 1 or 2 users) through a complex networked solution accessed by many users at varying locations. This variety of systems encompasses a wide spectrum of potential public health risks and different appropriate approaches to achieve the same level of residual risk.

As outlined in GAMP®5 (5) and ASTM E2500 (4) in defining the appropriate validation strategy, principles such as the following should be considered:

- Potential impact of computerized system on
  - Patient safety,
  - Product quality,
  - Data integrity
- System complexity
- System novelty/level of standardization
- Quality by design
- Critical aspects (critical quality attributes, critical process parameters)
- Vendor capability/potential for leveraging of vendor documentation and activities.

and a process of science-based quality risk management applied throughout the computerized system life cycle, as outlined below.

VALIDATION STRATEGY

As per recent FDA Draft Guidance (7) and also PIC/S good practices (23), the blood establishment computerized system is a composite consisting of the computer system hardware, software (operating system plus tools and applications), peripheral devices, personnel, and documentation.

In assessing the potential impact of a computerized system it may be helpful to categorize blood establishment processes/operations into

- Manual operations: No dependence on computerized system, for example, review of donor health and medical history questionnaire (paper record)
- IT-dependent operations: Limited dependence on computerized system, for example, review and approval of report of analysis results
- Automated operations: Use of computerized system for GxP decision/action, for example, release of blood/blood component, as suitable for transfusion or further processing.

The user requirements will confirm the level of required automation and, therefore, the scope of the IT-dependent and automated operations that are within the scope of the computerized system. The focus of a regulatory inspector will be to determine how the computerized system was validated and to examine objective evidence that it performs its intended functions reliably, consistently, and capably.

Appropriate Controls and a Structured Life Cycle Approach

A life cycle approach, for example, as outlined in GAMP®5 (5) to the specification, design, implementation, and testing of the computer system components will provide a sound basis for the validation of blood establishment database systems. The life cycle should extend from "cradle to grave" , that is, encompassing all phases

- Concept
- Project
- Operation
- Retirement

While there is increasing acceptance of the need to maintain the validated state throughout the operational phase of a computerized system, and to address system retirement and the management of associated GxP data, there is still a reluctance to apply risk management principles and involve potential computerized system vendors sufficiently early in the concept phase.

These days most data management systems are based on standard database packages such as Oracle, Sequel Server, etc. Although some data entry screen and/or reports may be customized, much of the user interface will be via configured routines and standard database queries. Completely customized in-house developed systems are, therefore, becoming more rare. A larger blood establishment may well, however, elect to develop its own system, in which case the blood establishment additionally assumes the responsibilities associated with these life cycle steps and should follow one of the industry standards for software development, for example, ISO90001:2000 TickIT and ISO 27001 for information security.

See Figure 36.2 for outline of controls appropriate for a medium/large blood establishment database system.

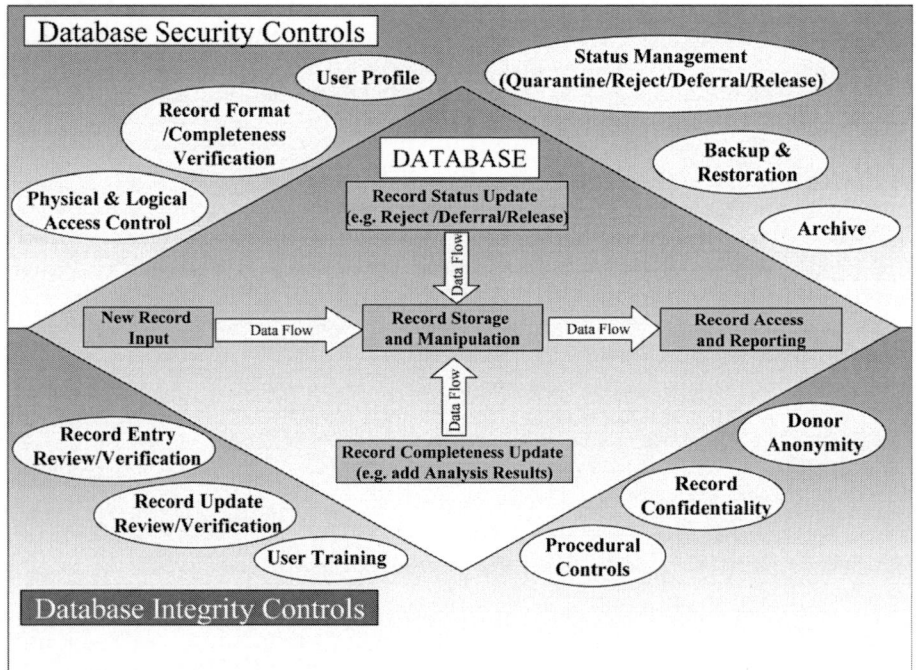

Figure 36.2   Blood product database record security and integrity.

Risk Management

It hardly needs repeating here that risk management is *the* "hot topic" of the regulators—at least thus far this century. Any blood establishment facing up to an inspection should be in no doubt that effective risk management based on a "science-based ... documented and justified risk assessment" should be the basis for CSV strategy.

In addition to a regulatory expectation that risk management has been applied to the determination, scoping, planning, and execution of CSV activities, the FDA is also on record as having observed failures such as "You have not performed a risk analysis for risks and hazards associated with...". Risk management should, therefore, be applied to all blood establishment computerized systems (BECS) and at all phases of their life cycle.

As outlined in ASTM E2500 (4), and also in Appendix M3 of the new GAMP®5 Guide (5) quality risk management for computerized systems should comprise a systematic process applied at appropriate stages. Properly performed, risk assessments provide a vehicle for the formulation and application of a scaleable approach to the documentation and "validation" of the computerized system.

In addition to considering the risks associated with patient safety, product quality and data integrity, the risks associated with the computerized system itself need to be considered. In so doing the risk reduction benefits of a standard solution, from a supplier with a proven track record in this field, of software that has been "proven in use" at other establishments, become apparent. A novel solution, developed by a team without previous experience in this field, may on paper appear advantageous in terms of perceived fit with user requirements, but must be assessed, categorized, and "managed" as a higher risk option.

As outlined in GAMP®5 (5), an initial "system level" risk assessment may be sufficient for a small database that is essentially being "configured" rather than "developed" to meet unique requirements. The results of such an assessment will be applied to the database as a whole and should determine-

- Level of initial vendor assessment and appropriate supervision throughout project
- Strategy for leveraging of vendor documentation/ level of project specific documentation
- Rigor of testing of system functions and reliance on vendor testing, etc.

At the other end of the scale, for a large networked solution, a phased approach will be appropriate as the initial risk assessment will be followed by more detailed studies of different computerized system components.

In embarking on these detailed assessments, serious consideration should be given as to how to split the computerized system in terms of processes/functions, and whether generic or specific hazards are to be assessed. It should also be realized that it is relatively quick and easy to associate risks with various system functions. A far greater challenge is to define and agree appropriate controls and to get those measures implemented in a timely manner. Effective risk management centers on follow-through to confirm that control measures have been implemented and are proven to be effective, this is where a number of projects struggle. The risk(s) associated with the system will have been appropriately "managed" when, and only when, these are mitigated, via successful implementation of defined control measures, to achieve an agreed lower residual risk.

Vendor Selection

One of the first decisions to be made by the blood establishment is as to whether the database system is to be

- Supplier development of an existing product,
- In-house development of an existing product,
- Supplier development of a new system, and
- In-house development of a new system.

Each of the above options presents a different risk profile, and this should be a key factor in making decisions on choice of software product and selection of vendor. In terms of software product, key considerations are degree of standardization of the proposed solution and the level of assurance provided by successful operation of similar solutions at other establishments. If a vendor can reference a number of clients running similar computerized systems, over a number of years, in terms of scope, complexity, size, and logistics, then not only will the risks associated with the software product be assessed as lower, but the ability of the vendor himself to meet expectations will be considered as having been, to some extent, proven in use.

If a third-party vendor is selected, the objective should be to make maximum use of vendor activities and vendor documentation. The first step in this strategy would be to assess the risks associated with the vendor to decide on the level of assessment/audit required to evaluate

- Quality management system (QMS)
- Technical capability
- Caliber of the project team being proposed
- Experience of development of similar systems.

In these days of increasingly global operations, it also behoves a cautious end user to confirm, which if any, system development activities are to be subcontracted to third-party suppliers or to internal "offshore" resources. The final assessment of the vendor, and associated risk, must reflect the totality of the supply chain.

It is unfortunately a fact that remote resources, while appearing to provide a lower cost solution, may well require more time and higher levels of supervision/deliverable review, and these factors need to be planned into the project organization, schedule, and budgeted costs.

Assuming that the vendor assessment is successful and the vendor is selected, then consideration should be given to using verification evidence from the vendor as proof that the computerized system is fit for use. For example, vendor test documentation could be integrated into the overall verification package removing the need for repeat testing on site/by the end user. As outlined in the preceding text, the decision as to whether to adopt such a strategy will not only be dependent on the competence and reliability of the vendor but also on the perceived GxP risks (patient safety, product quality, data integrity).

Personnel

As noted earlier, the capability and experience of vendor staff is a key factor in assessing and choosing a potential third-party vendor. The fact that the chosen vendor has successfully delivered a number of database systems to other blood establishments may be of little relevance if it becomes clear, early in the project, that staff assigned to a project have only recently been recruited and are not only grappling with the technical challenges of the project but also struggling to understand and apply the agreed QMS. A comprehensive vendor assessment should include the requirement for the vendor to nominate key project staff and provide details of experience and technical capability of these individuals.

Similar principles should also be applied to the selection of in-house staff for a computerized system project.

In outlining roles and responsibilities of individuals' key to achieving computerized system compliance, both GAMP®5 (5) and ASTM E2500 (4) feature the concept of "subject matter experts" (SMEs)—those possessing specialized expertise and/or responsibility, for example, business process, IT, quality, and validation.

ASTM E2500 (4) assigns these SMEs a lead role in performance of risk assessments, the definition of acceptance criteria, and verification activities. It also focuses review and approval of system documentation by the quality unit to critical system aspects.

This focus on patient safety, product quality, and data integrity is welcome, as is the recognition that technical aspects of system specification and performance are most efficiently and effectively defined and evaluated by the expert(s) in that field, as distinct from "committee decisions" without clarity as to individual responsibilities for review and approval.

While the new Draft Annex 11 (6) does lack some of the detail to be found in GAMP®5 (5), it does remind us of the most important consideration for any computerized system—the "closest cooperation between key personnel." It is no exaggeration to state that a successful computerized system project is one during which the appropriate key individuals are made available to the project and communicate effectively with one another. This will not only facilitate implementation of a compliant system but also is most likely to result in an on time and to cost solution that meets user and business requirements.

Planning

In its guidance on validation planning, GAMP®5 (5) now states that separate validation plans (VPs) may not be necessary to demonstrate the fitness for intended use, where the system can be regarded as one component of a wider manufacturing process or system, for example, in a quality by design environment or where process analytical technology (PAT) is adopted. However, this is unlikely to apply to in the context of a blood establishment database system, for which a "traditional" validation plan would still be the expected norm—as outlined in the U.S. Guidance for Industry (7). The referenced principles of software validation (24) are still valid as guidance on validation scope and activities, while GAMP®5 Appendix M1 (5) continues to outline a typical VP table of contents.

GAMP®5 (5) does, however, remind one that the level of detail provided in the VP should be based on the risk, size, complexity, and degree of standardization of the system. The focus on patient safety/ product quality/ data integrity risk is also reflected in the expanded section on validation strategy and the new section on quality risk management.

In summary, therefore, a blood establishment computerized system VP is still expected to cover the same general ground, but the validation activities described therein are expected to be scaled, based on quality risk management principles as applied to the computerized system in question.

Validation and Verification

While GAMP®5 (5), in its stated aim to be flexible, outlines the relationship of qualification terminology to "GAMP®5 verification activities," and is more sparing in its use of the term validation, ASTM E2500 (4) narrowly categorizes system validation under the "umbrella" of the verification phase. However, the U.S. Draft Guidance for Industry (7) in referring to variations in terminology between publications and organizations suggests a wider meaning to "software validation" than "software verification."

The EU draft Annex 11, its "acting sister" Annex 15 (6) are also still wedded to the more traditional use of validation to encompass the entire system life cycle.

Regardless of the terminology used, this area continues to be a focus of regulatory inspection, as demonstrated, for example, in an FDA-483 form for the American Red Cross, which references failures associated with validation of handheld computers/workstations.

Draft Annex 11 (6) (clause 3.7) states a number of requirements for database systems validation that would apply to most blood establishment systems. However, it should be viewed as a "starter for 10" rather than an exhaustive list. For example, the list is structured for a new system and does not encompass considerations, such as data cleansing, data migration, cutover, and requirements for parallel system running /comparative testing, etc.

Testing of the reliability and capability of search/filter tools and system reports could also have been highlighted.

The U.S. Draft Guidance for Industry (7) recommends that the output of reports be verified or validated and that data conversion from a legacy system to a new BECS be validated to avoid potential duplicate records.

The text in clause 3.7 of Draft Annex 11 (6) should, therefore, be used as a supplement to the rest of the guidance rather than as a comprehensive guide.

While the EU Draft Annex 11 (6) includes specific requirement for challenge testing, and the U.S. Draft Guidance for Industry (7) recommends boundary testing and testing at peak production times and with maximum number of users, the approach outlined in GAMP®5 Appendix D5 (5) where the need for this level of testing is evaluated and based on the system complexity, system novelty, and perceived risk is probably a more practical interpretation. The examples in Section 4.2.6 of GAMP®5 (5) provide a simple illustration of how the categorization of the software product/application can be applied in scaling validation activities, including the need for testing. Further GAMP guidance on testing can also be found in the Good Practice Guide for Testing of GxP Systems (25).

Support for the use of risk assessment as the basis for determining system test plan(s) and test cases is also to be found in the U.S. Draft Guidance for Industry (7).

Security

As noted in the new GAMP®5, security policies and procedures are typically generic in nature and for a global company, will apply across a wide range of systems, of different categorizations:-

- Business critical
- Finance critical
- GxP
- Noncritical

Recent guidance, therefore, (e.g., EU Draft Annex 11/ PIC/S PI 011) assumes that an information security management system (ISMS) is in place and provides a framework in which specific controls are applied to the computerized system, depending on the criticality of the data and the risk assessment of the system.

Despite the fact that ISMS may have been designed and implemented by the IT community to address business continuity or financial compliance concerns, it should provide a suitable framework for GxP systems. In fact, the fundamentals and drivers behind, for example, Sarbanes-Oxley compliance, are the same principles of data security and assurance of data integrity.

GAMP®5 Appendices O11 (security management) and O12 (system administration) (5) outline typical security management measures such as the information security policy, and a formal process for incident management.

While the EU Daft Annex 11 (6) lacks the detail of US 21 CFR Part 11 (17), it does highlight the need for built-in checks to ensure that data is correctly and securely entered, and the requirement for recording of the identity of the individual performing critical data entry.

Both the EU Draft Annex 11 (6) and the U.S. Draft Guidance (7) refer to testing of database security measures. The extent and rigor of this testing should be determined by risk assessment; for example, the controls of "superuser" functions/privileges would be expected to be more extensively tested due to the greater potential impact on the system/data.

User Access Management

The synergy between the ISMS processes and the security measures required for a GxP system is apparent. Key controls for a GxP system such as

- Formal identification of key system roles (system/business process owner and system administrator);
- All user access requests documented, reviewed, and approved by system or business process owner;
- Agreement of system access based on agreed system roles and permissions;
- Implementation of user access request by independent system administration function;
- Completion of successful user training before access to "live" system; and
- Periodic review of all system access

should be included in the ISMS user access management process and can be readily applied to a blood establishment computerized system to ensure that system security measures are formally documented and that the risks of unauthorized access or modification to data are minimized.

Access to database tables/transactions should be based strictly on the requirements for the defined role assigned to the individual user. Read/write access should only be granted if it could be demonstrated as a requirement for the execution of the role.

Note that system access management extends to remote access by third parties, for example, for system troubleshooting. The U.S. Draft Guidance for Industry (7) recommends maintaining a log of remote system access, including name, date, time, and reason for system access.

Change Management

Even the most cursory glance at inspection reports will confirm that change management continues to be a hot topic and a focus of regulatory inspection. FDA inspections, for example, have identified, not only deficiencies in modification to/releases of new software versions, but also multiple separate observations on "failures to maintain or review records of investigations of software problems" and "failures to investigate computer software deficiencies."

These observations have been prompted by the failure of more than one national blood establishment facility to manage its BECS via an effective operational process for change control. The opening sentences of GAMP®5 (5) Appendix O6 (operational change and configuration management) (5) contain a timely reminder that change should not be executed for change's sake, but limited to system improvements, which must be managed so as to minimize risk to regulatory processes and records and potential inspection risk to the facility concerned.

For third-party supplied software, the contract between the vendor and the blood establishment should clearly state how system changes are to be managed. Before release of the system for GxP use, clarity is required on the following:

- What database changes are to be made by blood establishment staff (e.g., addition of new users) and which will require vendor support?
- How much is the vendor software subject to change and what new releases/upgrades are planned?
- How likely is it that the blood establishment will wish/need to accept this new software?
- What documentation is available from the vendor on the extent and impact of the changes made, and what support is available to agree required regression testing?
- What process is to be followed at the BECS to document, justify, approve, and implement changes (by the vendor or internally originated)?

The U.S. Draft Guidance highlights regulatory concerns that, due to the complexity of many BECS, even relatively small changes may have a significant overall system effect. To meet current expectations it would, therefore, be prudent to ensure that appropriate evidence is prepared and available for all system changes with respect to their potential impact on or risk to patient safety, product quality, and data integrity. In fact, the EU Draft Annex 11 (6) goes further, calling for periodic reviews to review the cumulative effect of incremental system changes, performance issues and/or regulatory developments, and whether any further work is appropriate to confirm validation or address data integrity issues.

As mentioned earlier, a significant change impacting database functionality (e.g., the way data is manipulated or interpreted) or its equivalence with a previously approved database system, may require a change to the submission for blood establishment computer software for CBER or other regulatory authorities.

Data Storage, Back Up, Archiving, Business Continuity

The requirement of EU directive 2005/61/EC (20) to "retain the data...for at least 30 years in an appropriate and readable storage medium to ensure traceability" has highlighted the need for effective data storage and data archiving. Those attending the MHRA blood seminars in September and October 2005 were left in no doubt that this topic will be a focus of regulatory scrutiny in the United Kingdom.

The EU Draft Annex 11 (6) outlines the fundamental requirement for data storage to secure by physical and/or electronic means, the availability and integrity of data. This principle carries through into both backup and archival processes, although both the EU latest proposed amendment to Annex 11 (6) and GAMP®5 (5) distinguish between the two.

For both backup and archival, key requirements are as follows:

- Definition of roles and responsibilities
- Written procedures for the process to be followed
- Periodic verification of procedures and processes, for example, regular testing of backups to confirm compatibility with live system, checks on archived data to confirm accessibility, durability, accuracy, and completeness.

Documented procedures and verification should extend to third-party organizations where these are used for off-line data storage.

In line with science-based risk management principles, the frequency, complexity, and level of verification of the above processes should be based on a documented risk assessment.

In these days of networked solutions, it is anticipated that most databases will be "swept-up" so as to speak in server backups by the IT group. Written procedures for activities by third-party data storage/data management companies are also becoming more common (even if only prompted by commercial considerations and data privacy legislation). It is unfortunately still the case, however, that partial or full database restores may not have been adequately tested and proven.

Business continuity plans, also, may remain "plans" while both the EU Draft Annex 11 (6) and GAMP®5 (5) require evidence of rehearsal of the planned scenario(s) and training of personnel involved.

In addition to GAMP®5 Appendices O9 (backup and restore), O10(business continuity management), and O13 (archival and retrieval) please refer, for electronic data archiving, to the GAMP Good Practice Guide for Electronic Data Archiving (26).

Procedures for System Use/Operation and Training

While neither the EU Draft Annex 11 (6) nor the U.S. Draft Guidance for Industry (7) devotes many words to these topics, the requirements are specific and comprehensive; all users of a blood establishment computer system must be trained, and all their activities must be governed by written and approved procedures. It is interesting to note the requirement [US Draft Guidance, Section III, F.8 (7)] that training should encompass an assessment of the user's ability to understand and use the system correctly, including the ability to respond appropriately to unexpected system events such as alarms.

The importance attached by the regulators to effective training is illustrated by the fact that the topic appears so frequently in FDA-483 forms, Warning Letters, and consent decree notifications. In fact, the American National Red Cross Consent Decree Notification of April 2003 contains a whole section of training deficiencies, and also observes, under item 1, that "training records were falsified." SOPs are also a focus, as illustrated by FDA Warning Letter observations that "the blood bank did not follow its SOP." and "the current and previous versions of SOP ... are inadequate."

The operation appendices from GAMP®5 (5) provide a useful starting point for system SOPs. It should be noted that risk management principles should extend through SOP creation. A complex facility management system with multiple users in many locations, accessed and maintained by a mix of staff, in-house contract and third-party personnel, will require a far more comprehensive SOP set than that appropriate for a small database with a couple of users.

CONCLUSION

While some owner/operators (e.g., those from U.K. National Health Service Trusts) may perceive an increased regulatory burden in the last few years; in general, blood establishments should take heart from recent significant developments in regulatory thinking.

The focus by the regulators on patient safety, product quality, and data integrity is highly supportive of a trend across the pharmaceutical sector away from paper-based systems and toward e-operations. We can begin to see the "virtuous circle" effect whereby increased application of blood establishment database systems have resulted in a greater number of more complex systems and a more experienced user and supplier community, alongside a growing body of evidence as to the robustness of these installations and the security and integrity of the donor/patient/blood product records that they contain.

While both the EU Draft Annex 11 (6) and the U.S. Draft Guidance for Industry (7) are recent publications, a close study confirms that they contain no new requirements. In fact, the U.S. Guidance (7), while stating that it, once finalized, will reflect the current thinking of FDA, constantly refers back to original predicate rule requirements contained in CFR 211 and CFR 606.

Recent industry guidance [e.g., ASTM E2500, Section 6.6 (4)] outlines a scaleable approach to system development and implementation on the basis of good engineering/good software development practice, supported by risk management focused on patient safety, product quality, and data integrity risks. The development of a risk-based approach to computer system validation does take time and require a fundamental understanding of the processes/controls being automated and the associated risks. However, this early investment not only ensures that validation effort and time are focused on mitigation of risk to public health (rather than generation of volume of evidence) but has been proven to be more cost-effective through both project and operation phases of the system life cycle.

REFERENCES

1. Pharmaceutical Development-Q8, International Conference on Harmonisation of Technical Requirements for Registration of Pharmaceuticals for Human Use, Available at: www.ich.org.
2. Quality Risk Management-Q9, International Conference on Harmonisation of Technical Requirements for Registration of Pharmaceuticals for Human Use, Available at: www.ich.org.
3. Pharmaceutical Quality System-Q10, International Conference on Harmonisation of Technical Requirements for Registration of Pharmaceuticals for Human Use, Available at: www.ich.org.
4. ASTM Standard E2500, 2007 "Standard Guide for Specification, Design and Verification of Pharmaceutical and Biopharmaceutical Manufacturing Systems and Equipment", Available at: www.astm.org.
5. ISPE GAMP®5. A Risk Based Approach to Compliant GxP Computerized Systems. Tampa, FL: ISPE, 2008.
6. Rules governing medicinal products in the European Community, Volume IV, Pharmaceutical legislation—Medicinal Products for Human and Veterinary Use—Good Manufacturing Practices, including Annex 4 and draft Annex 11, 1998 (Draft Annex 11 April 2008).
7. FDA/CBER Guidance for Industry "Blood Establishment Computer System Validation in the User's Facility" (October 2007).
8. Code of Federal Regulations Food and Drugs 21 Part 210—Current Good Manufacturing Practice in manufacturing, processing, packing, or holding of drugs; general.
9. Code of Federal Regulations Food and Drugs 21 Part 211—Current Good Manufacturing Practice for finished pharmaceuticals.
10. Code of Federal Regulations Food and Drugs 21 Part 800—Medical devices: General.
11. Code of Federal Regulations Food and Drugs 21 Part 820—Good Manufacturing Practice for medical devices: General.
12. US Public Health Service (PHS) Act.
13. Code of Federal Regulations Food and Drugs 21 Part 600 – Biological Products: General.
14. Code of Federal Regulations Food and Drugs 21 Part 606—Current Good Manufacturing Practice for blood and blood components.
15. Code of Federal Regulations Food and Drugs 21 Part 610—General biological products standards.
16. Code of Federal Regulations Food and Drugs 21 Part 640—Additional standards for human blood and blood products.
17. Code of Federal Regulations Food and Drugs 21 Part 11 – Electronic Records; Electronic Signatures.
18. EU Directive 2002/98/EC of 27 January 2003 setting standards of quality and safety for the collection, testing, processing, storage and distribution of human blood and blood components and amending Directive 2001/83/EC.
19. EU Directive 2004/33/EC of 22 March 2004 regarding technical requirements for blood and blood components.
20. EU Directive 2005/61/EC of 30 September 2005 regarding traceability requirements and notification of Serious Adverse Reactions and Events.
21. EU Directive 2005/62/EC of 30 September 2005 regarding Community standards and specifications relating to a quality system for Blood Establishments.
22. Blood Safety and Quality Regulations 2005 No 50 & Blood Safety and Quality (Amendment) (No 2) Regulations 2005 No 2898 (effective from November 8th 2005) Blood Safety and Quality (Amendment) Regulations 2006 No 2013 (effective from August 31st 2006).
23. PIC/S Guidance on Good Practices for Computerized Systems in Regulated "GxP" Environments (PI 011-3) September 2007.Available at: http://www.picscheme.org.
24. General principles of software validation; final guidance for industry and FDA staff, January 2002, FDA Center for Devices and Radiological Health, Center for Biologics Evaluation and Research.
25. ISPE. GAMP Good Practice Guide: Testing of GxP Systems. Tampa, FL: ISPE, 2006.
26. ISPE. GAMP Good Practice Guide: Electronic Data Archiving. Tampa, FL: ISPE, 2007.

37 | Case Study 19: Process Analytical Technology

Guy Wingate
GlaxoSmithKline

INTRODUCTION

Process analytical technologies[a] (PAT) provide "systems for analysis and control of manufacturing processes based on timely measurements during processing of critical quality parameters and performance attributes of raw and in-process materials and processes to assure acceptable end product quality at the completion of the process" (1). As such PAT offers higher manufacturing efficiency, but requires a change in the established quality paradigm from analytical quality control on finished goods to a practice of a true process quality assurance philosophy. This case study will look at the implications for supporting computer systems.

FOUNDING PRINCIPLES

Although not formally defined as such, PAT is founded on three basic principles concerning the following:

- Process analysis: registered critical control parameters (CCPs)
- Process control: control of product variation
- Process understanding: process validation and risk management

Process analysis should be based on moving away from testing to document quality toward "quality by design (QbD)." Product and process quality characteristics need to be scientifically designed to meet specific objectives, not merely empirically derived from performance of test batches. Current process characterization does not facilitate a smooth transition through drug development, registration, manufacturing scale-up, and subsequent process improvement. It is vital that critical process parameters (CPPs) registered with regulatory authorities support demonstrable process control but not impede ongoing improvements to process capability.

Process control should be based on continuous quality assurance—all key stages should be quality assured and monitored for acceptability. Figure 37.1 presents this control theory which shared the basic concepts of conventional process control using alert and alarm limits. Contemporaneous quality decisions are facilitated in anticipation of potential product failures if no corrective action was taken. The whole manufacturing process becomes more tightly coupled with product quality.

Process understanding is based on a significantly better understanding of the changing product characteristics during the manufacturing process. Understanding starts with a fundamental specification of critical quality attributes (CQAs) and definition of supporting CPPs. The process capability might even be registered including valid process variation. No extraneous measurements and data are registered, analyzed, or validated. Potential risks associated with change are thereby mitigated.

PAT removes redundant waiting time for finished goods testing, replaces it with a continuum of in-process sampling, and facilitates enhanced closed loop control of processes. Ultimately, if all process parameters are controlled through PAT, then manufacturing processes will become paperless workflows culminating in totally electronic batch records. Batch release could then be conducted in real-time through continuous quality verification with a final review of processing exceptions rather than as currently with a manual review and approval of the entire processing history of a batch covering in-specification information and deviations.

---

[a]Sometimes referred to as process assurance technology

(A)

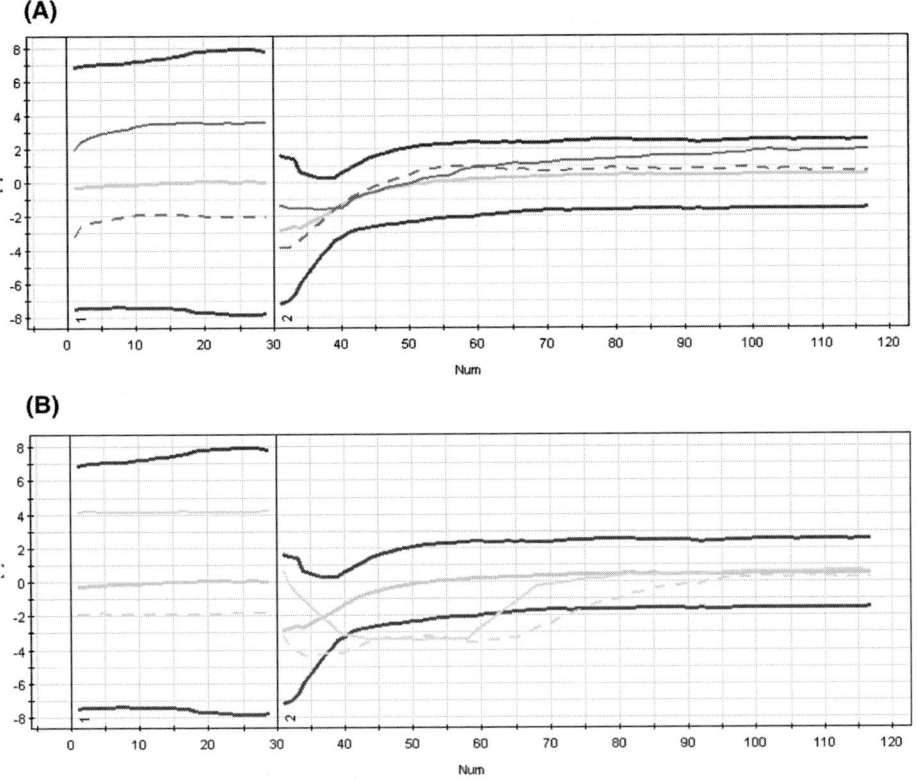

(B)

Figure 37.1 (**A**) Control model operating in design space. (**B**) Control model operating outside design space.

This new operating paradigm is not a silver bullet solving all product quality issues. It is a significant undertaking to try and establish a complete product and process understanding. Although products are manufactured to product specifications agreed with regulatory authorities deviations routinely occur leading to batch rejection, rework or in exceptional circumstances asking for regulatory relief to continue supply (e.g., danger to break in supply of medically critical product to which there is no alternative therapy). Conventional manufacturing specifications can be amended to reflect learning and improve product quality—but getting approval can be a slow and cumbersome process. In the new paradigm learning will also occur as for instance new critical control points are discovered (Fig. 37.2). The difference is that the amount of learning in the new operating paradigm should be dramatically less, reflecting the pay back from initially establishing a more robust product and process understanding. In theory the new paradigm offers much higher reliability in producing products that meet their specification despite the natural variation that occurs within materials and manufacturing processes.

PROCESS ANALYTICAL TECHNOLOGIES TOPOLOGY

The topology of PAT systems varies depending on the precise needs of the process it supports. An example topology is presented in Figure 37.3 for the purpose of discussion. It consists of the following:

- Data acquisition (process measurement)
- Chemometrics (multivariate data manipulation)
- Online prediction (product models)
- User reporting (contemporaneous reporting)
- Data historian (archiving)
- Interfacing to other systems

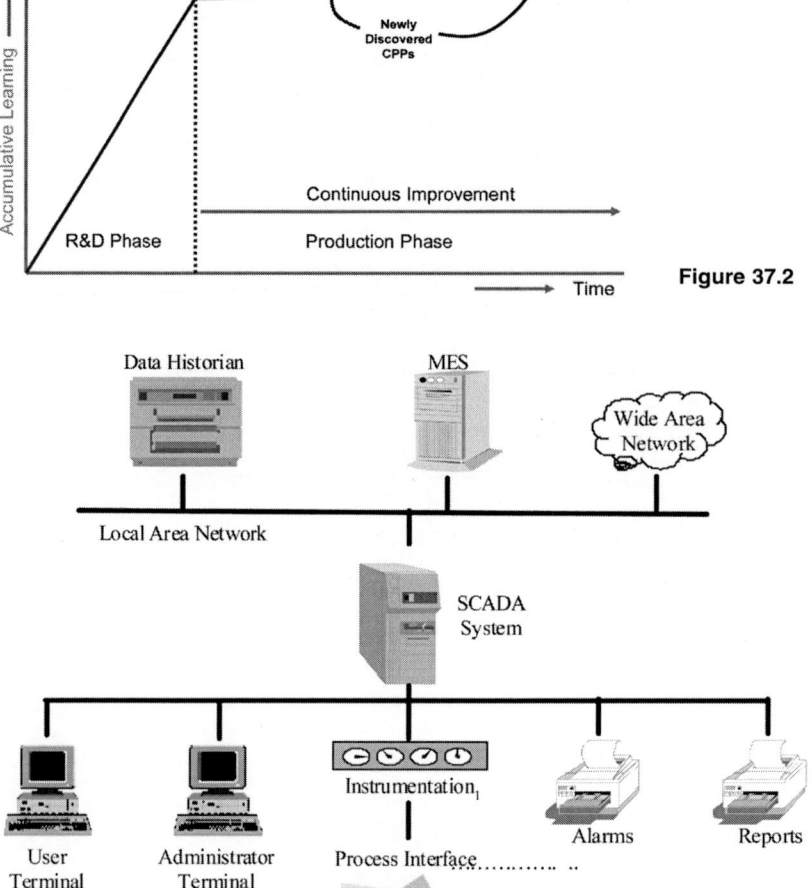

Figure 37.2   Learning life cycle.

Figure 37.3   Example PAT Topology.

Data acquisition is provided by instrumentation. Chemometrics, online prediction, and reporting is typically supported by some kind of supervisory control and data acquisition (SCADA) system and/or manufacturing execution system (MES). This may be a commercial off-the-shelf (COTS) product or a custom development. Data Historian is likely to be a separate interfaced application. To support this and exchange of information with other systems an "open systems" approach should be adopted.

Data Acquisition

The development of sensor technology to allow a range of nonintrusive measurements is a significant enabler of PAT. Examples of nonintrusive instruments (2) are as follows:

- Near infrared (NIR) spectroscopy
- FT-IR
- Raman spectroscopy
- UV/visible spectroscopy
- Acoustic emission spectroscopy

- Particle size characterization
- X-ray tomography
- NMR
- Mass spectrometry

The selection of instrumentation will be dependent of the chosen product characteristics being measured for the product form. Product characteristics can be categorized into physical structure, chemical identity, and homogeneity. Only CPPs affecting end product quality should be measured.

Chemometrics
Chemometrics provides a means of contemporaneously analyzing sample data to optimize processes. Data sources may be spectral, wet chemistry or a combination.

Chemometrics uses multivariate, multidimensional data to generate product specific models. These models are the basis against which future data can be compared to allow both qualitative and quantitative predications to be made.

Product specific models require management from initial creation, through, approval, use, refinement and eventual withdrawal and archiving. Data from different sensors must be correctly collated and built into the models for the products they support. As the body of knowledge concerning a process increases, the product-specific model can be refined and made ever more robust.

Online Prediction
Potential product rejects should be anticipated so that intervention can be prompted and corrective action taken in a timely manner to avoid final product rejection. Rejects may be due to product being out of specification (OOS), or the model being too sensitive. Data associated with the model being too sensitive should be analyzed and used to refine the model. Once the model has sufficient status it can be used as part of the registration of a product with a regulatory authority.

User Reporting
PAT operators do not solely react to alerts and warnings, but rather contemporaneously interact to what process is doing. Real-time reporting on product analysis is therefore required to enable immediate quality critical decisions to take place. It should be possible to configure standard reports for both on-screen display and printing. Trends will allow proactive management and optimization of in-process manufacturing.

Data Historian
Potentially huge volumes of data might need to be archived to satisfy regulatory electronic record requirements. Manufacturing processes might typically have between 10 and 1000 measurement points with individual measurements taken perhaps as regularly as every second. Not all this data will necessarily be handled by one PAT system, there could be many PAT systems deployed. A data strategy should be developed to map and define data flows and retention requirements.

DATA MANAGEMENT
The data processed by a PAT system can be thought of in terms of three categories of process parameters.

- General process parameters: measurements at particular points perhaps taken as much as every second
- Key process parameters: critical to process efficiency/effectiveness but not necessarily linked to CQAs
- Quality CPPs: a process parameter that influences a CQA and (as determined by risk assessment) presents a high risk to the process falling outside the design space

The flow of data is shown in Figure 37.4.

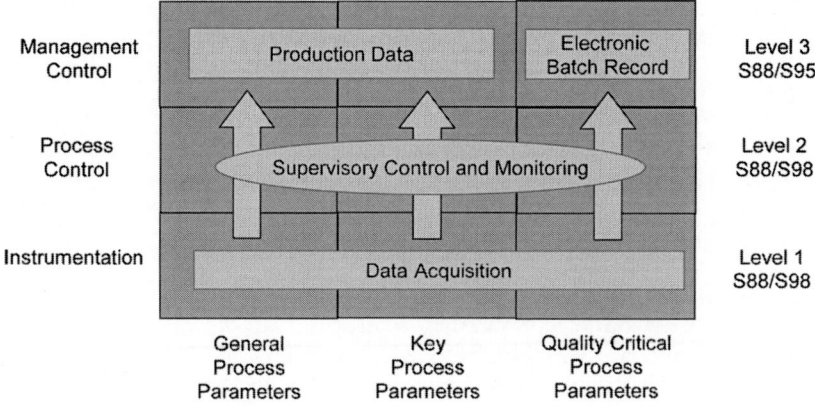

Figure 37.4 PAT data flow. From Ref. 3.

It is important to distinguish the use of information either as contributing to production data or the formal electronic batch record required by regulations to release product. By understanding what constitutes CPPs the amount of data requiring long-term storage the volume of data archiving can be significantly reduced without impact to retrospective product analysis.

The basic requirements for management and control of production data and batch record information are the same. Data integrity must be preserved (see chap. 11). Controls should include the following:

- System access controls (e.g., password, area token) to prevent unauthorized alteration
- Audit trail for authorized changes
- Data backups and archive copies
- Virus protection as appropriate

There are differences however when it comes to data retention. Production data is used for process efficiency/effectiveness and batch/campaign analysis. The likelihood of this data being used steadily decreases as illustrated in Figure 37.5. Consequently it is reasonable that production data is retained for life of batch to which it relates.

Retention requirements for batch records are very different. Batch record content should primarily consist of critical control points verifying CQAs and may be compressed (e.g., trends, means, calculations). Batch record content is used for batch release decisions and approvals. Batch records may be used over an extended period of time to support

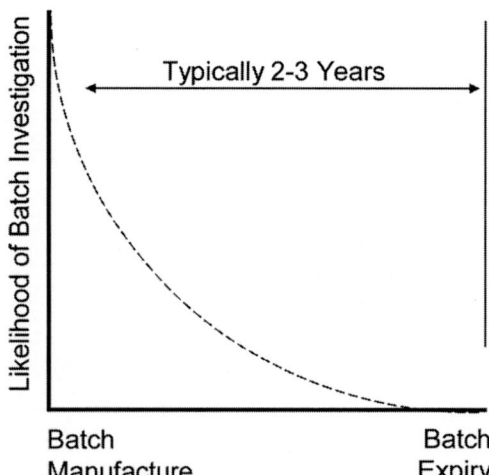

Figure 37.5 Use of production data. From Ref. 3.

investigations pertaining to product quality, efficacy and patient safety. Consequently this data should be retained for an extended period, typically for the shelf life of the batch to which it relates plus seven years.

COMPLIANCE STRATEGY

PAT systems do not require validation if they are used as part of a research project or proof of concept evaluation. The rationale behind this is that data is not used for GMP decisions and approvals. It is still assumed however that the PAT system is built in accordance with Good Practice. Data from the PAT system may later be used for GMP decisions and approvals in which case there would be a significant advantage if the PAT system were developed anticipating possible future validation with view to reducing later costs. PAT systems providing data for GMP decisions and approvals must be validated regardless of whether they are temporary or permanent systems.

VALIDATION AND VERIFICATION

Some companies are starting to use the term verification to encompass and replace the scope of validation. This change in terminology is supported by new industry guides such as ASTM E55 (4). International regulations however continue to use the term validation. The choice of terminology is at the user's discretion. This case study uses framework of validation plans and validation reports supported by specification and verification activities, which is consistent with GAMP guidance (5).

A validation master plan should be prepared. This may cover process and computer validation. Separate subordinate validation plans will normally be prepared for the MES and SCADA with concluding validation reports. The validation strategy for PAT should be clearly defined in these plans. The strategy should include risk assessments to focus PAT validation activities to best effect. Once validated all computer system elements should be maintained under change control and configuration management. More detail on validating different computer systems can be found in the other case studies available in this book.

Process Validation

Process validation will change dramatically compared with traditional concepts. Conventional validation consists of design qualification (DQ), followed by installation qualification (IQ), operational qualification (OQ), and performance qualification (PQ) where PQ is based on QC checks on a minimum of three consecutive batch runs. This approach works if conditions do not change but in reality they often do even with three consecutive batches let alone during the live of a products manufacture. Validating all possible process ranges can be cumbersome if not impossible. In these circumstances process validation does not help development of process capability, rather it often confirms a lack of a robust process.

PAT should bring about a fundamentally better understanding of processes and hence control. The challenge is to determine if advanced monitoring and control has supplanted the value of traditional process validation in assuring product quality (6). Validation in essence would be achieved through inherent process capability that is continually proven by successive successful product batches.

The rudimentary life cycle concept behind process validation remains unchanged although terminology may change. The basic life cycle should consist of the following:

- Planning and specification
- Design and build
- Testing and qualification

PAT systems can be validated within an integrated approach including facilities and equipment. Computerized systems do not have to be separated by default for the purpose of validation.

Many companies are merging traditional life cycle with QbD activities into a hybrid approach that maintains continuity with the past (preventing unnecessary rework of previous

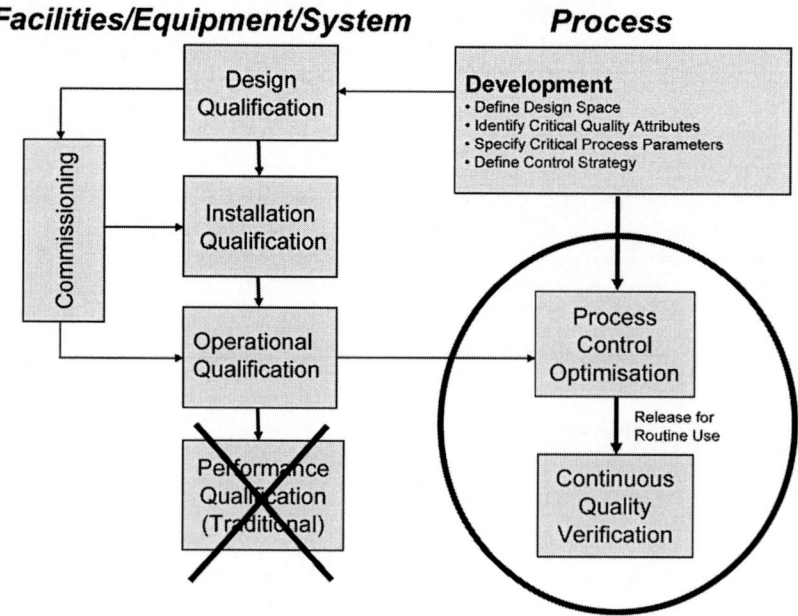

Figure 37.6 Process verification. From Ref. 3.

validation efforts) while fully facilitating the advantages offered by the QbD thinking. QbD activities should focus on the following:

- Definition of design space
- Identify CQAs
- Specific CPPs
- Validation of facilities/equipment/systems
- Define control strategy
- Verification of control model

Figure 37.6 illustrates the new approach. Once OQ is complete the process can be tuned to improve its performance. PQ does not exist. Because there is high process capability real-time batch release can be undertaken immediately underpinned by continuous quality verification.

Data Verification of Control Model
A new paradigm has emerged to verify the control model on the basis of the use of data rather than conventional testing of functional requirements. In essence a series of defined data sets are built to demonstrate correct operation in the design space. Data sets should individually or in combination examine the following:

- Fluctuating various control system inputs to fully explore the design space. This is particularly important in processes where inputs are subject to a high degree of variability.
- Forcing the process outside steady state control to demonstrate the control model's ability to respond and maintain the process within the design space.
- Excursions of the process outside the design space to ensure appropriate alarms and reports are generated.

Advanced control models may also be able to compare and contrast intermediate control points and sample measurements to identify inconsistencies indicative of a process problem.

ASTM recommends the amount of testing should be based on a risk assessment to determine an appropriate level of confidence required for a particular pharmaceutical or healthcare product being manufactured (7).

In adopting the data verification approach, it is important to remember key dependencies.

- Calibration of instrumentation providing data inputs to the PAT system is vital.
- Data sets must be controlled and managed. They must be secure and protected from change. Backups should be maintained and archive copies retained. Further information on this topic can be found in chapter 11.
- Other functions of the PAT system (e.g., security, alarms, and reporting) will still need to be validated using the conventional approach of testing functional requirements.

If all the above is done then the level of detail in specifications, testing documentation, and subsequent change control for the control system implementation of the control model can be relaxed because the data sets provide the primary verification of the control model.

Computer Validation

The basic computer validation requirements for computer validation are unchanged from the principles outlined earlier in this book. A life cycle approach should be adopted as discussed elsewhere (see Fig. 37.7).

The main computer elements in Figure 37.3 comprise

- Instruments,
- SCADA/MES systems, and
- Data historian.

The GAMP software categories should be applied to each of these elements to determine the basic validation requirements (5). Chapter 6 provides further details.

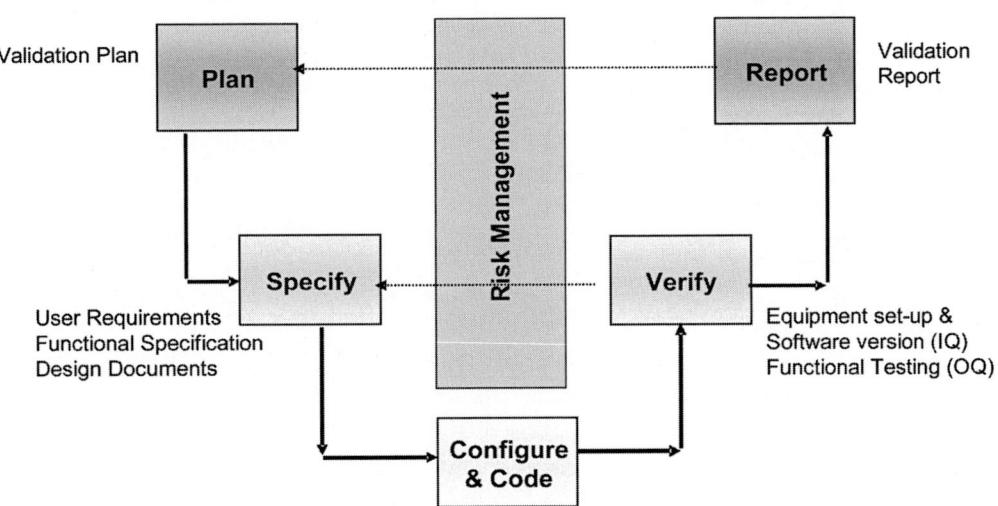

Maybe bespoke application or standard software.
Even if standard interfaces, calculations, reports
are often custom built.

Figure 37.7   Computer Validation Model.

Instruments should be calibrated and qualified. Firmware versions should be recorded with any configuration setup. Data transfer after acquisition should have integrity checking. This is normally facilitated through industry standard protocols. Supplier auditing might be appropriate for new instrument developments.

SCADA/MES should be validated. There is likely to be a mix of COTS software and bespoke code. The name and version of the operating system should be documented. The correct operation of statistical analysis software and predictive control software must be assured. Tests should challenge user programmed and configured functionality. User interfaces and user reports should receive particular attention. Custom programming such as macros should conform to good programming practices and be subject to source code reviews. Interfaces between systems should also be specified and tested. Supplier audits of SCADA and MES applications are typical conducted.

Data Historians should be validated. Data must be protected from unauthorized and unintentional modification. This might be achieved through locking down data (i.e., no subsequent write permissions given after data created). The requirement for any audit trails needs to be specified. Data retention requirements need to be defined in accordance with company policy and regulatory requirements. Software used to retrieve and analyze GMP data should be validated. Any dependencies on storage media need to be understood and managed. Backup and restore are key processes. System interfaces should also be specified and tested.

ELECTRONIC RECORDS AND ELECTRONIC SIGATURES

Electronic records should be identified on the basis of critical process control points and associated critical parameters. A defined process should be used to conduct this analysis, and one that builds on or is complementary to any assessment conducted as part of product registration. Consistency is key. There may be additional record requirements identified by predicate rules but care must be taken not to extend beyond these records (8). A risk analysis should be conducted to determine appropriate data management controls such as audit trail, archiving, and retention (9).

The electronic record strategy is likely to be focused on the role of the Data Historian. General audit trail requirements may be satisfied by the implementation of a transaction log. Particularly, critical records may warrant electronic audit trails implemented for individual records. Consideration should also be given to how copies of regulated records including audit trails may be provided during inspections and submissions to regulatory authorities.

Finally, a decision will need to be taken on how best to handle the application of signatures where this is required. FDA and other authorities will accept hybrid signatures applied to printed copies so long as it can be demonstrated that the equivalence of electronic and printed copies is maintained. Other systems might apply electronic signatures in which case there is no requirement to take a paper copy so long as electronic records with their signatures are secure and archived. The requirements for electronic records and electronic signatures are discussed more fully elsewhere in this book.

MAINTAINING COMPLIANCE

The compliance status of PAT systems as with any other regulated computerized system needs to be maintained. Key operational compliance activities include the following:

- Change control
- Configuration management
- Security management
- Patch updates
- Business continuity
- Disaster recovery

Further details on expectations for management and control during operation, maintenance, and final retirement can be found in chapters 11 and 12.

CONCLUSION

The principles extolled by PAT are not new and there are many examples from AstraZeneca, GlaxoSmithKline, and Pfizer of successful implementation over the past 10 to 20 years (2). The difference now is the recognition of the PAT philosophy as an approach, and the widespread available technology that can make this reality. The way product quality is addressed could be revolutionized, only time will tell. Continuing developments can be followed through ISPE PQLI publications (10).

This study is one of the few currently available for computer systems supporting PAT. The basic computer validation requirements for PAT are the same as any other computer system although their application needs to be tailored to meet the specific needs of PAT.

REFERENCES

1. U.S. Food and Drug Administration. Process Analytical Technologies Initiative, 2002.
2. Royal Pharmaceutical Society and American Association of Pharmaceutical Scientists. The key for achieving new standards of manufacturing excellence and regulatory compliance: process analytical technology. Eighth Arden House European Conference. London, March 24–26, 2003.
3. Wingate GAS. PAT & implications for GMP. Meeting of PIC/S Expert Circle on Computerised Systems. Dublin, October 2007.
4. American Society for Testing and Materials. E2500. Standard guide for specification, design, and verification of pharmaceutical and biopharmaceutical manufacturing systems and equipment, 2007.
5. International Society for Pharmaceutical Engineering. GAMP®5—A Risk-Based Approach to Compliant GxP Computerised Systems, 2008. Available at: www.ispe.org.
6. Pharmaceutical Research and Manufacturers of America. A risk-based approach to cGMPs, White Paper, 2003.
7. American Society for Testing and Materials. Guidance on the Verification of PAT Controls, WK9182, 2008.
8. Food and Drug Administration. U.S. Code of Federal Regulation, title 21, Part 11. Electronic signatures; electronic records. Rockville, MD.
9. Food and Drug Administration. Guidance for Industry, Part 11. Electronic Records; Electronic Signatures—Scope and Application, February 2003.
10. International Society for Pharmaceutical Engineering. Draft PQLI summary update report, version V04, September 14, 2007. Available at: www.ispe.com.

38 | Case Study 20: Computer Applications Supporting the Supply of Biotechnology Products

Guy Wingate

GlaxoSmithKline

Biotechnology is providing new and more effective therapies (i.e., biologicals, biopharmaceuticals, and vaccines) for many conditions such as leukemia and H1N1 "swine" flu and is the fastest growing sector in the pharmaceutical industry for new product licenses today. These products are often difficult to make because of their complexity and consequently carry a premium price. This makes them very attractive to pharmaceutical companies wanting to bolster their pipeline of new products. In many organizations, however, there is a steep learning curve to establish the necessary quality standards including those appropriate for computer applications.

The development, manufacture, and supply of biotechnology products are governed just like other medicinal products. Facilities are subject to the same basic Good Clinical Practices (GCP), Good Laboratory Practices (GLP), and Good Manufacturing Practice (GMP) requirements (1–3). In addition, there are particular regulatory expectations relating to biotechnology (4,5).

Computer applications are extensively used throughout development (including clinical trials), cell bank management, bulk drug substance manufacture, finished drug manufacturing, final packaging, analytical testing, and supply. Many products need to be kept frozen or at 2 to 8°C with only a limited time available at ambient temperature for processing. Computer systems are vital for monitoring and control of such cold chain conditions. Batch records are very important regulatory records with particular attention on time stamp information relating to processing steps and records of any deviations.

For biotech products there are no special requirements for computer validation over and above the basic regulatory requirements is specified in Refs. 6 to 8. The same position exists in regard to the use of electronic records and electronic signatures. Regulatory expectations of this are defined in Refs. 8–10.

Computer applications must be fit for their intended use and as such should be reliable, robust and secure. The high value of bulk drug substance and drug product means that these systems must not only protect the efficiency, quality and safety of products, they must also ensure manufacturing yields are under sufficient control to prevent unnecessary product loss. The timely feedback of process monitoring information [particularly critical process parameters (CPP) mapped to critical quality attributes (CQA)] for process control to assure product quality during production is vital (see also Case Study 19 on Process Analytical Technology). The value of a lost batch of product can far exceed the cost of validation. More importantly a lost batch may be very hard to recover in terms of continuity of supply of finished final product since cycle times to grow replacement cell cultured batches may be measured in terms of months instead of days. For a medically critical product this will be a key consideration.

Analytical methods for biologics can be complex and thus prone to greater error, which can impact the ease of validating supporting analytical equipment. The basic requirements of qualification (verifying operability) remain the same as used elsewhere in the pharmaceutical industry.

A Quality Management System (QMS) should be maintained that defines the approach being adopted. GAMP®5 (11) provides a framework on what is expected to be done. The principles presented in the main part of this book and the associated case studies for various computer systems are all applicable to facilities supporting biological, biopharmaceutical, and vaccine products. Practitioners validating computer applications are encouraged to refer to appropriate material.

Although not mandated a risk-based approach should be adopted to focus validation on critical aspects of the computer application that could impact product quality and patient safety. This is entirely consistent with the 21st century paradigm being promoted for Quality by Design (12,13) and should improve the effectiveness of validation while also reducing the overall costs.

Finally, remember that the practicalities of computer compliance are well understood in the wider pharmaceutical industry. It is important not to reinvent the wheel and inadvertently create bureaucratic ways of working that add questionable value. Compliance activities, if conducted correctly, should be a constructive and beneficial exercise.

REFERENCES

1. ICH Q7A (2000). Good Manufacturing Practice Guide for Active Pharmaceutical Ingredients, ICH Harmonised Tripartite Guideline, International Conference on Harmonisation of Technical Requirements for Registration of Pharmaceuticals for Human Use.
2. EU Directive 2001/83/EC amended by EU Directive 2003/94/EC and 2005/28/EC relating to investigational medicinal products for human use.
3. US Title 21 Code of Federal Regulation Parts 210 and 211, Current Good Manufacturing Practice in Manufacturing, Processing, Packaging or Holding of Drugs and Finished Pharmaceuticals.
4. US title 21 Code of Federal Regulations Parts 600, 601, and 610, Biological Products.
5. EU GMP Guidance for Manufacture of Biological Medicinal Products for Human Use, Annex 1 Manufacture of Sterile Products and Annex 2 Guidance for Biopharmaceuticals.
6. FDA (1983). *Guide to Inspection of Computerised Systems in Drug Processing*, Technical Report, Reference Materials and Training Aids for Investigators, Rockville, MD: Food and Drug Administration.
7. EU GMP Guidance for Manufacture of Biological Medicinal Products for Human Use, Annex 11 Computerised Systems and Annex 15 Process Validation.
8. Pharmaceutical Inspection Co-operation Scheme (2003). *Good Practices for Computerised Systems in Regulated GxP Environments*, Pharmaceutical Inspection Convention, PI 011-1, Geneva.
9. FDA. Preamble to Electronic Signatures and Electronic Records, Code of Federal Regulation Title 21, Part 11, 1997. Rockville, MD: Food and Drug Administration.
10. FDA. 21 CFR Part 11 Electronic Records; Electronic Signatures—Scope and Application, Guidance for Industry, 2003. Rockville, MD: Food and Drug Administration.
11. ISPE. GAMP®5: Risk-Based Approach to Compliant GxP Computerised Systems, 2008. Tampa, FL: International Society for Pharmaceutical Engineering. Available at: www.ispe.org.
12. FDA. Pharmaceutical CGMPs for The 21st Century: A Risk–Based Approach, 2004. Rockville, MD: Food and Drug Administration.
13. ASTM. E2500-07 Standard Guide for Specification, Design and Verification of Pharmaceutical and Biopharmaceutical manufacturing Systems and Equipment, American Society for Testing and Materials, 2007.

Glossary

Acceptance criteria ANSI/IEEE (1983): The criteria that a software product must meet to successfully complete a test phase or to achieve delivery requirements.

Actuator FDA (1995): A peripheral output device that translates electrical signals into mechanical actions, e.g., a stepper motor that acts on an electrical signal received from a computer system to turn its shaft a certain number of degrees or a certain number of rotations.

Alpha testing FDA (1995): Acceptance testing performed by the customer in a controlled environment at the developer's site. The software is used by the customer in a setting approximately the target environment with the developer observing and recording errors and usage problems.

As-built FDA (1995): Pertaining to an actual configuration of software code resulting from a software development project.

Assembly language FDA (1995): A low-level programming language that corresponds closely to the instruction set of a given computer, allows symbolic naming of operations and addresses, and usually results in a one-to-one translation of program instructions (mnemonics) into machine instructions.

Audit GMA-NAMUR (1996): An activity to determine through investigation the adequacy of, and adherence to, established procedures, instructions, specifications, codes, and standards or other applicable contractual and licensing requirements, and the effectiveness of implementation.

Garston-Smith (1997): An independent review for assessing compliance with software requirements, specifications, baselines, standards, or procedures.

Automation system A system based on a computer technology with input devices (e.g., sensors), output devices (e.g., actuators), and communication links (e.g., telemetry and cable networks) that are collectively designed to perform a specific function or group of functions (e.g., control, protection or monitoring). Automation systems may be linked into larger integrated systems. [Defined for this book]

Baseline FDA (1995): A specification or product that has been formally reviewed and agreed upon, that serves as the basis for further development, and that can be changed only through formal change control procedures.

Batch record IQA (1994): Documents (including those stored in photographic and electronic form) that record stages in the manufacture of a batch, details of ingredients and process equipment used, methods followed, in-process controls carried out, test results obtained, dates of manufacture, and testing and history of the storage of the pharmaceutical raw material.

Bespoke software GAMP (2008): A system produced for a customer, specifically to order, to meet a defined set of user requirements. Note: [Bespoke code includes so-called standard software where the version of the software to be used has not been market-tested over a period of time by other customers.]

Beta testing FDA (1995): Acceptance testing performed by the customer in a live application of the software at one or more end-user sites in an environment not controlled by the developer.

Black box testing See "Functional testing"

Bomb FDA (1995): A Trojan horse that attacks a computer system upon the occurrence of a specific logical event ("logic bomb"), the occurrence of a specific time-related logical event ("time bomb"), or something that is hidden in electronic mail or data triggered when read in a certain way ("letter bomb").

Bootstrap FDA (1995): A short computer program that is permanently resident or easily loaded into a computer and whose execution brings a larger program, such as an operating system or its loader, into memory.

Business continuity planning A documented process by which the recovery and continuation of critical business functions in the presence of events that significantly disrupt business operations. [Defined for this book]

Calibration FDA (1995): Ensuring continuous adequate performance of sensing, measurement and actuating equipment with regard to specified accuracy and precision requirements.

Certification FDA (1995): In computer systems, a technical evaluation, made as part to and in support of the accreditation process that establishes the extent to which a particular computer system or network design and implementation meet a prescribed set of requirements.

Change control FDA (1995): The processes, authorities for, and procedures to be used for all changes that are made to the computerized system and/or the system's data. Change control is a vital subset of the quality assurance program within an establishment and should be clearly described in the establishment's standard operating procedures.

GAMP (2008): A formal system by which qualified representatives of appropriate disciplines review proposed or actual changes that might affect a validated status. The intent is to determine the need for action that would ensure and document that the system is maintained in a validated state.

OECD (1995): Ongoing evaluation and documentation of system operations and changes to determine whether a validation process is necessary following any changes to the computerized system.

CHAZOP Computer HAZard and OPerability study to assess the threats and their control between automation systems, their users and operational environments, and the manufacturing process. CHAZOP studies for IT systems concentrate on the threats and their controls affecting data integrity [Defined for this book]

Client-server FDA (1995): A term used in a broad sense to describe the relationship between the received and the provider of a service ... a networked system where front-end applications, as the client, makes service requests on another networked system.

Code See "Software"

Comment FDA (1995): In programming languages, a language construct that allows explanatory text to be inserted into a program and that does not affect on the execution of the program.

Commercial off-the-shelf (COTS) products Versions of products that have been commercially available for at least six months and are widely used. Beta releases of products that are still under supplier evaluation and COTS products specifically customized than configured) for application are excluded from this definition. [Defined for this book]

Computer-aided software environments (CASE) Tools designed to support the analysis and design phase of the software development life cycle. The tools are usually orientated toward the support of graphical notations. [Defined for this book]

Computer system See "Automation system"

Computer virus FDA (1995): A program that secretly alters other programs to include a copy of itself and executes when the host program is executed. The execution of a virus program comprises a computer system by performing unwanted or unintended functions that may be destructive.

A program that alters other programs to include a copy of itself and executed when the host program is executed. The execution of a virus program compromises a computer system by performing unwanted or unintended functions that may be destructive. [Defined for this book]

Computerized system   A computer system plus the controlled process it operates [Defined for this book.]

Configuration   FDA (1995): The arrangement of a computer system or component as defined by the number, nature, and interconnection of its constituent parts.

Configuration parameters   Parameters that provide control values for computerized equipment. Configuration includes operating parameters (e.g., drug product manufacturing recipes, set points) and system environment parameters (e.g., file names, directory structures). Configuration provides a method to accomplish specific functionality without using a programming language [Defined for this book]

Control strategy   ISPE (2008): Comprehensive plan for ensuring that the final [pharmaceutical] product meets critical requirements, and therefore the needs of the patient.

Crash   FDA (1995): The sudden and complete failure of a computer system or component.

Critical impact   Computer systems have a critical impact if failure or latent design flaws can result in injury or illness to the consumer of the drug that

- Is life threatening
- Results in permanent impairment of body function
- Results in permanent damage to body structure
- Necessitates medical or surgical intervention to preclude the above

[Defined for this book]

Critical parameter   A process parameter that may cause significant variation in the quality of a finished product [Defined for this book]

Critical process   HPB (1998): A process that may cause significant variation in the quality of a finished product

Critical process parameter   ISPE (2008): A process parameter whose variability impacts a quality attribute and therefore needs to be controlled to ensure the process produces the desired quality. A critical process parameter remains critical even if it is controlled.

Critical quality attribute   ISPE (2008): A physical, chemical, biological or microbiological property or characteristic that needs to be controlled (directly or indirectly) to ensure [pharmaceutical] product quality.

Critical step   A step in a process that may cause significant variation in the quality of a finished product. [Defined for this book]

Data integrity   FDA (1995): The degree to which a collection of data is complete, consistent, and accurate.

Data validation   FDA (1995): A process used to determine if data are inaccurate, incomplete, or unreasonable. The process may include format checks, completeness checks, check—key tests, reasonableness checks, and limit checks.

Dead code   FDA (1995): Program code statements that can never execute during program execution. Such code can result from poor coding style or can be an artefact of previous versions or debugging efforts Dead code can be confusing, and is a potential source of erroneous software changes. Dead code is program logic that cannot execute because the program path does not permit the logic to be reached. Newly developed programs should be reviewed for the presence of dead code Dead code must be removed prior to compilation and submission for production implementation. In instances where program logic becomes dead code as a result of program modifications, the associated dead code should be removed from the program before recompilation and submission to the production implementation. Commented source code is not dead code because it is ignored by the compiler and does not become program logic. Code rendered inaccessible by configuration (e.g., switches, parameters, calls, etc.) is not dead code because this code is intended to be available for use depending on the need of a particular implementation. Similarly, code residing within a standard library, which is not accessed by the calling program, is not considered dead code because this code is intended to be available for use depending on the need of a particular implementation. Code that has been included for the purposes of testing or for later diagnosis during support work, and which can be configured "on" or "off' is not regarded as dead code. If the code is configurable for use in many different projects, each with a different configuration of options, the unused options should not be removed; however, source code and configuration review and testing processes must demonstrate that the correct options have been correctly deselected and do not function.

Debugging   FDA (1995): Determining the exact nature and location of a program error, and fixing the error.

Design   FDA (1995): The process of defining the architecture, components, interfaces, and other characteristics of an [automation] system or component.

Design qualification (DQ)   GAMP (2008): Formal and systematic verification that the requirements defined during specification are completely covered by the succeeding (design) specification or implementation.

GMA-NAMUR (1996): Formal and systematic verification that the requirements determined at the functional specification phase were completely met in the subsequent specification or implementation phase and that the higher authority of guidelines or laws have been taken into account.

Design review   Phrase synonymous with DQ used in relation to computer systems.

Desktop   GAMP (2001): Represents the end-user workstation and local software environment. Normally provides a Graphical User Interface (GUI) front-end menu providing users with access to required applications. Many desktop environments can be reconfigured by the end user.

Desktop build   Set of software on end-user workstations making up desktop environment. Also referred to as desktop configuration. [Defined for this book]

Disaster   A sudden, unplanned calamitous event that creates an inability on an orgait part to provide critical business functions for some period of time, which results in great damage or loss. [Defined for this book]

Electronic signature   OECD (1995): The entry in the form of magnetic impulses or computer data compilation of any symbol or series of symbols, executed, adapted, or authorized by a person to be equivalent to the person's handwritten signature.

Embedded system   GAMP (2008): A system, usually microprocessor or PLC based, whose sole purpose is to control a particular piece of automated equipment. This is contrasted with a stand-alone computer system.

Emulation   FDA (1995): A model that accepts the same inputs and produces the same outputs as a given system. To imitate one system with another.

Escrow   ACDM/PSI (1998): A legal term in Anglo-American law. A written agreement, constituting evidence between two or more parties (in this case the supplier and purchaser), that is given to a third party with instructions (in this case, to deliver source code and associated documentation) to be executed only on a future condition (in this case, the supplier going into receivership).

Failure mode effects analysis (FMEA)   A technique used to define, identify, and reduce known or potential failures to an acceptable level. [Defined for this book]

Firmware   FDA (1995): The combination of a hardware device, e.g., an integrated circuit, and computer instructions and data that reside as read-only software on that device. Such software cannot be modified by the computer during processing.

Functional specification   A written definition of the function that a system or system component can perform. [Defined for this book]

Functional testing   GAMP (2008): Also known as "Black Box" testing, since source code is not needed. This involves inputting normal information and abnormal test cases and then, evaluating outputs against those expected. Can apply to computer system or to a total system.

Good Clinical Practice (GCP)   The standard by which clinical trials are designed, implemented and reported so that there is public assurance that the data are credible, and that the rights, integrity, and confidentiality of subjects are protected. [Defined for this book]

Good Distribution Practice (GDP) EU (2007): GDP is the part of quality assurance that ensures that products are consistently stored, transported, and handled under suitable conditions.

Good Laboratory Practice (GLP) U.K. DoH (1995): GLP is concerned with the organizational processes and conditions under which studies are planned, performed, monitored, recorded, and reported to promote maintain the quality and reliability of the test data generated.

Good Manufacturing Practice (GMP) EU (2007): That part of quality assurance that ensures that products are consistently produced and controlled to the quality standards appropriate to their intended use.

IQA (1994): EU definition with I. It concerns production, quality control, and warehousing and distribution procedures.

GMP critical An aspect of the manufacturing process that if not properly managed can impact product quality. [Defined for this book]

Handshake FDA (1995): An interlocked sequence of signals connected components in which each component waits for the acknowledgment of its previous signal before proceeding with its action, such as data transfer.

Hardware FDA (1995): The physical equipment [making up a computer system], as opposed to programs, procedures, rules, and associated documentation. [Adapted]

Hardware design See "Design"

Hardware platform GAMP (2001): All computer hardware deployed to run software application programs. The definition covers servers, CPUs, memory devices, and peripheral controllers. [Adapted]

Hazard analysis See "CHAZOP"

Industry standard FDA (1995): Procedures or criteria recognized as acceptable practices by peer, professional, credentialing, or accrediting organizations.

Infrastructure GAMP (2001): All of the computer systems with their associated hardware, operating software (other than software applications), and networks used to run the business.

In-process control EU (2007): Checks performed during production to monitor and, if necessary, to adjust the process to ensure that the product conforms to its specification. The control of the environment or equipment tray also be regarded as part9f in-process control.

Installation qualification (IQ) FDA (1995): Establishing confidence that process equipment and ancillary systems [including computer systems] are compliant with appropriate codes and approved design intentions and that manufacturer's recommendations are suitably considered.

PMA (1990): Documented verification that all key aspects of hardware installation adhere to appropriate codes and approved design intentions and that the recommendations of the manufacturer have been suitably considered.

Integrated project support environment (IPSE) Tools supporting the configuration management of documentation and programming during the software development life cycle. [Defined for this book]

Major level of concern Computer system failure or latent design flaws potentially have a critical impact on the consumer of the drug product, operator, or bystander. [Defined for this book]

Metadata (DOD 5015.2-STD) Data describing stored data, that is, data describing the structure, data elements, interrelationships, and other characteristics of electronic records.

Minor level of concern Computer system failure or latent design flaws are not expected to result in any injury or illness to the consumer of the drug product, operator, or bystander. [Defined for this book]

Moderate level of concern Computer system failure or latent design flaws could result in injury or illness (but without a critical in on the consumer of the drug product, operator, or bystander. [Defined for this book]

Network GAMP (2001): A network is a data communications system that links two or more computers and peripheral devices. It consists of cabling, the network hardware, and communications software.

Object code A computer program that is the output of translated [assembler or complier] source code. [Defined for this book]

Operational qualification (OQ) FDA (1995): Establishing confidence that process equipment and subsystems [including computer systems] are capable of consistently operating within established limits and tolerances.

PMA (1990): Documented verification that the equipment-related system or sub performs as intended throughout all anticipated operating ranges.

Patch A change made directly to object code without retranslating [assembler or compiler source code. [Defined for this book]

Performance qualification (PQ) EPA (1995): Documented verification that the process-related system performs as intended throughout representative or anticipated operating ranges.

FDA (1995): Establishing confidence that the [manufacturing] process is effective and reproducible.

Peripherals GAMP (2001): Hardware deployed to extend the capability of the hardware platform. It includes printers, modems, keyboards, tape drives, screens, and scanners.

Platform FDA (1995): The hardware and software that must be present and functioning for an application program to perform as intended. A platform includes, but is not limited to the operating system or executive software, communication software, microprocessor, network, input/output hardware, any generic software libraries, database management user interface software, etc.

Procedures EU (2007): Description of the operations to be carried out, the taken and measures to be applied directly or indirectly related to the manufacture of a medicinal product.

Process qualification FDA (1995): Establishing confidence that a process is effective and reproducible.

Product qualification FDA (1995): Establishing confidence through appropriate testing that the finished product produced by a specified process meets all release requirements functionality and safety.

Production EU (2007): All operations involved in the preparation of a medicinal product, from receipt of material, through processing and packaging, to its completion as a finished product.

Program See "Software"

Project plan Similar to Quality Plan but including a detailed schedule of project activities and deliverables. [Defined for this book]

Prospective validation FDA (1995): Validation conducted prior to the distribution of either a new (drug) product, or product made under a revised manufacturing process, where the revisions may affect the [drug] product's characteristics.

GMA-NAMUR (1996): Documented evidence that a system does what it was planned to do before it is used in production. Also referred to as validating new systems.

Prototyping FDA (1995): An approach to accelerate the software development.

Process by facilitating the identification of required functionality during analysis and design phases. A limitation of this technique is the identification of system and software problems and hazards. [Adapted]

Quality   Garston-Smith (1997): The totality of features and characteristics of a product or service that bears on its ability to satisfy given needs.

Quality assurance (QA)   ANSI/IEEE (1983): A planned and systematic pattern of all actions necessary to provide adequate confidence that the item conforms to established technical requirements.

Quality by design   ISPE (2008): A systematic approach to development that begins with predefined objectives and emphasises product and process understanding based on sound science and quality risk management

Quality control (QC)   GAMP (2008): The regulatory process through which industry measures actual quality performance, compares it with standards, and acts on the differences.

Quality plan   GAMP (2008): A plan created by the supplier to define actions, deliverables, responsibilities, and procedures to satisfy the customer's quality and validation requirements.

Raw data   The first records of an action or observation saved onto a durable storage medium that are capable of being used immediately or as part of a further GMP decision or review process. Raw data may include photographs, microfilm or microfiche copies computer printouts, magnetic media, including dictated observations, and recorded data from automated systems. Raw data excludes transient electronic data. [Defined for this book]

GAMP (2008) Any worksheets, records, memoranda, notes, or exact copies thereof that are as a result of original observations and activities of, a study and are necessary for the reconstruction and evaluation of a work project, process or study report, etc. Raw data may be hard/paper copy or electronic but must be known and defined in the system procedures.

FDA (21 CFR 58): Raw data means any laboratory worksheets, records, memoranda, notes, or exact copies thereof that are the result of original observations and activities of a nonclinical laboratory study and are necessary for the reconstruction and evaluation of the report of that study. In the event that exact transcripts of raw data have been prepared (e.g., tapes that have been transcribed verbatim, dated, and verified accurate by signature), the exact copy or exact transcript may be substituted for the original source as raw data. Raw data may include photographs, microfilm or microfiche copies, computer printouts, magnetic media, including dictated observations, and recorded data from automated instruments.

Real time   FDA (1995) Pertaining to a system or mode of operation in which computation is performed during the actual time that an external process occurs, in order that the computation results can be used to control, monitor, or respond in a timely manner to the external process

Redundant code   See "Dead code"

Retrospective validation   FDA (1995): Validation of a process for a product already in distribution based on accumulated production, testing and control data.

GMA-NAMUR (1996): Documented evidence that a system does what it purports to do based on an analysis of historical information. Also referred to as validating existing systems.

Revalidation   FDA (1995): Relative to software changes, revalidation means validating the change itself, assessing the nature of the change to determine potential ripple effects, and performing the necessary regression testing.

GMA-NAMUR (1996) Repetition of the validation process or a specific portion of it [response to a change].

Review by exception   GAMP (2008): A systematic approach for screening data from manufacturing operations to create reports and disposition (i.e., release, quarantine, reject) that include critical process exceptions and reduce or eliminate the need for reviewing acceptable data or reports

Risk analysis   See "Risk assessment"

Risk assessment   A systematic approach to identifying the potential failures in processes and quantifying the risk, they present. [Defined for this book]

Risk management   ICH (Q8): The systematic application of quality management policies, procedures and practices to the tasks of assessing, controlling, communicating, and reviewing risk.

Security   OECD (1995): The protection of computer hardware and software from accidental or malicious access, use, modification, destruction or disclosure. Security also pertains to personnel data, communications, and physical protection of computer installations.

Sensor   FDA (1995): A peripheral input device that senses some variable in the system environment, such as temperature, and converts it to an electrical signal that can be further converted to a digital signal for processing by the computer.

Service level agreement   A formal agreement, possibly contract, defining the services to be provided by a supplier to a customer. [Defined for this book]

Simulation   FDA (1995): A model that behaves or operates like a given system when provided with a set of controlled inputs.

Software   GAMP (2008): A collection of programs, routines, and subroutines that controls the operation of a computer system.

Software design   See "Design"

Software inspection   FDA (1995): A manual testing technique in which program documents (including source code) are examined in a very formal and disciplined manner to discover errors, violations of standards, and other problems.

Source code   GAMP (2008): An original computer program expressed in human readable form (programming language) that must be translated into machine readable form before it can be executed by the computer.

Source code is the human readable form of program code, written in its original (source) programming language. Source code must be compiled, assembled or otherwise interpreted before it can be executed by a computer. The executable code is referred to as object code because it is not readily understandable because it exists as machine hexadecimal code. [Defined for this book]

Source code review   See "Software inspection"

Specification   IEEE (1998): A document that specifies in a complete, precise, verifiable manner the requirements, design, behavior, or other characteristics of a system or component, and often, the procedures for determining whether these provisions have been satisfied.

Standard   See "Industry standard"

Standard operating procedures   FDA (1995): Written procedures prescribing and describing the steps to be taken in normal and defined conditions that are necessary to assure control of production and processes
    See also "Procedures."

GMA-NAMUR (1996): Instruction that describes how something is to be accomplished. SOPs regulate the operation and maintenance of a computerized [system in order to use it in a correct way and also to fulfil its real purpose permanently. Structured, detailed instructions on how to do a task to ensure consistency and compliance. [Defined for this book]

Stepwise refinement   FDA (1995): A structured software design technique; c - steps are defined broadly at first, and then further defined with increasing detail.

Structural testing   GAMP (2008): Also known as "White Box" testing, it involves examining the internal structure of the source code. Includes low-level and high-level code review, path analysis, auditing of programming procedures, and standards actually used, inspection for extraneous "dead code," and boundary analysis and other techniques. Requires specific computer science and programming expertise. [Adapted]

Superfluous code   Software code that unnecessarily recalculates, rechecks, or re-performs calculations or actions that are unnecessary or that have already been done. [Defined for this book]

Supplier   GMA-NAMUR (1996): The company or group responsible for developing, constructing, and delivering a system or part of a system. A supplier can be an [equipment] vendor, a contractor [a system's application integrator], or a consultant.

GAMP (2008): Any organization of individuals contracted directly by the customer to supply a product.

TAG Unique label identifier given to instrumentation and/or equipment. [Defined for this book]

Testing FDA (1995): The process of operating a system or a component under specified conditions, observing or recording the results, and making an evaluation of some aspect of the system or component.

Transient data Data that have a temporary existence and is not retained. [Defined for this book]

Trojan horse FDA (1995): A method of attacking a computer system, typically by providing a useful program that contains code intended to compromise a computer system by secretly providing for unauthorized access, the unauthorized collection of privilege system and user data, the unauthorized reading or altering of files, the performance of unintended and unexpected functions, or the malicious destruction of software and hardware.

User GMA-NAMUR (1996): The company or group responsible for the operation of a system.

User requirements The customer's written functional needs with regard to a computer system. [Defined for this book]

Validation EU (2007): Action of proving, in accordance with the principles of Good Manufacturing Practice, that any procedure, process, equipment, material, activity, or system actually leads to the expected results.

FDA (1995): Establishing documented evidence that provides a high degree of assurance that a specific process will consistently produce a product meeting its predetermined specifications and quality attributes.

GMA-NAMUR (1996): Documented evidence that a specific process will consistently produce a product meeting its predetermined specification and quality attributes IQA (1994). The process of establishing documentary evidence that provides a high degree of assurance that any product, process, activity, procedure, system, equipment, or software used in the control or manufacture consistently meets its predetermined specification

OECD (1995): The demonstration that a computerized system is suitable for its intended purpose

Validation master plan A high-level plan coordinating a number of validation plans [Defined for this book]

Validation plan GMA-NAMUR (1996): A prospective plan of action whose implementation should produce formal and documented proof that the system is validated.

Vendor PICS (1999): A company or group responsible for developing, constructing, and delivering a system or part of a system
See also "Supplier"

Verification ISO (2005): Confirmation through provision of objective evidence that specified requirements has been fulfilled.

ASTM (2007): A systematic approach to verify that manufacturing systems, acting singly or in combination, are fit for intended use, have been properly installed, and are operating correctly. This is an umbrella term that encompasses all types of approaches to assuring systems are fit for use such as qualification, commissioning, verification and system validation, or other.

White box testing See "Structural testing"

White space Blank lines of code that have been purposely inserted into the software listing to make it easier to read. [Defined for this book]

Wireless device Devices, usually handheld, used for wireless data acquisition and communications. Examples include mobile phones and pagers. These devices can connect to intranet and internet services as well as facilitating dedicated communication links to host computer systems. [Defined for this book]

REFERENCE WEBSITES

http://www.21part11, Industry View of U.S. CFR Part 11
http://www.abpi.org.uk, Association of the British Pharmaceutical Industry
http://www.acdm.org.uk, Association of Clinical Data Management
http://www.barqa.com, British Association of Research Quality Assurance
http://www.bira.org.uk, The British Institute of Regulatory Affairs
http://www.bsi.com, British Standard Institute
http://www.computervalidation.com, Independent site
http://www.dashnet.com/acrpi, Association for Clinical Research in the Pharmaceutical Industry
http://www.diahome.org, Drug Information Association
http://www.eutha.org/emea.html, European Agency for the Evaluation of Medicinal Products
http://www.eudra.org, EMEA Home Page
http://www.fda.gov, FDA Home Page
http://www.ilpma.org/ichl.htnil, International Conference on Harmonization
http://www.ispe.org, ISPE Home Page (GAMP Forum)
http://www.ivthome.com, IVT Home Page
http://www.jettconsortium.org www, JETT Consortium Home Page
http://www.labcompliance.com, Ludwig Huber's Compliance Home Page
http://www.mhra.gov.uk, MHRA Home Page
http://www.pda.org, PDA Home Page
http://www.phrmafounthtion.org, PhRMA Home Page
http://www.picscheme.org, PlC/S Home Page
http://www.psiweb.org, Statisticians in the Pharmaceutical Industry

Index